Statistical Mechanics

Selecta of Elliott H. Lieb

ELLIOTT H. LIEB

# Statistical Mechanics

Selecta of Elliott H. Lieb

Edited by
B. Nachtergaele
J. P. Solovej
and J. Yngvason

Springer

Professor Elliott H. Lieb
Departments of Mathematics and Physics
Jadwin Hall
Princeton University
P.O. Box 708
Princeton, New Jersey 08544-0708, USA

Professor Bruno Nachtergaele
University of California
Department of Mathematics
One Shields Ave.
Davis, CA 95616-8633, USA

Professor Jan Philip Solovej
University of Copenhagen
Department of Mathematics
Universitetsparken 5
2100 Copenhagen, Denmark

Professor Jakob Yngvason
Universität Wien
Institut für Theoretische Physik
Boltzmanngasse 5
1090 Wien, Austria

ISBN 978-3-642-06092-2 ISBN 978-3-662-10018-9 (eBook)
DOI 10.1007/978-3-662-10018-9

Library of Congress Control Number: 2004108429



springeronline.com



Originally published by Springer-Verlag Berlin Heidelberg New York in 2004
Softcover reprint of the hardcover 1st edition 2004



Printed on acid-free paper 55/3141/xo 5 4 3 2 1 0

# Preface

This is the fourth Selecta of publications of Elliott Lieb, the first two being *Stability of Matter: From Atoms to Stars*, edited by Walter Thirring, and *Inequalities*, edited by Michael Loss and Mary Beth Ruskai.

A companion third Selecta on *Condensed Matter Physics and Exactly Soluble Models* is also edited by us.

The goal of statistical mechanics, whose foundations were laid in the second half of the 19th century by James Clerk Maxwell, Ludwig Boltzmann and Joshua Willard Gibbs, is a derivation of macroscopic properties of matter from assumptions about its microscopic constituents. Originally formulated in the context of classical mechanics, its basic principles survived the quantum revolution of the 20th century essentially unaltered. The famous formula, $S = k \ln W$, engraved on Boltzmann's tomb, is supposed to allow, at least in principle, the computation of all thermodynamic quantities such as specific and latent heats, thermal expansion coefficients and phase transition temperatures. Standard texts on statistical mechanics underpin this wide-ranging claim mainly by the study of simple models where interactions are absent or very weak and the thermodynamic potentials can be computed more or less explicitly. On the other hand, no one has so far been able to explain, not even qualitatively, such an ubiquitous phenomenon as the freezing of water and formation of ice crystals from the assumption that water consists of charged particles interacting by Coulomb forces. A rigorous derivation, from statistical mechanics, of thermodynamic properties of models with realistic forces is a formidable task where many basic questions are still unanswered.

Elliott Lieb is a mathematical physicist who meets the challenge of statistical mechanics head on, taking nothing for granted and not being content until the purported consequences have been shown, by rigorous analysis, to follow from the premises. The present volume contains a selection of his contributions to the field, in particular papers dealing with general properties of Coulomb systems, phase transitions in systems with a continuous symmetry, lattice crystals, and entropy inequalities. It includes also work on classical thermodynamics, a discipline that, despite many claims to the contrary, is logically independent of statistical mechanics and deserves a rigorous and unambiguous foundation of its own.

The subject of *Exactly Soluble Models* in statistical mechanics is such a large and important branch of Lieb's work that a separate volume of his Selecta is devoted to it, together with a selection of his papers on *Condensed Matter Physics*. The division of his work into separate volumes under different headings is, in fact, to some extent a matter of taste and convenience. His papers are masterpieces of mathematical physics that can all be studied for profit by any serious student of the field, whatever his or her specialization.

We thank Wolf Beiglböck for his support and encouragement of this project and Sabine Lehr, Brigitte Reichel-Mayer and Sandra Thoms of Springer-Verlag for their invaluable help with the production.

Davis, Princeton, Vienna, April 2004

*Bruno Nachtergaele*
*Jan Philip Solovej*
*Jakob Yngvason*

# Contents

## Part V. Classical Thermodynamics



## Part VI. Lattice Systems



## Part VII. Miscellaneous

# Commentaries

## A Survey by the Editors

The following is a brief description of Elliott Lieb's papers on *statistical mechanics*, excluding mostly the papers on exactly solvable models that are commented on in another volume of the Selecta. The numbers refer to the publication list of Elliott Lieb, which appears at the end of this volume. Some of the papers that are not included in this volume *Statistical Mechanics* appear in the other Selectas of Elliott Lieb, namely *The Stability of Matter: From Atoms to Stars*, *Inequalities*, and *Condensed Matter Physics and Exactly Soluble Models*. The publication list shows which papers appear in which Selecta. The papers in this Selecta are additionaly marked in boldface with their numbers as given in the table of contents. The numbers in square brackets refer to works by other authors, which are listed at the end of this survey.

### I. Thermodynamic Limit for Coulomb Systems

43 (**I.1**), 58 (**I.2**, 8 pages), 91

It was not always universally accepted that the Boltzmann/Gibbs partition function

$$Z = \text{Trace}(\exp[-H/kT]) \equiv \exp[-F(T)/kT]$$

defined a free energy $F(T)$ that correctly described all the physics of an equilibrium system of particles with Hamiltonian $H$ at temperature $T$. In particular, were phase transitions correctly described by singularities of this $F$? A more elementary question, but still far from an easy one, was the existence of the thermodynamic limit: Is it true that if one takes a system of $N$ particles in a box of volume $V$, and lets $V$ tend to $\infty$ with the density $\varrho = N/V$ fixed, then the free energy per particle, $F(T)/N$ converges to a function $f(\varrho, T)$? Obviously this limit will not exist if certain long-range forces are included, such as the universal gravitational attraction among particles.

This problem was attacked in its various aspects by several people in the 1960's (e.g., Fisher, Griffiths, Onsager, Ruelle, Van Hove [5]) and a strategy was found to prove the existence of the thermodynamic limit rigorously for systems with short-range forces. This excluded the real world situation of positive and negative particles, for which quantum mechanics was essential. Freeman Dyson and Andrew Lenard [4] showed that the ground state energy of such a quantum system was bounded below by a constant times $N$. However, it is essential that the negative particles are fermions; a system of charged bosons will have a ground state energy proportional to $-N^{7/5}$ (see 188 and 288 reprinted in the Stability of Matter Selecta). While this implied the existence of $F(T)$ it did not imply the existence of a thermodynamic limit for $F(T)$. The canonical method fails for Coulomb systems! In fact, the real problem turns out to be 'explosion' rather than 'implosion' (as occurs with gravitational attraction). To prevent explosion it is essential that the system be close to charge neutral.

The solution to the problem, obtained with Joel Lebowitz, was to decompose the volume into balls of decreasing radii instead of into 8 cubes of half the size, as had been done for the short-range case. This meant figuring out how to pack balls efficiently. Paper 43 (**I.1**) is an announcement of their result but it has the essentials. Paper 58 (**I.2**) has all the details and included short-range forces in addition to Coulomb forces. The first 7 pages are pedagogical and for that reason are included here. The proof, in various levels of completeness, can also be found in 51, 65, 66, 92, 99, 105 and 125.

The paper 91, with Heide Narnhofer, (reprinted in the Stability of Matter Selecta), proves the existence of the thermodynamic limit for *'jellium'*, i.e., the one-component charged gas with a neutralizing background. Oddly, this case turns out to be more complicated than the two-component gas.

## II. Hard Sphere Virial Coefficients
20 (**II.1**), 27 (**II.2**)

The second virial coefficient, $B$, of a dilute gas gives the leading departure of the equation of state from the ideal gas formula. In quantum mechanics, the $B$ for a fermion gas is different from the $B$ for a boson gas; the difference, $B_{\text{exch}}$ , is called the exchange second virial coefficient. Ignoring irrelevant constants, we can write

$$B_{\text{exch}} = \int G(\vec{x}, -\vec{x}; t) d^3 x$$

where $G(\vec{x}, \vec{y}; t) = \exp[(\Delta - V)t]$ is the heat kernel with two-body potential $V$ and with the time $t$ equal to $1/kT$. If $V$ is bounded then $B_{\text{exch}}$ goes as $t^3/2$ for small time (large $T$). For a hard core $V$, $B_{\text{exch}}$ is much smaller and goes, for large $T$, as $e^{-Ca^2/t}$, where $a$ is the radius of the hard core and $C$ is some constant. In 20 (**II.1**), with Sigurd Larsen, John Kirkpatrick and Harry Jordan, the fact that $G(\vec{x}, \vec{y}; t) < G_0(\vec{x}, \vec{y}; t) = \exp\{-|\vec{x} - \vec{y}|^2/4t\}$ was used to show that $C > 1$. In 27 (**II.2**), however, the correct asymptotic value $C = (\pi/2)^2$ was obtained using the path space Wiener integral and the Feynman-Kac formula. It is expected, and it is true, that for small $t$ $G(\vec{x}, -\vec{x}; t) \sim e^{-g(|x|)/4t}$, where $g(|x|)$ is the geodesic distance from $\vec{x}$ to $-\vec{x}$ going outside the ball of radius $a$. This involves complicated upper and lower bounds for the path space measure. Since then, similar results were found in the mathematics literature for the heat kernel in the presence of obstacles, but 27(**II.2**) may have been one of the first of this genre.

## III. Zeros of Partition Functions
49 (**III.1**), 60 (**III.3**), 61 (**III.2**), 133 (**III.4**)

Tsung-Dao Lee and Chen Ning Yang [10] discovered that the partition function, $Z$, of the Ising model of ferromagnetism, when viewed as a polynomial function of the quantity $z = e^{h/kT}$, with $h$ being the externally applied magnetic field, has the property that its zeros all lie on the unit circle $|z| = 1$. This fact has implications for the phase transition of the Ising model as $T$ and $h$ are varied. It also has applications to the existence of the thermodynamic limit, to mass gaps, and to correlation inequalities and inequalities for critical exponents. Zero theorems were later found for other systems; most notably, an idea of Taro Asano [1] permitted the extension to certain quantum systems, such as the Heisenberg model.

Paper 61 (**III.2**) shows that the corresponding zeros for the monomer-dimer problem lie on the imaginary $z$-axis. 49 (**III.1**) is a brief announcement. This result, with Ole Heilmann, concerns a general graph $G$; a dimer covering of $G$ is a (partial) matching of pairs of vertices of $G$. Unmatched vertices are called monomers. The partition function $Z$ is the sum over all partial matchings with a weight $z^m$, where $m$ is the number of monomers. More complicated results were also obtained for other quantities besides $Z$ and for the zeros as a function of other weights, such as edge weights. One interesting point is that some of the results can be obtained from the theory of domains of holomorphy in several complex variables – a subject that is often seen in quantum field theory but seldom in statistical mechanics. These results are obtained in sufficient generality that they have proved useful in combinatorial theory, where matchings play an important role.

No analog of the Lee-Yang theorem holds for the Ising antiferromagnet. The zeros in $z$ do not lie on simple curves. Nevertheless, paper 60 (**III.3**), with David Ruelle, sheds light on the absence of zeros for the antiferromagnet under the assumption that the corresponding ferromagnet has no zeros in the arc $|\arg z| < \theta$ for some $\theta > 0$. Then the partition function of the antiferromagnet is free of zeros in the disc that passes through the points $e^{\pm i\theta}$ and which is orthogonal to the unit disc at these two points. Deep theorems in the theory of several complex variables, such as the double cone theorem and calculation of domains of holomorphy for the product of polydiscs, are used here and appear to be needed.

Paper 133 (**III.4**), with Alan Sokal, contains a general theory of zeros of polynomials in several variables in which Grace's theorem plays a key role. Applications include 2-component classical ferromagnets.

## IV. Reflection Positivity

101 (**IV.1**), 104 (**IV.3**), 109 (**IV.2**), 110 (**IV.4**), 113, 116 (**IV.6**)
124 (**IV.5**), 183 (**IV.7**), 184 (**IV.8**), 194, 222, 227, 252, 253

Konrad Osterwalder and Robert Schrader [13] introduced reflection positivity into quantum field theory to show the equivalence, in certain cases, of the relativistic Minkowski theory with its more amenable 'Euclidean' counterpart. Later, Jürg Fröhlich, Barry Simon and Thomas Spencer [6] used the concept to prove the existence of a phase transition for the classical Heisenberg spin model. This was the first rigorous proof of a phase transition in a system with continuous symmetry.

Reflection positivity leads to 'chessboard estimates', which allow for the possibility of a 'Peierls type' argument for quantum spin systems. One of the first examples of this is 101 (**IV.1**) and 109 (**IV.2**), with Jürg Fröhlich, which show that the anisotropic antiferromagnetic Heisenberg model can have long range order in 2 and 3 dimensions. It was motivated by work of James Glimm, Arthur Jaffe and Thomas Spencer [8] in $\varphi^4$ field theory.

Paper 104 (**IV.3**), with Freeman Dyson and Barry Simon, uses reflection positivity to derive 'infrared bounds' that establish long range order for the isotropic Heisenberg antiferromagnet in 3 or more dimensions at positive temperature provided the individual spin value is large enough. This is the quantum generalization of the work [8] on classical spins and is the first proof of long range order for a quantum system with continuous symmetry. The method of infrared bounds is, so far, the only method to deal with continuous symmetry breaking, i.e., long range order.

Papers 183 (**IV.7**) and 184 (**IV.8**), with Tom Kennedy and Sriram Shastry, improve the infrared bounds of 104 (**IV.3**). For the $XY$ model the ferro- and antiferromagnets are mathematically identical on a bipartite lattice. Like the Ising model, one can be turned into the other by flipping the spins on one sublattice. In these papers and in 104 (**IV.3**) it is proved that the $XY$ model has long range order in 2 dimensions in its ground state and in 3 dimensions at positive temperature for all $S \geq 1/2$. (There can be no long range order in 2D for $T > 0$ by the Hohenberg-Mermin-Wagner theorem [11].) The *isotropic* Heisenberg model (sometimes called the XXX model) presents a different story. For the XXX ferromagnet the ground state has perfect order (even in dimension 1) but it is still an open problem to prove long range order rigorously for $T > 0$ in three or more dimensions. The antiferromagnetic XXX model is very different from the ferromagnet. One expects long range order in the ground state in 2D and for $T > 0$ for D$\geq$3. This is proved (partly in 104 (**IV.3**) and partly in 183 (**IV.7**)) in all cases except $S = 1/2$ in 2 dimensions, which is still open.

Two papers, 110 (**IV.4**) and 124 (**IV.5**), with Robert Israel, Jürg Fröhlich and Barry Simon, give the general theory of reflection positivity, infrared bounds, and chessboard estimates. Many models, classical and quantum, are treated. Paper 113 gives a brief review.

Reflection positivity also solves the flux phase problem for the Hubbard model on the hypercubic lattice. The optimum flux configuration is $\pi$ through every square face. This is in 222, reproduced in the Inequalities Selecta.

Another application of reflection positivity in 227 is to the proof, with Bruno Nachtergaele, of the Peierls instability for the Hubbard model and for the 'spin-Peierls' problem. This is discussed more in the accompanying Selecta volume on condensed matter and exactly soluble models.

An interesting problem, that does not concern spins but rather orientable molecules, is the liquid crystal problem. The goal is to find a model that can be shown to have a phase in which the spatial orientation of the molecules has long range order, but their centers of mass show no long range order. Paper 116 (**IV.6**), with Ole Heilmann, uses reflection positivity and a chessboard estimate to produce a Peierls type argument for the existence of *rotational* long range order in a lattice model of liquid crystals. Primitive as it is, this seems to be the only model with the desired property of orientational order at low temperature. It is 'evident' that the model has no long range translational order, but, this has never been proved rigorously. It is an open problem.

A twist on reflection positivity is to be found in paper 194 on electrons with spin moving on a lattice (the Hubbard model). Instead of reflecting a state of the system through a plane in ordinary space one reflects in spin space, i.e., the state is reflected between the up spins and the down spins. With this *spin-space* reflection positivity one can deduce that the ground state has zero total spin. This spin reflection has been used by Guang-Shan Tian [16] to deduce rigorously other interesting properties of correlated electrons. A variant on the theme in 226 (with James Freericks) shows that certain (continuum) electron-phonon systems have $S = 0$ in the ground state.

Papers 252 and 253, with Peter Schupp, use reflection positivity to treat certain *frustrated* two-dimensional spin systems (pyrochlore models) and prove that among the ground states there is one with total $S = 0$.

## V. Classical Thermodynamics

131 (**V.1**), 247 (**V.2**), 250, 259 (**V.3**), 266, 275, 289

With the exception of 131 (**V.1**), all these papers are about an attempt, with Jakob Yngvason, to understand the second law of thermodynamics. This law is interpreted to be *the entropy principle*, which is the existence of a real-valued, essentially unique function $S$ (the entropy) on the states of all equilibrium systems so that when a change (violent or gentle) is made on the system the value of $S$ is greater for the final state than it is for the initial state (provided that nothing else has changed in the universe except for the movement of a weight in a gravitational field). Lieb and Yngvason had always been puzzled by the usual derivations of entropy which seemed somewhat circular and to rely on idealized concepts that are hard to define exactly, like empirical temperatures, heat and Carnot-cycles.

This theory, which took more than half a decade to develop, is independent of statistical mechanics. It shows that the existence of entropy can be derived from about 16 axioms. This might seem to be many, but it is not really so. All of them are commonly accepted, but not mentioned explicitly in the usual derivations. The approach owes much to Constantin Carathéodory [3], and even more to Robin Giles [7]. The major contribution in the Lieb-Yngvason work is the promotion of the 'comparison hypothesis' of Giles from an axiom to a theorem.

The paper with all the details and a historical discussion is 250. It is too long to reprint here. (275 and 289, which are also long, contain updated and slightly simplified versions of 250.) Instead, 247 (**V.2**), which is a summary for mathematicians, and 259 (**V.3**), which is a summary for physicists, are reproduced. The paper 247 (**V.2**) was awarded the Conant prize for exposition by the American Mathematical Society.

Paper 131 (**V.1**), with Michael Aizenman, is about statistical mechanics. It addresses the question of the derivation of the third 'law' of thermodynamics from statistical mechanical models. This 'law' of Nernst [12], that the entropy is zero at $T = 0$, is not always satisfied. Some systems, such as ice and carbon monoxide, have a positive (or 'residual') entropy at $T = 0$. (One of the topics, paper 29, in the companion Selecta on exactly soluble models is a computation of the residual entropy of two-dimensional 'ice'.) Is the residual entropy defined as the zero-temperature limit of the entropy per site, which always exists, equal to the possibly ill-defined notion of ground state degeneracy per site? The answer is that this is true if one takes the maximum degeneracy with respect to all possible boundary conditions.

## VI. Lattice Systems

28 (**VI.1**), 86 (**VI.2**), 98 (**VI.3**), 132 (**VI.4**)

Various attempts were made to establish non-perturbative properties of lattice crystals, all with Joel Lebowitz.

The paper 28 (**VI.1**), with Zoltan Rieder, who was Lebowitz's graduate student at the time, is an attempt to verify Fourier's law that a crystal with different temperatures applied to two ends will have a linear distribution of temperature as we pass from one end to the other. This fails in this model of a harmonic, one-dimensional crystal; the temperature is constant throughout the crystal and makes a jump at both ends. Fourier's linear law remains to be explained in a rigorous way.

Paper 86 (**VI.2**), with Herm Jan Brascamp and Joel Lebowitz, addresses the question whether a Gibbs state (for the algebra of the difference variables) for a classical anharmonic crystal lattice exists for all dimensions. This is shown to be true under suitable convexity assumptions on the anharmonicity.

Paper 98 (**VI.3**), with Oscar Lanford, has a proof of a well defined time evolution for an infinite, classical anharmonic crystal under a wide variety of initial conditions.

Paper 132 (**VI.4**), with Jean Bricmont, Jean-Raymond Fontaine, and Thomas Spencer, considers a classical, ferromagnetic spin system with continuous symmetry, namely the $XY$ model in which the individual spins rotate only in the $XY$ plane. Can the Gibbs state at low temperature be derived by making an expansion about the fully magnetized state in which all the spins point in one direction? In the language of quantum mechanics this would be a spin-wave expansion. Because of the continuous symmetry this expansion is not at all obvious, but the answer is shown to be 'yes'. The proof fails for $XYZ$ spins (spins that can rotate on the unit sphere $S^2$) and this is taken to mean, by some, that such an expansion does not exist and this, in turn, is attributed to the non-abelian nature of the rotation group on the sphere.

## VII. Miscellaneous

63 (**VII.1**), 74 (**VII.2**), 114, 126 (**VII.3**), 203, 213 (**VII.4**), 219, 251

It is easy to prove, for infinite lattice spin-like systems, that the time evolution of a local operator $A$, given formally by $\tau_t(A) = e^{iHt}Ae^{-iHt}$ exists. If $x$ and $y$ are widely separated points, and $A(x)$ and $B(y)$ are localized near $x$ and $y$ respectively, we can look at the commutator $C(x, y; t) = [\tau_t(A(x)), B(y)]$ and ask whether the norm of this operator is small when $|x - y| > Vt$ for some universal number $V$, which we will identify as the velocity of sound. It is proved in 63 (**VII.1**) that there is a finite $V$ (depending only on the details of the system but not on $A$ and $B$) so that $\|C(x, y; t)\| < \exp[-|x - y|/Vt]$.

Paper 74 (**VII.2**), introduces Bloch coherent states to give upper and lower bounds to the free energy of quantum spin systems and, thereby, prove that as the spin $S$ goes to $\infty$ the classical spin limit is obtained. Berezin [2] had inequalities like these of a general nature, but did not apply them specifically. They are now known as Berezin-Lieb inequalities and, since then, show up in a variety of contexts, not just spin systems. Paper 203, with Jan Philip Solovej, introduces the notion of coherent *operators* as a generalization of coherent states. Some of the rigorous results obtained using coherent states are summarized in 219.

Wehrl introduced a *classical* entropy of quantum systems that remedies the defects of the usual classical Boltzmann-Gibbs prescription $S = -\int \varrho \log \varrho$ with

$\varrho = \exp[(-T + V)/kT]/\int \exp[(-T + V)/kT]$. For one thing, Wehrl's entropy is positive, and 114 solves the problem of its minimum value for Schrödinger-Glauber coherent states. Paper 114 also poses the same minimization problem for Bloch coherent spin states, which is still open. Paper 114 and other papers on coherent states are reproduced in the Inequalities Selecta.

Barry Simon [15] proved a useful correlation inequality for the Ising model and 126 (**VII.3**) carries the idea a bit further. A conjecture is made in 126 (**VII.3**) that the result also works for the $XY$ rotor, and this was proved by Vincent Rivasseau [14]. All three papers appear in the same issue of Commun. Math. Phys.

Paper 213 (**VII.4**), with Michael Loss, discusses what happens to the partition function, $Z$, for free fermions on a bipartite lattice with a 'magnetic field' $\theta$. That is, the lattice Laplacian $-\Delta f(x) = f(x \pm 1) - cf(x)$ is replaced by $e^{\pm i\theta} f(x \pm 1) - cf(x)$. Which value of $\theta$ will make $Z$ as large as possible? This problem (known as the 'flux phase problem'), and related questions are solved in some special cases. (The solution of the flux phase problem for the regular cubic lattice was given in 222.) The paper has several theorems about the closely related problem of enumerating dimer configurations on a lattice, including a direct proof of Kasteleyn's theorem [9] that does not have to make exceptions for special cases.

## Entropy Inequalities

47, 67, 68, 69, 80, 249

These papers are directed to the proof of *strong subadditivity of entropy* (SSA) for quantum mechanical density matrices, which was accomplished in 68 and 69 with Mary Beth Ruskai. The inequality is quite general, and even plays an important role in quantum information theory, Thus, it appears in the Selecta on *Inequalities*, but since its original motivation was in statistical mechanics it is mentioned here.

Let $\varrho_{1,2,3}$ be a density matrix on the product of three Hilbert spaces. Let $\varrho_{1,2}$ be the density matrix on the the first two obtained by taking the partial trace of $\varrho_{1,2,3}$ over the third space, and so forth. Let $S = -\mathrm{Trace}\varrho \log \varrho$ be the usual entropy of a density matrix. Then SSA is

$$S_{1,2,3} + S_2 \le S_{1,2} + S_{2,3}.$$

Paper 47, with Huzihiro Araki, has the partial result of subbaditivity: $S_{1,2} \le S_1 + S_2$ and also $S_{1,2} \ge |S_1 - S_2|$. Paper 67 contains inequalities needed in the proof of SSA, particularly the fact that the function $A \to \mathrm{Trace} \exp[K + \log A]$ is convex for $K$ hermitian and $A$ positive definite. Paper 80 is a review of subadditivity inequalities while 249, with Eric Carlen, contains another proof of SSA.

*Bruno Nachtergaele, Jan Philip Solovej, Jakob Yngvason*

## References


[1] T. Asano, *Theorems on the partition functions of the Heisenberg ferromagnets*, J. Phys. Soc. Japan **29** (1970), 350–359.

[2] F.A. Berezin, Izv. Akad. Nauk SSSR ser. Mat. **36**(5) (1972), 1134–1167. English translation: *Covariant and contravariant symbols of operators*, Math. USSR Izv. **6**(5) (1972), 1117–1151.

[3] C. Carathéodory, *Untersuchung über die Grundlagen der Thermodynamik*, Math. Annalen **67** (1909), 355–386; *Über die Bestimmung der Entropie und der absoluten Temperatur mit Hilfe von reversiblen Prozessen*, Sitzungsber. Preuss. Akad. Wiss., Phys. Math. Kl. (1925), 39–47.

[4] F.J. Dyson, A. Lenard, *Stability of Matter I and II*, J. Math. Phys, **8** (1967), 423–434; ibid **9** (1968), 698–711.

[5] M.E. Fisher, *The free energy of a macroscopic system*, Arch. Rational Mech. Anal. **17** (1964), 377–410; R.B. Griffiths, *Microcanonical Ensemble in Quantum Statistical Mechanics*, J. Math. Phys. **6** (1965), 1447–1461; M.E. Fisher and D. Ruelle, *The Stability of Many-particle Systems*, J. Math. Phys. **7** (1966), 260–270; L. Onsager, Electrostatic Interaction of Molecules, *Jour. Phys. Chem.* **43** (1939), 189–196; L. Van Hove, *Quelques proprietés generales de l'intégrale de configuration d'un système de particules avec interaction*, Physika **15** (1949), 951–961.

[6] J. Fröhlich, B. Simon, T. Spencer, *Phase Transitions and Continuous Symmetry Breaking*, Phys. Rev. Lett. **36** (1976), 804–806; *Infrared bounds, phase transitions and continuous symmetry breaking*, Commun. Math. Phys. **50** (1976), 79–85.

[7] R. Giles, *Mathematical Foundations of Thermodynamics*, Pergamon Press, Oxford, 1964.

[8] J. Glimm, A. Jaffe, and T. Spencer *Phase transitions for $\Phi_2^4$ quantum fields*, Commun. Math. Phys. **45** (1975), 203–216.

[9] P.W. Kasteleyn, *The statistics of dimers on a lattice I. The number of dimer arrangements on a quadratic lattice*, Physica **27** (1961), 1209–1225; *Graph theory and crystal physics*, in: *Graph Theory and Theoretical Physics*, F. Harary (ed.), Academic Press, 1967, pp. 44–110.

[10] T.D. Lee, C.N. Yang, *Statistical Theory of Equations of State and Phase Transitions. II. Lattice Gas and Ising Model*, Phys. Rev. **87** (1952), 410–419.

[11] M.D. Mermin, H. Wagner, *Absence of Ferromagnetism or Antiferromagnetism in One- or Two-Dimensional Isotropic Heisenberg Models*, Phys. Rev. Lett. **17** (1966), 1133–1136; P. Hohenberg, *Existence of Long-Range Order in One and Two Dimensions*, Phys. Rev. **158** (1967), 383–386.

[12] W. Nernst, *Über die Berechnung chemischer Gleichgewichte aus thermischen Messungen*, Nachr. Kgl. Ges. Wiss. Gött., 1906, No. 1, pp. 1–40.

[13] K. Osterwalder, R. Schrader, *Axioms for Euclidean Green's Functions*, Commun. Math. Phys. **31** (1973), 83–112; **42** (1975), 281–305.

[14] V. Rivasseau, *Lieb's Correlation Inequality for Plane Rotors*, Commun. Math. Phys. **77** (1980), 145–147.

[15] B. Simon, *Correlation inequalities and the decay of correlations in ferromagnets*, Commun. Math. Phys. **77** (1980), 111–126.

[16] G.-S. Tian, *Lieb's Spin-Reflection-Positivity Method and its Applications to Strongly Correlated Electron Systems*, J. Stat. Phys. **116** (2004), 629–680.

Part I

# Thermodynamic Limit for Coulomb Systems

VOLUME 22, NUMBER 13 PHYSICAL REVIEW LETTERS 31 MARCH 1969

# EXISTENCE OF THERMODYNAMICS FOR REAL MATTER WITH COULOMB FORCES

J. L. Lebowitz*
Belfer Graduate School of Science, Yeshiva University, New York, New York 10033

and

Elliott H. Lieb†
Department of Mathematics, Massachusetts Institute of Technology, Cambridge, Massachusetts 02139
(Received 3 February 1969)

It is shown that a system made up of nuclei and electrons, the constituents of ordinary matter, has a well-defined statistical-mechanically computed free energy per unit volume in the thermodynamic (bulk) limit. This proves that statistical mechanics, as developed by Gibbs, really leads to a proper thermodynamics for macroscopic systems.

In this note we wish to report the solution to a classic problem lying at the foundations of statistical mechanics.

Ever since the daring hypothesis of Gibbs and others that the equilibrium properties of matter could be completely described in terms of a phase-space average, or partition function, $Z = \mathrm{Tr}e^{-\beta H}$, it was realized that there were grave difficulties in justifying this assumption in terms of basic microscopic dynamics and that such delicate matters as the ergodic conjecture stood in the way. These questions have still not been satisfactorily resolved, but more recently still another problem about $Z$ began to receive attention: Assuming the validity of the partition function, is it true that the resulting properties of matter will be extensive and otherwise the same as those postulated in the science of thermodynamics? In particular, does the thermodynamic, or bulk, limit exist for the free energy derived from the partition function, and if so, does it have the appropriate convexity, i.e., stability properties?

To be precise, if $N_j$ are an unbounded, increasing sequence of particle numbers, and $\Omega_j$ a sequence of reasonable domains (or boxes) of volume $V_j$ such that $N_j/V_j \to \text{constant} = \rho$, does the free energy per unit volume

$$f_j = -kT(V_j)^{-1} \ln Z(\beta, N_j, \Omega_j) \tag{1}$$

approach a limit [called $f(\beta, \rho)$] as $j \to \infty$, and is this limit independent of the particular sequence and shape of the domains? If so, is $f$ convex in the density $\rho$ and concave in the temperature $\beta^{-1}$? Convexity is the same as thermodynamic stability (non-negative compressibility and specific heat).

Various authors have evolved a technique for proving the above,[1,2] but always with one severe drawback. It had to be assumed that the interparticle potentials were short range (in a manner to be described precisely later), thereby excluding the Coulomb potential which is the true potential relevant for real matter. In this note we will indicate the lines along which a proof for Coulomb forces can be and has been constructed. The proof itself, which is quite long, will be given elsewhere.[3] We will also list here some additional results for charged systems that go beyond the existence and convexity of the limiting free energy.

To begin with, a sine qua non for thermodynamics is the stability criterion on the $N$-body Hamiltonian $H = E_K + V$. It is that there exists a constant $B \geq 0$ such that for all $N$,

$$V(r_1, \cdots, r_N) > -BN \quad \text{(classical mechanics)}, \tag{2}$$

$$E_0 > -BN \quad \text{(quantum mechanics)}, \tag{3}$$

where $E_0$ is the ground-state energy in infinite space. (Classical stability implies quantum-mechanical stability, but not conversely.) Heuristically, stability insures against collapse. From the mathematical point of view, it provides a lower bound to $f_j$ in (1). We wish to emphasize that stability of the Hamiltonian ($H$ stability), while necessary, is insufficient for assuring the existence of thermodynamics. For example, it is trivial to prove $H$ stability for charged particles all of one sign, and it is equally obvious that the thermodynamic limit does not exist in this case.

It is not too difficult to prove classical and thus also quantum-mechanical $H$ stability for a wide variety of short-range potentials or for charged particles having a hard core.[2,4] But real charged particles require quantum mechanics and the recent proof of $H$ stability by Dyson and Lenard[5] is as difficult as it is elegant. They show that stability will hold for any set of charges and masses provided that the negative particles and/or the positive ones are fermions.

The second requirement in the canonical proofs[1] is that the potential be tempered, which is to say that there exist a fixed $r_0$ and constants $C \geq 0$ and $\epsilon > 0$ such that if two groups of $N_a$ and $N_b$ particles are separated by a distance $r > r_0$, their interparticle energy is bounded by

$$V(N_a \oplus N_b) - V(N_a) - V(N_b) \leq C r^{-(3+\epsilon)} N_a N_b. \tag{4}$$

Tempering is roughly the antithesis of stability because the requirements that the forces are not too repulsive at infinity insures against "explosion." Coulomb forces are obviously not tempered and for this reason the canonical proofs have to be altered. Our proof, however, is valid for a mixture of Coulomb and tempered potentials and this will always be understood in the theorems below. It is not altogether useless to include tempered potentials along with the true Coulomb potentials because one might wish to consider model systems in which ionized molecules are the elementary particles.

Prior to explaining how to overcome the lack of tempering we list the main theorems we are able to prove. These are true classically as well as quantum mechanically. But first three definitions are needed:

(D1) We consider $s$ species of particles with charges $e_i$, particle numbers $N^{(i)}$, and densities $\rho^{(i)}$. In the following $N$ and $\rho$ are a shorthand notation for $s$-fold multiplets of numbers. The conditions for $H$ stability (see above) are assumed to hold.

(D2) A neutral system is one for which $\sum_1^s N^{(i)} \times e_i = 0$, alternatively $\sum_1^s \rho^{(i)} e_i = 0$.

(D3) The ordinary $s$-species grand canonical partition function is

$$\sum_{N^{(s)}=0}^{\infty} \cdots \sum_{N^{(1)}=0}^{\infty} \prod_1^s z_i^{N^{(i)}} Z(N, \Omega). \tag{5}$$

The neutral grand canonical partition function is the same as (5) except that only neutral systems enter the sum.

The theorems are the following:

(T1) The canonical, thermodynamic limiting free energy per unit volume $f(\beta, \rho)$ exists for a neutral system and is independent of the shape of the domain for reasonable domains. Furthermore, $f(\beta, \rho^{(1)}, \rho^{(2)}, \cdots)$ is concave in $\beta^{-1}$ and jointly convex in the $s$ variables $(\rho^{(1)}, \cdots, \rho^{(s)})$.

(T2) The thermodynamic limiting microcanonical[6] entropy per unit volume exists for a neutral system and is a concave function of the energy per unit volume. It is also independent of domain shape for reasonable shapes and it is equal to the entropy computed from the canonical free energy.

(T3) The thermodynamic limiting free energy per unit volume exists for both the ordinary and the neutral grand canonical ensembles and are independent of domain shape for reasonable domains. Moreover, they are equal to each other

VOLUME 22, NUMBER 13 PHYSICAL REVIEW LETTERS 31 MARCH 1969

and to the neutral canonical free energy per unit volume.

Theorem 3 states that systems which are not charge neutral make a vanishingly small contribution to the grand canonical free energy. While this is quite reasonable physically, it does raise an interesting point about nonuniform convergence because the ordinary and neutral partition functions are definitely not equal if we switch off the charge before passing to the thermodynamic limit, whereas they are equal if the limits are taken in the reverse order.

An interesting question is how much can charge neutrality be nonconserved before the free energy per unit volume deviates appreciably from its neutral value? The answer is in theorem 4.

(T4) Consider the canonical free energy with a surplus (i.e., imbalance) of charge $Q$ and take the thermodynamic limit in either of three ways: (a) $QV^{-2/3}\to 0$; (b) $QV^{-2/3}\to\infty$; (c) $QV^{-2/3}\to$ const. In case (a) the limit is the same as for the neutral system while in case (b) the limit does not exist, i.e., $f\to\infty$. In case (c) the free energy approaches a limit equal to the neutral-system free energy plus the energy of a surface layer of charge $Q$ as given by elementary electrostatics.

We turn now to a sketch of the method of proof and will restrict ourselves here to the neutral canonical ensemble. As usual, one first proves the existence of the limit for a standard sequence of domains. The limit for an arbitrary domain is then easily arrived at by packing that domain with the standard ones. The basic inequality that is needed is that if a domain $\Omega$ containing $N$ particles is partitioned into $D$ domains $\Omega_1, \Omega_2, \cdots, \Omega_D$ containing $N_1, N_2, \cdots, N_D$ particles, respectively, and if the interdomain interaction be neglected, then

$$Z(N,\Omega) \geq \prod_1^D Z(N_i, \Omega_i). \tag{6}$$

If $\Omega$ is partitioned into subdomains, as above, plus "corridors" of thickness $>r_0$ which are devoid of particles, one can use (4) to obtain a useful bound on the tempered part of the omitted interdomain interaction energy. We will refer to these energies as surface terms.

The normal choice[1] for the standard domains are cubes $C_j$ containing $N_j$ particles, with $C_{j+1}$ being composed of eight copies of $C_j$ together with corridors, and with $N_{j+1}=8N_j$. Neglecting surface terms one would have from (6) and (1)

$$f_{j+1} \leq f_j. \tag{7}$$

Since $f_j$ is bounded below by $H$ stability, (7) implies the existence of a limit. To justify neglect of the surface terms one makes the corridors increase in thickness with increasing $j$; although $V_j^C$, the corridor volume, approaches $\infty$ one makes $V_j^C/V_j\to 0$ in order that the limiting density not vanish. The positive $\epsilon$ of (4) allows one to accomplish these desiderata.

Obviously, such a strategy will fail with Coulomb forces, but fortunately there is another way to bound the interdomain energy. The essential point is that it is not necessary to bound this energy for all possible states of the systems in the subdomains; it is only necessary to bound the "average" interaction between domains, which is much easier. This is expressed mathematically by using the Peierls-Bogoliubov inequality[7] to show that

$$Z(N,\Omega) \geq e^{-\beta U} \prod_1^D Z(N_i, \Omega_i), \tag{8}$$

where $U$ is the average interdomain energy in an ensemble where each domain is independent. $U$ consists of a Coulomb part, $U_C$, and a tempered part, $U_t$, which can be readily bounded.[1]

We now make the observation, which is one of the crucial steps in our proof, that independently of charge symmetry $U_C$ will vanish if the subdomains are spheres and are overall neutral. The rotation invariance of the Hamiltonian will produce a spherically symmetric charge distribution in each sphere and, as Newton[8] observed, two such spheres would then interact as though their total charges (which are zero) were concentrated at their centers.

With this in mind we choose spheres for our standard domains. Sphere $S_j$ will have radius $R_j = p^j$ with $p$ an integer. The price we pay for using spheres instead of cubes is that a given one, $S_k$, cannot be packed arbitrarily full with spheres $S_{k-1}$ only. We prove, however, that it can be packed arbitrarily closely (as $k\to\infty$) if we use all the previous spheres $S_{k-1}, S_{k-2}, \cdots S_0$. Indeed for the sequence of integers $n_1, n_2, \cdots, n_j = (p-1)^{j-1}p^{2j}$ we can show that we can simultaneously pack $n_j$ spheres $S_{k-j}$ into $S_k$ for $1\leq j\leq k$. The fractional volume of $S_k$ occupied by the $S_{k-j}$ spheres is $\varphi_j = p^{-3j}n_j$, and from (8) we then have

$$f_k \leq \varphi_1 f_{k-1} + \varphi_2 f_{k-2} + \cdots + \varphi_k f_0, \tag{9}$$

and

$$\sum_1^\infty \varphi_j = 1. \tag{10}$$

VOLUME 22, NUMBER 13 PHYSICAL REVIEW LETTERS 31 MARCH 1969

[Note that the inequality (6) is correct as it stands for pure Coulomb forces because $U_C$ in (8) is identically zero. If short-range potentials are included there will also be surface terms, as in the cube construction, but these present only a technical complication that can be handled in the same manner as before.[1]] While Eq. (9) is more complicated than (7), it is readily proven explicitly that $f_k$ approaches a limit as $k \to \infty$. [Indeed, it follows from the theory of the renewal equation[9] that (9) will have a limit if $\sum_1^\infty j\psi_j < \infty$.]

The possibility of packing spheres this way is provided by the following geometrical theorem which plays the key role in our analysis. We state it without proof, but we do so in $d$ dimensions generally and use the following notation: $\sigma_d$ = volume of a unit $d$-dimensional sphere $= \frac{4}{3}\pi$ in three dimensions and $\alpha_d = (2^d-1)2d^{\frac{1}{2}}$.

(T5) Let $p \geq \alpha_d + 2^d\sigma_d^{-1}$ be a positive integer. For all positive integers $j$, define radii $r_j = p^{-j}$ and integers $n_j = (p-1)^{j-1}p^{j(d-1)}$. Then it is possible to place simultaneously $\bigcup_j (n_j$ spheres of radius $r_j)$ into a unit $d$-dimensional sphere so that none of them overlap.

The minimum value of $p$ required by the theorem in three dimensions is 27.

Many of the ideas presented here had their genesis at the Symposium on Exact Results in Statistical Mechanics at Irvine, California, in 1968, and we should like to thank our colleagues for their encouragement and stimulation: M. E. Fisher, R. Griffiths, O. Lanford, M. Mayer, D. Ruelle, and especially A. Lenard.

---

*Work supported by Air Force Office of Scientific Research, U. S. Air Force under Grant No. AFOSR 68-1416.

†Work supported by National Science Foundation Grant No. GP-9414.

[1]These developments are clearly expounded in M. E. Fisher, Arch. Ratl. Mech. Anal. 17, 377 (1964); D. Ruelle, Statistical Mechanics (W. A. Benjamin, Inc., New York, 1969). For a synopsis, see also J. L. Lebowitz, Ann. Rev. Phys. Chem. 19, 389 (1968).

[2]R. B. Griffiths, Phys. Rev. 176, 655 (1968), and footnote 6a in A. Lenard and F. J. Dyson [J. Math. Phys. 9, 698 (1968)]; O. Penrose, in Statistical Mechanics, Foundations and Applications, edited by T. Bak (W. A. Benjamin, Inc., New York, 1967), p. 98.

[3]E. H. Lieb and J. L. Lebowitz, "The Constitution of Matter," to be published.

[4]L. Onsager, J. Phys. Chem. 43, 189 (1939); M. E. Fisher and D. Ruelle, J. Math. Phys. 7, 260 (1966).

[5]F. J. Dyson and A. Lenard, J. Math. Phys. 8, 423 (1967); A. Lenard and F. J. Dyson, J. Math. Phys. 9, 698 (1968); F. J. Dyson, J. Math. Phys. 8, 1538 (1967).

[6]R. B. Griffiths, J. Math. Phys. 6, 1447 (1965).

[7]K. Symanzik, J. Math. Phys. 6, 1155 (1965).

[8]I. Newton, in Mathematical Principles, translated by A. Motte, revised by F. Cajori (University of California Press, Berkeley, Calif., 1934), Book 1, p. 193, propositions 71, 76.

[9]W. Feller, An Introduction to Probability Theory and Its Applications (J. Wiley & Sons, New York, 1957), 2nd ed. Vol. 1, p. 290.

Reprinted from ADVANCES IN MATHEMATICS
All Rights Reserved by Academic Press, New York and London

Vol. 9, No. 3, December 1972
*Printed in Belgium*

# The Constitution of Matter: Existence of Thermodynamics for Systems Composed of Electrons and Nuclei

ELLIOTT H. LIEB*

*Department of Mathematics, Massachusetts Institute of Technology, Cambridge, Massachusetts 02139*

AND

JOEL L. LEBOWITZ†

*Belfer Graduate School of Science, Yeshiva University, New York, New York 10033*

We establish the existence of the infinite volume (thermodynamic) limit for the free energy density of a system of charged particles, e.g., electrons and nuclei. These particles, which are the elementary constituents of macroscopic matter, interact via Coulomb forces. The long range nature of this interaction necessitates the use of new methods for proving the existence of the limit. It is shown that the limit function has all the convexity (stability) properties required by macroscopic thermodynamics. For electrically neutral systems, the limit functions is domain-shape independent, while for systems having a net charge the thermodynamic free energy density is shape dependent in conformity with the well-known formula of classical electrostatics. The analysis is based on the statistical mechanics ensemble formalism of Gibbs and may be either classical or quantum mechanical. The equivalence of the microcanonical, canonical and grand canonical ensembles is demonstrated.

## Table of Contents



* Work supported by National Science Foundation Grant GP-31674X.
† Work supported by AFOSR Contract #F44620-71-C-0013.

## I. Introduction

In this paper we present a proof of the existence of the thermodynamic limit for Coulomb systems. A statement of the main results appeared in Lebowitz and Lieb (1969) and an outline of the proof is to be found in Lieb and Lebowitz (1972).

We start with a brief overview of the paper and defer precise definitions to later sections, mainly Section II.

### A. *Perspective*

Statistical Mechanics as developed by Gibbs and others rests on the hypothesis that the equilibrium properties of matter can be completely described in terms of a phase-space average, or canonical partition function $Z = \mathrm{Tr}\{\exp(-\beta H)\}$, with $H$ the Hamiltonian and $\beta$ the reciprocal temperature. It was realized early that there were grave difficulties in justifying this assumption in terms of basic microscopic dynamics. These questions, which involve the time evolution of macroscopic systems, have still not been satisfactorily resolved, but the great success of equilibrium statistical mechanics in offering qualitative and quantitative explanations of such varied phenomena as superconductivity, specific heats of crystals, chemical equilibrium constants, etc., have left little doubt about the essential correctness of the partition function method. However, since $Z$ cannot be evaluated explicitly for any reasonable physical Hamiltonian $H$, comparison with experiment always involves some uncontrolled approximations. Hence, the following problem deserves attention: Is it true that the thermal properties of matter obtained from an exact evaluation of the partition function would be extensive and otherwise have the same form as those postulated in the science of thermodynamics? In particular, does the thermodynamic, or bulk limit exist for the Helmholtz free energy/unit volume derived from the canonical partition function, and, if so, does it have the appropriate convexity, i.e., stability properties?

To be more precise: Let $\{\Lambda_j\}$ be a sequence of bounded open sets (domains) in $\mathbb{R}^d$ with $\Lambda_j$ becoming infinitely large as $j \to \infty$ in some "reasonable way" which will be specified later. [We shall be concerned primarily with $d = 3$ but many of our results are valid for all $d$. For some results on classical Coulomb systems in $d = 1$ and 2, cf. Lenard (1961), Hiis Hauge and Hemmer (1972).] The volume of $\Lambda_j$ will be denoted by $V(\Lambda_j)$ and $V(\Lambda_j) \to \infty$ as $j \to \infty$. Consider now a sequence of systems consisting of $S$ species of particles contained in the domains $\{\Lambda_j\}$. Let

$\mathbf{N}_j = (N_j^1, \ldots, N_j^S)$ be the particle number vector specifying the system in $\Lambda_j$, i.e., $N_j^i$ is a nonnegative integer and is the number of particles of species $i$ contained in $\Lambda_j$. The canonical partition function of the $j$-th system at reciprocal temperature $\beta$ is then given by

$$Z(\beta, \mathbf{N}_j\,;\Lambda_j) = \operatorname{Tr} e^{-\beta H} = \sum_{\alpha=0}^{\infty} \exp[-\beta E_\alpha(\mathbf{N}_j\,;\Lambda_j)]$$
$$\equiv \exp[V(\Lambda_j)\, g(\beta, \boldsymbol{\rho}_j\,;\Lambda_j)], \tag{1.1}$$

where $E_\alpha(\mathbf{N}_j; \Lambda_j)$ are the energy levels of the $j$-th system, $\boldsymbol{\rho}_j \equiv \mathbf{N}_j/V(\Lambda_j)$ is the particle density vector, and $-\beta^{-1}g(\beta, \boldsymbol{\rho}_j; \Lambda_j)$ is the Helmholtz free energy per unit volume of the $j$-th system. According to statistical mechanics, knowledge of $g$ determines all the equilibrium properties of this system. The question to be studied is the following: Given a sequence of particle density vectors $\{\boldsymbol{\rho}_j\}$ which approach a limit $\boldsymbol{\rho}$ as $j \to \infty$, does $g(\beta, \boldsymbol{\rho}_j; \Lambda_j)$ approach a limit $g(\beta, \boldsymbol{\rho})$ as $j \to \infty$, and is this limit independent in some sense of the particular sequence of domains $\{\Lambda_j\}$ and density vectors $\{\boldsymbol{\rho}_j\}$ used in going to the limit? If so, does the limiting free energy density have, as a function of $\boldsymbol{\rho}$ and $\beta$, the convexity properties required for thermodynamic stability, i.e., is $g(\beta, \boldsymbol{\rho})$ convex in $\beta$ and concave in $\boldsymbol{\rho}$? With regard to $\beta$, we see from (1.1) that each $g(\beta, \boldsymbol{\rho}_j; \Lambda_j)$ is convex in $\beta$. Therefore, if the limit $g(\beta, \boldsymbol{\rho})$ exists, it will automatically be convex in $\beta$. Consequently, we can set $\beta = 1$ and omit mention of $\beta$, and shall do so henceforth.

In addition to proving the above, one also wants to show that the "same" thermodynamic results are obtained from the microcanonical and grand canonical partition functions (to be defined later). This program is referred to as proving the existence of the thermodynamic limit.

Various authors have evolved a technique for establishing the existence of this limit for systems whose Hamiltonians satisfy certain conditions. [The different names associated with this development are: Van Hove, Lee and Yang, van Kampen, Wils, Mazur and van der Linden, Griffiths, Dobrushin, and in particular, Ruelle and Fisher. The reader is referred to Fisher (1964) and Ruelle (1969) for an exposition and references. For a synopsis and more references see also Lebowitz (1968) and Griffiths (1971).] These Hamiltonians are the sum of kinetic energies of the individual particles plus an interaction potential energy among the particles, the latter depending only on the particle coordinates. There are two basic conditions required by the above authors on the interaction among the particles constituting the microscopic units of macroscopic matter.

The first of these requirements is that the interaction be short-range or *tempered*. The requirement of tempering unfortunately excludes the Coulomb potential which is the true potential relevant for real matter. That thermodynamics is applicable to systems with Coulomb forces is a fact of common experience, but the proof that it does so is a much more subtle matter than for short-range forces. It is screening, brought about by the long-range nature of the Coulomb force itself, that causes the Coulomb force to behave as if it were short-range. This has the consequence, as we shall prove in this paper, that for a sequence of systems, each of which is *overall neutral*, the approach of $g(\boldsymbol{\rho}_j; \Lambda_j)$ to its limit $g(\boldsymbol{\rho})$ and the properties of $g(\boldsymbol{\rho})$ are the same as those obtained for systems with tempered interactions (except that the $\rho^i$, $i = 1, \ldots, S$ are constrained by the neutrality requirement). In particular, $g(\boldsymbol{\rho})$ is the same for different "shapes" of the domains $\{\Lambda_j\}$. This shape independence disappears when the constraint of charge neutrality is lifted and systems with a "nonnegligible" amount of net charge are considered. The true long-range nature of the Coulomb force now becomes manifest, leading in some cases to a shape dependent limit of the free energy density and in other cases (when the excess charge is too large) to an infinite limit.

The second basic requirement, which is essential also for Coulomb systems, is a stability criterion on the $N$-body Hamiltonian $H$. It is that there exists a constant $B < \infty$ such that for a system of $N$ particles, $H > -BN$. We shall refer to this condition as *H-stability*. Heuristically, $H$-stability insures against collapse of the system. Mathematically, it provides an upper bound to the sequence $\{g(\boldsymbol{\rho}_j; \Lambda_j)\}$ and this bound plays an essential role in the proof. It should be emphasized, however, that H-stability does not in itself imply a thermodynamic limit. As an example, it is trivial to prove $H$-stability for charged particles all of one sign, and it is equally obvious that the thermodynamic limit does not exist in that case. Since the kinetic energy is positive, it is clearly sufficient for $H$-stability that the interaction energy is $H$-stable by itself. For classical systems this is also a necessary condition since the kinetic energy can be arbitrarily small. While it is not too difficult to prove classical $H$-stability for a wide variety of interaction potentials [cf. Ruelle (1969)], it is clear that classical $H$-stability will not hold for a system composed of positive and negative point charges. Even for a single pair, the Coulomb energy in three dimensions, $-1/r$, is unbounded below. Interestingly, though, if the charged particles have *hard cores*, classical $H$-stability is satisfied, as shown by Onsager (1939).

Onsager's results were generalized by Fisher and Ruelle (1966). Their work, however, still left open the question of whether a quantum system of point Coulomb charges, which may be taken as the building blocks of real matter, is $H$-stable. Now when dealing with a quantum system of charges, the nonexistence of a lower bound to $-1/r$ might not appear as serious as in the classical case since we expect that the Heisenberg uncertainty principle, which prevents particles from having their positions "close to each other" without also having a large kinetic energy, will insure the existence of a lower bound to the Hamiltonian. This is indeed the case for any finite system, ($-13.5$ eV for a system composed of one electron and one proton), and generally $H > -\infty$ for any $N$ (cf. Simon, Appendix B to this paper). We need, however, a lower bound proportional to $N$ and this, it turns out, the uncertainty principle alone cannot provide. The required result was proved by Dyson and Lenard (1967, 1968), who showed that $H$-stability holds for a system of point charges in three dimensions when *all* species with negative and/or positive charges are fermions. This is happily the case in nature where the electrons are fermions. (When neither of the charges are fermions, Dyson (1967) found an *upper* bound to the ground state energy that is proportional to $-N^{7/5}$; hence such a system will not be thermodynamically stable). The Dyson–Lenard theorem is as fundamental as it is difficult.

We note here that Griffiths (1969), found a way to extend the "canonical" proof to electrically neutral systems with Coulomb forces under the restrictive assumption of complete charge symmetry, i.e., that positive and negative particles have the same mass, spin, etc., but this is clearly insufficient for nuclei and electrons. Also, in a recent paper, Penrose and Smith (1972) established the existence of the thermodynamic limit for classical systems with electromagnetic interactions (including external fields) when the systems are confined in superconducting like containers which modify the electromagnetic interaction among the constituent particles.

This paper deals with the general nonrelativistic, classical or quantum mechanical Coulomb system without restriction. We do not consider any relativistic effects, such as spin–spin and spin–orbit couplings; the simple spin–spin dipolar coupling containing an $r^{-3}$ interaction is not $H$-stable even for two particles. However, if the particles have a hard core, the dipolar interaction is $H$-stable and, although it is not tempered, it can be satisfactorily treated [Griffiths (1969); cf. also Remark (ii) after Theorem 2.6 in Section II].

Needless to say, we also do not deal with the strong (nuclear) and weak interactions. As pointed out by Dyson (1967), the magnitude of the nuclear forces is so large that they would give completely different binding energies for molecules and for crystals if they played any role in the thermal properties of ordinary matter. We are also neglecting gravitational forces which certainly are important for large aggregates of matter and thus might be thought important in the thermodynamic limit. To quote Onsager (1967), "The common concept of a homogeneous phase implies dimensions that are large compared to the molecules and small compared to the moon." When we speak of the thermodynamic limit, which is mathematically the infinite system limit, we have in mind its physical application to systems containing $10^{22} \sim 10^{28}$ particles, i.e., systems which are large enough for surface effects to be negligible and yet small enough for internal gravitational effects also to be completely negligible.

## B. *Outline and Summary of Results*

In Section II we establish the basic notation and definitions, and list some inequalities needed in the sequel. Here we rely heavily on Appendix B contributed by Simon to whom we are indebted. The proof of the existence of the thermodynamic limit proceeds, as in the tempered case, by first establishing the limit for a standard sequence of domains. The limit for an arbitrary sequence of domains is then easily arrived at by packing those domains with the standard ones. The usual choice for the standard domains is a sequence of cubes $\{\Gamma_j\}$ of sides essentially $2^j$. These have the desirable geometric property that $\Gamma_{j+1}$ can be packed with $2^d$ copies of $\Gamma_j$. For the Coulomb case we find it necessary to use balls $\{B_j\}$, and Section III is devoted to showing that the unit ball can be packed efficiently with a sequence of balls of decreasing diameter.

In Section IV we combine the results of Sections II and III to establish the existence of the thermodynamic limit of $g(\rho_j; B_j)$ defined in (1.1), when each system in the sequence is overall strictly neutral. Section V generalizes this result to arbitrary domains, while keeping the condition of strict neutrality. The limiting free energy $g(\rho)$ is found to be shape-independent.

Section VI is devoted to systems that are *not* overall neutral and we establish the fundamental fact of electrostatics that in the thermodynamic limit the free energy is the sum of the neutral system free energy and $\frac{1}{2}Q^2/C$, where $Q$ is the surplus charge and $C$ is the (shape-dependent) capacity. For technical reasons, we are able to do this only for a sequence of domains whose shapes are essentially ellipsoidal.

Section VII deals with the grand canonical ensemble. We prove the existence of the thermodynamic limit for the grand canonical pressure and show that the thermodynamic properties are the same as for the neutral canonical ensemble, i.e., nonneutral systems make a vanishingly small contribution to the grand canonical pressure regardless of the choice of the chemical potentials of the different species. This is a very special feature of the Coulomb potential.

The microcanonical ensemble is treated in Section VIII. For simplicity, and not for any reason of technical difficulty, we consider only neutral systems in balls. We make use of a microcanonical partition function that is a little different from the usual ones, but has the virtue of satisfying a minimax principle. This ensemble, the usual microcanonical ensembles, and the canonical ensemble are shown to have the same thermodynamic properties in the limit.

This reprint contains only the Introduction of the original paper.

Part II

# Hard Sphere Virial Coefficients

Reprinted from THE PHYSICAL REVIEW, Vol. 140, No. 1A, A129–A130, 4 October 1965
Printed in U. S. A.

# Suppression at High Temperature of Effects Due to Statistics in the Second Virial Coefficient of a Real Gas*

SIGURD YVES LARSEN†
*National Bureau of Standards, Washington, D. C.*

AND

JOHN E. KILPATRICK†
*Department of Chemistry, Rice University, Houston, Texas*

AND

ELLIOT H. LIEB‡
*Belfer Graduate School of Science, Yeshiva University, New York, New York*

AND

HARRY F. JORDAN§
*Los Alamos Scientific Laboratory, University of California, Los Alamos, New Mexico*

(Received 23 February 1965)

It is shown that the repulsive core present in realistic two-body potentials and in hard spheres leads to the rapid suppression of the effects of statistics in the second virial coefficient, except at very low temperatures. For hard spheres, an upper bound is obtained which goes down exponentially with temperature when the latter becomes large.

THE effects of quantum mechanics on the second virial coefficient may be formally separated into diffraction effects which obtain for a Boltzmann gas and exchange contributions associated with the Bose-Einstein or Fermi-Dirac character of the gas.[1] This separation arises very naturally in the formalism developed by Lee and Yang[2] and allows us to consider the virial as being the sum of a direct term

$$B_{\mathrm{direct}}=-(N/2)\int d\mathbf{r}[2^{3/2}\lambda_T{}^3\langle\mathbf{r}|e^{-\beta H_{\mathrm{rel}}}|\mathbf{r}\rangle-1],$$

which in the limit $h\to 0$ gives us the classical answer, and of an exchange term

$$B_{\mathrm{exch}}=\mp(N/2)[1/(2S+1)]\int d\mathbf{r}2^{3/2}\lambda_T{}^3\langle\mathbf{r}|e^{-\beta H_{\mathrm{rel}}}|-\mathbf{r}\rangle.$$

$H_{\mathrm{rel}}$ is the relative Hamiltonian, $\beta^{-1}$ is Boltzmann's constant times the temperature, $\lambda_T$ is the thermal wavelength defined as $h(2\pi mkt)^{-1/2}$, $N$ is Avogadro's constant, $S$ is the spin of the individual component, and the sign is negative for Bose-Einstein statistics and positive for Fermi-Dirac cases.

In the case of a perfect gas we have

$$B_{\mathrm{exch}}=\mp N(\lambda_T{}^3/2^{5/2})[1/(2S+1)].$$

At high temperatures this value is customarily[1] used to represent the quantum-mechanical effects due to statistics of a gas such as helium, while a Wigner-Kirkwood expansion is used to evaluate the direct term.

The purpose of this note is to point out that, in fact, for a real gas the presence of a strong repulsive core entails a drastic suppression of the exchange effect at high temperature.[3] We first show this to be the case for

* Work performed in part under the auspices of the U. S. Atomic Energy Commission.
† This work was completed at Los Alamos Scientific Laboratory while serving as consultant.
‡ This work was supported by Air Force Office of Scientific Research Grant No. AF-AFOSR-713-64.
§ Summer student from the Digital Computer Laboratory, University of Illinois, Urbana, Illinois.
[1] See J. O. Hirschfelder, C. F. Curtis, and R. B. Bird, *Molecular Theory of Gases and Liquids* (John Wiley & Sons, Inc., New York, 1954) with special reference to the article by J. deBoer and R. Byron Bird on the quantum theory and the equation of state.
[2] T. D. Lee and C. N. Yang, Phys. Rev. 113, 1165 (1959).
[3] Lloyd D. Fosdick has, independently, reached similar conclusions (private communication).

hard spheres and then consider more realistic potentials.

Introducing a complete set of eigenfunctions of the energy $\psi_n$, we can write

$$\langle \mathbf{r}|e^{-\beta H_{\mathrm{rel}}}|-\mathbf{r}\rangle=\sum_n \psi_n(\mathbf{r})\psi_n(-\mathbf{r})e^{-\beta E_n}.$$

Setting the collision diameter of the hard spheres at $r=\sigma$, we see that the matrix element is zero for $r<\sigma$ since the wave functions are zero inside this region. Next we show that for any $\mathbf{r}$ the matrix element for free particles is an upper bound to the exchange matrix element for particles subject to repulsive forces only. This result is immediate once we write the Wiener integral expression[4] for the exchange matrix element

$$\langle \mathbf{r}|e^{-\beta H_{\mathrm{rel}}}|-\mathbf{r}\rangle =\int_{C_{\beta;2\mathbf{r}}} \exp\left\{-\int_0^\beta d\tau V[\mathbf{X}(\tau)-\mathbf{r}]\right\} d_{\omega(\beta;2\mathbf{r})}\mathbf{X}, \quad (1)$$

which is less than or equal to

$$\langle \mathbf{r}|e^{-\beta T_{\mathrm{rel}}}|-\mathbf{r}\rangle=\int_{C_{\beta;2\mathbf{r}}} d_{\omega(\beta;2\mathbf{r})}\mathbf{X}, \quad (2)$$

since the exponential is less than 1. ($T_{\mathrm{rel}}$ is the relative kinetic energy.) In fact since paths passing through the sphere contribute for free particles and not for hard spheres the inequality obtains. Evaluating the exchange matrix element for the kinetic energy yields

$$\langle \mathbf{r}|e^{-\beta T_{\mathrm{rel}}}|-\mathbf{r}\rangle=(1/2^{3/2}\lambda_T^3)e^{-2\pi r^2/\lambda_T^2}. \quad (3)$$

We thus have

$$|B_{\mathrm{exch}}|<[2\pi N/(2S+1)] \times\int_\sigma^\infty dr\, r^2\langle \mathbf{r}|e^{-\beta T_{\mathrm{rel}}}|-\mathbf{r}\rangle 2^{3/2}\lambda_T^3,$$

which equals

$$[2\pi N/(2S+1)]\int_\sigma^\infty dr\, r^2 e^{-2\pi r^2/\lambda_T^2}.$$

At low temperatures ($\lambda_T$ large) this integral has for limiting value the free-particle result, while at high temperatures we obtain the asymptotic expansion

$$|B_{\mathrm{perf\ exchange}}|\times 2^{3/2}(\sigma/\lambda_T) \times e^{-2\pi(\sigma/\lambda_T)^2}[1+(1/4\pi)(\lambda_T/\sigma)^2+\cdots].$$

Since $\lambda_T$ is proportional to $T^{-1/2}$, we see that our upper bound goes down exponentially with temperature. In fact, if we set the collision diameter at about 2 Å and choose a value for the mass suitable for helium, we find that the dependence is roughly $e^{-T/2}$. Note that this precludes an asymptotic expansion in powers of $1/T$.

Physically, we can understand this formal result by noting that the free-particle exchange matrix element (Eq. 3) is highly peaked about $r=0$ and appreciable only for $r$ of the order of $\lambda_T/(2\pi)^{1/2}$ or less. In other words we see that the exchange is nontrivial only if the particles are allowed to come closer to each other than the thermal wavelength. If this is not possible, because of the presence of repulsive forces, the exchange is negligible. This is the case for hard spheres when the temperature is large enough so that the collision diameter $\sigma$ is greater than $\lambda_T$. In the example mentioned above $(\sigma/\lambda_T)\sim 1$ when $T$ is $\sim 16°$K. As the previous remark made on deriving the inequality (Eq. 1 $\leqslant$ Eq. 2) indicates, the matrix element outside the core will be smaller than the free-particle result and the consequent $B_{\mathrm{exch}}$ smaller for a given temperature than has been estimated. This point will not be considered further in this note.[5]

Turning our attention now to more realistic potentials, we note two differences. In the first place the intermolecular potentials have an attractive part. If $\epsilon$ represents the maximum well depth ($\epsilon/k\sim 10°$K for helium) then Eqs. (1) and (2) show that

$$e^{\beta\epsilon}\langle \mathbf{r}|e^{-\beta T_{\mathrm{rel}}}|-\mathbf{r}\rangle=e^{\beta\epsilon}(1/2^{3/2}\lambda_T^3)e^{-2\pi r^2/\lambda_T^2}$$

is an upper bound to the exchange matrix element for all $\mathbf{r}$. At high temperature $e^{\beta\epsilon}\to 1$ and we recover the free-particle result. Another difference is of course that though realistic potentials provide strong repulsive forces they lack the abrupt "all or nothing" character of hard spheres. Nevertheless, since the repulsion is so strong, the potential rising rapidly and reaching values many orders of magnitude larger than the maximum well depth, the wave functions are essentially zero for $r$'s within the core and so will be the exchange element.

We thus see again that at high temperature where the thermal wavelength is much smaller than the core radius, the exchange contribution to the virial will be completely negligible.

[4] S. G. Brush, Rev. Mod. Phys. 33, 79 (1961). Especially relevant is the discussion pertinent to and centered about Eq. (2.13); see also Eqs. (5.4) and (5.5).

[5] We hope to show in a subsequent paper that the leading term in the asymptotic form of the logarithm of $B_{\mathrm{exch}}$ is in fact proportional to $-\frac{1}{2}\pi^2(\sigma/\lambda_T)^2$.

J. Math. Phys. *8*, 43–52 (1967)

JOURNAL OF MATHEMATICAL PHYSICS VOLUME 8, NUMBER 1 JANUARY 1967

# Calculation of Exchange Second Virial Coefficient of a Hard-Sphere Gas by Path Integrals*

ELLIOTT H. LIEB

*Department of Physics, Northeastern University, Boston, Massachusetts*

(Received 21 February 1966)

By direct examination of the path (Wiener)-integral representation of the diffusion Green's function in the presence of an opaque sphere, we are able to obtain upper and lower bounds for that Green's function. These bounds are asymptotically correct for short-time, even in the shadow region. Essentially, we have succeeded in showing that diffusion probabilities for short-time intervals are concentrated mainly on the optical path. By integrating the Green's function, we obtain upper- and lower-bound estimates for the exchange part of the second virial coefficient of a hard-sphere gas. We can show that, for high temperature, it is asymptotically very small compared to the corresponding quantity for an ideal gas, viz.,

$$B_{\text{exch}}/B^0_{\text{exch}} = \exp\{-\tfrac{1}{2}\pi^3(a/\Lambda)^2 + O[(a/\Lambda)^{2/3}]\},$$

where $\Lambda$ is the thermal wavelength and $a$ is the hard-sphere radius. While it was known before that $B_{\text{exch}}/B^0_{\text{exch}}$ is exponentially small for high temperatures, this is the first time that a precise asymptotic formula is both proposed and proved to be correct.

## I. INTRODUCTION

FOR a gas of particles that interact via a two-body potential, the calculation of the second virial coefficient[1] involves an analysis of only a two-body problem. This simplification holds for quantum as well as for classical mechanics, but there the similarity between the two kinds of mechanics ends. Classically, the second virial coefficient depends neither on particle mass, $m$, nor on statistics and, for a one-component gas, is given by the simple configuration integral:

$$B_{\text{cl}}(T) = \tfrac{1}{2}N \int d\mathbf{r}\{1 - \exp[-\beta v(\mathbf{r})]\}, \tag{1.1}$$

where $v(\mathbf{r})$ is the pair potential, $N$ is Avogadro's number, and $\beta = (kT)^{-1}$.

Quantum-mechanically, no such simple formula as (1.1) exists, for the calculation of $B(T)$ requires either a detailed knowledge of the solutions of the two-particle Schrödinger equation at all energies, or, alternatively, a solution of the corresponding diffusion problem. Thus, while the problem of calculating the second virial coefficient may not be as profound as the original many-body problem from which it arose, it does require the answer to interesting questions about the classical analysis of the three-dimensional diffusion equation. To become familiar with the problem is to realize how difficult it is to calculate quantum corrections to (1.1).[2]

The true physicist will doubtless inquire whether quantum corrections to (1.1) are in fact significant, and the answer is that for helium they are quite important. Even for temperatures as high as 60°K, the quantum corrections in helium are about a third of the total.[3] For a hard-sphere gas, the quantum corrections do not drop to a tenth of the total until a temperature of about 1200°K is reached.[4] Since experimental values of the second virial coefficient are used in attempting to determine the effective inter-atomic helium potential, these quantum corrections are certainly worthy of consideration.

There is also[3] a pronounced difference between the second virial coefficient of $He^3$ and $He^4$, especially below 60°K. Assuming (as is always done) that the interaction potential is the same for the two isotopes, the difference could conceivably come from three sources: (a) the atomic mass difference; (b) the difference in nuclear spin which affects the statistical weights; and (c) the difference between Fermi–Dirac and Bose–Einstein statistics. For an ideal (noninteracting) quantum gas (b) and (c) are everything [see Eq. (1.11) below], and one might be tempted to conclude that, for helium too, the isotopic mass difference was relatively unimportant. Numerical calculations have, however, indicated the

* This paper was supported by the U. S. Air Force Office of Scientific Research under Grant No. 508-66 at Yeshiva University, New York.

[1] The $n$th-virial coefficient is the temperature-dependent coefficient of $v^{-n+1}$ in the series

$$Pv = RT[1 + B(T)v^{-1} + C(T)v^{-2} + \cdots].$$

Here, $P$ is the pressure, $T$ is the temperature, $R$ is the gas constant, and $v$ is the volume per mole of gas. In terms of $N$ (Avogadro's number), $v = N\rho^{-1}$ and $R = Nk$, where $\rho$ is the particle number density and $k$ is Boltzmann's constant.

[2] Hugh E. DeWitt, J. Math. Phys. **3**, 1003 (1962).

[3] J. Kilpatrick, W. Keller, E. Hammel, and N. Metropolis, Phys. Rev. **94**, 1103 (1954); J. Kilpatrick, W. Keller, and E. Hammel, *ibid.* **97**, 9 (1955).

[4] F. Mohling, Phys. Fluids **6**, 1097 (1963).

reverse. Above about 4°K, almost all of the difference in the two second virial coefficients is a mass effect.[3] This difference is about 10% at 60°K and drops only to the order of 5% at room temperature. In other words, on the one hand the mass effect is unusually large for helium, while on the other hand the effects due to statistics and spin decrease very rapidly with increasing temperature. For an ideal gas, these latter effects decrease as $T^{-\frac{3}{2}}$, but for helium the decrease is far more rapid. Under the assumption that the repulsive part of the helium interaction potential can be effectively replaced by a hard core, it has been *proved*[5] that the statistical and spin effects decrease *at least* exponentially fast with increasing temperature (for high temperatures). The suppression of exchange effects is so rapid that 20°K may be considered to be a high temperature for which asymptotic formulas are reasonably valid.

It is the purpose of this paper to prove that the exponential law for the hard-sphere gas mentioned above is more than just an upper bound, that it is in fact correct. The true coefficient appearing in the law [cf., (1.13) below] is, however, different from that of the bound given in Ref. 5, although the correct value was stated there, without proof, in a footnote.

To the casual reader, the problem must seem almost trivial. In the first place, we have eschewed calculating the true equation of state, and have, instead, contented ourselves with examining only the second virial coefficient—a simple matter of a two-body problem. Secondly, we are examining only the effects of spin and statistics. Thirdly, we are confining ourselves to high temperatures. That there is no simple perturbation theory for this problem must appear strange. But it is a fact that, in many respects, the problem is similar to the classical problem of diffraction of waves (of short wavelength) around a sphere into the dark zone, a problem which has exercised mathematicians for years.

The mathematical statement of the problem is as follows: The quantum-mechanical second virial coefficient may be written as the sum of a direct and an exchange part,

$$B = B_{\text{direct}} + B_{\text{exch}}, \tag{1.2}$$

where

$$B_{\text{direct}} = \tfrac{1}{2}N \int d\mathbf{r}\, [1 - 2^{\frac{3}{2}}\Lambda^3 G(\mathbf{r}, \mathbf{r}; \beta)], \tag{1.3}$$

$$B_{\text{exch}} = \mp\sqrt{2}\, \Lambda^3 N(2S+1)^{-1} \int d\mathbf{r}\, G(\mathbf{r}, -\mathbf{r}; \beta), \tag{1.4}$$

and

$$\Lambda^2 = 2\pi\hbar^2\beta/m. \tag{1.5}$$

In (1.4) the − sign is for bosons and the + sign is for fermions. $S$ is the total spin of the atom (the nuclear spin alone in the case of helium), it is to be noted that the spin enters only into $B_{\text{exch}}$. Thus, (b) and (c) mentioned above go together.

The function $G(\mathbf{r}, \mathbf{r}'; t)$ is the diffusion Green's function (also known as the Bloch function), and it satisfies

$$[-D\nabla_r^2 + v(\mathbf{r}) + \partial/\partial t]G(\mathbf{r}, \mathbf{r}', t) = 0, \quad (\text{for } t > 0) \tag{1.6}$$

with the initial condition

$$\lim_{t\to 0} G(\mathbf{r}, \mathbf{r}'; t) = \delta(\mathbf{r} - \mathbf{r}'). \tag{1.7}$$

In addition, $G$ satisfies appropriate boundary conditions in $\mathbf{r}$, such as vanishing on the walls of a box. In our case, we are interested in the limit of an infinite volume which means that $G$ satisfies (1.6) for all $\mathbf{r}$ but vanishes when $r \to \infty$. It is to be noted that boundary conditions need only be defined with respect to $\mathbf{r}$. Despite this fact, and despite the fact that (1.6) refers really only to $\mathbf{r}$, $G$ automatically turns out to be a symmetric function of $\mathbf{r}$ and $\mathbf{r}'$ for all $t$.

Equation (1.6) describes diffusion in a potential $v(\mathbf{r})$, with $\mathbf{r}'$ the source point, $t$ the elapsed time, and $D$ the diffusion constant. For quantum-mechanical purposes, $t$ is interpreted as $\beta$, $v$ is the interparticle potential, and $D$ is related to the mass of a single atom by

$$D = \hbar^2/m \equiv \tfrac{1}{4}\alpha. \tag{1.8}$$

Thus,

$$\Lambda^2 = 2\pi\, Dt = \tfrac{1}{2}\pi\alpha t. \tag{1.9}$$

In the case of *no interaction* ($v = 0$), $G$ is given by

$$G_0(\mathbf{r}, \mathbf{r}', t) = (\pi\alpha t)^{-\frac{3}{2}} \exp[-(\mathbf{r} - \mathbf{r}')^2/\alpha t], \tag{1.10}$$

and when this is inserted into (1.3) and (1.4), we obtain the result:

$$B^0_{\text{direct}} = 0, \tag{1.11a}$$

$$B^0_{\text{exch}} = \mp N\Lambda^3 2^{-5/2}(2S+1)^{-1}. \tag{1.11b}$$

[5] S. Larsen, J. Kilpatrick, E. Lieb, and H. Jordan, Phys. Rev. 140, A129 (1965). While it was realized in Ref. 4 that exchange effects are small at high temperatures, no proof of this assertion nor statement of its exponential character were offered. For further results on the hard sphere problem, see the following papers: M. Boyd, S. Larsen, and J. Kilpatrick, J. Chem. Phys. 45, 499 (1966); S. Larsen, K. Witte, and J. Kilpatrick, J. Chem. Phys. (to be published). Recently, J. B. Keller and R. A. Handelsman, Phys. Rev. 148, 94 (1966), have calculated the first few terms in a high-temperature power series for the *direct* second virial coefficient of a hard-sphere gas.

For a hard-sphere potential,

$$\begin{aligned} v(r) &= \infty, \quad \text{for } r \leq a, \\ &= 0, \quad \text{for } r > a, \end{aligned} \tag{1.12}$$

Eq. (1.11b) is a very misleading approximation to $B_{exch}$ for high temperatures. We are to *prove* that, for small $t$ or $\Lambda$,

$$\ln\left\{\frac{B_{exch}}{B^0_{exch}}\right\} = -\frac{\pi^3}{2}\left(\frac{a}{\Lambda}\right)^2 + O\left(\left(\frac{a}{\Lambda}\right)^{\frac{5}{2}}\right). \tag{1.13}$$

The proof consists in obtaining upper and lower bounds for $G(\mathbf{r}, \mathbf{r}'; t)$ by means of Wiener, or path integrals. These bounds are valid for all temperatures, and we could, in fact, give a more detailed estimate than is indicated in (1.13). The bounds are, however, complicated functions of $t$, and it seems neither necessary nor desirable to go beyond the asymptotic formula in (1.13).

Before giving the proof, it is worthwhile mentioning an alternative formulation of the problem which, at first sight, seems to offer an immediate solution. For a particle in a box, we can write

$$G(\mathbf{r}, \mathbf{r}'; t) = \sum_{n=1}^{\infty} \exp(-te_n)\psi_n(\mathbf{r})\psi_n^*(\mathbf{r}'), \tag{1.14}$$

where $e_n$ is the $n$th energy level and $\psi_n$ is the corresponding normalized eigenfunction. When (1.14) is inserted into (1.3) and (1.4), it is seen that knowledge of the energy levels alone is required. When the box is very large compared to the range of the potential, the virial coefficient can be expressed in terms of the bound-state energy levels (if any) and the scattering phase shifts of the potential, viz.

$$B_{direct} = -\sqrt{2}\, N\Lambda^3 \sum_{\text{all } l} (2l + 1)B_l, \tag{1.15a}$$

$$B_{exch} = B^0_{exch} \mp (2S + 1)^{-1} \sqrt{2}\, N\Lambda^3 \times \{\sum_{l \text{ even}} - \sum_{l \text{ odd}}\}(2l + 1)B_l, \tag{1.15b}$$

where

$$B_l = \sum_n \exp[-\beta e_n(l)] + (\Lambda^2/\pi^2)\int_0^\infty e^{-\Lambda^2 k^2/2\pi}\eta_l(k)k\, dk. \tag{1.16}$$

In (1.16) the sum is over negative energy levels (if any), while the integral contains the phase shift $\eta_l$—all for the appropriate angular momentum, $l$. For the case of no bound state, the above formula for the second virial coefficient in terms of the phase shifts was apparently first stated by Gropper and by Beth and Uhlenbeck,[6] and a derivation of it can be found in Ref. 3.

For the hard-sphere potential, there are no bound states, and it would appear that (1.16) and (1.15) should give the answer simply, especially as the phase shifts are given by the elementary formula

$$\eta_l(k) = -\tan^{-1}\{(-1)^l J_{l+\frac{1}{2}}(ka)/J_{-l-\frac{1}{2}}(ka)\}. \tag{1.17}$$

For small $\Lambda$, however, we see that large values of $k$ are important in (1.16). For very large $k$, the sum on $l$ in (1.15b) may be performed with the aid of Watson's transformation, and it is similar to the problem of diffraction around a sphere at short wavelength.[7] Apart from certain technical convergence difficulties connected with the fact that we are really interested in the diffracted field on a diameter (that is to say a caustic), there is another more important problem.[8] This problem is that there may also be contributions to (1.16) from small $k$, a region where Watson's transformation is not of great use. Finite $k$ contributions would, from (1.16), be expected to give a power series in $\Lambda$ for small $\Lambda$. But it is a fact that there is a remarkable cancellation between even and odd $l$ in (1.15b) so that *every* term in this power series vanishes. The final result, as shown in (1.13), is a function that vanishes faster than any power as $\Lambda \to 0$. If the potential were finite, instead of a hard core, this power series would *not* vanish. Thus, in summary, (1.16) and (1.15b) is a difficult starting point for hard spheres, despite the simplicity of the phase shifts and the existence of Watson's transformation.

Our approach is to go back to (1.4) and, as we mentioned before, to estimate $G(\mathbf{r}, -\mathbf{r}; t)$ directly through its expression in terms of a Wiener integral. Such integrals play an important theoretical role in analysis but, unless the integrand is Gaussian, it is difficult to obtain numerical answers from them. There have, of course, been rare exceptions such as Feynman's treatment of the Polaron problem.[9] Nevertheless, the analysis presented here is one of the very few cases, if not the only one, in which both an upper *and* a lower bound to a function is obtained with path integrals. The path integral approach also has the great virtue of transparency because it brings out the close connection between the diffusion equation, (1.6), and a random walk

[6] L. Gropper, Phys. Rev. 51, 1108 (1937); E. Beth and G. Uhlenbeck, Physica 4, 915 (1937); see also G. Uhlenbeck and E. Beth, *ibid.* 3, 729 (1936).
[7] B. Levy and J. Keller, Commun. Pure Appl. Math. 12, 159 (1959), where the relevent asymptotic formulas are given on p. 201. See also J. Keller, J. Opt. Soc. Am. 52, 116 (1962).
[8] I am indebted to Dr. S. Larsen for pointing this out to me.
[9] R. Feynman, Phys. Rev. 97, 660 (1955).

J. Math. Phys. *8*, 43–52 (1967)



problem. For these reasons, we believe the sequel might also possess an intrinsic mathematical value.

## II. LOWER BOUND BY PATH INTEGRALS

The solution to (1.6) and (1.7) is easily shown to be unique and to satisfy the relation

$$G(\mathbf{r}, \mathbf{r}'; t) = \int dz\, G(\mathbf{r}, \mathbf{z}; t_1) G(\mathbf{z}, \mathbf{r}'; t_2) \qquad (2.1)$$

for any positive $t_1$ and $t_2$ such that $t_1 + t_2 = t$. If the time interval $t$ is divided into $n + 1$ intervals of duration $\Delta$, so that $t = (n + 1)\Delta$, then, from (2.1),

$$\begin{aligned} G(\mathbf{r}, \mathbf{r}'; t) &= \lim_{n\to\infty} \int dZ\, G(\mathbf{r}, \mathbf{z}_1; \Delta) G(\mathbf{z}_1, \mathbf{z}_2; \Delta) \cdots \\ &\quad \times G(\mathbf{z}_{n-1}, \mathbf{z}_n; \Delta) G(\mathbf{z}_n, \mathbf{r}', \Delta) \\ &= \lim_{n\to\infty} \int dZ\, G_0(\mathbf{r}, \mathbf{z}_1; \Delta) e^{-\Delta v(\mathbf{z}_1)} \\ &\quad \times G_0(\mathbf{z}_1, \mathbf{z}_2; \Delta) e^{-\Delta v(\mathbf{z}_2)} \cdots \\ &\quad \times e^{-\Delta v(\mathbf{z}_{n-1})} G_0(\mathbf{z}_{n-1}, \mathbf{z}_n; \Delta) \\ &\quad \times e^{-\Delta v(\mathbf{z}_n)} G_0(\mathbf{z}_n, \mathbf{r}'; \Delta), \end{aligned} \qquad (2.2)$$

where $dZ = dz_1\, dz_2 \cdots dz_n$.

The heuristic justification for (2.2) is that, if $\alpha = 4D$ were zero, then

$$G(\mathbf{r}, \mathbf{r}'; \Delta) = \delta(\mathbf{r} - \mathbf{r}') \exp[-\Delta v(\mathbf{r})],$$

whereas if $v = 0$ then $G = G_0$, which is very nearly $\delta(\mathbf{r} - \mathbf{r}')$ for small $\Delta$. The combination

$$G_0(\mathbf{r}, \mathbf{r}'; \Delta) \exp[-\Delta v(\mathbf{r})]$$

is, hopefully, a good approximation to $G$ [at least as far as the integral in (2.2) is concerned] for very small $\Delta$. Formally, this combination satisfies (1.6) to leading order in $\Delta$ *for those values of* $\mathbf{r}$ *and* $\mathbf{r}'$ *such that* $G(\mathbf{r}, \mathbf{r}'; \Delta)$ *significantly contributes to* (2.2).

The fact that (2.2) is correct for a large class of bounded potentials has been known for some time. We are interested, however, in the hard-core potential [see Eq. (2.3) below] for which a special proof is apparently required. We remark that Ginibre has previously used (2.2) for the hard-core case, but without giving an explicit proof.[10]

I am indebted to Professor D. Babbitt for the proof in the hard-core case, which is outlined as the following. Take $D = \frac{1}{2}$ for convenience, and let $\Omega$ be the set of functions (paths) from $[0, \infty)$ into $\dot{\mathcal{R}}^3$, where $\dot{\mathcal{R}}^3$ is the one-point compactification of $\mathcal{R}^3$, the three-dimensional Euclidean space. Let $\{P_{\mathbf{r},\mathbf{r}';t};\ \mathbf{r}, \mathbf{r}' \in \mathcal{R}^3, t > 0\}$ denote the family of conditional Wiener measures on $\Omega$ as defined by Ginibre.[11] The crucial point to note is that $P_{\mathbf{r},\mathbf{r}';t}$ is concentrated on the paths that are bounded and continuous on $[0, t]$. Denote integration of $P_{\mathbf{r},\mathbf{r}';t}$ integrable functionals, $F$, on $\Omega$, by $\int F(\omega) P_{\mathbf{r},\mathbf{r}';t}(d\omega)$, where $\omega$ denotes a generic path in $\Omega$. Let

$$\Omega_t(\omega) = \begin{cases} 1 & \text{if } |\omega(\tau)| > a \quad \text{for all } 0 < \tau \le t, \\ 0 & \text{otherwise.} \end{cases}$$

Then $\Omega_t$ is $P_{\mathbf{r},\mathbf{r}';t}$ integrable and

$$G(\mathbf{r}, \mathbf{r}'; t) = \int \Omega_t(\omega) P_{\mathbf{r},\mathbf{r}';t}(d\omega).$$

This result is essentially given, with different notation, by Ray.[12] Since $P_{\mathbf{r},\mathbf{r}';t}$ is concentrated on the bounded, continuous paths on $[0, t]$, it follows that

$$\lim_{n\to\infty} \left\{ \prod_{k=1}^{n} \theta\left[\omega\left(\frac{kt}{n+1}\right)\right] \right\} = \Omega_t(\omega)$$

$P_{\mathbf{r},\mathbf{r}';t}$—almost everywhere on $\Omega$. The function $\theta$ is defined in Eq. (2.3) below. Hence, applying the dominated convergence theorem we have

$$\lim_{n\to\infty} \int \left\{ \prod_{k=1}^{n} \theta\left[\omega\left(\frac{kt}{n+1}\right)\right] \right\} P_{\mathbf{r},\mathbf{r}';t}(d\omega) = \int \Omega_t(\omega) P_{\mathbf{r},\mathbf{r}';t}(d\omega).$$

By definition of $P_{\mathbf{r},\mathbf{r}';t}$, the left side of this equation is identical to the right side of (2.2) for the hard-core case [cf. Eq. (2.3) below].

Having established (2.2), we use it as the rigorous starting point for our analysis. The limit $n \to \infty$ in (2.2) defines a conditional Wiener integral or path integral (conditional because both ends, $\mathbf{r}$ and $\mathbf{r}'$, are fixed). The $n$-fold integral in (2.2) bears to the path integral essentially the same relationship as a finite sum bears to the ordinary Riemann integral. Brush[13] has remarked that "it is usually impossible to do this" (evaluate the path integral) "by the direct method of finding an explicit formula for the finite dimensional integral and then passing to the limit of a continuous integral". Contrary to this dictum, we find, in fact, upper and lower bounds to the finite integral in (2.2) and then pass to the limit $n \to \infty$. In this way, we obtain upper and lower bounds to $G(\mathbf{r}, \mathbf{r}'; t)$.

We are interested in the case that $v$ is a hard core, (1.12), and hence the factor $\exp[(-\Delta)v(\mathbf{z})]$ in (2.2) is equal to the simpler expression

[10] J. Ginibre, J. Math. Phys. **6**, 1432 (1965). See especially Eqs. (A1.6)–(A1.10).

[11] J. Ginibre, J. Math. Phys. **6**, 238 (1965); see the Appendix.

[12] D. Ray, Trans. Am. Math. Soc. **77**, 299 (1954).

[13] S. Brush, Rev. Mod. Phys. **33**, 79 (1961).

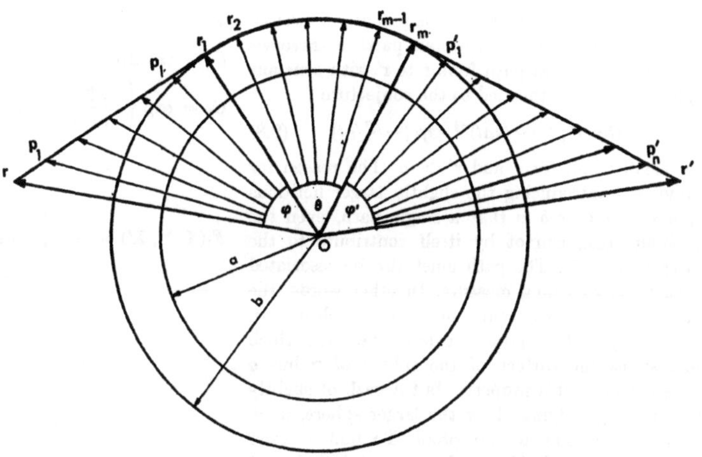


FIG. 1. Important quantities for calculating the path integral [cf. Eq. (2.9) et seq.]. The opaque sphere having radius $a$ is shown centered at the origin, $O$. A slightly larger, concentric sphere of radius $b$ is also shown. The vectors $\mathbf{r}$ and $\mathbf{r}'$ are the observation and source points, respectively, and the curve from $\mathbf{r}$ to $\mathbf{r}'$ via $\mathbf{r}_1$ and $\mathbf{r}_m$ is the shortest path from $\mathbf{r}$ to $\mathbf{r}'$ lying entirely outside the larger sphere. The straight line $\mathbf{r}_1 - \mathbf{r}$ is divided into $(l+1)$ equal parts by the vectors $\mathbf{p}_1, \ldots, \mathbf{p}_l$; the arc $\theta$ from $\mathbf{r}_1$ to $\mathbf{r}_m$ is divided into $(m-1)$ equal arcs by the vectors $\mathbf{r}_2, \ldots \mathbf{r}_{m-1}$; and the straight line $\mathbf{r}' - \mathbf{r}_m$ is divided to $(n+1)$ equal parts by the vectors $\mathbf{p}_1', \ldots, \mathbf{p}_n'$.

$$\begin{aligned}\theta(\mathbf{z}) &= 1, \quad \text{for } z > a \\ &= 0, \quad \text{for } z \leq a\end{aligned} \tag{2.3}$$

for all $\Delta$. Hence, the integrand in (2.2) is over a simple product of $G_0$ functions, but the integration range for each $\mathbf{z}_i$ is restricted to $z > a$. Such an integral is impossible to calculate. Since the integrand is positive, however, it is easy to obtain a lower bound to (2.2) by restricting the integration range still further, in such a manner that the restricted integral can be calculated exactly. To do this, we must define certain geometric quantities as shown in Fig. 1.

The plane of Fig. 1 is the $\mathbf{r}$, $\mathbf{r}'$ plane, and $\mathbf{0}$ is the center of the sphere of radius $a$. A larger, concentric sphere of radius $b > a$ is shown, and it is assumed that

$$b < \text{minimum}\ (r, r'). \tag{2.4}$$

The two straight lines, $(\mathbf{r}, \mathbf{r}_1)$ and $(\mathbf{r}_m, \mathbf{r}')$, together with the circular arc $(\mathbf{r}_1, \mathbf{r}_m)$ delineate the path which would be followed by a piece of string drawn taut between $\mathbf{r}$ and $\mathbf{r}'$. Thus, $\mathbf{r}_1 \cdot (\mathbf{r} - \mathbf{r}_1) = 0$ and $\mathbf{r}_m \cdot (\mathbf{r}' - \mathbf{r}_m) = 0$. The angles $\varphi$, $\theta$, and $\varphi'$ are the angles between $\mathbf{r}$ and $\mathbf{r}_1$, $\mathbf{r}_1$ and $\mathbf{r}_m$, and $\mathbf{r}_m$ and $\mathbf{r}'$, respectively, whence

$$\psi = \varphi + \theta + \varphi' \tag{2.5}$$

is the angle between $\mathbf{r}$ and $\mathbf{r}'$. Note that the angle $\theta$ may be zero and that the shortest path from $\mathbf{r}$ to $\mathbf{r}'$ may consist of only one straight line that does not touch the sphere of radius $b$. In that case $\mathbf{r}_1$ and $\mathbf{r}_n$ are not defined, but the subsequent analysis remains valid with trivial modifications. In any event,

$$S_b = r \sin \varphi + r' \sin \varphi' + b\theta \tag{2.6}$$

is the distance from $\mathbf{r}$ to $\mathbf{r}'$ along the shortest path lying outside of a sphere of radius $b$.

An intuitive discussion of (2.2) is useful at this point in order to motivate the subsequent analysis. This and the following paragraph are entirely heuristic and are not part of our proof. It will be recalled that we are interested in $G(\mathbf{r}, \mathbf{r}'; t)$ for small $t$. In this regime, the $G_0$ factors in (2.2) give a large weight to that "path" (or sequences of points $\mathbf{z}_1, \cdots, \mathbf{z}_n$) from $\mathbf{r}$ to $\mathbf{r}'$ which is of shortest length. That path is, moreover, traversed with constant speed (i.e., $|\mathbf{z}_{i+1} - \mathbf{z}_i|/\Delta = \text{const}$) and is, in fact, the path of classical geometrical optics. Alternatively, we may say that a Brownian particle, which is observed to go from $\mathbf{r}$ to $\mathbf{r}'$ in a short time, most likely went by way of the Newtonian, non-Brownian, trajectory. As the time increases, the optimum path ceases to have such a preponderant weight and other paths contribute more and more to (2.2). For the case of no interaction, however, we see from (1.10) that $G$ is *always* proportional to the maximum of the integrand, namely $\exp[-S^2/\alpha t]$, where $S$ is the distance from $\mathbf{r}$ to $\mathbf{r}'$. When $v \neq 0$, this simple relationship will not hold for all time, but for *short time* it is clear that the "optical" path is strongly preferred *if $v$ is finite*. Thus, for *finite $v$*, the factors $\exp[-\Delta v]$ in (2.2) contribute approximately the average potential along the optical path and

$$G(\mathbf{r}, \mathbf{r}'; t) \sim (\pi\alpha t)^{-\frac{3}{2}} \exp[-S^2/\alpha t] \times \exp\left[-t \int_0^1 v(\mathbf{r} + \mu(\mathbf{r}' - \mathbf{r})\, d\mu\right]. \tag{2.7}$$

For the hard-core case, (2.7) is patently nonsense. Instead, the fictitious Brownian particle traverses the shortest *allowed* path from $\mathbf{r}$ to $\mathbf{r}'$ with constant speed and we are thus led to the conjecture

$$G(\mathbf{r}, \mathbf{r}'; t) \sim (\pi\alpha t)^{-\frac{3}{2}} \exp\left[-S_a^2/\alpha t\right] \tag{2.8}$$

for small $t$ and for $r$ and $r' > a$. The reason for previously introducing the slightly larger fictitious sphere of radius $b$ is that a single path, even the optimum one, cannot by itself contribute to the integral in (2.2). The path must also be associated with a nonvanishing measure. In other words, the path must be at the center of a tube which in turn lies wholly in the allowed region. The path which just skims the surface of the sphere of radius $a$ does not have this property, but a path of slightly greater length, lying along the larger sphere, does.

We return now to our proof. To find a lower bound we now, divide the line $(\mathbf{r}, \mathbf{r}_1)$ into $l+1$ equal parts, designated by the vectors $\mathbf{p}_1, \cdots, \mathbf{p}_l$. Likewise, divide $(\mathbf{r}_m, \mathbf{r}')$ into $n+1$ equal parts, designated by $\mathbf{p}'_1, \cdots, \mathbf{p}'_n$. The arc $(\mathbf{r}_1, \mathbf{r}_m)$ is to be divided into $m-1$ equal arcs, of angle $\delta \doteq \theta/(m-1)$, and designated by $\mathbf{r}_2, \cdots, \mathbf{r}_{m-1}$. We define

$$S_b^m = r\sin\varphi + r'\sin\varphi' + b(m-1)\sin\delta, \tag{2.9}$$

so that

$$S_b = \lim_{m\to\infty} S_b^m.$$

Associated with these three divisions, we define the time intervals

$$\begin{aligned}\Delta_l &= tr\sin\varphi/(l+1)S_b^m,\\ \Delta_n &= tr'\sin\varphi'/(n+1)S_b^m,\\ \Delta_m &= tb\sin\delta/S_b^m,\end{aligned} \tag{2.10}$$

whence $(l+1)\Delta_l + (m-1)\Delta_m + (n+1)\Delta_n = t$. Furthermore, in (2.2) let there be $l+m+n$ variables of integration and we take the limit $l, m, n \to \infty$. We make the following changes from the $\mathbf{z}_i$ variables to $\mathbf{x}_i$, $\mathbf{y}_i$, and $\mathbf{x}'_i$:

$$\begin{aligned}\mathbf{z}_i &= \mathbf{p}_i + \mathbf{x}_i \quad (i = 1, \cdots, l),\\ \mathbf{z}_{i+l} &= \mathbf{r}_i + \mathbf{y}_i \quad (i = 1, \cdots, m),\\ \mathbf{z}_{i+l+m} &= \mathbf{p}'_i + \mathbf{x}'_i \quad (i = 1, \cdots, n).\end{aligned} \tag{2.11}$$

We also use the symbol $G_{lmn}$ to designate the integral in (2.2) before taking the limit on $l$, $m$, and $n$.

$$G_{lmn} = C_1C_2 \int d\mathbf{X}\, d\mathbf{Y}\, d\mathbf{X}'\, F_1(\mathbf{X}, \mathbf{Y}, \mathbf{X}')F_2(\mathbf{Y}), \tag{2.12}$$

where

$$C_1 = (\pi\alpha\,\Delta_l)^{-\frac{3}{2}(l+1)}(\pi\alpha\,\Delta_m)^{-\frac{3}{2}(m-1)}(\pi\alpha\,\Delta_n)^{-\frac{3}{2}(n+1)}$$

$$C_2 = \exp\left\{-\frac{S_b^m}{\alpha t}\left[r\sin\varphi + r'\sin\varphi' + 2b(m-1)\frac{1-\cos\delta}{\sin\delta}\right]\right\}, \tag{2.13}$$

$$\begin{aligned}F_1(\mathbf{X}, \mathbf{Y}, \mathbf{X}') = \exp\Bigg\{&-(\alpha\Delta_l)^{-1}\left[\sum_{j=2}^{l}|\mathbf{x}_j - \mathbf{x}_{j-1}|^2 + |\mathbf{x}_1|^2 + |\mathbf{y}_1 - \mathbf{x}_l|^2\right]\\ &- (\alpha\Delta_m)^{-1}\sum_{j=2}^{m}|\mathbf{y}_j - \mathbf{y}_{j-1}|^2\\ &- (\alpha\Delta_n)^{-1}\left[\sum_{j=2}^{n}|\mathbf{x}'_j - \mathbf{x}'_{j-1}|^2 + |\mathbf{x}'_m|^2 + |\mathbf{y}_n - \mathbf{x}'_1|^2\right]\Bigg\},\end{aligned} \tag{2.14}$$

$$F_2(\mathbf{Y}) = \exp\left\{-\frac{2}{\alpha\Delta_m}\sum_{j=1}^{m}\mathbf{y}_j\cdot\mathbf{u}_j\right\}, \tag{2.15}$$

with

$$\begin{aligned}\mathbf{u}_j &= 2\mathbf{r}_j - \mathbf{r}_{j-1} - \mathbf{r}_{j+1}, \quad \text{for } j = 2, \cdots, m-1,\\ \mathbf{u}_1 &= \mathbf{r}_1 - \mathbf{r}_2 + (\mathbf{r}_1 - \mathbf{r})b\sin\delta/r\sin\varphi,\\ \mathbf{u}_m &= \mathbf{r}_m - \mathbf{r}_{m-1} + (\mathbf{r}_m - \mathbf{r}')b\sin\delta/r'\sin\varphi'.\end{aligned} \tag{2.16}$$

We come now to the important point for which Eqs. (2.9)–(2.16) were preparations. From (2.11), it is clear that, by restricting the integration variables $\mathbf{x}_i$, $\mathbf{y}_i$, and $\mathbf{x}'_i$ to the regions

$$|\mathbf{x}_i| < c, \quad |\mathbf{y}_i| < c, \quad \text{and} \quad |\mathbf{x}'_i| < c, \tag{2.17}$$

where $c = b - a$, we can, on the one hand, satisfy the hard-sphere condition (2.3) and, on the other hand, obtain a lower bound for $G_{lmn}$. We also note that

$$\begin{aligned}|\mathbf{u}_i| &= 2b(1-\cos\delta), \quad \text{for } i = 2, \cdots, m-1\\ &= b(1-\cos\delta), \quad \text{for } i = 1 \text{ or } m.\end{aligned} \tag{2.18}$$

Thus, in the region, (2.17), we can replace the factor $F_2(\mathbf{Y})$ by the bound

$$\begin{aligned}F_2(\mathbf{Y}) &\geq \exp\left\{-2(\alpha\Delta_m)^{-1}\sum_{j=1}^{m}c\,|\mathbf{u}_j|\right\}\\ &= \exp\left\{-4c(m-1)(1-\cos\delta)S_b^m/\alpha t\sin\delta\right\}\\ &\underset{m\to\infty}{\longrightarrow} \exp\left\{-2S_bc\theta/\alpha t\right\} \equiv C_3.\end{aligned} \tag{2.19}$$

We also note that

$$\lim_{m\to\infty} C_2 = \exp\{-S_b^2/\alpha t\}. \tag{2.20}$$

We must now calculate the quantity (which is independent of $\mathbf{r}$ and $\mathbf{r}'$)

$$C_4 = C_1 \int d\mathbf{X}\, d\mathbf{Y}\, d\mathbf{X}'\, F_1(\mathbf{X}, \mathbf{Y}, \mathbf{X}'),$$

and we note that, in the limit $l, m, n \to \infty$, this is the Wiener integral for a well-known Green's function. Namely, consider the solution to (1.6) and (1.7) with zero potential but with $\mathbf{r}$ and $\mathbf{r}'$ in the *interior* of a sphere of radius $c$ and with $G = 0$ boundary conditions on the surface of the sphere. If we denote this Green's function by $G_c(\mathbf{r}, \mathbf{r}'; t)$ then, in the limit $l, m, n \to \infty$,

$$C_4 = G_c(0, 0; t). \tag{2.21}$$

To compute $G_c$, it is convenient to use the expansion (1.14). Each $\psi_n(\mathbf{r})$ is a spherical harmonic times a spherical Bessel function but, since we are interested only in the point $\mathbf{r} = \mathbf{r}' = 0$, only $S$-wave (spherically symmetric) solutions will be relevant. For $S$ waves, the normalized radial functions are simply $(2\pi c)^{-\frac{1}{2}} \sin kr/r$, the energies are $e(k) = \frac{1}{4}\alpha k^2$, and $k = n\pi/c$ with $n = 1, 2, 3, \cdots$. Thus,

$$C_4 = \frac{\pi}{2c^3} \sum_{n=1}^{\infty} n^2 \exp\left\{-\left(\frac{\alpha^{\frac{1}{2}}\pi}{2c}\right)^2 tn^2\right\} \tag{2.22}$$

$$> \frac{\pi}{2c^3} \exp\left\{-\left(\frac{\alpha^{\frac{1}{2}}\pi}{2c}\right)^2 t\right\}. \tag{2.23}$$

Our lower bound for $G(\mathbf{r}, \mathbf{r}'; t)$ is the product of $C_2$, $C_3$, and $C_4$, each of which depends on $\mathbf{r}$ and $\mathbf{r}'$ and/or the radius $b$ (or $c = b - a$):

$$G(\mathbf{r}, \mathbf{r}'; t)$$

$$> \frac{\pi}{2c^3} \exp\left\{-\frac{S_b^2 + 2S_b c\theta}{\alpha t} - \left(\frac{\alpha^{\frac{1}{2}}\pi}{2c}\right)^2 t\right\}. \tag{2.24}$$

The inequality (2.24) is generally valid, even if the geodesic from $\mathbf{r}$ to $\mathbf{r}'$ around the sphere of radius $b$ is a straight line. In that case the term $2S_b c\theta/\alpha t$ is to be omitted.

The next step is to determine $c$ so that the right-hand side of (2.24) is maximized. This is a tedious problem since the dependence of $S_b$ on $c$ is complicated. Furthermore, $b$ must always be less than $r$ and $r'$. To calculate $B_{\text{exch}}$, however, we are interested in having $\mathbf{r} = -\mathbf{r}'$ and, from (1.4), it is clear that $r \sim a$ is the important region to consider in the integral. For our purpose—the proof of (1.13)—it is sufficient, as well as legitimate, to take $c = r - a$. The distance $S_b$ is then simply $\pi r$, while $\theta$ is simply $\pi$ for all $r > a$.

Thus,

$$\frac{B_{\text{exch}}}{B^0_{\text{exch}}} = 8 \int G(\mathbf{r}, -\mathbf{r}; 2\Lambda^2/\pi\alpha)\, d\mathbf{r}$$

$$> 8 \int_a^{\infty} 4\pi r^2\, dr \frac{\pi}{2(r-a)^3}$$

$$\times \exp\left\{-\pi^3 \frac{r^2 + 2r(r-a)}{2\Lambda^2} - \frac{\pi\Lambda^2}{2(r-a)^2}\right\}$$

$$> 4\pi^2\Omega^4 e^{-\frac{1}{2}\pi^3(a/\Lambda)^2} \int_0^{\infty} \frac{dp}{p^3}$$

$$\times \exp\left\{-\frac{\pi^2}{2}\left[3\frac{p^2}{\Omega} + \Omega\left(2p + \frac{1}{p^2}\right)\right]\right\}, \tag{2.25}$$

where

$$\Omega = (4\pi)^{\frac{1}{3}}(a/\Lambda)^{\frac{2}{3}}. \tag{2.26}$$

The second inequality in (2.25) is obtained by noting that $r^2 \geq a^2$, and by changing variables to $p = (2\pi^2)^{1/3}\Lambda^{-4/3}a^{1/3}(r - a)$.

The inequality (2.25) is plainly of the form stated in (1.13). To make it more definite, however, we can obtain a lower bound to the integral in (2.25) in the following way: Replace the integration region by (0, 1) instead of (0, $\infty$); in this region, the terms $p^2$ and $p$ in the exponent may be replaced by unity. We are thus left with an integral of the form $\int_0^1 dp p^{-3} \cdot \exp(-\frac{1}{2}\pi^2\Omega p^{-2}) = (\pi^2\Omega)^{-1} \exp(-\frac{1}{2}\pi^2\Omega)$. Collecting the various factors, we obtain

$$\frac{B_{\text{exch}}}{B^0_{\text{exch}}} > \exp\left\{-\frac{\pi^3}{2}\left(\frac{a}{\Lambda}\right)^2 - \frac{3\pi^2}{2}\left(2\sqrt{\pi}\frac{a}{\Lambda}\right)^{\frac{2}{3}}\right.$$

$$\left. + 2\ln\left(4\sqrt{\pi}\frac{a}{\Lambda}\right) - \frac{3\pi^2}{2}\left(2\sqrt{\pi}\frac{a}{\Lambda}\right)^{-\frac{2}{3}}\right\} \tag{2.27}$$

as our final lower bound for $B_{\text{exch}}$.

## III. UPPER BOUND BY PATH INTEGRALS

We are interested in computing the path integral, (2.2), when the factors $\exp[-\Delta v(\mathbf{z})]$ are omitted, but when the integration ranges are restricted to $|z_i| > a$ for all $i$. The lower bound to (2.2) was obtained in Sec. II by restricting the integration range still further, namely, to a tube lying just outside the sphere. At first sight it would seem that the opposite procedure—integrating over too great

J. Math. Phys. 8, 43–52 (1967)



a region—should yield a suitable upper bound. Indeed, when $\mathbf{r}$ and $\mathbf{r}'$ are in each other's *line of sight* (i.e., when the straight line between the two points does not interest the sphere), then the simple expedient of integrating over *all* space yields an upper bound which is at once useful and accurate for small time (high temperature), viz:

$$G(\mathbf{r},\mathbf{r}';t) < G_0(\mathbf{r},\mathbf{r}';t). \tag{3.1}$$

While (3.1) is true for all $\mathbf{r}$ and $\mathbf{r}'$, it is quite misleading when the two points are in each other's *shadow*. A more sensitive extension of the integration range is required; but, unfortunately, allowing the paths to penetrate the sphere only slightly does not render the integral any more tractable than the original. In order to make the integration feasible, it appears to be necessary to extend the integrations to all space; but then the upper bound so obtained, (3.1), is virtually useless.

Our resolution of the dilemma is to integrate over all space, but at the same time to include an additional weight factor in the integrand of (2.2) so that paths which penetrate the sphere are effectively suppressed.

As in Sec. II, we consider the "taut string" shown in Fig. 1, except that this time we take $c = 0$ (i.e., radius $b$ = radius $a$). Otherwise, everything is the same as given in Eqs. (2.10)–(2.16). The first step in obtaining an upper bound is to integrate over the variables $\mathbf{X}$ and $\mathbf{X}'$ (alternatively, $\mathbf{z}_i$ for $i=1,\cdots,l$ and $i=l+m+1,\cdots,l+m+n$) over *all* space. We then pass to the limit $l$ and $n\to\infty$ and obtain

$$G(\mathbf{r},\mathbf{r}';t) < \lim_{m\to\infty} G_m(\mathbf{r},\mathbf{r}';t), \tag{3.2}$$

where

$$G_m(\mathbf{r},\mathbf{r}';t) = D_1C_2\int_R d\mathbf{Y}\, F_2(\mathbf{Y})F_3(\mathbf{Y}), \tag{3.3}$$

with

$$D_1 = (\pi\alpha t_1)^{-\frac{3}{2}}(\pi\alpha t_2)^{-\frac{3}{2}}(\pi\alpha\Delta_m)^{-\frac{3}{2}(m-1)}, \tag{3.4}$$

$$F_3(\mathbf{Y}) = \exp\left\{-(\alpha t_1)^{-1}|\mathbf{y}_1|^2 - (\alpha t_2)^{-1}|\mathbf{y}_m|^2 - (\alpha\Delta_m)^{-1}\sum_{i=2}^{m}|\mathbf{y}_i-\mathbf{y}_{i-1}|^2\right\}, \tag{3.5}$$

and

$$t_1 = (l+1)\,\Delta_l = tr\sin\varphi/S_a^m,\qquad t_2 = (n+1)\,\Delta_n = tr'\sin\varphi'/S_a^m. \tag{3.6}$$

The quantities $C_2$ and $F_2$ are as given in (2.13) and (2.15), respectively (with $b = a$, of course).

The integration range in (3.3) is

$$R{:}\ |\mathbf{y}_i + \mathbf{r}_i| > a,\quad \text{for}\quad i = 1,\cdots,m. \tag{3.7}$$

Since the $\mathbf{r}_i$ are different, one from another, the integration range for each $i$ is different. To overcome this complication, we integrate (3.3) over all space *after* first replacing the function $F_2(\mathbf{Y})$ by another positive function, $\bar{F}_2(\mathbf{Y})$, which has the property that $\bar{F}_2(\mathbf{Y}) \geq F_2(\mathbf{Y})$ for $\mathbf{Y}$ in the allowed region, $R$, while $\bar{F}_2(\mathbf{Y})$ is generally less than $F_2(\mathbf{Y})$ for paths which penetrate the sphere. First note that the vectors $\mathbf{u}_i$, given in (2.16), are parallel to $\mathbf{r}_i$:

$$\begin{aligned}\mathbf{u}_i &= 2(1-\cos\delta)\mathbf{r}_i, \quad \text{for}\quad i = 2,\cdots,m-1\\ &= (1-\cos\delta)\mathbf{r}_i, \quad \text{for}\quad i = 1 \ \text{or}\ m.\end{aligned} \tag{3.8}$$

In the allowed region, $R$, we have $a^2 \leq |\mathbf{y}_i + \mathbf{r}_i|^2 = |\mathbf{y}_i|^2 + 2\mathbf{y}_i\cdot\mathbf{r}_i + a^2$. Thus, in $R$,

$$\begin{aligned}\mathbf{y}_i\cdot\mathbf{u}_i &\geq -|\mathbf{y}_i|^2(1-\cos\delta), \quad \text{for}\quad i = 2,\cdots,m-1\\ &\geq -\tfrac{1}{2}|\mathbf{y}_i|^2(1-\cos\delta), \quad \text{for}\quad i = 1\ \text{or}\ m.\end{aligned} \tag{3.9}$$

Hence, in, $R$

$$F_2(\mathbf{Y}) \leq \bar{F}_2(\mathbf{Y}) = \exp\left\{\frac{1-\cos\delta}{\alpha\,\Delta_m}\times\left[|\mathbf{y}_1|^2 + |\mathbf{y}_m|^2 + 2\sum_{i=2}^{m-1}|\mathbf{y}_i|^2\right]\right\}. \tag{3.10}$$

Now, the integral over all space of the product $\bar{F}_2(\mathbf{Y})F_3(\mathbf{Y})$ is a simple $m$-dimensional Gaussian integral, which can be evaluated by using the well-known formula

$$\int_{-\infty}^{\infty}dx_1\cdots\int_{-\infty}^{\infty}dx_N\exp\left\{-\sum_{i,j=1}^{N}x_iA_{ij}x_j\right\} = \pi^{N/2}[\operatorname{Det}A]^{-\frac{1}{2}}, \tag{3.11}$$

for any symmetric, positive definite $N$-square matrix $A$. Applying this formula to $G_m$ (with $F_2$ replaced by $\bar{F}_2$), we obtain

$$G_m(\mathbf{r},\mathbf{r}';t) < C_2\left[\frac{\pi\,\alpha t_1t_2}{\Delta_m}|B^m|\right]^{-\frac{3}{2}}, \tag{3.12}$$

where $|B^m|$ is the determinant of the tri-diagonal $m$-square matrix

$$B^m = \begin{bmatrix} \frac{\Delta_m}{t_1} + \cos\delta & -1 & & & & & 0 \\ -1 & 2\cos\delta & -1 & & & & \\ & -1 & 2\cos\delta & -1 & & & \\ & & -1 & \ddots & \ddots & & \\ & & & \ddots & \ddots & -1 & \\ & & & & \ddots & 2\cos\delta & \\ 0 & & & & -1 & 2\cos\delta & -1 \\ & & & & & -1 & \frac{\Delta_m}{t_2} + \cos\delta \end{bmatrix} \tag{3.13}$$

The exponent $\frac{3}{2}$ in (3.12) instead of $\frac{1}{2}$ as in (3.11) comes about because each of the $m$ variables of integration is three dimensional.

In order for (3.12) to be valid, it is necessary that $B_m$ be positive definite. If $\delta = 0$, that criterion is surely satisfied and (by continuity) $B_m$ is positive definite for $0 < \delta < \bar{\delta}$, where $\bar{\delta}$ is the smallest value of $\delta$ for which $|B^m| = 0$.

To evaluate $|B^m|$, we expand in the first row and column as well as in the $m$th row and column and obtain

$$|B^m| = \left(\cos\delta + \frac{\Delta_m}{t_1}\right)\left(\cos\delta + \frac{\Delta_m}{t_2}\right)U_{m-2} - \left(2\cos\delta + \frac{\Delta_m}{t_1} + \frac{\Delta_m}{t_2}\right)U_{m-3} + U_{m-4}, \tag{3.14}$$

where $U_m$ is the $m$-square determinant

$$U_m = \mathrm{Det}\begin{bmatrix} 2\cos\delta & -1 & & & & 0 \\ -1 & 2\cos\delta & -1 & & & \\ & -1 & 2\cos\delta & -1 & & \\ & & -1 & \ddots & \ddots & \\ & & & \ddots & 2\cos\delta & -1 \\ 0 & & & & -1 & 2\cos\delta \end{bmatrix} \tag{3.15}$$

Since $U_m$ obviously satisfies the recursion relationship

$$U_m = 2\cos\delta U_{m-1} - U_{m-2}, \tag{3.16}$$

it follows that $U_m$ ($\cos\delta$) is the Chebyshev polynominal of the second kind[14] (in the variable $\cos\delta$), whence

$$U_m = \sin(m+1)\,\delta/\sin\delta. \tag{3.17}$$

Combining (3.17) with (3.14) and, recalling that $\theta = (m-1)\delta$, we obtain

$$|B^m| = \frac{\Delta_m^2 \sin\theta}{t_1 t_2 \sin\delta} - \Delta_m\left(\frac{1}{t_1} + \frac{1}{t_2}\right)\cos\theta - \sin\theta\sin\delta. \tag{3.18}$$

Now, recalling the definitions (2.10), (3.6) and the fact that $r\cos\varphi = a = r'\cos\varphi'$, (3.18) is equivalent to

$$\frac{t_1 t_2}{\Delta_m}|B^m| = \frac{ta\sin(\varphi + \varphi' + \theta)}{S_a^m \cos\varphi\cos\varphi'}. \tag{3.19}$$

But $\varphi + \varphi' + \theta = \psi =$ angle between $\mathbf{r}$ and $\mathbf{r}'$ [cf. (2.5)]. Thus, combining (3.19) with (2.13), (3.3), and (3.12) and passing to the limit $m \to \infty$, we have our upper bound

$$G(\mathbf{r}, \mathbf{r}'; t) < \left[\frac{S_a \cos\varphi\cos\varphi'}{\pi a t a \sin\psi}\right]^{\frac{3}{2}} \exp\left\{-\frac{S_a^2}{\alpha t}\right\}. \tag{3.20}$$

[14] A. Erdelyi, Ed., *Higher Transcendental Functions* (McGraw-Hill Book Co., Inc., New York, 1953), Vol. II, Chap. 10, p. 183.

J. Math. Phys. 8, 43–52 (1967)



Formula (3.20) has the essential feature that we have sought, namely, the factor exp $\{-(\text{shortest distance from } \mathbf{r} \text{ to } \mathbf{r}' \text{ around the sphere})^2/\alpha t\}$. It also has the factor $(\pi\alpha t)^{-\frac{3}{2}}$, characteristic of $G_0$. The factor $(S_a \cos\varphi \cos\varphi'/a \sin\psi)$, while it is usually of the order of unity, can be embarrassingly large when $\psi \sim \pi$. Unfortunately, it is precisely the case of diametric juxtaposition of $\mathbf{r}$ and $\mathbf{r}'$ that is of interest in calculating $B_{\text{exch}}$. Plainly, some slight improvement is required before inserting (3.20) into (1.4).

It is interesting to note, however, that the divergence in our upper bound at $\psi = \pi$ is not entirely unexpected. This is because many paths of the *same length* come together at that angle. In other words, $\psi = \pi$ can be regarded as a caustic. Our upper bound concentrated essentially on only one path around the sphere and, since that one path is not sufficient at $\psi = \pi$, difficulties were encountered there. It is noteworthy that precisely the same divergence is encountered in the classical asymptotic expansion for diffraction around a sphere.[7]

A simple artifice to overcome the annoying $(\sin\psi)^{-\frac{1}{2}}$ factor is the following: Let $\mathbf{O}Q$ be a vector of length $q < a$ perpendicular to $\mathbf{r}$ and let $s'$ be the sphere of radius $b = a - q$ centered at the point $Q$. This sphere is clearly tangent to the original sphere, $s$, (of radius $a$) at the single point $(a/q)\mathbf{O}Q$ and otherwise lies entirely inside the larger sphere, $s$. Also, let $G_{s'}(\mathbf{r}, \mathbf{r}'; t)$ be the Green's function for the exterior of $s'$, just as $G(\mathbf{r}, \mathbf{r}'; t)$ is the Green's function for the exterior of $s$. From (2.2), we see at once,

$$G(\mathbf{r}, \mathbf{r}'; t) < G_{s'}(\mathbf{r}, \mathbf{r}'; t) \qquad (3.21)$$

for all points $\mathbf{r}$ and $\mathbf{r}'$. We can, in turn, say that $G_{s'}$ is less than the right-hand side of (3.20), where the quantities $\varphi$, $\varphi'$, $\psi$, and $S$ are now measured relative to the sphere $s'$ centered at $Q$.

For our purposes, we want $\mathbf{r}' = -\mathbf{r}$ with $r > a$. Relative to the sphere $s'$, we have the following simple geometric inequalities for all $r > a$:

$$\pi(a - 3q) < S < \pi r, \qquad (3.22)$$
$$\sin\psi = \frac{2rq}{r^2 + q^2} > \frac{q}{r}.$$

In addition, $\cos\varphi \cos\varphi' < 1$, whence

$$G(\mathbf{r}, -\mathbf{r}; t) < r^3(\alpha t a q)^{-\frac{3}{2}}$$
$$\times \exp\{-\pi^2(a - 3q)^2/\alpha t\}, \qquad (3.23)$$

for any $0 < q < a/3$ and for all $r > a$.

We can now evaluate $B_{\text{exch}}$ as given by (1.4). To do so, we divide the integration range $\int_a^\infty dr$ into two parts: $\int_a^{2a} dr$ and $\int_{2a}^\infty dr$. In the former range, we use the bound (3.23), while in the later range, we use the very simple bound $G_0$ as in (3.1). Thus,

$$\frac{B_{\text{exch}}}{B^0_{\text{exch}}} = 8 \int G(\mathbf{r}, -\mathbf{r}; 2\Lambda^2/\pi\alpha)\, d\mathbf{r}$$
$$< 8 \int_a^{2a} dr\, 4\pi r^2 r^3 [2\Lambda^2 q a/\pi]^{-\frac{3}{2}}$$
$$\times \exp\{-\pi^3(a - 3q)^2/2\Lambda^2\}$$
$$+ 8 \int_{2a}^\infty dr\, 4\pi r^2 (2\Lambda^2)^{-\frac{3}{2}} \exp\{-2\pi r^2/\Lambda^2\}. \qquad (3.24)$$

In the first integral, take $q = \Lambda^2/(2\pi^3 a)$, assuming that $(\Lambda/a)^2 < 2\pi^3/3$. The second integral is clearly Order $\{\exp[-8\pi(a/\Lambda)^2]\}$ and is therefore exponentially small compared to $\exp[-\frac{1}{2}\pi^3(a/\Lambda)^2]$. While an upper bound to this second terms can be easily found, there is little point in doing so.

Evaluating the first integral in (3.24) and combining it with the second, we obtain our final upper bound:

$$\frac{B_{\text{exch}}}{B^0_{\text{exch}}} < \exp\left\{-\frac{\pi^3}{2}\left(\frac{a}{\Lambda}\right)^2 + \ln\left[\frac{1}{3}\, 2^{10}\pi^7\left(\frac{a}{\Lambda}\right)^6\right] + \frac{3}{2}\right.$$
$$\left. - \frac{9}{(2\pi)^3}\left(\frac{\Lambda}{a}\right)^2 + O\left[\exp\left(\tfrac{1}{2}\pi^3 - 8\pi\right)\left(\frac{a}{\Lambda}\right)^2\right]\right\}. \qquad (3.25)$$

## IV. CONCLUSIONS

By means of the discrete version of the Wiener integral, (2.2), we have obtained upper and lower bounds to the diffusion Green's function in the presence of an opaque sphere [Eqs. (2.24) and (3.20), respectively]. These bounds are useful for short time (high temperature), especially when the source point and the observation point are in each other's shadow.

The bounds enable us to calculate lower and upper bounds to the exchange part of the second virial coefficient of a hard-sphere gas. These bounds, respectively, given in (2.27) and (3.25), permit us to assert that the correct $B_{\text{exch}}$ diminishes with temperature much more rapidly than the noninteracting $B^0_{\text{exch}}$, in a manner given by the equation

$$\frac{B_{\text{exch}}}{B^0_{\text{exch}}} = \exp\left\{-\frac{\pi^3}{2}\left(\frac{a}{\Lambda}\right)^2 + O\left[\left(\frac{a}{\Lambda}\right)^{\frac{4}{3}}\right]\right\}.$$

## ACKNOWLEDGMENTS

The author thanks Dr. S. Larsen, Dr. J. Kilpatrick, and H. Jordan, who first stimulated my interest in the problem. Thanks are also due to Dr. E. Hammel for the hospitality of his department at Los Alamos, where these conversations occurred. I am also indebted to Dr. S. Larsen for many valuable comments during the course of this work.

# Part III

# Zeros of Partition Functions

VOLUME 24, NUMBER 25 PHYSICAL REVIEW LETTERS 22 JUNE 1970

# MONOMERS AND DIMERS

Ole J. Heilmann*†
Kemisk Laboratorium III, H. C. Orsted Institutet, University of Copenhagen, Denmark

and

Elliott H. Lieb*
Department of Mathematics, Massachusetts Institute of Technology, Cambridge, Massachusetts 02139
(Received 1 May 1970)

We prove that the free energy of an arbitrary monomer-dimer system is analytic in the density and temperature for nonzero density, and hence that the system has no phase transition. This result can also be used to locate the $z=e^{2\beta h}$ roots of the Heisenberg ferromagnet or antiferromagnet at high temperature.

From time to time the monomer-dimer (MD) system is used as a model of a physical system,[1] but primarily it is interesting as the phototypical lattice statistical mechanics problem. Although the pure dimer (PD) problem can be solved for planar lattices,[2] nothing was heretofore known rigorously about the MD problem because no theorems and no exact solutions were available (except in one dimension where the problem is not very interesting).

In this note we present the outline of a complete theory of the subject which allows us to answer most questions of physical interest. Essentially, the only question left unanswered is the nature of the singularity, if any, as we approach the PD limit, the answer to which is surely lattice dependent and, therefore, complicated. Gaunt's series expansions[3] offer nonrigorous but convincing evidence of the existence and lattice dependence of this singularity. Otherwise, our theorems show that the free energy is analytic in the monomer density $\rho$ and the temperature $T=(k\beta)^{-1}$. This had been conjectured before for specific lattices on the basis of numerical calculations.[3] We can also show that the correlation functions exist and enjoy the same analyticity properties. An appropriate variable in which to form a power series convergent for all $\rho>0$ is easily derived. In short, all that remains to be done in any specific problem is to use conventional graphical expansions to calculate coefficients on a computer. Admittedly, this procedure is likely to be impractical for small monomer densities unless one has a clear idea of the singularity at $\rho=0$.

We also show how the MD theory can be used to locate the $z=e^{2\beta h}$ roots of the Heisenberg and Ising ferromagnet and antiferromagnet at high temperature ($h$ is the magnetic field).

To formulate the problem consider a lattice $L$ consisting of $N$ vertices and a set of $\binom{N}{2}$ non-negative bond weights $w_{ij}$. We can introduce $T$ by setting $w_{ij}=\exp(-\beta J_{ij})$ for suitable real $J_{ij}$. $L$ is said to be articulated if the vertices can be numbered so that $w_{12}, w_{23}, w_{34}, \cdots, w_{N-1,N}\neq 0$. $L$ is said to be bounded by $W$ if $\sum_j w_{ij}\leq W$ for all $i$. Dimers can be placed on pairs of vertices so that each vertex has at most one dimer. The weight of a covering by $d$ dimers on $(a,b)$, $(c,d)$, $\cdots$ is $w_{ab}, w_{cd}, \cdots$, and $Z_d$ is the sum of these weights for all possible $d$-dimer coverings. Uncovered vertices are regarded as being occupied by monomers having an activity $x$, so that the total MD partition function is

$$P_L(x)=\sum_{d=0}^{M} Z_d x^{N-2d}, \tag{1}$$

where $M$ is the largest integer in $N/2$. Let $P_{L'}$ be the partition function for the lattice with vertex $N$ (and its edges) removed and let $P_{L'}{}^k$ be the same when vertices $N$ and $k$ are removed. The key equation is

$$P_L(x)=xP_{L'}(x)+\sum_{k=1}^{N-1} w_{k,N}P_{L'}{}^k(x). \tag{2}$$

For $N=1$ or 2 the roots of $P_L(x)=0$ are imaginary. We are led to the following theorem whose proof involves a simple modification of the classical inductive argument (on $N$) appropriate to a Sturm sequence and which can easily be supplied by the reader. The proof of the bound on the roots does not appear to be standard but it requires merely an addition to the inductive hypothesis which, in toto, reads, "For all lattices of order $N$, (a) Theorem 1 is true; (b) for $x=i\alpha$ and $\alpha\geq 2W^{1/2}$, $P_{L'}(x)/P_L(x)=i\delta$ with $\delta\geq -W^{-1/2}$."

Theorem 1.

$$P_L(x)=\prod_{j=1}^{M}(x^2+b_j),\ N \text{ even}$$

$$=x\prod_{j=1}^{M}(x^2+b_j),\ N \text{ odd}, \tag{3}$$

where $0 \le b_j < 4W$. The roots of $P_L = 0$ interlace those of $P_{L'} = 0$ and, if $L$ is articulated, the roots strictly interlace and hence are simple.

The significance of this theorem is that $\ln P_L(x)$ is analytic in the right-hand plane and hence that no phase transition can occur. The simplicity of the roots will be needed later in connection with the Heisenberg and Ising models.

Corollary 1.

$$2\ln Z_d \ge \ln Z_{d-1} + \ln Z_{d+1} + \ln\frac{(M-d+1)(d+1)}{(M-d)d}. \quad (4)$$

This is merely a statement of Newton's inequality.[4] Its meaning is that even for a finite system the free energy per unit volume is a strictly convex function of the monomer (or dimer) density.

The grand free energy per unit volume is

$$-\beta F = N^{-1}\ln P_L(x), \quad (5)$$

and it is clearly analytic away from the cut, $(-2iW^{1/2}, 2iW^{1/2})$. To expand $F$ near zero monomer density, the natural variable to use is $u = x^{-1}$. As $F$ has singularities in the $u$ plane along $(i/2W^{1/2}, i\infty)$ and $(-i/2W^{1/2}, -i\infty)$, a power series in $u$ will have only a finite radius of convergence. To remedy this use $s = (2W^{1/2}u)^{-1}[-1 + (1+4Wu^2)^{1/2}]$ or $W^{1/2}u = s(1-s^2)^{-1}$ which maps the unit $s$ disk conformally onto the $u$ plane less the cuts. Thus one constructs the usual power series in $u$, rearranges in powers of $s$, and convergence for all real $u$ is guaranteed.

The monomer density is defined as

$$\rho = xN^{-1}d\ln P_L(x)/dx. \quad (6)$$

It is easy to prove from (3) the following:

Theorem 2. The roots of $d\rho/dx = 0$ lie in

$$D = \{x: |x| < 2W^{1/2}, \pi/4 < \arg x < 3\pi/4 \text{ and } 5\pi/4 < \arg x < 7\pi/4\}.$$

The significance of this theorem is that the inverse function theorem guarantees a neighborhood of the positive $\rho$ axis in which $F$, considered as a function of $\rho$, is analytic.

The dimers can also be thought of as hard-core particles on the line (or covering) graph of $L$. For example, if $w_{ij} = 1$ on the edges of a planar hexagonal lattice and zero otherwise, the MD problem is the same as the nearest-neighbor exclusion problem on a Kagome lattice. Our theorem tells us that there is no phase transition on the Kagome lattice as there is for the square lattice.[5] Using this point of view, however, we can modify the Ginibre-Penrose method[6] to yield a lower bound on the compressibility,

$$\beta\chi^{-1} \equiv -\beta\rho\partial F/\partial\rho = \rho^2[x d\rho(x)/dx]^{-1}. \quad (7)$$

By this method one first derives the inequality

$$dZ_d^2 < Z_{d-1}\{(d+1)Z_{d+1} + 2WZ_d\}, \quad (8)$$

and then

$$\beta\chi^{-1} < \tfrac{1}{2}\rho^2[1 + 2Wx^{-2}]/(1-\rho). \quad (9)$$

(We wish to thank Professor J. Lebowitz for calling our attention to the fact that by treating the dimers as hard-core particles the theorems of Ref. 6 independently yield the same qualitative conclusion as Theorem 2, namely that $F$ is real analytic in $\rho$.)

Other bounds which can be derived directly from (3) are

$$\tfrac{1}{2}\rho/(1-\rho) \le \beta\chi^{-1} \le \tfrac{1}{2}\rho^2 W/(1-\rho)^2\chi^2, \quad (10)$$

$$\rho(x) \ge [1 + Wx^{-2}]^{-1}. \quad (11)$$

Turning now to a generalization of Theorem 1, we may consider a system in which placing a monomer at vertex $i$ entails a Boltzmann factor $m_i x_i$ (instead of merely $x$) where $m_i > 0$, all $i$, and are regarded as fixed and $x_i$ is the (variable) activity at site $i$. In this case we say that $L$ is bounded by $W$ if $m_i^{-1}\sum_j W_{ij} m_j^{-1} \le W$ for all $i$.

Theorem 3. If $x_i = x$ for all $i$ then Theorem 1 is still true. Otherwise, if $\mathrm{Re}(x_i) > 0$, all $i$, or $\mathrm{Re}(x_i) < 0$, all $i$, then $P_L(x_1, \cdots, x_N) \ne 0$. The proof is an adaptation of that for Theorem 1.

If $w_{ij}$ depends on temperature, as aforementioned, analyticity in $x$ does not trivially imply analyticity in $\beta$. The problem is similar to that for the Ising ferromagnet where the circle theorem holds.[7] Following the sophisticated analytic tour de force of Lebowitz and Penrose,[8] however, we can likewise show that for the MD problem there is analyticity in $(x, \beta)$ for $x$ in the right-hand plane and $\beta$ real and positive. As they did, we can also establish the existence of correlation functions. These statements are, of course, trivial for a finite system. The difficulty lies in proving them in the thermodynamic $(N \to \infty)$ limit. To our knowledge, no one has ever carried out the proof of the convergence of the virial series for the MD problem, but this can be done in a manner parallel to that for the Ising model. Hammersley has, however, proved the existence of the thermodynamic limit.[9]

The analogy with the Ising circle theorem is not fortuitous. Fisher[10] has shown how a zero-field Ising model can be put into one-one correspondence with a PD problem, and with non-neg-

VOLUME 24, NUMBER 25 PHYSICAL REVIEW LETTERS 22 JUNE 1970

ative bond weights in the case of a ferromagnet. He did not show how to include a magnetic field, but that lack is easily filled with the result that the Ising model becomes a MD problem with $x = (z-1)/(z+1)$, $z = e^{2\beta h}$. The only difference is that not all sites are allowed to have monomers (i.e., $m_i = 0$ or 1 in the above) so that Theorem 3 is called into play. In brief, our analysis starting with (3) and using the notion of a Sturm sequence provides a completely independent proof of the circle theorem (note that $|z|=1$ is equivalent to $x$ imaginary).

On the other hand, we can start with the circle theorem and derive Theorem 1, except for the statement about the simplicity of the roots. For generality, consider a spin-$\frac{1}{2}$ Heisenberg Hamiltonian: $H = H_0 + H_1$; $H_0 = \sum J_{ij} s_{iz} s_{jz}$; $H_1$ = arbitrary Hermitian quadratic form in $\{s_{ix}, s_{iy}\}$, and let

$$Z = (e^{\beta h} + e^{-\beta h})^{-N} \operatorname{Tr}[\exp(-\beta H) \exp(2\beta h \textstyle\sum s_{iz})].$$

Next, expand $e^{-\beta H}$ in a Taylor series, take the trace term by term, and express the result as an even polynomial of order $N$ in $y = (e^{2\beta h}-1)/(e^{2\beta h}+1)$. If $c_{2k}(\beta)$ be the coefficient of $y^{2k}$ in $Z$ then $c_{2k}(\beta) = \beta^k Z_k + r_{2k}(\beta)$ where $Z_k$ is the $k$-dimer partition function on a lattice in which $w_{ij} = J_{ij}$. The significant fact, which the reader can easily verify, is that $r_{2k}$, while complicated, is of higher order in $\beta$ than $k$. Hence,

$$Z = t^{-N} P_L(t) + R_L(t), \tag{12}$$

where $t^{-2} \equiv \beta y^2$ and $R_L(t)$ is a polynomial whose coefficients all vanish with $\beta$. If $H_1 = 0$ (Ising model) and $J_{ij} \geq 0$, the circle theorem tells us that the roots of $Z$ (in $t$) are imaginary for all positive $\beta$. It is easy to prove that the leading polynomial, $P_L(t)$, must necessarily also have this property and thus we have another proof of most of Theorem 1. Conversely, if the roots of $P_L(t)$ are imaginary and simple, $Z$ will also have this property for sufficiently small $\beta$. If we also note that negating the sign of all $J_{ij}$ is the same as changing $t$ to $it$ in $P_L$ we have the following:

Theorem 4. Let $H$ be the Hamiltonian of a Heisenberg ferromagnet ($J_{ij} \geq 0$) or antiferromagnet ($J_{ij} \leq 0$) such that the lattice of $\{J_{ij}\}$ is articulated. Then there exists a $\beta_0 > 0$ such that for $\beta < \beta_0$ the roots of $Z = 0$ in $e^{2\beta h}$ are (i) on the unit circle for a ferromagnet and (ii) on the negative real axis for an antiferromagnet.

Theorem 4 complements Suzuki's proof[11] of the circle theorem for sufficiently large $\beta$ but, unlike his hypothesis, we are not obliged to place any constraint on the off-diagonal part, $H_1$. As in Suzuki's case, we are obliged to state that we can give no bound for $\beta_0$ which is independent of $N$.

After this work was completed, we received a preprint from T. Asano[12] which contains a complete proof of the circle theorem for the anisotropic Heisenberg ferromagnet. Consequently, the ferromagnetic part of our Theorem 4 is obsolete.

---

*Work partially supported by National Science Foundation Grant No. GP-9414.

†Work partially supported by Statens Naturvidenskabelige Forskningsråd Grant No. 511-208/69.

[1]See, e.g., R. H. Fowler and G. S. Rushbrooke, Trans. Faraday Soc. 33, 1272 (1937).

[2]P. W. Kasteleyn, in *Graph Theory and Theoretical Physics*, edited by F. Harary (Academic, New York, 1967), p. 43. In certain cases the correlation functions for a few monomers in otherwise infinite sea of dimers can also be calculated: M. E. Fisher and J. Stephenson, Phys. Rev. 132, 1411 (1963); R. E. Hartwig, J. Math. Phys. 7, 286 (1966).

[3]D. S. Gaunt, Phys. Rev. 179, 174 (1969); L. K. Runnels, J. Math. Phys. 11, 849 (1970); J. F. Nagle, Phys. Rev. 152, 190 (1966); R. J. Baxter, J. Math. Phys. 9, 650 (1968).

[4]G. H. Hardy, J. E. Littlewood, and G. Polya, *Inequalities* (Cambridge Univ., Cambridge, England, 1959), Theorem 144.

[5]R. L. Dobrushin, Funktsional Analiz i Ego Prilozhen 2, 44 (1968) [Funct. Anal. Appl. 2, 302 (1968)].

[6]Cf. D. Ruelle, *Statistical Mechanics* (Benjamin, New York, 1969), Proposition (3.4.9).

[7]T. D. Lee and C. N. Yang, Phys. Rev. 87, 410 (1952).

[8]J. L. Lebowitz and O. Penrose, Commun. Math. Phys. 11, 99 (1968).

[9]J. M. Hammersley, in *Research Papers in Statistics; Festschrift in Honor of Jerzy Neyman*, edited by Florence Nightingale David and Evelyn Fix (Wiley, New York, 1966), p. 125.

[10]M. E. Fisher, J. Math. Phys. 7, 1776 (1966).

[11]M. Suzuki, Progr. Theoret. Phys. (Kyoto) 41, 1438 (1969).

[12]T. Asano, "The Rigorous Theorems for the Heisenberg Ferromagnet," (to be published).

Commun. math. Phys. 25, 190—232 (1972)


# Theory of Monomer-Dimer Systems*

OLE J. HEILMANN** and ELLIOTT H. LIEB

Department of Mathematics
Massachusetts Institute of Technology, Cambridge, Massachusetts, USA

Received November 30, 1971

**Abstract.** We investigate the general monomer-dimer partition function, $P(x)$, which is a polynomial in the monomer activity, $x$, with coefficients depending on the dimer activities. Our main result is that $P(x)$ has its zeros on the imaginary axis when the dimer activities are nonnegative. Therefore, no monomer-dimer system can have a phase transition as a function of monomer density except, possibly, when the monomer density is minimal (i.e. $x = 0$). Elaborating on this theme we prove the existence and analyticity of correlation functions (away from $x = 0$) in the thermodynamic limit. Among other things we obtain bounds on the compressibility and derive a new variable in which to make an expansion of the free energy that converges down to the minimal monomer density. We also relate the monomer-dimer problem to the Heisenberg and Ising models of a magnet and derive Christoffell-Darboux formulas for the monomer-dimer and Ising model partition functions. This casts the Ising model in a new light and provides an alternative proof of the Lee-Yang circle theorem. We also derive joint complex analyticity domains in the monomer and dimer activities. Our considerations are independent of geometry and hence are valid for any dimensionality.

## I. Introduction

A monomer-dimer system is specified by a graph, $G$ (also called a lattice in the physics literature), together with a family of weights (or Boltzmann factors) assigned to the edges of $G$. The precise definition of a weighted graph is given in Section II, but for the present we shall assume the reader is familiar with the concept. Dimers can be placed on the edges of $G$ so that no vertex has more than one dimer. Uncovered vertices are called monomers and have a fugacity which we call $x$. One can also define related problems, such as the monomer-trimer problem, and although the history of these various problems are intertwined we shall consider only the monomer-dimer problem in this paper.

We shall answer the question whether, as the monomer concentration is varied, a phase transition can occur for an infinite system. Our answer,

* Work supported by National Science Foundation Grant GP-26526.

** Permanent address: Kemisk Laboratorium III, the H.C. Ørsted Institute, University of Copenhagen, 2100 Copenhagen, Denmark.

which is independent of the geometry of the graph, is that the only place a singularity can possibly occur is at minimum monomer density (which is zero for lattices normally considered).

Although this paper is mathematical in nature and is not a survey paper, it might be useful to begin with a brief historical sketch (with no pretense of completeness, and therefore with apologies to the authors we have unintentionally omitted) to show how monomer-dimer systems are related to chemistry and physics.

The problem of placing nonoverlapping dimers on a lattice goes back at least to 1935 when Roberts [1] considered the problem of absorption of oxygen and hydrogen on a tungsten surface. His assumptions were based on kinetic considerations: If oxygen molecules are absorbed on the surface in a single layer with the two oxygen atoms of a molecule covering two neighboring tungsten atoms, how does the chance of finding two neighboring tungsten atoms both unoccupied depend on the density of absorbed oxygen and what is the average maximum density? Roberts attacked the problem by a straightforward Monte Carlo calculation [1–3] and later theoretically by applying the Bethe approximation [4]. More extensive Monte Carlo calculations were undertaken after the appearance of high speed computers [5, 6], and the problem has been given rigorous mathematical treatment in one dimension by McQuistan [7–10]. It should be noted that since it is assumed that molecules once absorbed do not move or leave the surface, then the statistics of this problem differs from the statistics of the ordinary, equilibrium monomer-dimer problem treated in the present paper and we shall deal no further with it. It is a pecularity of the Bethe approximation that the two problems are equivalent in first order.

The earliest treatment of the equilibrium monomer-dimer problem is due to Fowler and Rushbrooke [11], who took up the problem to settle a question raised earlier in 1937 by Guggenheim [12], namely whether deviations in the properties of a binary mixture from those of an ideal mixture could be caused solely by a difference in size of the two components. The Fowler-Rushbrooke paper discusses approximations valid at low dimer density and treats the pure dimer covering (i.e., no monomers) problem by finding eigenvalues for the transfer matrix for narrow strips.

The monomer-dimer problem was soon attacked in the Bethe approximation or random mixing approximation [4, 13–17], in which cases it is easy to include an interaction between neighboring vertices both of which are occupied by dimers. Almost simultaneously, the more general problem of monomer-polymer mixtures was attacked in the same approximations [18, 19, 21, 22] and in the slightly cruder approximations of the Flory-Huggins theory [23]. A description of these methods

14*

and the results can be found in Guggenheim's book on mixtures [24]. Rushbrooke *et al.* [25] attacked the monomer-polymer problem by series expansions valid at low polymer density.

Newer and more accurate results on the monomer-dimer problem have been obtained by series expansions valid at low dimer density [26–29] (the results of our paper show that the choice of expansion parameter in [27] actually makes the expansion valid at all dimer densities which are lower than the density of the pure dimer covering; for a proof of this fact see Chapter IX). Other results have been obtained by finding the largest eigenvalue of the transfer matrix either numerically for narrow strips [30] or by a variational procedure [31]. A numerical comparison of the entropy of mixing as calculated from the different approximations can be found in [32]. Monte Carlo calculations have also been performed [33].

The application of monomer-dimer systems as models for real, physical systems is rather limited. Fair agreement has been obtained for binary mixtures like benzene-diphenyl [34–36], benzene-diphenylmethane [35] and benzene-dibenzyl [35]. Mixtures like hexane-cetane [37], do not fit the model because of the flexibility of the "monomer" (hexane) [38]. Other possibilities are mixtures of monovalent and divalent ions; two examples are absorption of sodium and cupri-ions on a sterate film on water [13] and melted $Li_2O$–NiO mixtures [39]. The monomer-dimer theory has also found application in the cell-cluster theory [40] and in improvements of the Debye-Hückel theory when some of the ions are large [41].

The first rigorous results on the monomer-dimer problem were mostly of a negative nature [42, 43], and the earliest breakthrough came in the related dimer-covering problem (i.e., zero monomer density). The theory of this problem received a great impetus when the exact solution for planar lattices was discovered by Kasteleyn, Temperley and Fisher in 1961 [44–46]. We shall only mention the dimer-covering problem briefly in this paper, primarily because the nature of the pure dimer problem is dependent on the geometry of the lattice, while our main concern is with results independent of the structure of the lattice; part of the succeeding development of the dimer covering problem can be found in references [47–53].

The solution of the planar dimer-covering problem made it possible to calculate the monomer-monomer correlation function in the case of only two monomers [54, 55]. Other rigorous results on the monomer-dimer problems are lower bounds on the free energy obtained by Bondy and Welsh [56] and Hammersley and Menon [57, 58], and Hammersley's proof [59] of the existence of the thermodynamic limit in the sense of Van Hove for simple cubic lattices of arbitrary dimension.

Apart from the numerical computation of the monomer-dimer partition function for specific simple lattices, the central theoretical question to be answered is this: can a monomer-dimer system have a phase transition? The numerical work previously cited had led to the conjecture that a monomer-dimer phase transition did not occur on simple lattices but a proof was lacking. (It is worth pointing out that similar conclusions were also drawn about monomer-trimer systems, but we now have an example of a special monomer-trimer system that can be proved to have a phase transition [60].) It is the purpose of this paper to prove that a phase transition cannot occur for any lattice and, since our results are quite general, no mention is made of lattice geometry except insofar as it is necessary to prove the existence of the thermodynamic limit (Section VIII). (Our method of proof is different from Hammersley's [59].) In Section II we give the basic definitions and in Section III evaluate the monomer-dimer partition function for some simple lattices by way of illustration. In Section IV we give our main results on the zeros of the monomer-dimer partition function, when considered as a polynomial in $x$, for positive edge weights. The zeros lie on the imaginary axis. This result can be generalized to non-positive (complex) edge weights (Theorem 4.9). One practical consequence of locating the zeros is that by changing the variable from $x$ to some simple function of $x$ one can make a Taylor expansion of the free energy for all non-zero monomer densities and hence can undertake numerical calculations with greater confidence than before, including bounds on the error. This is shown in Section IX.

Section V relates the monomer-dimer problem to various magnetic systems and Section VI gives Christoffel-Darboux formulas for the monomer-dimer and magnetic system partition functions. In Section VII we use our knowledge of the location of the zeros to give bounds on the compressibility and show that these are stronger than the bounds derived using Ginibre's general method. In Section IX we prove the existence and analyticity of correlation functions and of the free energy.

A preliminary report of this work was given in Heilmann and Lieb [61]. Shortly thereafter Kunz [62] and Gruber and Kunz [63], in their work on the general monomer-polymer problem, announced several of the same theorems, notably the one on the location of the zeros.

To conclude this introduction we wish to draw attention to the fact that our monomer-dimer result is related to another problem in statistical mechanics – the nearest neighbor exclusion (or hard core lattice gas) model. Suppose that $G$ is an unweighted graph on whose vertices particles may be placed provided no edge has particles at both of its terminal vertices. We might expect a phase transition to occur as the particle density is varied. Indeed, Dobrushin [64] has shown this to be the case for the square and cubic lattice and Heilmann [60] has extended this conclusion to the planar triangular and face centered cubic lattices. If, however, $G$ is the line graph of a graph $G^*$ (i.e., the vertices of $G$ are the edges of $G^*$ and two vertices of $G$ are connected if the corresponding edges of $G^*$ have a common vertex) then the exclusion problem on $G$ is the same as the monomer-dimer problem (with unit edge weights) on $G^*$. Hence the exclusion problem on $G$ has no phase transition. An example of this is the Kagome lattice which is the line graph of the planar hexagonal lattice.

## II. Basic Definitions

The notation and terminology used in this paper will differ in certain respects from those employed in the previous paper [61], partly to make the notation more convenient and partly to bring the terminology into closer agreement with what is generally accepted.

*Graphs.* The basic terminology for unweighted graphs will be that suggested by Essam and Fisher [65] and we shall not repeat their definitions. Unfortunately, they did not suggest any terminology for *weighted graphs:* If $G$ is a graph then we shall associate a complex number, $W(i,j)$, called an *edge weight*, with each unordered pair of vertices, $[i,j]$, in $G$. Unless otherwise stated, $W(i,j)$ will be assumed non-negative for all pairs, $[i,j]$, and positive if there is an edge connecting $i$ and $j$. We shall further associate a complex number, $x_i$ (called a *vertex weight*), with each vertex in $G$. We shall use $N(G)$ (or just $N$ if no ambiguity is caused) for the number of vertices in $G$. We shall write $G-G'$ for the section graph of $G$ which is obtained by deleting the vertices in the subgraph, $G'$. If $G'$ consists only of the $i$'th vertex we shall write $G-i$.

*Definition 2.1.* A graph is said to be *Hamiltonian* if it contains a Hamilton walk, i.e. if it is possible to number the vertices in the graph such that

$$W(1,2)\,W(2,3)\ldots W(N-1,N)\neq 0\,. \tag{2.1}$$

*Dimers.* A *dimer* is a molecule which can be placed on the graph, $G$, such that it covers an edge and the two vertices on which the edge is incident. We shall use $\langle i,j\rangle$ for the dimer which covers vertices $i$ and $j$. A *dimer arrangement*, $D$, is a set of dimers placed on $G$ such that no vertex is covered by more than one dimer. The set of vertices of $G$ covered by the dimers in $D$ will be denoted by $[D]$. The set of all dimer arrangements will be denoted by $\mathcal{D}$. A dimer arrangement which covers all vertices in $G$ is called a *dimer covering*.

*Canonical Weights.* For a dimer arrangement, $D$, the canonical weight, $W(D)$, is given by

$$W(D)=\prod_{\langle i,j\rangle\in D} W(i,j)\,. \tag{2.2}$$

The *d-dimer partition function* (or *weight*) of all dimers arrangements with $d$ dimers is

$$Z_d = \sum_{\substack{D \in \mathscr{D} \\ \# D = d}} W(D), \tag{2.3}$$

where $\# D$ is the number of dimers in $D$.

If we denote the largest integer in $N/2$ by $M$, i.e.

$$M = [N/2], \tag{2.4}$$

then we can define the generating function for $Z_d$:

$$P(G; x) = \sum_{d=0}^{M} Z_d x^{N-2d}. \tag{2.5a}$$

Obviously $P(G; x)$ is a polynomial in $x$ of degree $N$ and will be called simply the *partition function* (of the weighted graph $G$). It will prove convenient to define $P(G; x)$ for a graph having no vertices by

$$P(\emptyset; x) = 1. \tag{2.5b}$$

*Remark.* If $W(i, j)$ is written as

$$W(i, j) = \exp(-\beta J_{ij}), \tag{2.6}$$

then $Z_d$ can be interpreted as the canonical partition function for $d$ dimers on the graph, $G$, if $J_{ij}$ is the energy gained by placing a dimer on the two vertices $i$ and $j$. Furthermore, if $x$ is taken to be the activity of a monomer, and if it is supposed that all vertices not covered by a dimer contain a monomer, then $P(G; x)$ is the grand canonical partition function for the distribution of monomers and dimers on $G$. Alternatively, $P(G; 1)$ can be considered to be the grand canonical partition function for the distribution of hard dimers on $G$; this analogy is particularly useful since much of the standard theory of statistical mechanics applies directly to this case.

*Further polynomials:* Besides $P(G; x)$ we shall also need two related polynomials, $Q(G; x)$ and $R(G; x)$:

$$Q(G; x) = i^{-N} P(G; ix), \tag{2.7}$$

$$R(G; x) = \begin{cases} P(G; \sqrt{x}), & N \text{ even}, \\ P(G; \sqrt{x})/\sqrt{x}, & N \text{ odd}. \end{cases} \tag{2.8}$$

*A Generating Function in N Variables.* Finally we shall need a more general form of $P(G; x)$ which includes the vertex weights:

$$P(G; x_1, x_2, \ldots, x_N) = \sum_{D \in \mathscr{D}} W(D) \prod_{i \in G - [D]} x_i, \tag{2.9}$$

where the sum runs over all dimer arrangements on $G$ and the product runs over all vertices of $G$ which are not in $[D]$.

One can easily derive the following relation:

$$P(G; m_1 x, m_2 x, \dots, m_N x) = \left(\prod_{i=1}^{N} m_i\right) P(G'; x), \tag{2.10}$$

where the graph $G'$ differs from $G$ only in the edge weights which are related by:

$$m_i m_j W'(i,j) = W(i,j). \tag{2.11}$$

In particular one has

$$P(G; x, x, \dots, x) = P(G; x). \tag{2.12}$$

An important property of $P(G; x_1, \dots, x_N)$ is that it is linear in each variable, i.e. considered as a function of $x_j$ it is a polynomial of degree one.

## III. Examples of Partition Functions

In this section we shall give closed form expressions for $Q(G, x)$ for some simple graphs. Most of these results are well known and, as the purpose of this section is primarily that of illustration, it may be omitted without any loss of continuity.

*Example 1. The Linear Chain.* In this case the weights are given by

$$\begin{aligned} W(i,j) &= 1, \quad \text{if} \quad |i-j| = 1 \\ &= 0, \quad \text{otherwise}. \end{aligned} \tag{3.1}$$

If we define $L(N)$ to be the linear chain with $N$ vertices and edge weights as given by (3.1), one has the following recursion formula for $N \geqq 2$:

$$Q(L(N); x) = xQ(L(N-1); x) - Q(L(N-2); x), \tag{3.2}$$

where the first term of the right hand side corresponds to a monomer at the $N$'th vertex and the second term corresponds to a dimer covering the vertices $N$ and $N-1$. Since one has

$$Q(L(0); x) = 1, \qquad Q(L(1); x) = x, \tag{3.3}$$

one easily find that

$$Q(L(N); x) = U_N(\tfrac{1}{2}x), \tag{3.4}$$

where $U_n(x)$ is the Chebyshev polynomial of the second kind of degree $n$ which is defined by

$$U_n(x) = \sin((n+1)\theta)/\sin\theta, \tag{3.5}$$

$$\cos\theta = x. \tag{3.6}$$

It is easily seen that the zeros of $Q(L(N); x)$ are real and lie in the interval $|x| < 2$. Consequently the zeros of $P(L(N); x)$ are pure imaginary and are limited by $|\mathrm{Im}(x)| < 2$. It should be noted that this bound on the roots is the best possible bound that is independent of $N$.

*Example 2. The Polygon.* In this case the weights are given by

$$W(i,j) = \begin{cases} 1 & \text{if } |i-j| = 1 \text{ or } (i,j) = (1, N), \\ 0 & \text{otherwise}. \end{cases} \tag{3.7}$$

If we use $P(N)$ for the polygon with $N$ vertices and all edge weights equal to one we have the following formula for $N \geqq 2$:

$$Q(P(N); x) = xQ(L(N-1); x) - 2Q(L(N-2); x), \tag{3.8}$$

where the first term of the right hand side corresponds to a monomer on the $N$'th vertex and the second term corresponds to a dimer which includes the $N$'th vertex (and which can be either $\langle N-1, N \rangle$ or $\langle 1, N \rangle$). Using Eq. (3.4) one finds

$$Q(P(N); x) = 2T_N\left(\frac{x}{2}\right), \tag{3.9}$$

where $T_n(x)$ is the Chebyshev polynomial of the first kind of degree $n$, defined by:

$$T_n(x) = \cos(n\theta), \qquad \cos\theta = x. \tag{3.10}$$

It is easily seen that the zeros have the same properties and the same bounds as those given for the linear chain.

*Example 3. The Complete Graph.* In this case the edge weights are given by

$$W(i,j) = 1, \quad \text{all } (i,j). \tag{3.11}$$

If we use $K(N)$ for the complete graph with $N$ vertices and edge weights equal to one we have the following recursion formula for $N \geqq 2$:

$$Q(K(N); x) = xQ(K(N-1); x) - (N-1)\,Q(K(N-2); x), \tag{3.12}$$

where again the first term of the right hand side corresponds to a monomer on the $N$'th vertex and the second term corresponds to a dimer including the $N$'th vertex. Since one has

$$Q(K(0); x) = 1, \qquad Q(K(1); x) = x, \tag{3.13}$$

it follows that

$$Q(K(N); x) = He_N(x), \tag{3.14}$$

where $He_n(x)$ is the Hermite polynomial of degree $n$ defined by

$$He_n(x) = (-1)^n e^{x^2/2} \frac{d^n}{dx^n} e^{-x^2/2}. \tag{3.15}$$

If one renormalizes the edge weights to

$$W'(i,j) = (N-1)^{-1}, \quad (N \geqq 2), \tag{3.16}$$

then one gets, with $K'(N)$ for the complete graph with edge weights normalized according to (3.16):

$$Q(K'(N); x) = (N-1)^{-N/2}\, He_N(x\sqrt{N-1}). \tag{3.17}$$

As before one finds that the zeros of $Q(K'(N); x)$ are real and the best $N$ independent bound is $|x| < 2$ [66].

*Example 4. The Bethe Graph.* We shall define *the rooted Bethe graph* of degree $d$ and order $n$, $B(d, n)$, as follows: $B(d, 1)$ consists of a single vertex called the root; $B(d, n)$ consists of a vertex called the root, which is connected by edges to the roots of $d-1$ graphs $B(d, n-1)$. With edge weights equal to one on all edges, one finds in the usual manner for $n \geqq 2$ that

$$\begin{aligned} Q(B(d,n); x) = {} & x[Q(B(d,n-1); x)]^{d-1} \\ & -(d-1)\,[Q(B(d,n-2); x)]^{d-1}\,[Q(B(d,n-1); x)]^{d-2}. \end{aligned} \tag{3.18}$$

Defining the quantity $\hat{Q}_n(x)$ for $n \geqq 2$ by

$$\hat{Q}_n(x) = Q(B(d,n); x)\left[\prod_{i=1}^{n-1} Q(B(d,i); x)\right]^{2-d}, \tag{3.19}$$

one gets for $n \geqq 3$

$$\hat{Q}_n(x) = x\hat{Q}_{n-1}(x) - (d-1)\,\hat{Q}_{n-2}(x), \tag{3.20}$$

with

$$\hat{Q}_1(x) = x, \quad \hat{Q}_2(x) = x^2 - (d-1). \tag{3.21}$$

Therefore, $\hat{Q}_n(x)$ is a polynomial and, by comparison with (3.2) and (3.4),

$$\hat{Q}_n(x) = (d-1)^{n/2}\, U_n(\tfrac{1}{2}x(d-1)^{-\frac{1}{2}}), \tag{3.22}$$

where $U_n(x)$ is again the Chebyshev polynomial defined in Eq. (3.5). The inversion of (3.19) is given by

$$Q(B(d,n); x) = \hat{Q}_n(x) \prod_{j=1}^{n-1} [\hat{Q}_j(x)]^{(d-2)(d-1)^{(n-1-j)}}, \tag{3.23a}$$

for $n > 1$, while

$$Q(B(d,1); x) = \hat{Q}_1(x) = x. \tag{3.23b}$$

The ($n$ independent) bound on the zeros of $Q(B(d,n); x)$ is seen to be $|x| < 2\sqrt{d-1}$.

## IV. Location of Zeros of Monomer Dimer Partition Functions

All the proofs in this section will be based on the two-step recurrence relation satisfied by the monomer-dimer polynomials. For $P(G;x)$ it reads (with $i$ being any vertex in $G$):

$$P(G;x) = xP(G-i;x) + \sum_{j\in G-i} W(i,j)\,P(G-i-j;x)\,. \tag{4.1}$$

The same is true for $Q(G;x)$ if the plus sign is replaced by a minus sign. The first term on the right is the contribution of a monomer placed on the $i$'th vertex, while the summation over $j$ gives the contribution of all the ways of placing a dimer at $i$. Obviously, if $x$ is a zero of $P(G;x)$ then $ix$ is a zero of $Q(G;x)$ and conversely.

**Lemma 4.1.** *If $G$ is a complete graph (i.e. $W(i,j)>0$ for all pairs $[i,j]$) then the zeros of $Q(G;x)$ are all real. Furthermore, if $i$ is any vertex in $G$ then the zeros of $Q(G;x)$, $a_1, a_2, \ldots, a_N$, and the zeros of $Q(G-i,x)$, $a'_1, a'_2, \ldots, a'_{N-1}$ obey a strict interlacing relation:*

$$a_1 < a'_1 < a_2 < a'_2 < \cdots < a'_{N-1} < a_N\,. \tag{4.2}$$

*Proof.* If $G$ is the empty graph and if $G$ contains only one vertex the lemma is trivially true. The lemma can then be proved for $N$ larger than one by induction, assuming that the lemma holds for all complete graphs, $G'$, with $N(G') \leqq N-1$. For $Q(G;x)$ the recurrence relation (4.1) takes the form

$$Q(G;x) = xQ(G-i;x) - \sum_{j\in G-i} W(i,j)\,Q(G-i-j;x)\,. \tag{4.3}$$

Here $Q(G-i;x)$ is polynomial of degree $N-1$ and the sum-polynomial

$$\sum_{j\in G-i} W(i,j)\,Q(G-i-j;x)\,, \tag{4.4}$$

is polynomial of degree $N-2$. From the induction assumption it follows that the zeros of $Q(G-i-j;x)$ interlace the zeros of $Q(G-i;x)$ for all $j\in G-i$. Since $W(i,j)$ is positive for all $j\in G-i$ then the zeros of the sum-polynomial (4.4) will also interlace the zeros of $Q(G-i,x)$. Now, considering the sign of the right-hand side of (4.3) as $x$ takes the values of the zeros of $Q(G-i;x)$ together with the sign as $x$ approaches $+\infty$ and $-\infty$ (see Fig. 1), one easily concludes that $Q(G;x)$ has $N$ real zeros which

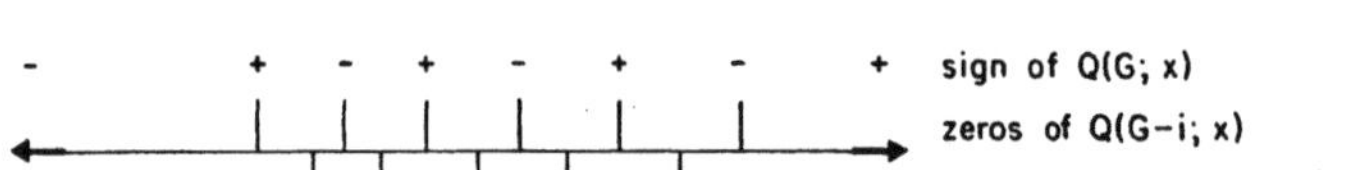


Fig. 1.

satisfy (4.2). Since $Q(G; x)$ is a polynomial of degree $N$ it does not have any other zeros.

By a continuity argument one can easily extend the lemma to include the case where some edge weights are zero and obtain the following theorem.

**Theorem 4.2.** *For any weighted graph, $G$, $Q(G; x)$ has only real zeros. Further, if $i$ is any vertex in $G$ and if $a_1, a_2, \ldots, a_N$ are the zeros of $Q(G; x)$ while the zeros of $Q(G-i; x)$ are $a'_1, a'_2, \ldots, a'_{N-1}$, then*

$$a_1 \leqq a'_1 \leqq a_2 \leqq a'_2 \leqq \cdots \leqq a'_{N-1} \leqq a_N . \qquad (4.5)$$

*Remark 4.1.* Theorem 4.2 does, of course, imply that the zeros of $P(G; x)$ are purely imaginary.

*Remark 4.2.* One might be interested in knowing when the inequalities in (4.5) will all be strict, especially since strict interlacing implies distinct zeros. Example 4 in Section III shows that connectedness is not sufficient to ensure distinct zeros. One can also construct counterexamples to the conjecture that the absence of articulation points would be sufficient to ensure distinct zeros (see Fig. 2). Instead we propose the following amendment which, although it does not exhaust all the possibilities, covers most cases of interest.

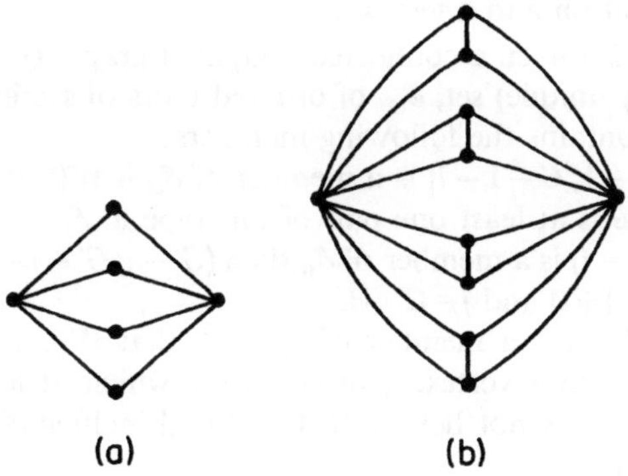


Fig. 2.

**Amendment to Theorem 4.2.** *If $G$ has a Hamilton walk which ends at the $i$'th vertex then all the inequalities in* (4.5) *are strict.*

*Proof.* As for Lemma 4.1, the proof inductively uses the recurrence relation (4.3). The important thing to notice is that if $G-i$ contains at least one vertex, $k$, such that $W(i, k) > 0$ and such that the zeros of $Q(G-i; x)$ and $Q(G-i-k; x)$ obey the strict interlacing property, then the zeros of $Q(G-i;x)$ and the sum-polynomial (4.4) will also obey the strict interlacing property. Consequently the strict interlacing property is retained along the Hamilton Walk.

*Definition 4.1.* For any graph, $G$, and any vertex, $j$, in $G$ we define the *vertex sum* $W_j$ by

$$W_j = \sum_{i\in G-j} W(i,j) - \min\{W(i,j): i\in G, W(i,j)>0\}. \tag{4.6}$$

We further define the numbers $B_1$, $B_1'$ and $B_1''$ by

$$B_1' = \max\{W_j : j\in G\}, \tag{4.7}$$

$$B_1'' = \tfrac{1}{4}\max\{W(i,j); i\in G, j\in G\} \tag{4.8}$$

and

$$B_1 = \max\{B_1', B_1''\}. \tag{4.9}$$

**Theorem 4.3.** *The (real) zeros of* $Q(G;x)$, $a_1, a_2, \ldots, a_N$, *satisfy*

$$|a_i| < 2\sqrt{B_1}, \text{ all } i. \tag{4.10}$$

*Proof.* If $G$ does not have a component which contains only two vertices then clearly $B_1'' < B_1'$. Since the bound on the zeros imposed by $B_1''$ is correct for a graph containing only two vertices we can concentrate on proving that $2\sqrt{B_1'}$ is a bound for the zeros when $G$ is a connected graph consisting of at least three vertices. For this purpose we shall need one more definition and a lemma:

*Definition 4.2.* Given a connected, weighted graph, $G$, we construct a (not necessarily unique) set, $\mathscr{E}_G$, of ordered pairs of section graphs of $G$ such that $\mathscr{E}_G$ contains the following members:
1) The pair $(G-1, G-1-i)$ is a member of $\mathscr{E}_G$ if $W(1,i)\neq 0$. Since $G$ is connected there is at least one pair of this type in $\mathscr{E}_G$.

2) If $(G', G'-i)$ is a member of $\mathscr{E}_G$ then $(G'-i, G'-i-j)$ is a member whenever $W(i,j)\neq 0$ and $j\in G'-i$.

3) If $(G', G'-i)$ is a member of $\mathscr{E}_G$ such that $W(i,j)=0$ for all $j$ in $G'-i$, then choose a vertex, $j$, in $G'-i$ for which at least one of the neighbors in $G$ does not belong to $G'-i$ and include $(G'-i, G'-i-j)$ as a member of $\mathscr{E}_G$.

**Lemma 4.4.** *If $G$ is a connected graph, $(G', G'-i)\in\mathscr{E}_G$ and $x\geqq 2\sqrt{B_1'}$ then $Q(G';x)/Q(G'-i;x)\geqq\sqrt{B_1'}$ and $Q(G',x)>0$, provided $G'$ is not $G$.*

*Proof.* Consider again the fundamental recurrence relation:

$$Q(G';x) = xQ(G'-i;x) - \sum_{j\in G'-i} W(i,j)\, Q(G'-i-j;x). \tag{4.11}$$

If $(G', G'-i)\in\mathscr{E}_G$ then it follows from the definition of $\mathscr{E}_G$ that

$$\sum_{j\in G'-i} W(i,j) \leqq B_1', \tag{4.12}$$

Table 1. Comparison between bound for the zeros (Theorem 4.3) and the actual value of the largest zero

| Lattice | Largest zero squared[a] | $4B_1$ |
|---|---|---|
| Simple quadratic | 11.24 | 12 |
| Plane triangular | 17.86 | 20 |
| Tetrahedral | 11.76 | 12 |
| Simple cubic | 19.23 | 20 |
| Body-centered cubic | 26.81 | 28 |
| Face-centered cubic | 41.32 | 44 |

[a] $-1/r$ as given by Eq. (3.1) in Gaunt [28].

and that the sum over $j$ in (4.11) can be confined to $j$'s for which $(G'-i, G'-i-j) \in \mathscr{E}_G$. Hence, if one divides Eq. (4.11) by $Q(G'-i; x)$ one can easily prove the lemma by induction.

*Proof of Theorem 4.3.* (cont.) The only step missing in the lemma above is the last step where $G'$ in Eq. (4.11) is identical to $G$. Also in this case one can confine the sum over $j$ to values for which $(G'-i, G'-i-j) \in \mathscr{E}_G$. Using

$$\sum_{j=2}^{N} W(1,j) \leqq 2B_1', \tag{4.13}$$

one can then furnish this final step in the proof by showing that Lemma 4.4 with $G' \equiv G-1$ implies that the right hand side of Eq. (4.3) is non-negative and strictly positive if $x > 2\sqrt{B_1}$.

*Remark 4.3.* It is seen that the bound on the zeros given by Theorem 4.3 is the best possible for the four examples considered in Section III. This cannot be expected to be true in general, but that the bound is rather good, at least for graphs with equal edge weights, can be seen from Table 1 where the bound given by Theorem 4.3 is compared with the value for the largest zero found numerically by Gaunt [28]. It is also interesting to compare the bound given by Theorem 4.3 with the bound which can be obtained by applying the general theory of Ruelle [67] to the system of hard dimers. $C(\beta)$ in Ruelle's notation is, in this case, given by

$$C(\beta) = 2B_4 \equiv \max_{\substack{[i,j]\in G \\ W(i,j)>0}} \left\{ \sum_{\substack{h \neq i,j \\ h \in G}} [W(i,h) + W(j,h) + W(i,j)] \right\},$$

while $B$ in Ruelle's notation is zero. One obtains $2eB_4$ as an upper bound on the square of the zeros as compared to $4B_1'$ with the present theory.

**Theorem 4.5.** *Assume that $G$ is a graph with complex edge weights and write $|G|$ for the graph obtained from $G$ by replacing the edge weights* *by their moduli. If $A$ is any upper bound on the zeros of $Q(|G|;x)$ then $P(G;x_1,x_2,\ldots,x_N)$ is not zero if $|x_j|>A$ for $1\leqq j\leqq N$.*

*Proof.* From Eq. (4.11) it follows that

$$\begin{aligned}&\left|\frac{P(G';x_1,x_2,\ldots)}{P(G'-i;x_1,x_2,\ldots)}\right|\\ &\geqq |x_i|-\sum_{j\in G'-i}|W(i,j)|\left|\frac{P(G'-i;x_1,x_2,\ldots)}{P(G'-i-j;x_1,x_2,\ldots)}\right|^{-1}.\end{aligned} \tag{4.14}$$

From this inequality one easily deduces inductively that

$$\left|\frac{P(G';x_1,x_2,\ldots)}{P(G'-i;x_1,x_2,\ldots)}\right|\geqq\frac{Q(|G'|;A)}{Q(|G'-i|;A)}, \tag{4.15}$$

if $x_1,x_2,\ldots,x_N$ and $A$ fullfill the condition of the theorem and $G'$ is a section graph of $G$. The inequality (4.15) implies the theorem.

**Theorem 4.6.** *Let $G$ be a weighted graph with $N$ vertices and $G'$ a subgraph. If $\operatorname{Re}(x_i)>0$ for all $i\in G'$ and $\operatorname{Re}(x_i)=0$ for all $i\in G-G'$ then the polynomial $P(G;x_1,\ldots,x_N)$ can only be zero if $P(G;x_1,\ldots,x_N)$ is identically zero when $P(G;x_1,\ldots,x_N)$ is considered as a polynomial in $\{x_i\}$, $i\in G'$, with the other vertex weights $\{x_i\}$, $i\in G-G'$, kept fixed at the given values. The same statement is true if $\operatorname{Re}(x_i)<0$ for all $i\in G'$. Here we return to the case of real, positive edge weights and $G'$ does not have to be a proper subgraph of $G$ because the only relevant fact about $G'$ is the specification of its vertices.*

*Proof.* Define the corresponding $Q$-polynomial:

$$Q(G;x_1,\ldots,x_N)=i^{-N}P(G;ix_1,\ldots,ix_N). \tag{4.16}$$

We shall then prove the theorem by proving the following two lemmas: The first, Lemma 4.7, implies that the theorem is true with $G'=G$, in which case the theorem simply says that $P(G;x_1,\ldots,x_N)$ is not zero when $\operatorname{Re}(x_i)>0$ for $i=1,\ldots,N$. The second, Lemma 4.8, then allows one to generalize the theorem as follows: Suppose that the vertices in $G'$ are $1,\ldots,k-1$ and that we wish to set $x_j=i\alpha_j$ for $j=k,\ldots,N$. Then the polynomial (and hence holomorphic function)

$$P(G;x_1,\ldots,x_{k-1},i\alpha_k,\ldots,i\alpha_N)$$

is equal to

$$\lim_{x_k\to i\alpha_k}\cdots\lim_{x_N\to i\alpha_N}P(G;x_1,\ldots,x_N),\ \operatorname{Re}(x_j)>0.$$

**Lemma 4.7.** *If* $\mathrm{Im}(x_j)<0$ *for* $1 \leqq j \leqq N$ *and if* $(G', G'-i) \in \mathscr{E}_G$ *then*

$$\mathrm{Im}\{Q(G'-i; x_1, x_2, \ldots)/Q(G'; x_1, x_2, \ldots)\} > 0\,, \tag{4.17}$$

*and*

$$Q(G'; x_1, x_2, \ldots) \neq 0\,. \tag{4.18}$$

*Proof.* If $G'$ contains only one vertex then the lemma clearly holds since one then has:

$$Q(G'-i; x_1, x_2, \ldots)/Q(G'; x_1, x_2, \ldots) = 1/x_i\,. \tag{4.19}$$

If $G'$ contains more than one vertex then we use the recurrence relation (4.11) to obtain

$$\begin{aligned}\mathrm{Im}\{Q(G'; x_1, x_2, \ldots)/Q(G'-i; x_1, x_2, \ldots)\} &= \mathrm{Im}\{x_i\} - \sum_{j \in G'-i} W(i,j) \\ &\cdot \mathrm{Im}\{Q(G'-i-j; x_1, x_2, \ldots)/Q(G'-i; x_1, x_2, \ldots)\}\,,\end{aligned} \tag{4.20}$$

and once again we can complete the proof by induction.

**Lemma 4.8.** *Let $D$ be a connected open set in* $\mathbb{C}^n$ *and let* $\{f_j\}$ *be a sequence of holomorphic functions on $D$ with the following properties:*

(i) *The* $f_j$ *are uniformly bounded on compact subsets of* $D$*;*

(ii) $\{f_j\}$ *converges to a function $f$ pointwise on* $D$*;*

(iii) *For each $j$ and each* $z \in D$, $f_j(z) \neq 0$.

*Then the convergence is uniform on compact subsets of $D$, $f$ is holomorphic, and either* $f \equiv 0$ *on $D$ or else* $f(z) \neq 0$ *for all* $z \in D$.

*Proof.* The uniform convergence on compacta and the analyticity of $f$ is Montel's Theorem. When $n=1$, the remainder of the lemma is Hurwitz's Theorem. If $n>1$ and if $f(z_0)=0$ for some $z_0 \in D$, then Hurwitz's Theorem states that $f \equiv 0$ on $P \cap D$, where $P$ is any one-dimensional hyperplane through $z_0$. Hence $f \equiv 0$ on some polydisc containing $z_0$ and, consequently, $f \equiv 0$ on $D$.

One might wonder how Theorem 4.6 would change if one allowed complex edge weights and whether there is a general theorem that combines Theorems 4.5 and 4.6. We first remark that if the edge weights are of the simple form

$$W(k,j) = |W(k,j)| \exp[i(\theta_k + \theta_j)]\,, \tag{4.21}$$

then we can use the correspondence given in Eqs. (2.10), (2.11) in reverse to obtain

$$P(G; x_1, \ldots, x_N) = P(|G|; x_1 \exp(-i\theta_1), \ldots, x_N \exp(-i\theta_N))\,, \tag{4.22}$$

where $|G|$ again means that we have replaced the edge weights by their moduli. Theorems 4.5 and 4.6 are directly applicable in this case which may suffice for many purposes. There exists, however, a general theorem which allows the edge weights to vary independently of each other and we shall conclude this section with that Theorem 4.9. Essentially it sums up all the preceding theorems. First we need a definition of the types of regions of the complex plane which we are going to consider.

*Definition 4.3*. The closed circular disk $D_-(A,\theta)$ in the complex plane is defined for $A$ real and non-negative and $0 \leqq \theta \leqq \pi/2$ as the closed subset of the complex plane:

$$D_-(A,\theta) = \{x : x \in \mathbb{C}, |x + A\cot\theta| \leqq A/\sin\theta\}\,. \tag{4.23}$$

Similarly $D_+(A,\theta)$ is defined by

$$D_+(A,\theta) = \{x : x \in \mathbb{C}, |x - A\cot\theta| \leqq A/\sin\theta\}\,. \tag{4.24}$$

$D_\pm(A,\theta)$ is defined as the intersection of $D_-(A,\theta)$ and $D_+(A,\theta)$, i.e.

$$\begin{aligned} D_\pm(A,\theta) = \{x : x \in \mathbb{C}, &|x + A\cot\theta| \leqq A/\sin\theta\,, \\ &|x - A\cot\theta| \leqq A/\sin\theta\}\,. \end{aligned} \tag{4.25}$$

The boundary of $D_-(A,\theta)$ and the boundary of $D_+(A,\theta)$ include the points $x = \pm iA$. We write $D_-^C(A,\theta)$ for the complement of $D_-(A,\theta)$ with respect to the complex plane, and similarly for $D_+^C(A,\theta)$. Clearly $0 \in D_\pm(A,\theta)$.

**Theorem 4.9.** *Let G be a graph with complex edge weights such that*

$$W(i,j) = -U(i,j)^2\,, \tag{4.26}$$

$$U(i,j) \in D_\pm(V(i,j),\theta), \quad \forall i,j \in G\,, \tag{4.27}$$

*with* $V(i,j)$ *real and non-negative and with* $0 \leqq \theta \leqq \pi/2$. *Let* $\bar{G}$ *be the graph with positive edge weights* $V(i,j)^2$. *Then* $P(G;x_1,\ldots,x_N)$ *is not zero if* $x_i \in D_-^C(A,\theta)$, *all* $i \in G$; *or if* $x_i \in D_+^C(A,\theta)$, *all* $i \in G$, *where A is the largest zero of* $Q(\bar{G};x)$.

*Proof*. The proof is by induction from the recurrence relations for a subgraph $G'$ in the form

$$\begin{aligned} &\frac{P(G';x_1,\ldots,x_N)}{P(G'-i;x_1,\ldots,x_N)} \\ &= x_i - \sum_{j \in G'-i} U(i,j)^2 \left[\frac{P(G'-i;x_1,\ldots,x_N)}{P(G'-i-j;x_1,\ldots,x_N)}\right]^{-1}. \end{aligned} \tag{4.28}$$

Choose $\delta > 0$ and assume that $x_i \in D_-^C(A+\delta,\theta)$ for all $i$. The induction assumption is that

$$\frac{P(G';x_1,\ldots,x_N)}{P(G'-i;x_1,\ldots,x_N)} \in D_-^C\left(\frac{Q(\bar{G}';A+\delta)}{Q(\bar{G}'-i;A+\delta)},\theta\right), \tag{4.29}$$

where $\bar{G}'$ is related to $\bar{G}$ as $G'$ is to $G$. The induction is easily established from the following three lemmas and since $\delta$ was arbitrary, the theorem is proved.

**Lemma 4.10.** *Let $A>0$ and $0\leqq\theta\leqq\pi/2$. Then $x\in D_-^C(A,\theta)$ if and only if $x^{-1}\in D_+(A^{-1},\theta)$. The same is true if $-$ and $+$ are interchanged.*

*Proof.* Trivial.

**Lemma 4.11.** *If $x\in D_\pm(A,\theta)$ and $y\in D_+(B,\theta)$ with $A\geqq 0$, $B\geqq 0$ and $0\leqq\theta\leqq\pi/2$, then $x^2y\in D_-(A^2B,\theta)$.*

*Proof.* By the maximum modulus principle it is sufficient to prove the lemma when $x$ and $y$ are on the boundaries of their respective domains. By symmetry it is also sufficient to consider $\mathrm{Re}(x)\geqq 0$. This means that we want to consider points in the complex plane of the form

$$x^2y=A^2B(-\cot\theta+e^{i\phi}/\sin\theta)^2(\cot\theta+e^{i\eta}/\sin\theta),\quad -\theta\leqq\phi\leqq\theta;\quad 0\leqq\eta\leqq 2\pi. \tag{4.30}$$

We want to prove that

$$|x^2y+A^2B\cot\theta|\leqq A^2B/\sin\theta, \tag{4.31}$$

when $x^2y$ is given by (4.30). The condition (4.31) is equivalent to

$$|(-\cos\theta+e^{i\phi})^2(\cos\theta+e^{i\eta})+\sin^2\theta\cos\theta|^2-\sin^4\theta\leqq 0. \tag{4.32}$$

By tedious computation one can transform the left side of (4.32) to

$$8\cos\theta(\cos\theta-\cos\phi)\left(\cos\theta\sin\frac{\phi-\eta}{2}+\sin\frac{\phi+\eta}{2}\right)^2.$$

Since $-\theta\leqq\phi\leqq\theta\leqq\pi/2$ implies that $\cos\theta\leqq\cos\phi$, (4.32) is satisfied.

**Lemma 4.12.** *If $x\in D_-^C(A,\theta)$ and $y\in D_-(B,\theta)$ with $A\geqq B\geqq 0$ and $0\leqq\theta\leqq\pi/2$, then $x-y\in D_-^C(A-B,\theta)$.*

*Proof.* Follows trivially from the triangle inequality.

### *An Alternative Proof of Theorem 4.9 Using Analytic Function Theory*

By the above, rather involved analysis of the recurrence relation we were able to prove Theorem 4.9 which can be thought of as interpolating between Theorems 4.5 and 4.6. It is a remarkable conclusion of the theory of analytic functions of several complex variables, however, that Theorems 4.5 and 4.6 automatically imply Theorem 4.9 without regard to the specific nature of the function $P(G;x_1,\ldots,x_N)$. Since the powerful concept of a domain of holomorphy in $\mathbb{C}^N$ has heretofore rarely been used in statistical mechanics, the following alternative proof of Theorem 4.9 may have some value.

Instead of proving that $P$ has no zeros we shall prove that $f\equiv 1/P$ is analytic. We suppose there are $M$ different edge weights which we label as $W_1, \ldots, W_M$ and define new variables $y_1, \ldots, y_M$ by

$$\exp(iy_j) = \frac{i - U_j/V_j}{i + U_j/V_j}, \quad j = 1, \ldots, M, \tag{4.33}$$

where the $U$'s and $V$'s are defined in (4.26) and (4.27). Clearly $y_j$ is holomorphic in $U_j/V_j$ except on the cuts $(i, i\infty)$ and $(-i, -i\infty)$. The curve in the $U_j$ plane defined by $\mathrm{Re}(y_j) = \text{constant}$ is a circular arc with end points $+iV_j$ (corresponding to $\mathrm{Im}(y_j) = \infty$) and $-iV_j$ (corresponding to $\mathrm{Im}(y_j) = -\infty$). In fact, for $0 \leqq \theta \leqq \pi/2$, the intersection of the boundary of the disc $D_+(V_j, \theta)$ with the right hand plane is the curve $\mathrm{Re}(y_j) = \pi - \theta$ while the intersection of the boundary of $D_+(V_j, \theta)$ with the left hand plane is the curve $\mathrm{Re}(y_j) = -\theta$ and similarly for $D_-(V_j, \theta)$. The domain $D_\pm(V_j, \theta)$ maps onto the domain $|\mathrm{Re}(y_j)| < \theta$.

For the activity variables we first make the replacement $x_j \to 1/t_j$, $j = 1, \ldots, N$ and then consider the polynomial

$$\tilde{P} \equiv \left[\prod_{j=1}^{N} t_j\right] P(G; t_1^{-1}, \ldots, t_N^{-1})$$

which is analytic in a neighborhood of $t_1 = t_2 = \cdots = t_N = 0$. Recalling that $A \geqq 0$ is the largest zero of $Q(\bar{G}; x)$ we define new variables $z_1, \ldots, z_N$ by

$$\exp(iz_j) = \frac{i - At_j}{i + At_j}, \quad j = 1, \ldots, N. \tag{4.34}$$

Let us consider the complex numbers $Y = (y_1, \ldots, y_M)$ together with $Z = (z_1, \ldots, z_N)$ as a point in $\mathbb{C}^{M+N}$ and define $f: \mathbb{C}^{M+N} \to \mathbb{C}$ by $f(Y, Z) \equiv 1/\tilde{P}$. Then Theorem 4.5 can be rephrased as:

$$\begin{aligned} &f \text{ is holomorphic on the tube, } T, \text{ with real base} \\ &|\mathrm{Re}(y_j)| < \pi/2, \quad j = 1, \ldots, M \quad \text{and} \\ &|\mathrm{Re}(z_j)| < \pi/2, \quad j = 1, \ldots, N. \end{aligned} \tag{4.35}$$

Likewise, Theorem 4.6 can be rephrased as:

$$\begin{aligned} &f \text{ is holomorphic on the tubes } T_+ \text{ and } T_- \text{ with real bases} \\ &\mathrm{Re}(y_j) = 0, \quad j = 1, \ldots, M \quad \text{and for} \\ &j = 1, \ldots, N, \quad 0 \leqq \mathrm{Re}(z_j) < \pi \quad \text{for } T_+ \text{ while} \\ &-\pi < \mathrm{Re}(z_j) \leqq 0 \quad \text{for } T_-. \end{aligned} \tag{4.36}$$

As $D = T \cup T_+ \cup T_-$ is connected, the *Tube Theorem* states that the envelope of holomorphy of $D$ is the domain $ch(D)$ where $ch(\;)$ means convex hull. That is, every function that satisfies (4.35) and (4.36) has a holomorphic extension to $ch(D)$. We claim that $ch(D)$ is precisely the domain given by Theorem 4.9, i.e.

$$ch(D) = \bigcup_{0 \leqq \theta \leqq \pi/2} [K_\theta^+ \cup K_\theta^-] \tag{4.37}$$

where

$$\begin{aligned} K_\theta^+ &= \{(Y, Z): |\mathrm{Re}(y_j)| < \theta, j = 1, \ldots, M, \; -\theta < \mathrm{Re}(z_j) < \pi - \theta, \\ &\quad j = 1, \ldots, N\} \\ K_\theta^- &= \{(Y, Z): |\mathrm{Re}(y_j)| < \theta, j = 1, \ldots, M, \theta - \pi < \mathrm{Re}(z_j) < \theta, j = 1, \ldots, N\}. \end{aligned} \tag{4.38}$$

To see that $K_\theta^+$, for example, is in $ch(D)$ we let $\alpha$ denote the subset of $\{1, \ldots, N\}$ for which $\mathrm{Re}(z_j) \leqq 0$ and let $\beta$ be the complement of $\alpha$. Then $(Y, Z) \in K_\theta^+$ means that $\mathrm{Re}(y_j) = \gamma_j \theta$, with $|\gamma_j| < 1$, $j = 1, \ldots, M$, $\mathrm{Re}(z_j) = \delta_j(\pi - \theta)$, $j \in \alpha$ and $\mathrm{Re}(z_j) = -\delta_j \theta$, $j \in \beta$ with $0 \leqq \delta_j < 1$. We then form a convex combination of a point in $T$ with a point in $T_+$, i.e.

$$(Y, Z) = \frac{2\theta}{\pi}(Y_1, Z_1) + \left(1 - \frac{2\theta}{\pi}\right)(Y_2, Z_2),$$

with

$$\begin{aligned} Y_1 &= \left(\gamma_1 \frac{\pi}{2}, \ldots, \gamma_M \frac{\pi}{2}\right) \\ Z_1 &= \left(S_1 \delta_1 \frac{\pi}{2}, \ldots, S_N \delta_N \frac{\pi}{2}\right), \; S_j = 1, \; j \in \alpha; \; S_j = -1, \; j \in \beta \\ Y_2 &= (0, \ldots, 0) \\ Z_2 &= (S_1 \delta_1 \pi, \ldots, S_N \delta_N \pi), \; S_j = 1, \; j \in \alpha; \; S_j = 0, \; j \in \beta. \end{aligned} \tag{4.39}$$

Thus, we have recovered Theorem (4.9). The only thing that remains to be shown is that $ch(D)$ is not larger than the domain given by the right side of (4.37). It is clear that $L_+ \equiv ch(T \cup T_+) = \bigcup_{0 \leqq \theta < \pi/2} K_\theta^+$ and similarly for $L_- \equiv ch(T \cup T_-)$. We claim that $ch(D) = L_+ \cup L_-$. Since $T$, $T_+$ and $T_-$ are all convex it follows easily that if there is a point in $ch(D)$ not in $L_+ \cup L_-$ then there must be a point in $ch(D)$ which is in $ch(T_+ \cup T_-)$ and which is not in $L_+ \cup L_-$. A point $(Y, Z)$ in $ch(T_+ \cup T_-)$ has the property that $Y = (0, \ldots, 0)$ and $|z_i - z_j| < \pi$ for all $i, j = 1, \ldots, N$. However, it is easy to see that $L_+ \cup L_-$ contains all such points, which completes the proof.

## V. Relationship of the Monomer-Dimer System to the Ising Model and to the Heisenberg Ferro- and Antiferromagnet

Fisher [68] has shown how the Ising model with zero magnetic field can be put into a one to one correspondence with the dimer coverings of a suitably chosen weighted graph. As we shall show, that method can easily be extended to the Ising model with non-zero magnetic field, which can be put into a one to one correspondence with the monomer-dimer problem in which some of the vertices have zero monomer weight. One proceeds as follows: The Ising partition function corresponding to a weighted graph, $\hat{G}$, with edge weights $\{K_{ij}\}$, and $N$ vertices can be written as

$$Z(\hat{G}; z_1, z_2, \ldots, z_N) = \sum_{s=\pm 1}^{\hat{G}} \exp\left[\sum_{[i,j]\in \hat{G}} K_{ij}(s_i s_j - 1)\right] \prod_{i=1}^{N} z_i^{s_i}. \tag{5.1}$$

To each vertex, $i$, is associated a spin-variable, $s_i$, and a fugacity, $z_i$. The first sum runs over all values of the spin variables, and it is assumed that different vertices have different fugacities. With $s_i$ and $s_j$ equal to $+1$ or $-1$ one has the identities:

$$\exp(K_{ij} s_i s_j) = \cosh(K_{ij})\,[1 + V_{ij} s_i s_j], \tag{5.2}$$

$$V_{ij} = \tanh(K_{ij}), \tag{5.3}$$

$$z_i^{s_i} = t_i(1 + y_i s_i), \tag{5.4}$$

$$t_i = \tfrac{1}{2}(z_i + z_i^{-1}), \tag{5.5}$$

$$y_i = (z_i - z_i^{-1})/(z_i + z_i^{-1}). \tag{5.6}$$

These can be used to obtain an alternative form of (5.1):

$$Z(\hat{G}; z_1, z_2, \ldots, z_N) = \left[\prod_{[i,j]\in \hat{G}} \cosh(K_{ij}) \exp(-K_{ij})\right] 2^N \prod_{i=1}^{N} t_i \cdot \Upsilon(G; y_1, y_2, \ldots, y_N), \tag{5.7}$$

$$\Upsilon(G; y_1, y_2, \ldots, y_N) = 2^{-N} \sum_{s=\pm 1}^{G} \left[\prod_{[i,j]\in G} (1 + V_{ij} s_i s_j)\right] \prod_{i=1}^{N} (1 + y_i s_i). \tag{5.8}$$

Here, $G$ is the weighted graph with edge weights $\{V_{ij}\}$, vertex weights $\{y_i\}$ and $N$ vertices. It is important to notice that $G$ has no negative edge weights if and only if $\hat{G}$ has the same property. This latter condition means that $\hat{G}$ is an Ising *ferromagnet*. $\Upsilon(G; y_1, y_2, \ldots, y_N)$ is essentially a partition function for walks on $G$ such that no edge is visited more than once.

The next step is to construct the expanded graph, $G^E$: For each vertex, $i$, in $G$ with coordination number, $q(i)$, larger than one, $q(i)$ new vertices, $i_1, i_2, \ldots, i_{q(i)}$, are substituted. The edges which were incident on $i$ in $G$ become edges in $G^E$ with the same weight and incident on $\{i_2, i_3, \ldots, i_{q(i)}\}$ such that $i_2, i_3, \ldots, i_{q(i)-1}$ get one edge each while $i_{q(i)}$ gets two edges. Finally edges with weight one are added between $[i_1, i_2]$,

$[i_2, i_3], \ldots, [i_{q(i)-1}, i_{q(1)}]$. The vertex weight of $i_1$ in $G^E$ is set equal to $y_i$ while $i_2, i_3, \ldots, i_{q(i)}$ are given vertex weight zero. Vertices in $G$ with coordination number one are transferred unaltered. Numbering the vertices in $G^E$ such that $i_1$ in the above notation gets number $i$ one finds

$$\Upsilon(G; y_1, \ldots, y_N) = \Upsilon(G^E; y_1, \ldots, y_N, 0, 0, \ldots, 0). \tag{5.9}$$

It is important that the vertices in $G^E$ have coordination number at most 3 and that vertices with non-zero value of the vertex weight have coordination number one.

The terminal graph, $(G^E)^T$, of $G^E$ can now be constructed: For each vertex, $i$, of $G^E$ with coordination number $q(i)$ a cluster of $q(i)$ new vertices is substituted; each of these is connected to the other $q(i)-1$ vertices by edges (called internal edges) with edge weight one. The vertex weight of the new vertices in $(G^E)^T$ are all taken to be equal to the vertex weight of the corresponding vertex in $G^E$. The edges which were incident on the vertex, $i$, in $G^E$ become edges incident on the vertices of the corresponding cluster in $(G^E)^T$, such that each of the $q(i)$ vertices gets connected by one edge. These edges are called external edges in $(G^E)^T$; their edge weights are taken to be the reciprocal of the edge weights of the corresponding edges in $G^E$.

Finally, it will be shown that $P((G^E)^T; y_1, \ldots, y_N, 0, \ldots, 0)$ and $\Upsilon(G^E; y_1, \ldots, y_N, 0, \ldots, 0)$ can be expanded in such a manner that a one to one correspondence between non-zero terms in the expansions of the two partition functions obtains and that, furthermore, corresponding terms only differ by a factor which is the same for all terms. $P((G^E)^T; y_1, \ldots, y_N, 0, \ldots, 0)$ is expanded as a sum over all dimer arrangements (see Eq. (2.9)). For $\Upsilon(G^E; y_1, \ldots, y_N, 0, \ldots, 0)$ one takes the definition (5.8), expands the two products, and sums over all values of the spin variables. Corresponding to the term 1 in the factor $(1 + V_{ij}s_is_j)$ in Eq. (5.8), a dimer is placed on the corresponding external edge in $(G^E)^T$, while the term $V_{ij}s_is_j$ corresponds to the external edge being free of a dimer. In the expansion of the second product in Eq. (5.8), the term $y_is_i$ corresponds to a monomer on the analogous vertex in $(G^E)^T$ which otherwise must be covered by a dimer. Having kept the important properties of $G^E$ in mind, it is easily perceived that

$$\begin{aligned}\Upsilon(G^E; y_1, \ldots, y_N, 0, \ldots, 0)& \\ &= \left[\prod_{\langle i,j\rangle \in G} V_{i,j}\right] P((G^E)^T; y_1, \ldots, y_N, 0, \ldots, 0).\end{aligned} \tag{5.10}$$

This last polynomial, considered as a polynomial in $(y_1, \ldots, y_N)$ is not identically zero since otherwise the original Ising polynomial, (5.7), would be identically zero. Thus, the conditions of Theorem 4.6 are satisfied and we can conclude that $P((G^E)^T; y_1, \ldots, y_N, 0, \ldots, 0)$, as a polynomial in $(y_1, \ldots, y_N)$, satisfies Theorem 4.6. Since Eqs. (5.6) implies that $|z_i| < 1$ is equivalent to $\mathrm{Re}(y_i) > 0$ while $|z_i| > 1$ is equivalent to $\mathrm{Re}(y_i) < 0$, then *Theorem 4.6 implies the Lee-Yang circle theorem for the Ising ferromagnet.*

On the other hand, by taking the high- temperature limit of the Ising-partition function, one can prove that *the Lee-Yang circle theorem implies the fact that the zeros of the monomer-dimer partition function are purely imaginary.* For convenience we write the Ising partition function as

$$Z(G; z) = (z + z^{-1})^{-N} \sum_{s=\pm 1}^{G} \exp\left(\beta \sum_{[i,j]\in G} J_{ij} s_i s_j\right) z^{\sum_{i=1}^{N} s_i}, \tag{5.11}$$

and, in contrast to what we did before, take the edge weights of the graph $G$ to be the $J_{ij}$'s ($J_{ij} \geqq 0$ corresponding to the ferromagnet). Introducing the variable $x$ defined by

$$x = \beta^{-\frac{1}{2}} (z + z^{-1})(z - z^{-1})^{-1}, \tag{5.12}$$

one obtains the relation

$$(z + z^{-1})^{-1} z^{s_i} = \frac{1}{2x} (x + s_i \beta^{-\frac{1}{2}}). \tag{5.13}$$

When this is substituted into Eq. (5.11) one gets

$$Z(G; z) = (2x)^{-N} \sum_{s=\pm 1}^{G} \exp\left(\beta \sum_{[i,j]\in G} J_{ij} s_i s_j\right) \prod_{i=1}^{N} (x + s_i \beta^{-\frac{1}{2}}). \tag{5.14}$$

$Z(G; z)$ is now expanded in powers of $\beta$. It is not difficult to see that the term of lowest order in $\beta$ is of zero'th order in $\beta$ and that it is precisely the monomer-dimer partition function for $G$. Consequently,

$$(2x)^N Z(G; z) = P(G; x) + 0(\beta). \tag{5.15}$$

Since $|z| = 1$ corresponds to purely imaginary $x$, the Lee-Yang circle theorem implies that the left-hand side of (5.15) considered as a polynomial in $x$ has purely imaginary zeros for all values of $\beta > 0$. If one assumed that $P(G; x)$ had a zero away from the imaginary axis one could easily prove from continuity of the zeros that for sufficiently small, but nonzero, $\beta$ the total right hand side of Eq. (5.15) would also have a zero outside the imaginary axis and one would thus have a contradiction. Consequently, the zeros of $P(G; x)$ are all purely imaginary. Thus, by starting with the Lee-Yang theorem, one can derive part of Theorem 4.2 but not the amendment.

The proof just given will also work the other way *provided* the zeros of $P(G;x)$ are all simple. This means that Theorem 4.1 implies directly that if $G$ has a Hamilton walk and if $\beta$ is sufficiently small, then the zeros of the corresponding Ising ferromagnet partition function lie on the unit circle, are simple, and fulfill an interlacing condition similar to Eq. (4.5).

One can easily convince oneself that (5.15) also holds when $Z(G;z)$ is a *Heisenberg* ferromagnet partition function of the form:

$$Z(G;z) = (z+z^{-1})^N \operatorname{Tr}\left\{\exp(\beta H)\, z^{\sum_{i=1}^{N} s_{iz}}\right\}, \tag{5.16}$$

$$\begin{aligned} H = &\sum_{[i,j]\in G} (J_{ij} s_{iz} s_{jz}) \\ &+ (\text{quadratic expression in } s_{1x}, s_{2x}, \ldots, s_{1y}, s_{2y}, \ldots)\,. \end{aligned} \tag{5.17}$$

Furthermore, if $J_{ij}$ is changed to $-J_{ij}$ in the formulas above one obtains an antiferromagnet in place of the ferromagnet and at the same time $P(G;x)$ is changed to $Q(G;x)$. Since $z^2$ negative and real corresponds to $x$ real, these considerations imply that *for an antiferromagnet the zeros in $z^2$ are all on the negative real axis for sufficiently small $\beta$ when $G$ has a Hamilton walk.* One should note that the considerations above do not yield a bound, independent of the size of $G$, on the range of $\beta$ in which the zeros have the stated properties.

Our result, that the zeros of the Heisenberg ferromagnet lie on the unit circle when $\beta$ is small, is overshadowed by the result of Asano [70] that the zeros lie on the unit circle for *all* $\beta > 0$. Asano's result has been extended by Suzuki and Fisher [71].

## VI. Christoffell-Darboux Type Formulas for Monomer-Dimer and Ising Systems

The recursion formula (4.1) suggests that it might be worthwhile to make an investigation of the general theory of orthogonal polynomials in order to obtain ideas which can be applied with success to the theory of monomer-dimer systems. Unfortunately, most results for orthogonal polynomials are not derived from the recursion formulae; the only important exception seems to be the Christoffel-Darboux Formula. This will be the main theme of the present section.

*Definition 6.1.* A *self-avoiding walk* on a weighted graph, $G$, is an ordered set of vertices in $G$, such that no vertex appears more than once. The *weight*, $W(S)$, associated with the walk

$$S = \{i_1, i_2, \ldots, i_{m+1}\}, \tag{6.1}$$

of length $m$, for $m>0$, is given by

$$W(S)=\prod_{j=1}^{m} W(i_j, i_{j+1}). \tag{6.2}$$

For $m=0$ we take $W(S)=1$. We shall write $[S]$ for the self-avoiding walk, $S$, considered as a subgraph of $G$, and the set of all self-avoiding walks on $G$ which start at the vertex $i$ and have non-zero weight will be denoted $\mathscr{S}_i$.

*Definition 6.2.* We define the kernel, $K(G,i;y,x)$ by

$$K(G,i;y,x)=\sum_{S\in\mathscr{S}_i} W(S)\,Q(G-[S];x)\,Q(G-[S];\bar{y}), \tag{6.3}$$

where $\bar{y}$ is the complex conjugate of $y$.

**Theorem 6.1.** *For every vertex, $i$, in a weighted graph, $G$,*

$$(x-\bar{y})\,K(G,i;y,x)=[Q(G;x)\,Q(G-i;\bar{y})-Q(G;\bar{y})\,Q(G-i;x)]. \tag{6.4}$$

*Proof.* By induction from the recursion formula (4.3).

Theorem 6.1 is the Christoffel-Darboux formula for the monomer-dimer partition function.

**Corollary 6.2.** *For every vertex, $i$, in a weighted graph, $G$,*

$$K(G,i;\bar{x},x)=Q'(G;x)\,Q(G-i;x)-Q(G;x)\,Q'(G-i;x), \tag{6.5}$$

*where the prime denotes differentiation with respect to $x$.*

*Proof.* Trivial.

*Remark 6.1.* Theorem 4.2 with the amendment can be obtained easily from Theorem 6.1 and Corollary 6.2: If $Q(G;x)$ had a complex zero, $a$, then $\bar{a}$ would also be a zero of $Q(G;x)$. Taking $y=x=a$ in Eq. (6.4) one arrives at a contradiction because every term in (6.3) is clearly nonnegative and real. If $G$ has a Hamilton walk, there is at least one nonzero term, namely when $[S]=G$. If $G$ does not have a Hamilton walk we can appeal to a continuity argument. The interlacing can then be proved from Eq. (6.5) by considering the sign of $Q(G-i;x)$ for $x$ equal to the zeros of $Q(G;x)$.

**Theorem 6.3.** *Let $H$ be a proper, non-empty subgraph of $G$, let $j$ be a vertex in $G-H$, and let $\mathscr{S}_j(H)$ be the set of all self-avoiding walks which start at the vertex $j$ and end at some vertex in $H$ without visiting $H$ before. Then*

$$\begin{aligned}&Q(G;x)\,Q(G-H-j;x)-Q(G-j;x)\,Q(G-H;x)\\&\quad=-\sum_{S\in\mathscr{S}_j(H)} W(S)\,Q(G-[S];x)\,Q(G-[S]-H;x),\end{aligned} \tag{6.6}$$

*where in $G-[S]-H$ it is understood that the vertex common to $[S]$ and $H$ is deleted only once.*

*Proof*. The proof again follows by a simple application of the recursion formula.

It is interesting that while the recursion formula for the monomer-dimer system has no interesting analog for the Ising model, the Christoffel-Darboux formula, (6.4), does. If $G$ is a weighted graph with positive edge weights, $\{V_{ij}\}$, and if $H$ is a subset of vertices in $G$, then we define

$$\Upsilon(G,H;z)=2^{-N(G)}\sum_{s=\pm 1}^{G}\left\{\left[\prod_{j\in H}(-is_j)\right]\cdot\left[\prod_{[j,k]\in G}(1+V_{jk}s_js_k)\right]\left[\prod_{j=1}^{N}z^{s_j}\right]\right\}, \tag{6.7}$$

where $i$ is the imaginary unit. We shall allow the same vertex of $G$ to occur repeatedly in $H$, in which case the corresponding factor in $\prod_{j\in H}(-is_j)$ should also be repeated. $\Upsilon(G,\emptyset;z)$ is seen to be a version of the Ising partition function (see Eqs. (5.3)–(5.8)). In the following, $H+j$ denotes the union of the set $H$ and the vertex $j$.

**Theorem 6.4.** *Let $G, H$ and $\Upsilon$ be as defined above and introduce*

$$\begin{aligned}\Psi(G,H,j;x,y)=\;&\Upsilon(G,H+j;x)\,\Upsilon(G,H;y)\\&-\Upsilon(G,H+j;y)\,\Upsilon(G,H;x).\end{aligned} \tag{6.8}$$

*Then there exists an expansion of $\Psi$:*

$$\begin{aligned}\Psi(G,H,j;x,y)=\frac{1}{2i}\left(\frac{x}{y}-\frac{y}{x}\right)\sum_{G'\subseteq G}\sum_{H'\subseteq G'}W(G,H,j;G',H')\\ \Upsilon(G',H';x)\,\Upsilon(G',H';y),\end{aligned} \tag{6.9}$$

*such that the numbers $W(G,H,j;G',H')$, are all non-negative when the edge weights $\{V_{jk}\}$ are all non-negative.*

*Proof*. Expanding the factor $(1+V_{jk}s_js_k)$ in Eq. (6.7) we obtain

$$\begin{aligned}\Upsilon(G,H+j;z)=\;&\Upsilon(G-V_{jk},H+j;z)\\&+V_{jk}\Upsilon(G-V_{jk},H+k;z),\end{aligned} \tag{6.10a}$$

$$\Upsilon(G,H;z)=\Upsilon(G-V_{jk},H;z)-V_{jk}\Upsilon(G-V_{jk},H+j+k;z), \tag{6.10b}$$

where we have used $G-V_{jk}$ to denote the graph $G$ with the edge $[j,k]$ deleted. Then,

$$\begin{aligned}\Psi(G,H,j;x,y)=\;&\Psi(G-V_{jk},H,j;x,y)\\&+V_{jk}\Psi(G-V_{jk},H,k;x,y)\\&+V_{jk}\Psi(G-V_{jk},H+j,k;x,y)\\&+V_{jk}^2\Psi(G-V_{jk},H+k,j;x,y).\end{aligned} \tag{6.11}$$

If $j$ is a vertex in a weighted graph, $G'$, which is not connected to any other vertex in $G'$ then, upon summing over $s_j = \pm 1$, one obtains:

$$\Psi(G', H, j; x, y) = \frac{1}{2i}\left(\frac{x}{y} - \frac{y}{x}\right) \Upsilon(G-j, H; x)\, \Upsilon(G-j, H; y). \tag{6.12}$$

This result holds even if $H$ contains $j$, but in that case one should delete $j$ from $H$ on the right side of (6.12). The theorem then follows by combining Eqs. (6.11) and (6.12).

Eq. (6.9) is the Christoffel-Darboux formula for the Ising model.

**Theorem 6.5.** *The zeros of $\Upsilon(G, H; z)$ lie on the unit circle (i.e. satisfy $|z| = 1$) for all G and H when all the $V_{ij}$ are non-negative (ferromagnetic case). If $e^{i\phi_1}, e^{i\phi_2}, \ldots, e^{i\phi_N}$ are the ordered zeros of $\Upsilon(G, H; z)$,*

$$-\pi < \phi_1 \leqq \phi_2 \leqq \cdots \leqq \phi_N \leqq \pi,$$

*and if $e^{i\psi_1}, e^{i\psi_2}, \ldots, e^{i\psi_N}$ are the zeros of $\Upsilon(G, H+j; z)$ ordered in the same manner, then either*

$$\phi_1 \leqq \psi_1 \leqq \phi_2 \leqq \psi_2 \leqq \cdots \leqq \phi_N \leqq \psi_N, \tag{6.13a}$$

*or*

$$\psi_1 \leqq \phi_1 \leqq \psi_2 \leqq \phi_2 \leqq \cdots \leqq \psi_N \leqq \phi_N. \tag{6.13b}$$

*If G is connected then all the inequalities in* (6.13) *are strict.*

*Proof.* From the definition (6.7)

$$\begin{aligned} \Upsilon(G, H; z^{-1}) &= (-1)^{N(H)}\, \Upsilon(G, H; z), \\ \Upsilon(G, H; \bar{z}) &= (-1)^{N(H)}\, \overline{\Upsilon(G, H; z)}. \end{aligned} \tag{6.14}$$

Let $x$ be a root of $\Upsilon(G, H; x) = 0$ and set $y = (\bar{x})^{-1}$ in (6.8). On the right side of (6.9) there will be at least one non-zero $W$, corresponding to $G' = G - j$ and $H' = H$, as can be seen from (6.11). Then, by using (6.14) and induction, the right side of (6.9) is non-zero unless $|x| = 1$. The proof of the interlacing parallels that in Remark 6.1.

## VII. Some Inequalities on the Compressibility and Other Quantities

In this section we display some inequalities that can be derived from the fact that the roots of $Q(G; x)$ are real and come in equal and opposite pairs (except for $x = 0$). Theorem 7.1 is of theoretical interest and will be used in Section VIII. Among other inequalities we derive a lower bound on the compressibility and show in Theorem 7.6 that this is stronger than the bound that can be derived using Ginibre's general method for repulsive potentials [72].

*Definition 7.1.* Let $a_1, a_2, \ldots, a_N$ be the (real) zeros of $Q(G; x)$ ordered such that

$$a_j = -a_{N+1-j}, \tag{7.1}$$

(which is possible since $Q(G; x)$ contains either only even powers or only odd powers of $x$). We then define $b_i \geqq 0$ $(i = 1, 2, \ldots, N)$ by

$$b_i = a_i^2 . \tag{7.2}$$

The numbers $-b_1, \ldots, -b_M$ (with $M = [N/2]$) are the zeros of $R(G; x)$.

**Theorem 7.1.** *The canonical partition function $Z_d$ satisfies the inequality:*

$$2\ln Z_d \geqq \ln Z_{d-1} + \ln Z_{d+1} + \ln \frac{(M-d+1)(d+1)}{(M-d)d}. \tag{7.3}$$

*Remark 7.1.* This inequality states that even for finite systems the free energy per unit volume is a strictly convex function of the dimer density.

*Proof.* If Newton's inequality is applied to $R(G; x)$ which is a polynomial of degree $M$ having $M$ real zeros (according to Theorem 4.2), then one gets

$$d(M-d)\, Z_d^2 - (d+1)(M-d+1)\, Z_{d-1} Z_{d+1} \geqq 0, \tag{7.4}$$

which trivially yields (7.1).

*Definition 7.2.* For a weighted graph, $G$, and activity, $x$, the *monomer density*, $\varrho_m$, is given by

$$\varrho_m = xN(G)^{-1}\, d\ln P(G; x)/dx = xN(G)^{-1}\, P'(G; x)/P(G; x). \tag{7.5}$$

**Theorem 7.2.** *If $G$ has at least one edge then the zeros of $d\varrho_m/dx$ satisfy*

$$|x^2 + b_1/2| \leqq b_1/2, \tag{7.6}$$

*where $-b_1$ is the smallest (largest modulus) zero of $R(G; x)$. If $G$ has no edges then $d\varrho_m/dx$ vanishes identically. In particular* $\mathrm{Re}(x^2) \leqq 0$ *and* $|x^2| \leqq b_1$.

*Proof.* With $y = x^2$ one has

$$\varrho_m = 1 - N^{-1} \sum_{j=1}^{N} b_j (y + b_j)^{-1}, \tag{7.7}$$

since $-b_j$, $j = 1, 2, \ldots, M$, are the zeros of $R(G; x)$. Differentiation gives

$$x \frac{d\varrho_m}{dx} = \frac{2y}{N} \sum_{j=1}^{N} \frac{b_j}{(y+b_j)^2}. \tag{7.8}$$

Setting $y = re^{i\theta} \neq 0$ one finds

$$\operatorname{Im}\left\{e^{2i\theta}\frac{N}{4x}\frac{d\varrho_m}{dx}\right\} = \sin\theta \sum_{j=1}^{N} \frac{b_j^2(r+b_j\cos\theta)}{|y+b_j|^4}. \tag{7.9}$$

If $x$ fails to satisfy condition (7.6) then $r + b_j \cos\theta$ is positive for all $j$. The theorem then follows because $b_j$ is real for all $j$ and nonzero for at least one value of $j$. (If $G$ does not contain any edges then $b_j = 0$ for all $j$ and $d\varrho_m/dx = 0$, identically.)

*Remark* 7.2. Clearly $\varrho_m$ is an increasing function of $x \in (0, \infty)$ and Theorem 7.1, together with the implicit function theorem, guarantees that $x$ is real analytic in $\varrho_m$. In the thermodynamic limit (whose existence we shall prove in sections VIII and IX) the zeros of $d\varrho_m/dx$ also satisfy Theorem 7.2 (by Vitali's theorem). *This means that the thermodynamic limit of $N(G)^{-1} \ln P(G; x)$ can be thought of as a real analytic function of $\varrho_m$*. The following Theorems 7.3 and 7.4 complement this assertion by giving an explicit lower bound to $d\varrho_m/dx$, but by themselves they do not obviously guarantee analyticity of $x$ in $\varrho_m$ since they are concerned solely with real $x$.

**Theorem 7.3.** *The following bounds on $\varrho_m$ and $x d\varrho_m/dx$ follow from Theorem 4.1:*

$$\varrho_m \geqq (1 + B_2 x^{-2})^{-1}, \tag{7.10}$$

$$x d\varrho_m/dx \geqq 2x^2(1-\varrho_m)^2/B_2, \tag{7.11}$$

$$x d\varrho_m/dx \leqq 2\varrho_m(1-\varrho_m), \tag{7.12}$$

*for real, positive $x$. $B_2$ is given by*

$$B_2 = 2N^{-1} \sum_{\langle i,j\rangle \in G} W(i,j). \tag{7.13}$$

*Proof*. Since $\frac{1}{2}NB_2$ is the coefficient of $x^{M-1}$ in $R(G; x)$ one has that

$$(1 + B_2 x^{-2}) = x^{-2} N^{-1} \sum_{j=1}^{N} (b_j + x^2). \tag{7.14}$$

Eq. (7.7) can alternatively be written

$$\varrho_m = x^2 N^{-1} \sum_{j=1}^{N} (b_j + x^2)^{-1}, \tag{7.15}$$

and

$$1 - \varrho_m = N^{-1} \sum_{j=1}^{N} \frac{b_j}{x^2 + b_j}. \tag{7.16}$$

The inequality (7.10) follows from (7.14) and (7.15) since

$$\left[N^{-1} \sum_{j=1}^{N} (b_j + x^2)\right]\left[N^{-1} \sum_{j=1}^{N} (b_j + x^2)^{-1}\right] \geqq 1. \tag{7.17}$$

The two inequalities:

$$(1-\varrho_m)^2 \leqq B_2 N^{-1} \sum_{j=1}^{N} \frac{b_j}{(x^2+b_j)^2}, \tag{7.18}$$

and

$$(1-\varrho_m)^2 \leqq N^{-1} \sum_{j=1}^{N} b_j^2 (x^2+b_j)^{-2}, \tag{7.19}$$

can be obtained from Eq. (7.16) by applying Cauchy's inequality. The inequality (7.11) then follows by substituting (7.8) into (7.18), while (7.12) can be obtained from (7.19) since

$$x\frac{d\varrho_m}{dx} = \frac{2}{N}\sum_{j=1}^{N} \frac{b_j}{(x^2+b_j)} - \frac{2}{N}\sum_{j=1}^{N} \frac{b_j^2}{(x^2+b_j)^2}. \tag{7.20}$$

A lower bound on $x d\varrho_m/dx$ can be obtained in an entirely different way by viewing the system as a hard core dimer gas. One can then prove

**Theorem 7.4.** $$x\frac{d\varrho_m}{dx} \geqq 2(1-\varrho_m)(1+2B_3 x^{-2})^{-1}, \tag{7.21}$$

*where*

$$B_3 = \max\left\{\sum_{j\in G-i} W(i,j) : i \in G\right\}.$$

Following Ginibre [72] one first proves

**Lemma 7.5.** $$dZ_d^2 \leqq Z_{d-1}[(d+1)Z_{d+1} + 2B_3 Z_d]. \tag{7.22}$$

*Proof of Lemma* 7.5. One has:

$$dZ_d = \sum_{\substack{D\in\mathscr{D}\\ \#D=d-1}} W(D) \sum_{\langle i,j\rangle \in G-[D]} W(i,j). \tag{7.23}$$

An application of Cauchy's inequality yields

$$\begin{aligned}(dZ_d)^2 &\leqq \left[\sum_{\substack{D\in\mathscr{D}\\ \#D=d-1}} W(D)\right]\left[\sum_{\substack{D\in\mathscr{D}\\ \#D=d-1}} W(D)\left\{\sum_{\langle i,j\rangle\in G-[D]} W(i,j)\right\}^2\right]\\ &\leqq Z_{d-1}\left[d(d+1)Z_{d+1} + \sum_{\substack{D\in\mathscr{D}\\ \#D=d-1}} W(D)\right.\\ &\qquad \left.\sum_{\langle i,j\rangle\in G-[D]} W(i,j)\left\{\sum_{k\in G}(W(i,k)+W(j,k))\right\}\right]\\ &\leqq Z_{d-1}[d(d+1)Z_{d+1} + dZ_d 2B_3],\end{aligned} \tag{7.24}$$

which proves the lemma.

*Proof of Theorem* 7.4. If one introduces the notation

$$\langle f(d)\rangle = \sum_{d=0}^{M} f(d)\, Z_d x^{N-2d}/P(G;x), \tag{7.25}$$

then one can write $\varrho_m = 1 - 2N^{-1}\langle d\rangle$ and

$$x\frac{d\varrho_m}{dx} = 4N^{-1}\langle (d-\langle d\rangle)^2\rangle . \tag{7.26}$$

If, furthermore, the inequality (7.22) is written as

$$\begin{aligned}&[(d+1)!\, Z_{d+1}x^{N-2(d+1)}]\,[d!\, Z_d x^{N-2d}]^{-1}\\ &\quad \geqq [d!\, Z_d x^{N-2d}]\,[(d-1)!\, Z_{d-1}x^{N-2(d-1)}]^{-1} - 2B_3 x^{-2},\end{aligned} \tag{7.27}$$

then Ginibre's main theorem [72] implies

$$\langle (d-\langle d\rangle)^2\rangle/\langle d\rangle \geqq (1+2B_3x^{-2})^{-1}, \tag{7.28}$$

which is equivalent to (7.21).

Theorem 7.4 is weaker than the lower bound on $x\frac{d\varrho_m}{dx}$ given in Theorem 7.3, Eq. (7.11). This can be seen from the following theorem:

**Theorem 7.6.** $$(1-\varrho_m) \geqq B_2(x^2+2B_3)^{-1}. \tag{7.29}$$

*Proof*. The inequality (7.29) may alternatively be written as

$$2B_3 \geqq B_2(1-\varrho_m)^{-1} - x^2 . \tag{7.30}$$

From the inequality (7.11) one obtains

$$\frac{d}{dx}(B_2(1-\varrho_m)^{-1}) \geqq 2x, \tag{7.31}$$

which implies that the right hand side of (7.30) is an increasing function of $x$. It will consequently be sufficient to prove (7.30) in the limit $x\to\infty$, i.e. to prove that

$$2B_3 \geqq \left(N^{-1}\sum_{j=1}^{N} b_j^2\right)\left(N^{-1}\sum_{j=1}^{N} b_j\right)^{-1}. \tag{7.32}$$

Lemma 7.5 gives for $d=1$

$$Z_1^2 < 2Z_2 + 2B_3 Z_1 . \tag{7.33}$$

Since

$$Z_1 = -\sum_{j=1}^{M}(-b_j) = \frac{1}{2}\sum_{j=1}^{N} b_j, \tag{7.34}$$

$$Z_2 = \sum_{j=1}^{M}\sum_{i=j+1}^{M}(-b_j)(-b_i) = \frac{1}{2}Z_1^2 - \frac{1}{4}\sum_{j=1}^{N} b_j^2, \tag{7.35}$$

then (7.32) follows from (7.33).

*Remark 7.3*. In this section and in Section IV we have introduced four different numbers $B_1, B_2, B_3$ and $B_4$ which all involve the sum of the weights of the edges incident on one vertex. $B_3$ is the maximum value of the sum; $B_2$ is the average value of the sum; $B_1$ is the maximum when one edge is deleted from the sum, and $B_4$ is somewhere between $B_3$ and $B_1$. If all vertices are equivalent except for boundary points then one obtains $B_2 = B_3$ in the limit of an infinite graph if the fraction of boundary vertices tends to zero. If, moreover, all edge weights are equal to $W$ and the coordination number is $q$ then in the limit of an infinite graph one has

$$B_2 = B_3 = qW; \quad B_1 = (q-1)W; \quad B_4 = (q-\tfrac{1}{2})W. \tag{7.36}$$

## VIII. The Thermodynamic Limit for Monomer-Dimer Systems: Basic Properties

Since the monomer-dimer problem can be considered as the problem of a hard core dimer gas, it is fairly obvious from the general results on thermodynamic limits that the limit exists when the weighted graph $G$ tends to infinity in a reasonable manner. We shall, nevertheless, give an explicit proof of the existence of the thermodynamic limit for the monomer-dimer problem, partly to obtain stronger results and partly to demonstrate a different method of proof.

*Definition 8.1*. A *Weighted Lattice*, (or simply *Lattice*), $L$, is an infinite, weighted graph imbedded in $\mathbb{R}^\nu$ such that if $\boldsymbol{a} = \{a_1, a_2, \ldots, a_\nu\}$ is any vector with integer components then $L$ is mapped onto itself if it is translated by $\boldsymbol{a}$. We also assume that every bounded subset of $\mathbb{R}^\nu$ contains only a finite number of vertices of $L$.

*Definition 8.2*. The lattice, $L$, will be said to have *compact interaction* if there exists a finite upper bound on the Euclidean length of the edges incident at any vertex.

*Remark 8.1*. In the following we shall assume that the lattice, $L$, has compact interaction and that the units of the Euclidean space are chosen such that if the edge $[i,j]$ is represented by a vector from the $i$'th vertex to the $j$'th vertex, and if $\boldsymbol{j}-\boldsymbol{i} = \{j_1 - i_1, j_2 - i_2, \ldots, j_\nu - i_\nu\}$,

$$\max_{W(i,j)>0} \left\{ \max_{1 \le k \le \nu} |j_k - i_k| \right\} < 1. \tag{8.1}$$

This will ensure that the interactions (nonzero edges) do not extend beyond the neighboring unit cells.

*Remark 8.2*. In the following we shall only consider subgraphs of the lattice $L$. If not otherwise specified all subgraphs will be assumed to be section graphs.

*Remark 8.3.* Unless otherwise specified, vectors will be assumed to have integer components.

*Definition 8.3. The section graph $\Lambda(\boldsymbol{a})$ includes every vertex with coordinates $x_1, x_2, \ldots, x_\nu$ which satisfies*

$$0 \leqq x_1 < a_1, \quad 0 \leqq x_2 < a_2, \ldots, 0 \leqq x_\nu < a_\nu . \tag{8.2}$$

*The section graph $\Lambda_{\boldsymbol{n}}(\boldsymbol{a})$ is the graph $\Lambda(\boldsymbol{a})$ translated by $\{n_1 a_1, n_2 a_2, \ldots, n_\nu a_\nu\}$.*

*Definition 8.4. Let $\Lambda$ be any section graph of L. The number $N_{\boldsymbol{a}}^+(\Lambda)$ is then defined as the number of graphs $\Lambda_{\boldsymbol{n}}(\boldsymbol{a})$ which, for $\boldsymbol{a}$ fixed and $\boldsymbol{n}$ running over all integer vectors, has at least one vertex in common with $\Lambda$. The number $N_{\boldsymbol{a}}^-(\Lambda)$ is defined similarly as the number of graphs $\Lambda_{\boldsymbol{n}}(\boldsymbol{a})$ which are section graphs of $\Lambda$.*

*Definition 8.5.* Let $\{\Lambda\}$ be a sequence of finite graphs such that $N(\Lambda) \to \infty$. If

$$\begin{aligned} &\lim N_{\boldsymbol{a}}^-(\Lambda) \to \infty , \\ &\lim N_{\boldsymbol{a}}^-(\Lambda)/N_{\boldsymbol{a}}^+(\Lambda) \to 1 , \end{aligned} \tag{8.3}$$

for all $\boldsymbol{a}$ then the sequence is said to *tend to infinity in the sense of Van Hove.*

*Remark 8.4.* Whenever a sequence of section graphs of $L$ is said to tend to infinity it should always be understood to be in the sense of Van Hove.

*Definition 8.6.* $A_0(\Lambda)$ is the modulus of the zero of maximum modulus of $P(\Lambda; x)$.

*Definition 8.7.* $\Lambda_n \equiv \Lambda(\{2^n - 1, 2^n - 1, \ldots, 2^n - 1\})$; $\Lambda_n + \boldsymbol{a}$ means the cube $\Lambda_n$ translated by the vector $\boldsymbol{a}$.

**Lemma 8.1.** *The following limit exists:*

$$\lim_{n \to \infty} A_0(\Lambda_n) = A_0 . \tag{8.4}$$

*Proof.* From Theorem 4.2, $A_0(\Lambda_n) \leqq A_0(\Lambda_{n+1})$. From Theorem 4.3, $A_0(\Lambda_n) < 2\sqrt{B_1}$.

**Theorem 8.2.** *If $\{\Lambda\}$ is a sequence of section graphs which tends to infinity then*

$$\lim_{\Lambda \to \infty} A_0(\Lambda) = A_0 . \tag{8.5}$$

*Proof.* Since $\Lambda$ is finite there exist an $n$ and an integer vector $\boldsymbol{a}$ such that $\Lambda \subset \Lambda_n + \boldsymbol{a}$. This proves that $\limsup A_0(\Lambda) \leqq A_0$. For any $n$ and sufficiently large $\Lambda$ there exists an integer vector $\boldsymbol{a}$ such that $\Lambda_n + \boldsymbol{a} \subset \Lambda$. This proves that $\liminf A_0(\Lambda) \geqq A_0$.

*Definition 8.8.* $N(\Lambda, r)$ is defined for $r > 0$ as the number of zeros of $P(\Lambda; x)$ with modulus less than $r$. For $r \leqq 0$, one defines $N(\Lambda, r) = 0$.

*Definition 8.9.* $n^*(\Lambda, r) \equiv N(\Lambda, r)/N(\Lambda)$, $n^*(n, r) \equiv n^*(\Lambda_n, r)$.

**Lemma 8.3.** *The following limit exists for all r:*

$$\lim_{n\to\infty} n^*(n,r) = n^*(r). \tag{8.6}$$

*The convergence is uniform in r; indeed for all r:*

$$|n^*(n,r) - n^*(r)| \leqq 4\nu 2^{-n}. \tag{8.7}$$

*Proof.* One has (with $N_1 = N(\Lambda_1)$):

$$N(\Lambda_n) = N_1 (2^n - 1)^\nu. \tag{8.8}$$

If one deletes the middle rows of unit cells in $\Lambda_n$, i.e. the vertices with $2^{n-1} - 1 \leqq x_j < 2^{n-1}$ for at least one of its coordinates, then $\Lambda_n$ is divided into $2^\nu$ disjoint cubes $\Lambda_{n-1}$. From the interlacing statement of Theorem 4.1 one then gets

$$|N(\Lambda_n, r) - 2^\nu N(\Lambda_{n-1}, r)| \leqq N(\Lambda_n) - 2^\nu N(\Lambda_{n-1}), \tag{8.9}$$

because the right side of (8.9) is the number of deleted vertices; as each vertex is deleted $N(\Lambda, r)$ either increases or decreases by exactly one. Since

$$|1 - 2^\nu N(\Lambda_{n-1})/N(\Lambda_n)| \leqq |1 - (1 - 2^{1-n})^\nu| \leqq 2\nu 2^{-n}, \tag{8.10}$$

one gets from Eq. (8.9):

$$|n^*(n,r) - n^*(n-1,r)| \leqq 4\nu 2^{-n}. \tag{8.11}$$

Eq. (8.7), and thereby the lemma, is an immediate consequence of Eq. (8.11).

**Theorem 8.4.** *If $\{\Lambda\}$ is a sequence of section graphs which tends to infinity, then the following limit is uniform in r*

$$\lim n^*(\Lambda, r) = n^*(r). \tag{8.12}$$

*Proof.* Take $\boldsymbol{a} = \{2^n + 1, 2^n + 1, \ldots, 2^n + 1\}$ and pack $\Lambda$ with $N_{\boldsymbol{a}}^-(\Lambda)$ copies of the cube $\Lambda(\boldsymbol{a})$. Next, delete the unit cells at the boundaries of the cubes $\Lambda(\boldsymbol{a})$. As above, one then gets

$$|N(\Lambda, r) - N_{\boldsymbol{a}}^-(\Lambda)\, N(\Lambda_n, r)| \leqq N(\Lambda) - N_{\boldsymbol{a}}^-(\Lambda)\, N(\Lambda_n). \tag{8.13}$$

Further, one has:

$$\begin{aligned} 0 &\leqq 1 - N_{\boldsymbol{a}}^-(\Lambda)\, N(\Lambda_n)/N(\Lambda) \\ &\leqq 1 - [N_{\boldsymbol{a}}^-(\Lambda)/N_{\boldsymbol{a}}^+(\Lambda)]\,[N(\Lambda_n)/N(\Lambda(\boldsymbol{a}))] \\ &\leqq 1 - [N_{\boldsymbol{a}}^-(\Lambda)/N_{\boldsymbol{a}}^+(\Lambda)]\,[1 - 2\nu 2^{-n}]. \end{aligned} \tag{8.14}$$

If $N_0$ is chosen such that

$$1 - N_{\boldsymbol{a}}^-(\Lambda)/N_{\boldsymbol{a}}^+(\Lambda) < 2^{-n+1} \tag{8.15}$$

when $N(\Lambda) > N_0$, then one has for $N(\Lambda) > N_0$

$$|n^*(\Lambda, r) - n^*(n, r)| \leqq 4(v+1)2^{-n}. \tag{8.16}$$

Using Eq. (8.7),

$$|n^*(\Lambda, r) - n^*(r)| \leqq 8(v+1)2^{-n}, \tag{8.17}$$

which proves the theorem.

*Remark 8.5.* If $\phi(r)$ and $\psi(r)$ are two non-decreasing functions which satisfy

$$\phi(r) = \psi(r), \quad \text{for} \quad r \leqq a \quad \text{and} \quad r \geqq b,$$

$$|\phi(r) - \psi(r)| < \varepsilon, \quad \text{for} \quad a \leqq r \leqq b,$$

and if $f(r)$ is continuous and non-decreasing in the closed interval $[a, b]$ and differentiable in the open interval $(a, b)$ then

$$\left| \int_{-\infty}^{\infty} f(r)\, d\phi(r) - f(r)\, d\psi(r) \right| < \varepsilon[f(b) - f(a)]. \tag{8.18}$$

If one defines a function

$$g(\Lambda) = \int_{-\infty}^{\infty} \gamma(r)\, dn^*(\Lambda, r) \tag{8.19}$$

for $\gamma$ continuous, then, as $N(\Lambda) \to \infty$,

$$\lim_{\Lambda \to \infty} g(\Lambda) = \int_{-\infty}^{\infty} \gamma(r)\, dn^*(r) = g. \tag{8.20}$$

If $\gamma(r)$ is differentiable and non-decreasing in $0 \leqq r \leqq A_0$ and if $|n^*(r) - n^*(\Lambda, r)| < \varepsilon$, then

$$|g(\Lambda) - g| < \varepsilon[\gamma(A_0) - \gamma(0)]. \tag{8.21}$$

This is a useful tool for proving uniformity of convergence.

*Definition 8.10.*

$$g(\Lambda; \mu) \equiv \tfrac{1}{2}\mu + N(\Lambda)^{-1} \ln P(\Lambda; e^{-\frac{1}{2}\mu}). \tag{8.22}$$

**Lemma 8.5.** $g(\Lambda, \mu) = \dfrac{1}{2}\mu + \dfrac{1}{2} \displaystyle\int_{-\infty}^{\infty} \ln(e^{-\mu} + r^2)\, dn^*(\Lambda, r).$ (8.23)

*Proof.* Trivial.

*Definition 8.11.*

$$g(\mu) \equiv \tfrac{1}{2}\mu + \tfrac{1}{2}\int \ln(e^{-\mu} + r^2)\, dn^*(r). \tag{8.24}$$

16*

**Theorem 8.6.** *If $\Lambda$ tends to infinity, then, for all $\mu$,*

$$\lim_{\Lambda\to\infty} g(\Lambda;\mu) = g(\mu). \tag{8.25}$$

*The convergence is uniform on subsets of the form $\mu \leqq \mu_0$.*

*Proof.* From Remark 8.5 it follows that

$$|g(\Lambda;\mu) - g(\mu)| < \varepsilon \ln(1 + A_0^2 e^{\mu}), \tag{8.26}$$

when $N(\Lambda) > N_0$, where $N_0$ is chosen to make

$$|n^*(\Lambda, r) - n^*(r)| < \varepsilon$$

for $N(\Lambda) > N_0$.

*Remark 8.6.* One can alternatively prove the existence of the grand canonical limit $g(\mu)$ directly, and then infer the existence of the distribution $n^*(r)$ from the power series expansion of $g(\mu)$ in $e^{\mu}$. This follows by the theory of the Hamburger moment problem from the fact that the domain of increase of $n^*(r)$ is finite by Theorem 8.4.

*Definition 8.12.* If, for a given $\Lambda$, $d_0$ is the maximum value of $d$ for which $Z_d$ is strictly positive, we define

$$\varrho_0(\Lambda) = d_0/N(\Lambda). \tag{8.27}$$

**Lemma 8.7.**

$$\varrho_0(\Lambda) = \lim_{\mu\to\infty} \frac{dg(\Lambda,\mu)}{d\mu}, \tag{8.28}$$

$$\lim_{\Lambda\to\infty} \varrho_0(\Lambda) = \lim_{\mu\to\infty} \frac{dg(\mu)}{d\mu} \equiv \varrho_0. \tag{8.29}$$

*Proof.* Eq. (8.28) is a trivial consequence of the fact that $\exp[g(\Lambda;\mu)]$ is a polynomial in $\exp(\frac{1}{2}\mu)$; the non-trivial part of the theorem is (8.29), for it says that we can interchange the limits $\mu\to\infty$ and $\Lambda\to\infty$. The left side of (8.29) is $\lim_{\Lambda\to\infty}\lim_{\mu\to\infty} \int r^2(e^{-\mu}+r^2)^{-1}\, dn^*(\Lambda, r)$ and the right side is the same with the limits reversed. On the right side we can take the limit $\Lambda\to\infty$ which, by Remark 8.5, means that $n^*(\Lambda, r)\to n^*(r)$. Now it is clear that for any increasing $m(r)$, with $m(\infty)=1$ and with $m(0)=0$ for $r<0$,

$$\lim_{\mu\to\infty} \int r^2(e^{-\mu}+r^2)^{-1}\, dm(r) = 1 - m(0+), \tag{8.30}$$

where $m(0+) = \lim_{r\to 0} m(r)$. Thus, to prove the theorem we require that $\lim_{\Lambda\to\infty} n^*(\Lambda, 0+) = n^*(0+)$. This, in turn, is a consequence of the uniform convergence of $n^*(\Lambda,\cdot)$ to $n^*(\cdot)$.

*Definition 8.13.* For $0 \leqq \varrho \leqq \varrho_0(\Lambda)$ we define the function $h(\Lambda,\varrho)$, for $\varrho N(\Lambda)$ integral, by

$$h(\Lambda,\varrho) = N(\Lambda)^{-1} \ln Z_{\varrho N(\Lambda)}. \tag{8.31}$$

For other values of $\varrho$ we define $h(\Lambda, \varrho)$ by linear interpolation between $\varrho_1$ and $\varrho_2$ which are chosen such that $\varrho_1 < \varrho < \varrho_2$, $N\varrho_1$ is integral and $N(\varrho_2 - \varrho_1) = 1$. For $\varrho > \varrho_0$ we define $h(\Lambda, \varrho) = -\infty$. It is a very important consequence of Newton's inequality (Theorem 7.1) that $h(\Lambda, \varrho)$ is concave in $\varrho$ for $0 \leqq \varrho \leqq 1$.

*Definition 8.14.* For $0 \leqq \varrho \leqq 1$

$$h(\varrho) \equiv \inf\{-\varrho\mu + g(\mu): -\infty < \mu < \infty\}. \tag{8.32}$$

**Theorem 8.8A.** $\limsup_{\Lambda\to\infty} h(\Lambda, \varrho) = h(\varrho), \quad 0 \leqq \varrho \leqq 1.$ (8.33)

*Proof.*

$$\begin{aligned} g(\Lambda, \mu) &\geqq \max\{\mu N(\Lambda)^{-1} d + N(\Lambda)^{-1} \ln Z_d : d \leqq N(\Lambda)\} \\ &= \sup\{\varrho\mu + h(\Lambda, \varrho) : 0 \leqq \varrho \leqq 1\}. \end{aligned} \tag{8.34}$$

Choose $N_0$ such that $|g(\Lambda, \mu) - g(\mu)| < \varepsilon$ for $N(\Lambda) > N_0$ and for $\mu \leqq \mu_0 < \infty$. Then

$$g(\mu) + \varepsilon \geqq \sup\{\varrho\mu + h(\Lambda, \varrho): 0 \leqq \varrho \leqq 1\},$$

i.e. for all $\varrho$ and $N(\Lambda) > N_0$ and $\mu \leqq \mu_0$

$$h(\Lambda, \varrho) \leqq g(\mu) - \varrho\mu + \varepsilon. \tag{8.35}$$

Therefore,

$$\limsup_{\Lambda\to\infty} h(\Lambda, \varrho) \leqq \inf\{g(\mu) - \varrho\mu : \mu \leqq \mu_0\} \tag{8.36}$$

for all $\mu_0$, and (8.33) follows.

**Theorem 8.8B.** *For* $0 \leqq \varrho < \varrho_0$,

$$\liminf_{\Lambda\to\infty} h(\Lambda, \varrho) = h(\varrho). \tag{8.37}$$

*Proof.* One has

$$\begin{aligned} g(\Lambda, \mu) &\leqq N(\Lambda)^{-1} \ln N(\Lambda) + \max\{\mu N(\Lambda)^{-1} d + N(\Lambda)^{-1} \ln Z_d : d \leqq N(\Lambda)\} \\ &= N(\Lambda)^{-1} \ln N(\Lambda) + \sup\{\varrho\mu + h(\Lambda, \varrho); 0 \leqq \varrho \leqq 1\}. \end{aligned} \tag{8.38}$$

Since $h(\Lambda, \varrho)$ is concave in $\varrho$,

$$h(\Lambda, \varrho) = \inf\{-\varrho\mu + g(\Lambda, \mu): -\infty < \mu < \infty\} - N(\Lambda)^{-1} \ln N(\Lambda). \tag{8.39}$$

Now choose $\mu_0$ such that

$$dg(\mu_0)/d\mu_0 = \tfrac{1}{2}(\varrho_0 + \varrho) \tag{8.40}$$

and then choose $N_0$ such that for $N(\Lambda) > N_0$

$$\begin{aligned} &N(\Lambda)^{-1} \ln N(\Lambda) < \tfrac{1}{2}\varepsilon, \\ &\partial g(\Lambda, \mu_0)/\partial\mu_0 > \varrho, \end{aligned} \tag{8.41}$$

and for $\mu \leqq \mu_0$

$$g(\Lambda, \mu) - g(\mu) > -\tfrac{1}{2}\varepsilon\,. \tag{8.42}$$

From (8.41) it follows that for $N(\Lambda) > N_0$

$$\inf\{-\varrho\mu + g(\Lambda, \mu): \infty < \mu < \infty\} = \inf\{-\varrho\mu + g(\Lambda, \mu): -\infty < \mu \leqq \mu_0\}\,.$$

From (8.42) it follows that for $N(\Lambda) > N_0$

$$\inf\{-\varrho\mu + g(\Lambda, \mu): -\infty < \mu \leqq \mu_0\} \geqq h(\varrho) - \tfrac{1}{2}\varepsilon\,.$$

Consequently, for $N(\Lambda) > N_0$

$$h(\Lambda, \varrho) \geqq h(\varrho) - \varepsilon\,, \tag{8.43}$$

which proves the theorem.

*Remark 8.7.* The limit

$$\lim_{\Lambda \to \infty} h(\Lambda, \varrho_0)$$

will generally depend on the specific boundary condition and does not need to exist. Hammersley [59] proved that for simple cubic lattices in any dimension and for $\{\Lambda\}$ being rectangular boxes with faces perpendicular to the directions of the edges and for an even number of vertices one has

$$\lim_{\Lambda \to \infty} h(\Lambda, \varrho_0) = h(\varrho_0)\,. \tag{8.44}$$

We conjecture that (8.44) holds generally for cyclic boundary conditions provided $\Lambda$ is chosen to have a dimer arrangement with the density $\varrho_0$.

## IX. Thermodynamic Limit for Monomer-Dimer Systems: Analyticity Properties

In this section we shall combine the results of Sections IV and VIII in order to furnish the general analyticity properties of the thermodynamic functions. We shall also prove the existence and analyticity of the infinite lattice correlation functions. We begin by restating the general analyticity condition of Theorem 4.9.

*Definition 9.1.* Let $G$ be a graph with general, complex edge weights. Write the edge weights as

$$W(i,j) = -U(i,j)^2 \tag{9.1}$$

and assume

$$U(i,j) \in D_\pm(V(i,j), \theta), \quad \text{all} \quad i,j \in G\,, \tag{9.2}$$

for some set of real, non-negative numbers $\{V(i,j)\}$ and some value of $\theta$, $0 \leqq \theta \leqq \pi/2$. Finally let $\bar{G}$ be the graph with the edge weights changed to $V(i,j)^2$. If $x$ is the monomer fugacity, we then say that $x \in \mathscr{A}(G)$ if $\theta$ and the set of numbers $\{V(i,j)\}$ can be chosen such that (9.2) is fulfilled and simultaneously

$$x \in D^C_\pm(A_0(\bar{G}), \theta) = D^C_+(A_0(\bar{G}), \theta) \cup D^C_-(A_0(\bar{G}), \theta). \tag{9.3}$$

If (9.3) is changed to

$$x \in D^C_\pm(A_0(\bar{G}) + \varepsilon, \theta) \tag{9.4}$$

then $x \in \mathscr{A}_\varepsilon(G)$.

Finally let $L$ be a lattice (Definition 8.1) with complex edge weights, let the edge weights be written in the form (9.1) and satisfy (9.2) with a set of weights $\{V(i,j)\}$ such that $\bar{L}$ (defined analogously to $\bar{G}$) is a lattice (i.e. that the translation invariance is preserved), and let $A_0$ be taken to be the value of $A_0$ obtained for $\bar{L}$ by the limit in Theorem 8.2. If $\theta$ and the set of numbers $\{V(i,j)\}$ can be chosen such that both (9.2) is satisfied and simultaneously

$$x \in D^C_\pm(A_0, \theta), \tag{9.5}$$

then we say that $x \in \mathscr{A}(L)$.

Next, we modify the definition of the free energy slightly from Definition 8.10.

*Definition 9.2.*

$$g_m(G; x) \equiv N(G)^{-1} \ln P(G; x) - \ln x \tag{9.6}$$

where the branch cuts of the right side of (9.6) are the $N(G)$ line segments $(z, 0)$, where $z$ is any one of the zeros of $P(G; x)$.

One then has the following lemma as a trivial corollary of Theorem 4.9.

**Lemma 9.1.** *If $G$ is a weighted graph and if $x \in \mathscr{A}_\varepsilon(G)$ then*

$$|g_m(G; x)| < |\ln \varepsilon| + 2\pi. \tag{9.7}$$

By application of Vitali's theorem one easily proves the following theorem from Lemma 9.1 and Theorem 8.6:

**Theorem 9.2.** *Let $L$ be a lattice with complex edge weights and let $\{\Lambda\}$ be a sequence of section graphs that tends to infinity in the sense of Van Hove. Then the limit*

$$\lim_{\Lambda \to \infty} g_m(\Lambda; x) = g_m(x) \tag{9.8}$$

*exists, is independent of the sequence, and is uniform on closed subsets of the following domains:*

(i) *If the edge weights are held fixed then $g_m(x)$ is analytic in $x \in \mathscr{A}(L)$.*

(ii) *If $g_m(x)$ is also considered as a function of the edge weights then it is analytic on the following domain:*

$$\bigcup_{\{V(i,j)\}} \bigcup_{0 \leqq \theta \leqq \pi/2} \{(x, \{W\}) : x \in D^C_\pm(A_0, \theta), W(i,j) = -U(i,j)^2 \text{ and } U(i,j) \in D_\pm(V(i,j), \theta)\}, \tag{9.9}$$

*where the allowed $\{V(i,j)\}$ are translation invariant.*

*Remark.* If all the edge weights are real and positive, then Theorem 9.2 states that the free energy is analytic in $x$ in the cut $x$-plane where the cut runs from $-iA_0$ to $iA_0$ along the imaginary axis. One can then obtain a natural expansion variable $s$ by the substitution

$$x = \tfrac{1}{2}A_0(1-s^2)/s \tag{9.10}$$

$$s = (x/A_0)\,[1 + A_0^2/x^2)^{\frac{1}{2}} - 1] \tag{9.11}$$

which maps the cut $x$-plane conformally onto the unit disk of the $s$-plane. A power series in $s$ will then converge in the whole physical region.

As mentioned in the introduction, the variable $s$ is essentially the same as the expansion variable, $y_2$, used by Nagle [27]. To establish the connection we note that if $A_0$ is replaced by some $A' > A_0$ in (9.10, 11), one obtains an expansion variable $s'$ with the same property that the analyticity region is mapped into the unit $s'$-disk. Naturally, the best $A'$ to use is $A_0$, but in the absence of accurate knowledge of $A_0$ we can use the bound given by Theorem 4.3. For a connected lattice, all of whose non-zero edge weights are a constant, $W$, and which has coordination number, $q$, the upper bound is

$$A' = [4W(q-1)]^{\frac{1}{2}}. \tag{9.12}$$

It is this $A'$ that appears in Nagle's $y_2$. We also note in passing that our previous report [61] contained an inferior estimate in which $(q-1)$ was replaced by $q$.

As the last point we turn to the correlation functions.

*Definition* 9.3. The correlation function $\varrho(G, S)$ is defined for subsets $S$ of the graph $G$ $(S \subset G)$ by

$$\varrho(G, S; x) = x^{N(S)} P(G-S; x)/P(G; x). \tag{9.13}$$

We first prove two lemmas which are the equivalent of Lemma 9.1 and Theorem 8.2 and then obtain the final theorem by an application of Vitali's theorem. For the sake of the proof of the lemmas it is more convenient to work with the function

$$f(G, S; x_1, \ldots, x_N) = P(G-S; x_1, \ldots, x_N)/P(G; x_1, \ldots, x_N). \tag{9.14}$$

**Lemma 9.3.** *If G is a weighted graph and if* $x \in \mathscr{A}_\varepsilon(G)$ *then*

$$|f(G, S; x, \ldots, x)| \leqq \varepsilon^{-N(S)}. \tag{9.15}$$

*Proof.* If we choose

$$x_j = \begin{cases} x' & j \in S \\ x'' & j \in G - S \end{cases} \tag{9.16}$$

then, in an obvious notation, we have

$$P(G; x', x'') = P(G - S; x'') \prod_{j=1}^{N(S)} [x' - a_j(x'')] \tag{9.17}$$

since the coefficient of $(x')^{N(S)}$ is $P(G-S; x'')$. Now $P(G-S; x'')$ is surely not zero if $x'' \in \mathscr{A}(G)$ and therefore

$$f(G, S; x', x'') = \prod_{j=1}^{N(S)} [x' - a_j(x'')]^{-1}. \tag{9.18}$$

It follows from Theorem 4.9 that if $x'' \in D_+^C(A_0(\bar{G}), \theta)$ then

$$a_j(x'') \in D_+(A_0(\bar{G}), \theta). \tag{9.19}$$

Consequently, if $x = x' = x'' \in \mathscr{A}_\varepsilon(G)$ then

$$|x' - a_j(x'')| > \varepsilon, \tag{9.20}$$

which proves the lemma.

**Lemma 9.4.** *If G is a weighted graph whose edge weights are real and positive,* $N(S) = N$, *and if x is purely imaginary and*

$$-i\bar{x} > A_0(G) \tag{9.21}$$

*then* $i^{N(S)} f(G, S; m_1 x, \ldots, m_N x)$ *is monotone decreasing in each of the variables,* $m_j (j \in G - S)$, *when* $m_k$ *is real and greater than one for* $k = 1, \ldots, N$.

*Proof.* Suppose $\{1\} \subset G - S$. Then if $\hat{G} = G - 1$

$$\begin{aligned} &\partial f(G, S; m_1 x, \ldots, m_N x)/\partial m_1 \\ &\quad = x[P(G; m_1 x, \ldots, m_N x)\, P(\hat{G} - S; m_2 x, \ldots, m_N x) \\ &\qquad - P(\hat{G}; m_2 x, \ldots, m_N x) \\ &\qquad \cdot P(G - S; m_1 x, \ldots, m_N x)]/P^2(G; m_1 x, \ldots, m_N x). \end{aligned} \tag{9.22}$$

Application of Eqs. (2.7), (2.10) and (2.11) then yields

$$\begin{aligned} &\Big(\prod_{j \in S} m_j\Big) i^{N(S)} \partial f(G, S; m_1 x, \ldots, m_N x)/\partial m_1 \\ &= y m_1^{-1} [Q(G'; y)\, Q((\hat{G} - S)'; y) - Q(G'; y)\, Q((G - S)'; y)]/Q^2(G'; y), \end{aligned} \tag{9.23}$$

with $y = -ix$. The monotonicity then follows from Theorem 6.3 since $y > A_0(G) \geqq A_0(G')$ by assumption.

**Theorem 9.5.** *Let $S \subset G_1 \subset G_2 \subset \cdots$ be an infinite sequence of section graphs with complex edge weights. Then the limit*

$$\lim_{i \to \infty} \varrho(G_i, S; x) = \varrho(S; x) \tag{9.24}$$

*exists uniformly on closed subsets of the following domains*

(i) *If the edge weights are held fixed then $\varrho(S; x)$ is analytic on* $\mathrm{Int}\left[\bigcap_i \mathscr{A}(G_i)\right]$.

(ii) *If $\varrho(S; x)$ is also considered as a function of any finite number of distinct edge weights then it is analytic on*

$$\begin{aligned}\mathrm{Int}\Big[\bigcap_i \Big(\bigcup_{\{V(i,j)\}} \bigcup_{0 \leq \theta \leq \pi/2} &\{(x, \{W\}) : x \in D_{\pm}^{C}(A_0(G_i), \theta),\\ &W(i,j) = -U(i,j)^2 \text{ and } U(i,j) \in D_{\pm}(V(i,j), \theta)\}\Big)\Big].\end{aligned} \tag{9.25}$$

*If the sequence is a sequence of section graphs of a lattice, L, and if the $\{V(i,j)\}$ in (9.25) are restricted to be translation invariant, then the limit is independent of the sequence when the sequence tends to infinity in the sense of Van Hove, provided the distance of S from the boundary of $G_i$ tends to infinity as $i \to \infty$.*

*Proof.* We notice that $G_i$ can be thought of as $G_{i+1}$ in which the vertex weights on the vertices belonging to $G_{i+1} - G_i$ are infinite. Consequently, if the edge weights are positive and if $x$ is sufficiently large and imaginary, Lemmas 9.3 and 9.4 tell us that $\{\varrho(G_i, S; x)\}$ is a monotone bounded sequence and hence has a limit. The rest follows from Vitali's theorem. The uniqueness of the limit in the case of a Van Hove sequence follows from the same cube packing arguments as in Section VIII.

## References


1. Roberts, J. K.: Proc. Roy. Soc. (London) A **152**, 469 (1935).
2. — Proc. Roy. Soc. (London) A **161**, 141 (1937).
3. — Proc. Cambridge Phil. Soc. **34**, 399 (1938).
4. — Miller, A. R.: Proc. Cambridge Phil. Soc. **35**, 293 (1939).
5. Readhead, P. A.: Trans. Faraday Soc. **57**, 641 (1961).
6. Rossington, D. R., Bost, R.: Surface Sci. **3**, 202 (1965).
7. Lichtman, D., McQuistan, R. B.: J. Math. Phys. **8**, 2441 (1967).
8. McQuistan, R. B., Lichtman, D.: J. Math. Phys. **9**, 1660 (1968).
9. — J. Math. Phys. **10**, 2205 (1969).
10. — Lichtman, S. J.: J. Math. Phys. **11**, 3095 (1970).
11. Fowler, R. H., Rushbrooke, G. S.: Trans. Faraday Soc. **33**, 1272 (1937).
12. Guggenheim, A.: Trans. Faraday Soc. **33**, 151 (1937).
13. Chang, T. S.: Proc. Roy. Soc. (London) A **169**, 512 (1939).
14. — Proc. Cambridge Phil. Soc. **35**, 265 (1939).
15. Miller, A. R.: Proc. Cambridge Phil. Soc. **38**, 109 (1942).

16. Orr, W. J. C.: Trans. Faraday Soc. **40**, 306 (1944).
17. McGlashan, M. L.: Trans. Faraday Soc. **47**, 1042 (1951).
18. Miller, A. R.: Proc. Cambridge Phil. Soc. **39**, 54 (1943).
19. — Proc. Cambridge Phil. Soc. **39**, 131 (1943).
20. Orr, W. J. C.: Trans. Faraday Soc. **40**, 320 (1944).
21. Guggenheim, E. A.: Proc. Roy. Soc. (London) A **183**, 203 (1944).
22. — Proc. Roy. Soc. (London) A **183**, 213 (1944).
23. We shall not attempt to give a complete bibliography of the Flory-Huggins theory; the reader is referred to standard textbooks. The earliest references seem to be P. J. Flory: J. Chem. Phys. **10**, 51 (1942) and Huggins: Ann. N. Y. Acad. Sci. **43**, 9 (1942).
24. Guggenheim, E. A.: Mixtures, Chapter X. Oxford: Claredon Press 1952.
25. Rushbrooke, G. S., Scoins, H. I., Wakefield, A. J.: Discussions Farad. Soc. **15**, 57 (1953).
26. Travena, D. H.: Proc. Phys. Soc. **84**, 969 (1964).
27. Nagle, J. F.: Phys. Rev. **152**, 190 (1966).
28. Gaunt, D. S.: Phys. Rev. **179**, 174 (1969).
29. Bellemans, A., Fuks, S.: Physica **50**, 348 (1970).
30. Runnels, L. K.: J. Math. Phys. **11**, 842 (1970).
31. Baxter, R. J.: J. Math. Phys. **9**, 650 (1968).
32. Craen, J. van, Bellemans, A.: Bull. Acad. Pol. Sci. **19**, 45 (1971).
33. Hammersley, J. M.: In: Proceedings of the 2nd Annual Conference on Computational Physics, pp. 1–8 (Institute of Physics and Physical Society, London (1970)).
34. Baxendale, J. H., Enüstün, B. V., Stern, J.: Phil. Trans. Roy. Soc. (London) A **243**, 169 (1951).
35. Everett, D. H., Penney, M. F.: Proc. Roy. Soc. (London) A **212**, 164 (1952).
36. Tompa, H.: J. Chem. Phys. **16**, 292 (1948).
37. Brøndsted, J. N., Koefoed, J.: Kgl. Danske Videnskab. Selskob. Mat-Fys. Medd. **22**, No. 17 (1946).
38. Tompa, H.: Trans. Faraday Soc. **45**, 101 (1949).
39. Pizzini, S., Morlotti, R., Wagner, V.: J. Electrochem. Soc. **116**, 915 (1969).
40. Cohen, E. G. D., De Boer, J., Salsburg, Z. W.: Physica **21**, 137 (1955).
41. Conway, B. E., Verall, R. E.: J. Phys. Chem. **70**, 1473 (1966).
42. Fisher, M. E., Temperley, H. N. V.: Rev. Mod. Phys. **32**, 1029 (1960).
43. Katsura, S., Inawashiro, S.: Rev. Mod. Phys. **32**, 1031 (1960).
44. Kasteleyn, P. W.: Physica, Grav. **27**, 1209 (1961).
45. Temperley, H. N. V., Fisher, M. E.: Phil. Mag. Serie 8 **6**, 1061 (1961).
46. Fisher, M. E.: Phys. Rev. **124**, 1664 (1961).
47. Kasteleyn, P. W.: J. Math. Phys. **4**, 287 (1963).
48. Montroll, E. W.: In: Applied combinatorial mathematics (Ed. F. Beckenbach). New York: J. Wiley & Sons, 1964.
49. Lieb, E. H.: J. Math. Phys. **8**, 2339 (1967).
50. Gibberd, R. W.: Can. J. Phys. **46**, 1681 (1968).
51. Wu, T. T.: J. Math. Phys. **3**, 1265 (1962).
52. Ferdinand, A. E.: J. Math. Phys. **8**, 2332 (1967).
53. Hammersley, J. M., Feuerverger, A., Izenman, A., Mahani, S.: J. Math. Phys. **10**, 443 (1969).
54. Fisher, M. E., Stephenson, J.: Phys. Rev. **132**, 1411 (1963).
55. Hartwig, R. E.: J. Math. Phys. **7**, 286 (1966).
56. Bondy, J. A., Welsh, D. J. A.: Proc. Cambridge Phil. Soc. Math. Phys. Sci. **62**, 503 (1966).
57. Hammersley, J. M.: Proc. Cambridge Phil. Soc. Math. Phys. Sci. **64**, 455 (1968).
58. — Menon, V. V.: J. Inst. Math. Appl. **6**, 341 (1970).
59. — In: Research papers in statistics. Festschrift für J. Neyman, p. 125 (Editor, F. N. David). New York: John Wiley & Sons 1966.

60. Heilmann, O. J.: Existence of phase transitions in certain lattice gases with repulsive potentials (to be published).
61. — Lieb, E. H.: Phys. Rev. Letters **24**, 1412 (1970).
62. Kunz, H.: Phys. Letters **32** A, 311 (1970).
63. Gruber, C., Kunz, H.: Commun. math. Phys. **22**, 133 (1971).
64. Dobrushin, R. L.: Funct. Anal. Appl. **2**, No. 4, 44 (1968), (English translation **2**, 302 (1968)).
65. Essam, J. W., Fisher, M. E.: Rev. Mod. Phys. **42**, 271 (1970).
66. Szegö, G.: Orthogonal polynomials (American Mathematical Society, Colloquium Publications Vol. XXIII, Providence 1939).
67. Ruelle, D.: Statistical mechanics. New York: W. A. Benjamin 1969.
68. Fisher, M. E.: J. Math. Phys. **7**, 1776 (1966).
69. Lee, T. D., Yang, C. N.: Phys. Rev. **87**, 410 (1952).
70. Asano, T.: J. Phys. Soc. Japan **29**, 350 (1970); Phys. Rev. Letters **24**, 1409 (1970).
71. Suzuki, M., Fisher, M. E.: J. Math. Phys. **12**, 235 (1971).
72. Ginibre, J.: Phys. Letters **24** A, 223 (1967).

E. H. Lieb
Department of Mathematics, 2—375
Massachusetts Institute of Technology
Cambridge, Mass. 02139, USA

# Theory of Monomer-Dimer Systems

O. J. Heilmann and E. H. Lieb

Commun. math. Phys. **25**, 190–232 (1972)

The legends to the figures were inadvertently omitted; they should read as follows:

Fig. 1. The sign variation of $Q(G; x)$ for $N(G) = 7$ as $x$ goes from $-\infty$ to $+\infty$.

Fig. 2. Two graphs for which $Q(G; x)$ has two identical zeros. (a) gives two zeros at $x = 0$ independent of the edge weights. In (b) if the weights on all four vertical edges are $a^2$ there will be double zeros at $x = \pm a$.

# A Property of Zeros of the Partition Function for Ising Spin Systems

Elliott H. Lieb*
*Department of Mathematics, Massachusetts Institute of Technology, Cambridge, Massachusetts 02139*
and
David Ruelle
*Institut des Hautes Etudes Scientifiques, 91 Bures-sur-Yvette, France*
(Received 22 November 1971)

Given an Ising antiferromagnet on a lattice with an $AB$ substructure (bipartite lattice), one can consider the associated ferromagnet in which all the exchange constants are negated. Suppose the ferromagnet is above its critical temperature in the sense that there is an arc $(-\theta, \theta)$ of the unit circle on which the partition function has no zeros in $z = \exp(2\beta H)$. We prove that the original antiferromagnet partition function will have no zeros in $z$ in the disc orthogonal to the unit disc and passing through the two end points of the arc. In other words, the antiferromagnet free energy is analytic in the magnetic field for small fields.

## 1. STATEMENT OF RESULTS

Let $Z^\nu$ be the lattice of points with $\nu$ integer coordinates. We assume that a function $\epsilon$ on $Z^\nu$ with values $\pm 1$ is given such that

(a) $\epsilon$ is not identically $+1$,

(b) $\epsilon(x)\epsilon(y) = \epsilon(x+y)$.

We can then say that the point $x$ is *even*, resp. *odd*, if $\epsilon(x) = +1$, resp. $-1$.

A spin variable with two possible values $\sigma_x = \pm 1$ is associated with each lattice site $x \in Z^\nu$ and we introduce a translation-invariant pair interaction $J$ such that $\sum_{x \in Z^\nu} |J(x)| < +\infty$. If $\Lambda = \{x_1, \ldots, x_m\}$ is a finite subset of $Z^\nu$, the energy of the spin configuration $\sigma = (\sigma_{x_1}, \ldots, \sigma_{x_m})$ is

$$E(\sigma) = \tfrac{1}{2} \sum_{x\in\Lambda} \sum_{y\in\Lambda} J(x-y)(1-\sigma_x\sigma_y) - H \sum_{x\in\Lambda} \sigma_x,$$

where $H$ is the magnetic field. The free energy per site at temperature $\beta^{-1}$ is then

$$\psi_J(\beta, H) = -\lim_{\Lambda\to\infty} \beta^{-1} |\Lambda|^{-1} \log \sum_\sigma \exp[-\beta E(\sigma)],$$

where $|\Lambda|$ is the number of points in $\Lambda$ and $\Lambda \to \infty$ may be taken to mean that $\Lambda$ is a rectangular box with all sides tending to infinity.

*Theorem 1:* (a) Let $J$ be ferromagnetic, i.e., $J \geq 0$, and let $\beta^{-1}$ be above the critical temperature in the sense that some neighborhood of $H = 0$ is free of zeros of the partition function $\sum_\sigma \exp[-\beta E(\sigma)]$. Then some neighborhood of $H = 0$ is also free of zeros of the partition function for the interaction $\epsilon J$. In particular $\psi_{\epsilon J}(\beta, H)$ is analytic around $H = 0$ [this can be applied to antiferromagnetic nearest neighbor (Ising) interactions because they are of the form $\epsilon J$].

(b) More precisely, let $z = e^{2\beta H}$. We shall show that if the partition function for the interaction $J$ does not vanish when $|z| = 1$, $|\arg z| < \theta$, then the partition function for the interaction $\epsilon J$ does not vanish when $z \in \Delta_\theta$. Here $\Delta_\theta$ is the open region containing the point 1, and bounded by the circle orthogonal to $\{z : |z| = 1\}$ going through $e^{\pm i\theta}$ (see Fig. 1). In particular, $\psi_{\epsilon J}$ is analytic in $\Delta_\theta$.

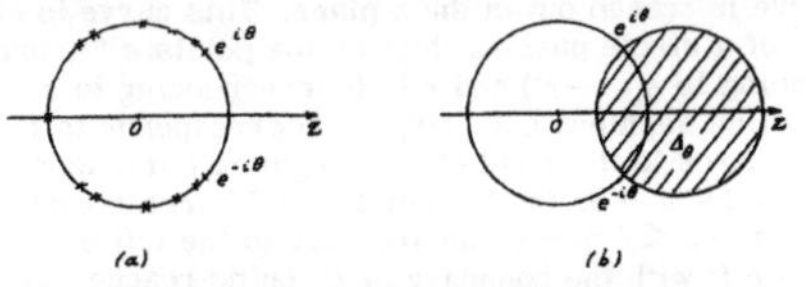


FIG. 1. (a) Interaction $J$: the crosses are zeros of the partition function. (b) Interaction $\epsilon J$: the cross-hatched region $\Delta_\theta$ is free of zeros of the partition function.

The proof uses the theory of analytic functions of several complex variables, and is accomplished in two steps. The first step (Sec. 2) is an application of Borchers' double cone theorem.[1] The second step (Sec. 3) is the computation of a holomorphy envelope (a limiting case of the holomorphy envelope of two polydiscs; see Ref. 2).

At the end of Sec. 3, some extensions of the above theorem are indicated.

## 2. PROOF OF PART (a) OF THEOREM 1

Let us fix $\beta > 0$ and write

$$F_J^\Lambda(H_1, H_2) = -\beta^{-1}|\Lambda|^{-1} \log \sum_\sigma \exp[-\beta E^*(\sigma)],$$

$$E^*(\sigma) = \tfrac{1}{2} \sum_{x\in\Lambda} \sum_{y\in\Lambda} J(x-y)(1-\sigma_x\sigma_y) - \sum_{x\in\Lambda} \{\tfrac{1}{2}[1+\epsilon(x)]H_1\sigma_x + \tfrac{1}{2}[1-\epsilon(x)]H_2\sigma_x\}.$$

It is easily seen that

$$\psi_J(\beta, H) = \lim_{\Lambda\to\infty} F_J^\Lambda(H, H),$$

$$\psi_{\epsilon J}(\beta, H) = -\beta^{-1} \sum_{x\in Z^\nu} \tfrac{1}{2}[\epsilon(x) - 1]\, J(x) + \lim_{\Lambda\to\infty} F_J^\Lambda(H; -H).$$

We shall now make use of a theorem of Borchers.[1,3]

*Theorem 2 (double cone theorem):* Let $C$ be an open convex cone in $\mathbf{R}^n$, with apex at the origin. Let the function $F$ be defined and analytic in a domain $D \subset \mathbf{C}^n = \mathbf{R}^n + i\mathbf{R}^n$ such that

(i) $D \supset (C + i\mathbf{R}^n) \cup (-C + i\mathbf{R}^n)$,

(ii) $D$ contains the points $i[\alpha x + (1-\alpha)y]$, where $x, y \in \mathbf{R}^n$, $x - y \in C$ and $\alpha$ runs between 0 and 1.

Then $F$ is analytic in a fixed complex neighborhood of the set $i[(x - C) \cap (y + C)]$

We apply this theorem to the function $F_J^\Lambda(H_1, H_2)$. Here $n = 2$ and $C$ consists of the points with coordinates $> 0$ in $\mathbf{R}^2$. Thus (i) means that $F_J^\Lambda$ is analytic for Re $H_1 > 0$ and Re $H_2 > 0$, or Re $H_1 < 0$ and Re $H_2 < 0$. Condition (i) is satisfied by $F_J^\Lambda$ as a consequence of the Lee–Yang circle theorem.[4] By assumption, (ii) is satisfied with

$$x = (K, K), \qquad y = (-K, -K)$$

where $2\beta K = \theta$ and $\theta$ is defined in part (b) of Theorem 1. Therefore, $F_\Lambda^J$ is analytic in a fixed complex neighborhood of the set

$$\{(H_1, H_2) \in i\mathbf{R}^2 : |H_1| < K, \ |H_2| < K\}.$$

In particular, when we take $H_1 = -H_2 = H$ and let $\Lambda \to \infty$, we obtain the result announced in part (a) of Theorem 1.

## 3. CALCULATION OF THE DOMAIN OF HOLOMORPHY

Consider a finite Ising spin system composed of $N$ sites and a collection of $\binom{N}{2}$ interaction constants $\{J_{ij}\}$ which are nonnegative. (Previously we assumed the system was translation invariant, but for this section there is no need to do so.) To each site we assign a magnetic field $H_i$, $i = 1, \ldots, N$. As was explained before, by letting $H_i = H$ for $i$ an even site and $H_i = -H$ for $i$ an odd site the system is, in fact, equivalent to an Ising antiferromagnet. In this section we shall allow the $\{H_i\}$ to be arbitrary. The advantage of doing so is that we can thereby consider a more general system in which the spin magnetic moment varies from site to site.

It is convenient to work with the "activity" variables

$$z_i = \exp(2\beta H_i), \qquad i = 1, \ldots, N, \tag{3.1}$$

so that the partitions function can be expressed as

$$Z(H_1, \ldots, H_N) = \exp\left(-\beta \sum_{i=1}^{N} H_i\right) P(z), \tag{3.2}$$

where $P$ is a polynomial and $z = (z_1, \ldots, z_N)$ is regarded as a point in $\mathbf{C}^N$.

*Notation:* (a) $D$ denotes the open unit disc in $\mathbf{C}$:

$$D = \{z \in \mathbf{C} : |z| < 1\}.$$

(b) $D^N$ denotes the symmetric unit polydisc in $\mathbf{C}^N$:

$$D^N = \{z \in \mathbf{C}^N : |z_i| < 1, \ i = 1, \ldots, N\}.$$

(c) $E^N$ denotes $[\mathrm{Int}\,(\sim D)]^N$:

$$E^N = \{z \in \mathbf{C}^N : |z_i| > 1, \ i = 1, \ldots, N\}$$

(d) For $0 < \theta < \pi$, $A_\theta \subset \mathbf{C}$ is the open arc:

$$A_\theta = \{z \in \mathbf{C} : z = e^{i\psi}, \ -\theta < \psi < \theta\}$$

(e) $B_\theta = \partial D \backslash A_\theta$ is the (closed) arc complementary to $A_\theta$.

(f) $A_\theta^N \subset \mathbf{C}^N$ is

$$A_\theta^N = \{z \in \mathbf{C}^N : z_i \in A_\theta, \ i = 1, \ldots, N\}.$$

Given that $P(z)$ has no zeros in $D^N \cup E^N$ (the Lee–Yang theorem) and given that, for $z \in \mathbf{C}$, $P(z, z, \ldots, z)$ has no zeros in $A_\theta$, we have established previously that there exists a complex neighborhood $O$ of $A_0^N$ in which $P(z)$ is free of zeros. In this section we address ourselves to obtaining a minimal estimate of $O$ by calculating the envelope of holomorphy of $D_N \cup E_N \cup O$. As will be seen, this envelope is, in fact, a domain in $\mathbf{C}^N$. In this domain, $f(z) = 1/P(z)$ is holomorphic and hence $P(z)$ has no zeros.

For $z \in \mathbf{C}$ we change variables $z \to \psi$ as follows:

$$\exp(\psi + i\theta) = -(z - e^{i\theta})/(z - e^{-i\theta}) \tag{3.3}$$

so that

$$\psi = i(\pi - \theta) + \ln[(z - e^{i\theta})/(z - e^{-i\theta})] \tag{3.4}$$

with the logarithm defined to be holomorphic in $\mathbf{C}\backslash B_\theta$ and $\to 1$ as $z \to \infty$.

Alternatively,

$$z = \cosh \tfrac{1}{2}(\psi - i\theta)/\cosh \tfrac{1}{2}(\psi + i\theta) \tag{3.5}$$

so that $\psi \to -\psi$ is equivalent to $z \to 1/z$.

The mapping (3.3) maps $\mathbf{C}\backslash B_\theta$ conformally onto

$$K_\theta = \{\psi = s + it : -\infty < s < \infty, \ -\pi < t < \pi\} \backslash \{i(\pi - \theta)\}. \tag{3.6}$$

To describe the mapping, we refer to Figs. 2 and 3. If we fix $t \neq \pi - \theta$, then as $s$ goes from $-\infty$ to $\infty$ a curve is traced out in the $z$ plane. This curve is an arc of a circle passing through the points $e^{i\theta}$ (corresponding to $s = -\infty$) and $e^{-i\theta}$ (corresponding to $s = \infty$). The line segment $L = [e^{i\theta}, e^{-i\theta}]$ corresponds to $t = -\theta$. Arcs to the right of this segment correspond to $-\theta < t < \pi - \theta$. In particular, $t = 0$ corresponds to $A_\theta$. If $-\pi < t < -\theta$, the arcs are to the left of $L$ and inside $D$ with the boundary of $B_\theta$ being reached as $t \downarrow -\pi$. If $\pi - \theta < t < \pi$, the arcs are to the left of $L$ and outside $D$ with $B_\theta$ being reached as $t \uparrow \pi$. The exceptional case is $t = \pi - \theta$, where $s \in \{-\infty, 0\}$ corresponds to the vertical line $\{e^{i\theta}, e^{i\theta} + i\infty\}$ while $s \in \{0, \infty\}$ corresponds to the vertical line $\{e^{-i\theta} - i\infty, e^{-i\theta}\}$. A complete circle through the points $e^{i\theta}$ and $e^{-i\theta}$ is always composed of two arcs whose $t$ values, $t$ and $t'$, satisfy $|t - t'| = \pi$. If a value of $t$ (and hence an arc) is fixed, then for every point on the arc the angle subtended by the line segment $L$ is $|\pi - |t + \theta||$. Among the circles through $\{e^{i\theta}, e^{-i\theta}\}$ there is a particularly important one to which we shall return later, namely the circle orthogonal to $\partial D$. It has the property that it is invariant under inversion: $z \to 1/z$. The arc of this circle inside $D$ corresponds to $t = -\pi/2$ while the arc outside $D$ corresponds to $t = \pi/2$.

Except when $\theta = \pi/2$, this circle has a radius $|\tan \theta|$ and its center is at $z = (1/\cos \theta, 0)$. When $\theta = \pi/2$ the "circle" is the imaginary axis.

Returning to $\mathbb{C}^N$, we apply the conformal map (3.3) to each $z_i$, i.e., $z_i \to \psi_i$, $i = 1, \ldots, N$, so that $1/P(z) = f(z) \to h(\psi)$ with $\psi = (\psi_1, \ldots, \psi_N)$. $D^N$ is mapped onto

$$\delta^N = \{\psi : \psi_j = s_j + it_j, \; -\infty < s_j < \infty, \; -\pi < t_j < 0, \; j = 1, \ldots, N\}$$

while $E^N$ is mapped onto $\epsilon^N = \mu^N \backslash \gamma^N$, where

$$\mu^N = \{\psi : \psi_j = s_j + it_j, \; -\infty < s_j < \infty, \; 0 < t_j < \pi, \; j = 1, \ldots, N\}$$

and

$$\gamma^N = \{\psi : \psi_j = s_j + it_j, \; -\infty < s_j < \infty, \; 0 < t_j < \pi, \; j = 1, \ldots, N$$

and $\psi_j = i(\pi - \theta)$ for some $j\}$.

The set $A_\theta^N$ is mapped onto

$$\alpha^N = \{\psi : \psi_j = s_j + it_j, \; -\infty < s_j < \infty, \; t_j = 0, \; j = 1, \ldots, N\}.$$

Since each $z_i$ is a holomorphic function of $\psi_i$ in $K_\theta$, $h(\psi)$ is holomorphic in $\delta^N \cup \epsilon^N \cup \Omega^N$, where $\Omega^N$ is some complex neighborhood of $\alpha_\theta^N$. This is a consequence of the double cone theorem stated in Sec. 2.

The next step is to establish analyticity of $h(\psi)$ in some neighborhood of $\gamma^N$ so that we can replace $\epsilon^N$ by $\mu^N$. Let $G^N$ be the inverse image of $\gamma^N$ in $\hat{\mathbb{C}}^N$ so that $z \in G^N$ means that some $z_i$ is infinite. The mere fact that $P$ is a polynomial does not guarantee that $1/P$ is holomorphic in a neighborhood of $G^N$.

[*Remark*: The fact that $P$ comes from an Ising ferromagnetic spin system does guarantee this analyticity, however, because the coefficient $Q$ of $\prod_{j=1}^n z_i$ (for example) in $P$, when considered as a polynomial in $z_{n+1}, \ldots, z_N$, is nonzero in $E^{N-n}$. This is so because $Q$ is the polynomial of an Ising ferromagnetic spin system of $N - n$ spins which interact with $n$ other spins fixed in the $+1$ direction. Hence $1/P = 0$ on $G^N$.]

To remedy the lack of analyticity in the general case, we can replace $f(z)$ by $F(z) = f(z)\,\gamma(z)$, where $\gamma(z) = \prod_{j=1}^N \rho(z_j)$ and $\rho(z) = \sin^M\theta[z^2 + 1 - 2z\cos\theta]^{-M/2}$ for a sufficiently large positive integer $M$. Clearly $\rho(z)$ is holomorphic away from the cut $B_\theta$ *and is never zero* except at $z = \infty$, so that $F$ has the same analyticity properties as $f$. However, $F = 0$ on $G^N$ and is analytic in some neighborhood of $G^N$. In terms of $\psi$, the equivalent substitution is

$$h(\psi) \to H(\psi) = h(\psi) \prod_{j=1}^{N} \cosh \tfrac{1}{2}(\psi + i\theta)^M.$$

Hence, $H$ is holomorphic on some neighborhood of the tube $T = \delta^N \cup \mu^N \cup \alpha^N$. Since any neighborhood of $T$ contains a connected component which contains $\delta^N \cup \mu^N$, we can use the tube theorem[5] to assert that $H$ is also holomorphic on the convex hull of $T$ which we shall call $T'$. It is easy to see that

$$T' = \bigcup_{0 \le a \le \pi} T_a,$$

where $T_a$ is the tube with imaginary base $\{(t_1, \ldots, t_N) : -\pi + a < t_j < a, j = 1, \ldots, N\}$. I.e., $T'$ is the union of translates of $\delta^N$ (or $\mu^N$). The inverse image of $T_a$ is a symmetric domain $D_a^N \subset \mathbb{C}^N$ where $\partial D_a \subset \mathbb{C}$ is a circle through the points $e^{i\theta}$ and $e^{-i\theta}$. As $a$ goes from 0 to $\pi$, all circles are reached; however, $D_a$ is either the disc interior to the circle or else the exterior of the circle and is chosen such that $D_a$ contains $A_\theta$.

A more geometric interpretation of our conclusion is this:

*Theorem 3:* Let $P(z_1, \ldots, z_N)$ be the polynomial of a ferromagnetic Ising spin system of $N$ spins and suppose that $P(z, \ldots, z) \neq 0$ when $z \in A_\theta = \{e^{i\varphi} : -\theta < \varphi < \theta\}$ for some $\theta \in (0, \pi)$. Let $C$ be any circle in $\mathbb{C}$ that includes the points $e^{i\theta}$ and $e^{-i\theta}$ and let $\Delta$ be the interior or exterior or $C$ chosen such that $\Delta \supset A_\theta$. Then $P(z_1, \ldots, z_N) \neq 0$ if $z_i \in \Delta$ for $i = 1, \ldots, N$.

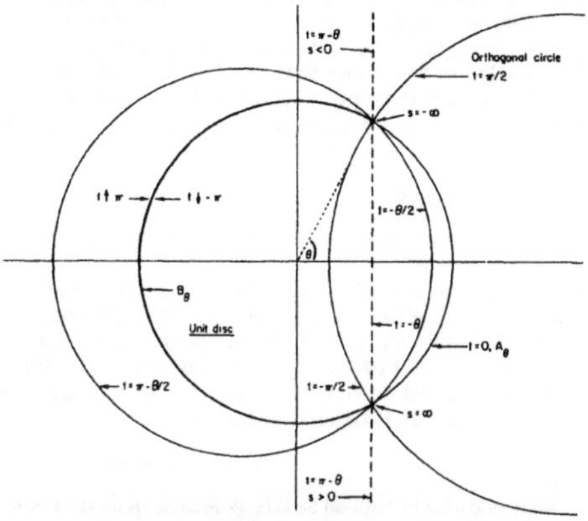


FIG. 2. The unit disc and several circles through the points $e^{i\theta}$ and $e^{-i\theta}$ are shown for the case $\theta < \pi/2$. If $\psi = s + it$, each arc corresponds to some fixed $t$ value (as shown) and $-\infty < s < \infty$. The Ising antiferromagnetic partition function has no zeros in the *interior* of the orthogonal circle.

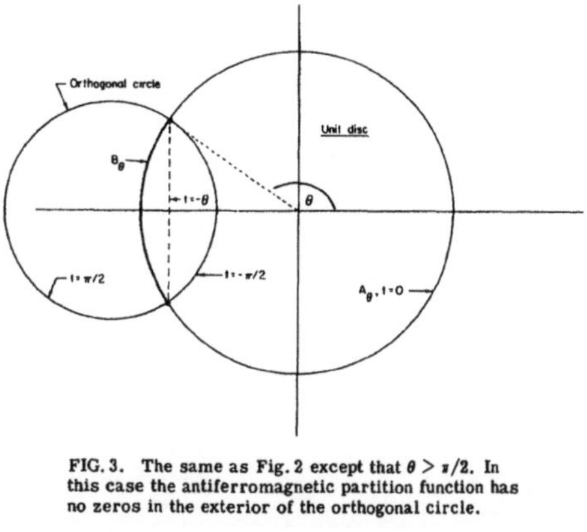


FIG. 3. The same as Fig. 2 except that $\theta > \pi/2$. In this case the antiferromagnetic partition function has no zeros in the exterior of the orthogonal circle.

To apply Theorem 3 to the antiferromagnet considered in the previous section, we must set $z_i = z = \exp(2\beta H)$ for $i$ an even site and $z_i = 1/z$ for $i$ an odd site. Let $A_\theta$ be the arc in which the ferromagnetic partition function has no zeros and let $C_\theta$ be the circle through the points $e^{i\theta}$ and $e^{-i\theta}$ that is *orthogonal* to the unit circle. Let $\Delta_\theta$ be the *interior* of $C_\theta$ when $0 < \theta < \pi/2$ and the *exterior* of $C_\theta$ when $\pi/2 < \theta < \pi$. When $\theta = \pi/2$, $\Delta_{\pi/2}$ is the right-hand complex plane, $\Delta_{\pi/2} = \{z : \mathrm{Re}(z) > 0\}$. Then the antiferromagnetic partition function is nonzero when $z \in \Delta_\theta$.

*Remarks:* (i) Theorem 3 can be applied to any system that has the Lee-Yang property such as the Heisenberg ferromagnet. It is only for the Ising model, however, that the replacement of $z$ by $1/z$ on odd sites is equivalent to the replacement of the ferromagnet by an antiferromagnet. In particular, Theorem 3 can be applied to an Ising model of arbitrary spin, not just spin $\frac{1}{2}$.

(ii) Instead of using the variable $z = \exp(2\beta H)$ one could have used $\xi = \exp(\beta H)$ so that the angle $\theta$ would be replaced by $\theta/2$. The new domain $\Delta'$ would be smaller in the sense that $(\Delta')^2 \subset \Delta$. In other words, the fact that the polynomial is a function of $z = \xi^2$ is a stronger hypothesis and naturally leads to a larger domain in which there are no zeros.

## 4. EXTENSIONS

Consider an assembly of spins with ferromagnetic pair interaction $J$ in a magnetic field $H$, and assume that the grand partition function has no zeros for $\mathrm{Re}\, H = 0$, $|\mathrm{Im}\, H| < K$. Suppose now that different magnetic fields $H_1, \ldots, H_n$ act on different classes of spins, and let $F(H_1, \ldots, H_n)$ be the corresponding free energy. We deduce as in Sec. 2 that $F$ is analytic in a complex neighborhood of the set

$$\{(H_1, \ldots, H_n) \in i\mathbf{R}^n : |H_1| < K, \ldots, |H_n| < K\}.$$

Let $z_i = e^{2\beta H_i}$ and $f(z_1, \ldots, z_n) = F(H_1, \ldots, H_n)$. The argument in Sec. 3 shows that $f$ is analytic when all $z_i$ are in the same $D_\alpha$, $D_\alpha$ being any open region containing the point 1 and bounded by a circle through the points $e^{\pm 2i\beta K}$.

The free energy for various (not all!) spin systems with interactions of antiferromagnetic or other type and magnetic moment varying from site to site may be written (up to constants) as

$$F(\epsilon_1 H, \ldots, \epsilon_n H) = f(z^{\epsilon_1}, \ldots, z^{\epsilon_n}),$$

where $\epsilon_1, \ldots, \epsilon_n$ are some real numbers. The above results prove analyticity in a neighborhood of $z = 1$, more precisely for all $z$ such that

$$1, z^{\epsilon_1}, \ldots, z^{\epsilon_n}$$

are all inside (or outside) the same circle through $e^{\pm 2i\beta K}$. However, if the region of analyticity for $z$ contains a piece of the negative real axis, it will in general be necessary to exclude it (introduce a cut). This is so because, if $\epsilon_i$ is not a positive integer, the functions $z^{\epsilon_i}$ are not holomorphic at 0 and $\infty$. Even if all $\epsilon_i$ are integers, it is still necessary to introduce a cut from $-1$ to $-\infty$.

## ACKNOWLEDGMENTS

This paper was written while the authors participated in the 1971 Battelle "Recontre" at the Battelle Seattle Research Center. We are grateful for the hospitality shown us. E. L. would like to thank R. Stora for introducing him to the mapping (3.3) and its relevance to the tube theorem. The same mapping has been used[2] to find the envelope of holomorphy of the union of two polydiscs.

* Work supported by National Science Foundation Grant GP 26526.

1 H. J. Borchers, Nuovo Cimento **19**, 787 (1961).
2 V. Glaser and J. Bros, "L'envelope d'holomorphie de l'union de deux polycercles," CERN preprint (1961).
3 H. Epstein, thesis (to appear).
4 T. D. Lee and C. N. Yang, Phys. Rev. **87**, 410 (1952).
5 S. Bochner and W. T. Martin, *Several Complex Variables* (Princeton U.P., Princeton, N. J., 1948).

Commun. Math. Phys. 80, 153–179 (1981)

Communications in Mathematical Physics


# A General Lee–Yang Theorem for One-Component and Multicomponent Ferromagnets

Elliott H. Lieb[1*] and Alan D. Sokal[2**]

1 Departments of Mathematics and Physics, Princeton University, Princeton, NJ 08544, USA
2 Department of Physics, Princeton University, Princeton, NJ 08544, USA

**Abstract.** We show that any measure on $\mathbb{R}^n$ possessing the Lee–Yang property retains that property when multiplied by a ferromagnetic pair interaction. Newman's Lee–Yang theorem for one-component ferromagnets with general single-spin measure is an immediate consequence. We also prove an analogous result for two-component ferromagnets. For $N$-component ferromagnets ($N \geqq 3$), we prove a Lee–Yang theorem when the interaction is sufficiently anisotropic.

## 1. Introduction

The Lee–Yang theorem on the zeros of the partition function is an important tool in the rigorous study of phase transitions in lattice spin systems [1]. In addition, it has applications to the proof of existence of the infinite-volume limit [2] and of a mass gap [3, 4], and to the proof of correlation inequalities [5, 6] and inequalities for critical exponents [4, 7, 8].

In this paper we shall give a new proof of a generalized Lee–Yang Theorem. Our methods lead to an essentially complete result for one-component and two-component (classical) ferromagnets with quite general single-spin measures. We have also some promising partial results for $N$-component ferromagnets ($N \geqq 3$). We end the paper with some conjectures.

Consider, for purposes of orientation, the model of one-component "spins" $\phi_i$ defined by the partition function

$$Z = \int \exp\left[\sum_{i,j=1}^{n} J_{ij}\varphi_i\varphi_j + \sum_{i=1}^{n} h_i\varphi_i\right]\prod_{i=1}^{n} d\nu_i(\varphi_i). \tag{1.1}$$

Here the $d\nu_i$ are suitable probability measures on the real line; the pair interaction coefficients $J_{ij}$ are nonnegative ("ferromagnetic"); and the magnetic fields $h_i$ are allowed to take arbitrary complex values. The Lee–Yang theorem then states

* Research supported in part by NSF grant PHY 78-25390 A01
** Research supported in part by NSF grant PHY 78-23952

0010-3616/81/0080/0153/$05.40

that, for suitable measures $d\nu_i$, the partition function $Z(h_1, \ldots, h_n)$ is nonzero whenever $\operatorname{Re} h_i > 0$ for all $i$. The theorem was originally proven by Lee and Yang [9] only for the spin-$\frac{1}{2}$ model

$$d\nu_i(\varphi) = \tfrac{1}{2}[\delta(\varphi - 1) + \delta(\varphi + 1)]d\varphi, \text{ all } i. \tag{1.2}$$

Subsequently, numerous alternate proofs for the spin-$\frac{1}{2}$ case were found [10–17, 41], and the theorem was also extended to more general single-spin measures $d\nu_i$ [18, 19, 15, 16, 42, 43]. The best result is that of Newman [15], which allows arbitrary even measures $d\nu_i$ with the property that

$$\int e^{h\varphi} d\nu_i(\varphi) \neq 0 \text{ for } \operatorname{Re} h \neq 0, \text{ all } i. \tag{1.3}$$

This result is essentially the best possible: it states that the Lee–Yang property holds for all $J_{ij} \geqq 0$ if and only if it holds for $J_{ij} \equiv 0$. But while the condition (1.3) is exceedingly natural, Newman's method of proof is quite indirect: he shows that (1.3) is a necessary and sufficient condition for the model (1.1) to be approximable in a certain sense by spin-$\frac{1}{2}$ models; and then he appeals to the already proven Lee–Yang theorem for the spin-$\frac{1}{2}$ case. The original motivation of the present work, therefore, was to find a *direct* proof of Newman's result, utilizing directly the condition (1.3). We did discover a rather elementary such proof; it is given in Appendix A. But we also discovered a far-reaching generalization of Newman's theorem, one which we believe clarifies the underlying structure of the Lee–Yang theorem.

Our method is based on the identity

$$Z(h_1, \ldots, h_n) = \exp\left[\sum_{i,j=1}^{n} J_{ij} \frac{\partial}{\partial h_i} \frac{\partial}{\partial h_j}\right] Z_0(h_1, \ldots, h_n) \tag{1.4}$$

where

$$Z_0(h_1, \ldots, h_n) = \int \exp\left[\sum_{i=1}^{n} h_i \varphi_i\right] \prod_{i=1}^{n} d\nu_i(\varphi_i). \tag{1.5}$$

Now the hypothesis (1.3) ensures precisely that $Z_0$ has the Lee–Yang property; so what we need to show is that this property is preserved by a certain (infinite-order) differential operator. Noting additionally the identity

$$\exp\left[\sum_{i,j=1}^{n} J_{ij} z_i z_j\right] = \lim_{k \to \infty} \prod_{i,j=1}^{n} (1 + k^{-1} J_{ij} z_i z_j)^k, \tag{1.6}$$

and taking account of the hypothesis $J_{ij} \geqq 0$, the Lee–Yang theorem is then reduced (modulo the approximation of entire functions by polynomials) to the following proposition about polynomials: if $P(z_1, \ldots, z_n)$ and $Q(z_1, \ldots, z_n)$ are polynomials which are nonvanishing when $\operatorname{Re} z_i > 0$ for all $i$, then the polynomial $S(z_1, \ldots, z_n) = P(\partial/\partial z_1, \ldots, \partial/\partial z_n)\ Q(z_1, \ldots, z_n)$ also has this property (or else is identically zero). Now this is a well-known result in the case $n = 1$ (Proposition 2.1); but it is also true in general, as we demonstrate (Proposition 2.2).

In fact, we deduce immediately the following generalization of Newman's result: Let $d\mu_0$ be any measure on $\mathbb{R}^n$ (not necessarily a product measure) possessing the "Lee–Yang property" (defined precisely in Sect. 3); then, for any set of $J_{ij} \geqq 0$, the measure

$$d\mu(\varphi) = \exp\left[\sum_{i,j=1}^{n} J_{ij}\varphi_i\varphi_j\right] d\mu_0(\varphi) \tag{1.7}$$

also has the Lee–Yang property. In other words, ferromagnetic pair interactions (among others) are "universal multipliers for Lee–Yang measures".

Similar considerations yield a general Lee–Yang theorem for two-component ferromagnets: Let

$$Z = \int \exp\left[\sum_{i,j=1}^{n}\sum_{\alpha=1}^{2} J_{ij}^{(\alpha)}\varphi_i^{(\alpha)}\varphi_j^{(\alpha)} + \sum_{i=1}^{n}\sum_{\alpha=1}^{2} h_i^{(\alpha)}\varphi_i^{(\alpha)}\right]\prod_{i=1}^{n} d\nu_i(\varphi_i), \tag{1.8}$$

where $J_{ij}^{(1)} \geqq |J_{ij}^{(2)}|$ for all $i, j$, and each $d\nu_i$ is a rotationally symmetric measure on $\mathbb{R}^2$ whose projection onto one of the coordinates has the Lee–Yang property. Then $Z \neq 0$ whenever $\operatorname{Re} h_i^{(1)} > |\operatorname{Im} h_i^{(2)}|$ for all $i$. This generalizes a result obtained by Dunlop [20] for the plane-rotator model

$$d\nu_i(\varphi) = \delta(|\varphi|^2 - 1)d\varphi, \text{ all } i \tag{1.9}$$

by infinitely more complicated (though intriguing) methods.

Sadly, we are unable to give a similarly complete solution of the Lee–Yang problem for $N$-component ferromagnets with $N \geqq 3$. At present, we have only the following partial result: in the obvious generalization of (1.8), one has $Z \neq 0$ whenever

$$\operatorname{Re} h_i^{(1)} > \left[\sum_{\alpha=2}^{N} (\operatorname{Im} h_i^{(\alpha)})^2\right]^{1/2} \text{ for all } i, \tag{1.10}$$

provided that

$$J_{ij}^{(1)} \geqq \sum_{\alpha=2}^{N} |J_{ij}^{(\alpha)}| \text{ for all } i, j. \tag{1.11}$$

This is a Lee–Yang theorem for highly anisotropic $N$-component ferromagnets, the first such result (known to us) for $N > 3$. On the other hand, it is clearly unsatisfactory: the condition (1.11) ought to be replaced by

$$J_{ij}^{(1)} \geqq \max_{2 \leqq \alpha \leqq N} |J_{ij}^{(\alpha)}| \text{ for all } i, j, \tag{1.12}$$

as is known by entirely different methods [10, 11, 21, 20] for $N = 3$ (with a restricted class of single-spin measures). This result (for all $N$) would indeed follow by an extension of our methods, as we indicate in Sect. 5, *provided* that an as-yet-unproven generalization of Proposition 2.2 is true. But we are unable to find a proof—we hope that others will be more clever!

## 2. General Theorems

We begin with a result about polynomials of a single complex variable, which gives the flavor of our methods.

**Proposition 2.1.** *Let $P$ and $Q$ be polynomials in a single complex variable, with the property that $P(z) \neq 0$ whenever $\operatorname{Re} z > 0$, and $Q(z) \neq 0$ whenever $\operatorname{Re} z > c$ ($c$ real). Then $S(z) \equiv P(d/dz)Q(z)$ is either nonvanishing whenever $\operatorname{Re} z > c$ or else is identically zero. Moreover, $S(z) \equiv 0$ if and only if $P(z)$ has a zero at $z = 0$ of order $m > \deg Q$.*

*Proof.* $P$ can be factored as

$$P(z) = a \prod_{i=1}^{\deg P} (z - \alpha_i)$$

with $a \neq 0$ and $\operatorname{Re} \alpha_i \leqq 0$ for all $i$. Hence it suffices to prove the proposition for $P(z) = z - \alpha$, $\operatorname{Re} \alpha \leqq 0$; the general case follows by repeated application of this special case. Now $Q$ can be factored as

$$Q(z) = b \prod_{j=1}^{\deg Q} (z - \beta_j)$$

with $b \neq 0$ and $\operatorname{Re} \beta_j \leqq c$ for all $j$. Then

$$\frac{Q'(z)}{Q(z)} = \sum_{j=1}^{\deg Q} \frac{1}{z - \beta_j},$$

and this has strictly positive real part whenever $\operatorname{Re} z > c$ (unless $\deg Q = 0$, in which case it is identically zero). Hence $Q'(z)/Q(z) \neq \alpha$ whenever $\operatorname{Re} z > c$ (unless $\deg Q = 0$ and $\alpha = 0$); that is, $P(d/dz)\, Q(z) = Q'(z) - \alpha Q(z) \neq 0$ for $\operatorname{Re} z > c$. The last assertion of the proposition is easily verified. QED.

*Remarks.* 1. Proposition 2.1 is actually a special case of a much more general result of Takagi [22] (see Marden [23, pp. 82–84]). The proof given here is a simplification of the method of Benz [32]; it is modeled on the standard proof of the Gauss–Lucas theorem [23, p. 22].

2. The arbitrariness of $c$ is a trivial consequence of invariance under translation of the variable associated with $Q$. Note, however, that the variable associated with $P$ *cannot* be translated; here zero is a distinguished point.

3. Proposition 2.1 was implicitly noted by Newman [15] in the course of the proof of an intermediate result (his Proposition 2.4). It was our attempt to understand the role of this proposition in the proof of the Lee–Yang theorem that led to the present work.

Proposition 2.1 is already sufficient, together with the approximation theorems given later in this section, to prove Newman's version of the Lee–Yang theorem; this proof is given in Appendix A. But it is in fact possible to prove a yet more general result which makes clear (we believe) what is really going on in the Lee–Yang theorem. To do this, we need a multi-variable generalization of Proposition 2.1.

*Notation.* If $z = (z_1, \ldots, z_n) \in \mathbb{C}^n$ and $c = (c_1, \ldots, c_n) \in \mathbb{R}^n$, then $\operatorname{Re} z > c$ means that $\operatorname{Re} z_j > c_j$ for all $j$; analogously for $\operatorname{Re} z \geqq c$. $\partial/\partial z$ means the $n$-tuple $(\partial/\partial z_1, \ldots, \partial/\partial z_n)$.

**Proposition 2.2.** *Let $P_i$ and $Q_i (1 \leqq i \leqq k)$ be polynomials in n complex variables, and define*

$$R(v, w) = \sum_{i=1}^{k} P_i(v) Q_i(w)$$

*and*

$$S(z) = \sum_{i=1}^{k} P_i(\partial/\partial z) Q_i(z).$$

(a) *If* $R(v, w) \neq 0$ *whenever* $\mathrm{Re}\, v \geqq 0$ *and* $\mathrm{Re}\, w \geqq c (c \in \mathbb{R}^n)$, *then* $S(z) \neq 0$ *whenever* $\mathrm{Re}\, z \geqq c$.

(b) *If* $R(v, w) \neq 0$ *whenever* $\mathrm{Re}\, v > 0$ *and* $\mathrm{Re}\, w > c$, *then either* $S(z) \neq 0$ *whenever* $\mathrm{Re}\, z > c$, *or else* $S(z)$ *is identically zero.*

This Theorem generalizes Proposition 2.1 in two major ways: first, the single complex variable is replaced by $n$ complex variables; and second, the single product $PQ$ is replaced by a sum of $k$ such terms. The second generalization can be thought about as follows: write $R(v, w)$ as a sum of monomials with all variables $v$ standing to the left of all variables $w$; then $S(z)$ is obtained by replacing each $v_j (1 \leqq j \leqq n)$ by $\partial/\partial z_j$, and each $w_j$ by $z_j$. This representation makes clear that $S$ depends only on $R$, not on the particular decomposition of $R$ into $\sum P_i Q_i$. This second generalization is of no particular interest for the application we have in mind, but it turns our to be quite natural for the proof of the propositions.

Note first that by translation invariance (Remark 2 above), we can take $c = 0$. We then proceed in a series of lemmas:

**Lemma 2.3.** *Let $Q_0$ and $Q_1$ be polynomials in a single complex variable, and assume that* $R(v, w) \equiv Q_0(w) + vQ_1(w) \neq 0$ *whenever* $\mathrm{Re}\, v \geqq 0$ *and* $\mathrm{Re}\, w \geqq 0$. *Then* $S(z) \equiv Q_0(z) + Q_1'(z) \neq 0$ *whenever* $\mathrm{Re}\, z \geqq 0$.

*Proof.* Setting $v = 0$, we find that $Q_0(z) \neq 0$ whenever $\mathrm{Re}\, z \geqq 0$. If $Q_1 \equiv 0$, this completes the proof; so assume that $Q_1 \not\equiv 0$. Then, letting $c \to +\infty$, we find that $Q_1(z) \neq 0$ whenever $\mathrm{Re}\, z > 0$ (for otherwise, by Hurtwitz' Theorem [23, p. 4] applied to $v^{-1}Q_0 + Q_1$, there would exist zeros of $R(v, w)$ with $\mathrm{Re}\, w > 0$ for any sufficiently large $|v|$). Moreover, if $Q_1(z_0) = 0$ and $\mathrm{Re}\, z_0 = 0$, then $Q_1'(z_0)/Q_0(z_0)$ is real and nonnegative (for otherwise, by the implicit function theorem, there would exists zeros of $R(v, w)$ with $w$ near $z_0$ and $\mathrm{Re}\, w > 0$ for suitable (large) $v$ with $\mathrm{Re}\, v > 0$). Finally, we note that $\mathrm{Re}[Q_0(z)/Q_1(z)] > 0$ whenever $\mathrm{Re}\, z \geqq 0$ and $Q_1(z) \neq 0$ (for otherwise there would exist a zero of $R(v, w)$ with $\mathrm{Re}\, v \geqq 0$ and $\mathrm{Re}\, w \geqq 0$).

Now $Q_1$ can be factored as

$$Q_1(z) = b \prod_{j=1}^{\deg Q_1} (z - \beta_j)$$

with $b \neq 0$ and $\mathrm{Re}\, \beta_j \leqq 0$ for all $j$. Then

$$\frac{Q_1'(z)}{Q_1(z)} = \sum_{j=1}^{\deg Q_1} \frac{1}{z - \beta_j},$$

and this has nonnegative real part for Re $z \geqq 0$ (except at the zeros of $Q_1$, where it is undefined). Hence

$$Q_0(z) + Q_1'(z) = Q_1(z)\left[\frac{Q_0(z)}{Q_1(z)} + \frac{Q_1'(z)}{Q_1(z)}\right] \neq 0$$

whenever Re $z \geqq 0$ and $Q_1(z) \neq 0$. On the other hand, if Re $z = 0$ and $Q_1(z) = 0$, then

$$Q_0(z) + Q_1'(z) = Q_0(z)\left[1 + \frac{Q_1'(z)}{Q_0(z)}\right] \neq 0.$$

This completes the proof.

*Remark 1.* It is indeed possible for $Q_1$ to have zeros on the imaginary axis: consider, for example, $Q_0(w) = 1 + w$ and $Q_1(w) = w$.

*Remark 2.* A related result has been obtained by Dieudonne [24].

**Lemma 2.4.** (*Grace* [25]). *Let $K \subset \mathbb{C}$ be a circular region (i.e. a closed disc, the closed exterior of a disc, or a closed half-plane), and let*

$$F(x) = \sum_{m=0}^{N} a_m x^m$$

*be a polynomial which is nonvanishing whenever $x \in K$. Next let $x_1, \ldots, x_N$ be complex variables, and let $E_0, \ldots, E_N$ be the elementary symmetric functions of the $\{x_i\}$,* i.e. $E_0 = 1$, $E_1 = \sum_i x_i$, *and*

$$E_m = \sum_{i_1 < i_2 < \ldots < i_m} x_{i_1} x_{i_2} \ldots x_{i_m}.$$

*Then the polynomial*

$$\tilde{F}(x_1, \ldots, x_N) = \sum_{m=0}^{N} a_m \binom{N}{m}^{-1} E_m(x_1, \ldots, x_N)$$

*is nonvanishing whenever $x_1, \ldots, x_N$ are all in K.*

*Proof.* See Obreschkoff [26, *pp.* 23–24] or Marden [23, pp. 62–63].

*Proof of Proposition 2.2.* Let $N$ be any integer $\geqq$ the maximal degree of $R(v, w)$ in any of the variables $v_j$; and introduce new variables $v_j^{(k)}$, $1 \leqq k \leqq N$. Now let $\tilde{R}(V, w)$ be the polynomial obtained by expanding $R(v, w)$ as a sum of monimials and replacing each factor $v_j^m$ by

$$\binom{N}{m}^{-1} E_m(v_j^{(1)}, \ldots, v_j^{(N)}).$$

By repeated application of Lemma 2.4, $\tilde{R}(V, w)$ is nonvanishing whenever Re $v_j^{(k)} \geqq 0$ for all $j, k$ and Re $w_j \geqq 0$ for all $j$. Now $\tilde{R}(V, w)$ is of degree at most 1 in each variable $v_j^{(k)}$; so we can repeatedly apply Lemma 2.3 to convert each $v_j^{(k)}$ into $\partial/\partial w_j$, while all other variables are fixed in the closed right half-plane. The result of this process is easily seen to be $S(w)$. This proves (*a*).

To prove (*b*), let $\varepsilon > 0$ and define $P_i^{(\varepsilon)}(v) = P_i(v_1 + \varepsilon, \ldots, v_n + \varepsilon)$, and likewise for $Q_i^{(\varepsilon)}$; and define $R^{(\varepsilon)}$ and $S^{(\varepsilon)}$ in the obvious way involving $P_i^{(\varepsilon)}$ and $Q_i^{(\varepsilon)}$. Now clearly $R^{(\varepsilon)}(v, w) \neq 0$ whenever $\operatorname{Re} z \geqq 0$ and $\operatorname{Re} w \geqq 0$, so by part (a), $S^{(\varepsilon)}(z) \neq 0$ whenever $\operatorname{Re} z \geqq 0$. But $S^{(\varepsilon)}(z)$ converges to $S(z)$ as $\varepsilon \downarrow 0$, uniformly on compacts; so Hurwitz' theorem on $\mathbb{C}^n$ implies that either $S(z)$ is nonvanishing on the open set $\operatorname{Re} z > 0$, or else $S(z)$ is identically zero. This completes the proof.

*Remarks.* 1. Part (b) of Proposition 2.2 can also be proven by an "elementary" argument (i.e. one avoiding Grace's Theorem); the proof is based on the identity

$$S(w) = \left[\prod_{j=1}^{n} \exp\left(\frac{\partial}{\partial v_j}\frac{\partial}{\partial w_j}\right)\right] R(v, w)\bigg|_{v=0}.$$

More precisely, we define $R^{(\varepsilon)}$ as above, and note that

$$S(w) = \lim_{\varepsilon \downarrow 0} \lim_{m \to \infty} \left(\left[\prod_{j=1}^{n}\left(1 + m^{-1}\left(\frac{\partial}{\partial v_j} + \varepsilon\right)\left(\frac{\partial}{\partial w_j} + \varepsilon\right)\right)^m\right] R^{(\varepsilon)}(v, w)\bigg|_{v=0}\right)$$

uniformly on compacts. Now the differential operator in brackets is a product of polynomials each of which is of degree 1 in each variable and which is nonvanishing when the real parts of all variables are nonnegative; so it follows from Lemma 2.3, by a repetitive argument similar to that used above, that this operator preserves the nonvanishing of $R^{(\varepsilon)}$ for $\operatorname{Re} v \geqq 0, \operatorname{Re} w \geqq 0$. The conclusion of part (b) then follows by Hurwitz' theorem.

2. If the coefficients in $P_i$ and $Q_i$ are allowed to depend analytically on an auxiliary variable $\zeta$ varying in a domain $D \subset \mathbb{C}^r$ and the hypothesis of the proposition holds for all $\zeta \in D$, then in part (b) of the proposition, $S(z; \zeta)$ can vanish identically for one value of $\zeta$ only if it does so for *all* $\zeta \in D$. This is an immediate consequence of including the variable $\zeta$ in the Hurwitz argument.

3. Grace's theorem has been employed in a similar way by Millard and Viswanathan [27].

Our next goal is to extend Propositions 2.1 and 2.2 to suitable classes of entire functions. If $f$ is an entire function $\mathbb{C}^n$, and $b > 0$, we define

$$\| f \|_b = \sup_{z \in \mathbb{C}^n}\left[\exp\left(-b \sum_{i=1}^{n} |z_i|^2\right)|f(z)|\right]. \tag{2.1}$$

Then, for each $a \geqq 0$, let $\mathscr{A}^n_{a+}$ be the space of entire functions $f$ such that $\|f\|_b < \infty$ for all $b > a$. That is, $\mathscr{A}^n_{a+}$ is the space of entire functions of exponential order strictly less than 2, or of order 2 and type at most $a$. We equip $\mathscr{A}^n_{a+}$ with the family of norms $\|\cdot\|_b, b > a$ (or equivalently, the countable family $\|\cdot\|_{a+1/k}, k$ integer); then $\mathscr{A}^n_{a+}$ is a Fréchet space. Note also that $\mathscr{A}^n_{a+}$ is closed under differentiation; this is a simple consequence of the Cauchy integral formula. Finally, we note two other elementary facts about $\mathscr{A}^n_{a+}$ [28]:

1. A *bounded* sequence (or net) in $\mathscr{A}^n_{a+}$ converges in the topology of $\mathscr{A}^n_{a+}$ if and only if it converges pointwise on $\mathbb{C}^n$ (or even on an arbitrarily small nonempty open subset of $\mathbb{C}^n$).

2. For any $f \in \mathscr{A}^n_{a+}$, the partial sums of the Taylor series of $f$ converge to $f$ in the topology of $\mathscr{A}^n_{a+}$. Hence the polynomials are dense in $\mathscr{A}^n_{a+}$.

$$\text{Let } f(z) = \sum_{\mathbf{m}} \alpha_{\mathbf{m}} z^{\mathbf{m}} \text{ and } g(z) = \sum_{\mathbf{m}} \beta_{\mathbf{m}} z^{\mathbf{m}}$$

be entire functions on $\mathbb{C}^n$; then we can define the formal power series

$$[f(\partial)g](z) = \sum_{\mathbf{k}} \sum_{\mathbf{m}} \alpha_{\mathbf{k}} \beta_{\mathbf{m}} (\partial/\partial z^{\mathbf{k}}) z^{\mathbf{m}}. \tag{2.2}$$

[Here $\mathbf{m} = (m_1, \dots, m_n)$ is a multi-index, $z^{\mathbf{m}} = \prod_{i=1}^{n} z_i^{m_i}$ and $\partial/\partial z^{\mathbf{k}} = \partial/\partial z_1^{k_1} \dots \partial z_n^{k_n}$.]

For suitable $f$ and $g$, we can actually make sense of (2.2):

**Proposition 2.5.** *Let $a, b > 0$ with $ab < \frac{1}{4}$, and let $c > b/(1 - 4ab)$. Let $f, g$ be entire functions on $\mathbb{C}^n$ with $\|f\|_a < \infty$, $\|g\|_b < \infty$. Then the series (2.2) is absolutely convergent for all z, and defines an entire function such that*

$$\|f(\partial)g\|_c \leqq K_{abc}^n \|f\|_a \|g\|_b \tag{2.3}$$

*for some $K_{abc}^n < \infty$ independent of $f$ and $g$. It follows that*

$$(f, g) \mapsto f(\partial)g$$

*is a continuous bilinear map from $\mathscr{A}_{a+}^n \times \mathscr{A}_{b+}^n$ into $\mathscr{A}_{b/(1-4ab)+}^n$, for any $a, b \geqq 0$ with $ab < \frac{1}{4}$.*

*Proof.* By a simple estimate using the Cauchy integral formula,

$$|\alpha_{\mathbf{k}}| \leqq \|f\|_a \prod_{i=1}^{n} (2ea/k_i)^{k_i/2}$$

$$|\beta_{\mathbf{m}}| \leqq \|g\|_b \prod_{i=1}^{n} (2eb/m_i)^{m_i/2}$$

(with $0^0 \equiv 1$). Since $(k/2e)^{k/2} \geqq C_1 \Gamma((k+1)/2)$ with $C_1 > 0$, it follows that

$$|\alpha_{\mathbf{k}}| \leqq C_2 \|f\|_a \prod_{i=1}^{n} a^{k_i/2}/A(k_i)$$

$$|\beta_{\mathbf{m}}| \leqq C_2 \|g\|_b \prod_{i=1}^{n} b^{m_i/2}/A(m_i) \tag{2.4}$$

where we have defined $A(0) = 1$, $A(2s+1) = A(2s+2) = s!$ for $s = 0, 1, 2, \dots$. Therefore the proposition reduces to the case $n = 1$ with

$$f(z) = 1 + (z + z^2)\exp(az^2)$$
$$g(z) = 1 + (z + z^2)\exp(bz^2)$$

and $z$ real and positive. Clearly the terms 1 are unimportant. The evaluation of the double series (2.2) is then a combinatorial problem that can be handled as follows: For $x$ real, write

$$\exp(ax^2) = C_3(a) \int_{-\infty}^{\infty} \exp(-t^2/a + 2tx)dt \tag{2.5}$$

and use this (formally) with $x = \partial/\partial z$. Since

$$\exp(2t\partial/\partial z)g(z) = g(z + 2t),$$

we have

$$\begin{aligned}&\exp(a\,\partial^2/\partial z^2)[(z+z^2)\exp(bz^2)]\\&\quad = C_3(a)\int_{-\infty}^{\infty}[(z+2t)+(z+2t)^2]\exp[-t^2/a+b(z+2t)^2]dt\\&\quad = P_1(a,b;z)\exp(c'z^2)\end{aligned}$$

where $c'=b/(1-4ab)$ and $P_1$ is a quadratic polynomial in $z$ whose coefficients depend only on $a$ and $b$. Then

$$\begin{aligned}&(\partial/\partial z+\partial^2/\partial z^2)\exp(a\partial^2/\partial z^2)[(z+z^2)\exp(bz^2)]\\&\quad = P_2(a,b;z)\exp(c'z^2)\end{aligned}$$

where $P_2$ is a quartic polynomial in $z$ whose coefficients depend only on $a$ and $b$. The rest of the proof is easy.

*Remarks.* 1. A partially alternate proof can be based on the methods of [29, Theorem 7] or [30, Lemma 14.1.1.].

2. We do not know whether the estimate (2.3) is true for $c=b/(1-4ab)$. We suspect that it is not, but we have no counterexample.

For any set $A\subset\mathbb{C}^n$, let $\mathscr{P}^n(A)$ be the set of polynomials on $\mathbb{C}^n$ which are nonvanishing on $A$ and let $\overline{\mathscr{P}^n_{a+}}(A)$ be the closure of $\mathscr{P}^n(A)$ in $\mathscr{A}^n_{a+}$. It follows immediately by Hurwitz' theorem that any $f\in\overline{\mathscr{P}^n_{a+}}(A)$ is either identically zero or else is nonvanishing in the interior of $A$. However, the converse is *not* true: as we shall see shortly (Proposition 2.7), there exist entire functions $f\in\mathscr{A}^n_{a+}$, nonvanishing on $A$, which are not approximable by polynomials nonvanishing on $A$.

Let $D^n$ denote the set $\{z\in\mathbb{C}^n:\operatorname{Re} z>0\}$. Then Propositions 2.2 and 2.5 immediately imply:

**Proposition 2.6.** *Let $a,b\geqq 0$ with $ab<\frac{1}{4}$, and let $f\in\overline{\mathscr{P}^n_{a+}}(D^n)$ and $g\in\overline{\mathscr{P}^n_{b+}}(D^n)$. Then $h(z)=f(\partial/\partial z)\,g(z)$ is in $\overline{\mathscr{P}^n_{b/(1-4ab)+}}(D^n)$.*

For a partial converse to Proposition 2.6, see [31, 32] and [33, Sect. IX.6].

For our application we shall need to know which "pair interactions" lie in $\overline{\mathscr{P}^n_{a+}}(D^n)$. The criterion is simple:

**Proposition 2.7.** *Let $B$ be a (complex) $n\times n$ symmetric matrix, and let $f(z)=\exp(\sum_{i,j}B_{ij}z_iz_j)$. Then the following are equivalent:*

(a) *$B_{ij}\geqq 0$ for all $i,j$.*

(b) *$f\in\overline{\mathscr{P}^n_{\|B\|+}}(D^n)$, where $\|B\|$ is the norm of $B$ considered as a bilinear form on $\mathbb{C}^n$ (or $\mathbb{R}^n$) equipped with the Euclidean norm.*

(c) *There exist polynomials $\{P_i\}$ in $\mathscr{P}^n(D^n)$ converging pointwise to $f$.*

*Proof.* To prove (a)$\Rightarrow$(b), note that

$$f(z)=\lim_{k\to\infty}\prod_{i,j}(1+k^{-1}B_{ij}z_iz_j)^k;$$

and since $B_{ij}\geqq 0$, the polynomials on the right are all nonvanishing in $D^n$. Moreover,

$$\left|\prod_{i,j}(1+k^{-1}B_{ij}z_iz_j)^k\right| \leqq \exp\left(\sum_{i,j}B_{ij}|z_i||z_j|\right)$$

$$\leqq \exp\left(\|B\|\sum_{i=1}^{n}|z_i|^2\right),$$

from which the convergence in $\mathscr{A}^n_{\|B\|+}$ easily follows [28]. Clearly (b) implies (c). Finally, note that (c) implies that for fixed $(z_2,\ldots,z_n)\in D^{n-1}$, there exist polynomials $\{Q_i\}$ in $\mathscr{P}^1(D^1)$ converging pointwise to

$$\hat{f}(z)=\exp\left[B_{11}z^2+\left(2\sum_{j=2}^{n}B_{1j}z_j\right)z\right].$$

Now $Q_i\in\mathscr{P}^1(D^1)$ implies that $|Q_i(z)|\geqq|Q_i(z')|$ whenever $\operatorname{Re}z\geqq|\operatorname{Re}z'|$ and $\operatorname{Im}z=\operatorname{Im}z'$, and this inequality clearly carries over to $\hat{f}$. But it is not hard to see that this implies $B_{11}\geqq 0$ and $\operatorname{Re}\sum_{j=2}^{n}B_{1j}z_j\geqq 0$. Since this holds for all $(z_2,\ldots,z_n)\in D^{n-1}$, we must have $B_{1j}\geqq 0$ for $2\leqq j\leqq n$. Analogously one shows that $B_{ij}\geqq 0$ for all $i,j$; hence (c) implies (a). QED.

*Remarks.* 1. $\|B\|\leqq\max_i\sum_j|B_{ij}|$, by a simple argument using the Riesz–Thorin interpolation theorem (or Hölder's inequality).

2. Polya [31] and Obrechkoff [34] have shown that $f\in\overline{\mathscr{P}^1_{a+}}(D^1)$ if and only if

$$f(z)=\mathrm{K}e^{\beta z^2+\gamma z}\prod_j\left(1-\frac{z}{\alpha_j}\right)e^{a/\alpha_j} \tag{2.6}$$

with $0\leqq\beta\leqq a$, $\operatorname{Re}\alpha_j\leqq 0$ for all $j$, $\sum_j|\alpha_j|^{-2}<\infty$, and $\operatorname{Re}\gamma\geqq-\sum_j\operatorname{Re}\alpha_j^{-1}$. For a proof, see Levin [33, Sect. VIII.1]. Analogous results exist for various other regions in $\mathbb{C}^1$ [33, 35].

For each $a>0$, let $\mathscr{T}^n_a$ be the space of tempered distributions $T$ on $\mathbb{R}^n$ such that

$$T(x)=\exp\left[-a\sum_{i=1}^{n}x_i^2\right]T_a(x) \tag{2.7}$$

for some tempered distribution $T_a$. We equip $\mathscr{T}^n_a$ with the weak topology generated by the test functions

$$f(x)=\exp\left[a\sum_{i=1}^{n}x_i^2\right]f_a(x) \tag{2.8}$$

with $f_a\in\mathscr{S}(\mathbb{R}^n)$. That is, a sequence (or net) of distributions $T^{(j)}\in\mathscr{T}^n_a$ converges to $T\in\mathscr{T}^n_a$ if and only if the distributions $T^{(j)}_a$ [defined as in (2.7)] converge to $T_a$ in the usual (weak) topology of $\mathscr{S}'(\mathbb{R}^n)$. Also, we define $\mathscr{T}^n_\infty=\bigcap_{a>0}\mathscr{T}_a$, equipped with the obvious topology.

**Lemma 2.8.** *Let* $0 < a \leqq \infty$. *Then the Laplace transform* $T \mapsto \hat{T}$ *defined by*

$$\hat{T}(x) = \int e^{z \cdot x} T(x) dx \tag{2.9}$$

*is a sequentially continuous linear map of* $\mathscr{T}^n_a$ *into* $\mathscr{A}^n_{1/4a+}$. *That is, if a sequence* $T^{(j)}$ *converges to* $T$ *in* $\mathscr{T}^n_a$, *then* $\hat{T}^{(j)}$ *converges to* $\hat{T}$ *in* $\mathscr{A}^n_{1/4a+}$.

*Proof.* Assume first that $0 < a < \infty$. Then $\hat{T}(z) = T_a(f_z)$, where $f_z(x) = \exp(z \cdot x - ax^2)$. Since $T_a \in \mathscr{S}'(\mathbb{R}^n)$, we have.

$$|T_a(f)| \leqq K \sup_x (1 + |x|^M) \sum_{|\alpha| \leqq N} |\partial^\alpha f(x)| \tag{2.10}$$

for some $K, M, N$. Now

$$\sup_x (1 + |x|^M) \sum_{|\alpha| \leqq N} |\partial^\alpha f_z(x)| \leqq C(1 + |z|^{M+N}) \exp(|z|^2/4a) \tag{2.11}$$

for a suitable constant $C$ (depending on $M, N, a$). Hence $T \in \mathscr{A}^n_{1/4a+}$. Now fix $\beta > 1/4a$ and let $g_z(x) = \exp(z \cdot x - ax^2 - \beta |z|^2)$. Then it follows from (2.11) that $\{g_z\}_{z \in \mathbb{C}^n}$ is a *bounded* family in $\mathscr{S}(\mathbb{R}^n)$. Hence, since weak and strong convergence are equivalent for *sequences* in $\mathscr{S}'(\mathbb{R}^n)$ [36, pp. 74, 238], $T^{(j)}_a(g_z)$ converges to $T_a(g_z)$ *uniformly* for $z \in \mathbb{C}^n$; in other words, $\hat{T}^{(j)}$ converges to $\hat{T}$ in $\mathscr{A}^n_{1/4a+}$.

The case $a = \infty$ follows immediately from the foregoing, since convergence in $\mathscr{T}^n_\infty$ [resp. $\mathscr{A}^n_{0+}$] is equivalent to convergence in $\mathscr{T}^n_a$ for all $a < \infty$ [resp. in $\mathscr{A}^n_{\varepsilon+}$ for all $\varepsilon > 0$]. QED.

*Remark.* We would get continuity instead of just sequential continuity if we had equipped $\mathscr{T}^n_a$ with the *strong* topology.

**Proposition 2.9.** *Let* $0 \leqq \alpha < \beta \leqq \infty$; *let* $T$ *be a distribution in* $\mathscr{T}^n_\beta$ *whose Laplace transform* $\hat{T}$ *lies in* $\overline{\mathscr{P}^n_{1/4\beta+}}(D^n)$; *and let* $f \in \overline{\mathscr{P}^n_{\alpha+}}(D^n)$. *Then, for every* $\gamma < \beta - \alpha$ [*and for* $\gamma = \infty$ *if* $\beta = \infty$], *the distribution* $fT$ *lies in* $\mathscr{T}^n_\gamma$ *and its Laplace transform* $\widehat{fT}$ *lies in* $\overline{\mathscr{P}^n_{1/4\gamma+}}(D^n)$.

*Proof.* Clearly $fT \in \mathscr{T}^n_\gamma$. To prove the statement about the Laplace transform, assume first that $f$ is a polynomial, i.e. $f \in \mathscr{P}^n(D^n)$. Then clearly

$$\widehat{fT}(z) = f(\partial/\partial z)\hat{T}(z),$$

so by Proposition 2.6 [with $a = \alpha, b = 1/4\beta$] we have $\widehat{fT} \in \overline{\mathscr{P}^n_{1/4(\beta-\alpha)+}}(D^n) \subset \overline{\mathscr{P}^n_{1/4\gamma+}}(D^n)$. For general $f \in \overline{\mathscr{P}^n_{\alpha+}}(D^n)$, let $\{f_j\}$ be a sequence in $\mathscr{P}^n(D^n)$. For general $f \in \overline{\mathscr{P}^n_{\alpha+}}(D^n)$, let $\{f_j\}$ be a sequence in $\mathscr{P}^n(D^n)$ converging to $f$ in $\mathscr{A}^n_{\alpha+}$. Then $\{f_j T\}$ converges to $fT$ in $\mathscr{T}^n_\gamma$, so by Lemma 2.8, $\{\widehat{f_j T}\}$ converges to $\widehat{fT}$ in $\mathscr{A}^n_{1/4\gamma+}$. Since $\overline{\mathscr{P}^n_{1/4\gamma+}}(D^n)$ is closed in $\mathscr{A}^n_{1/4\gamma+}$, this proves the proposition.

Finally, let us append a remark which clarifies the "strong Lee–Yang theorem" of Newman [15, section 3]:

$$\partial^{\mathbf{m}}_x |Z(x+iy)|^2 > 0 \text{ whenever } x \in (0, \infty)^n \text{ and } y \in \mathbb{R}^n,$$
$$\text{for every multi-index } \mathbf{m}. \tag{2.12}$$

(Here we write $x = \operatorname{Re} h$, $y = \operatorname{Im} h$. The ordinary Lee–Yang theorem is just the case $\mathbf{m} = 0$.) The point is that (2.12) is actually a *consequence* of our form of the Lee–Yang Theorem:

**Proposition 2.10.** *Let $f$ be an analytic function on $D^n$ which is a limit, uniformly on compacts, of polynomials $P \in \mathscr{P}^n(D^n)$. [In particular, $f \in \overline{\mathscr{P}^n_{\alpha+}}(D^n)$ for some $\alpha$ suffices.] For $x \in (0, \infty)^n$ and $y \in \mathbb{R}^n$, let $F(x, y) = |f(x+iy)|^2$ and let $G_{\mathbf{m}}(x, y) = \partial^{\mathbf{m}} F/\partial x^{\mathbf{m}}$. Then, for each multi-index $\mathbf{m}$, we have either*
(i) *$G_{\mathbf{m}}(x, y) > 0$ for all $x \in (0, \infty)^n$ and all $y \in \mathbb{R}^n$ or else*
(ii) *$G_{\mathbf{m}}(x, y) = 0$ for all $x \in (0, \infty)^n$ and all $y \in \mathbb{R}^n$.*

*Proof.* Consider first the case $f = P \in \mathscr{P}^n(D^n)$. Let $Q(z) = \overline{P(\bar{z})}$; clearly $Q \in \mathscr{P}^n(D^n)$. Now define

$$R(z, z') = P(z + iz')Q(z - iz'). \tag{2.13}$$

Note that if $x$ and $y$ are real, then $R(x, y) = F(x, y)$. Moreover, $R$ is a polynomial which is nonvanishing on the open set

$$\Omega_a = \{(z, z') : \operatorname{Re} z_i > a, |\operatorname{Im} z'_i| < a \text{ for } 1 \leq i \leq n\}$$

for each $a > 0$. By Proposition 2.2 (and Remark 2 following its proof), we have for each multi-index $\mathbf{m}$ *either*
(i) $\partial^{\mathbf{m}} R/\partial z^{\mathbf{m}} \neq 0$ in $\Omega_a$

or else

(ii) $\partial^{\mathbf{m}} R/\partial z^{\mathbf{m}} \equiv 0$ in $\Omega_a$.

Moreover, if (ii) holds for one value of $a$ then it holds for all $a$, by analytic continuation; and $\bigcup_{a>0} \Omega_a$ contains the set $(0, \infty)^n \times \mathbb{R}^n$ of real points. Thus, to complete the proof for the case $f = P$, we need only determine the sign of $G_{\mathbf{m}}(x, y)$ in case (i); we use induction on each component of $\mathbf{m}$. Clearly $G_{\mathbf{0}}(x, y) \geq 0$. Suppose that $G_{\mathbf{m}}(x, y) \geq 0$ for all $x \in (0, \infty)^n$ and all $y \in \mathbb{R}^n$, but that $G_{\mathbf{r}}(x', y') < 0$ for some $x' \in (0, \infty)^n$ and $y' \in \mathbb{R}^n$, with $\mathbf{r} = \mathbf{m} + (1, 0, \ldots, 0)$. Then, by the above, $G_{\mathbf{r}}(x, y) < 0$ for all $x \in (0, \infty^n)$ and $y \in \mathbb{R}^n$. Fix $w = (x_2, \ldots, x_n) \in (0, \infty)^{n-1}$ and $y \in \mathbb{R}^n$, and consider $G_{\mathbf{m}}(x_1, w, y)$ and $G_{\mathbf{r}}(x_1, w, y) = \partial G_{\mathbf{m}}/\partial x_1$ as polynomials in $x_1$. Let $G_{\mathbf{m}}(x_1) = \sum_{k=0}^{K} c_k c_1^k$ with $c_K \neq 0$ (we suppress the dependence on $w$ and $y$, which are fixed once and for all). Clearly $c_K > 0$, since otherwise $G_{\mathbf{m}}(x_1) \to -\infty$ as $x_1 \to +\infty$, contrary to the hypothesis on $G_{\mathbf{m}}$. Therefore either $G_{\mathbf{r}}(x_1) \equiv 0$ for all $x_1$ (if $K = 0$) or else $G_{\mathbf{r}}(x_1) > 0$ for $x_1 \to +\infty$. But either possibility contradicts $G_{\mathbf{r}}(x, y) < 0$ for all $x \in (0, \infty)^n$ and $y \in \mathbb{R}^n$. This completes the proof in the special case $f = P$.

Now let $f = \lim_{j \to \infty} P_j$ with each $P_j \in \mathscr{P}^n(D^n)$. Form $R_j$ from $P_j$ as before; since the convergence is uniform on compacts, all derivatives converge as well, so we have

$$G_{\mathbf{m}}(x, y) = \lim_{j \to \infty} \partial_z^{\mathbf{m}} R_j(z, z')|_{z=x,\, z'=y}$$

for each multi-index $\mathbf{m}$. Fix $\mathbf{m}$. Then, by the above, each $\partial_z^{\mathbf{m}} R_j$ is either strictly positive on $\bigcup_{a>0} \Omega_a$ or else identically zero there. It is then easy to see, using Hurwitz' theorem and perhaps passing to a subsequence, that $\lim_{j\to\infty} \partial_z^{\mathbf{m}} R_j$ is either strictly positive on the connected open set $\bigcup_{a>0} \Omega_a$ or else is identically zero there. This completes the proof.

*Remarks.* 1. If $G_{\mathbf{m}} \equiv 0$ for some $\mathbf{m}$, then obviously $G_{\mathbf{m}'} \equiv 0$ for all $\mathbf{m}' \geqq \mathbf{m}$; moreover, $F(x, y)$ must be a polynomial in $x$ of degree less than $\mathbf{m}$ (with coefficients depending on $y$). This happens, of course, if $f$ is a polynomial; but it also happens in other cases, e.g. $f(x) = \exp(iz)$ with $n = 1, m = 1$.

2. If $f$ is the Laplace transform of an even, positive measure $\mu$ not supported at the origin, then case (ii) of the proposition cannot occur for any multi-index $\mathbf{m}$, since $F(x, 0)$ must increase at least exponentially rapidly as $x \to \infty$ in a suitable direction in $(0, \infty)^n$. This observation, combined with our form of the Lee–Yang theorem, will immediately imply (2.12), Newman's strong Lee–Yang theorem.

## 3. One-Component Models

The proof of a very general Lee–Yang Theorem for one-component ferromagnets is now essentially complete; all we have to do is to collect the pieces from the preceding section.

*Definition 3.1.* A finite (positive) measure $\mu$ on $\mathbb{R}^n (\mu \not\equiv 0)$ is said to have the *Lee–Yang property (with falloff $\beta$) if $\mu \in \mathcal{T}^n_\beta$ and $\hat{\mu} \in \mathcal{P}^n_{1/4\beta+}(D^n)$.*

Since $\mu \not\equiv 0$ implies that $\hat{\mu} \not\equiv 0$, it follows that $\hat{\mu}(z) \neq 0$ for $\operatorname{Re} z > 0$; this is the usual conclusion of the Lee–Yang theorem. Note, however, that $\hat{\mu} \in \mathcal{P}^n_{1/4\beta+}(D^n)$ is a *stronger* hypothesis: it says not only that $\hat{\mu}$ is nonvanishing in $D^n$ but that it is approximable by polynomials with this property.

**Theorem 3.2.** *Let $\mu_0$ have the Lee–Yang property with falloff $\beta$; and let $f \in \mathcal{P}^n_{\alpha+}(D^n)[\alpha < \beta]$ be nonnegative on the support of $\mu_0$, and strictly positive on a set of nonzero $\mu_0$-measure. Then $\mu = f\mu_0$ has the Lee–Yang property with falloff $\gamma$, for every $\gamma < \beta - \alpha$ [and $\gamma = \infty$ if $\beta = \infty$]. In particular, we can take*

$$f(\varphi) = \exp\left[\sum_{i,j=1}^{n} J_{ij}\varphi_i\varphi_j\right]$$

*with all $J_{ij} \geqq 0$, provided that $\alpha = \|J\| < \beta$. (Here $\|J\|$ is the norm of $J$ considered as a bilinear form on $\mathbb{R}^n$ equipped with the Euclidean norm.)*

Theorem 3.2 follows immediately from Propositions 2.9 and 2.7; the positivity conditions on $f$ are needed only to ensure that $\mu \geqq 0$ and $\mu \not\equiv 0$.

**Corollary 3.3.** *Let $\{v_i\}_{1 \leqq i \leqq n}$ be measures on $\mathbb{R}^1$, each having the Lee–Yang property with falloff $\beta$; and let $J$ be a symmetric $n \times n$ matrix with nonnegative entries [$\|J\| < \beta$]. Then the measure $\mu$ on $\mathbb{R}^n$ given by*

$$d\mu(\varphi) = \exp\left[\sum_{i,j=1}^{n} J_{ij}\varphi_i\varphi_j\right]\prod_{i=1}^{n} dv_i(\varphi_i) \tag{3.1}$$

*has the Lee–Yang property with falloff $\gamma$, for every $\gamma < \beta - \|J\|$ [and $\gamma = \infty$ if*

$\beta = \infty$]. *In particular, we can let each* $\nu_i$ *be an* even *measure in* $\mathscr{T}^1_\beta$ *satisfying condition (1.3).*

*Proof.* Only the last sentence (which is Newman's [15] Lee–Yang Theorem) requires further explanation; it is a consequence of the following lemma:

**Lemma 3.4.** *Let* $\nu$ *be an even measure in* $\mathscr{T}^1_\beta$ *satisfying condition* (1.3). *Then* $\nu \in \overline{\mathscr{P}^1_{1/4\beta+}}(D^1)$, *and*

$$\hat{\nu}(h) = \mathrm{K}e^{bh^2}\prod_j\left(1+\frac{h^2}{\alpha_j^2}\right) \tag{3.2}$$

*with* $K>0, 0 \leqq b \leqq 1/4\beta$ *and* $0<\alpha_1 \leqq \alpha_2 \leqq \ldots$, *with* $\sum_j \alpha_j^{-2} < \infty$*; here the sequence* $\{\alpha_j\}$ *may be empty, finite or infinite.*

*Proof* [15, 5]. Since $\hat{\nu} \in \mathscr{T}^1_\beta$, $\hat{\nu}$ is an entire function either of order strictly less than 2, or of order 2 and type at most $1/4\beta$. Moreover, $\hat{\nu}$ is even and has only pure imaginary zeros, which we shall denote $\pm i\alpha_j$. If $\hat{\nu}$ is of order $\rho < 2$, then (3.2) [with $b=0$] follows from the Hadamard factorization theorem [33, Sect. I.10] after grouping conjugate pairs of factors. If $\hat{\nu}$ is of order 2 and finite type, then (3.2) follows similarly from Lindelof's extension of the Hadamard factorization Theorem [33, Sect. I.11]; we must have $b$ real, since $\hat{\nu}(h)$ is real for $h$ real, and we must have $b \geqq 0$, since otherwise $\hat{\nu}(h)$ would vanish as $h \to \pm\infty$, which is impossible for the Laplace transform of a measure. It easily follows [28] from (3.2) that $\hat{\nu} \in \overline{\mathscr{P}^1_{1/4\beta+}}(D^1)$, since the obvious approximating polynomials form a bounded sequence in $\mathscr{A}^1_{1/4\beta+}$ [they are all bounded in absolute value by $\hat{\nu}(|h|)$] which is pointwise convergent to $\hat{\nu}$. QED.

*Remarks.* 1. For non-even measures $\nu$, condition (1.3) is *not* in general sufficient to imply $\hat{\nu} \in \overline{\mathscr{P}^1_{1/4\beta+}}(D^1)$. For example, consider $\nu = \delta_a$, so that $\hat{\nu}(h) = \exp(ah)$. Clearly this satisfies (1.3); but $\hat{\nu} \in \overline{\mathscr{P}^1_{1/4\beta+}}(D^1)$ only if $a \geqq 0$. In general one must test whether $\hat{\nu}$ is of the form (2.6).

2. Lemma 3.4 may also be extended to even measures $\nu$ satisfying a weakened form of (1.3):

$$\int e^{h\varphi}d\nu(\varphi) \neq 0 \text{ for } |\mathrm{Re}\, h| > c, \tag{3.3}$$

for some $c$. Then the $\alpha_j$ in (3.2) no longer need be real, but come in complex-conjugate pairs and satisfy $|\mathrm{Im}\,\alpha_j| \leqq c$. Still, it is easy to show that $\sum_j |\alpha_j|^{-2} < \infty$. Moreover, we have $\hat{\nu} \in \overline{\mathscr{P}^1_{1/4\beta+}}(D_c)$. where $D_c = \{z : \mathrm{Re}\, z > c\}$. As an example of a model for which this extension is useful, consider the spin-1 measure

$$\nu = \kappa\delta_0 + \frac{1-\kappa}{2}(\delta_1 + \delta_{-1}) \tag{3.4}$$

with $0 \leqq \kappa \leqq 1$; here

$$\hat{\nu}(h) = \kappa + (1-\kappa)\cosh h. \tag{3.5}$$

For $0 \leqq \kappa \leqq \frac{1}{2}$, $\nu$ has the ordinary Lee–Yang property (1.3). But even for $\frac{1}{2} < \kappa < 1$, $\nu$ satisfies (3.3) with $c = \cosh^{-1}(\kappa/(1-\kappa))$. Hence we can apply the obvious generalization of Corollary 3.3 in which the region $D^n$ is replaced by $D^n_c$. (This is the reason for the otherwise pedantic insistence on arbitrary $c$ in Propositions 2.1 and 2.2.) The physical consequences of this theorem is that a ferromagnetic model with single-spin measure (3.4) is free of phase translations in the region $h > c$ (and by symmetry $h < -c$). Of course, for $\kappa > \frac{1}{2}$ there will in general be phase transitions at $h \neq 0$; indeed, at suitable temperature one expects the appearance of *three* distinct phase as $h$ is varied [37, 38].

In Appendix $B$ we compare our approach to the Lee–Yang Theorem with the Asano contraction method [10–14, 43], and give an "explanation" from within our own approach of why the Asano method works.

## 4. Two-Component Models

We now begin the application of our methods to Lee–Yang theorems for $N$-component classical ferromagnets ($N \geqq 2$). First we must determine the zero-free region for the Laplace transform of the single-spin measure: this is the largest region for which one can even hope for a Lee–Yang theorem.

**Proposition 4.1.** *Let* $\nu \in \mathscr{T}^N_\beta$ *be a rotationally invariant measure on* $\mathbb{R}^N$ $(N \geqq 2)$ *satisfying*

$$\int e^{h\varphi^{(1)}} d\nu(\varphi) \neq 0 \text{ for } \operatorname{Re} h \neq 0. \tag{4.1}$$

*Then the Laplace transform* $\hat{\nu}$ *is of the form*

$$\hat{\nu}(h) = F\left( \sum_{\alpha=1}^{N} h^{(\alpha)2} \right), \tag{4.2}$$

*where F is an entire function (of order at most 1) with only real negative zeros. More precisely,*

$$\hat{\nu}(\mathbf{h}) = a\, e^{b\zeta} \prod_j \left( 1 + \frac{\zeta}{\alpha_j^2} \right) \tag{4.3}$$

*with* $\zeta = \sum_{\alpha=1}^{N} h^{(\alpha)^2}$, $a > 0$, $0 \leqq b \leqq 1/4\beta$ *and* $0 < \alpha_1 \leqq \alpha_2 \leqq \ldots$, *with* $\sum_j \alpha_j^{-2} < \infty$; *here the sequence* $\{\alpha_j\}$ *may be empty, finite or infinite. Finally,* $\hat{\nu} \in \mathscr{P}^N_{1/4\beta+}(L_N)$, *where*

$$L_N = L_N^+ \cup L_N^- \tag{4.4}$$

*and*

$$L_N^{\pm} = \left\{ \mathbf{h} : \pm \operatorname{Re} h^{(1)} > \left[ \sum_{\alpha=2}^{N} (\operatorname{Im} h^{(\alpha)})^2 \right]^{1/2} \right\}. \tag{4.5}$$

*Remarks.* 1. We indicate the components of a spin by Greek superscripts in parentheses, running from 1 to $N$; we label the spins on a lattice by lower-case Latin subscripts, running from 1 to $n$.

2. (4.1) says that the projection of $\nu$ onto the first coordinate has the Lee–Yang

property as a measure on $\mathbb{R}^1$. It does *not* say that $v$ has the Lee–Yang property as a measure on $\mathbb{R}^N$. Indeed, (4.2) implies that $v$ *cannot* have the Lee–Yang property as a measure on $\mathbb{R}^N$ (unless $v$ is Gaussian, so that $\{\alpha_j\}$ is empty): for one can always find $\mathbf{h}\in\mathbb{R}^N$ with $\operatorname{Re} h^{(\alpha)} > 0$ for all $\alpha$ such that $\zeta = \sum_\alpha h^{(\alpha)^2}$ takes an arbitrary negative real value.

*Proof.* By rotational invariance, $\hat{v}(\mathbf{h})$ is a function only of $\zeta = \sum_\alpha h^{(\alpha)^2}$. By hypothesis (4.1), $\hat{v}(\mathbf{h})$ has only pure imaginary zeros in $h^{(1)}$ when $h^{(2)} = \ldots = h^{(N)} = 0$. The representation (4.2) follows. Indeed, Lemma 3.4 implies the representation (4.3) Then $\hat{v}\in\mathscr{P}^N_{1/4\beta+}(L_N)$ follows as in Lemma 3.4, as a result of the following lemma.

**Lemma 4.2.** *Let $L_N$ be defined by (4.4) and (4.5) and let $\mathbf{h}\in L_N$. Then $\zeta = \sum_{\alpha=1}^{N} h^{(\alpha)^2}$ is never real and negative*

*Proof.* Write $\mathbf{h} = (x^{(1)} + iy^{(1)}, \mathbf{x} + i\mathbf{y})$ with $\mathbf{x}, \mathbf{y}\in\mathbb{R}^{N-1}$. Then $\operatorname{Re}\zeta = x^{(1)^2} - y^{(1)^2} + |\mathbf{x}|^2 - |\mathbf{y}|^2$ and $\operatorname{Im}\zeta = 2(x^{(1)}y^{(1)} + \mathbf{x}\cdot\mathbf{y})$. If $\operatorname{Im}\zeta = 0$, then $|\mathbf{y}| \geqq |x^{(1)}||y^{(1)}|/|\mathbf{x}|$ by the Schwarz inequality. But $\mathbf{h}\in L_N$ means that $|x^{(1)}| > |\mathbf{y}|$; hence $|\mathbf{x}| > |y^{(1)}|$. It follows that $\operatorname{Re}\zeta > 0$. QED.

It is convenient, following Dunlop [20], to introduce the variable $\tilde{\mathbf{h}} = (h^{(1)}, ih^{(2)}, \ldots, ih^{(N)}$. Then the set $\mathbf{h}\in L_N$ becomes the *tube*

$$\operatorname{Re}\tilde{\mathbf{h}}\in\Gamma_+ \cup \Gamma_-, \tag{4.6}$$

where

$$\Gamma_\pm = \left\{\mathbf{x} : \pm x^{(1)} > \left[\sum_{\alpha=2}^{N} x^{(\alpha)^2}\right]^{1/2}\right\} \tag{4.7}$$

are the forward and backward light cones. Our ultimate goal is a Lee–Yang theorem for the region (1.10), that is, for the tube

$$\operatorname{Re}\tilde{\mathbf{h}}_i\in\Gamma_+ \text{ for all } i. \tag{4.8}$$

We study first the case $N = 2$. This case is particularly simple because the tube (4.8) is equivalent by linear transformation to a product of half-planes. That is, introducing the new variables

$$h^\pm = 2^{-1/2}(\tilde{h}^{(1)} \pm \tilde{h}^{(2)}) = 2^{-1/2}(h^{(1)} \pm ih^{(2)}), \tag{4.9}$$

the tube (4.8) becomes

$$\operatorname{Re} h_i^+ > 0, \operatorname{Re} h_i^- > 0 \text{ for all } i. \tag{4.10}$$

We are then precisely in the situation studied in Sect. 2; the Lee–Yang Theorems of Sect. 3 carry over immediately. We need only note that

$$\exp[h^{(1)}\varphi^{(1)} + h^{(2)}\varphi^{(2)}] = \exp[h^+\varphi^- + h^-\varphi^+], \tag{4.11}$$

so that differentiation of the partition function with respect to $h^\pm$ brings down a factor of $\varphi^\mp$. Thus, a "ferromagnetic pair interaction" is an entire function of the form

$$f(\varphi)=\exp\left[\sum_{i,j=1}^{n}(J_{ij}^{++}\varphi_i^+\varphi_j^+ + J_{ij}^{+-}\varphi_i^+\varphi_j^- + J_{ij}^{-+}\varphi_i^-\varphi_j^+ + J_{ij}^{--}\varphi_i^-\varphi_j^-)\right] \tag{4.12}$$

with all coefficients $J_{ij}^{\pm\pm} \geqq 0$. Rewriting this using (4.9), we find that

$$f(\varphi)=\exp\left[\sum_{i,j=1}^{n}\sum_{\alpha,\beta=1}^{2} J_{ij}^{(\alpha\beta)}\varphi_i^{(\alpha)}\varphi_j^{(\beta)}\right] \tag{4.13}$$

subject to the conditions [20]

$$\begin{cases} J_{ij}^{(11)} \text{ and } J_{ij}^{(22)} \text{ are real} \\ J_{ij}^{(12)} \text{ and } J_{ij}^{(21)} \text{ are pure imaginary} \\ J_{ij}^{(11)} + J_{ij}^{(22)} \geqq |J_{ij}^{(12)} - J_{ij}^{(21)}| \\ J_{ij}^{(11)} - J_{ij}^{(22)} \geqq |J_{ij}^{(12)} + J_{ij}^{(21)}| \end{cases} \tag{4.14}$$

for all $i, j$. In particular, in the usual case in which

$$J_{ij}^{(12)} = J_{ij}^{(21)} = 0 \text{ for all } i,j, \tag{4.15}$$

we recover the well-known [21] condition

$$J_{ij}^{(11)} \geqq |J_{ij}^{(22)}| \text{ for all } i,j. \tag{4.16}$$

We then have the following immediate analogues of Theorem 3.2 and Corollary 3.3:

**Theorem 4.3.** *Let $\mu_0$ be a finite (positive) measure on $\mathbb{R}^{2n}(\equiv(\mathbb{R}^2)^n)$ with $\mu_0 \not\equiv 0$, $\mu_0 \in \mathcal{T}_\beta^{2n}$ and $\hat\mu_0 \in \overline{\mathcal{P}^{2n}_{1/4\beta+}}((L_2^+)^n)$; and let $f \in \overline{\mathcal{P}^{2n}_{\alpha+}}((L_2^+)^n)[\alpha<\beta]$ be nonnegative on the support of $\mu_0$, and strictly positive on a set of nonzero $\mu_0$-measure. Then $\mu = f\mu_0$ is a finite (positive) measure with $\mu \not\equiv 0$, $\mu \in \mathcal{T}_\gamma^{2n}$ and $\hat\mu \in \overline{\mathcal{P}^{2n}_{1/4\gamma+}}((L_2^+)^n)$, for every $\gamma < \beta - \alpha$ [and $\gamma = \infty$ if $\beta = \infty$]. In particular, we can take $f$ of the form (4.13) – (4.16), provided that $\alpha = \|J\| < \beta$. (Here $\|J\|$ is the norm of $J$ considered as a bilinear form on $\mathbb{R}^{2n}$ equipped with the Euclidean norm.)*

**Corollary 4.4.** *For $1 \leqq i \leqq n$, let $\nu_i \in \mathcal{T}_\beta^2$ be a rotationally invariant measure on $\mathbb{R}^2$ satisfying condition (4.1); and let $J$ be a symmetric real $2n \times 2n$ matrix satisfying (4.5) – (4.16) [$\|J\| < \beta$]. Then the measure $\mu$ on $\mathbb{R}^{2n}$ given by*

$$d\mu(\varphi)=\exp\left[\sum_{i,j=1}^{n}\sum_{\alpha=1}^{2} J_{ij}^{(\alpha\alpha)}\varphi_i^{(\alpha)}\varphi_j^{(\alpha)}\right]\prod_{i=1}^{n} d\nu_i(\varphi_i) \tag{4.17}$$

*has $\mu \not\equiv 0$, $\mu \in \mathcal{T}_\gamma^{2n}$ and $\hat\mu \in \overline{\mathcal{P}^{2n}_{1/4\gamma+}}((L_2^+)^n)$, for every $\gamma < \beta - \|J\|$ [and $\gamma = \infty$ if $\beta = \infty$]. In particular, the partition function*

$$\hat\mu(\mathbf{h}) = \int \exp\left[\sum_{i=1}^{n}\sum_{\alpha=1}^{2} h_i^{(\alpha)}\varphi_i^{(\alpha)}\right] d\mu(\varphi) \tag{4.18}$$

*is nonvanishing if*

$$\operatorname{Re} h_i^{(1)} > |\operatorname{Im} h_i^{(2)}| \text{ for all } i. \tag{4.19}$$

*Remarks.* 1. The last sentence of Corollary 4.4 has been proven by Dunlop [20] for the special case of the plane rotator (1.9), by quite different methods. His proof also extends [21] to the two-component $|\varphi|^4$ model.

2. The zero-free region obtained in Corollary 4.4 can in some cases be extended by exploiting the covariance (or invariance) of the partition function under (complex) rotations [20, Theorems 1 and 3]. For example, if the interaction is isotropic ($J_{ij}^{(11)} = J_{ij}^{(22)}$ for all $i,j$), then the partition function is invariant under the simultaneous rotation of all spins; hence the region (4.19) can be extended to

$$\bigcup_{\substack{\mathbf{u}\in\mathbb{R}^2 \\ |\mathbf{u}|=1}} \{\mathbf{h} : \mathbf{u}\cdot \mathrm{Re}\,\mathbf{h}_i > |\mathbf{u} \times \mathrm{Im}\,\mathbf{h}_i| \text{ for all } i\} \tag{4.20}$$

## 5. N-Component Models (N ≧ 3)

The main result of this section is Corollary 5.5, a Lee–Yang Theorem for $N$-component ferromagnets in which the interaction is sufficiently *anisotropic* [see (1.11)]. As explained in the Introduction, this result—unlike those in the previous sections—is not "best possible". But it is the only Lee–Yang Theorem we know of, for $N > 3$! Moreover, we believe that our methods can probably be extended to derive a "best possible" Lee–Yang theorem for general $N$.

The case $N \geqq 3$ is considerably more difficult than the case $N = 2$, because the tube (4.8) is no longer equivalent to a product of half-planes. The trouble is that, in three or more dimensions, the light cone is round! As a result, our fundamental theorems—Propositions 2.2 and 2.7—are no longer adequate. Rather, we require *generalizations* of these propositions to tubes more general than products of half-planes. We state these conjectured generalizations in the form of two questions, to which we can provide at present only some partial answers.

*Question 5.1.* Let $\Gamma \subset \mathbb{R}^n$ be a closed convex cone, and let

$$\Gamma^* = \{x \in \mathbb{R}^n : x \cdot y \geqq 0 \text{ for all } y \in \Gamma\} \tag{5.1}$$

be its dual cone. Let $R(v, w)$ and $S(z)$ be defined as in Proposition 2.2. Now assume that $R(v, w) \neq 0$ whenever $\mathrm{Re}\, v \in \Gamma^*$ and $\mathrm{Re}\, w \in \Gamma + c$. Does it follow that $S(z) \neq 0$ whenever $\mathrm{Re}\, z \in \Gamma + c$? If not, for which cones $\Gamma$ and which polynomials $R$ is it true?

*Question 5.2.* (a) Let $\Gamma_1 \subset \mathbb{R}^{n_1}$ and $\Gamma_2 \subset \mathbb{R}^{n_2}$ be open convex cones, and let $B$ be a (real or complex) $n_1 \times n_2$ matrix. For which $B$ is the function

$$f(w, z) = \exp\left( \sum_{i=1}^{n_1} \sum_{j=1}^{n_2} B_{ij} w_i z_j \right) \tag{5.2}$$

approximable by polynomials nonvanishing in the set

$$T_{\Gamma_1} \times T_{\Gamma_2} \{(w, z) : \mathrm{Re}\, w \in \Gamma_1, \mathrm{Re}\, z \in \Gamma_2\}\,? \tag{5.3}$$

(b) [Restricted form] In the above, let $n_1 = n_2 = n$ and $\Gamma_1 = \Gamma_2 = \Gamma$, and let $B$

be symmetric. For which $B$ is the function

$$f(z) = \exp\left(\sum_{i,j=1}^{n} B_{ij} z_i z_j\right) \tag{5.4}$$

approximable by polynomials nonvanishing in the set

$$T_\Gamma = \{z : \mathrm{Re}\, z \in \Gamma\}\ ? \tag{5.5}$$

With regard to Question to Question 5.1, we have the following result, applicable to *special* polynomials $P_i$:

**Proposition 5.3.** *Assume that there exist vectors $\lambda^{(1)}, \ldots, \lambda^{(M)} \in \Gamma$ and polynomials $\tilde{P}_i (1 \leqq i \leqq k)$ in $M$ complex variables, such that*

$$P_i(v) = \tilde{P}_i(\lambda^{(1)} \cdot v, \ldots, \lambda^{(M)} \cdot v) \text{ for all } i, \tag{5.6}$$

*and such that*

$$\tilde{R}(\tilde{v}, w) \equiv \sum_{i=1}^{k} \tilde{P}_i(\tilde{v}) Q_i(w) \tag{5.7}$$

*is nonvanishing whenever* $\mathrm{Re}\, \tilde{v} \geqq 0$ *and* $\mathrm{Re}\, w \in \Gamma + c$. *Then*

$$S(z) \equiv \sum_{i=1}^{k} P_i(\partial/\partial z) Q_i(z) \tag{5.8}$$

*is nonvanishing whenever* $\mathrm{Re}\, z \in \Gamma + c$.

To understand this proposition, note that the vectors $\lambda^{(1)}, \ldots, \lambda^{(M)} \in \Gamma$ define "supporting hyperplanes" for the cone $\Gamma^*$. Thus, if $\mathrm{Re}\, v \in \Gamma^*$, then $\mathrm{Re}\, \lambda^{(\ell)} \cdot v \geqq 0$ for $1 \leqq \ell \leqq M$. Hence the hypothesis in $\tilde{R}$ implies the usual hypothesis on $R$. In fact (and this is the weakness of the proposition), it is stronger: it implies that $R(v, w)$ is nonvanishing whenever $\mathrm{Re}\, w \in \Gamma + c$ and $\mathrm{Re}\, v \in \{\lambda^{(1)}, \ldots, \lambda^{(M)}\}^*$; and this latter cone is in general *larger* than $\Gamma^*$. (If $\Gamma^*$ is "round", it is necessarily larger.) Hence not all $R$ can be accommodated.

*Examples.* 1. Take $k = 1$; then the hypothesis on $\tilde{R}$ reduces simply to the requirement that $\tilde{P}(\tilde{v}) \neq 0$ whenever $\mathrm{Re}\, \tilde{v} \geqq 0$. But by (5.6) this implies that $P(v) \neq 0$ whenever $\mathrm{Re}\, v \in \{\lambda^{(1)}, \ldots, \lambda^{(M)}\}^*$. As an example of a polynomial $P(v)$ which is nonvanishing for $\mathrm{Re}\, v$ in $\Gamma^*$ but not in this larger set, let $\Gamma = \Gamma^* = \bar{\Gamma}_+$, the closed forward light cone in $\mathbb{R}^n (n \geqq 3)$; and let

$$P(v) = 1 + (v_1 + \alpha)^2 - \sum_{j=2}^{n} v_j^2 \tag{5.9}$$

with $\alpha > 0$. By Lemma 4.2, $P(v)$ is nonvanishing if $\mathrm{Re}\, v \in \Gamma^*$. But for $v$ real, the zero manifold of $P$ is a spacelike hyperboloid which is asymptotic to the cone $\Gamma^*$. Hence, if $\tilde{\Gamma}$ is a convex cone strictly larger than $\Gamma^*$, $P$ necessarily has a zero in $\tilde{\Gamma}$

2. Again let $\Gamma = \Gamma^* = \bar{\Gamma}_+$, and take $P$ of the *special* form

$$P(v) = \prod_{j=2}^{n} \tilde{P}_j(v_1, v_j). \tag{5.10}$$

Then the hypothesis of the proposition *can* be satisfied: take $M = 2n - 2$; $\lambda^{(1)} = e^{(1)} + e^{(2)}$, $\lambda^{(2)} = e^{(1)} - e^{(2)}$, etc. [the $e^{(j)}$ $(1 \leqq j \leqq n)$ are the unit vectors]; and take

$$\tilde{P}(\tilde{v}) = \prod_{j=2}^{n} \bar{P}_j\left(\frac{\tilde{v}_{2j-3} + \tilde{v}_{2j-2}}{2}, \frac{\tilde{v}_{2j-3} - \tilde{v}_{2j-2}}{2}\right). \tag{5.11}$$

But this is possible because the special form (5.10) implies that $P(v)$ is nonvanishing not only for Re $v \in \Gamma^*$ but also for Re $v$ in the larger cone

$$\tilde{\Gamma} = \{\lambda^{(1)}, \ldots, \lambda^{(M)}\}^* = \{x : x_1 \geqq \max_{2 \leqq j \leqq n} |x_j|\}. \tag{5.12}$$

*Proof of Proposition 5.3.* Introduce also the new variables $\tilde{w}^{(1)}, \ldots, \tilde{w}^{(M)} \in \mathbb{C}$, and define

$$\hat{R}(\tilde{v}, \tilde{w}, w) = \sum_{i=1}^{k} \tilde{P}_i(\tilde{v}) Q_i\left(w + \sum_{\ell=1}^{M} \tilde{w}^{(\ell)} \lambda^{(\ell)}\right). \tag{5.13}$$

Now $\hat{R}$ is a polynomial which is nonvanishing whenever Re $\tilde{v} \geqq 0$, Re $\tilde{w} \geqq 0$ and Re $w \in \Gamma + c$. So fix $w$ with Re $w \in \Gamma + c$, and apply Proposition 2.2. It follows that

$$\hat{S}(\tilde{w}, w) = \sum_{i=1}^{k} \tilde{P}_i(\partial/\partial\tilde{w}) Q_i\left(w + \sum_{\ell=1}^{M} \tilde{w}^{(\ell)} \lambda^{(\ell)}\right) \tag{5.14}$$

is nonvanishing whenever Re $\tilde{w} \geqq 0$ and Re $w \in \Gamma + c$. But

$$\begin{aligned} &\tilde{P}_i(\partial/\partial\tilde{w}) Q_i\left(w + \sum_{\ell=1}^{M} \tilde{w}^{(\ell)} \lambda^{(\ell)}\right) \\ &= \tilde{P}_i\left(\lambda^{(1)} \cdot \frac{\partial}{\partial w}, \ldots, \lambda^{(M)} \cdot \frac{\partial}{\partial w}\right) Q_i\left(w + \sum_{\ell=1}^{M} \tilde{w}^{(\ell)} \lambda^{(\ell)}\right) \\ &= P_i(\partial/\partial w) Q_i\left(w + \sum_{\ell=1}^{M} \tilde{w}^{(\ell)} \lambda^{(\ell)}\right) \end{aligned} \tag{5.15}$$

by (5.6). Setting $\tilde{w} = 0$ completes the proof.

*Remark.* There is also an analogue here of Proposition 2.2 (b); its proof is essentially identical.

With regard to Question 5.2, we are able to make only a few remarks:

1. We state Question 5.2. in two forms—the general form and the "on-diagonal" form—because here, unlike in Proposition 2.7, the two forms are not *obviously* equivalent. (They may be nonobviously equivalent.)

2. A *necessary* condition in case (a) is that

$$\sum_{i=1}^{n_1} \sum_{j=1}^{n_2} B_{ij} u_i x_j \geqq 0 \text{ for all } u \in \Gamma_1, x \in \Gamma_2. \tag{5.16}$$

This follows by taking $w = \alpha u$, $z = \alpha x$ with $\alpha \in \mathbb{C}$ and applying the $n = 1$ case of Proposition 2.7. *Perhaps* (5.16) is also a sufficient condition. This is certainly the case if $\Gamma_1$ and $\Gamma_2$ are the "positive hyperoctants"

$$\Gamma_1 = \{x \in \mathbb{R}^{n_1} : x_i > 0 \text{ for all } i\}, \text{etc.}, \tag{5.17}$$

since in this case (5.16) reduces to the requirement $B_{ij} \geqq 0$ for all $i, j$, and the result is true by a minor modification of the proof of Proposition 2.7. However, we have been unable to prove this conjecture for more general $\Gamma_1, \Gamma_2$.

3. In case (a) we do have the following *sufficient* condition:

**Proposition 5.4.** *Let $\mu^{(1)}, \ldots, \mu^{(M)} \in \Gamma_1^*$ and $\lambda^{(1)}, \ldots, \lambda^{(M)} \in \Gamma_2^*$. Now define*

$$P_k(w, z) = \prod_{\ell=1}^{M} ((1 + k^{-1}(\mu^{(\ell)} \cdot w)(\lambda^{(\ell)} \cdot z))^k.$$

*Then $P_k$ is nonvanishing in the set (5.3), and*

$$\lim_{k \to \infty} P_k(w, z) = f(w, z), \tag{5.19}$$

*where $f$ is defined by (5.2) and*

$$B_{ij} = \sum_{\ell=1}^{M} \mu_i^{(\ell)} \lambda_j^{(\ell)}. \tag{5.20}$$

*Moreover, the limit (5.19) holds in the topology of $\mathscr{A}_{a+}^{n_1+n_2}$ for sufficiently large a.*

*Proof.* Since $\Gamma_1$ is open, we have

$$\Gamma_1^* = \{0\} \bigcup \{x : x \cdot y > 0 \text{ for all } y \in \Gamma_1\}.$$

Hence, for $\operatorname{Re} w \in \Gamma_1$, we have either $\mu^{(\ell)} \cdot w = 0$ or $\operatorname{Re} \mu^{(\ell)} \cdot w > 0$; and likewise for $\lambda^{(\ell)} \cdot z$. It follows that $P_k$ is nonvanishing in the set (5.3). The limit (5.19) follows from

$$\lim_{k \to \infty} (1 + k^{-1}x)^k = \exp(x).$$

The converenge in $\mathscr{A}_{a+}^{n_1+n_2}$ is proved as in Proposition 2.7.

Note that every matrix $B$ of the form (5.20) satisfies (5.16) [as it must!]; but the converse is *not* in general true. For example, let $\Gamma_1 = \Gamma_2 = \Gamma_+$, the forward light cone, with $n_1 = n_2 = n \geqq 3$. Then for a diagonal matrix $B$, (5.16) is equivalent to

$$B_{11} \geqq \max_{2 \leqq j \leqq n} |B_{jj}|, \tag{5.21}$$

while (5.20) is equivalent to

$$B_{11} \geqq \sum_{j=2}^{n} |B_{jj}|. \tag{5.22}$$

[To see that (5.22) is a necessary condition, note that (5.20) implies $\operatorname{tr}(BG) \geqq 0$ for every diagonal matrix $G$ with $G_{11} = 1$ and $G_{jj} = \pm 1 (2 \leqq j \leqq n)$. It is easy to see that (5.22) is a sufficient condition: it suffices to take $M = n - 1$, for each $j (2 \leqq j \leqq n)$ to take

$$\begin{aligned} \mu^{(j-1)} &= |B_{jj}|^{1/2} (\alpha_j e^{(1)} + e^{(j)}) \\ \lambda^{(j-1)} &= |B_{jj}|^{1/2} (\alpha_j e^{(1)} + \operatorname{sign}(B_{jj}) e^{(j)}) \end{aligned} \tag{5.23}$$

with suitable $\alpha_j \geqq 1$.]

These remarks imply a Lee–Yang theorem for $N$-component ferromagnets with *highly anisotropic* interactions:

**Corollary 5.5.** *For $1 \leqq i \leqq n$, let $v_i \in \mathcal{T}_\beta^N$ be a rotationally invariant measure on $\mathbb{R}^N$ satisfying condition (4.1); and let $J_{ij}^{(\alpha)}[1 \leqq i, j \leqq n, 1 \leqq \alpha \leqq N]$ be real numbers satisfying*

$$J_{ij}^{(1)} \geqq \sum_{\alpha=1}^{N} |J_{ij}^{(\alpha)}| \text{ for all } i, j. \tag{5.24}$$

*Then the measure $\mu$ on $\mathbb{R}^{Nn}$ given by*

$$d\mu(\varphi) = \exp\left( \sum_{i,j=1}^{n} \sum_{\alpha=1}^{N} J_{ij}^{(\alpha)} \varphi_i^{(\alpha)} \varphi_j^{(\alpha)} \right) \prod_{i=1}^{n} dv_i(\varphi_i) \tag{5.25}$$

*has $\mu \not\equiv 0$, $\mu \in \mathcal{T}_\gamma^{Nn}$ and $\hat{\mu} \in \overline{\mathcal{P}_{1/4\gamma+}^{Nn}}((L_N^+)^n)$, provided that $\beta$ is sufficiently large (how large depends on $J$ and $\gamma$). In particular, the partition function*

$$\hat{\mu}(\mathbf{h}) = \int \exp\left( \sum_{i=1}^{n} \sum_{\alpha=1}^{N} h_i^{(\alpha)} \varphi_i^{(\alpha)} \right) d\mu(\varphi) \tag{5.26}$$

*is nonvanishing if*

$$\operatorname{Re} h_i^{(1)} > \left( \sum_{\alpha=1}^{N} (\operatorname{Im} h_i^{(\alpha)})^2 \right)^{1/2} \text{ for all } i. \tag{5.27}$$

*Proof.* This follows from Proposition 5.4, the above remarks, and Proposition 5.3 in the accustomed manner [note that (5.18) satisfies the hypotheses of Proposition 5.3. for essentially the same reason as (5.10)]. QED

*Remarks (continued).* 4. In case (b) of Question 5.2, a *necessary* condition [analogous to (5.16)] is that

$$\sum_{i,j=1}^{n} B_{ij} x_i x_j \geqq 0 \text{ for all } x \in \Gamma. \tag{5.28}$$

However, this is clearly *not* a sufficient condition: (5.28) would allow any positive definite matrix $B$, for any $\Gamma$, which clearly contradicts Proposition 2.7. We suspect that (5.16) may be a necessary condition in case (b) as well as in case (a).

5. For the special case $\Gamma = \Gamma_+$, the forward light cone, (5.21) is a *sufficient* condition for case (b): for by a minor modification of Lemma 4.2, (5.21) implies that $\sum_{i,j=1}^{n} B_{ij} z_i z_j = \sum_{i=1}^{n} B_{ii} z_i^2$ is never real and negative, for $\operatorname{Re} z \in \Gamma$; so we can write

$$f(z) = \exp\left( \sum_{i,j=1}^{n} B_{ij} z_i z_j \right) = \lim_{k \to \infty} \left( 1 + k^{-1} \sum_{i,j=1}^{n} B_{ij} z_i z_j \right)^k, \tag{5.29}$$

a limit of polynomials nonvanishing in $\operatorname{Re} z \in \Gamma$.

6. The preceding remark would allow us to handle rotators with *isotropic* interactions, for arbitrary $N$, provided that we could find a satisfactory answer to Question 5.1 in this case. This is so because we can write

$$\exp[J_{ij} \varphi_i \cdot \varphi_j] = \exp[\tfrac{1}{2} J_{ij} (\varphi_i + \varphi_j)^2] \exp[-\tfrac{1}{2} J_{ij} (\varphi_i^2 + \varphi_j^2)], \tag{5.30}$$

and the last factor is a harmless constant for the plane rotator measure (1.9). The same remark applies to single-spin measures which, though not of the form (1.9), satisfy the following strong version of (4.1):

$$\int e^{h\varphi^{(1)}-c|\varphi|^2}\,d\nu(\varphi) \neq 0 \text{ for } \operatorname{Re} h \neq 0, \text{ for } all\ c \geqq 0 \tag{5.31}$$

—for in this case we can absorb the last factor in (5.30) into the single-spin measure. Newman [39] (see also the remarks at the end of [40]) has found all rotationally invariant measures satisfying (5.31); aside from (1.9), they are

$$d\nu(\varphi) = C|\varphi|^{2m}\exp(-a|\varphi|^4 - b|\varphi|^2)\prod_j (1+|\varphi|^2/\alpha_j^2)\exp(-|\varphi|^2/\alpha_j^2) \tag{5.32}$$

with $C > 0$, $m \geqq 0$ integral, $a \geqq 0$, $b$ real, and $\alpha_j > 0$ with $\sum_j \alpha_j^{-4} < \infty$; here the sequence $\{\alpha_j\}$ may be empty, finite or infinite. In particular, by taking $m = 0$ and $\{\alpha_j\}$ empty, we obtain the $N$-component $|\varphi|^4$ lattice field theory. Finally, note that this idea also handles *some* anisotropic interactions, in particular those which can be written as a product of terms each of which looks like (5.30) except that $(\varphi_i + \varphi_j)^2$ is replaced by $\hat{\varphi}_{ij}^2$, with

$$\hat{\varphi}_{ij} = (\varphi_i^{(1)} + \varphi_j^{(1)}, \varphi_i^{(2)} \pm \varphi_j^{(2)}, \ldots, \varphi_i^{(N)} \pm \varphi_j^{(N)}) \tag{5.33}$$

for some sequence of $\pm$ signs. This allows some (but not all) interactions of the form (1.12).

The moral of this rather long story is that the $N$-component Lee–Yang theorem ($N \geqq 3$) rides on finding a satisfactory answer to Question 5.1, for the case where $\Gamma$ is a product of forward light cones. But this we must leave as an exercise for the ambitious reader.

## Appendix A: Alternate Proof of Newman's Lee–Yang Theorem

In this Appendix we shall give an alternate proof of Corollary 3.3—which is a slight generalization of Newman's [15] Lee–Yang Theorem—based on the elementary Proposition 2.1 instead of the more difficult Proposition 2.2. (Actually, we shall prove only the $\beta = \infty$ case of Corollary 3.3; see Remark 1 following the proof.)

Note first that since $J_{ii} \geqq 0$, we can absorb the factor $\exp(J_{ii}\varphi_i^2)$ into $d\nu_i(\varphi_i)$ and preserve the Lee–Yang property of the latter [15, Proposition 2.4]; this follows from the $n = 1$ case of Proposition 2.7. and 2.9. Hence we can assume that $J_{ii} = 0$.

The proof is now by induction on $n$. By hypothesis the theorem is true for $n = 1$. So assume that it is true for $n = N - 1$, that is, assume that the function $\hat{\mu}_{N-1}$ defined by

$$\hat{\mu}_{N-1}(h_1, \ldots, h_{N-1}) = \int \exp\left(\sum_{i,j=1}^{N-1} J_{ij}\varphi_i\varphi_j + \sum_{i=1}^{N-1} h_i\varphi_i\right)\prod_{i=1}^{N-1} d\nu_i(\varphi_i) \tag{A.1}$$

lies in $\mathscr{P}_{0+}^{N-1}(D^{N-1})$. Now by definition of $\hat{\mu}_N$ (and Fubini's theorem), we have

$$\hat{\mu}_N(h_1, \ldots, h_N) = \int \hat{\mu}_{N-1}(h_1 + \tilde{J}_1\varphi_N, \ldots, h_{N-1} + \tilde{J}_{N-1}\varphi_N)e^{h_N\varphi_N}d\nu_N(\varphi_N) \tag{A.2}$$

with

$$\tilde{J}_i = J_{iN} + J_{Ni} = 2J_{iN}(1 \leqq i \leqq N-1). \tag{A.3}$$

Now the point is that since $\tilde{J}_i \geqq 0$, the function

$$g(\varphi_N) = \hat{\mu}_{N-1}(h_1 + \tilde{J}_1\varphi_N, \ldots, h_{N-1} + \tilde{J}_{N-1}\varphi_N) \tag{A.4}$$

lies in $\overline{\mathscr{P}^1_{0+}}(D^1)$, for each fixed $(h_1, \ldots, h_{N-1}) \in D^{N-1}$; and so, by the $n = 1$ case of Proposition 2.9, $\hat{\mu}_N$ lies in $\overline{\mathscr{P}^1_{0+}}(D^1)$ as a function of $h_N$ for each fixed $(h_1, \ldots, h_{N-1}) \in D^{N-1}$. Of course, this is not quite what we need to prove (though it is the essential idea of the proof). To complete the rigorous proof, let $\{f_j\}$ be a sequence in $\mathscr{P}^{N-1}(D^{N-1})$ converging to $\hat{\mu}_{N-1}$ in $\mathscr{A}^{N-1}_{0+}$; that is, for each $\varepsilon > 0$ we have

$$\begin{aligned} |\hat{\mu}_{N-1}(\ldots) - f_j(\ldots)| &\leqq c_j^{(\varepsilon)} \exp\left[\varepsilon \sum_{i=1}^{N-1} |h_i + \tilde{J}_i\varphi_N|^2\right] \\ &\leqq c_j^{(\varepsilon)} \exp\left[2\varepsilon \sum_{i=1}^{N-1} (|h_i|^2 + \tilde{J}_i^2\varphi_N^2)\right] \end{aligned} \tag{A.5}$$

for some sequence of constants $\{c_j^{(\varepsilon)}\}$ converging to zero. Inserting this into (A.2), and using the hypothesis

$$\int \exp(b\varphi_N^2)\,d\nu_N(\varphi) < \infty \text{ for all } b, \tag{A.6}$$

we find easily that

$$\left|\hat{\mu}_N(\ldots) - \int f_j(\ldots)e^{h_N\varphi_N}d\nu_N(\varphi_N)\right| \leqq Kc_j^{(\varepsilon)} \exp\left[2\varepsilon \sum_{i=1}^{N} |h_i|^2\right] \tag{A.7}$$

for some constant $K < \infty$. Hence the integral converges to $\hat{\mu}_N$ in $\mathscr{A}^N_{0+}$. But the integral equals

$$\begin{aligned} &f_j\left(h_1 + \tilde{J}\frac{\partial}{\partial h_N}, \ldots, h_{N-1} + \tilde{J}_{N-1}\frac{\partial}{\partial h_N}\right)\int e^{h_N\varphi_N}d\nu_N(\varphi_N) \\ &\equiv g_j\left(\frac{\partial}{\partial h_N}\right)\int e^{h_N\varphi_N}d\nu_N(\varphi_N) \end{aligned} \tag{A.8}$$

(since $f_j$ is a polynomial, this equality is trivial), and $g_j \in \mathscr{P}^1(D^1)$ whenever $(h_1, \ldots, h_{N-1}) \in D^{N-1}$. Now let $\{p_k\}$ be a sequence in $\mathscr{P}^1(D^1)$ converging to $\hat{\nu}_N$ in $\mathscr{A}^1_{0+}$. Then, by Proposition 2.1, $g_j(\partial/\partial h_N)p_k(h_N)$ is a polynomial in $(h_1, \ldots, h_N)$ which is nonvanishing in $D^N$ (or else is identically zero[1]). But by Proposition 2.5 and an easy estimate, $g_j(\partial/\partial h_N)p_k(h_N)$ converges in $\mathscr{A}^N_{0+}$ to (A.8) as $k \to \infty$. Hence the function (A.8) is in $\overline{\mathscr{P}^N_{0+}}(D^N)$; so by (A.7), $\hat{\mu}_N \in \overline{P^N_{0+}}(D^N)$ as well. This completes the proof.

1 We use Remark 2 following Proposition 2.2 (or an equivalent argument based on the last sentence of Proposition 2.1) to ensure that $g_j(\partial/\partial h_N)P_k(h_N)$ vanishes identically in $h_N$ for one value of $(h_1, \ldots, h_{N-1})$ only it does so for *all* $(h_1, \ldots, h_{N-1})$

*Remarks.* 1. By keeping careful track of the rate of Gaussian falloff in the above proof, one can also handle $\beta$ finite but sufficiently large (depending on the matrix $J$). But the inductive structure of the proof, which treats the $n$ spins asymmetrically, is unlikely to allow the optimal result $\|J\| < \beta$ proven in Corollary 3.3.

2. The inductive idea—considering the spin $\varphi_N$ as a "magnetic field" acting on the spins $\varphi_1, \ldots, \varphi_{N-1}$—is also the basis of the proofs of the (spin-$\frac{1}{2}$) Lee–Yang Theorem due to Newman [15, Theorem 3.1] and Sherman [14].

**Appendix B: Comparison with the Asano Contraction Method**

The present approach to the Lee–Yang theorem is based on the idea that certain functions $F$—namely, $F \in \overline{\mathscr{P}^n_{\alpha+}}(D^n)$—are "universal multipliers" for Lee–Yang measures: that is, whenever $d\mu_0(\varphi)$ has the Lee–Yang property, so does $F(\varphi)d\mu_0(\varphi)$.

The Asano contraction method [10–14, 43], by contrast, is based on the idea that certain *measures* $\mu_0$ have the following property: if $F_1(\varphi)d\mu_0(\varphi)$ and $F_2(\varphi)d\mu_0(\varphi)$ have the Lee–Yang property, then so does $F_1(\varphi)F_2(\varphi)d\mu_0(\varphi)$. This idea is extremely powerful, since it allows one to prove the Lee–Yang theorem for a large model simply by verifying it for each elementary interaction, and this is often a trivial computation. Unfortunately, however, the only base-measure $\mu_0$ for which this idea is known to work is the uncoupled spin-$\frac{1}{2}$ Ising measure

$$d\mu_0(\varphi) = \prod_{i=1}^{n} \tfrac{1}{2}[\delta(\varphi_i - a_i) + \delta(\varphi_i + a_i)]d\varphi_i. \tag{B.1}$$

That the Asano property is *not* a general property of Lee–Yang measures can be seen from two simple examples in $n = 1$:

1. Let $\mu_0$ be the usual spin-1 measure

$$\mu_0 = \tfrac{1}{3}[\delta_{-1} + \delta_0 + \delta_1]. \tag{B.2}$$

Then $\exp(-b\varphi^2)d\mu_0(\varphi)$ has the Lee–Yang property if and only if $b \leqq \log 2$. So take $F_1(\varphi) = F_2(\varphi) = \exp(-b\varphi^2)$ with $\frac{1}{2}\log 2 < b \leqq \log 2$; then the Asano property fails.

2. Let $\mu_0$ be a spin-$\frac{1}{2}$ Ising measure in a positive magnetic field:

$$d\mu_0(\varphi) = e^{h_0\varphi}[\delta(\varphi - 1) + \delta(\varphi + 1)]\, d\varphi \tag{B.3}$$

with $h_0 > 0$. Now let $F_1(\varphi) = F_2(\varphi) = \exp(-h_1\varphi)$ with $h_0/2 < h_1 \leqq h_0$; the Asano property again fails.

On the other hand, we can "explain" in terms of our own approach why the Asano property does hold for the measure (B.1): the point is that if $F_1(\varphi)d\mu_0(\varphi)$ has the Lee–Yang property, then there exists a function $P_1(\varphi)$, equal to $F_1(\varphi)$ on the support of $\mu_0$, which is a *universal multiplier* [in fact, $P_1 \in \mathscr{P}^n(D^n)$]; hence

$$F_1(\varphi)F_2(\varphi)d\mu_0(\varphi) = P_1(\varphi)F_2(\varphi)d\mu_0(\varphi)$$

has the Lee–Yang property. To be explicit:

**Proposition B.1.** *Let $F(\sigma_1, \ldots, \sigma_n)$ be defined for $\{\sigma_1, \ldots, \sigma_n\} = \pm 1$, and let $P(\varphi_1, \ldots, \varphi_n)$ be the unique polynomial of degree at most 1 in each variable which* *coincides with F, i.e.*

$$P(\varphi_1,\ldots,\varphi_n)=\sum_{\sigma_1=\pm 1}\cdots\sum_{\sigma_n=\pm 1}F(\sigma_1,\ldots,\sigma_n)\prod_{i=1}^{n}\frac{1+\sigma_1\varphi_i}{2}. \tag{B.4}$$

*Then the following are equivalent:*

$$\text{(a)}\ Q(z_1,\ldots,z_n)\equiv\sum_{\sigma_1=\pm 1}\cdots\sum_{\sigma_n=\pm 1}F(\sigma_1,\ldots,\sigma_n)\prod_{i=1}^{n}z_i^{(1/2)(1-\sigma_i)} \tag{B.5}$$

*is nonvanishing if all* $|z_i|<1$ [*this is the Lee–Yang property in the activity variables* $z_i=\exp(-2h_i)$];

(b) $P(\varphi_1,\ldots,\varphi_n)$ *is nonvanishing if all* Re $\varphi_i>0$.

*Proof.* The transformation

$$z_i=(1-\varphi_i)/(1+\varphi_i) \tag{B.6}$$

maps $|z_i|<1$ onto Re $\varphi_i>0$; and moreover

$$P(\varphi_1,\ldots,\varphi_n)=Q(z_1,\ldots,z_n)\prod_{i=1}^{n}\frac{1+\varphi_i}{2}. \tag{B.7}$$

Since $(1+\varphi_i)/2\neq 0$ for Re $\varphi_i>0$, this completes the proof.

*Remarks.* 1. Although example 1 above shows that the Asano property does not hold for the spin-1 measure, a *modified* Asano property does hold [27] for spin 1 and in fact for all the classical discrete spins: there is a weight function $G(\varphi)$ [certain binomial coefficients] such that if $F_1(\varphi)d\mu_0(\varphi)$ and $F_2(\varphi)d\mu_0(\varphi)$ have the Lee–Yang property, then so does $G(\varphi)F_1(\varphi)F_2(\varphi)d\mu_0(\varphi)$. It would be interesting to have a deeper understanding of this phenomenon. Is there any direct generalization, for example, to classical Heisenberg spins?

2. The Asano contraction method has the advantage over the method of the present paper in that it is suited to studying zero-free regions other than half-planes (or circles in the activity variables) [12, 13].

*Acknowledgement.* One of us (A.S.) wishes to thank Barry Simon for a helpful conversation.

## References

1. Ruelle, D.: Statistical mechanics—rigourous results. Chap. 5. New York: Benjamin 1969
2. Fröhlich, J.: Poetic phenomena in (two dimensional) quantum field theory: non-uniqueness of the vacuum, the solitons and all that. In: Les méthodes mathématiques de la théorie quantique des champs (1975 Marseille conference). Paris: CNRS 1976
3. Penrose, O., Lebowitz, J. L.: Commun. Math. Phys. **39** 165–184 (1974)
4. Guerra, F., Rosen, L., Simon, B.: Commun. Math. Phys. **41**, 19–32 (1975)
5. Newman, C. M.: Commun. Math. Phys. **41**, 1–9 (1975)
6. Dunlop, F.: J. Stat. Phys. **21**, 561–572 (1979)
7. Baker, G. A. Jr.: Phys. Rev. Lett. **20**, 990–992 (1968)
8. Sokal, A. D.: J. Stat. Phys. **25**, 25–50 (1981)

9. Lee, T. D., Yang, C. N. : Phys. Rev. **87**, 410–419 (1952)
10. Asano, T. : J. Phys. Soc. Jpn **29**, 350–359 (1970)
11. Suzuki, M., Fisher, M. E. : J. Math. Phys (N. Y.) **12**, 235–246 (1971)
12. Ruelle, D. : Phys. Rev. Lett. **26**, 303–304, 870 (E) (1971)
13. Ruelle, D. : Commun. Math. Phys. **31**, 265–277 (1973)
14. Simon, B. : The P $(\varphi)_2$ Euclidean (quantum) field theory. Sect. IX.3. Princeton: Princeton University Press 1974
15. Newman, C. M. : Commun. Pure Appl. Math. **27**, 143–159 (1974)
16. Dunlop, F. : J. Stat. Phys. **17**, 215–228 (1977)
17. Sylvester, G. S. : A note on the Lee–Yang Theorem. Oklahoma State University preprint (1980)
18. Griffiths, R. B. : J. Math. Phys. **10**, 1559–1565 (1969)
19. Simon, B., Griffiths, R. B. : Commun. Math. Phys. **33**, 145–164 (1973)
20. Dunlop, F. : Commun. Math. Phys. **69**, 81–88 (1979)
21. Dunlop, F., Newman, C. M. : Commun. Math. Phys. **44**, 223–235 (1975)
22. Takagi, T. : Proc. Physico-Math. Soc. Japan **3**, 175–179 (1921). The collected papers of T. Takagi Kuroda, S. (ed.) Tokyo: Iwanami Shoten Publishers 1973
23. Marden, M. : Geometry of polynomials (2nd edit). Providence: American Mathematical Society 1966
24. Dieudonné, J. : C. R. Acad. Sci. Paris **199**, 999–1001 (1934)
25. Grace, J. H. : Proc. Cambridge Philos. Soc. **11**, 352–357 (1902)
26. Obreschkoff, N. : Verteilung und Berechung der Nullstellen reeller Polynome. Berlin: VEB Deutscher Verlag der Wissenschaften 1963
27. Millard, K. Y., Viswanathan, K. S. : J. Math. Phys. **15**, 1821–1825 (1974)
28. Taylor, B. A. : Some locally convex spaces of entire functions. In: Entire functions and related parts of analysis. Korevaar, J. (ed.). Proceedings of symposia in pure mathematics, Vol. XI. Providence: American Mathematical Society 1968
29. Sikkema, P. C. : Differential operators and differential equations of infinite order with constant coefficients. Groningen–Djakarta: Noordhoff 1953
30. Hille, E. : Analytic function theory, Vol. II. Boston: Ginn 1962
31. Pólya, G. : Sur les opérations fonctionelles linéaires, échangeables avec la dérivation et sur les zéros des polynômes. C. R. Acad. Sci. (Paris) **183**, 413–414 (1926). Collected papers, Vol. II Boas, R. P. (ed.). pp. 261–262, 427. Cambridge: MIT Press 1974
32. Benz, E. : Comments. Math. Helv. **7**, 243–289 (1935)
33. Levin, B. Ja : Distribution of zeros of entire functions. Providence: American Mathematical Society 1964 (Translated from Russian)
34. Obrechkoff, N. : Quelques classes de fonctions entières limites de polynômes et de fonctions méromorphes limites de fractions rationelles. Jn: Actualités Sci. et Ind. Vol. 891 pp. 24–25. Paris: Hermann 1941
35. Korevaar, J. : Limits of polynomials whose zeros lie in a given set. In: Entire functions and related parts of analysis (Korevaar, J. (ed.). Proceedings of symposia in pure mathematics, Vol. XI. Providence: American Mathematical Society 1968
36. Schwartz, L : Théorie des distributions (nouv. éd.). Paris: Hermann 1966
37. Pirogov, S. A., Sinai, Ya. G. : Teor. Mat. Fiz. **25**, 358–369 (1975); **26**, 61–76 (1976) Theor. Math. Phys. (USSR) **25**, 1185–1192 (1975); **26**, 39–49 (1976)
38. Gawedzki, K. : Commun. Math. Phys. **59**, 117–142 (1978)
39. Newman, C. M. : Proc. Am. Math. Soc. **61**, 245–251 (1976)
40. Newman, C. M. : J. Stat. Phys. **15**, 399–406 (1976)
41. Heilmann, O. J., Lieb, E. H. : Commun. Math. Phys. **25**, 190–232 (1972); Erratum **27**, 166 (1972)
42. Dunlop, F. : Zeros of the partition function for some generalized ising models. Esztergom Colloquium. IHES Preprint 1979
43. Runnels, L. K., Lebowitz, J. L. : J. Stat. Phys. **23**, 1–10 (1980)

Communicated by A. Jaffe

Received October 27, 1980

Part IV

# Reflection Positivity

VOLUME 38, NUMBER 8 PHYSICAL REVIEW LETTERS 21 FEBRUARY 1977

# Existence of Phase Transitions for Anisotropic Heisenberg Models

Jürg Fröhlich*
*Department of Mathematics, Princeton University, Princeton, New Jersey 08540*

and

Elliott H. Lieb†
*Department of Mathematics and Department of Physics, Princeton University, Princeton, New Jersey 08540*
(Received 21 December 1976)

The two-dimensional anisotropic, nearest-neighbor Heisenberg model on a square lattice, both quantum and classical, has been shown rigorously to have a phase transition in the sense that the spontaneous magnetization is positive at low temperatures. This is so for all anisotropies. An analogous result (staggered polarization) holds for the antiferromagnet in the classical case; in the quantum case it holds if the anisotropy is large enough (depending on the single-site spin).

In statistical mechanics and solid state physics, magnetic systems are often described by models of spins on a cubic lattice with nearest-neighbor quadratic interactions. Moreover, for spin $S=\frac{1}{2}$, these models yield idealized descriptions of hard-core Bose gases. A primary question is whether some of the equilibrium states of such models in the thermodynamic limit have long-range order at low (but nonzero) temperatures, $T$, i.e., whether there are first-order phase transitions in the magnetic field. To date, the rigorous results for Heisenberg-type models include the following: (i) In three or more dimensions the classical isotropic and anisotropic (ferromagnetic and antiferromagnetic) models have a first-order phase transition.[1] The same is true in the quantum case,[2] except that, for the antiferromagnet, the existence of a phase transition is only known for spin $S \geq 1$. (Here, $S$ is the spin at one site.) (ii) In two dimensions the Ising model (case of extreme anisotropy) has, at low temperatures, equilibrium states with long-range order,[3] but the isotropic Heisenberg ferromagnet or antiferromagnet (including the rotor model) does not.[4] It has been suggested[5] that the two-dimensional isotropic models might have phase transitions not accompanied by long-range order, but this problem will not concern us here.

It is therefore of interest to know whether the two-dimensional anisotropic models have a first-order phase transition, for all anisotropies, i.e., whether they behave more like the Ising model than the isotropic models. An affirmative answer was recently given for the classical ferromagnetic and antiferromagnetic models.[6] Previous work had shown that long-range order exists at low temperatures if the anisotropy is large (in both the quantum[7,8] and classical[9] cases). The open question was whether the quantum fluctuations caused by the $xy$ part of the Hamiltonian could destroy the ordering for small anisotropy.

We have rigorously proved that, at low temperatures, the equilibrium states of the following models have long-range order: the quantum mechanical, ferromagnetic Heisenberg model with arbitrary anisotropy; the quantum mechanical antiferromagnet, provided $S$ is large enough, depending on the anisotropy. As a by-product we have found alternative proofs of the results of Ref. 6.

We consider three different models defined by the Hamiltonians

$$H_p = S^{-2}\{H_p^{z} + \alpha H_p^{xy}\}, \quad p=1,2,3,$$

where $0 \leq \alpha < 1$ is the anisotropy parameter. Model 1 is the usual ferromagnet:

$$H_1^{z} = -\sum S_i^{z} S_j^{z}, \quad H_1^{xy} = -\sum\{S_i^{x} S_j^{x} + S_i^{y} S_j^{y}\}$$

(with the sums taken over distinct nearest-neighbor pairs). Model 2 is a more subtle version of model 1:

$$H_2^{z} = \tfrac{1}{2}\sum (S_i^{z} - S_j^{z})^2,$$
$$H_2^{xy} = \tfrac{1}{2}\sum\{(S_i^{x} - S_j^{x})^2 + (S_i^{y} - S_j^{y})^2\}.$$

Note that

$$H_2 = H_1 + \frac{1-\alpha}{2S^2}\sum\{(S_i^{z})^2 + (S_j^{z})^2\} + C,$$

where $C$ is a trivial constant proportional to $1 + S^{-1}$. For $S > \frac{1}{2}$, the distinction between $H_2$ and $H_1$ is interesting: In the quantum case the ground state of $H_1$ *and* $H_2$ is doubly degenerate and has all spins maximally oriented in the $z$ direction. In the classical limit (where the spins become unit vectors and the trace is replaced by integra-

VOLUME 38, NUMBER 8 PHYSICAL REVIEW LETTERS 21 FEBRUARY 1977

tions over unit spheres[10]) the ground state of $H_2$ is infinitely degenerate: There is no preferred direction. Thus, in model 2 long-range order is a more subtle phenomenon than in model 1, and this fact causes our estimate for $T_c$ to go to zero as $S\to\infty$, for each fixed $\alpha<1$, in model 2.[11] In model 1, $T_c$ is bounded away from 0, uniformly in $S$, for all $\alpha<1$. Finally, model 3 is the usual antiferromagnet:

$$H_3=-H_1.$$

To state our theorems we define (for spin $S$ and temperature $T$)

$$\sigma(S,T)=S^{-2}\lim_{|j|\to\infty}\epsilon_j\langle S_0^z S_j^z\rangle$$

to be the long-range-order parameter for the equilibrium state $\langle\text{-}\rangle$ obtained as a thermodynamic limit of Gibbs states $\langle A\rangle_\Lambda=Z_\Lambda^{-1}\mathrm{Tr}[A\exp(-H_p^{(\Lambda)}/kT)]$ with periodic boundary conditions at the boundary of the rectangles $\Lambda$. In the ferromagnet $\epsilon_j=1$, whereas in the antiferromagnet $\epsilon_j=(-1)^{|j|}$.

*Model 1*: For each $\alpha<1$ there is a $T_c>0$ and a function $\mu(T)$, both independent of $S$, such that $\sigma(S,T)\geq\mu(T)>0$ when $T<T_c$. Moreover, $\mu(T)\to1$ as $T\to0$.

*Model 2*: For each $\alpha<1$ there is a $T_c(S)>0$ and a function $\mu(S,T)$ such that $\sigma(S,T)\geq\mu(S,T)>0$ when $T<T_c(S)$. $\mu(S,T)\to1$ as $T\to0$, but $T_c(S)\to0$ as $S\to\infty$.

*Model 3*: For each $S$ there is an $\alpha_c(S)$, with $0<\alpha_c(S)<1$, such that, for fixed $\alpha<\alpha_c(S)$, there is a function $\mu(S,T)$ and a $T_c$ with the property that $\sigma(S,T)\geq\mu(S,T)>0$ when $T<T_c$. Moreover $\alpha_c(S)\to1$ as $S\to\infty$, and $\mu(S,0)\geq\frac{1}{3}$ with $\mu(S,0)\to1$ as $S\to\infty$. Note that long-range order $[\sigma(S,T)>0]$ implies that there are pure phases with spontaneous magnetization (or staggered polarization).

There are two principal parts to our proof. As in the Peierls argument[3] we want to show that the probability of a contour $\gamma$ is small, i.e.,

$$D(\gamma)\equiv\langle P_\gamma\rangle$$

is small. Here $P_\gamma$ is the projection onto the subspace in which all spins just inside (outside) $\gamma$ have negative (positive) $z$ component. Our first step is to relate $D(\gamma)$, which is a number of order unity, to an extensive quantity. The inequality is

$$D(\gamma)\leq\overline{\lim_{\Lambda\uparrow Z^2}}\langle P\rangle_\Lambda^{|\gamma|/4N},$$

where $|\gamma|$ is the length of $\gamma$ and $N$ is the number of sites in $\Lambda$. $P$ is the projection onto the subspace of states in which every spin has either a positive or negative $z$ component, and, in cases 1 and 2, the first two rows of spins must have positive $z$ component, the next two rows have negative $z$ component and so on, alternately. In case 3 the signs are reversed on one of the two sublattices.

The second step is to show that

$$D\equiv\overline{\lim_{\Lambda\uparrow Z^2}}\langle P\rangle_\Lambda^{1/N}$$

is small when $T$ is small. We write

$$\langle P\rangle_\Lambda=\textstyle\sum_j\langle\varphi_j,P\varphi_j\rangle\exp(-e_j/kT)/Z_\Lambda,$$

$$Z_\Lambda=\textstyle\sum_j\exp(-e_j/kT),$$

where $\varphi_j$ and $e_j$ are the eigenvectors and eigenvalues of $H^{(\Lambda)}$. If $e_0$ is the ground-state energy, the terms in the numerator in which $e_j>e_0+\epsilon N$, for $\epsilon>0$, are easy to dispose of, because they have a small Boltzmann factor, i.e., they give a small contribution for energetic reasons. The terms with $e_j$ close to $e_0$ present a difficulty. Classically they *vanish*, but in the quantum case one needs an estimate showing that states of low energy have small projections onto states which have a large $H^z$ energy. This cannot be inferred by energetic considerations alone. We have proved a theorem which we call "exponential localization of eigenvectors." In the case at hand it says roughly that when $e_j<e_0+\epsilon N$ then $\langle\varphi_j,P\varphi_j\rangle\leq a^d$, where $a<1$ and $d$ is the number of times $H^{xy}$ must be applied to a state specified by $P$ such that the resulting state has $H^z$ energy less than $e_0+\gamma N$, with $\gamma$ small. Using this, it is possible to show that $D\to0$ as $T\to0$.

To summarize our results, we have established that quantum fluctuations are unimportant for the anisotropic ferromagnet and hence that some kind of spin-wave theory may be applicable. In the antiferromagnetic case, quantum fluctuations are unimportant if $S$ is large enough (depending on $\alpha$), but the situation for small $S$ has not been decided rigorously.

---

*A. P. Sloan Foundation Fellow; partially supported by U. S. National Science Foundation, Grant No. MPS75-11864.

†Work partially supported by U. S. National Science Foundation, Grant No. MCS 75-21684 A01.

[1]J. Fröhlich, B. Simon, and T. Spencer, Commun. Math. Phys. 50, 79 (1976).

[2]F. Dyson, E. Lieb, and B. Simon, "Phase Transitions in Quantum Spin Systems with Isotropic and Non-Isotropic Interactions" (to be published). See also Phys. Rev. Lett. 37, 120 (1976).

[3]R. E. Peierls, Proc. Cambridge Philos. Soc. 32, 477 (1936); see also R. Griffiths, in *Statistical Mechan-*

VOLUME 38, NUMBER 8 PHYSICAL REVIEW LETTERS 21 FEBRUARY 1977

*ics and Quantum Field Theory*, edited by C. DeWitt and R. Stora (Gordon and Breach, New York, 1971), and references therein.

[4]N. D. Mermin and H. Wagner, Phys. Rev. Lett. 17, 1133 (1966); N. D. Mérin, J. Math. Phys. (N.Y.) 8, 1061 (1967); R. L. Dobrushin and S. B. Shlosman, Commun. Math. Phys. 42, 31 (1975).

[5]H. E. Stanley and T. A. Kaplan, Phys. Rev. Lett. 17, 913 (1966).

[6]S. Malyshev, Commun. Math. Phys. 40, 75 (1975).

[7]J. Ginibre, Commun. Math. Phys. 14, 205 (1969).

[8]D. Robinson, Commun. Math. Phys. 14, 195 (1969).

[9]A. Bortz and R. Griffiths, Commun. Math. Phys. 26, 102 (1972).

[10]E. Lieb, Commun. Math. Phys. 31, 327 (1973).

[11]Absence of a phase transition in the classical model for $\alpha = 0$ follows from a result of O. McBryan and T. Spencer (to be published).

Existence of Phase Transitions for Anisotropic Heisenberg Models

PHYSICAL REVIEW LETTERS

Commun. math. Phys. 60, 233—267 (1978)

Communications in
**Mathematical Physics**


# Phase Transitions in Anisotropic Lattice Spin Systems

Jürg Fröhlich[1]* and Elliott H. Lieb[2]**

**Department of Mathematics[1] and Departments of Mathematics and Physics[2], Princeton University, Princeton, New Jersey 08540, USA**

**Abstract.** A general method for proving the existence of phase transitions is presented and applied to six nearest neighbor models, both classical and quantum mechanical, on the two dimensional square lattice. Included are some two dimensional Heisenberg models. All models are anisotropic in the sense that the groundstate is only finitely degenerate. Using our method which combines a Peierls argument with reflection positivity, i.e. chessboard estimates, and the principle of exponential localization we show that five of them have long range order at sufficiently low temperature. A possible exception is the quantum mechanical, anisotropic Heisenberg ferromagnet for which reflection positivity is *not* proved, but for which the rest of the proof is valid.

## I. Summary of Results and General Strategy of Proofs

One of the main purposes of this paper is to explain a general method for proving the existence of phase transitions, in the sense of long range order at sufficiently low temperatures, in classical and quantum lattice systems. In principle, our method can be applied to arbitrary lattice systems satisfying *reflection positivity* (a condition closely related to the existence of a self-adjoint positive definite transfer matrix), the groundstates of which are essentially *finitely degenerate* (e.g. the space of groundstates decomposes into finitely many subspaces labelled by a discrete order parameter, sometimes related to a broken discrete symmetry group).

Our method is inspired by recent work of Glimm, Jaffe and Spencer concerning phase transitions in the $(\lambda\phi^4)_2$ quantum field model, [16]. In this paper their ideas are extended in two ways:

1. We systematize the use of *reflection positivity* and *chessboard estimates*

---

* Present address: Institut des Hautes Etudes Scientifiques, F-91440 Bures-sur-Yvette, France.

A Sloan Foundation fellow. Work partially supported by U.S. National Science Foundation grant no. MPS 75-11864

** Work partially supported by U.S. National Science Foundation grant no. MCS 75-21684 A01

0010-3616/78/0060/0233/$07.00

in obtaining upper bounds on the statistical weight of contours arising in a Peierls argument and we show how to apply these methods to *quantum* lattice systems. This reduces the proof of long range order to estimating a ratio between a *constrained* partition function and the usual partition function. (This is basically a *thermodynamic* estimate).

2. We introduce the *principle of exponential localization* in order to derive upper bounds on constrained partition functions. This principle is particularly useful in the analysis of quantum lattice systems.

Reflection positivity, originally inspired by work of Osterwalder and Schrader [18], and the principle of exponential localization are useful tools in contexts other than the theory of phase transitions.

In Section I.A we introduce six different classical and quantum mechanical models on the two dimensional square lattice in terms of which we develop and illustrate our general method. A summary of our main results concludes that section.

In Section I.B we recall the connections between phase transitions and the occurrence of various forms of long range order (LRO) at sufficiently low temperatures.

In Section I.C, D and E we present the main ideas behind our general method; (Section I.C contains a convenient variant of the Peierls argument, essentially identical to the one of [16]; see also [10]).

In Section II we establish, (or review) reflection positivity for five of our six models, the exception being the quantum Heisenberg ferromagnet. We prove a generalization of the Hölder inequality for traces which, when combined with reflection positivity, yields the *chessboard estimates*. They extend constructive field theory estimates of [19].

In Section III we introduce the *principle of exponential localization* and apply it to our models for the purpose of estimating constrained partition functions. This is an expansion of the idea used in [22].

In Section IV the proofs of our main results are completed by combining the estimates of Sections I.C, II and III. Sections II and III contain results which are of some interest in their own right: Theorems 2.1, 2.2, 3.1 and Corollary 3.2. The reader can understand their statements and proofs without being familiar with the rest of this paper.

Next, we describe the models studied in this paper in general terms and recall some typical aspects of two dimensional lattice systems.

Two facts are well established about two-dimensional (quantum or classical) lattice spin systems with short range interactions:

(i) The Ising model has a first order phase transition (i.e. long range order for large $\beta = (kT)^{-1}$); for all values $S = 1/2, 1, \ldots$ of the spin.

(ii) Models with *continuous* symmetry (e.g. the isotropic Heisenberg models) have *no* such ordering. The proof of this is due to Mermin-Wagner [1], Mermin [2] and Hohenberg [3], (MWH). Thus, a natural question is whether the anisotropic models have LRO for all values of the anisotropy parameter, $\alpha$, with $0 < \alpha < 1$. For the *classical* Heisenberg (H) model this was proved recently by Malyshev [4]. Kunz, Pfister and Vuillermot [5] later gave a simplified proof for the planar rotator. Ginibre [6] and Robinson [7] proved LRO for the quantum Heisenberg *ferromagnet* for very small $\alpha$.

In [8] we announced proofs of LRO for a variety of anisotropic models: in particular for the quantum ferromagnetic H model, for all $\alpha < 1$. Subsequently we became aware of a flaw in one of the lemmas for the ferromagnetic model. This is basically the same flaw as in the announced Dyson, Lieb and Simon [9] proof of LRO for the *three* dimensional H model. The other results stated in [8] are correct. Here we will present the details of the proofs, including the part of the proof for the ferromagnetic H model that is correct. It is hoped that before very long the missing piece of the puzzle will be filled in.

An obvious remark has to be made: All the models we consider have *no* LRO at high temperature, a fact which can be proved by high temperature expansions, for example. Since LRO implies the existence of a spontaneously polarized state, our proof of LRO at low temperature implies the existence of a phase transition.

The models discussed here are all two-dimensional but, as in the usual Peierls argument [23], all our results and methods can be extended to higher dimensions. They can also be extended to some other lattices, e.g. the honeycomb lattice; see [12].

Some of our results were reviewed in [10] and [11]. Additional applications of the ideas presented here are to be found [10], [12] and [13].

### *I.A. Description of the Models and Main Results*

All models are on the square lattice $\mathbb{Z}^2$ and have only nearest neighbor interactions. Thus $\sum_{\langle i,j\rangle}$ means a sum over nearest neighbors, each term being included once; $H$ is the Hamiltonian.

(1) Classical $N$-Vector Model ($N > 1$)

$$H = -\sum_{\langle i,j\rangle}\left\{S_i^1 S_j^1 + \alpha \sum_{k=2}^{N} S_i^k S_j^k\right\}. \tag{1.1}$$

Each $S_i$ is a unit vector in $\mathbb{R}^N$, uniformly distributed on the sphere. (More general rotationally symmetric spin distributions could be accomodated by our methods.) Note that in this classical case, the ferromagnet (minus sign in (1.1)) is equivalent to the antiferromagnet (plus sign) by reversing the spins on the odd sublattice. This is *not* true in quantum models.

Our result in this case is that for every $\alpha < 1$ there is LRO at low $T$. In other words for every $\alpha < 1$ there is a $\beta_c(\alpha)$ such that there is LRO for $\beta > \beta_c(\alpha)$. Our estimate on $\beta_c(\alpha)$ is

$$\beta_c(\alpha) = O((1-\alpha)^{-1}). \tag{1.2}$$

The MWH result is that $\beta_c(\alpha = 1) = \infty$. Our proof is simpler than Malyshev's [4].

(2) Classical Anharmonic Crystal (AC Model)

The Hamiltonian of this model is given by

$$H = \sum_{\langle i,j \rangle} \phi(x_i, x_j),$$

where $x_i$ is the coordinate of an $N$-vector classical oscillator bound to site $i$, with apriori distribution $d^N x$; $\phi$ is some *continuous*, anisotropic interaction potential,

$$\phi(x, y) = \phi_1(x) + \phi_1(y) + \phi_2(x, y),$$

where $\phi_1 \geqq 0$ is a one body potential, and $\phi_2$ is a two body potential.

In other words we are considering some sort of anharmonic, anisotropic classical crystal (resp. a Euclidean lattice field theory). We will prove LRO at high $\beta$ under the following assumptions on $\phi$:

$$\min_{x,y} \phi(x, y) = \varepsilon_0$$

occurs for $x$ and $y$ in the *same* direction, (Typically at $x = y = x_0$, for some $x_0 \neq 0$). But if $x^1$ and $y^1$ (the 1-components of $x$, resp. $y$) have *opposite* signs

$$\phi(x, y) \geqq \varepsilon_0 + \alpha + \lambda(\phi_1(x) + \phi_1(y)),$$

for some $\alpha > 0$ and some $\lambda > 0$ with the property that for sufficiently large $\beta$

$$\int e^{-\beta\lambda\phi_1(x)} d^N x < \infty .$$

Examples of such potentials are:

1. $\phi_1(x)$ e.g. $gx^4 - \frac{1}{4}(x^1)^2 + (64g)^{-1}$, for some $g > 0$, $\phi_2(x, y) = V(x - y)$, where $V(x)$ is some strictly convex function with minimum at $x = 0$.
2. $\phi_1(x)$ e.g. $\gamma x^2, \gamma \ll 1$, $\phi_1(x, y) = V(x - y)$, with $V(x)$ e.g. $gx^4 - \frac{1}{4}(x^1)^2$, for some $g > 0$, (or $V$ an arbitrary continuous function with two sharp minima at $x = \pm(x^{01}, 0, \ldots, 0)$).
3. $\phi_1(x) = \gamma \log(|x| + 1)$, $\phi_2$ as in example 2.

Examples 2 and 3 (of anti-ferromagnetic type) are *not* of the general form of model AC, but can be brought into this form by replacing $x_i$ by $-x_i$ on one of the sublattices.

*Remarks.* It is of interest to consider also the case where $\phi_1(x)$ is replaced by const. $\beta^{-1}\phi_1(x)$. Then these models certainly do *not* have LRO for large $\beta$, as can be shown by a high temperature expansion.

The symmetry $\phi(x, y) = \phi(-x, -y)$ is *not* crucial for our arguments; see also [10, 12]. The main point of the study of model AC is that $\exp[-\beta\phi_2(x, y)]$ is *not* required to be of positive type. Next nearest neighbor interactions (coupling $x_{(m,n)}$ with $x_{(m\pm 1, n\pm 1)}$) could be included.

Physically more interesting models of an anharmonic crystal would be obtained by setting $\phi_1(x) = 0$ and assuming that $\phi_2$ is translation invariant. Our methods do *not* apply to such models.

(3) Quantum Antiferromagnetic Heisenberg Model

$$H = H^a = S^{-2}[H^z + \alpha H^{xy}] \tag{1.3}$$

$$H^z = \sum_{\langle i,j\rangle} S_i^z S_j^z$$

$$H^{xy} = \sum_{\langle i,j\rangle} \{S_i^x S_j^x + S_i^y S_j^y\}\,; \tag{1.4}$$

$S = 1/2, 1, 3/2, \ldots$ is the total spin at each site. We will prove that there is LRO at sufficiently large $\beta$ and small $\alpha$: For each $S$ there is an $\alpha(S)$ and $\beta_c(\alpha)$ such that for $\alpha < \alpha(S)$ and $\beta > \beta_c(\alpha)$ there is LRO. As $S \to \infty, \alpha(S) \to 1$. We do not know if there is LRO for all $\alpha < 1$ when $S$ is finite. This is an *open problem*. Because the $S \to \infty$ limit is the same as the classical model [14], we have here a generalization of the Malyshev result.

There is an equivalent form for (1.4) which is more convenient for our purposes, namely

$$H^z = -\sum_{\langle i,j\rangle} S_i^z S_j^z$$

$$H^{xy} = -\sum_{\langle i,j\rangle} \{S_i^x S_j^x + (iS_i^y)(iS_j^y)\}\,. \tag{1.4a}$$

This is obtained by making a rotation of $\pi$ about the $y$-axis for the spins on one of the two sublattices; for such spin operators $S^z \to -S^z, S^x \to -S^x, S^y \to +S^y$. In this representation *all* the terms in (1.3) are then of the form—(real matrix at $i$) (real matrix at $j$). Then reflection positivity, as discussed in Section II.A, holds. See [9] for more details.

(4) Quantum Ferromagnetic Heisenberg Model

$$H = H^f = -H^a. \tag{1.5}$$

The announced result [8] was that there is LRO for all $\alpha < 1$ when $\beta$ is large enough (uniformly in $S$). Unfortunately we cannot prove this because the proof of reflection positivity (Section II) is missing, but the second stage of the proof is correct and is given in Section III.

(5) The Two Quantum Models Can Be Modified as Follows

$$H^z = \sum_{\langle i,j\rangle} (S_i^z - S_j^z)^2$$

$$H^{xy} = \sum_{\langle i,j\rangle} \{(S_i^x - S_j^x)^2 + (S_i^y - S_j^y)^2\}.$$

$$H = S^{-2}[H^z + \alpha H^{xy}]. \tag{1.6}$$

This was mentioned in [8]. We will not give the details of the proofs here which are straight forward variants of the ones for (3), (4). This model is, however, interesting for the following reasons:

First, consider this model classically. When $\alpha = 0$ there is no LRO for any

$\beta$ by the Brascamp-Lieb argument [15]. Refined statements about exponential clustering were proved in [21]. When $\alpha = 1$ there is no LRO by MWH. We expect that there is no LRO for any $0 < \alpha < 1$ and any $\beta$.

However, the quantum model has a phase transition. In view of the foregoing remark, it is not surprising that our method yields the following in the ferromagnetic case (assuming reflection positivity): For $\alpha < 1$ there is a $\beta_c(\alpha, S)$, with LRO when $\beta > \beta_c(\alpha, S)$. However, $\beta_c(\alpha, S) \to \infty$ as $S \to \infty$ or $\alpha \to 1$.

(6) Quantum and Classical $xy$ Model

For convenience we take this model in the form

$$H = -S^{-2} \sum_{\langle i,j \rangle} \{S_i^z S_j^z + \alpha S_i^x S_j^x\}. \tag{1.7}$$

This is the *ferromagnet*. However by making a notation by $\pi$ about the $y$-axis for all spins on one sublattice (as in model (3)), we see that the antiferromagnet (defined with a + sign in (1.7)) is *equivalent* to the ferromagnet. See [9] for further details. For this model, as given by (1.7), reflection positivity does hold: (see Section II.A and use the standard representation in which $S^z$ and $S^x$ are real matrices).

Since the results and proofs for this model are the same as for the antiferromagnet (model (3)), resp. for model (1), we will not give further details.

### *I.B. Remarks about Long Range Order*

Let $\langle - \rangle_\Lambda$ be the Gibbs state of a system in a bounded rectangle $\Lambda \subset \mathbb{Z}^2$ with periodic boundary conditions, at inverse temperature $\beta$. The system in the thermodynamic limit, $\Lambda \uparrow \mathbb{Z}^2$, is said to have LRO if

$$\sigma(\beta) \equiv \lim_{\Lambda \uparrow \mathbb{Z}^2} \langle m_\Lambda^2 \rangle_\Lambda > 0, \tag{1.8}$$

where $m_\Lambda = \frac{1}{|\Lambda|} \sum_{i \in \Lambda} m_i$ is the magnetization, and $m_i$ is defined, in the different models, by

(1) $m_i = S_i^1$
(2) $m_i = x_i^1$
(3) $m_i = S^{-1}(-1)^{i_1+i_2} S_i^z$
(this is the staggered magnetisation)
(4), (5), (6) $m_i = S^{-1} S_i^z$.

The inequality $\sigma(\beta) > 0$ implies that there is spontaneous magnetization; see e.g. [9]. It is well known that $\sigma(\beta) \geqq M^2 > 0$ is implied by

$$\langle m_0 m_j \rangle_\Lambda \geqq M^2 > 0, \tag{1.9}$$

uniformly in $\Lambda$ and $j$. We will establish (1.9) at small temperature.

For this purpose, define $P_i^{\pm\delta}$ to be the projection operator onto all configurations satisfying $m_1 \geqq \delta$, resp. $m_i \leqq -\delta$. Moreover

$$P_i^{<\delta} = 1 - P_i^{+\delta} - P_i^{-\delta} \tag{1.10}$$

is the projection onto all configurations for which $|m_i| < \delta$. Finally, $P_i(\lambda) = P_i^{-\lambda} + P_i^{<\lambda}$ is the projection onto all configurations for which $m_i < \lambda$. For all models, except the AC model, $|m_i| \leqq 1$. Then

$$\begin{aligned}\langle m_0 m_j\rangle_\Lambda &= \int \lambda\lambda' \langle dP_0(\lambda) dP_j(\lambda')\rangle_\Lambda \\ &\geqq \delta^2\{\langle P_0^{+\delta} P_j^{+\delta}\rangle_\Lambda + \langle P_0^{-\delta} P_j^{-\delta}\rangle_\Lambda\} \\ &\quad - \{\langle P_0^{+\delta} P_j^{-\delta}\rangle_\Lambda + \langle P_0^{-\delta} P_j^{+\delta}\rangle_\Lambda\} \\ &\quad - \delta^2 \langle P_0^{<\delta} P_j^{<\delta}\rangle_\Lambda. \end{aligned} \tag{1.11}$$

The three terms on the right side of (1.11) are labelled I, II, III.

First we discuss II. Since, in all models, $m_0$ and $m_j$ commute, for all $j$, $P_0^{+\delta} P_j^{-\delta} \leqq P_0^+ P_j^-$, for $\delta > 0$, with

$$\begin{aligned} P_j^+ &= P_j^{+(\delta=0)} = 1 - P_j(0), \\ P_j^- &= P_j(0). \end{aligned} \tag{1.12}$$

Therefore

$$\langle P_0^{+\delta} P_j^{-\delta}\rangle_\Lambda \leqq \langle P_0^+ P_j^-\rangle_\Lambda. \tag{1.13}$$

The right side of (1.13) will be estimated by means of a new version of the Peierls argument inspired by work of Glimm, Jaffe and Spencer [16], and will be shown to be small, for large $\beta$, in the following sense which depends on the model: For some $\varepsilon > 0$ and $\beta$ large enough

$$\langle P_0^+ P_j^-\rangle_\Lambda < \frac{\varepsilon}{2}, \tag{1.14}$$

uniformly in $\Lambda$ and $j$. Thus,

$$\mathrm{II} > -\varepsilon. \tag{1.15}$$

Next we discuss term III on the right side of (1.11). By the Schwarz inequality for the state $\langle - \rangle_\Lambda$

$$\langle P_0^{<\delta} P_j^{<\delta}\rangle_\Lambda \leqq \langle P_0^{<\delta}\rangle_\Lambda, \tag{1.16}$$

and we have used

$$(P_j^{<\delta})^2 = P_j^{<\delta}$$

and

$$\langle P_j^{<\delta}\rangle_\Lambda = \langle P_0^{<\delta}\rangle_\Lambda,$$

which follows from the translation invariance of $\langle - \rangle_\Lambda$. We will prove by purely thermodynamic considerations that for some $\varepsilon > 0$ and sufficiently large $\beta$ (depending on the model)

$$\langle P_0^{<\delta}\rangle_\Lambda < \varepsilon. \tag{1.17}$$

Therefore

$$\mathrm{III} > -\varepsilon\delta^2 > -\varepsilon. \tag{1.18}$$

Finally we discuss term I on the right side of (1.11).

$$\begin{aligned}\langle P_0^{+\delta}P_j^{+\delta}\rangle_\Lambda &= \langle P_0^{+\delta}(1-P_j^{-\delta}-P_j^{<\delta})\rangle_\Lambda \\ &= \langle P_0^{+\delta}\rangle_\Lambda - \langle P_0^{+\delta}P_j^{<\delta}\rangle_\Lambda - \langle P_0^{+\delta}P_j^{-\delta}\rangle_\Lambda \\ &\geqq \langle P_0^{+\delta}\rangle_\Lambda - \langle P_0^{<\delta}\rangle_\Lambda - \langle P_0^{+\delta}P_j^{-\delta}\rangle_\Lambda . \end{aligned} \tag{1.19}$$

In all the models considered in this paper there is a symmetry taking $m_j$ to $-m_j$, for all $j \in \Lambda$. Therefore

$$\langle P_0^{-\delta}\rangle_\Lambda = \langle P_0^{+\delta}\rangle_\Lambda$$

so that

$$\langle P_0^{+\delta}\rangle_\Lambda = \tfrac{1}{2} - \tfrac{1}{2}\langle P_0^{<\delta}\rangle_\Lambda . \tag{1.20}$$

Combination of (1.14), (1.17), (1.19) and (1.20) yields

$$\langle P_0^{+\delta}P_j^{+\delta}\rangle_\Lambda > \tfrac{1}{2} - 2\varepsilon , \tag{1.21}$$

uniformly in $\Lambda$ and $j$, (provided $\beta$ is large enough, depending on the model). Therefore

$$\mathrm{I} \geqq \delta^2 - 4\delta^2\varepsilon .^{1} \tag{1.22}$$

Insertion of (1.15), (1.18) and (1.22) into inequality (1.11) gives

$$\langle m_0 m_j\rangle_\Lambda \geqq \delta^2 - 4\delta^2\varepsilon - 2\varepsilon > \delta^2 - 6\varepsilon , \tag{1.23}$$

uniformly in $\Lambda$ and $j$. Therefore

$$\sigma(\beta) > \delta^2 - 6\varepsilon . \tag{1.24}$$

In each model we will choose $\delta$ and $\varepsilon$ to depend on $\beta$ in such a way that, for sufficiently large $\beta$, $\delta^2 - 6\varepsilon > 0$.

The most difficult inequality to prove is (1.14). The strategy will be explained in three steps, $C$, $D$ and $E$ below. The inequality (1.17) is relatively simple and will be given in Section IV.

### *I.C. The Peierls Argument*

In this section we describe a general form of the Peierls argument. We consider a finite classical or quantum lattice system in a square $\Lambda \subset \mathbb{Z}^2$. For convenience we wrap $\Lambda$ on a torus, but this is inessential for this part of the argument. At each site $i \in \Lambda$ we are given two orthogonal projection operators, $P_j^{\pm}$, with

$$P_j^+ + P_j^- = 1, \text{ for all } j. \tag{1.25}$$

In the following $\langle - \rangle \equiv \langle - \rangle_\Lambda$. We propose to derive an upper bound on $\langle P_m^+ P_n^- \rangle$, where $m$ and $n$ are arbitrary, fixed sites in $\Lambda$, and $m \neq n$. The first step is the trivial identity

$$\langle P_m^+ P_n^- \rangle = \left\langle P_m^+ P_n^- \prod_{\substack{j\in\Lambda \\ m\neq j\neq n}} (P_j^+ + P_j^-)\right\rangle , \tag{1.26}$$

[1] This also holds when $\langle P_0^{+\delta}\rangle_\Lambda \neq \langle P_0^{-\delta}\rangle_\Lambda$

an immediate consequence of (1.25). We now expand the product on the right side of (1.26).

*Definition 1.* A *configuration* $c$ is a function on $\Lambda$ with values in $\{+,-\}$, and $c(m)=+, c(n)=-$. A *contour* $\gamma\subset\Lambda$ is a family of *nearest neighbor pairs* $\{\langle i_1, j_1\rangle,\dots,\langle i_l,j_l\rangle : l=4,6,\dots\}$ which decomposes $\Lambda$ into *precisely two* disjoint subsets

$$\Lambda_m=\Lambda_m(\gamma)\supset\{i_1,\dots,i_l,m\},\text{ and}$$

$$\Lambda_n=\Lambda_n(\gamma)\supset\{j_1,\dots,j_l,n\}\text{ with}$$

$$\Lambda_m\cup\Lambda_n=\Lambda.$$

Given a configuration $c$, we let $\Gamma(c)$ denote that class of all contours $\gamma=\{\langle i_1, j_1\rangle,\dots,\langle i_l,j_l\rangle\}$ with $c(i_k)=+, i_k\in\Lambda_m(\gamma), c(j_k)=-, j_k\in\Lambda_n(\gamma), k=1,\dots,l$. Since, for any configuration $c, c(m)=+, c(n)=-$, we conclude that, given an arbitrary $c$, there exists a contour $\gamma(c)\in\Gamma(c)$ with the property that there exists a connected set $\Lambda_c\subset\Lambda_m(\gamma(c))$ such that $m\in\Lambda_c, c(i)=+$, for all $i\in\Lambda_c, \{i_1,\dots,i_l\}\subseteq\Lambda_c$. (A set $X$ is connected if any two sites $i,j$ in $X$ belong to a chain $\{i=i_0,i_1,\dots,i_k,i_{k+1}=j\}\subseteq X$ such that $i_l$ and $i_{l+1}$ are nearest neighbors, $l=0,\dots,k$). Using Definition 1 we get from (1.26) by expanding

$$\langle P_m^+P_n^-\rangle=\sum_c\left\langle\prod_{j\in\Lambda}P_j^{c(j)}\right\rangle$$

$$=\sum_\gamma\sum_{\{c,\gamma(c)=\gamma\}}\left\langle\prod_{j\in\Lambda}P_j^{c(j)}\right\rangle. \tag{1.27}$$

Next, we note that

$$0\leqq P_i^\pm\leqq 1, [P_i^{c(i)},P_j^{c(j)}]=0,$$

for $i\neq j$, arbitrary $c$. Hence, for $Y\subseteq X$,

$$0\leqq\prod_{j\in X}P_j^{c(j)}\leqq\prod_{j\in Y}P_j^{c(j)},\text{ so that}$$

$$0\leqq\left\langle\prod_{j\in X}P_j^{c(j)}\right\rangle\leqq\left\langle\prod_{j\in Y}P_j^{c(j)}\right\rangle\leqq 1, \tag{1.28}$$

for all $c$. Therefore

$$\sum_{\{c:\gamma(c)=\gamma\}}\left\langle\prod_{j\in\Lambda}P_j^{c(j)}\right\rangle$$

$$\leqq\sum_{\{c,\Gamma(c)\ni\gamma\}}\left\langle\prod_{j\in\Lambda}P_j^{c(j)}\right\rangle$$

$$=\left\langle P_m^+P_n^-\prod_{\langle i,j\rangle\in\gamma}P_i^+P_j^-\right\rangle$$

$$<\left\langle\prod_{\langle i,j\rangle\in\gamma}P_i^+P_j^-\right\rangle. \tag{1.29}$$

Therefore we have the inequality

$$\langle P_m^+ P_n^- \rangle \leqq \sum_{\gamma} \left\langle \prod_{\langle i,j \rangle \in \gamma} P_i^+ P_j^- \right\rangle . \tag{1.30}$$

Let $|\gamma|$ denote the number of nearest neighbor pairs in $\gamma$, (the "length" of $\gamma$).

**Theorem 1.1.** (*Peierls Argument*). *Suppose that* (*for large enough* $|\Lambda|$)

$$\left\langle \prod_{\langle i,j \rangle \in \gamma} P_i^+ P_j^- \right\rangle \leqq e^{-K|\gamma|}, \tag{1.31}$$

*for some constant* $K > \ln 3$ (*independent of* $\Lambda$). *Then*

$$\langle P_m^+ P_n^- \rangle \leqq \sum_{l=2}^{\infty} 2l 3^{2l-2} e^{-2lK} < \infty , \tag{1.32}$$

*for arbitrary m and n in* $\Lambda$ (*and all sufficiently large squares* $\Lambda$).

*Remark.* The assertions of Theorem 1.1 do not depend on the size of $\Lambda$ and extend without change to the infinite system $\Lambda = \mathbb{Z}^2$; (see the subsequent proof and [10, Section 3]).

*Proof.* By Definition 1 the length of a contour is always *even*. The smallest contours are $\{\langle m, j_1 \rangle, \langle m, j_2 \rangle, \langle m, j_3 \rangle, \langle m, j_4 \rangle\}$ and $\{\langle i_1, n \rangle, \ldots, \langle i_4, n \rangle\}$, i.e. have length 4. Hence, by (1.30),

$$\langle P_m^+ P_n^- \rangle \leqq \sum_{l=2}^{\infty} \sum_{\{\gamma : |\gamma| = 2l\}} \left\langle \prod_{\langle i,j \rangle \in \gamma} P_i^+ P_j^- \right\rangle . \tag{1.33}$$

(When $\Lambda$ is finite these sums are finite). Given some fixed length $2l$, well known combinatorics shows that there are no more than $2(l-1)3^{2l-2}$ contours of length $2l$, provided $\Lambda$ is large enough, depending on $m$ and $n$. (The factor $3^{2l-2}$ comes from a standard 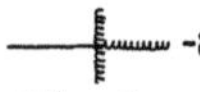 -argument and the fact that all contours consist of one or two *closed* pieces. The factor $2(l-2)$ comes from the fact that each contour must separate $m$ from $n$). Theorem 1.1 now follows from (1.33) and the inequality $K > \ln 3$ which guarantees that the series $\sum_{l=2}^{\infty} 2(l-1)3^{2l-2}e^{-2lK}$ *converges*.

Q.E.D.

Theorem 1.1 has the following

**Corollary 1.2.** *Given* $\varepsilon > 0$, *there exists some finite* $K(\varepsilon)$ *such that, for all* $K \geqq K(\varepsilon)$,

$$\langle P_m^+ P_n^- \rangle < \frac{\varepsilon}{2} .$$

*Remarks.* 1. The relevance of Corollary 1.2 for the proof of long range order has been explained in Section 1.B.

2. Theorem 1.1 and Corollary 1.2 can easily be generalized to the case of more than two positive operators (e.g. projections) $P_i^1, \ldots, P_i^M$ say, with

$$\sum_{l=1}^{M} P_i^l = 1 .$$

We could apply this more refined Peierls argument to the quantum ferromagnet, model (4), with $P_i^1 = P_i^{+\delta}$, $P_i^2 = P_i^{<\delta}$, $P_i^3 = P_i^{-\delta}$. This extension is important in models with more complicated phase diagrams involving at least $M > 2$ pure phases; (see e.g. [10], Sections 3 and 8).

3. Clearly these techniques extend to arbitrary dimensions $\geqq 2$ and other than simple, cubic lattices. See also [12].

### I.D. *Reflection Positivity and Chessboard Estimate*

In Section I.C we have seen that in order to prove

$$\langle P_m^+ P_n^- \rangle < \varepsilon,$$

uniformly in $m$ and $n$, it is sufficient to show

$$\left\langle \prod_{\langle i,j\rangle \in \gamma} P_i^+ P_j^- \right\rangle \leqq e^{-K|\gamma|} \tag{1.34}$$

for some constant $K = K(\varepsilon) \gg \ln 3$. Here we want to sketch how (1.34) can be reduced to a purely thermodynamic estimate.

Let $\Lambda$ be a square with sides of length $N = 4M, M = 1, 2, 3, \ldots$. We define a "universal projection"

$$P_\Lambda = \prod_{m=0}^{M-1} \left[ \prod_{n=0}^{N-1} P^+_{(4m,n)} P^-_{(4m+1,n)} P^-_{(4m+2,n)} P^+_{(4m+3,n)} \right]. \tag{1.35}$$

The following self explanatory figure illustrates Equation (1.35).

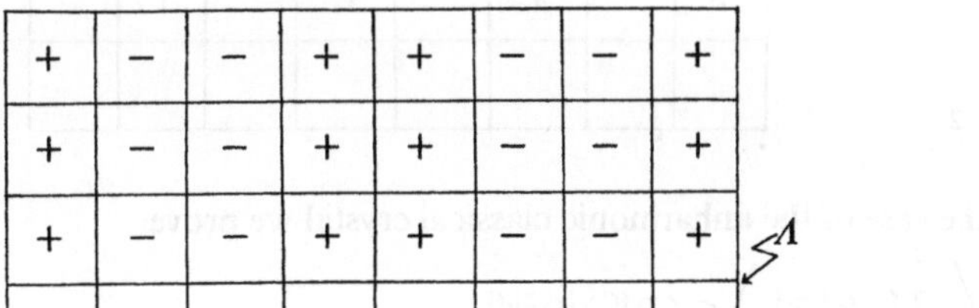


Fig. 1

One of the *key estimates* in our approach to proving LRO is the inequality

$$\left\langle \prod_{\langle i,j\rangle \in \gamma} P_i^+ P_j^- \right\rangle \leqq \langle P_\Lambda \rangle^{|\gamma|/2|\Lambda|} \tag{1.36}$$

which we shall prove for models (1), (3), (5) (antiferromagnetic case) and (6), i.e. all models except the quantum ferromagnets and the anharmonic classical crystal, model (2). For the former, we believe that (1.36) holds but we have no proof; (1.36) will be assumed to hold in the sequel.

For the AC model, the definition of the universal projection has to be modified: Let $\Lambda$ be a square with sides of even length $N=2M$. We define

$$P_\Lambda^{AC} = \prod_{m=0}^{M-1}\left[\prod_{n=0}^{M-1} P^+_{(2n,2m)}P^-_{(2n+1,2m)}\right]. \tag{1.37}$$

The following figure explains the definition of $P_\Lambda^{AC}$.

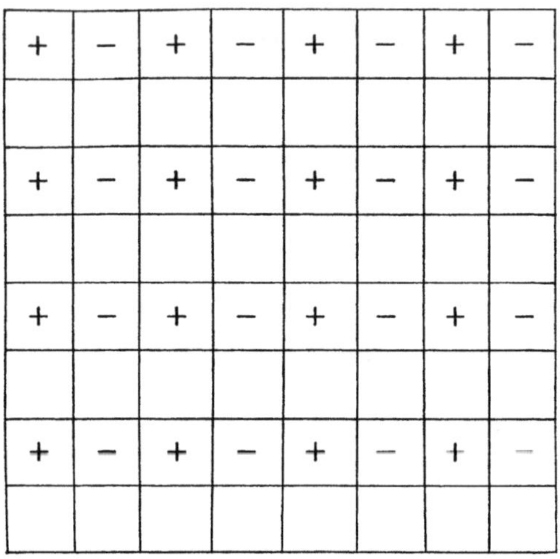


$N=8, M=4$

**Fig. 2**

In the case of the anharmonic classical crystal we prove

$$\left\langle \prod_{\langle i,j\rangle \in \gamma} P_i^+P_j^-\right\rangle \leqq \langle P_\Lambda^{AC}\rangle^{|\gamma|/2|\Lambda|}. \tag{1.38}$$

(This inequality *also* holds for the classical $N$-vector model, model (1)). Our proofs of inequalities (1.36) and (1.38) are based on the notion of *reflection positivity* (or $O-S$ positivity) which we now explain: We choose a pair of lines $l$ parallel to one of the coordinate axes, cutting $\Lambda$ into two congruent pieces $\Lambda_+$ and $\Lambda_-$ (Note: $l$ is a pair of lines, because $\Lambda$ is wrapped on a torus). In models (1), (3), (4) and (5) the lines $l$ are *between* two lattice lines, so that $\Lambda_+ \cap \Lambda_- = \emptyset$, whereas in model (2) $l$ consists of two *lattice lines*, and $\Lambda_+ \cap \Lambda_- = l$. Let $\theta_l$ be the reflection at $l$. Let $F=F(m)_{\Lambda_+}$ be a complex-valued function of all the $m_i$'s (see Section 1.B), with $i\in\Lambda_+$. We define $\theta_l F = \theta_l F(m)_{\Lambda_-}$ to be the function obtained from $F$ by substituting $m_{\theta_l j}$ for $m_j$. *Reflection positivity* is the inequality

$$\langle F(\theta_l \bar{F})\rangle \geqq 0, \tag{1.39}$$

where $\bar{F}$ is the *complex conjugate* of $F$. A somewhat more general inequality (also called reflection positivity) is discussed in Section II; as an example we mention that in the $N$-vector model, (1.39) is true for arbitrary, complex-valued functions $F$ of $\{\mathbf{S}_i : i \in \Lambda_+\}$ and both choices of $l$ (between two lattice lines or coinciding with a lattice line).

Reflection positivity (1.39) yields the following Schwarz inequality: If $F$ and $G$ are functions of $(m)_{\Lambda_+}$ then

$$|\langle F(\theta_l \bar{G})\rangle|^2 \leqq \langle F(\theta_l \bar{F})\rangle \langle G(\theta_l \bar{G})\rangle. \tag{1.40}$$

Next, we indicate how (1.36), resp. (1.38) follow from (1.40). Let $\gamma_h$ be all pairs $\langle i,j\rangle$ of the contour $\gamma$ for which $j-i$ points in the 1-direction. Such pairs are called "horizontal". Furthermore $\gamma_V = \gamma \backslash \gamma_h$ denotes all "vertical" pairs in $\gamma$. For $\langle i,j\rangle \in \gamma_h$, let $i \wedge j$ denote the site with smaller 1-coordinate; for $\langle i,j\rangle \in \gamma_V$, $i \wedge j$ is the site with smaller 2-coordinate. Suppose that reflection positivity (1.39) holds for reflections $\theta_l$ at lines $l$ *between* two lattice lines. Then we define

$$\gamma_{h,e} = \{\langle i,j\rangle \in \gamma_h : i \wedge j \text{ even}\}$$
$$\gamma_{h,0} = \{\langle i,j\rangle \in \gamma_h : i \wedge j \text{ odd}\} = \gamma_h \backslash \gamma_{h,e}.$$

Similarly $\gamma_{V,e}$ and $\gamma_{V,0}$ are defined. By the standard Schwarz inequality for $\langle - \rangle$ we have

$$\left\langle \prod_{\langle i,j\rangle \in \gamma} P_i^+ P_j^- \right\rangle \leqq \left\langle \prod_{\langle i,j\rangle \in \gamma_h} P_i^+ P_j^- \right\rangle^{1/2} \left\langle \prod_{\langle i,j\rangle \in \gamma_V} P_i^+ P_j^- \right\rangle^{1/2}$$
$$\leqq \prod_{\substack{\alpha = h,V \\ \beta = e,0}} \left\langle \prod_{\langle i,j\rangle \in \gamma_{\alpha,\beta}} P_i^+ P_j^- \right\rangle^{1/4} \tag{1.41}$$

To each factor on the right side we now apply reflection positivity (1:39) and inequality (1.40) repeatedly, for many different choices of $l$. This yields

$$\left\langle \prod_{\langle i,j\rangle \in \gamma_{\alpha,\beta}} P_i^+ P_j^- \right\rangle \leqq \langle P_\Lambda \rangle^{2|\gamma_{\alpha,\beta}|/|\Lambda|}. \tag{1.42}$$

This inequality is a special case of a general corollary of reflection positivity, called *chessboard estimate* [19], which we prove in Section 2. Clearly, inequalities (1.41) and (1.42) yield the *key inequality* (1.36).

In the classical, anharmonic crystal, model (2), we first decompose $\gamma$ into two pieces, $\gamma_h$ consisting of horizontal and $\gamma_V$ consisting of vertical pairs. For $\langle i,j\rangle \in \gamma_h$, let $(ij)_2$ denote the 2-coordinate of both $i$ and $j$, for $\langle i,j\rangle \in \gamma_V$, let $(ij)_1$ be the 1-coordinate of $i$ and $j$. We define

$$\gamma_{h,e} = \{\langle i,j\rangle \in \gamma_h : (ij)_2 \text{ even}\}$$
$$\gamma_{h,0} = \{\langle i,j\rangle \in \gamma_h : (ij)_2 \text{ odd}\}$$
$$\gamma_{V,e} = \{\langle i,j\rangle \in \gamma_V : (ij)_1 \text{ even}\}$$
$$\gamma_{V,0} = \{\langle i,j\rangle \in \gamma_V : (ij)_1 \text{ odd}\}.$$

Applying again the standard Schwarz inequality for $\langle - \rangle$, we obtain

$$\left\langle \prod_{\langle i,j\rangle \in \gamma} P_i^+ P_j^- \right\rangle \leqq \prod_{\substack{\alpha = h,V \\ \beta = e,0}} \left\langle \prod_{\langle i,j\rangle \in \gamma_{\alpha,\beta}} P_i^+ P_j^- \right\rangle^{1/4}. \tag{1.43}$$

To each term on the r.s. of (1.43) we apply the Schwarz inequality (1.40) repeatedly, for all allowed choices of reflections $\theta_l$ at *lattice lines* $l$. This yields (see Section II)

$$\left\langle \prod_{\langle i,j\rangle \in \gamma_{\alpha,\beta}} P_i^+ P_j^- \right\rangle \leqq \langle P_A^{AC} \rangle^{2|\gamma_{\alpha,\beta}|/|A|}. \tag{1.44}$$

The *key inequality* (1.38) follows from inequalities (1.43) and (1.44). Further details concerning reflection positivity and the chessboard estimates (1.42) and (1.44) are given in Section II.

## *I.E. Estimate of $\langle P_A \rangle$ and Exponential Localization*

In this section we sketch the main ideas of how to estimate

$R_A(\beta) \equiv \langle P_A \rangle$ and $R_A^{AC}(\beta) \equiv \langle P_A^{AC} \rangle$.

By definition of $\langle - \rangle$,

$$R_A^{(AC)}(\beta) \equiv \frac{\mathrm{Tr}(\exp[-\beta H_A] P_A^{(AC)})}{\mathrm{Tr}(\exp[-\beta H_A])}. \tag{1.45}$$

where $R_A^{(AC)}$ means either $R_A$ or $R_A^{AC}$. Here $H_A$ is the Hamiltonian of the model under consideration, and Tr is the usual trace in the quantum mechanical models, and, in the classical models, an integral with measure the product of the single spin distributions over all sites in $A$. Let $E_A(de)$ denote the spectral measure of the Hamiltonian $H_A$. By the spectral theorem

$$\mathrm{Tr}(\exp[-\beta H_A]C) = \int_{e_0}^{\infty} e^{-\beta e} \, \mathrm{Tr}(E_A(de)C), \tag{1.46}$$

where $e_0 \equiv e_0(A) = \mathrm{infspec}\, H_A$ is the groundstate energy, and $C$ is an arbitrary operator, resp. function. We will choose some positive number $\Delta = \Delta(\beta)$, depending on the model under consideration, and decompose $R_A^{(AC)}(\beta)$ into two pieces

$$R_-^{(AC)}(\beta, \Delta) = Z_A(\beta)^{-1} \int_{e_0}^{e_0 + \Delta|A|} e^{-\beta e} \, \mathrm{Tr}(E_A(de) P_A^{(AC)})$$

$$R_+^{(AC)}(\beta, \Delta) = Z_A(\beta)^{-1} \int_{e_0 + \Delta|A|}^{\infty} e^{-\beta e} \, \mathrm{Tr}(E_A(de) P_A^{(AC)}), \tag{1.47}$$

where

$$Z_A(\beta) = \mathrm{Tr}(\exp[-\beta H_A]) = \int_{e_0}^{\infty} e^{-\beta e} \, \mathrm{Tr}(E_A(de)) \tag{1.48}$$

is the partition function. We estimate $R_+^{(AC)}(\beta, \Delta)$ by

$$R_+^{(AC)}(\beta,\Delta) \leqq Z_\Lambda(\beta)^{-1} \exp\{-\beta[e_0 + \Delta|\Lambda|]\} \int_{e_0+\Delta|\Lambda|}^{\infty} \mathrm{Tr}(E_\Lambda(de))$$
$$\leqq \exp\{-\beta[e_0+\Delta|\Lambda|]\}\{\mathrm{Tr}(1)Z_\Lambda(\beta)^{-1}\}. \quad (1.49)$$

The Peierls-Bogoliubov inequality will be shown to give

$$\mathrm{Tr}(1)/Z_\Lambda(\beta) \leqq \exp \beta[e_0 + \tfrac{1}{2}\Delta|\Lambda|], \quad (1.50)$$

for $\beta$ sufficiently large. Thus

$$R_+^{(AC)}(\beta,\Delta) \leqq \exp\{-\beta\Delta|\Lambda|/2\}. \quad (1.51)$$

Next we consider $R_-^{(AC)}(\beta, \Delta)$. In the classical cases this will vanish for the following reason: $\Delta$ will be chosen sufficiently small so that $P_\Lambda^{(AC)}$ will be a projection onto configurations with energy greater than $e_0 + \Delta|\Lambda|$. Thus the integral for $R_-^{(AC)}(\beta)$ will vanish identically.

In the quantum cases, the situation is more complicated. Although $P_\Lambda$ will be a projection onto states whose *average energy* exceeds $e_0 + \Delta|\Lambda|$, the integral does *not* vanish, because $P_\Lambda$ will have nonvanishing matrix elements in eigenstates of $H_\Lambda$ with energy $< e_0 + \Delta|\Lambda|$. To be explicit, let $e_0 \leqq e_1 \leqq$ be the eigenvalues of $H_\Lambda$ with eigenvectors $\phi_0, \phi_1, \ldots$. Then

$$R_-(\beta,\Delta) = Z_\Lambda(\beta)^{-1} \sum_{i\Delta}{}' C_i \exp[-\beta e_i] \quad (1.52)$$

where $\sum'$ means a sum over $i$ such that $e_i \leqq e_0 + \Delta|\Lambda|$, and

$$C_i \stackrel{\Delta}{\equiv} (\phi_i, P_\Lambda \phi_i). \quad (1.53)$$

Now $C_i$ is independent of $\beta$, and terefore $R_-(\beta, \Delta)$ does not necessarily vanish as $\exp[-\beta(\text{const.})]$. What we have to show is that $C_i \to 0$ sufficiently fast as $i \to 0$. Then we can hope that $R_-(\beta, \Delta)$ goes to zero sufficiently fast as $\beta \to \infty$, for a suitable choice of $\Delta$.

The estimate on $C_i$, carried out in Section III, comes about in the following way: We write $H_\Lambda = A_\Lambda + B_\Lambda$, where $B_\Lambda$ is suitably small compared to $A_\Lambda$, and such that $P_\Lambda$ is an eigenprojection for $A_\Lambda$ onto $A_\Lambda$ eigenvectors having energy $> e_0 + n\Delta|\Lambda|$ for some $n$. In model (3), for example, $A = S^{-2}H^z$. If $B_\Lambda$ were zero then $C_i = 0$ for $e_i \leqq e_0 + \Delta|\Lambda|$. The *principle of exponential localization* will tell us that $A_\Lambda$ eigenvectors of $A_\Lambda$ energy greater than $e_0 + n\Delta|\Lambda|, (n \geqq 2)$ when expanded in the $\phi_i$, are strongly (indeed, *exponentially well* in $|\Lambda|$) localized around $\phi_i$ with $e_i > e_0 + \Delta|\Lambda|$. This, in turn, will lead to $C_i$ being *small* for $e_i \approx e_0$.

*Acknowledgements.* We thank B. Simon for some very useful suggestions.

## II. Reflection Positivity and Chessboard Estimates

### II.A. Reflection Positivity

In this section we recall the proofs of reflection positivity, inequality (1.39), for the models studied in this paper. For the classical $N$-vector models, reflection positivity is shown in [20]. In terms of a transfer matrix formalism it is used in [10]. For the quantum anti-ferromagnet and the quantum mechanical $xy$ model (models (3) and (6)) reflection positivity was discovered in [9]. The proof given there also applies to the classical $N$-vector models.

First we consider the classical, anharmonic crystal, model (2), for which (1.39) is new. We choose a pair of lattice lines $l$ cutting $\Lambda$ into two congruent pieces, $\Lambda_+$ and $\Lambda_-$, with $\Lambda_+ \cap \Lambda_- = l$. Let $\tilde{\Lambda}_\pm = \Lambda_\pm \backslash l$. The $N$-vector oscillators attached to sites in $\tilde{\Lambda}_\pm$ have coordinates $(y)_\pm \equiv \{y_j \in \mathbb{R}^N : j \in \tilde{\Lambda}_\pm\}$. The coordinates of the $N$-vector oscillators attached to sites in $l$ are denoted by

$$(z) \equiv \{z_j \in \mathbb{R}^N : j \in l\}.$$

Given a function $F$ of $(y)_+, (z)$, we define $\theta_l F$ to be the function of $(y)_-, (z)$ obtained by substituting $y_{\theta_l j}$ for $y_j$, for all $j \in \tilde{\Lambda}_+$, i.e. $\theta_l F$ is the reflection of $F$ in the lines $l$. The Hamilton function $H_\Lambda$ of the AC model is given by

$$\begin{aligned} H_\Lambda &= \sum_{\langle i,j\rangle \subset \Lambda} \phi(x_i, x_j) \\ &= \sum_{\langle i,j\rangle \subset \Lambda_+} \phi(x_i, x_j) + \sum_{\langle i,j\rangle \subset \Lambda_-} \phi(x_i, x_j) \\ &= \sum_{\langle i,j\rangle \subset \Lambda_+} \{\phi(x_i, x_j) + \phi(x_{\theta_l i}, x_{\theta_l j})\} \\ &\equiv B((y)_+, (z)) + (\theta_l B)((y)_-, (z)). \end{aligned}$$

Let $dx$ be the a priori distribution of a single oscillator, and set

$$d(y)_\pm = \prod_{j \in \tilde{\Lambda}_\pm} d^N y_j$$
$$d(z) = \prod_{j \in l} d^N z_j.$$

Let $F = F((y)_+, (z))$ be an arbitrary function localized on $\Lambda_+$. Then

$$\begin{aligned} \langle F(\overline{\theta_l F})\rangle &= Z_\Lambda(\beta)^{-1} \int d(z) d(y)_+ d(y)_- e^{-\beta H_\Lambda} F((y)_+, (z)) \overline{\theta_l F((y)_-, (z))} \\ &= Z_\Lambda(\beta)^{-1} \int d(z) \{\textstyle\int d(y)_+ e^{-\beta B((y)_+, (z))} F((y)_+, (z))\} \\ &\quad \times \{\textstyle\int d(y)_- e^{-\beta(\theta_l B)((y)_-, (z))} \overline{(\theta_l F)((y)_-, (z))}\} \\ &= Z_\Lambda(\beta)^{-1} \int d(z) |\textstyle\int d(y)_+ \\ &\quad \times e^{-\beta B((y)_+, (z))} F((y)_+, (z))|^2, \end{aligned} \tag{2.1}$$

i.e.

$$\langle F(\overline{\theta_l F})\rangle \geqq 0, \tag{2.2}$$

which is reflection positivity. Clearly, this form of reflection positivity also holds for the classical $N$-vector models. Next, we consider the quantum mechanical models and the classical $N$-vector models. We let $l$ be a pair of lines between lattice lines cutting $\Lambda$ into two *disjoint*, congruent pieces, $\Lambda_+$ and $\Lambda_-$. Let $\mathfrak{A}_j$ denote the family of all bounded functions of the spin $\mathbf{S}_j$ ("algebra of observables" at site $j$). We define

$$\mathfrak{A}_\pm = \bigotimes_{j \in \Lambda_\pm} \mathfrak{A}_j,$$

and

$$\mathfrak{A} = \mathfrak{A}_+ \otimes \mathfrak{A}_- .$$

Given some $B \in \mathfrak{A}$, we define $\theta B \equiv \theta_l B$ by

$$(\theta B)((\mathbf{S})_\Lambda) = B((\theta \mathbf{S})_\Lambda), \tag{2.3}$$

where

$$(\mathbf{S})_\Lambda = \{\mathbf{S}_j : j \in \Lambda\}, \text{ and}$$
$$(\theta\mathbf{S})_\Lambda = \{\mathbf{S}_{\theta_l j} : j \in \Lambda\},$$

i.e. $\theta B$ is obtained from $B$ by substituting $\mathbf{S}_{\theta_l j}$ for $\mathbf{S}_j$, all $j \in \Lambda$. Clearly $\theta$ defines an isomorphism from $\mathfrak{A}_+$ onto $\mathfrak{A}_-$ (and conversely). Furthermore we define

$$\bar{B} = (B^T)^* \tag{2.4}$$

to be the *complex conjugate* (*not* the adjoint) of $B$, for arbitrary $B \in \mathfrak{A}$. Following [9] we study Hamiltonians of the following general form:

$$H = B + \theta(\bar{B}) - \sum_i C_i \theta(\bar{C}_i), \tag{2.5}$$

where $B, C_1, \ldots, C_k, \ldots$ are in $\mathfrak{A}_+$, (and $B = B^*, C_i = \pm C_i^*$, for all $i$, so that $H$ is selfadjoint). The following result is a slight variation of Theorem E.1 of [9].

**Theorem 2.1.** (*Reflection Positivity*). *Let* $F \in \mathfrak{A}_+$. *Then*

$$\langle F(\overline{\theta F}) \rangle \equiv \frac{\mathrm{Tr}(e^{-\beta H} F(\overline{\theta F}))}{\mathrm{Tr}(e^{-\beta H})} \geqq 0,$$

*where* "Tr" *means the usual trace in the quantum case and an integral in the classical case.*

*Proof.* It clearly suffices to prove that $\mathrm{Tr}(e^{-\beta H} F(\overline{\theta F})) \geqq 0$. By the Trotter product formula,

$$e^{-\beta H} = \lim_{n \to \infty} G_n, \qquad \text{where}$$

$$G_n = \left( e^{-(\beta/n)B} e^{-(\beta/n)\theta\bar{B}} \left[ 1 + \frac{\beta}{n} \sum_i C_i \theta \bar{C}_i \right] \right)^n. \tag{2.6}$$

Thus, Theorem 2.1 is proved if

$$\mathrm{Tr}(G_n F(\overline{\theta F})) \geqq 0, \text{ for all } n. \tag{2.7}$$

To prove (2.7), note that all elements in $\mathfrak{A}_+$ commute with all elements in $\mathfrak{A}_-$. In (2.7) all elements with a $\theta$ (which are in $\mathfrak{A}_-$) can therefore be moved to the right of all elements without a $\theta$ (which are in $\mathfrak{A}_+$). This shows that $\mathrm{Tr}(G_n F(\overline{\theta F}))$ is a sum of terms of the form

$$\begin{aligned} \mathrm{Tr}(D_1 \ldots D_m F \theta \bar{D}_1 \ldots \theta \bar{D}_m \theta \bar{F}) &= \mathrm{Tr}(D_1 \ldots D_m F \theta(\bar{D}_1 \ldots \bar{D}_m \bar{F})) \\ &= \mathrm{Tr}(D_1 \ldots D_m F)\, \mathrm{Tr}(\bar{D}_1 \ldots \bar{D}_m \bar{F}), \end{aligned}$$

with $D_1, \ldots, D_m$ in $\mathfrak{A}_+$. Here we have used the obvious facts that $\mathrm{Tr}(AB) = \mathrm{Tr}(A)\,\mathrm{Tr}(B)$, for $A \in \mathfrak{A}_+, B \in \mathfrak{A}_-$, and $\mathrm{Tr}(\theta A) = \mathrm{Tr}(A)$, for all $A \in \mathfrak{A}_+$. Finally

$$\mathrm{Tr}(D_1 \ldots D_m F)\,\mathrm{Tr}(\bar{D}_1 \ldots \bar{D}_m \bar{F}) = |\mathrm{Tr}(D_1 \ldots D_m F)|^2 \geqq 0,$$

by definition of complex conjugation ($B \mapsto \bar{B}$). Q.E.D.

We leave it to the reader to check that the Hamiltonians $H_\Lambda$ of models (1), (3) and (6) are of the form (2.5). See also [9]. Hence Theorem 2.1 proves *reflection positivity*, inequality (1.39), for these models. However, for the quantum ferromagnet, models (4) and (5) (ferromagnetic case), $H_\Lambda$ is *not* of the form (2.2) (because of the $S_i^y S_j^y$ terms), and the proof of Theorem 2.1 breaks down. At present, *no* useful form of reflection positivity is known for these models. In the sequel, we will assume that inequality (1.36), which follows from reflection positivity (as shown in Section II.B) *does* hold for the ferromagnetic models, even though we have no proof of it.

### *II.B. Chessboard Estimate*

Our goal in this subsection is to use reflection positivity to prove inequalities (1.42) and (1.44) (chessboard estimate). We prove a general theorem that includes (1.42) as a special case.

**Theorem 2.2.** (*Generalized Hölder Inequality*). *Let $\mathfrak{A}$ be a vector space with antilinear involution $J$ (to be thought of as complex conjugation). Let $\omega$ be a multilinear functional on $\mathfrak{A}^{\times 2M}$, for some integer $M > 0$, with the properties*

*(C) $\omega(A_1, \ldots, A_{2M}) = \omega(A_2, \ldots, A_{2M}, A_1)$ (cyclicity), and*

*($\theta$) The matrix $K$ whose matrix elements $K_{ij}$ are given by*

$$K_{ij} = \omega(JA_{i,1}, \ldots, JA_{i,M}, A_{j,1}, \ldots, A_{j,M}),$$

*with $A_{l,m}$ an arbitrary vector in $\mathfrak{A}$, for all $l = 1, \ldots, n, m = 1, \ldots M$, is a positive semi-definite $n \times n$ matrix, for all $n = 1, 2, \ldots$; (Reflection Positivity). Then*

$$(1)\quad |\omega(A_1, \ldots, A_{2M})| \leqq \prod_{j=1}^{2M} \omega(JA_j, A_j, \ldots, JA_j, A_j)^{1/2M},$$

*(chessboard estimate)*

*and*

$$(2)\quad \|A\|_{2M} \equiv \omega(JA, A, \ldots, JA, A)^{1/2M}$$

*is a semi-norm on $\mathfrak{A}$.*

*Proof.* 1. *A Schwarz Inequality.* Let $\mathscr{L}(\mathfrak{A}^{\times M})$ be the vector space over the complex numbers spanned by all elements in $\mathfrak{A}^{\times M}$. Hypothesis ($\theta$) tells us precisely that $\omega$ defines an *inner product* on $\mathscr{L}(\mathfrak{A}^{\times M})$. As a special consequence of the *Schwarz inequality* for this inner product we have

$$|\omega(A_1, \ldots, A_{2M})| \leqq \omega(A_1, \ldots, A_M, JA_M, \ldots, JA_1)^{1/2} \cdot \omega(JA_{2M}, \ldots, JA_{M+1}, A_{M+1}, \ldots, A_{2M})^{1/2}. \quad (2.8)$$

2. *Proof of Theorem* 2.2 *for* $M=2$. This serves to exhibit the main ideas behind the proof of the general case. By (2.8) and hypothesis (C),

$$\begin{aligned}|\omega(A,B,C,D)| &\leqq \omega(A,B,JB,JA)^{1/2}\omega(JD,JC,C,D)^{1/2}\\ &= \omega(B,JB,JA,A)^{1/2}\omega(JC,C,D,JD)^{1/2}\\ &\leqq \omega(B,JB,B,JB)^{1/4}\omega(JA,A,JA,A)^{1/4}\\ &\quad \cdot\omega(JC,C,JC,C)^{1/4}\omega(D,JD,D,JD)^{1/4}\\ &= \omega(JA,A,JA,A)^{1/4}\omega(JB,B,JB,B)^{1/4}\\ &\quad \cdot\omega(JC,C,JC,C)^{1/4}\omega(JD,D,JD,D)^{1/4}\end{aligned}$$

which is (1); (2) follows from the multilinearity of $\omega$ and (1).

3. *The General Case*. Since $\omega$ is multi-linear and

$$\omega(JA_j,A_j,\dots,JA_j,A_j)=\omega(A_j,JA_j,\dots,A_j,JA_j),$$

by hypothesis (C), we may assume that

$$\omega(JA_j,A_j,\dots,JA_j,A_j)=1, \tag{2.9}$$

for all $j=1,\dots,2M$; (if not, replace $A_j$ by $\omega(JA_j,A_j,\dots,JA_j,A_j)^{-1/2M}\cdot A_j$). We set $JA_j\equiv A_{j+2M}, j=1,\dots,2M$. A *configuration* $c$ is a function on $\{1,\dots,2M\}$ with values in $\{1,\dots,4M\}$. Let $z\equiv \max_c|\omega(A_{c(1)},A_{c(2)},\dots,A_{c(2M)})|$, i.e.

$$z\geqq|\omega(A_{c(1)},\dots,A_{c(2M)})|, \text{for all } c. \tag{2.10}$$

**Lemma.** $z=1$.

*Proof.* For $c$ defined by

$$c(2m-1)=j+2M, c(2m)=j,$$

$m=1,\dots,M$,

$$\omega(A_{c(1)},\dots,A_{c(2M)})=1,$$

by (2.9). Hence $z\geqq 1$. Thus, it suffices to show $z\leqq 1$. Let $\tilde c$ be a configuration for which

$$|\omega(A_{\tilde c(1)},\dots,A_{\tilde c(2M)})|=z.$$

Let $c(M+1)\equiv j$. Then, by the Schwarz inequality (2.8),

$$\begin{aligned}z&=|\omega(A_{\tilde c(1)},\dots,A_{\tilde c(2M)}|\\ &\leqq \omega(A_{\tilde c(1)},\dots,A_{\tilde c(M)},JA_{\tilde c(M)},\dots,JA_{\tilde c(1)})^{1/2}\\ &\quad\cdot\omega(JA_{\tilde c(2M)},\dots,JA_j,A_j,\dots,A_{\tilde c(2M)})^{1/2}\\ &\leqq z^{1/2}\omega(JA_{\tilde c(2M)},\dots,JA_j,A_j,\dots,A_{\tilde c(2M)})^{1/2}, \text{ by (2.10)}\\ &= z^{1/2}\omega(JA_{\tilde c(2M-1)},\dots,JA_j,A_j,\dots,A_{\tilde c(2M)},JA_{\tilde c(2M)})^{1/2}, \text{ by hypothesis (C)}\\ &\leqq z^{3/4}\omega(JA_{\tilde c(2M-1)},\dots,JA_j,A_j,JA_j,A_j,\dots,A_{\tilde c(2M-1)})^{1/4}, \text{ by (2.8) and (2.10)}\end{aligned}$$

$$\leqq \dots$$

$$\leqq z^{1-2^{-(m-1)}} \omega(\underbrace{JA_j, A_j, \dots, JA_j, A_j}_{M}, \underbrace{*, *, \dots}_{M})^{2^{-(m-1)}}$$

$$\leqq z^{1-2^{-m}} \omega(JA_j, A_j, \dots, JA_j, A_j)^{2^{-m}}, \text{ by (2.8) and (2.10)}$$

$$= z^{1-2^{-m}}, \text{ by (2.9)},$$

for some $m$ with $2^{m-1} \geqq M > 2^{m-2}$. Hence $z^{2^{-m}} \leqq 1$, i.e. $z \leqq 1$.

Q.E.D.

To prove Theorem 2.1, (1), let $c$ be given by $c(j) = j, j = 1, \dots, 2M$. By (2.10) and the Lemma,

$$|\omega(A_{c(1)}, \dots, A_{c(2M)})| = |\omega(A_1, \dots, A_{2M})| \leqq z = 1. \tag{2.11}$$

The multilinearity of $\omega$ and (2.11) completes the proof of (1). Theorem 2.2, (2) follows from the multilinearity of $\omega$ and hypothesis $(\theta)$ (which imply $\|A\|_{2M} \geqq 0$ and $\|\lambda A\|_{2M} = |\lambda| \, \|A\|_{2M}$) and from (1) (which implies that $\|A + B\|_{2M} \leqq \|A\|_{2M} + \|B\|_{2M}$). Q.E.D.

To apply Theorem 2.2 to the proof of estimates (1.42), resp. (1.44), one makes the following identifications:

$$\omega(\cdot) \mapsto \langle \cdot \rangle$$

$$A_j \mapsto P_i^+ P_j^-, \text{ with } i,j \text{ nearest neighbors;}$$

Theorem 2.2 must be applied twice, once in the *vertical direction* and once in the *horizontal direction*. This gives (1.42), resp. (1.44). We now must check that $\omega(\cdot) = \langle \cdot \rangle$ satisfies the hypothesis of Theorem 2.2: Clearly $\left\langle \prod_{j \in A} B_j \right\rangle$ is linear in each $B_j$, yielding multi-linearity of $\omega$.

Since we have wrapped $\Lambda$ on a torus (periodic boundary conditions),

$$\left\langle \prod_{j \in \Lambda} B_j \right\rangle = \left\langle \prod_{j \in \Lambda} B_{j+a} \right\rangle,$$

for arbitrary $a \in \Lambda$. This shows that $\omega$ satisfies hypothesis (C) in both, the vertical and the horizontal directions. Finally, hypothesis $(\theta)$ of Theorem 2.2 in both, the vertical and the horizontal directions, is an immediate consequence of *reflection positivity* (inequality (2.1), resp. Theorem 2.1). A more direct proof of inequalities (1.42) and (1.44) proceeds as follows; (we sketch the argument leading to (1.42); the case of the anharmonic crystal is treated similarly). Let $B_{h,e}$ denote all pairs of horizontal nearest neighbors $\langle i,j \rangle$ (directed, "horizontal bonds") with $i \wedge j$ even. Let $\mathcal{O}$ be an arbitrary, *non empty* subset of $B_{h,e}$. Let $|\mathcal{O}|$ denote the number of horizontal bonds in $\mathcal{O}$. We consider the family

$$\left\{ \left\langle \prod_{\langle i,j \rangle} P_i^+ P_j^- \right\rangle^{1/2|\mathcal{O}|} : \mathcal{O} \subseteq B_{h,e} \right\}.$$

Let

$$z = \max_{\mathcal{O}} \left\{ \left\langle \prod_{\langle i,j \rangle \in \mathcal{O}} P_i^+ P_j^- \right\rangle^{1/2|\mathcal{O}|} \right\},$$

and let $\tilde{\mathcal{O}}$ be some subset of directed, horizontal bonds on which the maximum $z$ is taken. Using translation invariance of $\langle - \rangle$ (corresponding to hypothesis (C) of Theorem 2.1) and reflection positivity of $\langle - \rangle$ (corresponding to ($\theta$)) and applying the Schwarz inequality (corresponding to (2.8)) repeatedly, as in inequality (2.11), in the horizontal and vertical direction, we obtain

$$z \leqq \langle P_\Lambda \rangle^{1/k|\Lambda|} z^{1-1/k},$$

for some integer $k > 0$. Hence $z \leqq \langle P_\Lambda \rangle^{1/|\Lambda|}$ from which we obtain (1.42). Finally we remark that Theorem 2.2 can be used to give alternate proofs of the general chessboard estimates of the last reference in [19] (Theorem 2.3, periodic boundary conditions) and of [10] (Lemma 4.5). Furthermore Theorem 2.2 implies the *Hölder inequality* for general traces and the *Peierls-Bogolubov* and *Golden-Thompson* inequalities.

## III. Exponential Localization

In this section we explain the difficult part in the required estimate of $R_\Lambda(\beta) = \langle P_\Lambda \rangle$, defined in (1.45), for the quantum mechanical models. We recall that in Section I.E. we have split $R_\Lambda(\beta)$ into two pieces

$$R_\Lambda(\beta) = R_-(\beta, \Delta) + R_+(\beta, \Delta), \tag{3.1}$$

where

$$R_-(\beta, \Delta) = Z_\Lambda(\beta)^{-1} \sum_{i\Delta}{}' C_i \exp[-\beta e_i]; \tag{3.2}$$

here $\sum_{i\Delta}'$ means a sum over all $i$ such that $e_i \leqq e_0 + \Delta |\Lambda|$, and

$$C_i \equiv (\phi_i, P_\Lambda \phi_i). \tag{3.3}$$

The easy estimate of $R_+(\beta, \Delta)$ is postponed to Section IV. In this section we prove upper bounds on $R_-(\beta, \Delta)$ for models (1)–(6). We claim that, for the classical models (1), (2) and (6) (classical case),

$$R_-(\beta, \Delta) = 0 \tag{3.4}$$

for sufficiently small $\Delta$. To show this we first estimate the minimum $\mathcal{E}(P_\Lambda^{(AC)})$ of the Hamilton function $H_\Lambda$ restricted to the configurations

$\{\mathbf{S} : \mathbf{S} \in P_\Lambda\}$ (models (1), (6)),

resp. $\{x : x \in P_\Lambda^{AC}\}$ (model (2)).

For models (1) and (6)

$$\mathcal{E}(P_\Lambda) \geqq -\tfrac{3}{2}|\Lambda| - \frac{\alpha}{2}|\Lambda|. \tag{3.5}$$

For model (2)

$$\mathscr{E}(P_\Lambda^{AC}) \geqq 2\varepsilon_0|\Lambda| + \frac{\alpha}{2}|\Lambda|. \tag{3.6}$$

Therefore

if $\Delta < (1-\alpha)/2$ (models (1), (6)) (3.7)

resp. $\Delta < \alpha/2$ (model (2)) (3.8)

then $R_-(\beta, \Delta) = 0$

which proves our contention.

As already noted in Section I.E, (3.4) is *false* for the quantum mechanical models, and we have to work much harder in order to obtain a good upper bound on $R_-(\beta, \Delta)$. The idea is to show that $C_i = (\phi_i, P_\Lambda \phi_i)$ is very small for eigenvalues $e_i$ of $H_\Lambda$ close to the ground state energy $e_0$. Although $P_\Lambda$ is a projection onto states of relatively high $H^z$-energy, $(\phi_i, P_\Lambda \phi_i)$ does *not* vanish, even for $e_i$ very close to $e_0$, as it does in the classical case.

## *III.A. Principle of Exponential Localization*

The following general result will be crucial for our analysis.

**Theorem 3.1.** (*Exponential Localization of Eigenvectors*). *Let A and B be selfadjoint operators (typically finite, hermitean matrices) on a Hilbert space $\mathscr{H}$ such that*

(*i*) $A \geqq 0$

(*ii*) $\pm B \leqq \varepsilon A$,

*with* $0 \leqq \varepsilon < 1$. *Suppose that*

$$(A+B)\psi = \lambda\psi, \ \|\psi\| = 1.$$

*Choose some* $\rho > \lambda \geqq 0$ *such that*

$$\sigma \equiv \varepsilon\rho(\rho - \lambda)^{-1} < 1. \tag{3.9}$$

*Let $P_\rho$ be the spectral projection of A corresponding to $[\rho, \infty)$, and $M_\rho \equiv P_\rho \mathscr{H}$, (all "eigenvectors" of A corresponding to eigenvalues $\geqq \rho$). Note that $(A-\lambda)$ restricted to $M_\rho > 0$. Finally, let $\phi \in M_\rho$ be a unit vector with the property*

(*iii*) $\{B(A-\lambda)^{-1}\}^j \phi \in M_\rho$,

*for* $j = 0, 1, \ldots, d-1$, *with* $d \geqq 1$.

*Then* $|(\phi, \psi)| \leqq \sigma^d$. (3.10)

*Remarks.* 1. Since $B \geqq -\varepsilon A$, by (ii),

$$A + B \geqq (1-\varepsilon)A \geqq 0, \tag{3.11}$$

so that all eigenvalues $\lambda$ of $A+B$ are nonnegative.

2. Clearly the condition $|B| \leqq \varepsilon A$ implies (ii), but the converse is *false*, as the example

$$A = \begin{pmatrix} 2 & \frac{3}{2} \\ \frac{3}{2} & 2 \end{pmatrix}, B = \begin{pmatrix} 1 & 0 \\ 0 & -1 \end{pmatrix}, |B| = 1, \varepsilon = 1$$

shows. Hypothesis (ii) is all we need to prove (3.10).

*Proof.* By hypothesis,

$$(A+B)\psi = \lambda\psi, \text{ i.e. } (A-\lambda)\psi = B\psi.$$

Thus, for some $\delta \geqq 0$,

$$\psi = (A - \lambda + i\delta)^{-1}(B\psi + i\delta\psi),$$

so that

$$\begin{aligned}|(\phi,\psi)| &= |(\phi,(A-\lambda+i\delta)^{-1}(B\psi+i\delta\psi))| \\ &= |((A-\lambda-i\delta)^{-1}\phi, B\psi+i\delta\psi)|.\end{aligned}$$

Since $\phi \in M_\rho$ and $\lambda < \rho$, by hypothesis,

$$\lim_{\delta\downarrow 0}(A-\lambda-i\delta)^{-1}\phi = (A-\lambda)^{-1}\phi,$$

hence

$$|(\phi,\psi)| = |(B(A-\lambda)^{-1}\phi,\psi)|. \tag{3.12}$$

By hypothesis (iii), $\{B(A-\lambda)^{-1}\}^j\phi \in M_\rho$, for $j = 0, 1, \ldots, d-1$. Therefore, for $d > 1$,

$$|(\phi,\psi)| = |(P_\rho B(A-\lambda)^{-1}\phi,\psi)|, \tag{3.12'}$$

and we can iterate (3.12′) $d-1$ times and then apply (3.12). This yields

$$\begin{aligned}|(\phi,\psi)| &= |(B(A-\lambda)^{-1}\{P_\rho B(A-\lambda)^{-1}P_\rho\}^{d-1}\phi,\psi)| \\ &= |(A^{-1/2}B(A-\lambda)^{-1}\{P_\rho B(A-\lambda)^{-1}P_\rho\}^{d-1}\phi, A^{1/2}\psi)| \\ &\leqq \|A^{-1/2}B(A-\lambda)^{-1/2}P_\rho\| \, \|P_\rho(A-\lambda)^{-1/2}B(A-\lambda)^{-1/2}P_\rho\|^{d-1} \\ &\quad \cdot \|(A-\lambda)^{-1/2}\phi\| \, \|A^{1/2}\psi\|,\end{aligned} \tag{3.13}$$

where we have used that $[A, P_\rho] = 0$. Now

$$\begin{aligned}\|A^{-1/2}B(A-\lambda)^{-1/2}P_\rho\| &\leqq \|A^{-1/2}BA^{-1/2}\| \, \|A^{1/2}(A-\lambda)^{-1/2}P_\rho\| \\ &\leqq \varepsilon\rho^{1/2}(\rho-\lambda)^{-1/2} = (\varepsilon\sigma)^{1/2},\end{aligned} \tag{3.14}$$

$$\begin{aligned}\|P_\rho(A-\lambda)^{-1/2}B(A-\lambda)^{-1/2}P_\rho\| &\leqq \|P_\rho(A-\lambda)^{-1/2}A^{1/2}\|^2 \, \|A^{-1/2}BA^{-1/2}\| \\ &\leqq \rho(\rho-\lambda)^{-1}\varepsilon = \sigma,\end{aligned} \tag{3.15}$$

and we have used the definition of $P_\rho$ and hypothesis (ii). Finally

$$\|(A-\lambda)^{-1/2}\phi\| = \|(A-\lambda)^{-1/2}P_\rho\phi\| \leqq (\rho-\lambda)^{-1/2}\|\phi\| \tag{3.16}$$

and

$$\|A^{1/2}\psi\| = [(\psi, A\psi)]^{1/2}$$
$$\leqq (1-\varepsilon)^{-1/2}[(\psi,(A+B)\psi)]^{1/2}, \text{ by (3.11)}$$
$$= [(1-\varepsilon)^{-1}\lambda]^{1/2} < \rho^{1/2}, \tag{3.17}$$

since $\sigma \equiv \varepsilon\rho(\rho-\lambda)^{-1} < 1$, i.e. $\lambda < \rho(1-\varepsilon)$. If we combine (3.13)–(3.17) we find $|(\phi,\psi)| \leqq \sigma^{d-1/2}\varepsilon^{1/2}(\rho-\lambda)^{-1/2}\rho^{1/2} = \sigma^d$. Q.E.D.

**Corollary 3.2.** *Suppose $N \subset M_\rho$ is a subspace of $M_\rho$ such that each $\phi \in N$ satisfies hypothesis (iii) of Theorem 3.1. If $P$ is the projection onto $N$ then (in the notations of Theorem 3.1)*

$$\langle \psi, P\psi \rangle \leqq \sigma^{2d}.$$

The proof is essentially identical to the one of Theorem 3.1. We now apply Corollary 3.2 to estimating the overlap of the universal projection $P_A$ with the low lying eigenstates of $H_A$, i.e. the numbers $C_i = (\phi_i, P_A\phi_i)$, when the eigenvalues $e_i \leqq e_0 + \Delta|A|$, for models (3) and (4), (quantum mechanical antiferromagnet, resp. ferromagnet. The case of the $xy$ model is similar to the antiferromagnet). For this purpose we identify

$$P = P_A \tag{3.18}$$

$$A = S^{-2}H^z - e_0(\alpha = 1), \tag{3.19}$$

where $e_0(\alpha=1)$ is the groundstate energy of the *isotropic* Hamiltonian,

$$B = \alpha S^{-2}H^{xy}. \tag{3.20}$$

In all the models discussed here, the groundstate energy $e_0(\alpha=1)$ of the Hamiltonian $H = S^{-2}(H^z + H^{xy})$ is bounded above by the groundstate energy $e_0^z$ of $S^{-2}H^z$. Therefore

$$A \geqq 0. \tag{3.21}$$

Furthermore

$$A + \frac{1}{\alpha}B = S^{-2}(H^z + H^{xy}) - e_0(1) \geqq 0. \tag{3.22}$$

If we rotate all spins on one of the sublattices by an angle $\pi$ around the $z$-axis we see that $A + \frac{1}{\alpha}B$ is unitarily equivalent to $A - \frac{1}{\alpha}B$, as $H^{xy}$ is taken into $-H^{xy}$ under this unitary transformation, but $H^z$ is unchanged. Therefore

$$\pm B \leqq \alpha A, \text{ i.e. } \varepsilon = \alpha. \tag{3.23}$$

Finally we set

$$\rho = e_0^z - e_0(\alpha=1) + n\Delta|A|, \tag{3.24}$$

where $\Delta$ and $n$ will be chosen to be dependent on the model. Thus the hypotheses of Theorem 3.1 and Corollary 3.2 are satisfied.

*III.B. Estimates for the Antiferromagnet*

Next we consider the quantum mechanical antiferromagnet, model (3), in detail. We shall estimate the overlap coefficients

$$C_i = (\phi_i, P_\Lambda \phi_i),$$

for all eigenvectors $\phi_i$ of the Hamiltonian $H_\Lambda$ corresponding to eigenvalues $e_i$ with

$$e_i \leqq e_0^z + \Delta |\Lambda|. \tag{3.25}$$

From (3.19) and (3.20) we infer that

$$A + B = H_\Lambda - e_0(\alpha = 1).$$

Therefore the eigenvalue $\lambda$ of $A + B$ introduced in Theorem 3.1, Corollary 3.2 satisfies

$$\lambda \leqq \delta |\Lambda| + \Delta |\Lambda|, \tag{3.26}$$

where

$$\delta \equiv \frac{1}{|\Lambda|}(e_0^z - e_0(\alpha = 1)).$$

It is shown in [17] that

$$e_0(\alpha = 1) \geqq -\left(1 + \frac{1}{4S}\right) 2|\Lambda| \tag{3.27}$$

so that

$$\delta \leqq (2S)^{-1}. \tag{3.28}$$

Combining (3.23) and (3.24) with (3.26) and (3.28) we arrive at the following estimate for $\sigma$:

$$\sigma \equiv \varepsilon \frac{\rho}{\rho - \lambda} \leqq \varepsilon \frac{(\delta + n\Delta)|\Lambda|}{(n-1)\Delta |\Lambda|}$$

$$\leqq \alpha \left(1 + \frac{1 + 2S\Delta}{2S\Delta(n-1)}\right). \tag{3.29}$$

Let

$$\mathscr{E}^z(P_\Lambda) \equiv \inf \operatorname{spec}(P_\Lambda H^z P_\Lambda) - e_0(\alpha = 1)$$

be the minimal $A$-energy of any state in $N \equiv P_\Lambda \mathscr{H}$. Recalling definition (1.35) of the universal projection $P_\Lambda$ we see that

$$\mathscr{E}^z(P_\Lambda) \geqq e_0^z - e_0(\alpha = 1) + \tfrac{1}{2}|\Lambda|,$$

so that

$$\mathscr{E}^z(P_\Lambda) - \rho \geqq (\tfrac{1}{2} - n\Delta)|\Lambda|. \tag{3.30}$$

Since $P_\Lambda$ plays the role of the projection $P$ introduced in Corollary 3.2, we have the constraint

$$\tfrac{1}{2} - n\Delta > 0. \tag{3.31}$$

**Lemma 3.3.** (*Estimate on d for Antiferromagnet*). *Let* $A \equiv S^{-2}H^z - e_0(\alpha = 1)$
(1) *Let* $\psi$ *be a vector of A-energy at least e, i.e.* $(1 - P_{\rho=e})\psi = 0$. *Then the A-energy of* $H^{xy}\psi$ *is at least* $e - 8S^{-1}$, *i.e.* $(1 - P_{e-8S^{-1}})H^{xy}\psi = 0$.
(2) $d \geqq [\frac{1}{16}(1 - 2n\Delta)S|\Lambda|]$, *for* $2n\Delta < 1$, *where* $[a]$ *is the largest integer* $\leqq a$.

*Proof.* In our representation (1.40) of the antiferromagnet

$$\begin{aligned} H^{xy} &= -S^{-2} \sum_{\langle i,j\rangle \subset \Lambda} \{S_i^x S_j^x + (iS_i^y)(iS_j^y)\} \\ &= -\frac{1}{2S^2} \sum_{\langle i,j\rangle \subset \Lambda} \{S_i^+ S_j^+ + S_i^- S_j^-\}, \end{aligned} \tag{3.32}$$

where $S^+, S^-$ are the spin-raising, resp. spin-lowering operators. Using (3.32) we see that one application of $H^{xy}$ to a vector $\psi$ can raise (resp. lower) the $z$-components of the spins of *one* nearest neighbor pair $\langle i,j\rangle \subset \Lambda$ by 1. Clearly, this cannot change the minimal $A$-energy of $\psi$ by more than $S^{-2}\cdot 8S = 8S^{-1}$, as a minute of reflection shows. More precisely,

$$(1 - P_{e-8/S})H^{xy}P_e = 0. \tag{3.33}$$

This completes the proof of (1). The proof of (2) is an immediate consequence of the definition of $d$, of inequality (3.30) and of part (1). Q.E.D.

**Proposition 3.4.** $R_- \leqq \sigma^{2d}$, *where*

$$\sigma = \alpha(1 + (S\eta)^{-1} + 0(\beta^{-\xi})) \text{ and}$$

$$d \geqq \left[\frac{(1-\eta)S}{16}|\Lambda|\right]$$

*for arbitrary* $\xi < 1$ *and* $\eta < 1$.

*Proof.* We choose

$$\Delta = \beta^{-\xi} \quad \text{and} \quad n = \tfrac{1}{2}\eta\beta^{\xi}. \tag{3.34}$$

Then $\frac{1}{2} - n\Delta = \frac{1}{2}(1 - \eta) > 0$, for $\eta < 1$, so that the constraint (3.31) is fulfilled. By Equation (3.2), (3.3),

$$\begin{aligned} R_- = R_-(\beta, \Delta) &= \frac{\sum_{e_i < e_0 + \Delta|\Lambda|} (\phi_i, P_\Lambda \phi_i)e^{-\beta e_i}}{\sum e^{-\beta e_i}} \\ &\leqq \max_{e_i < e_0 + \Delta|\Lambda|} (\phi_i, P_\Lambda \phi_i). \end{aligned} \tag{3.35}$$

Suppose the maximum on the right side of (3.35) occurs for $i = i_0, e_{i_0} < e_0 + \Delta|\Lambda|$. We set $\phi_{i_0} \equiv \psi, P_\Lambda \equiv P$ and apply Corollary 3.2. This gives

$$|(\phi_{i_0}, P_\Lambda \phi_{i_0})| \leqq \sigma^{2d}, \text{ i.e. } R_- \leqq \sigma^{2d}.$$

By (3.34) and (3.29)

$$\begin{aligned}\sigma &\leqq \alpha\left(1+\frac{1+2S\Delta}{2S\Delta(n-1)}\right)\\ &=\alpha\left[1+\left(1-\frac{2}{\eta\beta^{\xi}}\right)^{-1}\left(\frac{1+2S\beta^{-\xi}}{\eta S}\right)\right]\\ &=\alpha(1+(\eta S)^{-1}+O(\beta^{-\xi})).\end{aligned}$$

Furthermore, by Lemma 3.3, (2) and (3.34)

$$\begin{aligned}d &\geqq [\tfrac{1}{16}(1-2n\Delta)S|\Lambda|]\\ &=\left[\frac{1-\eta}{16}S|\Lambda|\right].\end{aligned}$$

Q.E.D.

*Remark.* The dependence of $\sigma$ and $d$ on the total spin $S$ will permit us to show that the critical anisotropy $\alpha_c(S)$, below which a phase transition occurs, tends to 1 as $S \to \infty$.

Our estimate for $d$ is not very good and can be improved; we illustrate how to do so for spin 1/2. We claim $d \geqq |\Lambda|(1-n\Delta)/4$ instead of $|\Lambda|(1-2n\Delta)S/8 = |\Lambda|(1-2n\Delta)/16$; $P_\Lambda$ (Fig. 1) now means a projection onto a definite pattern of up or down spins; $\mathscr{E}^z(P_\Lambda) = -|\Lambda| + \delta$, and we wish to lower it to an $A$-energy of $-2|\Lambda| + n\Delta|\Lambda| + \delta$. Let $e_h$ be the $H^z$-energy of the horizontal bonds. Initially, $e_h = 0$; finally $e_h \leqq -|\Lambda| + n\Delta|\Lambda|$, since the vertical energy $\geqq -|\Lambda|$. Also, $e_h = -|\Lambda| + 2b$ where $b$ is the number of bad (i.e. $+-$ or $-+$) horizontal bonds. At least $k = |\Lambda|(1-n\Delta)/2$ bad horizontal bonds must be removed; $d \geqq d' =$ numbers of steps to do this, while $d' \geqq d''/2$, where $d''$ is the number of *single* spin flips required to do the same thing. Since the initial horizontal pattern in in each row is *bgbgb*... ($g$ = good bond), it is easy to see that $d'' = k$. These arguments give the following improved estimates for $S = 1/2$:

$$\sigma = \alpha\left(1+\frac{2}{\eta}+0(\beta^{-\xi})\right), \tag{3.36}$$

$$d \geqq [\tfrac{1}{4}(1-\eta/2)|\Lambda|]. \tag{3.37}$$

### *III.C. Estimates for the Ferromagnet*

It is well known that in the quantum mechanical ferromagnet (model (4))

$$e_0(\alpha) = e_0^z, \text{ for all } |\alpha| \leqq 1; \tag{3.38}$$

in fact, the groundstates for $|\alpha| < 1$ are identical with the two groundstates of $H^z$. Therefore

$$\rho = n\Delta|\Lambda|, \text{ by (3.24), and}$$

$$A = S^{-2}H^z - e_0^z, B = \alpha S^{-2}H^{xy}.$$

We estimate the overlap coefficients $(\phi_i, P_\Lambda\phi_i)$ for all eigenvectors of the Hamiltonian $H_\Lambda$ corresponding to eigenvalues $e_i$ with $e_i \leqq e_0^z + \Delta|\Lambda|$. Thus the eigenvalue $\lambda$ of $A+B$ introduced in Theorem 3.1 and Corollary 3.2 must satisfy $\lambda \leqq \Delta|\Lambda|$. Therefore

$$\sigma \equiv \varepsilon \frac{\rho}{\rho-\lambda} = \alpha \frac{n}{n-1}. \tag{3.39}$$

As in the antiferromagnet one shows that

$$R_- \leqq \sigma^{2d}; \tag{3.40}$$

see the proof of Proposition 3.4. We are left with estimating the "distance" $d$ on the right side of (3.40).

*Estimate on d.* Let $l$ be an integer such that $|\Lambda|^{1/2}/l$ is an integer. We decompose $\Lambda$ into $|\Lambda|/l^2$ disjoint, congruent squares, $b(=\text{boxes})$, with sides of length $l$. Let $\phi$ be an eigenvector of $\{S_i^z : i \in \Lambda\}$. Clearly $\phi$ is also an eigenvector of $A$. For $\phi$, a *perfect square* is defined to be a square $b=b_\phi$ such that $S_i^z\phi = \sigma_i\phi$ and one of the following two properties holds:

(i) $\sigma_i \geqq (0.9)S$ for all $i \in b_\phi$

(ii) $\sigma_i \leqq -(0.9)S$ for all $i \in b_\phi$.

Suppose now that the $A$-energy of $\phi$ is $\leqq n\Delta|\Lambda|$. We propose to estimate the minimal number, $k$, of perfect squares $b_\phi$ for this $\phi$. For this purpose we assign an $A$-energy to every $b$ square in $\Lambda$ in such a way that the sum of the energies assigned to all squares in $\Lambda$ is $\leqq$ the $A$-energy of $\phi$. The $A$-energy of a perfect square is zero. Therefore, to a square which is *not* perfect, an $A$-energy of at least 2(0.1) must be assigned. There are $(|\Lambda|l^{-2})-k$ squares which are not perfect. Since the $A$ energy of $\phi$ is $\leqq n\Delta|\Lambda|$, we obtain the inequality

$$(|\Lambda|l^{-2}-k)(0.2) \leqq n\Delta|\Lambda|,$$

i.e.

$$k \geqq |\Lambda|(l^{-2}-5n\Delta). \tag{3.41}$$

Since $l \geqq 2$, we require that $n\Delta < 1/40$. Let $\psi$ be an arbitrary vector in the range of $P_\Lambda$, i.e. $P_\Lambda\psi = \psi$. Define $d$ (see Theorem 3.1) by the condition

$$(1-P_{n\Delta|\Lambda|})[B(A-\lambda)^{-1}]^d\psi \neq 0,$$

but (3.42)

$$(1-P_{n\Delta|\Lambda|})[B(A-\lambda)^{-1}]^j\psi = 0,$$

for all $j<d$.

We expand $[B(A-\lambda)^{-1}]^d\psi$ in terms of eigenvectors $\phi_{j,\psi}^z$ of $\{S_i^z : i \in \Lambda\}$. Let $\phi = \phi_{j_0,\psi}^z$ be a vector of $A$-energy $\leqq n\Delta$. By (3.42) such a $\phi \neq 0$ exists. By (3.41) $\phi$ has $k \geqq |\Lambda|(l^{-2}-5n\Delta)$ perfect squares. In order to obtain a perfect square by repeated application of $B(A-\lambda)^{-1}$ to $\psi, B(A-\lambda)^{-1}$ has to be applied to $\psi$

at least $m$ times, where

$$m \geqq (0.9) \frac{l^2}{2} \cdot S \cdot \frac{l}{4}. \tag{3.43}$$

This is so, because $\psi$ is an eigenvector of $P_A$, so that the $z$-components of $\frac{l^2}{2}$ spins in a box $b$ have to be raised from $S_z \leqq 0$ to $S^z = (0.9)S$, resp. lowered from $S^z \geqq 0$ to $S^z = -(0.9)S$, in order to convert $b$ into a perfect box. (Recall that $P_A$ is pictorially given by Figure 1, Section I.D).

For the quantum mechanical ferromagnet

$$\begin{aligned} B &= -\sum_{\langle i,j\rangle \subset A} \{S_i^x S_j^x + S_i^y S_j^y\} \\ &= -\tfrac{1}{2} \sum_{\langle i,j\rangle \subset A} \{S_i^+ S_j^- + S_i^- S_j^+\}. \end{aligned} \tag{3.44}$$

Equation (3.44) shows that when the $z$-component of a spin at some site is raised (lowered) the $z$-component of a spin at a nearest neighbor site is lowered (raised). Thus, in order to raise the $z$-component of a spin at some site $i \in b$ from $S^z \leqq 0$ to $S^z = (0.9)S$ without lowering the $z$-components of other spins in $b$, $B$ has to be applied

$$(0.9)\,\mathrm{dist}(i, \text{boundary of } b) \cdot S$$

times; hence, on the average, $(0.9)S \cdot \frac{l}{4}$ times. This completes the proof of (3.43).

If we combine (3.41) with (3.43) we obtain

$$d \geqq m \cdot k \geqq |A|(l^{-2} - 5n\Delta) \frac{l^3 S}{8} (0.9).$$

Choosing $l = [(10n\Delta)^{-1/2}] (\geqq 2)$ yields

**Proposition 3.5.** *Provided* $n\Delta < 1/40$

$$R_- \leqq \sigma^{2d},$$

*where*

$$\sigma = \alpha \frac{n}{n-1}, \text{ and}$$

$$d \geqq (0.9)|A| S [16 (10 n\Delta)^{1/2}]^{-1}.$$

*Remark.* The estimate on $d$ obtained in Proposition 3.5 for the ferromagnet is vastly superior to the estimates on $d$ obtained for the antiferromagnet (Proposition 3.4 and (3.37)). This will become apparent in the next section where we will allow $n\Delta$ to go to zero as $\beta \to \infty$. Then $d \to \infty$ for the ferromagnet, but *not* for the antiferromagnet. Finally we note that the general methods developed in this section can be applied in other contexts than the one considered here in order to get bounds on expectations of global observables in equilibrium states.

## IV. Estimates on $R_+$ and Completion of the Proof

### *IV.A. Summary of Previous Results*

Recall that our proof of LRO at low temperatures is completed by showing that

$$R_\Lambda(\beta) = \langle P_\Lambda \rangle$$

is "small" for large $\beta$, namely we require that

$$\sum_{n=2}^{\infty} 2n 3^{2n-2} R_\Lambda^{n/|\Lambda|} < \tfrac{1}{4}; \tag{4.1}$$

see Section I.C, Theorem 1.1 and Section I.D, inequalities (1.34), (1.36) and (1.38). In Section III we decomposed $R_\Lambda(\beta)$ into two parts,

$$R_\Lambda(\beta) = R_-(\beta, \Delta) + R_+(\beta, \Delta), \tag{4.2}$$

and we have established upper bounds on $R_-(\beta, \Delta)$, namely:

(a) In models (1) and (6) (classical case), i.e. the classical $N$-vector models:

$$R_-(\beta, \Delta) = 0, \text{ for } \Delta \leqq \tfrac{1}{2}(1-\alpha); \tag{4.3}$$

see (3.7).

(b) In model (2), the classical, anharmonic crystal:

$$R_-(\beta, \Delta) = 0, \text{ for } \Delta \leqq \alpha/2; \tag{4.4}$$

see (3.8).

(c) In model (3), the quantum antiferromagnet:

$$R_-(\beta, \Delta) \leqq \sigma^{2d}, \text{ for } \Delta = \beta^{-\xi}, \tag{4.5}$$

where

$$\sigma = \alpha(1 + (S\eta)^{-1} + 0(\beta^{-\xi})),$$

and

$$d \geqq \left[\frac{(1-\eta)S}{16} |\Lambda|\right],$$

resp.

$$d \geqq [\tfrac{1}{4}(1-\eta/2)|\Lambda|], \text{ for } S = 1/2,$$

with

$$0 < \xi < 1, 0 < \eta < 1$$

(to be chosen later). See Propositions 3.4 and (3.37). The estimates for model (6) (quantum $xy$ model) are identical.

(d) In model (4) (quantum ferromagnet)

$$R_-(\beta, \Delta) \leqq \sigma^{2d}, \tag{4.6}$$

where

$$\sigma = \alpha \frac{n}{n-1}, \tag{4.7}$$

and

$$d \geqq (0.9)|\Lambda| S[16(10n\Delta)^{1/2}]^{-1}, \tag{4.8}$$

with $n > 1$ and $\Delta > 0$ to be chosen later. We require $n\Delta < 1/40$.

### *IV.B. The $R_+$ Estimate*

We now estimate $R_+(\beta, \Delta)$ for these models.

(*a'*) *Models* (*1*) *and* (*6 Classical*). Let $P_\Lambda^{>\delta}$ be the subset of configurations such that $m_i \equiv S^{-1}S_i^z \geqq (1-\delta)^{1/2}$, for all $i \in \Lambda$. Then (with Tr defined by the usual normalized integral, i.e. $\mathrm{Tr}(1) = 1$)

$$Z_\Lambda(\beta) \geqq \mathrm{Tr}\{P_\Lambda^{>\delta} \exp[-\beta H_\Lambda]\}$$
$$\geqq \{\mathrm{Tr}\, P_\Lambda^{>\delta}\} \exp\{-\beta\, \mathrm{Tr}[P_\Lambda^{>\delta} H_\Lambda]/\mathrm{Tr}\, P_\Lambda^{>\delta}\}$$

by Jensen's inequality. By symmetry, the term proportional to $\alpha$ vanishes in $\mathrm{Tr}\, P_\Lambda^{>\delta} H_\Lambda$. Furthermore, $H_\Lambda \leqq -2|\Lambda|(1-\delta)$ whenever $P_\Lambda^{>\delta} \neq 0$. Moreover,

$$\mathrm{Tr}(P_\Lambda^{>\delta}) = \{\tfrac{1}{2}[1-(1-\delta)^{1/2}]\}^{|\Lambda|}. \tag{4.9}$$

Hence, choosing $\delta = \Delta/4 = (1-\alpha)/8$, we obtain

$$R_+(\beta, \Delta) \leqq \exp[\beta|\Lambda|(2-\Delta)] Z_\Lambda^{-1} \leqq e^{-(\beta/4)(1-\alpha)|\Lambda|} e^{c(\alpha)|\Lambda|}, \tag{4.10}$$

where $c(\alpha) \propto \alpha - \ln(1-\alpha)$ for $\alpha \approx 1$, is *independent* of $\beta$. Thus

$$\langle P_\Lambda \rangle^{1/|\Lambda|} = R_\Lambda(\beta)^{1/|\Lambda|} = R_+(\beta, \Delta)^{1/|\Lambda|} \leqq e^{-(\beta/4)(1-\alpha)+c(\alpha)}, \tag{4.11}$$

which tends to 0, as $\beta \to \infty$. *This completes the proof of* LRO *for models* (1) and (6, classical) *for large enough* $\beta$. An estimate on the spontaneous magnetization $\langle m_0 \rangle = S^{-1}\langle S_0^z \rangle$, resp. $\sigma(\beta)$ (see Section I.B) as a function of $\beta$ is given later.

(*b'*) *Model* (2). By definition of model (2) (anharmonic crystal, Section I.A),

$$\min \phi(x, y) = \varepsilon_0,$$

occurs when $x$ and $y$ have the same direction. Without loss of generality we may assume that there exist some $x_0 \neq 0$ such that

$$\phi(x_0, x_0) = \varepsilon_0. \tag{4.12}$$

But when $x^1$ and $y^1$ (the 1-components of $x$, resp. $y$) have opposite sign

$$\phi(x, y) \geqq \varepsilon_0 + \alpha + \lambda(\phi_1(x) + \phi_1(y)), \tag{4.13}$$

for some $\lambda > 0$; see Section I.A. We now choose $\delta > 0$ such that $x_0^1 - \delta \geqq 0$ and

$$\phi(x, y) \leqq \varepsilon_0 + \alpha/8,$$

for all $x$ and $y$ in a ball of radius $\delta$ centered at $x_0$. We can do so, since the interaction potential $\phi$ is by assumption *continuous*
Hence

$$Z_\Lambda(\beta) \geqq e^{-2\beta(\varepsilon_0+\alpha/8)|\Lambda|}(v_N(\delta))^{|\Lambda|}, \tag{4.14}$$

where $v_N(\delta)$ is the volume of a ball of radius $\delta$ in $\mathbb{R}^N$. Furthermore, for $\Delta = \alpha/2$ i.e. $R_-^{AC}(\beta, \Delta) = 0$ (see (4.4)),

$$R_+^{AC}(\beta) = R_+^{AC}(\beta, \Delta = \alpha/2)$$
$$\leqq \exp[-2\beta(\varepsilon_0 + \alpha/4)|\Lambda|]g(\beta)^{|\Lambda|}Z_\Lambda(\beta)^{-1}, \tag{4.15}$$

where

$$g(\beta) \equiv \int e^{-4\beta\lambda\phi_1(x)}d^N x.$$

This is an immediate consequence of (4.13); (see also inequality (1.49) of Section I.E). Combination of (4.14) and (4.15) yields

$$R_+^{AC}(\beta, \alpha/2) \leqq \exp[-(\beta\alpha/4)|\Lambda|](g(\beta)/v_N(\delta))^{|\Lambda|}.$$

By definition of the AC model (model (2), Section I.A.) there exists some finite $\beta_0$ such that for all $\beta \geqq \beta_0$

$$g(\beta) = \int e^{-4\beta\lambda\phi_1(x)}d^N x < \infty.$$

Obviously $g(\beta)$ is monotone decreasing in $\beta$, as $\phi_1$ is positive. Thus there exists a finite constant $c$ such that

$$g(\beta)/v_N(\delta) \leqq e^c, \text{ for all } \beta \geqq \beta_0.$$

Hence

$$R^{AC}(\beta) = R_+^{AC}(\beta, \alpha/2) \leqq \exp[-\beta\alpha/4 + c]|\Lambda|, \tag{4.16}$$

which tends to 0 as $\beta \to \infty$. Recalling condition (4.1) (resp. Theorem 1.1 of Section I.C and Section I.D, inequality (1.38)) we observe that *inequality* (4.16) *completes the proof of* LRO *for the* AC *model for large enough* $\beta$.

(c′) *Models* (3) *and* (4) (*Quantum Heisenberg Models*). In order to estimate $R_+(\beta, \Delta)$ we need a lower bound on the partition function $Z_\Lambda(\beta)$. This is done by comparing it with the partition function of the corresponding spin $S$ Ising model (anisotropy $\alpha = 0$) by means of the *Peierls-Bogolyubov inequality.*

**Lemma 4.1.** *For models* (3) *and* (4) *the partition function satisfies*

$$Z_\Lambda \geqq Z_\Lambda^I. \tag{4.17}$$

*Where* $Z_\Lambda^I$ *is the partition function of the spin S Ising model* (*i.e.* $\alpha = 0$ *in* (*1.3*) *and* (*1.5*)).

*Proof.* By the Peierls-Bogolyubov inequality,

$$Z_\Lambda \geqq \sum_j \exp[-\beta(\psi_j, H\psi_j)] \tag{4.18}$$

Phase Transitions in Anisotropic Lattice Spin Systems 265

for any set $\{\psi_j\}$ of orthonormal vectors. Choose the $\psi_j$ to be eigenvectors of all the $S_i^z, i\in\Lambda$. Then the right side of (4.18) is precisely $Z_\Lambda^I$ because $(\psi_j, H^{xy}\psi_j)=0$ for all $j$.

**Lemma 4.2.** *For models* (3) *and* (4)

$$Z_\Lambda \geqq [(\delta/8)(2S+1)]^{|\Lambda|}\exp\{2\beta|\Lambda|(1-\delta)\}$$

*for any* $0\leqq\delta\leqq 1$.

*Proof.* Using Lemma 4.1,

$$Z_\Lambda \geqq Z_\Lambda^I \geqq \sum{}' \exp(-\beta S^{-2}H^z)$$

where $\sum'$ means a restricted summation in which each $S_i^z \geqq S(1-\delta)^{1/2}$. (Note: the partition function for the Ising ferro and antiferromagnet are identical.) Then $H^z \leqq -2|\Lambda|S^2(1-\delta)$. To complete the proof we have to bound $\sum' 1 \equiv \mu^{|\Lambda|}$.

$$\mu = [S - S(1-\delta)^{1/2}+1]_+ \geqq [1+S\delta/2]_+ \geqq S\delta/2 \geqq (2S+1)\delta/8$$

for $S\geqq 1/2$, and where $[\ ]_+$ means integral part. To complete the bound on $R_+$ we use the fact that $\mathrm{Tr}\,1=(2S+1)^{|\Lambda|}$. For the ferromagnet, $e_0=-2|\Lambda|$. Thus, provided

$$\Delta > 4\delta \text{ (ferromagnet)}, \tag{4.19}$$

(1.50) and (1.51) are established for $\beta$ sufficiently large. For the antiferromagnet, $e_0 > -2|\Lambda|(1+1/4S)$. Thus, provided

$$\Delta > 4\delta + S^{-1} \text{ (antiferromagnet)}, \tag{4.20}$$

(1.50) and (1.51) are established for $\beta$ sufficiently large.

The final estimate for the ferromagnet is obtained from (4.6) and (1.51). Choose $n=(1+\alpha)/(1-\alpha)<2/(1-\alpha)$. Thus $\sigma<(1+\alpha)/2<1$. Choose $\Delta = K\beta^{-2/3}$ where $K$ is chosen such that $\sigma^{(1,8)S(10nK)^{-1/2}/16} < e^{-K/2}$. This can be done uniformly in $S>1/2$. For sufficiently large $\beta, n\Delta<1/40$. Furthermore, with this choice of $\Delta, R_+ > R_-$. Hence

$$\lim_{\Lambda\to\infty} \langle P_\Lambda\rangle^{1/|\Lambda|} \leqq \exp(-K\beta^{1/3}/2) \tag{4.21}$$

which tends to zero as $\beta$ tends to infinity. Note that there is a $\beta^{1/3}$, instead of a $\beta$ dependence in (4.21). This completes the proof of LRO for the quantum ferromagnet, except for the assumption that the chessboard estimate holds.

The calculation for the antiferromagnet is more complicated. We have to combine (4.5) with (1.51). As $\beta\to\infty, R_+(\beta,\Delta)^{1/|\Lambda|}\to 0$, provided $\xi<1$, by (1.51) and (4.5). The problem resides in $R_-(\beta,\Delta)$. This will *not* go to zero as $\beta\to\infty$, but for small enough $\alpha$ (depending on $S$), which we call $\alpha_c(S)$, we can make $R_-(\beta, \Delta)^{1/|\Lambda|}$ smaller than any given number, say $\mu$. Choose $\mu$ such that (4.1) is satisfied. We omit details, but note that $\alpha_c(S)$ tends to 1 as $S$ tends to infinity.

### *IV.C. Estimate of the Spontaneous Magnetization*

Consider the order parameter which satisfies the previously derived inequality

$$\sigma(\beta) > \delta^2 - 6\varepsilon \tag{1.24}$$

provided $\langle P_0^+ P_j^- \rangle < \varepsilon/2$ and $\langle P_0^{<\delta} \rangle < \varepsilon$. In the classical models (1) and (6) these inequalities hold for all $\varepsilon > 0$ and $\delta < 1$ if $\beta$ is large enough. This follows from chessboard estimates applied to $\langle P_0^{<\delta} \rangle$, and the results of Section IV. Thus $\sigma(\beta) \to 1$ as $\beta \to \infty$.

For the quantum antiferromagnets (3), (6), an estimate on $\langle P_0^{<\delta} \rangle$ can be obtained using chessboard estimates and exponential localization, as before, with the following result: Given $\varepsilon > 0$, $\delta < 1$ and $\alpha < 1$ there exists an $S(\varepsilon, \delta, \alpha) < \infty$ such that (1.24) holds as $\beta \to \infty$ for $S > S(\varepsilon, \delta, \alpha)$. For the ferromagnet (4), chessboard estimates, if they could be shown to be true, would easily yield $\sigma(\beta) \to 1$ as $\beta \to \infty$ for all $S$ and all $\alpha < 1$. Without using chessboard estimates, we can show that $\langle P_0^{<\delta} \rangle \to 0$ as $\beta \to \infty$ for all $\delta < 1$ and all $\alpha < 1$. This is proved by means of the following thermodynamic argument: It is sufficient to show

$$\langle (S_0^z)^2 \rangle \to 1 \text{ as } \beta \to \infty .$$

By the Schwarz inequality and translation invariance

$$S^{-2} \langle (S_0^z)^2 \rangle \geqq \tfrac{1}{4} S^{-2} \sum_{|i-0|=1} \langle S_0^z S_i^z \rangle .$$

This latter quantity is half the $H^z$-energy per site. The ground state has the property that $S^{-2} \langle S_i^z S_j^z \rangle = 1$ for all $i, j$. If $S^{-2} \langle S_0^z S_i^z \rangle \not\to 1$ as $\beta \to \infty$, for $|i-0| = 1$, then the free energy would not approach the ground state energy as $\beta \to \infty$. This, it is easy to see by the previous arguments, would be a contradiction.

## References


1. Mermin, N., Wagner, H.: Phys. Rev. Letters **17**, 113 (1966)
2. Mermin, N.: J. Math. Phys. **8**, 1061 (1967)
3. Hohenberg, P.: Phys. Rev. **158**, 383 (1967)
4. Malyshev, S.: Commun. math. Phys. **40**, 75 (1967)
5. Kunz, H., Pfister, C., Vuillermot, P.: J. Phys. A **9**, 1673 (1976); Phys. Lett. **54**A, 428 (1975)
6. Ginibre, J.: Commun. math. Phys. **14**, 205 (1969)
7. Robinson, D.: Commun. math. Phys. **14**, 195 (1969)
8. Fröhlich, J., Lieb, E.: Phys. Rev. Letters **38**, 440 (1977)
9. Dyson, F., Lieb, E., Simon, B.: Phys. Rev. Letters **37**, 120 (1976). The details are in: Phase transitions in quantum spin systems with isotropic and nonisotropic interactions. J. Stat. Phys. (in press)
10. Fröhlich, J.: Acta Phys. Austriaca, Suppl. **15**, 133 (1976)
11. Lieb, E.: New proofs of Long Range Order, Proceedings of the International Conference on the Mathematical Problems in Theoretical Physics, Rome 1977. Lecture notes in physics. Berlin-Heidelberg-New York: Springer (in press)
12. Fröhlich, J., Israel, R., Lieb, E., Simon, B.: (papers in preparation)
13. Heilmann, O., Lieb, E.: Lattice models of liquid crystals (in preparation)
14. Lieb, E.: Commun. math. Phys. **31**, 327 (1973)
15. Brascamp, H., Lieb, E.: Some inequalities for Gaussian measures and the long-range order of the one-dimensional plasma. In: Functional integration and its applications (ed. A. M. Arthurs), pp. 1–14. Oxford: Clarendon Press 1975

16. Glimm, J., Jaffe, A., Spencer, T.: Commun. math. Phys. **45**, 203 (1975)
17. Anderson, P. W. : Phys. Rev. **83**, 1260 (1951). See also the second paper cited in Ref. [9]
18. Osterwalder, K., Schrader, R.: Commun. math. Phys. **31**, 83 (1973); **42**, 281 (1975)
19. Fröhlich, J.: Helv. Phys. Acta **47**, 265 (1974)
Seiler, E., Simon, B.: Ann. Phys. (N.Y.) **97**, 470 (1976)
Fröhlich, J., Simon, B.: Ann. Math **105**, 493 (1977)
20. Fröhlich, J., Spencer, T.: In: New developments in quantum field theory and statistical mechanics (eds. M. Lévy, P. Mitter). New York: Plenum Publ. Corp. 1977; (Theorem 5.2, Lemma 5.3). The basic idea is due to:
Fröhlich, J., Simon, B., Spencer, T.: Commun. math. Phys. **50**, 79 (1976)
21. McBryan, O., Spencer, T.: Commun. math. Phys. **53**, 299 (1977)
22. Hepp, K., Lieb, E.H.: Ann. Phys. (N.Y.) **76**, 360 (1973), (Theorem 3.15)
23. Peierls, R.: Proc. Cambridge Phil. Soc. **32**, 477 (1936)
Griffiths, R.: Phys. Rev. **136A**, 437 (1964)
Dobrushin, R.: Dokl. Akad. Nauk SSR **160**, 1046 (1965)

Communicated by J. Glimm

Received December 29, 1977

**Note Added in Proof**

We are informed that an inequality similar to Theorem 2.2 was also proved independently by R. F. Streater and E. B. Davies (unpublished). Their proof is similar to ours.

*Journal of Statistical Physics, Vol. 18, No. 4, 1978*

# Phase Transitions in Quantum Spin Systems with Isotropic and Nonisotropic Interactions

**Freeman J. Dyson,[1] Elliott H. Lieb,[2] and Barry Simon[2,3]**

*Received November 1, 1976*

We prove the existence of spontaneous magnetization at sufficiently low temperature, and hence of a phase transition, in a variety of quantum spin systems in three or more dimensions. The isotropic spin 1/2 $x$–$y$ model and the Heisenberg antiferromagnet with spin 1, 3/2,...and with nearest neighbor interactions on a simple cubic lattice are included.

**KEY WORDS:** Phase transitions; Heisenberg ferromagnet.

## 1. INTRODUCTION

A basic subtlety in the study of statistical mechanics is the following: In nature, we observe abrupt changes in certain basic physical quantities, such as the magnetization of a magnet, but the statistical mechanics of systems with finitely many degrees of freedom is typically real analytic in all external variables. The resolution of this apparent paradox is that the abrupt changes are only approximately abrupt: True discontinuities only occur in the limit of an infinite system. For this reason, one must expect the problem of rigorously proving the existence of phase transitions to be a difficult one even for systems for which there is considerable numerical evidence or even a heuristic explanation for such a transition. In fact, until recently, the only general method available for directly proving the existence of phase transitions was Peierls' method, developed by Dobrushin[(8)] and Griffiths[(15)] on the basis of original ideas of Peierls.[(35)] The method was developed originally

Research supported by U.S. National Science Foundation under grants GP-40768X (F.J.D.), MCS 75-21684 (E.H.L.), and GP-39048 (B.S.).

[1] Institute for Advanced Study, Princeton, New Jersey.

[2] Departments of Mathematics and Physics, Princeton University, Princeton, New Jersey.

[3] Alfred Sloan Fellow.



0022-4715/78/0400-0335$05.00/0 

for classical spin systems, but it has been extended by Ginibre[13] and Robinson[39] to treat highly anisotropic (Heisenberg) quantum models and by Glimm *et al.*[14] to treat certain quantum field theories. The quantum spin systems were handled as perturbations of the classical (Ising) model, so that this method seems to be restricted to the highly anisotropic regime; the quantum field theories are treated by going to the Euclidean region, where they become essentially *classical* models.[19],4

One way of describing the limitations of the Peierls argument is in terms of the broken symmetries that often accompany phase transitions—the simplest physical example is the occurrence of a spontaneous magnetization in some direction in the absence of an external field. In all cases of the Peierls argument accompanied by a broken symmetry, this symmetry has been a discrete (finite) symmetry group. Until recently, one has been unable to prove rigorously the existence of phase transitions in systems with a continuous symmetry group, such as the classical Heisenberg model ("classical" spins with values on the unit sphere in $\mathbb{R}^3$) or the quantum Heisenberg model. This situation has been changed by recent work of Fröhlich, Simon, and Spencer[12] (henceforth FSS), who prove the existence of phase transitions in a variety of classical spin systems, including certain classical Heisenberg models. It is our goal in this paper to provide the first proof of phase transitions in any kind of quantum spin system with continuous symmetry—in particular, we will prove that such transitions occur in the spin-1 nearest neighbor, quantum Heisenberg antiferromagnet on a simple cubic lattice in three or more space dimensions. It is well known that a phase transition accompanied by a spontaneous magnetization cannot occur for this model in one or two dimensions.[21,28]

Since we use some of the ideas of FSS as an important element of our proof, it is useful to recall them. While FSS deal directly with infinite volume expectations, it is useful, for our purposes, to rephrase their results in terms of finite volume statements. Let $\Lambda$ be a parallelepiped in the simple $\nu$-dimensional cubic lattice $Z^\nu$ of the form

$$\Lambda = \{\boldsymbol{\alpha} | 0 \leqslant \alpha_1 \leqslant L_1 - 1, ..., 0 \leqslant \alpha_\nu \leqslant L_\nu - 1\}$$

We shall refer to $\Lambda$ as the standard $L_1 \times \cdots \times L_\nu$ box. In the classical model, one has a "spin" $\mathbf{S}_\alpha$ for each $\boldsymbol{\alpha} \in \Lambda$, where $\mathbf{S}$ has three components $S^{(j)}$, $j = 1, 2, 3$. These classical spins will be normalized by

$$\mathbf{S}_\alpha \cdot \mathbf{S}_\alpha \equiv \sum_j (S_\alpha^{(j)})^2 = 1 \tag{1}$$

and distributed according to the isotropic spherical distribution denoted by

4 The Peierls argument has recently been extended to the entire anisotropic regime in the classical case[48] and to a large regime for the quantum anisotropic ferromagnet.[49]

$d\lambda(\mathbf{S})$. The basic Hamiltonian is

$$H = -\sum_{\alpha,i} \mathbf{S}_\alpha \cdot \mathbf{S}_{\alpha+\delta_i} \tag{2}$$

where $i$ is summed from 1 to $\nu$, and $\boldsymbol{\alpha}$ is summed over $\Lambda$. Here $\boldsymbol{\delta}_i$ is the unit vector whose $i$th component is 1 and we use the convention that if $\alpha_i = L_i - 1$, then $(\boldsymbol{\alpha} + \boldsymbol{\delta}_i)_i = 0$, i.e., $H$ has periodic boundary conditions. It is important that changing the boundary conditions does not affect the existence or nonexistence of a phase transition, so that the imposition of periodic boundary conditions is purely a matter of mathematical convenience. $H$ has each pair of nearest neighbors (in $\Lambda$, viewed as a torus) interacting once with coupling 1. Since we will have the inverse temperature $\beta$ as a free variable, we do not add an additional factor $J$ in front of (2). We will occasionally use $\mathbf{S}_\alpha \cdot \mathbf{S}_\alpha = 1$ to rewrite $H$ as

$$H = \text{const} + \tfrac{1}{2}\sum_{\alpha,i} (S_\alpha - S_{\alpha+\delta_i})^2 \tag{2$'$}$$

The partition function $Z$ is defined by

$$Z = \int \exp[-\beta H(\mathbf{S})] \prod_{\alpha\in\Lambda} d\lambda\,(\mathbf{S}_\alpha) \tag{3a}$$

and thermal expectations by

$$\langle f(\mathbf{S})\rangle_{\Lambda,\beta} = Z^{-1} \int f(\mathbf{S}) \exp[-\beta H(\mathbf{S})] \prod_{\alpha\in\Lambda} d\lambda\,(\mathbf{S}_\alpha) \tag{3b}$$

Our translation of FSS is that their basic result is a proof that

$$\lim_{\Lambda\to\infty|} \left\langle \sum_{j=1}^{3} \left( |\Lambda|^{-1} \sum_{\alpha\in\Lambda} S_\alpha^{(j)} \right)^2 \right\rangle_{\Lambda,\beta} \neq 0 \tag{4}$$

for $\beta$ sufficiently large. Intuitively, (4) corresponds to macroscopic fluctuations in the bulk magnetization (since $\langle S_\alpha^{(j)}\rangle = 0$ by symmetry) and hence to the presence of a multiplicity of phases. We return to the question of relating (4) to other notions of phase transition at the conclusion of this section.

We introduce the Fourier variables $\hat{\mathbf{S}}_\mathbf{p}$ by

$$\hat{\mathbf{S}}_\mathbf{p} = \frac{1}{\sqrt{|\Lambda|}} \sum_{\alpha\in\Lambda} [\exp(-i\mathbf{p}\cdot\boldsymbol{\alpha})]\mathbf{S}_\alpha \tag{5}$$

where $\mathbf{p}$ runs through the dual lattice $\Lambda^*$, i.e., $p_j = 2\pi n_j/L_j$; $n_j = -\tfrac{1}{2}L_j + 1,\ldots, \tfrac{1}{2}L_j$ ($L_j$ even) or $-\tfrac{1}{2}(L_j - 1),\ldots, \tfrac{1}{2}(L_j - 1)$ ($L_j$ odd). In terms of these Fourier variables, $H$ has the form

$$H = \text{const} + \sum E_\mathbf{p} \hat{\mathbf{S}}_\mathbf{p} \cdot \hat{\mathbf{S}}_{-\mathbf{p}} \tag{2$''$}$$

where

$$E_\mathbf{p} = \tfrac{1}{2} \sum_{|\delta|=1} [1 - \exp(i\mathbf{p}\cdot\boldsymbol{\delta})] = \nu - \sum_{i=1}^{\nu} \cos p_i \geqslant 0 \tag{6}$$

and with the sum over $|\boldsymbol{\delta}|$ over the $2\nu$ vectors $\pm\boldsymbol{\delta}_i$. For a quantum system in which each spin has angular momentum $S$, $2SE_{\mathbf{p}}$ is the energy of a momentum-$\mathbf{p}$ spin wave.[4,9,10] For later purposes we note that

$$E_{\mathbf{p}} \sim |p|^2/2 \quad \text{for } |p| \text{ small} \tag{7}$$

In terms of the Fourier variables, (4) has the form

$$\lim_{\Lambda\to\infty} |\Lambda|^{-1} g_{\mathbf{p}=0} \neq 0 \tag{4'}$$

where

$$g_{\mathbf{p}} = \langle \hat{\mathbf{S}}_{\mathbf{p}} \cdot \hat{\mathbf{S}}_{-\mathbf{p}} \rangle \tag{8}$$

The FSS proof of (4′) comes from two bounds. The first is the Plancherel relation (which is completely trivial in this case since finite sums are involved), yielding a sum rule

$$\frac{1}{|\Lambda|} \sum_{\mathbf{p}\in\Lambda^*} g_{\mathbf{p}} = \langle (\mathbf{S}_{\boldsymbol{\alpha}})^2 \rangle = 1 \tag{9}$$

(any value of $\boldsymbol{\alpha}$ can be used). The second is the basic bound

$$g_{\mathbf{p}} \leqslant 3/(2\beta E_{\mathbf{p}}); \qquad \mathbf{p} \neq 0 \tag{10}$$

proven by FSS. The condition (10) has the physical interpretation that the average energy $E_{\mathbf{p}}\langle \hat{\mathbf{S}}_{\mathbf{p}}^{(j)} \cdot \hat{\mathbf{S}}_{-\mathbf{p}}^{(j)} \rangle$ per mode is dominated by its equipartition value of $\frac{1}{2}kT$ per degree of freedom, *counting each value of j as a separate degree of freedom*. The bound (10) implies that

$$\lim_{|\Lambda|\to\infty} |\Lambda|^{-1} \sum_{\mathbf{p}\neq 0} g_{\mathbf{p}} \leqslant (3/2\beta) G_\nu(0)$$

where

$$G_\nu(0) = (2\pi)^{-\nu} \int_{|p_i|\leq\pi} (E_{\mathbf{p}})^{-1}\, d^\nu p \tag{11}$$

and we have obtained the Fourier integral (11) as a limit of Fourier sums. $G_\nu(0)$ is finite when $\nu \geqslant 3$ by (7). The sum rule (9) implies that (4′) holds so long as

$$(3/2\beta) G_\nu(0) < 1$$

i.e.,

$$\beta > \tfrac{3}{2} G_\nu(0) \equiv \beta_c^{\mathrm{FSS}} \tag{12}$$

This method relies on the fact that the bound (10) and the sum rule (9) force a macroscopic occupation in the $\mathbf{p} = 0$ mode. This is a kind of spin-wave Bose condensation. The above discussion is for the ferromagnet, but the same obviously applies to the antiferromagnet with $\mathbf{p}$ replaced by $(\pi,\ldots,\pi) - \mathbf{p}$ in suitable places.

To explain the problems that have to be overcome in extending the FSS results to the quantum case, we must describe the model. Let $S$ be a fixed number chosen from 1/2, 1, 3/2,.... Each site $\boldsymbol{\alpha} \in \Lambda$ has associated with it a $(2S + 1)$-dimensional space $\mathscr{H}_{\boldsymbol{\alpha}} \cong \mathbb{C}^{2S+1}$ and three self-adjoint *operators* $\mathbf{S}_{\boldsymbol{\alpha}}$ $(S_{\boldsymbol{\alpha}}^{(j)}, j = 1, 2, 3)$ obeying the usual commutation relations (summation convention used on Latin indices):

$$[S_{\boldsymbol{\alpha}}^{(j)}, S_{\boldsymbol{\alpha}}^{(k)}] = i\epsilon_{jkl}S_{\boldsymbol{\alpha}}^{(l)} \tag{13}$$

However, (1) is replaced by

$$\mathbf{S}_{\boldsymbol{\alpha}}^{2} = \sum_{j} (S_{\boldsymbol{\alpha}}^{(j)})^2 = S(S + 1) \tag{1Q}$$

Since dim $\mathscr{H}_{\boldsymbol{\alpha}} = 2S + 1$, (13) and (1$Q$) essentially determine $\mathbf{S}_{\boldsymbol{\alpha}}$ uniquely. In volume $\Lambda$, the basic Hilbert space is $\mathscr{H}_{\Lambda} = \otimes_{\boldsymbol{\alpha}\in\Lambda} \mathscr{H}_{\boldsymbol{\alpha}} \cong \mathbb{C}^{(2S+1)|\Lambda|}$. We abuse notation by letting $\mathbf{S}_{\boldsymbol{\alpha}}$ stand for the triplet of operators on $\mathscr{H}_{\Lambda}$ that are the tensor product of 1 on each $\mathscr{H}_{\boldsymbol{\gamma}}$ for $\boldsymbol{\gamma} \neq \boldsymbol{\alpha}$ and $\mathbf{S}_{\boldsymbol{\alpha}}$ on $\mathscr{H}_{\boldsymbol{\alpha}}$. The basic Hamiltonian is still given by (2) [or (2′)] but now $Z$ and thermal expectations are given by

$$Z = \mathrm{Tr}_{\mathscr{H}_{\Lambda}}[\exp(-\beta H_{\Lambda})] \tag{3Qa}$$

$$\langle A\rangle_{\Lambda,\beta} = Z^{-1}\,\mathrm{Tr}[A \exp(-\beta H_{\Lambda})] \tag{3Qb}$$

We still define operators $\hat{\mathbf{S}}_{\mathbf{p}}$ by (5). Due to the commutation relations

$$[S_{\boldsymbol{\alpha}}^{(j)}, S_{\boldsymbol{\beta}}^{(k)}] = i\,\delta_{\boldsymbol{\alpha\beta}}\epsilon_{jkl}S_{\boldsymbol{\alpha}}^{(l)} \tag{13′}$$

one has that

$$[\hat{S}_{\mathbf{p}}^{(j)}, \hat{S}_{\mathbf{q}}^{(k)}] = |\Lambda|^{-1/2}(i\epsilon_{jkl}\hat{S}_{\mathbf{p}+\mathbf{q}}^{(l)}) \tag{13″}$$

In particular, $\hat{S}_{\mathbf{p}}^{(j)}$ commutes with its adjoint $\hat{S}_{\mathbf{p}}^{(j)*} = \hat{S}_{-\mathbf{p}}^{(j)}$ and, with the definition of $g_{\mathbf{p}}$ by (8), we have $g_{\mathbf{p}} = g_{-\mathbf{p}}$. The expression (2″) still holds [we caution the reader that the constants in (2′) and (2″) are different from those in the classical case]. The sum rule (9) is replaced by

$$\frac{1}{|\Lambda|}\sum_{\mathbf{p}\in\Lambda^*} g_{\mathbf{p}} = S(S + 1) \tag{9Q}$$

The difficult problem to overcome is that (10) cannot be true in the quantum case! For, if (10) holds, then as $\beta \to \infty$ with $\Lambda$ fixed and finite, $g_{\mathbf{p}}$ would approach zero and this, in turn, would imply that $\mathbf{S}_{\boldsymbol{\alpha}}\cdot\mathbf{S}_{\boldsymbol{\gamma}} \equiv f_{\boldsymbol{\alpha\gamma}}$ approaches a constant as $\beta \to \infty$. But as $\beta \to \infty$, $f_{\boldsymbol{\alpha\gamma}} \to S^2$ for $\boldsymbol{\alpha} \neq \boldsymbol{\gamma}$ and $S(S + 1)$ for $\boldsymbol{\alpha} = \boldsymbol{\gamma}$. Therefore, because of the values of $f_{\boldsymbol{\alpha\gamma}}$ in the ferromagnetic ground state, (10) is false in the quantum case. We believe that the following is true for the ferromagnet:

$$g_{\mathbf{p}} \leqslant \sqrt{\tfrac{3}{2}}S\coth(\sqrt{\tfrac{2}{3}}S\beta E_{\mathbf{p}}) \tag{10Q}$$

We will prove the analog of (10$Q$) for the antiferromagnet in a two-step process. First, we will prove (10) with $g_{\mathbf{p}}$ replaced by a "Duhamel two-point function," $b_{\mathbf{p}}$. Second, we will obtain bounds relating $b_{\mathbf{p}}$ and $g_{\mathbf{p}}$. The second step in the argument carries over to the ferromagnet. Our proof of the first part does not. We believe that the first bound will be true in that case, but are unable to prove it.

In Section 2, we discuss this Duhamel two-point function, which is not new (see the discussion in that section), and in Section 3 we prove a basic bound relating $b_{\mathbf{p}}$ to $g_{\mathbf{p}}$ (involving also the double commutator $\sum_{j=1}^{3} [[\hat{S}_{\mathbf{p}}^{(j)}, H], \hat{S}_{-\mathbf{p}}^{(j)}]$). Section 4 contains a proof of the analog of (10) with $g_{\mathbf{p}}$ replaced by $b_{\mathbf{p}}$. In Section 5 we put everything together to prove a phase transition in the sense of (4). In Sections 6 and 7 and the appendices we discuss the Heisenberg antiferromagnet and additional results.

We should emphasize here that certain aspects of our argument are very general. The bounds in Section 3 are "operator theoretic" in the sense that they depend on no special properties of the Hamiltonian. As we will explain, we believe that the bounds in Section 4 have an extension to any antiferromagnetic quantum lattice system, but this part of our proof only works for nearest neighbor interactions on a simple cubic lattice (or a rather small class of other lattices that does *not* include face- or body-centered cubic lattices). In any event, the bounds in Section 4 depend neither on algebraic properties of the spins nor on the norms of the spin operators. It is only in combining the bounds from Sections 3 and 4 with a sum rule of the type of ($9Q$) that these detailed properties of the spin enter. It is here that the $S$ dependence of our critical temperature bounds arises.

We also note that modulo a factor of 3/2, the bounds that would follow if ($10Q$) could be proved for the ferromagnet, have the interpretation of making rigorous certain elements of spin-wave theory.[4,9,10] We discuss this point further in Appendix B.

Finally, we turn to relating the criterion (4) for phase transitions to other criteria. This is a problem discussed already by Griffiths[16] (see Hepp and Lieb[20] for related results). Let us begin by giving an abstract version of Griffiths' main theorem and corollary (Ref. 16, §II.f) motivated by the form of the results given in the appendix to Ref. 20.

To motivate the following theorem, it is useful to think of $n$ as parametrizing the size of a magnetic system, $x$ as the magnetization, $y$ as a magnetic field, and $\alpha$ as the partition function.

**Theorem 1.1 (Griffiths[16]).** Let $\mu_n$ be a sequence of probability measures (i.e., $\int d\mu_n = 1$) on the real line such that, for $y$ in an interval $[-c, c]$ about 0,

$$\alpha(y) = \lim_{n\to\infty} \frac{1}{n} \ln \int e^{yx}\, d\mu_n(x)$$

exists (and is finite). We assume $c > 0$. Let

$$a_{\pm} = \pm \lim_{y\downarrow 0} \{y^{-1}[\alpha(\pm y) - \alpha(0)]\}$$

(which exists since $\alpha$ is convex). Let $\delta > 0$ be fixed. Then

$$\overline{\lim} \left[\int_{(a_+ + \delta)n}^{\infty} d\mu_n(x)\right]^{1/n} < 1; \qquad \overline{\lim} \left[\int_{-\infty}^{(a_- - \delta)n} d\mu_n(x)\right]^{1/n} < 1$$

*Proof.* We prove the first statement. Let

$$b_n = \int_{(a_+ + \delta)n}^{\infty} d\mu_n(x)$$

If the first statement is false, then $\overline{\lim}(1/n) \ln b_n \geqslant 0$. Clearly, for $y \geqslant 0$

$$\int e^{yx}\, d\mu_n(x) \geqslant b_n e^{n(a_+ + \delta)y}$$

so that, if $\overline{\lim}(1/n) \ln b_n \geqslant 0$,

$$\alpha(y) \geqslant (a_+ + \delta)y$$

Since $\alpha(0) = 0$,

$$\lim_{y\downarrow 0} y^{-1}[\alpha(y) - \alpha(0)] \geqslant a_+ + \delta$$

which is impossible. ■

This theorem says that

$$\int_{(a_- - \delta)n}^{(a_+ + \delta)n} d\mu_n(x)$$

goes to 1 with exponentially small error.

**Corollary 1.1.** Under the hypothesis of Theorem 1.1, if $a_+ = a_- = a$ [i.e., if $\alpha(y)$ is differentiable at $y = 0$], then for functions $f$ obeying $|f(x)| \leqslant Ae^{B|x|}$ for some $B < c$ (defined in Theorem 1.1)

$$\lim_{n\to\infty} \int f(x/n)\, d\mu_n(x) = f(a)$$

*Remark.* For $f(x) = x^k$, this is a result of Hepp and Lieb.[20]

*Proof.* For any $f$ that is bounded, this follows immediately from the theorem. Fix $\beta > 0$. Then

$$\begin{aligned}\int_{x\geq\beta n} |f(x/n)|\, d\mu_n(x) &\leqslant A \int_{x\geq\beta n} e^{Bx/n}\, d\mu_n(x) \\ &\leqslant Ae^{(B-c)\beta} \int e^{cx/n}\, d\mu_n(x) \\ &\leqslant Ae^{(B-c)\beta} \left[\int e^{cx}\, d\mu_n(x)\right]^{1/n}\end{aligned}$$

by Hölder's inequality. Thus

$$\overline{\lim} \int_{|x| \geq \beta n} |f(x/n)| \, d\mu_n(x) \leqslant Ae^{(B-c)\beta}[\alpha(c) + \alpha(-c)]$$

so that, given $\epsilon$, we can find $\beta_0$ such that

$$\overline{\lim_{n \to \infty}} \int_{|x| \geq \beta_0 n} |f(x/n)| \, d\mu_n(x) \leqslant \epsilon$$

Since $f(\cdot/n)$ is bounded on $(-\beta_0 n, \beta_0 n)$, we can use the remark at the beginning of the proof to conclude that

$$f(a) - \epsilon \leqslant \underline{\lim} \int f(x/n) \, d\mu_n(x) \leqslant \overline{\lim} \int f(x/n) \, d\mu_n(x) \leqslant f(a) + \epsilon$$

Since $\epsilon$ is arbitrary, the result follows. ■

**Corollary 1.2.** Under the hypotheses of Theorem 1.1, for any function $f$ obeying $|f(x)| \leqslant Ae^{B|x|}$ for some $B < c$

$$\overline{\lim_{n \to \infty}} \int f(x/n) \, d\mu_n(x) \leqslant \max_{a_- \leq y \leq a_+} f(y)$$

In particular, if $f(y) = y^{2k}$ ($k$ an integer) and $a_- = -a_+$, then

$$a_+ \geqslant \overline{\lim_{n \to \infty}} \left[ \int (x/n)^{2k} \, d\mu_n(x) \right]^{1/2k}$$

*Remark.* The case $f(y) = y^{2k}$ with $x/n$ bounded on supp $\mu_n$ is the main result of Griffiths,[16] whose proof is abstracted above.

*Proof.* For $f$ bounded, the result follows from Theorem 1.1. The general case follows as in Corollary 1.1. ■

Following Griffiths,[16] we can apply Corollary 1.2 to prove that long-range order in the sense of (4) implies a spontaneous magnetization. Let $H_\Lambda$ be the Hamiltonian of a system in a box with periodic boundary conditions (but with no restriction on the form of the interaction). Let $A_\alpha$ be an operator at site $\alpha$. Define $m(A)$ by

$$m(A) = \lim_{\mu \downarrow 0} \frac{d}{d\mu} \left( \lim_{\Lambda \to \infty} |\Lambda|^{-1} \ln\left\{ \mathrm{Tr}\left[ \exp\left( -\beta H_\Lambda + \mu \sum A_\alpha \right) \right] \right\} \right)$$

As an immediate consequence of Corollary 1.2, we have the following result:

**Theorem 1.2.** Under the above conditions, if $\sum_{\alpha \in \Lambda} A_\alpha$ commutes with $H_\Lambda$, then

$$m(A) \geqslant \overline{\lim_{\Lambda \to \infty}} \left\langle \left( |\Lambda|^{-1} \sum_{\alpha \in \Lambda} A_\alpha \right)^{2k} \right\rangle_\Lambda^{1/2k} \tag{14}$$

In particular, if there is a unitary operator leaving $H_\Lambda$ invariant but taking $A_\alpha$ to $-A_\alpha$ and if the right side of (14) is nonzero for some $k$, then there is a phase transition in the sense that $\lim_{\Lambda\to\infty} |\Lambda|^{-1} \ln\{\mathrm{Tr}[\exp(-\beta H_\Lambda + \mu \sum A_\alpha)]\}$ is nondifferentiable at $\mu = 0$.

*Remark 1.* It is well known[41] that the nondifferentiability of the free energy implies multiple "phases" in the sense of several equilibrium states.

*Remark 2.* If $\mathrm{Tr}[\exp(-\beta H_\Lambda + \mu \sum A_\alpha)]$ is replaced by $\mathrm{Tr}[\exp(-\beta H_\Lambda) \exp(\mu \sum A_\alpha)]$, the commutation condition is not needed; the commutation condition is only used to obtain the physically relevant object. The commutativity unfortunately fails in several cases of interest, notably the $x$–$y$ and the antiferromagnetic models. In Section 5, after Theorem 5.2, we develop a different strategy for proving the existence of a phase transition in the noncommutative case, and apply it there to the $x$–$y$ model. In Section 6 we use it again for the antiferromagnet.

Theorem 1.2 with $k = 1$ shows that (4) implies there is a phase transition in general systems, but one should expect that in the isotropic Heisenberg model it yields a lower bound on $m(S^{(3)})$ which is too small by a factor of three. For, in the isotropic model, $\langle(\sum S_\alpha^{(3)})^2\rangle = \frac{1}{3}\langle|\sum \mathbf{S}_\alpha|^2\rangle$ by symmetry, but as soon as an external field in the other direction is turned on, the Lee–Yang theorem[2] implies that $\langle(|\Lambda|^{-1} \sum S_\alpha^{(i)})^2\rangle \to 0$ as $|\Lambda| \to \infty$ for $i = 1, 2$. This can be remedied by the use of some angular momentum theory.

**Theorem 1.3.** In the isotropic Heisenberg model, the spontaneous magnetization $m(S^{(3)})$ obeys

$$m(S^{(3)})^2 \geqslant \lim_{\Lambda\to\infty} \left\langle \left| |\Lambda|^{-1} \sum_{\alpha\in\Lambda} \mathbf{S}_\alpha \right|^2 \right\rangle_\Lambda = 3 \lim_{\Lambda\to\infty} \left\langle \left| |\Lambda|^{-1} \sum S_\alpha^{(3)} \right|^2 \right\rangle$$

*Remark.* The only restriction on the interaction is its isotropy in spin space, i.e., that $H$ commutes with simultaneous rotations of all spins.

*Proof.* Let $J_{(i)} = \sum_{\alpha\in\Lambda} S_\alpha^{(i)}$ and define

$$|J| = (J_{(1)}^2 + J_{(2)}^2 + J_{(3)}^2 + \tfrac{1}{4})^{1/2} - \tfrac{1}{2}$$

Letting

$$a(j, k) = (2j + 1)^{-1} \sum_{n=-j}^{j} n^k$$

we have by the isotropy of $H$ that

$$\langle J_{(3)}^{2k}\rangle = \sum_j a(j, 2k)\, \mathrm{prob}(|J| = j)$$

For simplicity of notation, consider the case where $|\Lambda|$ is even, so that $|J|$ has integral eigenvalues. For $j$ integral

$$a(j, 2k) \geqslant \frac{2}{2j+1}\sum_{n=1}^{j} n^{2k} \geqslant \frac{2}{2j+1}\int_0^j x^{2k}\,dx$$

$$= \frac{j^{2k}}{2k+1}\frac{2j}{2j+1} \geqslant \frac{2}{3}\frac{j^{2k}}{2k+1}, \qquad j = 0, 1,\ldots; \quad k \geqslant 1$$

Therefore, for $k \geqslant 1$,

$$\langle J_{(3)}^{2k}\rangle \geqslant \frac{2}{3(2k+1)}\langle |J|^{2k}\rangle \geqslant \frac{2}{3(2k+1)}\langle |J|^2\rangle^k \tag{15}$$

Using the definition of $|J|$, it is not hard to see that

$$\lim_{\Lambda\to\infty}\left\langle \left| |\Lambda|^{-1}\sum_{\alpha\in\Lambda} \mathbf{S}_\alpha \right|^2 \right\rangle = \lim_{\Lambda\to\infty}|\Lambda|^{-2}\langle |J|^2\rangle$$

so that the result follows by taking $k \to \infty$ in (15) and applying Theorem 1.2. ∎

## 2. THE DUHAMEL TWO-POINT FUNCTION

For quantum systems in finite volume with partition function $Z = \mathrm{Tr}(e^{-\beta H})$, we define the Duhamel two-point function (DTF) by

$$(A, B) = Z^{-1}\int_0^1 \mathrm{Tr}(e^{-x\beta H}Ae^{-(1-x)\beta H}B)\,dx \tag{16}$$

One expression of the naturalness of this object is that it has been introduced and discussed by a variety of authors, e.g., Bogoliubov,[5] Kubo,[24] Hohenberg,[21] Mermin and Wagner,[28] Mori,[29] Naudts *et al.*,[30,31] Powers,[36] and Roepstorff.[40] Fröhlich[46] has independently noted that it is likely to be useful in finding quantum generalizations of (10). We warn the reader that (16) may differ by factors of $\beta$ and by adjoints from the conventions of the above authors.

The name we have chosen comes from the fact that $(1/2)\mu^2(A, A)Z$ is the second-order term in a perturbation expansion, first derived by Duhamel, for $\mathrm{Tr}[\exp(-\beta H + \mu A)]$, i.e.,

$$(A, B)Z = \frac{\partial^2}{\partial\mu\,\partial\lambda}\mathrm{Tr}[\exp(-\beta H + \mu A + \lambda B)] \tag{17}$$

From (17), or from the definition (16), it is obvious that

$$(A, B) = (B, A) \tag{18}$$

so that, in particular, if $A = A_r + iA_i$ with $A_r^* = A_r$, $A_i^* = A_i$, then

$$(A^*, A) = (A_r, A_r) + (A_i, A_i) \tag{19}$$

If $\langle B\rangle_\mu \equiv \{\mathrm{Tr}[\exp(-\beta H + \mu A)]\}^{-1}\,\mathrm{Tr}[B\exp(-\beta H + \mu A)]$, then

$$\left.\frac{\partial\langle B\rangle_\mu}{\partial\mu}\right|_{\mu=0} = (A, B) - \langle A\rangle\langle B\rangle = (A - \langle A\rangle, B - \langle B\rangle) \tag{20}$$

so that, for example, any kind of generalization of Griffiths' second inequality[17] to quantum systems would involve the DTF and not the thermal two-point function, $\langle AB\rangle \equiv Z^{-1}\,\mathrm{Tr}[AB\exp(-\beta H)]$. Unlike $(A, B)$, $\langle AB\rangle$ is not symmetric in $A$ and $B$. If $H$ has a complete set of eigenfunctions $\phi_i$ with $H\phi_i = \epsilon_i\phi_i$ and $a_{ij} = (\phi_i, A\phi_j)$, $b_{ij} = (\phi_i, B\phi_j)$, then

$$\begin{aligned}(A, B) &= Z^{-1}\int_0^1 \sum_{i,j} a_{ij}b_{ji}e^{-x\beta\epsilon_i}e^{-(1-x)\beta\epsilon_j}\,dx \\ &= (\beta Z)^{-1}\sum_{i,j} a_{ij}b_{ji}(e^{-\beta\epsilon_i} - e^{-\beta\epsilon_j})/(\epsilon_j - \epsilon_i)\end{aligned} \tag{21}$$

Up to factors of $\beta$ and $Z$, the reader will recognize the formula from Ruelle's book[41] for the inner product he uses in his proof of Bogoliubov's inequality. From either (21) or directly from (16) [writing $\mathrm{Tr}(e^{-x\beta H}A^*e^{-(1-x)\beta H}A) = \mathrm{Tr}(C_x^*C_x)$, with $C_x = e^{-(1-x)\beta H/2}Ae^{-x\beta H/2}$], one sees that

$$(A^*, A) \geqslant 0 \tag{22}$$

so that we have a Schwarz inequality [using $(A^*, A) = (A, A^*)$]

$$|(A, B)| \leqslant (A^*, A)^{1/2}(B^*, B)^{1/2} \tag{22$'$}$$

Since the thermal expectation $\langle AB\rangle$ is not symmetric in $A$ and $B$, it is not clear which "two-point function" is closest to its classical analog. Some insight into this is obtained by looking at harmonic oscillators with variable $\hbar$. Let $A = a\sqrt{\hbar}$, with $a$ being the "usual" creation operator. Then $H = \omega A^*A$. As can be seen by direct calculation (Appendix B), $(A^*, A)$ is independent of $\hbar$ but $(1/2)\langle A^*A + AA^*\rangle$ is not. In this sense, $(A^*, A)$ is the most classical two-point function that can be constructed, and from this point of view it is not surprising that the classical bound (10) also holds for the corresponding DTF.

From knowledge of the DTF for all pairs $A$, $B$, one can recover the thermal expectations via the trivial identity

$$\langle A\rangle = (A, 1) \tag{23}$$

Conversely, one can recover the DTF in finite volume from thermal expectations and the action of the group of time automorphisms

$$\alpha_t(A) = e^{itH}Ae^{-itH}$$

To do so, one can use the function

$$f(z) = \langle B\alpha_{z\beta}(A)\rangle \tag{24a}$$

which is defined a priori for $z$ real. It has an analytic continuation to the strip $\operatorname{Im} z \leqslant 1$ with

$$f(z + i) = \langle A\alpha_{-z\beta}(B)\rangle \tag{24b}$$

and

$$(A, B) = \int_0^1 f(ix)\, dx \tag{24c}$$

This connection between the DTF and the KMS boundary condition is not new; see, e.g., Refs. 31, 36, and 40. It leads quite easily[40] to the bound

$$(A^*, A) \leqslant \tfrac{1}{2}\langle A^*A + AA^*\rangle \tag{25}$$

for, since $f$ with $B = A^*$ is analytic, the three-line lemma implies that

$$|f(ix)| \leqslant a^x b^{1-x} \leqslant xa + (1 - x)b \tag{26}$$

where $a = \sup_{z\in\mathbb{R}}|f(i + z)|$; $b = \sup_{z\in\mathbb{R}}|f(z)|$. But, for $z \in \mathbb{R}$, $|f(z + i)| \leqslant |\langle A\alpha_{-z\beta}(A^*)\rangle| \leqslant \langle AA^*\rangle$ and $|f(z)| \leqslant \langle A^*A\rangle$ by the Schwarz inequality. Thus $a = \langle AA^*\rangle$ and $b = \langle A^*A\rangle$, whence (25) follows from (26) by integration. The connection with the KMS boundary condition is also the key to extending many of the above considerations to infinite volume.[31]

The remainder of this section is not needed for the argument of the paper but is included to give the reader a source for all the main lore about the DTF. First, we want to prove (following Powers[36]) Bogoliubov's inequality[28] using the representation (16). We will use (22′), (25), and the formula

$$\langle [A, B]\rangle = ([A, \beta H], B) \tag{27}$$

Now, (27) follows by noting that

$$\operatorname{Tr}(e^{-x\beta H}[A, \beta H]e^{-(1-x)\beta H}B) = \frac{d}{dx}\operatorname{Tr}(e^{-x\beta H}Ae^{-(1-x)\beta H}B)$$

so that (16) can be directly integrated. Thus

$$\begin{aligned}
|\langle [A, B]\rangle|^2 &\leqslant |([A, \beta H], B)|^2 && \text{[by (27)]}\\
&\leqslant (B^*, B)([A, \beta H]^*, [A, \beta H]) && \text{[by (22')]}\\
&\leqslant (\tfrac{1}{2}\langle B^*B + BB^*\rangle)([A^*, \beta H], [\beta H, A]) && \text{[by (25)]}\\
&\leqslant \tfrac{1}{2}\langle B^*B + BB^*\rangle\langle [A^*, [\beta H, A]]\rangle && \text{[by (27)]}
\end{aligned}$$

which is Bogoliubov's inequality,

$$|\langle[A, B]\rangle|^2 \leqslant \langle[A^*, [\beta H, A]]\rangle(\tfrac{1}{2}\langle B^*B + BB^*\rangle) \tag{28}$$

We would like to note two properties of the thermal expectation of the double commutator. First, $\langle[A^*, [\beta H, A]]\rangle \geqslant 0$. This follows from (27) or by an eigenfunction expansion

$$\begin{aligned}\langle[A^*, [\beta H, A]]\rangle &= \beta \sum_{i,j} (|a_{ij}|^2 + |a_{ji}|^2)(\epsilon_i - \epsilon_j)e^{-\beta\epsilon_j} \\ &= \tfrac{1}{2}\beta \sum_{i,j} (|a_{ij}|^2 + |a_{ji}|^2)(\epsilon_i - \epsilon_j)(e^{-\beta\epsilon_j} - e^{\beta\epsilon_i}) \geqslant 0\end{aligned}$$

since $(x - y)(e^{-y} - e^{-x}) \geqslant 0$ for all $x$ and $y$. Second, if $A = A_r + iA_i$ with $A_r$ and $A_i$ self-adjoint, then

$$\langle[A^*, [\beta H, A]]\rangle = \langle[A_r, [\beta H, A_r]]\rangle + \langle[A_i, [\beta H, A_i]]\rangle \tag{29}$$

Finally, we want to say a few words about the infinite-volume DTF.

**Proposition 2.1.** If the finite-volume DTFs converge through some sequence of volumes, so do the ordinary thermal expectations. Conversely, if the ordinary finite-volume thermal expectations converge and the finite-volume time automorphisms converge, so do the DTFs.

*Proof.* The proposition is a consequence of formulas (23) and (24) relating ordinary thermal expectations and DTFs. ■

*Remark 1.* In infinite volume, one cannot define the DTF by (16), but (24) still holds. From this realization[31] follow all the relations we use in the proof of the next section, so our results there hold directly in any KMS state without reference to proving the result in finite volume and making a limiting argument.

*Remark 2.* For discussion of the limit of finite-volume time automorphisms for spin systems, see Streater,[44] Robinson,[38] and Lieb and Robinson.[27]

**Definition.** We say the (infinite-volume) DTF clusters if and only if $\lim_{|\mathbf{a}|\to\infty}(A, \tau_{\mathbf{a}}(B)) = \langle A\rangle\langle B\rangle$, where $\tau_{\mathbf{a}}$ is the space translation.

**Theorem 2.1.** The DTF clusters if the ordinary thermal function clusters, i.e., $\lim_{|\mathbf{a}|\to\infty}\langle A\tau_{\mathbf{a}}(B)\rangle = \langle A\rangle\langle B\rangle$.

*Proof.* Suppose that the thermal function clusters. Let $f_{\mathbf{a}}(z) = \langle\tau_{\mathbf{a}}(B)\alpha_{z\beta}(A)\rangle$. Then, by hypotheses, $f_{\mathbf{a}}(z) - \langle A\rangle\langle B\rangle \to 0$ as $\mathbf{a} \to \infty$ for $z$ real or $z = i +$ real. Moreover, for all $z$ and $\mathbf{a}$, $|f_{\mathbf{a}}(z)| \leqslant \|A\|\,\|B\|$. Thus, by a simple complex variable argument, $f_{\mathbf{a}}(z) \to \langle A\rangle\langle B\rangle$ for all $z$, so $(A, \tau_{\mathbf{a}}(B)) = \int_0^1 f_{\mathbf{a}}(ix)\,dx \to \langle A\rangle\langle B\rangle$. ■

## 3. LOWER BOUNDS ON THE DUHAMEL TWO-POINT FUNCTION

As explained in the introduction, the natural quantum extension of (10) involves the DTF, while the sum rule involves the thermal two-point function. To put the two together, we need a lower bound on the DTF in terms of the thermal two-point function. We have already seen that the easy bound (25) goes in the other direction. The lower bound will involve the function $f$ from $[0, \infty)$ to $[0, 1)$ defined implicitly by the relation

$$f(x \tanh x) = x^{-1} \tanh x \tag{30}$$

and plotted in Fig. 1 (for which we thank J. F. Barnes). We will need:

**Lemma 3.1.** The function $f$ given by (30) is convex.

This lemma is proved in Appendix A. By absorbing $\beta$ into $H$ and adding a constant to $H$ so that $\mathrm{Tr}(e^{-H}) = 1$, we can always deal with thermal expectations defined by $\langle B \rangle = \mathrm{Tr}(Be^{-H})$. We now define

$$g(A) = \tfrac{1}{2}\langle A^*A + AA^* \rangle = \tfrac{1}{2}\mathrm{Tr}[(A^*A + AA^*)e^{-H}] \tag{31}$$

$$b(A) = (A^*, A) = \int_0^1 \mathrm{Tr}(A^*e^{-xH}Ae^{-(1-x)H})\, dx \tag{32}$$

$$c(A) = \langle [A^*, [H, A]] \rangle = \mathrm{Tr}([A^*, [H, A]]e^{-H}) \tag{33}$$

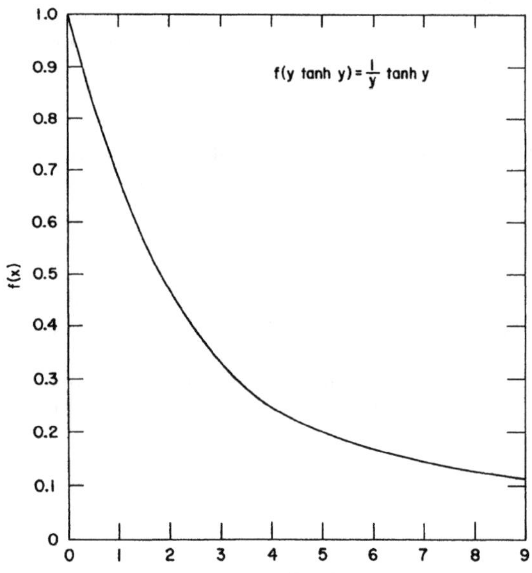


Fig. 1. A graph of the convex function $f$ defined for $x \geq 0$ by $f(y \tanh y) = y^{-1} \tanh y$. This function appears in the inequality, Theorem 3.1, between the Duhamel and ordinary two-point functions. For $x \geq 6$, $f(x) = x^{-1}$ to five-place accuracy. (Values computed by J. F. Barnes.)

The following basic bound is a generalization and improvement of a bound of Roepstorff[40]—we shall discuss the precise connection after its proof.[5]

**Theorem 3.1.** Let $f$ be the function given by (30). Then for any $A$ and self-adjoint $H$

$$b(A) \geqslant g(A)f(c(A)/4g(A)) \tag{34}$$

*Proof.* Let

$$h(x) = \mathrm{Tr}(A^* e^{-xH} A e^{-(1-x)H})$$

Then

$$g(A) = \tfrac{1}{2}[h(0) + h(1)], \qquad b(A) = \int_0^1 h(x)\,dx, \qquad c(A) = h'(1) - h'(0)$$

Moreover, if $H\phi_n = E_n\phi_n$ is an eigenfunction expansion of $H$,

$$h(x) = \sum_{n,m} |a_{nm}|^2 e^{-E_m} e^{+x(E_m - E_n)} \tag{35}$$

so that $h$ is the Laplace transform of a positive measure. As a result, inequality (34) clearly follows from Proposition 3.1 below. ■

**Proposition 3.1.** Let $f$ be the function given by (30). Let

$$h(x) = \int e^{xt}\,d\mu\,(t)$$

for some positive measure $\mu$. Then $b \geqslant gf(c/4g)$, where $b = \int_0^1 h(x)\,dx$, $g = \tfrac{1}{2}[h(0) + h(1)]$, $c = h'(1) - h'(0)$.

*Proof.* Let $d\nu$ be the measure

$$d\nu(t) = \tfrac{1}{2}(e^t + 1)\,d\mu(t)$$

Then

$$b = \int t^{-1}(e^t - 1)\,d\mu(t) = \int \frac{2}{t}\tanh\frac{t}{2}\,d\nu(t)$$

$$g = \int \tfrac{1}{2}(e^t + 1)\,d\mu(t) = \int d\nu(t)$$

$$c = \int t(e^t - 1)\,d\mu(t) = 4\int \tfrac{1}{2}t\tanh\tfrac{1}{2}t\,d\nu(t)$$

[5] Unfortunately we were unaware that Falk and Bruch[47] had previously proved our Theorem 3.1 [middle inequality in their Eq. (8)] and our Theorem A.4 [first inequality in their Eq. (8)].

Since $d\omega(t) = g^{-1}\, d\nu(t)$ is a probability measure and $f$ is convex (Lemma 3.1), we have by Jensen's inequality

$$\begin{aligned} gf(c/4g) &= gf\Big(\int \tfrac{1}{2}t \tanh \tfrac{1}{2}t \, d\omega\Big) \\ &\leqslant g\int f(\tfrac{1}{2}t \tanh \tfrac{1}{2}t)\, d\omega \\ &= g\int (2/t)\tanh \tfrac{1}{2}t\, d\omega = b \quad \blacksquare \end{aligned}$$

*Remark 1.* The strict convexity of $f$ and the above proof make it clear that equality holds in (34) if and only if $d\mu$ is a measure concentrated at a single point. From (35), this clearly holds if $A$ is a creation operator and $H$ is the Hamiltonian of a harmonic oscillator; this also follows from explicit calculation (see Appendix B).

*Remark 2.* Since one can shift the position of the point mass in Remark 1 by replacing the harmonic oscillator Hamiltonian $H$ by $\omega H$, we can adjust $\omega$ so that $c/4g$, the argument of $f$, has any preassigned positive value. Since equality always holds in (34) for the harmonic oscillator, it follows that the function $f$ in (34) is the best possible.

*Remark 3.* In Ref. 40, Roepstorff proved a bound of the form of (34) with two changes. First, he required that $A = A^*$, and second, he used the function

$$f_R(x) = x^{-1}(1 - e^{-x}) \tag{36}$$

in place of $f$. Since one can prove directly that

$$f(x) \geqslant f_R(x), \qquad \text{all } x \tag{37}$$

(see Appendix A), our inequality is stronger in two ways. Since we wish to use the inequality for $A$'s with $A \neq A^*$, the former change is more significant. Roepstorff's inequality, had he proved it for all $A$, would lead to a phase transition in all systems where we prove one, albeit only at a lower transition temperature than ours.

*Remark 4.* As we will see shortly, there is an abstract method of extending inequalities like (34) or Roepstorff's inequality from Hermitian $A$ to all $A$. Given this fact, (37) follows also from the best possible nature of our inequalities. (See Remark 2 above.)

**Lemma 3.2.** Let $c_1,..., c_n$, $b_1,..., b_n$, $g_1,..., g_n$ be real numbers and let $c = \sum_{i=1}^n c_i$, $b = \sum_{i=1}^n b_i$, $g = \sum_{i=1}^n g_i$. Assume that $g_i \geqslant 0$ and $g > 0$, and

$g_i = 0 \Rightarrow b_i \geqslant 0$ and $c_i = 0$. Let $F$ be *any* convex function on an interval $I$ that includes all $c_i/4g_i$ where $g_i > 0$ and such that

$$b_i \geqslant g_i F(c_i/4g_i)$$

when $g_i > 0$. Then $c/4g$ lies in the interval $I$ and

$$b \geqslant gF(c/4g)$$

*Proof.* Let $p_i \equiv g_i/g \geqslant 0$, so that $\sum_{i=1}^{n} p_i = 1$. Then

$$c/4g = \sum{}' (c_i/4g_i)p_i$$

where $\sum'$ means the summation over those $i$ with $p_i > 0$. Thus $c/4g$ is a convex combination of points in $I$ and is therefore in $I$. Moreover,

$$\begin{aligned} F(c/4g) &= F\Big(\sum{}' p_i(c_i/4g_i)\Big) \\ &\leqslant \sum{}' p_i F(c_i/4g_i) \qquad \text{(by convexity)} \\ &= g^{-1} \sum{}' g_i F(c_i/4g_i) \leqslant b/g \quad \blacksquare \end{aligned}$$

**Corollary 3.1.** An estimate of the form (34) need only be proved for $A$ self-adjoint (and in particular, Roepstorff's estimate for $A = A^*$ implies his estimate for all $A$).

*Remark.* Since our estimate is proved directly for all $A$ and is best possible, this corollary is of academic interest only.

*Proof.* Let $A = A_r + iA_i$ with $A_r = A_r^*$, $A_i = A_i^*$. We have already seen [Eqs. (19) and (29)] that $c(A) = c(A_r) + c(A_i)$, and $b(A) = b(A_r) + b(A_i)$. That $g(A) = g(A_r) + g(A_i)$ is trivial. $\blacksquare$

**Corollary 3.2.** Let $\mathbf{A} = (A_1, \ldots, A_n)$ be an $n$-tuple of operators, let $g(\mathbf{A}) = \sum_i g(A_i)$, etc., and let $f$ be given by (30). Then

$$b(\mathbf{A}) \geqslant g(\mathbf{A}) f(c(\mathbf{A})/4g(\mathbf{A})) \tag{38}$$

*Remark.* One can also prove this corollary by mimicking the proof of Theorem 3.1, using the fact that the sum of Laplace transforms of positive measures is again a positive measure.

In our applications, we want to go from upper bounds on $b$ and $c$ to one on $g$. Theorem 3.2 is the perfect vehicle for this.

**Theorem 3.2.** Suppose that $b \geqslant gf(c/4g)$, where $f$ is given by (30) and $b, g, c \geqslant 0$. Suppose that $b \leqslant b_0$ and $c \leqslant c_0$. Then $g \leqslant g_0$, where

$$g_0 = \tfrac{1}{2}(c_0 b_0)^{1/2} \coth x_0 \tag{39}$$

$$x_0{}^2 = c_0/4b_0 \tag{40}$$

*Proof.* First note that $g_0$ is just chosen so that $g_0 f(c_0/4g_0) = b_0$. Suppose that $g > g_0$. Then $c/4g < c_0/4g_0$ and since $f$ is monotone decreasing (see Appendix A),

$$gf(c/4g) \geqslant g_0 f(c_0/4g_0) = b_0 \geqslant b$$

thereby violating the hypothesis that $b \geqslant gf(c/4g)$. ∎

The following will not be needed in our applications but we think it of sufficient interest to mention it.

**Theorem 3.3.** Under the hypothesis of Theorem 3.2,

$$b - g \geqslant b_0 - g_0 = \frac{1}{2}(c_0 b_0)^{1/2}\left(\frac{1}{x_0} - \frac{1}{\tanh x_0}\right)$$

*Remark.* In the cases of interest, we emphasize that $b - g \leqslant 0$.

*Proof.* The function $g(f(c/4g) - 1)$ is seen to be monotone decreasing in both $g$ and $c$ (see Appendix A). Since $g \leqslant g_0$ and $c \leqslant c_0$,

$$b - g \geqslant g(f(c/4g) - 1) \geqslant g_0(f(c_0/4g_0) - 1) = b_0 - g_0 \quad ∎$$

## 4. GAUSSIAN DOMINATION—THE QUANTUM CASE

Our goal in this section is to prove the quantum analog of (10). We will succeed only if all the matrices can be simultaneously chosen to be real. As will be shown later, the antiferromagnet can be accommodated, even though it is essentially complex. This is discussed further in Ref. 50. We will restrict ourselves to nearest neighbor interactions on a simple cubic lattice, but no special commutation properties are required. Thus for each $\boldsymbol{\alpha}$ we choose a copy $\mathscr{H}_\alpha$ of the same Hilbert space and copies of $n + 1$ basic operators denoted by $S_\alpha^{(1)},\ldots, S_\alpha^{(n)}, \tilde{A}_\alpha$. To avoid unnecessary technical complications, we suppose $\dim \mathscr{H}_\alpha < \infty$, but it is clear that various unbounded operators on infinite-dimensional spaces could be accommodated. The basic Hamiltonian in $\Lambda$ is

$$H = \sum_{\alpha\in\Lambda}\left(\tilde{A}_\alpha - \sum_{m=1}^{\nu}\sum_{j=1}^{n} S_\alpha^{(j)} S_{\alpha+\delta_m}^{(j)}\right) \tag{41a}$$

$$= \sum_{\alpha\in\Lambda}\left[A_\alpha + \tfrac{1}{2}\sum_{m=1}^{\nu}(\mathbf{S}_\alpha - \mathbf{S}_{\alpha+\delta_m})^2\right] \tag{41b}$$

where $A_\alpha = \tilde{A}_\alpha - \nu\mathbf{S}_\alpha^2$. We define $\hat{\mathbf{S}}_\mathbf{p}$ by (5) and set

$$b_\mathbf{p}^{(j)} = (\hat{S}_\mathbf{p}^{(j)}, \hat{S}_{-\mathbf{p}}^{(j)})$$

the Duhamel two-point function. $E_\mathbf{p}$ is given by (6). Then we will prove the following result below:

**Theorem 4.1.** For Hamiltonians of the form (41) in boxes $\Lambda$ of sides $L_1 \times \cdots \times L_\nu$ with each an $L_j$ even integer and such that Theorem 4.2 holds,

$$b_\mathbf{p}^{(j)} \leqslant (2\beta E_\mathbf{p})^{-1} \tag{42}$$

As in the FSS proof of the classical case, we prove (42) on the basis of the following "Gaussian domination" estimate (Theorem 2.1 of Ref. 12). We thank J. Fröhlich for having suggested the following result to us as a conjecture:

**Theorem 4.2.** Let $H$ be a Hamiltonian of the form (41), in which all the matrices are real. Let $\{\mathbf{h}_i(\boldsymbol{\alpha}) | \boldsymbol{\alpha} \in \Lambda, i = 1,..., \nu\}$ be $\nu|\Lambda|$ vectors in $\mathbb{R}^n$. Let $\partial_j \mathbf{h}_i(\boldsymbol{\alpha}) \equiv \mathbf{h}_i(\boldsymbol{\alpha} + \boldsymbol{\delta}_j) - \mathbf{h}_i(\boldsymbol{\alpha})$ and $\sigma(\mathbf{h}) = \sum_{\boldsymbol{\alpha}} \mathbf{h}(\boldsymbol{\alpha}) \cdot \mathbf{S}_{\boldsymbol{\alpha}}$. Let $\Lambda$ be $L_1 \times \cdots \times L_\nu$ with each $L_i$ even. Then

$$\frac{\operatorname{Tr}\{\exp[-\beta H + \sigma(\sum \partial_i \mathbf{h}_i)]\}}{\operatorname{Tr}(\exp - \beta H)} \leqslant \exp \frac{\|h\|^2}{2\beta} \tag{43}$$

where $\|h\|^2 = \sum_{i,\boldsymbol{\alpha}} |\mathbf{h}_i(\boldsymbol{\alpha})|^2$.

*Proof of Theorem 4.1 Given Theorem 4.2.* One can follow FSS; we provide an essentially equivalent proof for the reader's convenience—since the proof below uses no operator theory, it will no doubt be more attractive to some, less attractive to others. Taking $h_i \to \lambda h_i$ in (43), subtracting 1 from both sides, dividing by $\lambda^2$, and taking $\lambda$ to zero, we find

$$\left(\overline{\sigma\left(\sum_i \partial_i \mathbf{h}_i\right)}, \sigma\left(\sum_i \partial_i \mathbf{h}_i\right)\right) \leqslant \beta^{-1} \sum_{i,\boldsymbol{\alpha}} |\mathbf{h}_i(\boldsymbol{\alpha})|^2 \tag{44}$$

This equation has just been proven for $h$ real-valued, but by (19) it extends to complex-valued $h$'s. Fix $\mathbf{p} \neq 0$ and $j$ in $\{1,..., n\}$. Choose now

$$[\mathbf{h}_i(\boldsymbol{\alpha})]_k = \delta_{jk}\{\exp[i\mathbf{p} \cdot (\boldsymbol{\alpha} - \boldsymbol{\delta}_i)] - \exp(i\mathbf{p} \cdot \boldsymbol{\alpha})\}|\Lambda|^{-1/2}$$

where the subscript $k$ labels the $n$ components of $\mathbf{h}$. We have

$$\sum_{i,\boldsymbol{\alpha}} |h_i(\boldsymbol{\alpha})|^2 = \sum_i |\exp(-i\mathbf{p} \cdot \boldsymbol{\delta}_i) - 1|^2 = 2E_{\mathbf{p}}$$

while

$$\left[\sum_i (\partial_i \mathbf{h}_i)(\boldsymbol{\alpha})\right]_k = \delta_{jk}|\Lambda|^{-1/2}(2E_{\mathbf{p}}) \exp(i\mathbf{p} \cdot \boldsymbol{\alpha})$$

so that (44) becomes

$$4E_{\mathbf{p}}^2(\hat{\sigma}_{\mathbf{p}}^{(j)}, \hat{\sigma}_{-\mathbf{p}}^{(j)}) \leqslant (2E_{\mathbf{p}})\beta^{-1}$$

which is (42). ■

We next turn to the proof of Theorem 4.2. In the FSS proof of (10) a critical role was played by the inequality

$$\left\{\int \exp[-\tfrac{1}{2}(x - y - a)^2]\, d\mu(x)\, d\nu(y)\right\}^2$$

$$\leqslant \int \exp[-\tfrac{1}{2}(x - y)^2]\, d\mu(x)\, d\mu(y) \int \exp[-\tfrac{1}{2}(x - y)^2]\, d\nu(x)\, d\nu(y)$$

for any measures $d\mu$, $d\nu$ on $\mathbb{R}^n$ and any $n$-tuple of reals $a$. We begin the proof of Theorem 4.2 with a quantum analog of this fact.

**Lemma 4.1.** Let $\mathscr{H}_1$ be a finite-dimensional vector space and let $\mathscr{H} = \mathscr{H}_1 \otimes \mathscr{H}_1$. If $A, B,...$ are operators on $\mathscr{H}_1$, we use the symbols $A, B,...$ for the operators $A \otimes 1$, $B \otimes 1,...$, and the symbols $\tilde{A}, \tilde{B},...$ for $1 \otimes A$, $1 \otimes B,...$. Then for any self-adjoint operators $A, B, C_1,..., C_l$ with real matrix representations and real numbers $h_1,..., h_l$

$$\left(\mathrm{Tr}\left\{\exp\left[A + \tilde{B} - \sum_{i=1}^{l} (C_i - \tilde{C}_i - h_i)^2\right]\right\}\right)^2$$

$$\leqslant \mathrm{Tr}\left\{\exp\left[A + \tilde{A} - \sum_{i=1}^{l} (C_i - \tilde{C}_i)^2\right]\right\} \mathrm{Tr}\left\{\exp\left[B + \tilde{B} - \sum_{i=1}^{l} (C_i - \tilde{C}_i)^2\right]\right\} \tag{45}$$

*Proof.* Let $\alpha$ denote the quantity being squared on the left-hand side of (45). By the Trotter product formula (see, e.g., Section VIII.8 of Ref. 37), we have that $\alpha = \lim_{n\to\infty} \alpha_n$, where

$$\alpha_n = \mathrm{Tr}(\{\exp(A/n)\exp(\tilde{B}/n)\exp[-(C_1 - \tilde{C}_1 - h_1)^2/n] \cdots\}^n)$$

Using the operator identity

$$\exp(-D^2) = (4\pi)^{-1/2} \int \exp(ikD)\exp(-k^2/4)\, dk$$

we have that

$$\begin{aligned}\alpha_n = (4\pi)^{-nl/2} \int d^{nl}k \\ \times \mathrm{Tr}\{\exp(A/n)\exp(\tilde{B}/n)\exp[ik_1(C_1 - \tilde{C}_1)/n^{1/2}] \cdots\} \\ \times \exp(-k^2/4)\exp(-ik_1h_1 + \cdots)\end{aligned} \tag{46}$$

The operators $A, B$, etc., can be thought of as matrices. We have assumed that they are all real. Then

$$\mathrm{Tr}\{\exp(A/n)\exp(\tilde{B}/n)\exp[ik_1(C_1 - \tilde{C}_1)/n^{1/2}]\cdots\}$$

$$= \mathrm{Tr}[\exp(A/n)\exp(ik_1C_1/n^{1/2})\cdots]\,\overline{\mathrm{Tr}[\exp(B/n)\exp(ik_1C_1/n^{1/2})\cdots]} \tag{47}$$

where we have used the reality of the matrices $A, B$, etc., to take the complex conjugate *without* reversing the order of the factors. We have also used the fact that $\tilde{D}$ and $F$ commute for any two operators $D$ and $F$. Using the

Schwarz inequality on the $dk$ integration in (46) and then using (47) with $B = A$, we obtain

$$|\alpha_n|^2 \leqslant \left((4\pi)^{-nl/2}\int d^{nl}k\ \mathrm{Tr}\{\exp(A/n)\exp(\tilde{A}/n)\right.$$

$$\left.\times \exp[ik_1(C_1 - \tilde{C}_1)/n^{1/2}]\cdots\}\exp(-k^2/4)\right)$$

$$\times \left((4\pi)^{-nl/2}\int d^{nl}k\ \mathrm{Tr}\{\exp(B/n)\exp(\tilde{B}/n)\right.$$

$$\left.\times \exp[ik(C_1 - \tilde{C}_1)/n^{1/2}]\cdots\}\exp(-k^2/4)\right)$$

Reversing the steps at the start of the proof, we obtain (45). ■

*Remark.* In our original announcement[51] we claimed that we could prove a phase transition for the ferromagnet. At the time we believed that Theorem 4.2 held for complex matrices by virtue of the following stratagem: Complex matrices can be made real by doubling the size of the representation. Unfortunately, if one doubles $\mathscr{H}$ then the reflection structure is destroyed. If one doubles $\mathscr{H}_1$ then the trace can be changed in a nontrivial way. An illustration of what can happen is provided by the fact that

$$\mathrm{Tr}\Big\{\Big(\sum_i A_i\tilde{A}_i\Big)^3\Big\} \geq 0$$

for real $A_i$, whereas for spin matrices,

$$\mathrm{Tr}(\sigma\cdot\sigma)^3 < 0$$

*Proof of Theorem 4.2.* Define

$$Z(\{\mathbf{h}_i(\alpha)\}) = \mathrm{Tr}\left[\exp\left(-\sum_{\alpha\in\Lambda}\left\{\beta A_\alpha + \frac{\beta}{2}\sum_{m=1}^{\nu}[\mathbf{S}_\alpha - S_{\alpha+\delta_m} + \beta^{-1}\mathbf{h}_m(\alpha)]^2\right\}\right)\right]$$

Then (43) is easily seen to be equivalent to

$$Z(\{\mathbf{h}_i(\alpha)\}) \leqslant Z(\{0\}) \tag{48}$$

for all real $\{\mathbf{h}_i(\alpha)\}$. Since $Z$ is continuous in the $h$'s and goes to zero as any $\mathbf{h}_i(\alpha) \to \infty$, it takes its maximum value $Z_0$ at some set of $\mathbf{h}$'s, say $\bar{\mathbf{h}}_i(\alpha)$. If this maximum value is taken at more than one point, choose a point with the largest number of $h$'s equal to zero. Thus, we must show $\bar{\mathbf{h}}_i(\alpha) = 0$ for all $\alpha$, $i$. If not, by relabeling we can suppose that $\bar{\mathbf{h}}_i(\alpha) \neq 0$ for $i = 1$ and $\alpha = (L_1 - 1, 0,..., 0)$. Let $\mathscr{H}_1$ be the tensor product of all the $\mathscr{H}_\gamma$ such that

$\gamma = (\gamma_1,..., \gamma_\nu)$ with $0 \leqslant \gamma_1 \leqslant \frac{1}{2}L_1 - 1$ and $\gamma_2,..., \gamma_\nu$ arbitrary. Then $\mathscr{H}_\Lambda = \mathscr{H}_1 \otimes \tilde{\mathscr{H}}_1$ in such a way that $\tilde{S}_\gamma = S_\gamma$, where $\tilde{\gamma}_2 = \gamma_2,..., \gamma_\nu = \tilde{\gamma}_\nu, \tilde{\gamma}_1 = L_1 - 1 - \gamma_1$. With this representation

$$Z(\{\mathbf{h}_i(\alpha)\}) = \operatorname{Tr}\left\{\exp\left[D + \tilde{D} - \sum_{i=1}^{l} (C_i - \tilde{C}_i - y_i)^2\right]\right\}$$

where $D$ is all "interactions" between $\mathscr{H}_1$-spins, the $C_i$ represent the spins at sites $(0, \gamma_2,..., \gamma_\nu)$ and $(\frac{1}{2}L_1 - 1, \gamma_2,..., \gamma_\nu)$, and the $y$'s are the corresponding $h$'s up to factors of $\beta$.

By Lemma 4.1, we conclude that

$$Z(\{\mathbf{h}_i(\alpha)\})^2 \leqslant Z(\{\mathbf{h}_i^{(1)}(\alpha)\})Z(\{\mathbf{h}_i^{(2)}(\alpha)\})$$

where $\mathbf{h}^{(1)}$ (resp. $\mathbf{h}^{(2)}$) is a set of $h$'s invariant under the $\gamma \to \tilde{\gamma}$ reflection and equal to the $\bar{\mathbf{h}}$'s on the $\mathscr{H}_1$ (resp. $\tilde{\mathscr{H}}_1$) spins and zero on the bonds between $\mathscr{H}_1$ and $\tilde{\mathscr{H}}_1$.

Now, on the one hand either $\{\mathbf{h}_i^{(1)}(\alpha)\}$ or $\{\mathbf{h}_i^{(2)}(\alpha)\}$ must contain strictly more zero elements than $\{\bar{\mathbf{h}}_i(\alpha)\}$ and on the other hand since $Z(\{\bar{\mathbf{h}}_i(\alpha)\}) = Z_0$ and $Z(\{\mathbf{h}_i(\alpha)\}) \leqslant Z_0$, we must have $Z(\{\mathbf{h}_i(\alpha)\}) = Z_0$. This contradicts the fact that the $\bar{h}$ has a maximal number of zeros, so it must be that all $\bar{h}$'s are zero. ∎

We conclude this section with a few remarks on the restrictions we have placed on the interaction. It is our belief that the basic bound (48), suitably generalized to allow $h$'s on each bond, holds for any ferromagnetic interaction, but we are in the unhappy situation of not being able to prove it (or phase transitions) even for face-centered or body-centered cubic lattices. Our proof is restricted to lattices with the following property: The perpendicular bisector of any bond contains no sites and the lattice is reflection symmetric about that bisector. For example, the two-dimensional honeycomb (hexagonal) lattice can be handled.

## 5. PHASE TRANSITIONS: THE FERROMAGNETIC CASE

In this section we want to put the results of Sections 3 and 4 together with explicit calculations of the double commutator to prove that phase transitions occur in the spin 1/2 $x$–$y$ model, where the matrices can be chosen simultaneously real, namely

$$S_x = 1/2\begin{pmatrix} 0 & 1 \\ 1 & 0 \end{pmatrix}, \qquad S_y = 1/2\begin{pmatrix} 1 & 0 \\ 0 & -1 \end{pmatrix}$$

This is Example 4 below. We also present the consequences, in terms of phase transitions, that would follow if Theorem 4.2 held for the ferromagnet; these are contained in Examples 1, 2, and 3.

We first rephrase the abstract result, Theorem 3.2. Suppose that we have a lattice system on $\mathbb{Z}^\nu$ (with no a priori restriction on the kind of interactions) with operators $S_\alpha^{(j)}$ at each site. Define $\hat{S}_{\mathbf{p}}^{(j)}$ in the usual way and

$$d^{(j)} = \langle S_\alpha^{(j)} S_\alpha^{(j)} \rangle; \qquad d = \sum_j d^{(j)}$$

$$b_{\mathbf{p}}^{(j)} = (\hat{S}_{\mathbf{p}}^{(j)}, \hat{S}_{-\mathbf{p}}^{(j)}); \qquad b_{\mathbf{p}} = \sum_j b_{\mathbf{p}}^{(j)}$$

$$g_{\mathbf{p}}^{(j)} = \langle \hat{S}_{\mathbf{p}}^{(j)} \hat{S}_{-\mathbf{p}}^{(j)} \rangle; \qquad g_{\mathbf{p}} = \sum_j g_{\mathbf{p}}^{(j)}$$

$$c_{\mathbf{p}}^{(j)} = \langle [\hat{S}_{\mathbf{p}}^{(j)}, [\beta H, \hat{S}_{-\mathbf{p}}^{(j)}]] \rangle \geqslant 0; \qquad c_{\mathbf{p}} = \sum_j c_{\mathbf{p}}^{(j)}$$

(Note that $\hat{S}_{\mathbf{p}}^{(j)}$ and $\hat{S}_{\mathbf{q}}^{(j)}$ commute for any $\mathbf{p}$ and $\mathbf{q}$.)

From the sum rule (Plancherel)

$$\frac{1}{|\Lambda|} \sum_{\mathbf{p} \in \Lambda^*} g_{\mathbf{p}}^{(j)} = d^{(j)}$$

and Theorem 3.2, one immediately concludes that:

**Theorem 5.1.** Suppose that there exist fixed measurable functions $B_{\mathbf{p}}^{(j)}$, $C_{\mathbf{p}}^{(j)}$ of $\mathbf{p}$ and a function $D^{(j)}(\beta)$ such that for every finite system $L_1 \cdots L_\nu$ with $L_i$ even, one has the bounds

(a) $d^{(j)} \geqslant D^{(j)}(\beta)$ and $\lim_{\beta \to \infty} D^{(j)}(\beta) \geqslant D_\infty^{(j)}$.
(b) $b_{\mathbf{p}}^{(j)} \leqslant \beta^{-1} B_{\mathbf{p}}^{(j)}$; $B_{\mathbf{p}} < \infty$ for $p \neq 0$.
(c) $c_{\mathbf{p}}^{(j)} \leqslant \beta C_{\mathbf{p}}^{(j)}$.

Further assume that

$$\lim_{|\Lambda| \to \infty} |\Lambda|^{-1} \sum_{\substack{\mathbf{p} \neq 0 \\ \mathbf{p} \in \Lambda^*}} B_{\mathbf{p}} = \int B_{\mathbf{p}} \, d^\nu p/(2\pi)^\nu$$

Then there is long-range order at some finite $\beta$ whenever (49) and (50) hold for some $j$:

$$D_\infty^{(j)} > (2\pi)^{-\nu} \int_{|p_i| \leqslant \pi} \tfrac{1}{2}(B_{\mathbf{p}}^{(j)} C_{\mathbf{p}}^{(j)})^{1/2} \, d^\nu p \tag{49}$$

$$\int_{|p_i| \leqslant \pi} B_{\mathbf{p}}^{(j)} \, d^\nu p < \infty; \qquad D_\infty^{(j)} < \infty \tag{50}$$

There is long-range order for any $\beta$ such that

$$D^{(j)}(\beta) > (2\pi)^{-\nu} \int_{|p_i| \leqslant \pi} \tfrac{1}{2}(B_{\mathbf{p}}^{(j)} C_{\mathbf{p}}^{(j)})^{1/2} \coth[\tfrac{1}{2}\beta(C_{\mathbf{p}}^{(j)}/B_{\mathbf{p}}^{(j)})^{1/2}] \, d^\nu p \tag{51}$$

In particular, if $D^{(j)}(\beta)$ is monotone nondecreasing in $\beta$ and (49) and (50) hold, then there is a long-range order for $\beta > \tilde{\beta}_c$, where $\tilde{\beta}_c$ is the unique solution of

$$D^{(j)}(\tilde{\beta}_c) = (2\pi)^{-\nu} \int_{|p_i| \leqslant \pi} \tfrac{1}{2}(B_{\mathbf{p}}^{(j)} C_{\mathbf{p}}^{(j)})^{1/2} \coth[\tfrac{1}{2}\tilde{\beta}_c (C_{\mathbf{p}}^{(j)}/B_{\mathbf{p}}^{(j)})^{1/2}] \, d^\nu p \tag{52}$$

*Proof.* Theorem 3.2 and the stated sum rule yield (4) whenever (51) holds. When (49) and (50) hold, (51) must hold for $\beta$ large. To prove this, (50) is used together with the bound (see Appendix A)

$$\coth x \leqslant x^{-1} + 1, \qquad x \geqslant 0$$

and the dominated convergence theorem to obtain (49) as the $\beta \to \infty$ limit of (51). Equation (52) has a unique solution since the right side of (51) decreases strictly monotonically from $\infty$ to

$$(2\pi)^{-\nu} \int_{|p_i| \leqslant \pi} \tfrac{1}{2}(B_{\mathbf{p}}^{(j)} C_{\mathbf{p}}^{(j)})^{1/2} \, d^\nu p \qquad \text{as} \quad \beta \to \infty \quad \blacksquare$$

*Remark.* If all superscripts $j$ are replaced by dot products (i.e., $\sum_{j=1}^{3}$ in the usual Heisenberg case), the result still holds.

**Example 1.** Usual [isotropic, $O(3)$] Heisenberg model of spin $S$. This is the model described in Section 1. For nearest neighbor simple cubic coupling, we consider the consequence of supposing that we may take $B_{\mathbf{p}} = 3/(2E_{\mathbf{p}})$. By a simple calculation (see below), we can take $C_{\mathbf{p}} = 4SE_{\mathbf{p}}$. Finally, $D(\beta) = S(S+1)$ can obviously be taken. Thus (50) holds as long as $\nu \geqslant 3$ and (49) becomes

$$S(S+1) > S\sqrt{\tfrac{3}{2}}$$

which holds for $S = 1/2, 1, 3/2$, etc. In order to solve (52), one needs the function (in $\nu = 3$ dimensions)

$$H_3(x) = (2\pi)^{-3} \int_{|p_i| \leqslant \pi} \coth(xE_{\mathbf{p}}) \, d^3 p$$

which is tabulated in Table I and graphed in Fig. 2 (we owe these to J. F. Barnes). The solutions of

$$S(S+1) = S\sqrt{\tfrac{3}{2}} H_3(\beta S \sqrt{\tfrac{8}{3}}) \tag{52H}$$

are shown in Table II (also due to J. F. Barnes).

For use below, we do the double commutator calculation in the general case. Let $H$ be given by (41) and define

$$iQ_\alpha^{ij} = [S_\alpha^{(i)}, S_\alpha^{(j)}], \qquad iP_\alpha^{k;ij} = [S_\alpha^{(k)}, Q_\alpha^{ij}], \qquad R_\alpha^{ij} = [S_\alpha^{(i)}, [A_\alpha, S_\alpha^{(j)}]]$$

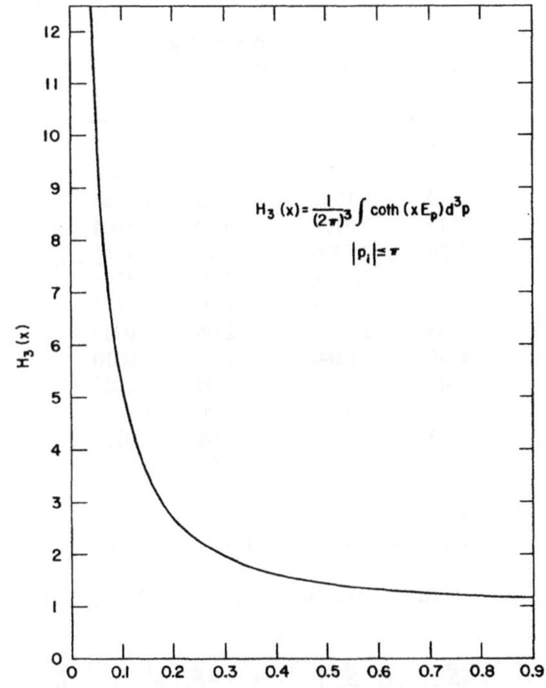


Fig. 2. Graph of the integral appearing in (52*H*) for three dimensions. The large-$x$ asymptotic form of $H_\nu(x)$ in $\nu$ dimensions is given by $H_\nu(x) \sim 1 + B_\nu x^{-\nu/2}$ and $B_\nu = 2^{-\nu+1}\pi^{-\nu/2}\zeta(\nu/2)$, with $\zeta$ being the Riemann zeta function. $B_3 = 0.117$. (Values computed by J. F. Barnes.)

Then, by a direct calculation,

$$[\hat{S}^{(i)}_{-\mathbf{p}}, [H, \hat{S}^{(i)}_{\mathbf{p}}]]$$
$$= -\frac{1}{\Lambda}\sum_{\alpha}\left\{R^{ii}_{\alpha} + \sum_{m=1}^{\nu}\sum_{j=1}^{n}[P^{iij}_{\alpha}S^{(j)}_{\alpha+\delta_m} + S^{(j)}_{\alpha}P^{iij}_{\alpha+\delta_m} + 2\cos(\mathbf{p}\cdot\boldsymbol{\delta}_m)\, Q^{ij}_{\alpha}Q^{ij}_{\alpha+\delta_m}]\right\} \tag{53}$$

In the case of Example 1, $R = 0$, $Q^{ij} = \epsilon^{ijk}S^{(k)}$, $\sum_i P^{iij} = -2S^{(j)}$, from which one concludes that

$$\sum_i [\hat{S}^{(i)}_{-\mathbf{p}}, [H, \hat{S}^{(i)}_{\mathbf{p}}]] = \frac{1}{|\Lambda|}\sum_{\alpha\in\Lambda}\sum_{m=1}^{\nu} 4(1 - \cos\mathbf{p}\cdot\boldsymbol{\delta}_m)\mathbf{S}_\alpha\mathbf{S}_{\alpha+\delta_m}$$

In particular, as an operator,

$$\sum_i [\hat{S}^{(i)}_{-\mathbf{p}}, [H, \hat{S}^{(i)}_{\mathbf{p}}]] \leqslant 4E_{\mathbf{p}}S^2 \tag{54}$$

**Table I. A Table of Values of the Function Shown in Fig. 2**

| $y$ | $H_3(y)$ | $y$ | $H_3(y)$ |
|---|---|---|---|
| 0.01 | 50.5562 | 0.75 | 1.2182 |
| 0.05 | 10.1591 | 0.80 | 1.1956 |
| 0.10 | 5.1537 | 0.85 | 1.1766 |
| 0.15 | 3.5168 | 0.90 | 1.1604 |
| 0.20 | 2.7206 | 0.95 | 1.1465 |
| 0.25 | 2.2591 | 1.00 | 1.1345 |
| 0.30 | 1.9637 | 1.50 | 1.0693 |
| 0.35 | 1.7621 | 2.00 | 1.0439 |
| 0.40 | 1.6180 | 2.50 | 1.0310 |
| 0.45 | 1.5114 | 3.00 | 1.0234 |
| 0.50 | 1.4305 | 4.00 | 1.0150 |
| 0.55 | 1.3676 | 5.00 | 1.0107 |
| 0.60 | 1.3178 | 6.00 | 1.0081 |
| 0.65 | 1.2778 | 8.00 | 1.0052 |
| 0.70 | 1.2451 | 10.00 | 1.0037 |

since $\mathbf{S}_\alpha \cdot \mathbf{S}_\beta \leqslant S^2$. In any state such that $\langle \mathbf{S}_\alpha \cdot \mathbf{S}_{\alpha+\delta_m} \rangle$ is independent of $\alpha$ and $m$,

$$\left\langle \sum_i [\hat{S}^{(i)}_{-\mathbf{p}}, [H, \hat{S}^{(i)}_{\mathbf{p}}]] \right\rangle = -4E_{\mathbf{p}} |\Lambda|^{-1} \nu^{-1} \langle H \rangle \tag{55}$$

The fact that the $P \cdot S$ and $Q \cdot Q$ terms combine so nicely is coincidental (see Examples 3 and 4 below).

**Example 2.** $O(n)$ Heisenberg model. Instead of three spin operators $S^{(i)}_\alpha$ ($i = 1, 2, 3$) obeying the $O(3)$ commutation relations, we have $n(n-1)/2$ spin operators obeying the $O(n)$ commutation relations with the same irreducible representation $R$ at each point.

The commutation relations have each spin operator commuting with $(n-2)(n-3)/2$ *other* spin operators. The other $2(n-2)$ spin operators

**Table II. Lower Bounds on the Critical β in Three Dimensions [Solutions of (52*H*)]**

| $S$ | $\beta_c(S)S(S+1)$ | $S$ | $\beta_c(S)S(S+1)$ |
|---|---|---|---|
| $\frac{1}{2}$ | 1.354 | $\frac{7}{2}$ | 0.788 |
| 1 | 0.965 | 4 | 0.783 |
| $\frac{3}{2}$ | 0.872 | $\frac{9}{2}$ | 0.778 |
| 2 | 0.832 | 5 | 0.775 |
| $\frac{5}{2}$ | 0.810 | $\infty$ | 0.758 |
| 3 | 0.797 | | |

break into $n - 2$ pairs, which, along with the given spin, form $n - 2$ copies of $O(3)$. Using these facts, one quickly computes as above that one can take as a bound on the double commutator

$$C_p = 4(n-2)\tilde{K}E_p$$

where $\tilde{K}$, depending on the representation $R$, is the maximum eigenvalue of $\mathbf{S}_\alpha \cdot \mathbf{S}_\beta$ (for $\alpha \neq \beta$). The putative Duhamel function bound $B_p$ is $n(n-1)[4E_p]^{-1}$. Finally, let $K$ be the value, depending on $R$, of $\sum (S^{(i)})^2$. Then the condition for a phase transition in $\nu \geqslant 3$ dimensions would be

$$K > \tilde{K}^{1/2}[\tfrac{1}{4}n(n-1)(n-2)]^{1/2} \tag{56}$$

if one could extend Theorem 4.2. We consider the following examples.

(*i*) *Identity representation.* The "natural" representation of $O(n)$ on $\mathbb{C}^n$ by real rotations (spin 1 for $n = 3$) has $K = n - 1$ [each $S^{(i)}$ obeys $\operatorname{Tr}(S^{(i)})^2 = 2$]. Moreover, $\tilde{K} = 1$—because the theory of weights[42] implies that the maximum value of $\mathbf{S}_\alpha + \mathbf{S}_\beta$ occurs in a space including a vector of the form $u \otimes v$, so that the maximum value is of the form $(u, S_\alpha u) \cdot (v, S_\beta v)$ and since $\operatorname{Tr}(\mathbf{S}_\alpha \cdot \mathbf{a}) = 0$, $\operatorname{Tr}[(S_\alpha \cdot \mathbf{a})^2] = 2|\mathbf{a}|^2$ for any $\frac{1}{2}n(n-1)$-vector $\mathbf{a}$, $(u, \mathbf{S}_\alpha u)$ is always a vector of norm less than or equal to 1. The condition (56) holds for $n = 3, 4, 5$ but not for $n \geqslant 6$.

(*ii*) $n = 4$. It is well known[42] that as a Lie algebra $O(4) = O(3) \oplus O(3)$. Thus representations $R$ are labeled by pairs $(j, k)$ of half-integers. By direct computation, $K = 2[j(j+1) + k(k+1)]$ and $\tilde{K} = 2[j^2 + k^2]$. One can see that (56) holds for all values of $(j, k)$ except for the values $(j, k) = (0, 0)$ and $(0, 1/2)$.

The moral of these computations is the following: When putting together the bounds of Sections 3 and 4, detailed estimates on algebraic properties are needed, which one can sometimes prove and sometimes not. It is fortunate that for the usual ferromagnetic $O(3)$ model one can always verify the algebraic conditions. We should emphasize that we believe phase transitions occur in *any* quantum system of the type discussed in Section 4 (with not all the $S_\alpha$ constant); it is our method that has limitations.

**Example 3.** Anisotropic Heisenberg model. The anisotropic model is like the isotropic model of Section 1, except that

$$H = -\sum_{\alpha \in \Lambda} \sum_{m=1}^{\nu} \left( \sum_{i=1}^{3} J_i S_\alpha^{(i)} S_{\alpha+\delta_m}^{(i)} \right)$$

We want to consider the prolate case $J_3 = 1$, $J_1 = J_2$ smaller than 1 (but *positive*). This is a generalization of the case treated by Ginibre[13] and Robinson[39] and if $\nu \geqslant 3$, we would, modulo the proof of Theorem 4.2, recover their results. In fact, our method is ideal for discussing quantum models of the sort that are small perturbations of classical models. For, if we try to prove (49) with $S^{(j)}$ being the "classical" spin about which one is perturbing, in this case $j = 3$, then the left side of (49) will have a nonzero limit as the perturbation parameter goes to zero, and the right side will go to zero. In the case at hand, we first argue that

$$\lim_{\beta\to\infty} D^{(3)}(\beta) = S^2 \tag{57}$$

and that (in *infinite* volume)

$$\lim_{\beta\to\infty} \langle S_\alpha^{(i)} S_{\alpha+\delta_m}^{(i)}\rangle = S^2\delta_{i3} \tag{58}$$

Clearly, (57) follows from (58) and the Schwarz inequality

$$\langle S_\alpha^{(3)} S_{\alpha+\delta_m}^{(3)}\rangle \leqslant \langle S_\alpha^{(3)} S_\alpha^{(3)}\rangle \leqslant S^2$$

in translation-invariant states. To prove (58), note that $S_\alpha \cdot \mathbf{S}_\beta \leqslant S^2$ and $S_\alpha^{(3)} S_\beta^{(3)} \leqslant S^2$ imply that $\langle H\rangle/|\Lambda| \leqslant \nu S^2$, while the bound $Z_\Lambda \geqslant \exp(\beta\nu S^2)$ (from the all-spin-up state) plus a standard convexity argument implies that

$$\lim_{\beta\to\infty} \sum_{m=1}^{\nu} [J_1\langle \mathbf{S}_\alpha \cdot \mathbf{S}_{\alpha+\delta_m}\rangle + (1 - J_1)\langle S_\alpha^{(3)} S_{\alpha+\delta_m}^{(3)}\rangle] = \nu S^2$$

From this it follows that $\langle S_\alpha^{(3)} S_{\alpha+\delta_m}^{(3)}\rangle$ and $\langle S_\alpha \cdot S_{\alpha+\delta_m}\rangle$ both go to $S^2$, which implies (58).

Now, suppose we can take $B_\mathbf{p}^{(3)} = 1/2E_\mathbf{p}$. Using (58) and (53),

$$\lim_{\beta\to\infty} \langle[\hat{S}_{-\mathbf{p}}^{(3)}, [H, \hat{S}_\mathbf{p}^{(3)}]]\rangle = 0$$

Thus, a simple modification of (49) (allowing $C_\mathbf{p}$ to be $\beta$ dependent) clearly holds. We conclude that in $\nu = 3$ dimensions the prolate anisotropic model ($J_3 > J_1 = J_2$) would have a phase transition for arbitrary spin if Theorem 4.2 extends to this case.

**Example 4.** The $x$–$y$ model. This is the highly oblate anisotropic case ($J_1 = J_2 = 1$, $J_3 = 0$). We will verify (49) in this case without the superscripts, i.e., sum over $j = 1$ and 2. Clearly $D(\beta) = \frac{1}{2}$ can be taken for all $\beta$ and $B_\mathbf{p} = 1/E_\mathbf{p}$. Thus, if $\nu \geqslant 3$, (50) holds. By the Schwarz inequality, (49) then holds if $\nu \geqslant 3$ and

$$\left[\int C_\mathbf{p}\, d^\nu p\, (2\pi)^\nu\right] G_\nu(0) < 1 \tag{59}$$

where $G_\nu(0) = (2\pi)^{-\nu} \int E_\mathbf{p}^{-1}\, d^\nu p$. By (53),

$$\begin{aligned}(2\pi)^{-\nu}\int C_\mathbf{p}\, d^\nu p &= \sum_{i=1}^{2}\sum_{m=1}^{\nu}\sum_{j=1}^{2} \langle P_\alpha^{iij} S_{\alpha+\delta_m}^{(j)} + S_\alpha^{(j)} P_{\alpha+\delta_m}^{iij}\rangle \\ &= \sum_{m=1}^{\nu}\sum_{\gamma=\pm 1} \langle S_\alpha^{(x)} S_{\alpha+\gamma\delta_m}^{(x)} + S_\alpha^{(y)} S_{\alpha+\gamma\delta_m}^{(y)}\rangle \\ &\leqslant \tfrac{1}{2}[\nu(\nu+1)]^{1/2}\end{aligned}$$

where the last inequality follows from a bound in Appendix C on the ground-state energy in the $x$–$y$ model. Thus (59) follows from

$$\tfrac{1}{2}[\nu G_\nu(0)](1 + \nu^{-1})^{1/2} < 1 \tag{60}$$

We will prove in Section 7 that $\nu G_\nu(0)$ is monotone decreasing in $\nu$, so the largest value of (60) with $\nu \geqslant 3$ occurs when $\nu = 3$, where (60) is seen to be true using Watson's value[45] of $G_3(0) = 0.505 \cdots$. We conclude that (49) holds in the $x$–$y$ model in any dimension $\nu \geqslant 3$.

We have thus verified (49) and therefore (4) for the $x$–$y$ model. By what we have done so far this does *not* imply a phase transition. Theorem 1.2 is not applicable because $\sum_{\alpha\in\Lambda} S_\alpha^{(1)}$ does not commute with $H$. Fortunately, the following alternative argument leading to a phase transition is available. Instead of proving that

$$\lim_{\Lambda\to\infty} \left\langle \left( \left|\Lambda\right|^{-1} \sum_{\alpha\in\Lambda} \mathbf{S}_\alpha \right)^2 \right\rangle_\Lambda \neq 0$$

we will in essence prove that

$$\lim_{\Lambda'\to\infty} \left( \lim_{\Lambda\to\infty} \left\langle \left( \left|\Lambda'\right|^{-1} \sum_{\alpha\in\Lambda'} S_\alpha \right)^2 \right\rangle_\Lambda \right) \neq 0$$

which directly implies nonclustering of the infinite-volume state and thus the existence of multiple phases. To prove this last result requires us to transfer our bound on $g_\mathbf{p}$ to infinite volume, which we will do by some standard manipulations of smearing spins with nice $x$-space functions.

Let $f$ be a function of compact support on $Z^\nu$. Define $g^\Lambda(f)$ by

$$\begin{aligned} g^\Lambda(f) &= \left\langle \left| \sum_{\alpha\in\Lambda} S_\alpha^{(1)} f(\alpha) \right|^2 \right\rangle_\Lambda \\ &= \sum_{\alpha,\beta\in\Lambda} \overline{f(\alpha)} f(\beta) \langle S_\alpha^{(1)} S_\beta^{(1)} \rangle_\Lambda \\ &= \frac{1}{|\Lambda|} \sum_{\mathbf{p}\in\Lambda^*} |\hat{f}^{(\Lambda)}(\mathbf{p})|^2 g_\mathbf{p}^{(\Lambda)} \end{aligned} \tag{61}$$

where $g_\mathbf{p}^{(\Lambda)} = \langle \hat{S}_\mathbf{p}^{(1)} \hat{S}_{-\mathbf{p}}^{(1)} \rangle_\Lambda$ and

$$\hat{f}^{(\Lambda)}(\mathbf{p}) = \sum_{\alpha\in\Lambda} f(\alpha) \exp(+i\mathbf{p}\cdot\alpha)$$

Now, for $\mathbf{p} \neq 0$

$$g_\mathbf{p}^{(\Lambda)} \leqslant G_\mathbf{p} = \tfrac{1}{2}(B_\mathbf{p}^{(1)} C_\mathbf{p}^{(1)})^{1/2} \coth[\tfrac{1}{2}\beta(C_\mathbf{p}^{(1)}/B_\mathbf{p}^{(1)})^{1/2}]$$

so that, picking a subsequence of $\Lambda$ for which the state $\langle \cdots \rangle$ converges,

$$\lim_{\Lambda\to\infty} g^\Lambda(f) \leqslant (2\pi)^{-\nu} \int_{|p_i|\leqslant\pi} |\hat{f}(\mathbf{p})|^2 G_\mathbf{p}\, d^\nu p$$

whenever $\hat{f}(\mathbf{p} = 0) = 0$. [The sum converges to the integral because $f$ has compact support, so $|\hat{f}(\mathbf{p})| \leqslant C|\mathbf{p}|$ if $\hat{f}(0) = 0$.] Let $F$ be defined by

$$F(\boldsymbol{\alpha} - \boldsymbol{\beta}) = \langle S_{\boldsymbol{\alpha}}^{(1)} S_{\boldsymbol{\beta}}^{(1)} \rangle_{\infty}$$

and let $\hat{F}$ be the measure given by the Fourier transform of $F$, i.e., formally

$$\hat{F}(\mathbf{p}) = \sum_{\boldsymbol{\alpha}} [\exp(i\mathbf{p} \cdot \boldsymbol{\alpha})] F(\boldsymbol{\alpha})$$

Thus, since the sum in (61) is finite

$$\int |\hat{f}(\mathbf{p})|^2 \hat{F}(\mathbf{p})\, d^\nu p = \sum_{\boldsymbol{\alpha},\boldsymbol{\beta}} \overline{f(\boldsymbol{\alpha})} f(\boldsymbol{\beta}) F(\boldsymbol{\alpha} - \boldsymbol{\beta})$$

$$= \lim_{\Lambda \to \infty} g^{\Lambda}(f) \leqslant \int |\hat{f}(\mathbf{p})|^2 G_{\mathbf{p}}\, d^\nu p$$

whenever $\hat{f}(0) = 0$. It follows by letting $\hat{f}$ approach a $\delta$-function at $\mathbf{p} \neq 0$ that

$$\hat{F}(\mathbf{p}) = C\, \delta(\mathbf{p}) + \text{absolutely continuous in } p$$

with $|\hat{F}(\mathbf{p})| \leqslant G_{\mathbf{p}}$ for $\mathbf{p} \neq 0$. By the Plancherel relation,

$$\lim_{\beta \to \infty} C \geqslant D_{\infty}^{(1)} - (2\pi)^{-\nu} \int_{|p_i| \leq \pi} \tfrac{1}{2} (B_{\mathbf{p}}^{(1)} C_{\mathbf{p}}^{(1)})^{1/2}\, d^\nu p$$

so that (49) implies that $\langle \cdots \rangle_{\infty}$ does not cluster and there is a multiplicity of phases.[41]

We summarize Example 4 in a theorem:

**Theorem 5.2.** The spin $1/2$ $x$–$y$ model with nearest neighbor interactions on a simple cubic lattice in $\nu \geqslant 3$ dimensions possesses a phase transition at sufficiently low temperatures.

As a final remark, we note that even if we could complete the proof of the existence of the phase transition in the isotropic model, we presumably do not get the correct value for the spontaneous magnetization as $T \to 0$. For, we would obtain (from Theorem 1.3)

$$\lim_{T \to 0} m(S^{(3)})^2 \geqslant S(S + 1 - \sqrt{\tfrac{3}{2}})$$

while the correct value as $T \to 0$ is presumably $S^2$.

## 6. PHASE TRANSITIONS: THE ANTIFERROMAGNETIC CASE

In the classical case, once one proves phase transitions for simple cubic, nearest neighbor ferromagnets, one automatically has them also for the antiferromagnets since the symmetry $\mathbf{S}_\alpha \to (-1)^{|\alpha|}\mathbf{S}_\alpha(|\alpha| = \sum_{i=1}^{\nu} \alpha_i)$ takes one Hamiltonian into the other. A similar argument works for certain quantum systems, e.g., the $x$–$y$ model, since there is a unitary operator that takes $\mathbf{S}_\alpha^{(i)} \to (-1)^{|\alpha|}\mathbf{S}_\alpha^{(i)}$ for $i = 1, 2$ (namely rotation by $\pi$ about the $z$ axis of those $\alpha$ with $|\alpha|$ odd). However, for the isotropic quantum Heisenberg model no such symmetry exists: Even for two spins, the smallest eigenvalue of $-\mathbf{S}_\alpha \cdot \mathbf{S}_\beta$ ($\alpha \neq \beta$) (which is $-S^2$) is different from the smallest eigenvalue of $\mathbf{S}_\alpha \cdot \mathbf{S}_\beta$ ($\alpha \neq \beta$) [which is $-S(S + 1)$]. In this section we shall prove that the antiferromagnet has phase transitions if either $S$ or $\nu$ is not very small.

We still use Theorem 5.1 (with some obvious modifications). To complete the proof, we need a modified version of the argument of Example 4 of Section 5, since the staggered magnetization does not commute with $H$. We have already computed $\sum_i [S_\alpha^{(i)}, [S_\alpha^{(i)}, H]]$ since the ferromagnetic and antiferromagnetic Hamiltonians differ only by a sign. Knowing this, we will turn below to estimating $C_{\mathbf{p}}$. Obviously we can take $D = S(S + 1)$. Thus our main problem will be to estimate $B_{\mathbf{p}}$, to which we turn first. Theorem 4.2, suitably modified, can be extended to the case of ferromagnetically coupled real matrices and antiferromagnetically coupled imaginary matrices. By a local rotation, the quantum antiferromagnet can be brought into this form.

**Theorem 6.1.** Let $H$ be the Hamiltonian of the simple cubic Heisenberg antiferromagnet with nearest neighbor interactions. Then

$$\sum_{j=1}^{3} b_{\mathbf{p}}^{(j)} \leqslant 3/2E_{\mathbf{p}}'$$

where

$$E_{\mathbf{p}}' = \nu + \sum_{i=1}^{\nu} \cos p_i$$

The proof is similar to that of Theorem 4.1, so we only sketch the details, emphasizing the differences. Instead of Lemma 4.1, we need the following result:

**Lemma 6.1.** Let $\mathscr{H}_1$ be a finite-dimensional vector space with a distinguished complex conjugation and let $\mathscr{H} = \mathscr{H}_1 \otimes \mathscr{H}_1$. If $A$ is an operator on $\mathscr{H}_1$, we use $A$ for $A \otimes 1$ and $\tilde{A}$ for $1 \otimes A$. If $A$, $B$, $C_1,\ldots, C_l$ are *real* self-adjoint operators and $D_1,\ldots, D_k$ are imaginary self-adjoint operators and $h_1,\ldots, h_l$ are real numbers, then

$$\{\mathrm{Tr}[\exp(X)]\}^2 \leqslant \mathrm{Tr}[\exp(Y)]\,\mathrm{Tr}[\exp(Z)]$$

where

$$X = A + \tilde{B} - \sum_{i=1} (C_i - \tilde{C}_i - h_i)^2 + \sum_{j=1}^{k} (D_j - \tilde{D}_j)^2$$

$$Y = A + \tilde{A} - \sum_{i=1}^{l} (C_i - \tilde{C}_i)^2 + \sum_{j=1}^{k} (D_j - \tilde{D}_j)^2$$

$$Z = B + \tilde{B} - \sum_{i=1}^{l} (C_i - \tilde{C}_i)^2 + \sum_{j=1}^{k} (D_j - \tilde{D}_j)^2$$

*Proof.* This follows the proof of Lemma 4.1, except that we use

$$\exp(F^2) = (4\pi)^{-1/2} \int \exp(-\tfrac{1}{4}k^2 + kF)\, dk$$

to "linearize" the $\exp(D - \tilde{D})^2$ factors. ∎

*Proof of Theorem 6.1.* Let $\mathbf{p}'$ be given by $p_i' = \pi - p_i$ so that $E_{\mathbf{p}}' = E_{\mathbf{p}'}$. Define $\mathbf{S}_\alpha'$ by

$$(S_\alpha')^{(i)} = (-1)^{|\alpha|} S_\alpha^{(i)}; \qquad i = 1, 3$$

$$(S_\alpha')^{(2)} = S_\alpha^{(2)}$$

There is a local unitary operator taking the $\mathbf{S}_\alpha$ into the $\mathbf{S}_\alpha'$ (rotations by $\pi$ about the $y$ axis for $|\alpha|$ odd). Moreover, the Hamiltonian for the $S'$ system is ferromagnetic in the 1 and 3 variables, which have real representatives, and antiferromagnetic in the 2 variable, which has an imaginary representative. Using Lemma 6.1 and the method of the proof of Theorems 4.1 and 4.2, we see that

$$(b_{\mathbf{p}}')^{(i)} \leqslant 1/2E_{\mathbf{p}}$$

for $i = 1, 3$. But $(b_{\mathbf{p}}')^{(i)} = b_{\mathbf{p}}^{(i)}$ for $i = 1, 3$, so we have that

$$b_{\mathbf{p}}^{(3)} \leqslant 1/2E_{\mathbf{p}}'$$

By symmetry

$$\sum_i b_{\mathbf{p}}^{(i)} \leqslant 3/2E_{\mathbf{p}}' \quad ∎$$

To estimate $C_{\mathbf{p}}$, we introduce a basic constant $\rho_\nu$ as follows. Let $H_\Lambda$ be the Hamiltonian of the antiferromagnet with periodic boundary conditions. We define

$$\rho_\nu = - \lim_{\Lambda \to \infty} (\nu|\Lambda|)^{-1} \inf \operatorname{spec}(H_\Lambda)$$

where one can show the limit exists as $\Lambda \to \infty$ in a suitable (e.g., van Hove) sense by mimicking the standard arguments on the existence of the thermodynamic limit for the free energy.[41] By (55) we have that

$$C_{\mathbf{p}} \leqslant 4E_{\mathbf{p}}\rho_\nu$$

Thus (49) becomes

$$S(S+1) > \sqrt{\tfrac{3}{2}}(\rho_\nu)^{1/2}(2\pi)^{-\nu}\int_{|p_i|\leq\pi}(E_{\mathbf{p}}/E_{\mathbf{p}}')^{1/2}\,d^\nu p \tag{62}$$

If $\nu \geqslant 3$ [so that (50) holds] and (62) holds, then a lower bound on the critical $\beta$ is given by the solution of

$$S(S+1) = (2\pi)^{-\nu}\int_{|p_i|\leq\pi}(E_{\mathbf{p}}/E_{\mathbf{p}}')^{1/2}(3\rho_\nu/2)^{1/2}\coth[\beta(2\rho_\nu E_{\mathbf{p}}E_{\mathbf{p}}'/3)^{1/2}]\,d\nu_p \tag{63}$$

We conclude this section by demonstrating that (62) holds for $\nu \geqslant 3$, $S \geqslant 1$; and when $S = 1/2$, for $\nu$ sufficiently large. To do this we use the bound of Anderson[1] (reproduced in Appendix C):

$$\rho_\nu \leqslant S[S+(2\nu)^{-1}] \tag{64}$$

Moreover, by the Schwarz inequality

$$\nu K_\nu = (2\pi)^{-\nu}\int_{|p_i|\leq\pi}(E_{\mathbf{p}}/E_{\mathbf{p}}')^{1/2}\,d^\nu p \leqslant [\nu G_\nu(0)]^{1/2} \tag{65}$$

[where $G_\nu(0)$ is given by (11)], so that (62) certainly holds if

$$S(S+1)^2 > \tfrac{3}{2}[S+(2\nu)^{-1}]\nu G_\nu(0) \tag{66}$$

Using the fact that $\nu G_\nu(0)$ is monotone decreasing in $\nu$ (see the next section) and that $G_3(0) = 0.505\ldots$,[45] we see that (66) certainly holds if $S \geqslant 1$ and $\nu \geqslant 3$. Moreover, since $\nu G_\nu(0) \to 1$ as $\nu \to \infty$, even for $S = 1/2$, (66) holds for $\nu$ sufficiently large—in fact, (62) is so close to holding for $\nu = 3$, $S = 1/2$ when (64) is used that we expect that numerical analysis of $\nu K_\nu$ for $\nu \geqslant 4$ would show that, using (64) to bound $\rho_\nu$, (62) holds for $\nu \geqslant 4$, $S = 1/2$. It is also true that $\nu K_\nu$ is monotone decreasing in $\nu$ by the method of Theorem 7.1, because $f(x) = [(1+x)/(1-x)]^{1/2}$ is convex in $[0, 1)$.

M. L. Glasser has kindly evaluated the integral in (62) for us and found that

$$3K_3 = 1.157$$

[to be compared with the bound 1.23 of (65)]. Thus for (62) to hold with $S = 1/2$, $\nu = 3$ one needs that $\rho_3 < 0.28$. The best rigorous bounds [from (64) and the trivial $\rho_\nu \geqslant S^2$] are $0.25 < \rho_3 < 0.33$. The best numerical estimate on $\rho_3$ we can find is $\rho_3 \cong 0.3$,[22] but there is a need for better rigorous bounds to be certain that (62) fails for $\nu = 3$, $S = 1/2$ and to check the $\nu$ for which (62) holds when $S = 1/2$. We summarize with:

**Theorem 6.2.** The nearest neighbor, simple cubic antiferromagnet has a phase transition at sufficiently low temperature if $\nu \geqslant 3$, $S = 1, 3/2,\ldots$ or if $S = 1/2$ and $\nu$ is sufficiently large.

## 7. DIMENSIONAL DEPENDENCE OF THE CRITICAL TEMPERATURE

It is a well-known element of folklore that as $\nu \to \infty$, transition temperatures approach those of the mean field approximation. Our goal in this final section is to note the extent to which this piece of folklore is proven by the upper bounds on the transition $\beta$ of FSS for the classical $N$-vector model and our putative bounds for the spin-$S$ quantum Heisenberg ferromagnet. For the classical case the results are summarized in Table III. In the third column we give Griffiths'[18] rigorous lower bound for the Ising-model critical $\beta$ and in the fourth column we give the lower bound obtained by applying the method of Brascamp and Lieb[6] to the case at hand (see Appendix D). The table should be supplemented with the following result:

**Theorem 7.1.** $\nu G_\nu(0)$ is monotone decreasing and approaches 1 as $\nu \to \infty$.

*Proof.* By the convexity of the function $g(x) = x^{-1}$ for $x \geqslant 0$, we have that for $f_1,\ldots,f_n \geqslant 0$ and $F = \sum_{j=1}^{n} f_j$

$$F^{-1} \leqslant \sum_{j=1}^{n} n^{-2}(n-1)(F-f_j)^{-1}$$

since

$$F = \sum_{j=1}^{n} \frac{1}{n}\frac{n}{n-1}(F-f_j)$$

**Table III. Transition β in N-Vector Model**

| $N$ | FSS upper bound | Griffiths' lower bound | BL lower bound | Mean field |
|---|---|---|---|---|
| 1 | $\frac{1}{2}G_\nu(0)$ | $\tanh^{-1}(2\nu)^{-1}$ | $1/4\nu$ | $1/2\nu$ |
| $\geqslant 2$ | $\frac{1}{2}NG_\nu(0)$ | — | $N/4\nu$ | $N/2\nu$ |

Taking $f_j = 1 - \cos p_j$ and integrating, we obtain

$$G_n(0) \leqslant n^{-1}(n-1)G_{n-1}(0)$$

proving the required monotonicity.

That the limit is 1 follows from the lemma below. ■

**Lemma 7.1.** Suppose that $f_1,\ldots,f_n,\ldots$ are identically distributed, independent, *nonnegative* random variables and that $F$ is a function on $(0,\infty)$ such that:

(i) $|F(x+\epsilon) - F(x)| \leqslant C\epsilon/x^k$, all $x > 0$, and $0 < \epsilon < 1$, for some $C, k > 0$.

(ii) $\mathrm{Exp}[(\sum_{i=1}^m f_i)^{-k}] < \infty$ for some $m$, where Exp = "expectation."

Then

$$\lim_{n\to\infty} \mathrm{Exp}\left[F\left(\frac{1}{n}\sum_{i=1}^n f_i\right)\right] = F(\mathrm{Exp}(f))$$

*Proof.* By the convexity of $1/x^k$ and the argument above, $\mathrm{Exp}\{[1/n \sum_{i=1}^n f_i]^{-k}\}$ is monotone decreasing in $n$. Thus, for $n \geqslant m$ and $0 < \epsilon < 1$

$$\left|\mathrm{Exp}\left[F\left(\frac{1}{n}\sum_{i=1}^n (f_i+\epsilon)\right)\right] - \mathrm{Exp}\left[F\left(\frac{1}{n}\sum_{i=1}^n f_i\right)\right]\right|$$

$$\leqslant C\epsilon\, \mathrm{Exp}\left[\left(\sum_{i=1}^m f_i\right)^{-k} m^k\right]$$

Moreover, by the usual strong law of large numbers[7]

$$\mathrm{Exp}\left[F\left(\frac{1}{n}\sum_{i=1}^n (f_i+\epsilon)\right)\right] \to F(\mathrm{Exp}(f)+\epsilon)$$

These two formulas and the continuity of $F$ imply the result. ■

Given this theorem, we see that the ratio of the FSS bound to the mean field theory bound approaches 1. Since the Griffiths bound also approaches mean field theory, we have the following theorem:

**Theorem 7.2.** Let $T_c^\nu$ be the true transition temperature of the simple cubic, $\nu$-dimensional Ising model. Then

$$T_c^\nu/2\nu \to 1$$

as $\nu \to \infty$.

Unfortunately (for reasons we explain in Appendix D), the Brascamp–Lieb bound appears to be off by a factor of two in the nearest neighbor case, so for $N \geqslant 2$, where the only upper bound we have on $T_c$ is theirs, we cannot prove that $NT_c^{\nu}/2\nu$ converges to 1. We do know, however, that its $\underline{\lim}$ is $\geqslant 1$ and its $\overline{\lim}$ is $\leqslant 2$, so that the FSS bound certainly has the right $\nu$ dependence and if mean field theory is asymptotically correct, then FSS is asymptotically correct.

To discuss the $\nu$ dependence of our presumed upper bound on the critical $\beta$ in the spin-$S$ quantum Heisenberg model, we abstract the argument of Theorem 7.1:

**Theorem 7.3.** Let $f_1,\ldots,f_n,\ldots$ be identically distributed, independent random variables. Let $F$ be a nonnegative convex (respectively concave) function on the convex hull of Ran $f_1$. Then $\mathrm{Exp}[F((1/n)\sum_{i=1}^{n} f_i)]$ is monotone decreasing (respectively increasing) in $n$.

*Proof.* Use the convexity and

$$\frac{1}{n}\sum_{i=1}^{n} f_i = \frac{1}{n}\sum_{j=1}^{n}\left(\frac{1}{n-1}\sum_{\substack{i\neq j\\ 1\leqslant i\leqslant n}} f_i\right) \quad \blacksquare$$

**Theorem 7.4.** Let $\tilde{\beta}_c(S,\nu)$ be the solution of (52*H*) and $\zeta(S,\nu) = \nu\tilde{\beta}_c(S,\nu)$. Let $\zeta(S,\infty)$ be the solution of the equation

$$S + 1 = \sqrt{\tfrac{3}{2}}\coth[\sqrt{\tfrac{3}{2}}S\zeta(S,\infty)] \tag{67}$$

(tabulated in Table IV). Then $\zeta(S,\nu)$ is monotone decreasing to $\zeta(S,\infty)$ as $\nu\to\infty$.

*Proof.* Let $f_i$ $(i = 1, 2,\ldots)$ be the function $1 - \cos p_i$ on $[-\pi,\pi]^\infty$ with the natural product probability measure $X\, dp_i/2\pi$. Let $F$ be the function

$$F(x) = \sqrt{\tfrac{3}{2}}\coth(\sqrt{\tfrac{3}{2}}x)$$

**Table IV. Values of[a] $\zeta(S,\infty) = \lim_{\nu\to\infty} \nu\tilde{\beta}_c(S,\nu)$**

| $S$ | $S^2\zeta(S,\infty)$ | $S(S+1)\zeta(S,\infty)$ |
|---|---|---|
| $\frac{1}{2}$ | 0.702 | 2.11 |
| 1 | 0.873 | 1.75 |
| $\frac{3}{2}$ | 0.985 | 1.64 |
| 2 | 1.06 | 1.59 |
| $\frac{5}{2}$ | 1.12 | 1.57 |
| 3 | 1.16 | 1.55 |
| $\frac{7}{2}$ | 1.20 | 1.54 |
| 4 | 1.22 | 1.53 |
| 15 | 1.41 | 1.50 |
| 50 | 1.47 | 1.50 |
| $\infty$ | 1.50 | 1.50 |

[a] $\nu$ is the lattice dimension; $S$ is the spin.

Then $\zeta(S, \nu)$ solves

$$S + 1 = \operatorname{Exp}\left[F\left(S\zeta(S, \nu)\nu^{-1} \sum_{i=1}^{\nu} f_i\right)\right] \tag{68}$$

Now $F$ is convex and strictly monotone decreasing (see Appendix A). So, using Theorem 7.3,

$$\begin{aligned}
&\operatorname{Exp}\left[F\left(S\zeta(S, \nu - 1)\nu^{-1} \sum_{i=1}^{\nu} f_i\right)\right] \\
&\quad \leqslant \operatorname{Exp}\left[F\left(S\zeta(S, \nu - 1)(\nu - 1)^{-1} \sum_{i=1}^{\nu-1} f_i\right)\right] \\
&\quad = S + 1 = \operatorname{Exp}\left[F\left(S\zeta(S, \nu)\nu^{-1} \sum_{i=1}^{\nu} f_i\right)\right]
\end{aligned}$$

By the monotonicity of $F$, $\zeta(S, \nu - 1) \geqslant \zeta(S, \nu)$. Thus $\zeta$ approaches a limit $\zeta_\infty$. By a simple extension of Lemma 7.1, $\zeta_\infty$ solves (67) and so equals $\zeta(S, \infty)$. ∎

Mean field theory gives the same transition temperature for the spin-$S$ Heisenberg model and the spin-$S$ Ising model. Thus the $\zeta(S, \infty)$ result does not agree with mean field theory, but at least the gross (i.e., $\nu^{-1}$) structure of mean field theory and our bound agree.

## APPENDIX A. A GARDEN OF COTH AND TANH

In this appendix, we collect some basic properties of the functions $\coth x$ and $f$ given by $f(x \tanh x) = x^{-1} \tanh x$.

**Theorem A.1.** The function

$$g(x) = \coth x$$

is strictly monotone decreasing and convex on $[0, \infty)$. The function

$$h(x) = g(x) - x^{-1}$$

is strictly monotone increasing and concave on $[0, \infty)$ with $h(0) = 0$. The function $g(x)$ obeys

$$x^{-1} < g(x) < 1 + x^{-1}, \qquad x > 0 \tag{A1}$$

$$|g(x + \epsilon) - g(x)| \leqslant \epsilon/x^2, \qquad 0 < \epsilon, \quad 0 < x \tag{A2}$$

*Proof.* By direct calculation, $g'(x) = -(\sinh x)^{-2}$ and $g''(x) = 2(\cosh x)(\sinh x)^{-3}$. The first sentence is then obvious. The bound $\sinh x > x$ (which comes from the power series for sinh) then yields $0 \geqslant g'(x) > -x^{-2}$, from which (A2) follows, as does $h' > 0$. Now $\lim_{x\to 0} h(x) = 0$ is obvious from the power series expansion of coth and sinh. Condition (A1) just says $h(0) < h(x) < h(\infty)$. This leaves the concavity of $h$. Given $g''$ above, we need that

$$x^3 \cosh x \leqslant \sinh^3 x$$

or equivalently that

$$\frac{\sinh 2x}{2x} \leqslant \left(\frac{\sinh x}{x}\right)^4 \tag{A3}$$

or

$$\prod_{n=1}^{\infty} [1 + (2x/n\pi)^2] \leqslant \prod_{n=1}^{\infty} [1 + (x/n\pi)^2]^4 \tag{A4}$$

which is obvious. ■

*Remark.* The convexity of $g$ and concavity of $h$ are useful in obtaining quick (i.e., hand calculator!) estimates on the function

$$H_\nu(x) = (2\pi)^{-\nu} \int_{|p_i|\leqslant\pi} \coth(xE_{\mathbf{p}})\, d^\nu p$$

For example, they lead to the bounds

$$\coth(\nu x) \leqslant H_\nu(x) \leqslant \coth(\nu x) + (\nu x)^{-1}[\nu G_\nu(0) - 1] \tag{A5}$$

via Jensen's inequality and the fact that $(2\pi)^{-\nu} \int_{|p_i|\leqslant\pi} E_{\mathbf{p}}\, d^\nu p = \nu$.

Now we turn to the study of the function given by

$$f(x \tanh x) = x^{-1} \tanh x \tag{A6}$$

**Theorem A.2.** $f$ is a well-defined, strictly monotone-decreasing, convex function of $(0, \infty)$ to $(0, 1)$ with $\lim_{x\to 0} f(x) = 1$ and $\lim_{x\to\infty} f(x) = 0$.

*Proof.* Let $\alpha(x) = x \tanh x$. Since $\alpha$ is strictly monotone increasing from $(0, \infty)$ to $(0, \infty)$, it has an inverse function $\beta(x)$ which is also strictly monotone increasing. Let $\gamma(x) = x^{-1} \tanh x$. Then $\gamma'(x) = x^{-1}(\cosh x)^{-2} - x^{-2} \tanh x = (x \cosh x)^{-1}[(\cosh x)^{-1} - x^{-1} \sinh x] < 0$ since $(\cosh x)^{-1} < 1$ and $x^{-1} \sinh x > 1$. Thus $f = \gamma \circ \beta$ is strictly monotone decreasing. Since $\beta(0) = 0$, $\beta(\infty) = \infty$ and $\gamma(0) = 1$, $\gamma(\infty) = 0$, we have the limiting statements.

This leaves the proof of the convexity of $f$. Our original proof involved a straightforward but rather brutal computation. V. Bargmann provided us with the following proof, which, while not free of complexity, is a considerable improvement. We first note that for a function $G$ on $(0, \infty)$ to be convex, it is necessary and sufficient that for any $0 < x < y < z$

$$\det\begin{vmatrix} 1 & x & G(x) \\ 1 & y & G(y) \\ 1 & z & G(z) \end{vmatrix} \geqslant 0 \tag{A7}$$

for, if $H$ is the linear function with $H(x) = G(x)$, $H(z) = G(z)$, the determinant in (A7) is $[H(y) - G(y)](z - x)$ and convexity is that $G(y) < H(y)$. Thus convexity of $f$ is equivalent to

$$\det\begin{vmatrix} 1 & x\tanh x & x^{-1}\tanh x \\ 1 & y\tanh y & y^{-1}\tanh y \\ 1 & z\tanh z & z^{-1}\tanh z \end{vmatrix} \geqslant 0 \tag{A8}$$

for $x \leqslant y \leqslant z$, where we have used the monotonicity of $\alpha$. Multiplying by $xyz(\coth x)(\coth y)(\coth z)$ and interchanging columns, we find that (A8) is equivalent to

$$\det\begin{vmatrix} 1 & x^2 & x\coth x \\ 1 & y^2 & y\coth y \\ 1 & z^2 & z\coth z \end{vmatrix} \leqslant 0 \tag{A9}$$

Again using the criterion (A7), we see that (A9) is equivalent to the concavity of the function $P(x) = x^{1/2}\coth(x^{1/2})$.

We have therefore shown the convexity of $f$ is equivalent to the concavity of $P$. By a straightforward calculation,

$$\begin{aligned} 4y^3(\sinh^3 y)P''(y^2) &= 2y^2\cosh y - \cosh y\sinh^2 y - y\sinh y \\ &= y(y\cosh y - \sinh y) + (\cosh y)(y^2 - \sinh^2 y) \\ &\leqslant y(y\cosh y - \sinh y) + (\cosh y)\left(-\frac{y^4}{3}\right) \\ &= \sum_{n=0}^{\infty} y^{2n+4}\left[\frac{1}{(2n+1)(2n+3)} - \frac{1}{3}\right]\frac{1}{(2n)!} \leqslant 0 \end{aligned}$$

In the third step, we used

$$-(y^2 - \sinh^2 y) = (y + \sinh y)(\sinh y - y) \geqslant (2y)(y^3/6). \quad \blacksquare$$

**Theorem A.3.** Let $F$ be any convex function on $[0, \infty)$. The function $H(y, c) = y[F(cy^{-1}) - 1]$ for $y, c > 0$ is jointly convex in $(y, c)$. If $F$ is monotone decreasing, then $H$ is monotone decreasing in both $y$ and $c$.

*Proof.* The second derivative matrix

$$F''\left(\frac{c}{y}\right)\begin{pmatrix} c^2y^{-3} & -cy^{-2} \\ -cy^{-2} & y^{-1} \end{pmatrix}$$

is positive semidefinite. For $y$ fixed, $H$ is monotone decreasing as $c$ increases since $F$ is monotone decreasing. For $c$ fixed, $\partial H/\partial y = F(cy^{-1}) - 1 - cy^{-1}F'(cy^{-1})$ is negative by the convexity of $F$ [which implies that $(0, F(0))$ must lie above the tangent to $F$ at the point $(cy^{-1}, F(cy^{-1}))$]. ■

**Theorem A.4.**[6]

$$f(x) \geqslant x^{-1}(1 - e^{-x}) \tag{A10}$$

*Remark.* As we explained in Section 3, this can also be proved by appealing to the best possible nature of our bound there and to the bound of Roepstorff.

*Proof.* In (A10) replace $x$ by $(x/2)\tanh(x/2) = (x/2)(e^x - 1)/(e^x + 1)$. Then (A10) is equivalent to

$$\frac{2}{x}\frac{e^x - 1}{e^x + 1} \geqslant \frac{2}{x}\frac{e^x + 1}{e^x - 1}\left[1 - \exp\left(-\frac{x}{2}\frac{e^x - 1}{e^x + 1}\right)\right]$$

or

$$4e^x(e^x + 1)^{-2} \leqslant \exp\left(-\frac{x}{2}\frac{e^x - 1}{e^x + 1}\right)$$

or

$$x - 2\ln\frac{e^x + 1}{2} \leqslant -\frac{x}{2}\frac{e^x - 1}{e^x + 1}$$

or

$$Q(x) \equiv x\frac{3e^x + 1}{e^x + 1} - 4\ln\frac{e^x + 1}{2} \leqslant 0 \tag{A11}$$

Now, by a simple computation, $Q(0) = 0$ and

$$Q'(x) = 2e^x(e^x + 1)^{-2}(x - \sinh x) < 0$$

so (A11) holds. ■

## APPENDIX B. COMPUTATIONS WITH HARMONIC OSCILLATORS

Let $H = \omega A^*A$ with $[A, A^*] = s$, where $s > 0$ is the $c$-number commutator, in a space where there is a vector $\psi_0$ with $A\psi_0 = 0$ and $\{(A^*)^n\psi_0\}$ a spanning set. Then there is an orthonormal basis $\psi_n$ with $H\psi_n = n\omega s\psi_n$

[6] See footnote 5, p. 349.

and $(\psi_m, A\psi_n) = \delta_{m,n-1}(ns)^{1/2}$. Thus, using (21) when $\beta = 1$, it is easy to verify that

$$(A^*, A) = \omega^{-1} \tag{B1}$$

Moreover,

$$\tfrac{1}{2}\langle(A^*A + AA^*)\rangle = \langle A^*A + \tfrac{1}{2}s\rangle = \tfrac{1}{2}s + \omega^{-1}\langle H\rangle = \tfrac{1}{2}s\coth(\tfrac{1}{2}\omega s) \tag{B2}$$

Finally,

$$[[A^*, H], A] = \omega s^2 \quad (c\text{-number}) \tag{B3}$$

so, in particular,

$$\frac{\langle[[A^*, H], A]\rangle}{4(\frac{1}{2}\langle A^*A + AA^*\rangle)} = \frac{\omega s}{2}\tanh\left(\frac{\omega s}{2}\right) \tag{B3'}$$

There are three important things to notice about the formulas (B1)–(B3'):

(i) Of the various quantities, only the Duhamel two-point function is independent of the commutator parameter $s$. This adds to the evidence elsewhere in this paper that the DTF is "closer" to the classical two-point function than the usual thermal two-point function.

(ii) The inequality of Section 3

$$(A^*, A) \geqslant \tfrac{1}{2}\langle A^*A + AA^*\rangle f(\langle[[A^*, H], A]\rangle/(4 \cdot \tfrac{1}{2}\langle A^*A + AA^*\rangle))$$

is saturated in the case at hand, and, as $\omega$ and/or $s$ vary, the argument of $f$ runs through all values from 0 to $\infty$. This implies that $f$ is the best possible function.

(iii) The Bogoliubov inequality in its strong form ($\beta = 1$)

$$|\langle[A, C]\rangle|^2 \leqslant (A^*, A)\langle[C^*, [H, C]]\rangle$$

is saturated in the example at hand if $C$ is chosen to be $A^*$.

Not only are the inequalities of Section 3 saturated by a single harmonic oscillator, but those of Section 4 are saturated by a system of coupled harmonic oscillators. This is not surprising since the classical analogs of these inequalities[12] are saturated by classical harmonic oscillators. For, in a $\nu$-dimensional box $\Lambda$, let

$$H = \tfrac{1}{2}\sum_{\alpha\in\Lambda}(p_\alpha{}^2 + x_\alpha{}^2) - J\sum_{|\alpha-\beta|=1} x_\alpha x_\beta$$

on $L^2(\mathbb{R}^{|\Lambda|})$ with $p_\alpha = (1/i)\,\partial/\partial x_\alpha$. By using normal modes (which are just Fourier series), it is easy to see that for $H$ to be positive one needs

$$2J\nu < 1 \tag{B4}$$

[for the frequencies obey $\mu(\mathbf{k}) = 1 - 2J\sum_{i=1}^{\nu}\cos k_i$] and

$$(\hat{x}(\mathbf{k})^*, \hat{x}(\mathbf{k})) = [\beta\mu(\mathbf{k})]^{-1} \tag{B5}$$

As $J$ approaches $(2\nu)^{-1}$ [the largest value allowed by (B4)], the inequality of Section 4 [in which the right side of (B5) is replaced by $(2JE_{\mathbf{k}})^{-1}$] is saturated.

The saturation of the inequalities from Sections 3 and 4 means the presumed lower bound on the transition temperature given by (52$H$) will agree with the transition temperature in a spin wave theory (which is essentially a harmonic approximation) with *three* degrees of freedom at each site instead of the usual two.

## APPENDIX C. LOWER BOUNDS ON THE GROUND-STATE ENERGIES OF SOME QUANTUM SPIN HAMILTONIANS

In this appendix, we wish to derive lower bounds on the ground-state energy of the Hamiltonian of the $x$–$y$ model and of the isotropic Heisenberg antiferromagnet. These bounds are needed, respectively, in Sections 5 and 6 to obtain upper bounds on $|\Lambda|^{-1}\langle -H_\Lambda\rangle$. At first sight one might expect the $x$–$y$ model to look more like the Heisenberg ferromagnet, which has an exactly calculable ground state, rather than the antiferromagnet. This expectation is wrong, as can be understood by making a 180° rotation about the $z$ axis at each site in one of the natural sublattices of the $x$–$y$ ferromagnet.

We shall prove the following result:

**Theorem C.1.** For $n + 1$ independent spin-1/2 spins $\mathbf{S}_0,\ldots,\mathbf{S}_n$ the maximum eigenvalue of

$$A_{n+1} = S_0^{(x)} \sum_{i=1}^{n} S_i^{(x)} + S_0^{(y)} \sum_{i=1}^{n} S_i^{(y)}$$

is $\frac{1}{2}l$ if $n = 2l - 1$ and $\frac{1}{2}[l(l+1)]^{1/2}$ if $n = 2l$.

**Theorem C.2 (Anderson[1]).** For $n + 1$ independent spin-$S$ spins $\mathbf{S}_0,\ldots,\mathbf{S}_n$, the maximum eigenvalue of

$$B_{n+1} = -\mathbf{S}_0 \cdot \sum_{i=1}^{n} \mathbf{S}_i$$

is $S(nS + 1)$.

Theorem C.2 and the proof we give here for the reader's convenience are due to Anderson.[1] Both lead directly to upper bounds on $|\Lambda|^{-1}\langle -H_\Lambda\rangle$ since $-H_\Lambda$ can be written as a sum of $|\Lambda|$ operators unitarily equivalent to $\frac{1}{2}A_{n+1}$ (resp. $\frac{1}{2}B_{n+1}$) where $n = 2\nu$.

*Proof of Theorem C.1.* We first note that for two spins, with the usual up–down notation,

$$A_2|++\rangle = 0, \qquad A_2|--\rangle = 0$$
$$A_2|+-\rangle = \tfrac{1}{2}|-+\rangle, \qquad A_2|-+\rangle = \tfrac{1}{2}|+-\rangle$$

Since $A_n$ is a sum of pair operators equivalent to $A_2$, $A_n$ is also a matrix with nonnegative elements. By the Perron–Frobenius theorem, the largest eigenvalue has an associated eigenvector with nonnegative coefficients in the up–down basis (see Lieb and Mattis[25,26]). Moreover, if we decompose the Hilbert space into $n + 2$ pieces $\mathscr{H}_j$ corresponding to $j$ up spins, $A_n$ leaves each piece invariant and the matrix is ergodic on each $\mathscr{H}_j$. Thus, on $\mathscr{H}_j$ the largest eigenvalue is simple and has strictly positive coefficients. In $\mathscr{H}_j$, there are $\binom{n}{j-1}$ basis vectors with $S_0^{(3)} = +\frac{1}{2}$ (call their sum $\phi$) and $\binom{n}{j}$ with $S_0^{(3)} = -\frac{1}{2}$ (call their sum $\psi$). By the above remarks, the eigenvector corresponding to the maximum eigenvalue $\lambda_j$ is of the form $\eta = a\phi + b\psi$.

Now, given the action of $A_2$, $A_n$ applied to a basis vector summand in $\phi$ yields $\frac{1}{2}$ times the sum of $n - j + 1$ vectors, which are summands in $\psi$. By symmetry it follows that

$$A_n\phi = \frac{1}{2}(n - j + 1)\left[\binom{n}{j-1}\Big/\binom{n}{j}\right]\psi = \frac{1}{2}j\psi$$

and similarly

$$A_n\psi = \frac{1}{2}j\left[\binom{n}{j}\Big/\binom{n}{j-1}\right] = \frac{1}{2}(n + 1 - j)\phi$$

The equation $A_n\eta = \lambda_j\eta$ thus becomes $\frac{1}{2}ja = \lambda_j b$ and $\frac{1}{2}(n + 1 - j)b = \lambda_j a$, so that $\lambda_j^2 = \frac{1}{4}(n + 1 - j)j$. The maximum value of $\lambda_j$ clearly occurs for $j = l$ when $n = 2l$ or $2l - 1$. ■

*Proof of Theorem C.2 (following Anderson*[1]*).* This is an exercise in addition of angular momentum since $B_{n+1} = -S_0 \cdot J$ with $J = \sum_{i=1}^{n} \mathbf{S}_i$. If $j$ is the eigenvalue of $J$ and $\alpha = \min(S, j)$, then the maximum value of $B_{n+1}$ is

$$-\tfrac{1}{2}[|S - j|(|S - j| + 1) - S(S + 1) - j(j + 1)] = jS + \alpha$$

Among the values $j = 0,\ldots,nS$, the maximum value of this occurs when $j = nS$. ■

## APPENDIX D. ON THE BRASCAMP–LIEB UPPER BOUND

In Ref. 6, Brascamp and Lieb applied a basic theorem on log concave function to give upper bounds on transition temperatures in pair interacting Ising models (with *no* restriction that the interaction be either ferromagnetic or nearest neighbor). In this appendix, we extend this result to the classical $N$-vector model. We begin with an abstraction of the Brascamp–Lieb argument:

**Theorem D.1.** Let $d\mu$ be a measure on $\mathbb{R}^N$ with support on $\{\mathbf{s} \mid |\mathbf{s}| = 1\}$. Define the convex function $f$ on $\mathbb{R}^N$ by

$$f(\mathbf{x}) = \log \int \exp(\mathbf{x} \cdot \mathbf{s})\, d\mu(\mathbf{s})$$

Let $a = \max_x(\partial^2 f/\partial x_i\, \partial x_j)_\infty$, where $(A_{ij})_\infty$ for a matrix $A_{ij}$ indicates its largest eigenvalue. Hence $-\frac{1}{2}a\mathbf{x}^2 + f(\mathbf{x})$ is concave in $\mathbb{R}^N$. Let $\Gamma$ be a lattice in $\mathbb{R}^\nu$ (i.e., discrete subgroup of $\mathbb{R}^\nu$) and let $J$ be a function on $\Gamma$ with $\sum_{\alpha\in\Gamma} |J(\boldsymbol{\alpha})| < \infty$ and $J(-\boldsymbol{\alpha}) = J(\boldsymbol{\alpha})$. Let $J_{\max}$ (resp. $J_{\min}$) denote the largest (resp. smallest) point of the spectrum of the operator

$$(\mathscr{J}x)_\alpha = \sum_\beta J_{\alpha-\beta} x_\beta \tag{D1}$$

on $l^2(\Gamma)$. Then, if

$$\beta(J_{\max} - J_{\min}) < a^{-1} \tag{D2}$$

there is no long-range order in the model with finite-volume partition function:

$$Z_\Lambda = \int \prod_{\alpha\in\Lambda} d\mu(\mathbf{s}_\alpha) \exp(\tfrac{1}{2}\beta \sum_{\alpha,\beta\in\Lambda} J_{\alpha\beta} \mathbf{s}_\alpha \cdot \mathbf{s}_\beta)$$

*Proof.* Without loss, we can take $\beta = 1$ by absorbing it into $J$. Let $\mathscr{J}^\Lambda$ denote the operator on $l^2(\Lambda)$ obtained from (D1) by restricting the sum to $\Lambda$. Choose $\epsilon$ small and $\Lambda$ large so that

$$\epsilon + J^\Lambda_{\max} - J^\Lambda_{\min} < a^{-1} \tag{D3}$$

Let

$$A^\Lambda_{\alpha\beta} = (\epsilon + J^\Lambda_{\max})\delta_{\alpha\beta} - J_{\alpha-\beta}$$

which is positive definite on $l^2(\Lambda)$. Moreover, $\exp[-\frac{1}{2}a\mathbf{x}^2 + f(\mathbf{x})]$ is a log concave function, so, by the argument in Ref. 6,

$$\sum_{\alpha,\beta\in\Lambda} z_\alpha z_\beta \langle s_\alpha s_\beta\rangle_\Lambda \leqslant \sum_{\alpha,\beta} z_\alpha z_\beta Q_{\alpha\beta}, \qquad Q = a(1 - aA^\Lambda)^{-1}$$

for any $\mathbf{z}$ [by (D3), $1 - aA^\Lambda$ is strictly positive definite]. Taking $\Lambda \to \infty$, we conclude the absence of long-range order. ■

*Remark 1.* By taking an Ising model with $J_{\alpha-\beta} = 1$ if $|\boldsymbol{\alpha} - \boldsymbol{\beta}| \leqslant n$ and taking $n \to \infty$ ("mean field model"), one sees that (D2) is best possible in that $\beta(J_{\max} - J_{\min}) = (1 + \epsilon)a^{-1}$ occurs with long-range order. However, for *ferromagnetic* Ising systems where $a = 1$, $\beta \sum_{\alpha\neq 0} J_\alpha < 1$ has no long-range order by a result of Griffiths.[18] If $J$ is normalized so that $J(0) = 0$, then $\sum_{\alpha\neq 0} |J_\alpha| = J_{\max}$ in the ferromagnetic case. For the nearest neighbor, simple cubic case, $J_{\max} - J_{\min} = 4\nu$ while $\sum_{\alpha\neq 0} |J_\alpha| = 2\nu$. As a result, (D2) is off by a factor of 2 at least in the Ising model, and we presume in general models, in this nearest neighbor case.

*Remark 2.* The restriction that $d\mu$ have support on the sphere is easily removed by replacing $f$ by $\log \int \exp(\mathbf{x} \cdot \mathbf{s} + \beta J_{\max}\mathbf{s}^2)\, d\mu(\mathbf{s})$.

**Theorem D.2.** Let

$$f(\mathbf{h}) = \log \int_{\substack{|\mathbf{s}|=1 \\ \mathbf{s}\in\mathbb{R}^N}} \exp(\mathbf{h} \cdot \mathbf{s})\, d\Omega_s$$

Then $\max_{\mathbf{h}} (\partial^2 f/\partial h_i\, \partial h_j)_\infty = 1/N$.

*Proof.* When $\mathbf{h} = 0$, $(\partial^2 f/\partial h_i\, \partial h_j)_\infty = 1/N$ by inspection. By symmetry, $f(\mathbf{h}) = F(h)$ is spherically symmetric. For $\mathbf{h} \neq 0$, it is easily seen that $\partial^2 f/\partial h_i\, \partial h_j$ has eigenvalue $F''(h)$ once and $F'(h)/h$, $N - 1$ times. Clearly $F'(0) = 0$ by symmetry, so $F'(h)/h \leqslant \max_{0<x<h} F''(x)$ and thus

$$\max_{\mathbf{h}} (\partial^2 f/\partial h_i\, \partial h_j)_\infty = \max_{h} F''(h)$$

We will prove that $F'''(h) \leqslant 0$ for $h \geqslant 0$ so that $\max F''(h) = F''(0) = 1/N$ [see, e.g., (D5) below]. We remark that the concavity of $F'$ for $N \geqslant 3$ follows[32] from general criteria of Ellis *et al.*[11] for GHS inequalities [the distribution for $s_1$ is a limit of distributions of the form $\exp(-V)$ with $V = a_2x^2 + a_4x^4 + \cdots$, with $a_4, a_6, \ldots \geqslant 0$].

In terms of the obvious probability measure,

$$\partial^2 f/\partial h_i\, \partial h_j = \langle s_i s_j\rangle - \langle s_i\rangle\langle s_j\rangle, \qquad \partial f/\partial h_i = \langle s_i\rangle$$

Picking $\mathbf{h} = (h, 0, \ldots)$,

$$\langle s_1^2\rangle = d^2F/dh^2 + (dF/dh)^2, \qquad \langle s_i^2\rangle = h^{-1}\, dF/dh, \quad i \geqslant 2$$

We conclude from $\langle \sum s_i^2\rangle = 1$ that $g(h) \equiv dF/dh$ obeys the differential equation

$$g'(h) + g(h)^2 + (N-1)h^{-1}g(h) = 1 \tag{D4}$$

From (D4) and the fact that $g$ is odd we see that

$$g(h) = N^{-1}h - N^{-2}(N+2)^{-1}h^3 + O(h^5) \tag{D5}$$

so that $g''(h) < 0$ for $h$ small. If $g''(h) < 0$ is not true for all $h > 0$, let $h_0$ be the first positive zero of $g''$. Now $g'$ is positive ($f$ is convex by Jensen's inequality) and $g$ is thus positive for all $h > 0$. Multiplying (D4) by $h$ and taking two derivatives and using the fact that $g, g' \geqslant 0$, we get

$$hg''' + [(N+1) + 2hg]g'' < 0$$

Since $hg$ is monotone and $g'' < 0$ on $(0, h_0)$, we have

$$hg''' + \alpha g'' \leqslant 0, \quad 0 < h < h_0; \qquad \alpha = N + 1 + 2h_0 g(h_0)$$

or

$$(h^\alpha g'')' \leqslant 0, \qquad 0 < h < h_0$$

It follows that

$$g''(h_0) \leqslant h_0^{-\alpha} h_1^{\alpha} g''(h_1) < 0$$

where $h_1$ is chosen so small that $g''(h_1) < 0$, $h_1 < h_0$. This contradiction proves that $F'''(h) < 0$, so that $\max F''(h) = F''(0) = 1/N$. ■

**Corollary D.1.** In the nearest-neighbor, $N$-vector, $\nu$-dimensional ferromagnet, there is no long-range order in the two-point function if $\beta \leqslant N/4\nu$.

## APPENDIX E. TRANSFER MATRICES IN QUANTUM SPIN SYSTEMS

The transfer matrix is a useful technique in classical spin systems, both as a calculational tool in one-dimensional systems and in the Onsager solution,[33,43] and as a general theoretical tool (e.g., the appendix of Ref. 12). It is a well-known folk theorem that quantum spin systems do not possess transfer matrices. In this appendix, we want to show that this folk theorem is wrong, although the definition of our transfer matrix is sufficiently abstract that it is unlikely to be a useful calculational tool; it may turn out to be a useful theoretical tool. Our construction is borrowed from axiomatic and constructive quantum field theory, where a particular positivity condition has been emphasized by Osterwalder and Schrader[34] in recovering the Hamiltonian semigroup from the Euclidean Green's functions. Klein[23] has emphasized the idea of exploiting the notion of Osterwalder–Schrader positivity in more general contexts.

It should be emphasized that our transfer matrix differs from the usual one in an important way: Our basic inner product will have a lattice spacing built into it; that is, if $A$, $B$ are operators at a single site and $T$ is the transfer matrix, then the matrix element $(A, T^n(B))$ will involve a thermal expectation of operators at sites a distance $n + 1$ from each other.

We shall consider general Hamiltonians of the form (41) with real matrices. Consider a volume $\Lambda$ with $L_1 = 2m_1$, $\alpha_1 = -m_1 + 1, -m_1 + 2, \ldots, m_1$. Define an automorphism $R$ from operators on the sites with $\alpha_1 \geqslant 1$ to the operators on the sites with $\alpha_1 \leqslant 0$ by reflecting about the plane $\alpha_1 = 1/2$, e.g.,

$$R(\mathbf{S}_\alpha) = \mathbf{S}_{\alpha'}$$

with $\alpha_1' = 1 - \alpha_1$, $\alpha_2' = \alpha_2, \ldots, \alpha_\nu' = \alpha_\nu$. Thus $H_\Lambda$ has the form

$$H_\Lambda = B + R(B) - \sum_{i=1}^{k} C_i R(C_i) \tag{E1}$$

where $B, C_1, \ldots, C_k$ are operators on the sites with $\alpha_1 \geqslant 1$. The first (resp. second) term represents the interactions to the right (resp. left) of $\alpha_1 = 1/2$, while the third term is the interaction across the plane $\alpha_1 = 1/2$. The minus sign in (E1), which restricts us to ferromagnets, will be crucial.

**Theorem E.1.** Suppose all matrices are real. For any operator $D$ on the sites with $\alpha_1 \geq 1$,

$$\langle R(D)D \rangle \geqslant 0 \tag{E2}$$

*Proof*

$$\langle R(D)D \rangle = [\mathrm{Tr}(e^{-H})]^{-1} \lim_{n\to\infty} \mathrm{Tr}\{R(D)D \exp(-B/n) \times \exp[-R(B/n)] \exp[C_1 R(C_1)/n] \cdots\}$$

by the Trotter product formula. Now expand each factor $\exp[C_i R(C_i)/n]$ in a power series. Then $\langle R(D)D \rangle$ is a sum of terms of the form $\mathrm{Tr}[R(Q_1)Q_1 \cdots R(Q_n)Q_n] = \mathrm{Tr}[R(Q_1 \cdots Q_n)]\ \mathrm{Tr}(Q_1 \cdots Q_n) = \mathrm{Tr}(Q_1 \cdots Q_n)^2 \geqslant 0$ since $Q_1 \cdots Q_n$ are real matrices. ■

Now let $\langle \cdots \rangle_\infty$ denote an infinite-volume state obtained as a limit point of the states in volumes with $L_1$ even. The state $\langle \cdots \rangle_\infty$ is automatically translation invariant since each finite-volume state is. Let $R$ be defined as above, and define an inner product $((\cdot, \cdot))$ on the operators on the sites with $\alpha_1 \geqslant 1$ by $((A, B)) = \langle R(A^*)B \rangle$. Let $\mathscr{H}$ be the Hilbert space obtained by completing in this inner product. Following the idea of Osterwalder and Schrader,[34] we have the result:

**Theorem E.2.** Let $\tau_n$ be the automorphism on the infinite-volume quasi-local algebra obtained by translating $n$ units to the right. Then, there is a self-adjoint contraction $T$ on $\mathscr{H}$ such that for any operators $A, B$ on the sites with $\alpha_1 \geqslant 1$

$$((A, \tau_n(B))) = ((A, T^n(B))) \tag{E3}$$

*Remark.* $T$ is an abstract version of the transfer matrix with the important difference of $n + 1$ spacing mentioned above.

*Proof.* We begin by proving that

$$|((A, \tau_1(B)))| \leqslant ((A, A))^{1/2}((B, B))^{1/2} \tag{E4}$$

for $A, B$ bounded. Clearly, $|((A, \tau_n(B)))| \leqslant \|A\|\,\|B\|$. Moreover,

$$\begin{aligned} ((A, \tau_1(B))) &\leqslant ((A, A))^{1/2}((\tau_1(B), \tau_1(B)))^{1/2} = ((A, A))^{1/2}((B, \tau_2(B)))^{1/2} \\ &\leqslant ((A, A))^{1/2}((B, B))^{1/4}((B, \tau_4(B)))^{1/4} \cdots \\ &\leqslant ((A, A))^{1/2}((B, B))^{(1-2^{-n})/2}((B, \tau_{2^{n+1}}(B)))^{2^{-n-1}} \end{aligned}$$

Taking $n$ to $\infty$, (E4) results. Thus, there is a contraction $T$ such that $((A, \tau_1(B))) = ((A, T(B))$. Since $((\tau_1(A), B)) = ((A, \tau_1(B)))$, $T$ is self-adjoint. Moreover, since $\tau_n = (\tau_1)^n$, we have (E3). ■

**Corollary E.1.** Fix $\alpha_2,\ldots, \alpha_\nu$ and an operator $B$ at a single site. Let $g(\alpha_1) = \langle B^*_{(0,\alpha_2,\ldots,\alpha_\nu)} B_{(\alpha_1,\alpha_2,\ldots,\alpha_\nu)}\rangle$ for $\alpha_1 \geqslant 1$. Then there is a positive measure $d\mu_B$ on $[-1, 1]$ such that

$$g(\alpha) = \int_{-1}^{1} \lambda^{\alpha-1}\, d\mu_B(\lambda), \qquad \alpha = 1, 2,\ldots \tag{E5}$$

*Proof.* Since $g(\alpha_1) = ((B_{(1,\alpha_2,\ldots,\alpha_\nu)}, T^{\alpha_1-1}B_{(1,\alpha_2,\ldots,\alpha_\nu)}))$, (E5) follows by the spectral theorem. ■

There is at this point one important difference from the Euclidean field theory case. A semigroup indexed by ℝ of self-adjoint operators is automatically a semigroup of positive operators. This is not automatic in the discrete case. It is natural to ask if this is so, i.e., if $d\mu_B$ in (E5) is supported on [0, 1].

## ACKNOWLEDGMENTS

It is a pleasure to thank several individuals for valuable aid: J. F. Barnes and M. L. Glasser for helping find numerical values of various integrals, V. Bargmann for simplifying one of our technical proofs (Theorem A.2), P. A. Vuillermot for valuable discussions, and J. Fröhlich for a variety of useful comments and, in particular, for communicating to us a conjectured bound which we proved (Theorem 4.2) in the real case and used as an important part of our argument. We also thank him for pointing out a serious error in our original manuscript, namely that our "proof" of Theorem 4.2 was wrong in the complex case.

## REFERENCES

1. P. W. Anderson, *Phys. Rev.* **83**:1260 (1951).
2. T. Asano, *J. Phys. Soc. Japan* **29**:350–359 (1970).
3. D. Bessis, P. Moussa, and M. Villani, *J. Math. Phys.* **16**:2318 (1975).
4. F. Bloch, *Z. Physik* **61**:206 (1930).
5. N. Bogoliubov, *Phys. Abh. S.U.* **1**:113–229 (1962).
6. H. Brascamp and E. H. Lieb, in *Functional Integration and Its Applications*, A. N. Arthurs, ed., Clarendon Press, Oxford (1975).
7. L. Breiman, *Probability*, Addison-Wesley (1967).
8. R. L. Dobrushin, *Teorija Verojatn i ee Prim.* **10**:209–230 (1965).
9. F. J. Dyson, *Phys. Rev.* **102**:1217 (1956).
10. F. J. Dyson, *Phys. Rev.* **102**:1230 (1956).

11. R. Ellis, J. Monroe, and C. Newman, *Comm. Math. Phys.* **46**:167–182 (1976).
12. J. Fröhlich, B. Simon, and T. Spencer, *Comm. Math. Phys.*, to appear.
13. J. Ginibre, *Comm. Math. Phys.* **14**:205 (1969).
14. J. Glimm, A. Jaffe, and T. Spencer, *Comm. Math. Phys.* **45**:203 (1975).
15. R. Griffiths, *Phys. Rev.* **136A**:437–439 (1964).
16. R. Griffiths, *Phys. Rev.* **152**:240–246 (1966).
17. R. Griffiths, *J. Math. Phys.* **8**:478–483 (1967).
18. R. Griffiths, *Comm. Math. Phys.* **6**:121–127 (1967).
19. F. Guerra, L. Rosen, and B. Simon, *Ann. Math.* **101**:111–259 (1975).
20. K. Hepp and E. H. Lieb, *Phys. Rev. A* **8**:2517–2525 (1973).
21. P. Hohenberg, *Phys. Rev.* **158**:383 (1967).
22. F. Keffer, in *Handbuch der Physik*, Band XVIII/2, Springer (1966), pp. 1–273.
23. A. Klein, *Bull. Am. Math. Soc.* **82**:762 (1976).
24. R. Kubo, *J. Phys. Soc. Japan* **12**:570 (1957).
25. E. H. Lieb and D. Mattis, *J. Math. Phys.* **3**:749 (1962).
26. E. H. Lieb and D. Mattis, *Phys. Rev.* **125**:164 (1962).
27. E. H. Lieb and D. W. Robinson, *Comm. Math. Phys.* **28**:251–257 (1972).
28. N. Mermin and H. Wagner, *Phys. Rev. Lett.* **17**:1133 (1966).
29. H. Mori, *Prog. Theor. Phys.* **33**:423 (1965).
30. J. Naudts and A. Verbeure, *J. Math. Phys.* **17**:419–423 (1976).
31. J. Naudts, A. Verbeure, and R. Weder, *Comm. Math. Phys.* **44**:87–99 (1975).
32. C. Newman, in *Proc. 1975 Marseille Conference on Quantum Field Theory and Statistical Mechanics.*
33. L. Onsager, *Phys. Rev.* **65**:117 (1944).
34. K. Osterwalder and R. Schrader, *Comm. Math. Phys.* **31**:83 (1974).
35. R. Peierls, *Proc. Camb. Phil. Soc.* **32**:477–481 (1936).
36. R. Powers, U.C. Berkeley Preprint (1976).
37. M. Reed and B. Simon, *Methods of Modern Mathematical Physics, I, Functional Analysis*, Academic Press (1972).
38. D. Robinson, *Comm. Math. Phys.* **7**:337 (1968).
39. D. Robinson, *Comm. Math. Phys.* **14**:195 (1969).
40. G. Roepstorff, *Comm. Math. Phys.* **46**:253–262 (1976).
41. D. Ruelle, *Statistical Mechanics*, Benjamin (1969).
42. H. Samelson, *Notes on Lie Algebras*, Van Nostrand (1969).
43. T. Schultz, D. Mattis, and E. H. Lieb, *Rev. Mod. Phys.* **36**:856 (1964).
44. R. Streater, *Comm. Math. Phys.* **6**:233 (1967).
45. G. N. Watson, *Quart. J. Math.* **10**:266 (1939).
46. J. Fröhlich, private communication.
47. H. Falk and L. W. Bruch, *Phys. Rev.* **180**:442 (1969).
48. S. Malyshev, *Comm. Math. Phys.* **40**:75 (1975).
49. J. Fröhlich and E. H. Lieb, *Phys. Rev. Lett.* **38**:440–442 (1977); J. Fröhlich and E. H. Lieb, *Comm. Math. Phys.*, in press.
50. J. Fröhlich, R. Israel, E. H. Lieb, and B. Simon, papers in preparation.
51. F. J. Dyson, E. H. Lieb, and B. Simon, *Phys. Rev. Lett.* **37**:120–123 (1976).

Commun. math. Phys. 62, 1—34 (1978)

Communications in Mathematical Physics



# Phase Transitions and Reflection Positivity. I. General Theory and Long Range Lattice Models

Jürg Fröhlich[1]* **, Robert Israel[2]***, Elliot H. Lieb[3]†, and Barry Simon[3]**

[1] Department of Mathematics, Princeton University, Princeton, NJ 08540, USA
[2] Department of Mathematics, University of British Columbia, Vancouver, B.C., Canada
[3] Departments of Mathematics and Physics, Princeton University, Princeton, NJ 08540, USA

**Abstract.** We systematize the study of reflection positivity in statistical mechanical models, and thereby two techniques in the theory of phase transitions: the method of *infrared bounds* and the chessboard method of estimating contour probabilities in Peierls arguments. We illustrate the ideas by applying them to models with long range interactions in one and two dimensions. Additional applications are discussed in a second paper.

## 1. Introduction

Among the recent developments in the rigorous theory of phase transitions have been the introduction of two powerful techniques motivated in part by ideas from constructive quantum field theory: the method of infrared bounds [10, 4] which provides the only presently available tool for proving that phase transitions occur in situations where a continuous symmetry is broken, and the chessboard estimate method of estimating contour probabilities in a Peierls' argument [14, 9]. This is the first of three papers systematizing, extending and applying these methods. In this paper, we present the general theory and illustrate it by considering phase transitions in one and two dimensional models with long range interactions. In II, [7], we will consider a large number of applications to lattice models and in III, [8] some continuous models including Euclidean quantum field theories. Reviews of some of our ideas and those in [4, 9, 10, 14] can be found in [5, 6, 23, 27, 43]. An application can be found in [19].

Three themes are particularly emphasized in these papers. The first, §§2—4, is the presentation of a somewhat abstract framework, partly for clarification (e.g.

* Present address: Institute des Hautes Études Scientifiques, 35, Route de Chartres, F-91440 Bures-sur-Yvette, France

** Research partially supported by US National Science Foundation under Grant MPS-75-11864

*** Research partially supported by Canadian National Research Council under Grant A4015

† Research partially supported by US National Science Foundation under Grant MCS-75-21684-A01

0010-3616/78/0062/0001/$06.80

the tricks in [4] to handle the quantum antiferromagnet may appear more natural in the light of §§2, 3 below) but mainly for the extensions of the theory thereby suggested (e.g. the second theme below and the use, for classical systems, of reflections in planes containing sites: this idea, occurring already in [9], will be critical for many of our applications, e.g. to the classical antiferromagnets in external field). The abstract framework also clarifies various limitations of the theory such as its present inapplicability to the quantum Heisenberg ferromagnets and its restriction to reflections in planes *between* lattice planes for quantum systems. The second theme is the extension of the methods beyond the nearest neighbor simple cubic models emphasized in [10, 4, 9]. It will turn out (§3) that rather few additional short range interactions can be accomodated but that a larger variety of long range interactions can be treated. This extension will allow us (§5) to recover and extend to suitable quantum models the results of Dyson [3] (resp. Kunz-Pfister [26]) on long range one (resp. two) dimensional systems. It will also allow us (see II) to discuss a number of lattice Coulomb gases: for example, a "hard core model" where each site can have charge 0, $+1$ or $-1$ will have two "crystal phases" for sufficiently low temperatures and large fugacity and, for sufficiently low temperatures and suitable fugacity, a third phase which can be thought of as a "plasma" or "gas" phase. Finally it will allow us to construct (see III) a *two* dimensional quantum field theory (a $\phi^4$ perturbation of a generalized free field) with a spontaneously broken continuous symmetry.

For pair interactions, Hegerfeldt and Nappi [18] have proposed our sufficient condition for reflection positivity but they did not discuss the connection with phase transitions or the quantum case; see also their footnote on p. 4 of their paper.

The final theme involves the development of an idea in [10, 5] for proving that phase transitions occur in a situation where there is no symmetry broken and thus no a priori clear value of external field or fugacity for the multiple phase point. In all cases, the value can be computed for zero-temperature and one shows that there are multiple phases at some nearby value for low temperature, although our methods do not appear to specify the value by any computationally explicit procedure. This technique, which we do not discuss until Paper II, allows us in particular to recover some results of Pirogov-Sinai [33–35] including the occurrence of transitions in the triangle model (ordinary Ising ferromagnet in external field but with an additional interaction $K\sum\sigma_i\sigma_j\sigma_k$ over all triples $ijk$ where $i$ and $k$ are nearest neighbors of $j$ in orthogonal directions) and the occurrence of three phases in the Fisher stabilized antiferromagnet in suitable magnetic field (ordinary Ising antiferromagnet but with additional next nearest neighbor ferromagnetic coupling). As another example we mention an analysis of some models of Ginibre, discussed by Kim-Thompson [32] in the mean field approximation, with the property that at low temperatures there are an *infinite* number of external field values with multiple phases.

Next we want to make some remarks on the limitations, advantages and disadvantages of the reflection positivity (RP) methods. As regards the chessboard Peierls argument, it is useful to compare it with the most sophisticated Peierls type method that we know of, that of Pirogov-Sinai (PS method) [33–35, 20] (a comparison with the "naive" Peierls argument can be found in [27]):

1) The most serious defect in the RP method is that the requirement of reflection positivity places rather strong restrictions on the interactions, especially for finite range interactions. For example, the PS analysis of the Fisher antiferromagnet would not be affected if one added an additional ferromagnetic coupling $\sigma_i\sigma_j$ for pairs $ij$ with $i-j=(8,10)$ (for example) while our argument would be destroyed no matter how small the coupling! More significantly, the RP analysis in this case requires that $\sigma_{(0,0)}\sigma_{(1,1)}$ and $\sigma_{(0,0)}\sigma_{(1,-1)}$ have equal couplings; PS does not. Similarly in the triangle model, an RP argument requires the four kinds of triangles to have equal couplings while PS does not.

2) RP can handle certain, admittedly special, long range couplings, among them interactions of physical interest such as Coulomb monopole and dipole couplings. PS in its present form is restricted to finite range interactions.

3) Inherent in the PS method is the notion that one is looking at a system with a "finitely degenerate ground state". This is not inherent in the RP method: all that is important is that a *finite* number of *specific* periodic states have a larger internal energy per unit volume than the true ground states. In some cases, e.g. the antiferromagnet without Fisher stabilization, there is no practical difference since the finite number of states of importance in RP are among the infinitely many ground states that prevent the application of PS. However, there is a model (of a liquid crystal) with an infinitely degenerate ground state to which Heilmann and Lieb [19] have applied the RP method with success. This model has only two ground states in finite volume with suitable boundary conditions, but infinitely many ground states in the PS sense in infinite volume.

4) The PS method gives much more detailed information than the RP method on the manifold of coexisting phases. For example in the Fisher antiferromagnet, there is, for $T$ small, an external field, $\mu(T)$, near the computable number $\mu(0)$, so that there are three (or more) phases at that value of $T$ and $\mu$. PS obtain continuity of $\mu(T)$ in $T$ while RP does not, but shows only that $\mu(T)\to\mu(0)$ as $T\to 0$.

5) While neither PS nor we have tried hard to optimize the lower bounds on transition temperatures, it seems reasonably clear that RP methods would produce better bounds.

6) PS require the number of values that a given spin takes to be finite. RP methods effortlessly extend to models like the anisotropic classical Heisenberg model (see [9]).

7) PS can only handle classical models, at least in its present version. RP methods can handle certain quantum models quite efficiently (see [9]).

8) RP works most naturally for states with periodic boundary conditions. This can occasionally be awkward.

9) PS obtain the exact number of phases at the maximum phase points while RP only yields a lower bound. This difference is probably not intrinsic, and RP methods could probably be combined with [11] to yield the exact number of phases.

10) To our, admittedly biased, tastes the RP method seems considerably simpler than the PS method.

As regards the infrared bounds method, there is no comparable method with which to compare it, but we note it is most unfortunate that the only available method for proving phase transitions depends so strongly on reflection positivity. We mention two examples to illustrate this remark:

1) In [10], it is proven that the classical Heisenberg ferromagnet with nearest neighbor interaction has a phase transition for a simple cubic lattice. The methods of §§2–4 easily extend this result to face centered cubic and many other lattices, but *not* to the body centered cubic lattice. This remains an open problem.

2) There has been some discussion recently (see [36] and references therein) of an intriguing model, originally due to Elliott [28], which should have "helical" long range order: consider a one dimensional plane rotor or $N$-vector, $N \geqq 3$ model with nearest neighbor ferromagnet coupling, $J$, and somewhat stronger second neighbor antiferromagnet coupling, $K$. It will have a helical ground state, i.e. in a ground state $\sigma_i \cdot \sigma_{i+1} = \cos\theta$ for some $\theta \neq 0, \pi$ depending on the exact value of $J/K$. Of course, this helical ordering won't persist to finite temperature in the one dimensional case, but if one adds two more dimensions with conventional nearest neighbor ferromagnetic couplings one expects helical order will persist. We do not see how to prove this with RP methods; indeed, infrared bounds obtained by RP methods always seem to blow up at a single $p$ while at least two $p$'s are involved here due to the evenness of the function $E_p$. We note that if one could prove an infrared bound, helical order would be proven since $E_p$ vanishes at precisely two $p$'s with a zero of order $p^2$.

Finally, we summarize the contents of the remaining sections. In §2, we present an abstract framework for reflection positivity and provide the basic perturbation criteria which allow one to go from reflection positivity for uncoupled spins to reflection positivity for suitably coupled spins. In §3, we specialize to spin systems and examine two questions: about what kinds of planes does one have reflection positivity for the system of uncoupled spins, and what kinds of interactions obey the basic perturbation criteria of §2? In §4, we review and describe the two basic RP methods of proving phase transitions when one has reflection positivity about the large family of planes obtained by translating a basic family of planes. In §5, we discuss the applications to recover the Dyson and Kunz-Pfister results already mentioned.

## 2. Abstract Theory of Reflection Positivity

Reflection positivity was introduced in quantum field theory by Osterwalder and Schrader [30] and it has continued to play an important role there. Its significance in the study of phase transitions for lattice gases was realized in [10, 5, 9], although we must emphasize that transfer matrix ideas are intimately connected with reflection positivity. Klein [25] has considered other abstractions in somewhat different contexts.

To understand the framework we are about to describe, it is useful to keep in mind a particular example, describing a chain of Ising spins, that is essentially that given in [10, 9] (we describe the example after the basic framework).

$\mathfrak{A}$ will be a real algebra (with unit) of observables. (We note that to say $\mathfrak{A}$ is a real algebra does not preclude $\mathfrak{A}$ from being, say, an algebra of complex valued functions: "real" means that we only suppose that one can multiply by real scalars.) Below we will freely use and expand exponentials and use the Trotter-product formula (in cases where $\mathfrak{A}$ is non-abelian). In most applications these manipulations present no problem since $\mathfrak{A}$ is usually finite dimensional. In III, we will deal with some unbounded operators and exercise some care on this point. We suppose we are given a linear functional $A \to \langle A \rangle_0$ on $\mathfrak{A}$ with $\langle 1 \rangle_0 = 1$. Given $H \in \mathfrak{A}$, we define

$$\langle A \rangle_H = \langle A e^{-H} \rangle_0 / \langle e^{-H} \rangle_0 . \tag{2.1}$$

Moreover, we suppose $\mathfrak{A}$ contains two subalgebras $\mathfrak{A}_+$ and $\mathfrak{A}_-$ and a real linear morphism $\theta : \mathfrak{A}_+ \to \mathfrak{A}_-$. [The phrase "real linear" does not preclude $\theta$ from being complex linear or complex antilinear; morphism means $\theta(AB) = \theta(A)\theta(B)$. In most examples, $\theta$ has an extension to $\mathfrak{A}_+ \cup \mathfrak{A}_-$ obeying $\theta^2 = 1$, but this property plays no role in our considerations below.]

The example to keep in mind involves $2n$ spin 1/2-Ising spins $\sigma_{-n+1}, \sigma_{-n+2}, \ldots, \sigma_n$. Then $\mathfrak{A}$ is the family of polynomials in all the $\sigma$'s, $\mathfrak{A}_+$ (resp. $\mathfrak{A}_-$) the polynomials in $\sigma_1, \ldots, \sigma_n$ (resp. $\sigma_0, \sigma_{-1} \ldots \sigma_{-n+1}$), and $\theta$ is defined so that $\theta(\sigma_i) = \sigma_{-i+1}$; $\langle A(\sigma) \rangle_0 = \frac{1}{4^n} \sum_{\sigma_i = \pm 1} A(\sigma_i)$. Although $\mathfrak{A}_+$ and $\mathfrak{A}_-$ have trivial intersection in this example, we will not suppose this to be true in the abstract setting; we will not even suppose that $\mathfrak{A}_+$ and $\mathfrak{A}_-$ commute with each other, although it will turn out that there are no cases for which we can prove perturbed reflection positivity with non-mutually-commuting $\mathfrak{A}_+$ and $\mathfrak{A}_-$ (with the exception of some Fermion systems).

*Definition.* A real linear functional $\langle \cdot \rangle$ on $\mathfrak{A}$ is called *reflection positive* (RP) if and only if $\langle A\theta(A) \rangle \geqq 0$ for all $A \in \mathfrak{A}_+$.

The reader should check RP and GRP (defined below) for the functional $\langle \cdot \rangle_0$ in the example. Unfortunately, we know of *no* abstract perturbation theory for functionals satisfying RP in the fully non-commutative setting, but a slightly stronger notion is preserved under suitable perturbations:

*Definition.* $\langle \cdot \rangle$ is called *generalized reflection positive* (GRP) if and only if

$$\langle A_1 \theta(A_1) \ldots A_m \theta(A_m) \rangle \geqq 0$$

for all $A_1, \ldots, A_m \in \mathfrak{A}_+$.

**Theorem 2.1.** *If* $-H = B + \theta(B) + \sum_{j=1}^{k} C_i \theta(C_i)$ *(or more generally* $B + \theta(B) + \int C(x) \theta[C(x)] d\varrho(x)$ *for a positive measure* $d\varrho$*) with* $B, C_i \in \mathfrak{A}_+$ *and if* $\langle \cdot \rangle_0$ *is GRP, then* $\langle \cdot \rangle_H$*, defined in* (2.1) *is GRP.*

*Proof.* For simplicity, let us consider first the case where $\mathfrak{A}$ is abelian even though it is a special case of the general situation we then discuss. Then, since $\theta$ is a morphism

$$e^{-H} = e^B \theta(e^B) e^{\Sigma C_i \theta(C_i)} .$$

Expanding the exponential, we see that

$$e^{-H} = \text{sum of terms of the form } (D_1 \theta(D_1) \ldots D_j \theta(D_j)),$$

so that by GRP for $\langle \cdot \rangle_0$, $\langle e^{-H} \rangle_0 \geqq 0$ and $\langle e^{-H} A_1 \theta(A_1) \ldots A_m \theta(A_m) \rangle_0 \geqq 0$.

For the general non-abelian case, we first use the Trotter product formula to write

$$e^{-H} = \lim_{k\to\infty} \left[ e^{B/k}\theta(e^{B/k}) \prod_i e^{C_i\theta(C_i)/k} \right]^k$$

and then expand to get $e^{-H}$ as a limit of sums of $\pi[D_j\theta(D_j)]$. □

In the next section, we will give a relevant example (Example 6) of a situation with $\langle\cdot\rangle_0$ RP but not GRP. There is one case where RP implies GRP (this, in fact, is the only case for which we know how to prove GRP!):

**Theorem 2.2.** *If $\mathfrak{A}_+$ and $\mathfrak{A}_-$ commute with each other, a linear functional is RP if and only if it is GRP.*

*Proof.* $\pi A_i\theta(A_i) = (\pi A_i)\theta(\pi A_i)$ since the $A_j$ and $\theta(A_i)$ commute and $\theta$ is a morphism. □

We will also need:

**Theorem 2.3.** *If $\mathfrak{A}_+$ and $\mathfrak{A}_-$ commute with each other and if $\langle\cdot\rangle_0$ is RP, then for any $A, B, C_i, D_i \in \mathfrak{A}_+$:*

$$|\langle e^{A+\theta B+\Sigma C_i\theta D_i}\rangle_0|^2 \leqq \langle e^{A+\theta A+\Sigma C_i\theta C_i}\rangle_0 \langle e^{B+\theta B+\Sigma D_i\theta(D_i)}\rangle_0 .$$

*Proof.* For simplicity of notation we suppose that $\mathfrak{A}$ is abelian. The general case follows by using the Trotter formula as in the proof of Theorem 2.1. Since $\langle\cdot\rangle_0$ is RP, we have a Schwarz inequality $|\langle A\theta B\rangle_0|^2 \leqq \langle A\theta A\rangle_0 \langle B\theta B\rangle_0$ and so (here we use that $\mathfrak{A}_+$ and $\mathfrak{A}_-$ commute)

$$\begin{aligned} &|\langle A_1\theta(B_1)\dots A_j\theta(B_j)\rangle_0|^2 \\ &\leqq \langle A_1\theta(A_1)\dots A_j\theta(A_j)\rangle_0 \langle B_1\theta(B_1)\dots B_j\theta(B_j)\rangle_0 . \end{aligned} \tag{2.2}$$

Now

$$\alpha \equiv e^{A+\theta B+\Sigma C_i\theta(D_i)} = e^A\theta(e^B)e^{\Sigma C_i\theta(D_i)},$$

so expanding the sums we can write it as sum of terms of the form $E_1\theta(F_1)\dots E_l\theta(F_l)$. Using (2.2), we see that

$$|\langle\alpha\rangle_0| \leqq \sum \langle \pi E_i\theta(E_i)\rangle_0^{1/2} \langle \pi F_i\theta(F_i)\rangle_0^{1/2},$$

so using the Schwarz inequality for sums

$$|\langle\alpha\rangle_0|^2 \leqq [\sum \langle \pi E_i\theta(E_i)\rangle_0][\sum \langle \pi F_i\theta(F_i)\rangle_0] .$$

We can now resum the exponential and so obtain the desired result. □

*Remarks.* Notice that only (2.2) was needed to obtain the result, so we could have paralleled the discussion of GRP and given (2.2) a name. We only know how to prove (2.2) when $\mathfrak{A}_+$ and $\mathfrak{A}_-$ commute.

The theorems in this section are only mild abstractions of ideas in [10, 4]. In fact, [4] already noted the importance of inequalities like those in Theorem 2.3 and of Hamiltonians of the form singled out in Theorem 2.1.

*Remark.* Independently, Osterwalder and Seiler have discussed RP for Euclidean Fermi lattice field theories [31] using ideas similar to ours.

There is a generalization of Theorem 2.3, which, while it will not be used in the sequel, is potentially of interest.

**Theorem 2.4.** *If* $\mathfrak{A}_+$ *and* $\mathfrak{A}_-$ *commute with each other and* $\langle \cdot \rangle_0$ *is RP, then for any* $C_i, D_i \in \mathfrak{A}_+$

$$\left|\left\langle e^{\sum_i C_i \theta D_i} - 1\right\rangle_0\right|^2 \leqq \left\langle\left(e^{\sum_i C_i \theta C_i} - 1\right)\right\rangle_0 \langle (e^{\sum D_i \theta D_i} - 1)\rangle_0 .$$

*Proof.* The same as for Theorem 2.3. One merely has to notice that the first term (namely 1) in the expansion of the exponential cancels. □

*Remark.* Theorem 2.3 is a Corollary of Theorem 2.4. Merely add $(\lambda^{-1}A+\lambda)\times(\lambda^{-1}\theta B+\lambda)$ to the exponential in Theorem 2.4 and then let $\lambda\to\infty$.

## 3. Reflections in a Single Plane

In this section, we consider the case where $\mathfrak{A}$ is an algebra of observables for a classical or quantum spin system on a lattice, $\langle \cdot \rangle_0$ is an uncoupled expectation and $\theta$ is a reflection in a plane. We concentrate on two distinct questions which are connected with our discussion in the last section: a) When is $\langle \cdot \rangle_0$ RP and/or GRP? b) What interactions lead to a Hamiltonian with $-H=B+\theta B+\sum C_i\theta C_i$? We discuss the first question in a series of examples.

### *1) Reflections in a Plane Without Sites-Classical Case*

We imagine the finite lattice $\Lambda$ (which may be a torus) being divided by a plane $\pi$ into two subsets $\Lambda_+$ (to the "right" of $\pi$) and $\Lambda_-$, with no sites on $\pi$. There is some "reflection" $r$ on $\Lambda$ such that $r$ takes $\Lambda_+$ into $\Lambda_-$ and $r^2=1$. The "spin" at each site is a random variable taking values in a compact set $K$ with some "a priori" Borel probability distribution $d\varrho$. Let $K_\Lambda=\prod_{i\in\Lambda} K_i$ and $K_\pm=\prod_{i\in\Lambda\pm} K_i$ (where each $K_i$ is a copy of $K$). For $x\in K_-$, define $\theta_* x\in K_+$ by $(\theta_* x)_i = x_{r(i)}$. We take $\mathfrak{A}$ to be all real-valued continuous functions on $K_\Lambda$ with $\mathfrak{A}_\pm$ the subalgebras of functions depending only on the spins in $\Lambda_\pm$. Define $\theta: \mathfrak{A}_+\to\mathfrak{A}_-$ by

$$(\theta F)(x)=F(\theta_* x).$$

Finally, we let $\langle F\rangle_0=\int_{K_\Lambda} F(x)\prod_{i\in\Lambda} d\varrho(x_i)$. Then $\langle \cdot \rangle_0$ is RP since

$$\begin{aligned}\langle F(\theta F)\rangle_0 &= \int_{K_-}\int_{K_+} F(x)F(\theta_* y)\prod_{i\in\Lambda_+} d\varrho(x_i)\prod_{j\in\Lambda_-} d\varrho(y_j)\\ &=\left[\int_{K_+} F(x)\prod_{i\in\Lambda_+} d\varrho(x_i)\right]^2 \geqq 0 .\end{aligned}$$

Since $\mathfrak{A}$ is abelian, $\langle \cdot \rangle_0$ is GRP. This example includes the kind of classical system in [10]. Alternatively, we could allow $\mathfrak{A}$, $\mathfrak{A}_\pm$ to be complex valued and then define $(\theta F)(x)=\overline{F(\theta_* x)}$.

*2) Reflections in a Plane Without Sites-"Real" Quantum Case*

The setup is very similar to 1) but now for each $i \in \Lambda$, we take a copy $\mathscr{H}_i$ of $\mathbb{R}^m$ with the natural inner product. One defines $\mathscr{H} = \bigotimes_{i \in \Lambda} \mathscr{H}_i$ and $\mathscr{H}_-$ (resp. $\mathscr{H}_+$) as the tensor product of the spaces associated with sites in $\Lambda_-$ (resp. $\Lambda_+$). $\mathfrak{A}$ is now all matrices on $\mathscr{H}$ and $\langle A \rangle_0 = \mathrm{Tr}_{\mathscr{H}}(A)/\mathrm{Tr}_{\mathscr{H}}(1)$. $\mathfrak{A}_+$ (resp. $\mathfrak{A}_-$) consists of all operators of the form $1 \otimes A$ (resp. $A \otimes 1$) under the tensor decomposition $\mathscr{H} = \mathscr{H}_- \otimes \mathscr{H}_+$. Finally $\theta(1 \otimes A) = A \otimes 1$. Then for $B = 1 \otimes A$

$$\mathrm{Tr}(B\theta B) = \mathrm{Tr}_{\mathscr{H}}(A \otimes A) = \mathrm{Tr}_{\mathscr{H}_+}(A)^2 \geqq 0$$

*since* $\mathrm{Tr}(A)$ *is real.* Thus $\langle \cdot \rangle_0$ is RP and, since $\mathfrak{A}_+$ and $\mathfrak{A}_-$ commute, GRP. This example includes the quantum $xy$ model [4] in the realization $\sigma_x = \begin{pmatrix} 0 & 1 \\ 1 & 0 \end{pmatrix}$, $\sigma_y = \begin{pmatrix} 1 & 0 \\ 0 & -1 \end{pmatrix}$. Alternatively, we could take $\mathscr{H}_i = \mathbb{C}^m$ and $\theta(1 \otimes A) = \bar{A} \otimes 1$ where $^-$ is complex conjugation.

*3) Reflections in a Plane Without Sites-General Quantum Case*

This is identical to the setup in (2) except for the fact that $\mathscr{H}_i$ is a copy of $\mathbb{C}^m$. If we take $\theta(1 \otimes A) = A \otimes 1$, then $\langle \cdot \rangle_0$ is *not* RP since $\mathrm{Tr}(A)$ may not be real. Indeed if $\mathfrak{A}$ and $\theta$ are chosen in some other way so that Tr is GRP, then the ferromagnetic Heisenberg Hamiltonian will *not* be expressible as $-H = B + \theta B + \sum C_i \theta C_i$, since $\mathrm{Tr}(\sigma_1 \cdot \sigma_0)^3 < 0$, while $(\sigma_1 \cdot \sigma_0)^3$ is a sum of $A_1 \theta A_1 \ldots A_3 \theta A_3$. Of course, if one takes $\theta_1(1 \otimes A) = \bar{A} \otimes 1$ where $^-$ is ordinary matrix complex conjugation, then for $B = 1 \otimes A$

$$\mathrm{Tr}(B\theta_1 B) = \mathrm{Tr}_{\mathscr{H}}(\bar{A} \otimes A) = |\mathrm{Tr}_{\mathscr{H}_+}(A)|^2 \geqq 0 .$$

So one recovers RP and GRP, but the usual Heisenberg ferromagnet is no longer of the form $\sum C_i \theta_1 C_i$, since $\boldsymbol{\sigma}_1 \cdot \theta_1 \boldsymbol{\sigma}_1 = \sigma_{1x}\sigma_{0x} + \sigma_{1z}\sigma_{0z} - \sigma_{1y}\sigma_{0y}$ in the usual realization of the $\sigma$'s.

The fact that $\langle \cdot \rangle_0$ is not RP does not stop it from being RP on a subalgebra; indeed in the Heisenberg case, for functions of $\sigma_z$'s alone, it is RP. It could happen that for the usual (anisotropic) Heisenberg case, $\langle \cdot \rangle_H$ is also RP on this subalgebra and this would lead to phase transitions in the two dimensional anisotropic case [9]. However, the failure of full GRP implies that our simple perturbation scheme of §2 will not yield a proof of this type of restricted RP.

*4) Twisted Reflections in a Plane Without Sites*

It is sometimes useful to define $\theta$ with a "twist". For example, in the setup of 3), take $m = 2S + 1$ and take $\sigma_x$, $\sigma_y$, $\sigma_z$ as the usual spin $S$ spins; i.e. $\sigma_z$ is diagonal and $\sigma_x \pm i\sigma_y$ are raising and lowering operators. Thus $\sigma_x$, $\sigma_z$ are real and $\sigma_y$ is pure imaginary. Let $U$ be the operator on $\mathscr{H}_-$ which rotates about the $y$ axis by 180° at each site. Let

$$\theta(1 \otimes A) = (\overline{UAU^{-1}}) \otimes 1 .$$

Then for $B=1\otimes A$

$$\begin{aligned}\operatorname{Tr}(B\theta B)&=\operatorname{Tr}(\overline{UAU^{-1}}\otimes A)=\overline{\operatorname{Tr}(UAU^{-1})}\operatorname{Tr}(A)\\&=|\operatorname{Tr}(A)|^2\geqq 0.\end{aligned}$$

So $\langle\cdot\rangle_0$ is RP and GRP. Moreover, $\theta(\sigma_j)=-\sigma_{r(j)}$ so that the *antiferromagnet* $-H=-\sum_{\langle ij\rangle}\sigma_i\cdot\sigma_j$ with a sum over nearest neighbors, is of the form $B+\theta B+\sum C_i\theta C_i$. This is essentially the method [4] used to discuss the antiferromagnet.

*5) Reflections in a Plane Containing Sites-Classical Case*

The setup is very similar to 1), but now there may be sites on $\pi$. Therefore we break up $\Lambda$ into three pieces, $\Lambda_-$, $\Lambda_0$, $\Lambda_+$ corresponding to sites to the "left" of $\pi$, on $\pi$, and to the right of $\pi$. $r$ now maps $\Lambda_+$ to $\Lambda_-$ and leaves $\Lambda_0$ invariant. $\mathfrak{A}_+$ (resp. $\mathfrak{A}_-$) is the family of all functions of the spins in $\Lambda_0\cup\Lambda_+$ (resp. $\Lambda_-\cup\Lambda_0$) and for $x=\{x_i\}_{i\in\Lambda_-\cup\Lambda_0}$, $\theta_* x=x_{r(i)}\in K_+\times K_0$. As before $\langle G\rangle_0=\int G\prod_{i\in\Lambda}d\varrho(x_i)$ and $(\theta F)(x)=F(\theta_* x)$. Then writing $(x,y,z)$ according to the decomposition $K_-\times K_0\times K_+$:

$$\begin{aligned}\langle F\theta F\rangle_0&=\int F(y,z)F(\theta_*(x,y))\prod_{i\in\Lambda_-}d\varrho(x_i)\prod_{j\in\Lambda_0}d\varrho(y_j)\prod_{k\in\Lambda_+}d\varrho(z_k)\\&=\int_{K_0}\prod_{j\in\Lambda_0}d\varrho(y_j)\left|\int F(y,z)\prod_{k\in\Lambda_+}d\varrho(z_i)\right|^2\geqq 0.\end{aligned}\tag{3.1}$$

Thus we have RP and GRP since $\mathfrak{A}$ is abelian. This kind of reflection is mentioned in [9] and will play a major role in many of the examples in II.

*6) Reflections in a Plane Containing Sites-"Real" Quantum Case*

The setup is as in 2) but with the modifications in 5). Thus $\mathscr{H}=\mathscr{H}_-\otimes\mathscr{H}_0\otimes\mathscr{H}_+$, $\mathfrak{A}_+$ is the linear span of the $1\otimes A\otimes B$, and $\mathfrak{A}_-$ the one of the $B\otimes A\otimes 1$. We take $\theta(1\otimes A\otimes B)=B\otimes A\otimes 1$. Noticing that for $C$, an operator on $\mathscr{H}_0\otimes\mathscr{H}_+$ [the analog of (3.1)]:

$$\operatorname{Tr}(C\theta C)=\operatorname{Tr}_{\mathscr{H}_0}([\operatorname{Tr}^{(P)}_{\mathscr{H}_+}(C)]^2)\geqq 0,$$

where $\operatorname{Tr}^{(P)}_{\mathscr{H}_+}$ is the partial trace on $\mathscr{H}_+$, we see that $\langle\cdot\rangle_0$ is RP. In this case $\mathfrak{A}_+$ and $\mathfrak{A}_-$ are not mutually commuting so that GRP is not automatic; indeed it is *false*. For let $\mathscr{H}_+=\mathscr{H}_-=\mathscr{H}_0=\mathbb{C}^2$ and let

$$\theta C=\sigma_x\otimes(1+\sigma_z)\otimes 1+\sigma_z\otimes(1+\sigma_x)\otimes 1$$

$$\theta D=\sigma_x\otimes(1-\sigma_z)\otimes 1+\sigma_z\otimes(1-\sigma_x)\otimes 1$$

in terms of the usual Pauli matrices. Then:

$$\begin{aligned}\operatorname{Tr}(C(\theta C)D(\theta D))&=8\operatorname{Tr}((1+\sigma_z)(1+\sigma_x)(1-\sigma_z)(1-\sigma_x))\\&=-32<0.\end{aligned}$$

Since this example is not so far from what could arise when expanding realistic spin systems, we conclude that reflections in planes containing sites are not likely to be permitted for quantum spin systems, even "real" ones.

We summarize the above examples in:

**Theorem 3.1.** $\langle \cdot \rangle_0$ *is GRP for conventional reflections in planes without sites for classical and simultaneously real quantum systems and for reflections in planes with sites (lattice planes) for classical systems.*

Now we turn to the question of which interactions lead to Hamiltonians of the form

$$-H = \theta B + B + \int C(x)\theta[C(x)]d\varrho(x). \tag{3.2}$$

To illustrate the ideas, we will first consider the case of pair interactions in one dimension and then more general cases. The main result is that the interaction has to be "reflection positive" for (3.2) to hold. The net result of the analysis and Theorem 2.1 is that $\langle \cdot \rangle_H$ is RP if and only if the interaction is reflection positive. This is very reminiscent of theorems of Schoenberg [40] (see also [2, 12, 38]) relating positive definiteness of $e^{+tF}$ to (conditional) positive definiteness of $F$, and, indeed, our results can be viewed as a special case of that circle of ideas (see Theorem 3.5).

We begin with consideration of spins $\sigma_{-n+1}, \ldots, \sigma_n$.

*Definition.* A function $(J(j))_{j \geqq 1}$ will be called *reflection positive* if and only if for all positive integers $m$ and $z_1, \ldots, z_m \in \mathbb{C}$:

$$\sum_{i,j \geqq 1} \bar{z}_i z_j J(i+j-1) \geqq 0. \tag{3.3}$$

If we know a priori that $J$ is real-valued [it is by (3.3)] (3.3) need only be checked for $z$ real. In this case the left side of (3.3) can be viewed as the interaction *between* spins at sites $1, \ldots, m$ with values $z_1, \ldots, z_m$ and the reflections of these spins at $j = \frac{1}{2}$ if the basic interaction is $\sum_{\alpha,\beta} J(\alpha - \beta)\sigma_\alpha \sigma_\beta$. This explains the name given.

The following comes from the realization of (3.3) as the condition of solvability of the Hamburger moment problem. For the readers ease, we sketch a standard proof ([37]):

**Proposition 3.2.** *Let* $(J(j))_{j \geqq 1}$ *be a real-valued bounded function. Then* (3.3) *holds if and only if*

$$J(j) = c\delta_{j1} + \int_{-1}^{1} \lambda^{j-1} d\varrho(\lambda) \tag{3.4}$$

*for a positive measure* $d\varrho$ *and* $c \geqq 0$.

*Remark.* If we interpret $0^{j-1}$ as $\delta_{j1}$, then $c\delta_{j1}$ is just the contribution of a $\delta(\lambda)$ piece of $d\varrho$. We write it as $c\delta_{j1}$ to be explicit.

*Proof.* If (3.4) holds, then

$$\sum_{i,j \geqq 1} \bar{z}_i z_j J(i+j-1) = c|z_1|^2 + \int_{-1}^{1} \left| \sum_{i=1}^{m} \lambda^{i-1} z_i \right|^2 d\varrho(\lambda) \geqq 0$$

so (3.3) holds. Conversely, if (3.3) holds, form a Hilbert space, $\mathscr{H}$, by starting with finite sequence $(z_1, \ldots, z_m)$ (arbitrary $m$) and letting

$$\langle (z), (w) \rangle = \sum \bar{z}_i w_j J(i+j-1)$$

and then dividing out by $z$'s with $\langle (z), (z) \rangle = 0$ and completing. For a finite sequence $(z_1, \ldots, z_m)$, let $A(z_1, \ldots, z_m) = (0, z_1, \ldots, z_m)$ and note that by repeated use of the Schwarz inequality:

$$\|Az\| \leqq \|z\|^{1/2} \|A^2 z\|^{1/2} \leqq \|z\|^{1-1/2^n} \|A^{2^n} z\|^{1/2^n}.$$

But

$$\|A^{2^n} z\|^2 = \sum \bar{z}_i z_j J(i+j+2^{n+1}-1)$$
$$\leqq \left(\sum |z_i|\right)^2 \sup |J(j)|$$

so, $\overline{\lim} \|A^{2^n} z\|^{1/2^n} \leqq 1$ as $n \to \infty$. We conclude that $\|Az\| \leqq \|z\|$, so $A$ extends to a map of $\mathscr{H}$ to $\mathscr{H}$. Moreover, by a direct calculation $(z, Az) = (Az, w)$. We conclude that $A$ is self-adjoint. Thus for any $z$

$$(z, A^{j-1} z) = \int_{-1}^{1} \lambda^{j-1} d\varrho_z(\lambda)$$

by the spectral theorem, where $0^{j-1} = \delta_{j1}$. Let $z = (1, 0, \ldots)$ so that $(z, A^{j-1} z) = J(j)$ and (3.4) holds. □

We want to emphasize two features of (3.4). First $J \geqq 0$ is *not* required. Secondly only the function $J(j) = c\delta_{j1}$ obeys (3.4) and has bounded support.

In order to obtain the simplest result relating (3.2) to (3.3) we consider free boundary conditions:

**Proposition 3.3.** *Let $(J(j))_{j \geqq 1}$ be given. For each $m$, consider spin 1/2 Ising spins, $\sigma_{-m+1}, \ldots, \sigma_m$ and let $\theta\sigma_i = \sigma_{-i+1}$*

$$-H_m(\sigma) = \sum_{\substack{i,j=-m+1 \\ i<j}}^{m} J(i-j)\sigma_i \sigma_j.$$

*Then $H_m$ has the form* (3.2) *for every $m$ if and only if $J$ obeys* (3.3).

*Remark.* One half of this theorem is also contained in Hegerfeldt and Nappi [18].

*Proof.* If $J$ obeys (3.3), then $J$ has a representation (3.4), so that

$$-H_m(\sigma) = B_m + \theta B_m + \int_{-1}^{1} C_m(\lambda) \theta [C_m(\lambda)] d\varrho(\lambda),$$

where $B_m = \sum_{1<i<j<m} J(i-j)\sigma_i \sigma_j$ and $C_m(\lambda) = \sum_{j=1}^{m} \lambda^{j-1} \sigma_j$. Conversely, suppose that $H_m$ has the form (3.2). Then $C(x) = \sum_{i=1}^{m} \mu_i(x) \sigma_i$ and so $\int C(x)[\theta C(x)] d\varrho(x)$

$= \sum_{j \leqq 0 < 1 \leqq i} J(i-j)\sigma_i \sigma_j$, where, for $1 \leqq i, j \leqq m$: $J(i+j-1) = \int \mu_i(x) \mu_j(x) d\varrho(x)$ because

if $F(\sigma) = \sum K_{ij} \sigma_i \sigma_j$, then $K_{ij}$ is unique. Thus $\sum_{1 \leqq i,j \leqq m} z_i \bar{z}_j J(i+j-1)$

$= \int \left| \sum_{i=1}^{m} z_i \mu_i(x) \right|^2 d\varrho(x)$ and therefore $J$ is reflection positive. □

This proposition is the basic result; we present a number of extensions and variations:

A) In applications, it is useful to know that periodic boundary consitions lead to a state obeying OS positivity. Given $m$ as above, we define for $i=1,2,\ldots,2m-1$.

$$J_m^P(i)=\sum_{k=-\infty}^{\infty} J(|i+2km|). \tag{3.5}$$

The Hamiltonian

$$-H_m^{\text{per}}=\sum_{-m+1\leqq i<j\leqq m} J_m^P(j-i)\sigma_j\sigma_i$$

is the Hamiltonian with periodic boundary conditions. If $J$ has the form (3.4), then

$$J_m^P(i)=c[\delta_{i1}+\delta_{i,2m-1}]+\int_{-1}^{1}[\lambda^{i-1}+\lambda^{-i+1}\lambda^{2m}](1-\lambda^{2m})^{-1}d\varrho(x)$$

so by the above arguments, $-H=B+\theta B+\int[C(x)\theta C(x)]d_\eta(x)$ for suitable $C$'s. We summarize in:

**Proposition 3.4.** *Under the hypothesis above, if $J$ obeys* (3.3), *then $H_m^{\text{per}}$ has the form* (3.2).

B) We could consider reflections about a plane containing a site. Then the above arguments imply that $J(1)$ is *arbitrary* and $J(i)=c\delta_{i2}+\int_{-1}^{1}\lambda^{i-2}d\varrho(x)$ for $i\geqq 2$. In particular, in that case, one can have second "linear" neighbor coupling.

C) If one considers a multidimensional cubic system and considers reflection in the plane $i_1=1/2$, the kind of analysis above shows that what one needs is that

$$\sum_{i_1,j_1\geqq 1}\bar{z}_i z_j J(i_1+j_1-1,i_2-j_2,\ldots,i_\nu-j_\nu)\geqq 0 \tag{3.6}$$

which leads to the requirement that for $i_1\geqq 1$

$$J(i_1,i_2,\ldots,i_\nu)=c_{i_2,\ldots,i_\nu}\delta_{i_1 1}+\int_{-1}^{1}\lambda^{i_1-1}d\varrho_{i_2,\ldots,i_\nu}(\lambda),$$

where $c_{i_2,\ldots,i_\nu}$ is a positive definite function on $\mathbb{Z}^{\nu-1}$ and $d\varrho$ obeys a similar condition. In particular, if

$$\begin{aligned} J(i)&=\alpha \quad \text{if} \quad |i_1|^2+\ldots+|i_\nu|^2=1\\ &=\beta \quad \text{if} \quad |i_1|^2+\ldots+|i_\nu|^2=2,\\ &=0 \quad \text{otherwise}\end{aligned}$$

(i.e. nearest neighbor coupling $\alpha$, next nearest $\beta$), then one will have RP about any plane bisecting a nearest neighbor bond as long as

$$\alpha-2|\beta|(\nu-1)\geqq 0. \tag{3.7}$$

In particular, $\beta$ can be negative. The case $\beta=-\alpha/2(\nu-1)$ is of some subtlety and is discussed in detail in Paper II. [To check (3.7) is equivalent to RP, we note that the function $c$, which has to be positive definite on $\mathbb{Z}^{\nu-1}$, has a Fourier transform $c(p)=\alpha-2\beta\sum_{j=1}^{\nu-1}\cos p_j$ so that the infimum occurs at $p_j=0$ (all $j$) if $\beta\geqq 0$ and at $p_j=\pi$ (all $j$) if $\beta\leqq 0$.]

D) Some clarity is obtained by considering a lattice gas in a very general language, i.e. by allowing multi-particle interactions. We will not explicitly use Theorem 2.1, and the connection with Schoenberg's work on conditionally positive definite functions will be manifest.

At each site $j\in\mathbb{Z}^\nu$ we are given a copy $K_j$ of some configuration space $K$ and a fixed probability measure $d\varrho(x_j)$ on $K_j$; $x_j$ denotes a point in $K_j$. (For the mathematically inclined reader we remark that $K$ is assumed to be a compact Hausdorff space, and $d\varrho$ is chosen to be a regular Borel measure. In fact all our spaces, resp. measures will have these properties.)

It helps one's intuition to imagine that $K$ is the two point set $\{1,-1\}$, and $d\varrho$ the measure assigning probability $\frac{1}{2}$ to 1 and $-1$. This will correspond to Ising models (see also Corollary 3.6, below).

Given a subset $X\subseteqq\mathbb{Z}^\nu$, we define

$$K^X=\mathop{\times}_{j\in X}K_j\quad\text{and}\quad K^\infty=K^{\mathbb{Z}^\nu}.$$

(Since $K$ is a compact Hausdorff space, so is $K^X$, for all $X\subseteqq\mathbb{Z}^\nu$.)

To each bounded subset $\Lambda\subset\mathbb{Z}^\nu$ there corresponds a finite system in $\Lambda$ with configuration space $K^\Lambda$, an algebra of "observables" $C(K^\Lambda)$, and whose states are the probability measures on $K^\Lambda$. [These are precisely the continuous, normalized, positive linear functionals on $C(K^\Lambda)$.]

We denote by tr the expectation on $C(K^\infty)$ given by the product measure $\prod_{j\in\mathbb{Z}^\nu}d\varrho(x_j)$. Clearly tr defines a state of the finite system in $\Lambda$, denoted $\mathrm{tr}_\Lambda$, by restriction to $C(K^\Lambda)$.

The dynamics of such systems is given in terms of an *interaction*, $\Phi$. This is a map from bounded subsets $X\subset\mathbb{Z}^\nu$ to $C(K^\infty)$ with the properties that

$$\Phi(X)\in C(K^X),\tag{3.8}$$

and

$$\mathrm{tr}_Y(\Phi(X))=\int\prod_{j\in Y}d\varrho(x_j)\Phi(X)(x)=0,\tag{3.9}$$

for all $Y$ with $Y\cap X\neq\emptyset$; $x=\{x_j\}_{j\in\mathbb{Z}^\nu}$.

Condition (3.9) is not loss of generality: given an arbitrary interaction $\tilde{\Phi}$ satisfying (3.8), one can always find a *physically equivalent* interaction $\Phi$ obeying (3.8) *and* (3.9)!

The Hamilton function of a finite system in $\Lambda$ with interaction $\Phi$ is given by

$$H_\Lambda^\Phi=\sum_{X\subset\Lambda}\Phi(X),$$

and the Gibbs equilibrium state with boundary condition $\varrho_{\partial\Lambda}\in L^1\left(K^\Lambda, \prod_{j\in\Lambda} d\varrho(x_j)\right)$, describing the interactions of the system in $\Lambda$ with its complement in $\Lambda^c$ (recall the Dobrushin-Lanford-Ruelle equations [39, 22]), is given by

$$\langle F\rangle(\Phi, \varrho_{\partial\Lambda}) = Z_\Lambda^{-1}\operatorname{tr}_\Lambda(Fe^{-H_\Lambda^\Phi}\varrho_{\partial\Lambda}), \tag{3.10}$$

for arbitrary $F\in C(K^\Lambda)$. Here

$$Z_\Lambda = \operatorname{tr}_\Lambda(e^{-H_\Lambda^\Phi}\varrho_{\partial\Lambda}).$$

We now consider a decomposition of $\mathbb{Z}^\nu$ into two disjoint sublattices $\Gamma_+, \Gamma_-$ (generally separated by a hyperplane); $r$ is the reflection taking $\Gamma_-$ to $\Gamma_+$ and $\theta_*$ the obvious reflection map from $K^{\Gamma_-}$ to $K^{\Gamma_+}$. For $F\in C(K^{\Gamma_+})$, we set

$$\theta F(x_-) = \overline{F(\theta_* x_-)},$$

where $x_\pm = \{x_j\}_{j\in\Gamma_\pm}$; we set $\Lambda_\pm = \Lambda\cap\Gamma_\pm$, and if $\Lambda_+ = r\Lambda_-$ we say that $\Lambda$ is *reflection symmetric* (RS).

Our previous notion of RP is equivalent to

$$\langle F\theta F\rangle(\Phi, \varrho_{\partial\Lambda}) \geqq 0, \tag{3.11}$$

for all $F\in C(K^{\Lambda_+})$. In this case $\langle - \rangle(\Phi, \varrho_{\partial\Lambda})$ is said to be RP.

We say that a b.c. $\varrho_{\partial\Lambda}$ satisfies RP iff $\operatorname{tr}_\Lambda(F\theta F\varrho_{\partial\Lambda}) \geqq 0$, (3.12)
for all $F\in C(K^{\Lambda_+})$.

Clearly there are b.c. $\varrho_{\partial\Lambda}$ which are not RP, but there are also plenty of b.c. which are $\left(\text{e.g. } \varrho_{\partial\Lambda} = \sum_k G_k\theta G_k, G_k\in C(K^{\Lambda_+}) \text{ for all } k\right)$!

*Remark.* Consider two b.c. $\varrho_{\partial\Lambda}$ and $\varrho'_{\partial\Lambda}$ such that

$$\varrho_{\partial\Lambda}\cdot\varrho'_{\partial\Lambda}\in L^1\left(K^\Lambda, \prod_{j\in\Lambda} d\varrho(x_j)\right).$$

If $\varrho_{\partial\Lambda}$ and $\varrho'_{\partial\Lambda}$ are RP then so is

$$\varrho''_{\partial\Lambda} = \varrho_{\partial\Lambda}\cdot\varrho'_{\partial\Lambda}, \tag{3.13}$$

by Schur's theorem.

From now on we shall always assume that $\Phi$ is reflection covariant, i.e.

$$\theta\Phi(X) = \Phi(rX), \tag{3.14}$$

for arbitrary $X\subset\Gamma_+$.

Our aim is to state and prove a *necessary and sufficient* condition on an interaction $\Phi$ such that $\langle - \rangle(\Phi, \varrho_{\partial\Lambda})$ is RP, for all RP b.c. $\varrho_{\partial\Lambda}$ and all bounded, RS regions $\Lambda$.

We call an interaction CRN (for "conditionally reflection negative") if and only if

$$\sum_{X\cap\Gamma_\pm \neq \emptyset} \operatorname{tr}(F\theta F\Phi(X)) \leqq 0, \tag{3.15}$$

for all $F\in C(K^Y)$, with $Y$ an arbitrary bounded subset of $\Gamma_+$, obeying $\operatorname{tr}(F) = 0$.

We call an interaction $\Phi$ RN (for "reflection negative") if and only if

$$\sum_{X\cap\Gamma_\pm \neq \emptyset} \operatorname{tr}(F\theta F\theta(X)) \leqq 0, \tag{3.16}$$

for all $F\in C(K^Y)$ and for arbitrary, bounded $Y\subset\Gamma_+$.

Let $\operatorname{diam}X=\max\{|i-j|:i,j\in X\}$, let $X+a$ denote the translate of $X$ by a vector $a\in\mathbb{Z}^\nu$, and let $\tau_a$ denote the natural isomorphism from $C(K^X)$ to $C(K^{X+a})$, for arbitrary $X$, i.e. $\{\tau_a\}$ are the translations. Finally, let $\|\cdot\|$ denote the supnorm on $C(K^\infty)$.

**Theorem 3.5.** 1) *The Gibbs state* $\langle - \rangle(\beta\Phi,\varrho_{\partial\Lambda})$ *is* RP, *for all inverse temperatures* $\beta\geqq 0$, *all* RP b.c. $\varrho_{\partial\Lambda}$ *and all* RS *regions* $\Lambda$ *if and only if* $\Phi$ *is* CRN.

2) *Suppose an interaction* $\Phi$ *fulfills* (3.9) *and has the property that*

$$\sup\{\|\Phi(X)\|:\operatorname{diam}X\geqq r\}\to 0, \tag{3.17}$$

*as* $r\to\infty$ *(this condition is fulfilled if* $\Phi$ *obeys any reasonable condition of thermodynamic stability!) Then* $\Phi$ *is* CRN *if and only if* $\Phi$ *is* RN.

3) *If* $\Phi$ *is* RN *and* $\Lambda$ *some* RS *bounded set then*

$$\sum_{X\cap\Lambda_\pm \neq \emptyset} \Phi(X)$$

*is a weak limit of functions of the form*

$$-\sum_k G_k^\Lambda \theta G_k^\Lambda,$$

*where* $G_k^\Lambda\in C(K^{\Lambda_+})$, *for all k. An analogous statement holds for* RP b.c. $\varrho_{\partial\Lambda}$.

*Remarks.* 1) The class of (C)RN interactions $\Phi$ forms a *convex cone.* An analogous statement holds for RP b.c. By (3.13), the convex cone of RP b.c. is *multiplicative.* Furthermore, note that RP is stable under taking the thermodynamic limit $\Lambda\uparrow\mathbb{Z}^\nu$ through a sequence of RS regions $\Lambda$, with RP b.c. $\varrho_{\partial\Lambda}$.

These facts and Theorem 3.5 represent a rather complete, mathematical characterization of RP Gibbs states in the classical case; see also Corollary 3.6.

2) Generally, CRN interactions and periodic b.c. lead to RP Gibbs states; (see also Proposition 3.4). If $\Phi$ obeys (3.17) and the periodic Gibbs states are RP, for all bounded hyper cubes $\Lambda$, then $\Phi$ *must* be RN.

Clearly, periodic b.c. lead to translation invariance, so that $\Lambda$ is RS with respect to many different pairs of hyperplanes, and—if $\Phi(X+a)=\tau_a(\Phi(X))$ (translation invariance)—the Gibbs state is translation invariant. For these reasons translation invariant $\Phi$'s and periodic b.c. play an (annoyingly) important role in our theory.

*Proof of Theorem 3.5.* 1) First we choose $\varrho_{\partial\Lambda}=1$. This b.c. is clearly RP. In this case, the Gibbs state $\langle - \rangle(\beta\Phi,1)$ is RP if and only if

$$R_\Lambda^{\beta\Phi}=\exp\Bigg[-\sum_{\substack{X\subset\Lambda\\ X\cap\Gamma_+\neq\emptyset}} \beta\Phi(X)\Bigg]$$

has the property

$$\operatorname{tr}(F\theta F R_\Lambda^{\beta\Phi})\geqq 0,$$

for all $F \in C(K^{\Lambda_+})$. This follows easily from (3.14) and the definition of the Gibbs state. If $R_\Lambda^{\beta\Phi}(x_+, x_-)$ denotes the integral kernel of $R_\Lambda^{\beta\Phi}$ the above inequality takes the form

$$\int \prod_{j \in \Gamma_+} d\varrho(x_j) d\varrho(y_j) F(x_+) \overline{F(y_+)} R_\Lambda^{\beta\Phi}(x_+, \theta_* y_+) \geqq 0 \tag{3.18}$$

for all $F \in C(K^{\Lambda_+})$.

Assuming that (3.18) holds for arbitrary RS regions $\Lambda$ and all $\beta \geqq 0$ and using a straight forward extension of Schoenberg's theorem [38] (Theorem XIII.52) we conclude that $\Phi$ must be CRN, i.e.

$$\sum_{X \cap \Gamma_\pm \neq \emptyset} \operatorname{tr}(F\theta F \Phi(X)) \leqq 0 .$$

for all $F \in C(K^{\Lambda_+})$ with $\operatorname{tr}(F) = 0$ and arbitrary, bounded $\Lambda_+ \subset \Gamma_+$. [Here we have used (3.9) to include regions $X \not\subset \Lambda$ in the summation. We recall that Schoenberg's theorem says that a matrix $(b_{ij})$ has the property that $(e^{\beta b_{ij}})$ is positive definite for all $\beta \geqq 0$ if and only if $\sum \bar{z}_i z_j b_{ij} \geqq 0$ for all z's with $\sum z_i = 0$.] This proves one direction of Theorem 3.5(1). Conversely suppose now that $\Phi$ is CRN. Then $\sum_{X \cap \Gamma_\pm \neq \emptyset} \operatorname{tr}(F\theta F \Phi(X)) \leqq 0$, for all $F \in C(K^{\Lambda_+})$ with $\operatorname{tr}(F) = 0$, for any RS region $\Lambda$. Now fix some RS, bounded $\Lambda$. By (3.9), it follows that

$$\sum_{X \cap \Lambda_\pm \neq \emptyset} \operatorname{tr}(F\theta F \Phi(X)) = \sum_{\substack{X \cap \Lambda_\pm \neq \emptyset \\ X \subseteq \Lambda}} \operatorname{tr}_\Lambda(F\theta F \Phi(X)) \leqq 0 ,$$

for all $F \in C(K^{\Lambda_+})$ with $\operatorname{tr}_\Lambda(F) = 0$. If we write this out as an integral and use Schoenberg's theorem in the other direction we immediately conclude that $R_\Lambda^{\beta\Phi}(x_+, \theta_* y_+)$ is a positive definite kernel.

Next, if $\varrho_{\partial\Lambda}$ is RP then the kernel of $\varrho_{\partial\Lambda}$, $\varrho_{\partial\Lambda}(x_+, \theta_* y_+)$ is positive definite. By Schur's theorem, $R_\Lambda^{\beta\Phi}(x_+, \theta_* y_+) \varrho_{\partial\Lambda}(x_+, \theta_* y_+)$ is positive definite, so that

$$\int \prod_{j \in \Lambda_+} d\varrho(x_j) d\varrho(y_j) F(x_+) \overline{F(y_+)} R_\Lambda^{\beta\Phi}(x_+, \theta_* y_+) \varrho_{\partial\Lambda}(x_+, \theta_* y_+)$$

$$= \operatorname{tr}_\Lambda(F\theta F R_\Lambda^{\beta\Phi} \varrho_{\partial\Lambda}) \geqq 0 ,$$

for all $F \in C(K^{\Lambda_+})$.

Since, by condition (3.14), $e^{-\beta H_\Lambda^\Phi} = e^{-H_\Lambda^{\beta\Phi}}$ is obviously of the form $G^\Lambda \theta G^\Lambda R_\Lambda^{\beta\Phi}$, with $G^\Lambda \in C(K^{\Lambda_+})$, Theorem 3.5.1) is now proven.

2) It is trivial that if $\Phi$ is RN then $\Phi$ is CRN. Therefore we must only show that if $\Phi$ is CRN and satisfies (3.9) and (3.17) then $\Phi$ is RN. For this purpose, let $F \in C(K^Y)$, for an arbitrary, but hence forth fixed $Y \subset \Gamma_+$. We define

$$G = F - \tau_a(F) ,$$

where a is a translation such that $Y+a\subset\Gamma_+$, i.e. $G\in C(K^{Y\cup Y+a})$ with $Y\cup Y+a\subset\Gamma_+$. Clearly $\mathrm{tr}(G)=\mathrm{tr}(F)-\mathrm{tr}(\tau_a(F))=\mathrm{tr}(F)-\mathrm{tr}(F)=0$. Hence if $\Phi$ is CRN then

$$\sum_{X\cap\Gamma_\pm\neq\emptyset}\mathrm{tr}(G\theta G\Phi(X))\leqq 0\,,\quad \text{i.e.}$$

$$\sum_{X\cap\Gamma_\pm\neq\emptyset}\mathrm{tr}(F\theta F\Phi(X))-\sum_{X_1\cap\Gamma_\pm\neq\emptyset}\mathrm{tr}(F\theta\tau_a(F)\Phi(X_1))$$

$$-\sum_{X_2\cap\Gamma_\pm\neq\emptyset}\mathrm{tr}(\tau_a(F)\theta F\Phi(X_2))$$

$$+\sum_{X_3\cap\Gamma_\pm\neq\emptyset}\mathrm{tr}(\tau_a(F)\theta\tau_a(F)\Phi(X_3))\leqq 0\,.$$

By condition (3.9), the only non-vanishing terms in the last three sums on the l.s. of this inequality fulfill the conditions $X_1\subset Y\cup r(Y+a)$, $X_2\subset Y+a\cup rY$ and $X_3\subset(Y+a)\cup r(Y+a)$. Moreover $X_j\cap\Gamma_\pm\neq\emptyset$, $j=1,2,3$. Applying now condition (3.17) we see that these three sums thend to 0 as a tends to $\infty$ in a direction for which $\Gamma_+ + a\subset\Gamma_+$, for all a of this direction. Thus

$$\sum_{X\cap\Gamma_\pm\neq\emptyset}\mathrm{tr}(F\theta F\Phi(X))\leqq 0\,,$$

for all $F\in C(K^Y)$. Since $Y$ is an arbitrary, bounded set in $\Gamma_+$, this proves Theorem 3.5(2).

3) Let $P$ be an orthogonal projection on $L^2_+=L^2\left(K^{\Lambda_+},\sum_{j\in\Lambda_+}d\varrho(x_j)\right)$. Then the distribution kernel of $P$, $P(x_+,y_+)$, is a weak limit of functions of the form

$$\sum_k\Psi_k(x_+)\overline{\Psi_k(y_+)}\,,\quad\text{where}\quad\Psi_k\in L^2_+\,,\qquad\text{for all } k.$$

This observation combined with the spectral theorem for negative, (resp. positive) bounded operators and the relation $\overline{\Psi_k(\theta_*y_-)}=(\theta\Psi_k)(y_-)$ clearly proves Theorem 3.5(3). □

As an application of this general theory we consider a classical spin system with many body interactions. The classical spin at site $i$ is denoted $\sigma_i$, and $\sigma_X=\prod_{i\in X}\sigma_i$. The expectation tr is chosen such that $\mathrm{tr}(\sigma_X)=0$ and $\mathrm{tr}(\sigma_X^2)>0$, for all non-empty $X$. The interaction $\Phi$ is given by

$$\Phi:X\to -J_X\sigma_X\,,\tag{3.19}$$

where $J=\{J_X\}$ is a family of real numbers indexed by the bounded subsets of $\mathbb{Z}^\nu$. The interaction $\Phi$ is translation invariant if $J_{X+a}=J_X$, for all $a\in\mathbb{Z}^\nu$, and reflection covariant, see (3.14), if $J_X=J_{rX}$, for all $X\subset\Gamma_+$.

*Example.* Ising model with multi-spin interactions.

*Definition.* We say that $J$ is RP if and only if

$$\sum_{X,Y\subset\Lambda_+}\bar z_X z_Y J_{X\cup rY}\geqq 0\,,\tag{3.20}$$

for arbitrary, finite sequences $\{z_X\}_{X\subset\Gamma_+}$ of complex numbers.

**Corollary 3.6.** 1) *Let $\Phi$ be given by* (3.19). *Then $\Phi$ is CRN if and only if $J$ is RP.*
2) *The family of all RP $J$'s forms a convex, multiplicative cone.*

*Proof.* 1) It is not hard to see that if $J$ is RP then $\Phi$, given by (3.19), is RN, thus CRN. Conversely, if $\Phi$ is CRN then, for an arbitrary function $F$ of $\{\sigma_j\}_{j\in\Gamma_+}$ with $\mathrm{tr}(F)=0$

$$\sum_{X,Y\subset\Gamma_+} J_{X\cup rY}\overline{\mathrm{tr}(F\sigma_X)}\mathrm{tr}(F\sigma_Y)\geqq 0. \tag{3.21}$$

Now choose $F=\sum\tilde{z}_X\sigma_X$, where $\tilde{z}_X=z_X\mathrm{tr}(\sigma_X^2)^{-1}$, and $\{z_X\}_{X\subset\Gamma_+}$ is a finite sequence of complex numbers. Then

$$\mathrm{tr}(F)=\sum\tilde{z}_X\mathrm{tr}(\sigma_X)=0,$$

and

$$\mathrm{tr}(F\sigma_X)=\sum_Y\tilde{z}_Y\mathrm{tr}(\sigma_Y\sigma_X)=\sum_Y\tilde{z}_Y\mathrm{tr}(\sigma^2_{Y\cap X})\mathrm{tr}(\sigma_{Y\Delta X})=\tilde{z}_X\mathrm{tr}(\sigma_X^2)=z_X, \tag{3.22}$$

so

$$\sum_{X,Y\subset\Gamma_+} J_{X\cup rY}\overline{\mathrm{tr}(F\sigma_X)}\mathrm{tr}(F\sigma_Y)=\sum_{X,Y\subset\Gamma_+}\bar{z}_X z_Y J_{X\cup rY},$$

and, by (3.21) and (3.22), this is non-negative. Since $\{z_X\}$ is arbitrary, it follows that $J$ is RP.

2) Convexity is obvious. Given $J$ and $J'$, both RP, we define $J''$ by

$$J''_X=J_X\cdot J'_X, \text{ for all } X.$$

By Schur's theorem $J''_X$ is then also RP. □

*Remark.* There are plenty of RP $J$'s with the property that $J_X\neq 0$, for subsets $X$ containing an arbitrarly large number of sites. (As an excercise we recommend that the reader construct some explicit examples of this type.) As a largely open problem we propose to investigate the detailed geometric properties of the cone of RN interaction within one of the standard Banach spaces of interactions, [39].

Theorem 3.5 and Corollary 3.6 provide a rather satisfactory, general theory of RP Gibbs states for classical systems. See also [6]. In the quantum case no complete characterization of RP Gibbs states is available, yet.

The reader can check that Theorem 3.5/Corollary 3.6 includes results in Proposition 3.3 and its consequences via Theorem 2.1 as a special case. In particular, the following should be noted. In Proposition 3.3, we assumed that $H$ has the form (3.2). This form was chosen so that the Gibbs state $\langle\cdot\rangle_{\beta H}$ is RP for *all* $\beta$. If, instead, one starts with the apparently weaker requirement that $\langle\cdot\rangle_{\beta H}$ is RP for *all* $\beta$, then Theorem 3.5.3) tells us that $H$ has to be of the form (3.2).

*Example.* Consider a two-dimensional Ising model with 2, 3, and 4 body interactions. Let

$$X=\sigma_{(0,0)}\sigma_{(0,1)}\sigma_{(1,1)}\sigma_{(1,0)},$$

$$Y=\sigma_{(0,0)}\sigma_{(1,0)}[\sigma_{(1,1)}+\sigma_{(0,1)}],$$

$$Z=\sigma_{(0,0)}\sigma_{(1,1)}.$$

Let $-H=\sum_{a\in\Lambda}\tau_a[JX+KY+LZ]$ where $J$, $K$, $L$ are numbers and $\tau_a$ represents translation by a unit. $H$ will be RN with reflection about the plane $i_1=1/2$ if $K^2=JL$ and $J,L\geqq 0$. To see this, note that in this case $-H$ has the form $B+\theta B+\sum_i C_i\theta C_i$, where $C\alpha\sigma_{(1,0)}\sigma_{(1,1)}+\beta\sigma_{(1,0)}$ and hence $C\theta C=\alpha^2 X+\alpha\beta Y+\beta^2 Z]$, and the sum on $i$ is over translations in the plane $i_1=1/2$.

## 4. Chessboard Estimates and Infrared Domination

In this section, we review, systematize and extend the basic methods of [10, 4, 14, 9] which are based on the use of RP about a large number of planes. For this reason, we will have to work with periodic boundary conditions or directly in infinite volume. We begin by describing "chessboard estimates", then mention the way these can be used in connection with a Peierls argument, and finally discuss the method of infrared bounds.

**Theorem 4.1.** *(Abstract Chessboard Estimates [9]). Let $\mathfrak{A}_0$ be a real vector space, let $r:\mathfrak{A}_0\to\mathfrak{A}_0$ be a real linear map with $r^2=1$ and let $F(a_1,\dots,a_{2n})$ be a complex-valued multilinear map obeying.*

$$F(a_1,\dots,a_{2n})=F(a_2,\dots,a_{2n},a_1) \tag{4.1}$$

*and*

$$|F(a_1,\dots,a_n,b_n,\dots,b_1)|^2 \leqq F(a_1,\dots,a_n,ra_n,\dots,ra_1)F(b_1,\dots,b_n,rb_n,\dots,rb_1). \tag{4.2}$$

*Then $\|a\|\equiv|F(a,ra,a,\dots,ra)|^{1/2n}$ is a semi-norm and*

$$|F(a_1,\dots,a_{2n})|\leqq\prod_{i=1}^{2n}\|a_i\|. \tag{4.3}$$

*Remarks.* 1. In the example of $2n$ spins on a line, one should think of $\mathfrak{A}_0$ as functions of a spin ata single site, and $F(a_1,\dots,a_{2n})=\left\langle\prod_{i=-n+1}^{n}a_{i+n}(\sigma_i)\right\rangle$; $r(a)=a$ (or $\bar a$ if we take complex valued functions) so that (4.1) is true if periodic boundary conditions are used and (4.2) is an expression of RP.

2. The statement and proof are patterned on [9]. For a discussion of its field theory forebears see [43]. For applications to Hölder's inequality for matrices, see [6].

3. It is a worthwhile exercise to prove this directly for the case $2n=4$, see [6, 43].

4. By (4.2) the $F(a_1,\dots,a_n,ra_n,\dots,ra_1)$ are either all $\geqq 0$ or all $\leqq 0$. We can suppose the former without loss.

*Proof.* We first prove (4.3) and then it follows that $\|\cdot\|$ is a semi-norm, since (4.3) implies the triangle inequality. Let $a_1,\dots,a_{2n}$ be given and suppose that $\|a_i\|\neq 0$ for all $i$. Let $b_1,\dots,b_{2n}$ be any $2n$ elements each of which is either an $a_i$ or an $r(a_i)$. Let

$$g(b_1,\dots,b_{2n})\equiv F(b_1,\dots,b_{2n})\Big/\prod_{i=1}^{2n}\|b_i\|$$

and let $g_0 = \max|g(b_i)|$ as the $b_i$ run through the $(4n)^{2n}$ possibilities. Among all choices with $|g(b_i)| = g_0$, pick one with the longest string of the form $a_i, r(a_i), a_i, \ldots, r(a_i)$ for $b_1, \ldots, b_{2l}$. Since (4.1) implies that $\|r(a_i)\| = \|a_i\|$, (4.2) shows that $g$ obeys the same Schwarz inequality as $F$. Thus, if $|g(b_1, \ldots, b_{2n})| = g_0$, we must have that $|g(b_1, \ldots, b_n, rb_n, \ldots, rb_1)| = g_0$. If $2l$ is not $2n$ in the above choice, let $b'_1, \ldots, b'_{2n}$ be a cyclic permutation of $b_1, \ldots, b_{2n}$ with $a_i, r(a_i), \ldots, a_i, r(a_i)$ occuring as $b'_{n-j}, \ldots, b'_n$ where $j = n-1$ if $2l > n$ and otherwise $j = 2l-1$. But then $b_1, \ldots, b_n, rb_n, \ldots, rb_1$ has a string of the form $a_i, r(a_i), \ldots$ of length $2j+2$. It follows that $g_0 = |g(a_i, r(a_i), \ldots, r(a_i))|$ for some $a_i$. But such a $g$ is always 1 so $g_0 \leqq 1$. This implies (4.3) if each $\|a_i\| \neq 0$.

If some $\|a_i\| = 0$, we claim that $F(a_i) = 0$. For, if not, let $b_1, \ldots, b_{2n}$ be a sequence with some $b_j = a_i$ so that the longest string $a_i, r(a_i), \ldots, r(a_i)$ occurs consistent with $F(b_j) \neq 0$. As above $b_1, \ldots, b_{2n}$ must be $a_i, r(a_i), \ldots, r(a_i)$ so there is a contradiction. □

Typical of the explicit versions of Theorem 4.1 are the following:

**Theorem 4.2.** *Let $\Lambda$ be a rectangular subset of $\mathbb{Z}^\nu$ with sides $2n_1 \times \ldots \times 2n_\nu$ ($n_1, \ldots, n_\nu$ positive integers). Let $\langle \cdot \rangle$ be an expectation value for a classical spin system which is invariant under translations* mod $n_i$ *(periodic boundary conditions) and which is RP with respect to (untwisted) reflections* (mod $n_i$) *in all planes perpendicular to coordinate axes running mid-way between neighboring points of $\Lambda$. Then for any functions $\{G_\alpha\}_{\alpha \in \Lambda}$:*

$$\left|\left\langle \prod_{\alpha \in \Lambda} G_\alpha(\sigma_\alpha) \right\rangle\right| \leqq \prod_{\alpha \in \Lambda} \left\langle \prod_{\beta \in \Lambda} G_\alpha(\sigma_\beta) \right\rangle^{1/|\Lambda|}. \tag{4.4}$$

*Proof.* Let $\mathfrak{A}_0$ be the functions of spins $\{\sigma_\alpha\}_{\alpha \in \Lambda; \alpha_1 = 1}$ and let

$$F(a_1, \ldots, a_{2n_1}) = \left\langle \prod_{j=1}^{2n_1} a_j(\{\sigma_\alpha\}_{\alpha_1 = j}) \right\rangle.$$

Using the assumed RP and Theorem 4.1, and setting $a_j = \prod_{\alpha_2, \ldots, \alpha_\nu} G_{j, \alpha_2, \ldots, \alpha_\nu}$, we obtain

$$\left|\left\langle \prod_{\alpha \in \Lambda} G_\alpha(\sigma_\alpha) \right\rangle\right| \leqq \prod_{j=1}^{2n_1} \left\langle \prod_{k=1}^{2n_1} \prod_{\alpha_2, \ldots, \alpha_\nu} G_{j, \alpha_2, \ldots, \alpha_\nu}(\sigma_{k, \alpha_2, \ldots, \alpha_\nu}) \right\rangle^{1/2n_1}.$$

Repeating the argument in the other $\nu - 1$ directions, (4.4) results. □

Now let $j$ be an element of the dual lattice, $\tilde{\Lambda}$, to $\Lambda$, i.e. $j$ is the center of a unit cube, $\Delta_j$ contained in $\Lambda$. Let $F$ be a function of the spins in $\Lambda$. We say that $F \in \Sigma_j$ if and only if $F$ is only a function of spins at the corners of $\Delta_j$. Given such an $F$ we set

$$\gamma(F) = \left\langle \prod_{i \in \tilde{\Lambda}} \tilde{F}_{(i)} \right\rangle^{1/|\Lambda|},$$

where $\tilde{F}_{(i)}$ is $F$ for $i = j$ and for nearest neighbor cubes $\Delta_i$ and $\Delta_{i'}$, $\tilde{F}_{(i)} = \theta_{ii'}[\tilde{F}_{(i)}]$ with $\theta_{ii'}$ untwisted reflection in the plane separating $\Delta_i$ and $\Delta_{i'}$. Thus, if $i - j$ has all even components, then $F_{(i)}$ is a translate of $F$ and if $i - j$ has $\nu_0$ odd components $F$ is a translate of $F$ reflected in $\nu_0$ orthogonal planes. The proof of Theorem 4.2 extends to:

**Theorem 4.3.** *If $\Lambda$ is the set in Theorem 4.2, $\langle\cdot\rangle$ is translation invariant and RP with respect to planes perpendicular to the coordinate axes but through the sites then*

$$\left\langle\prod_{i\in\Lambda} F_i\right\rangle \leqq \prod_{i\in\Lambda}\gamma(F_i)$$

*for $F_i \in \Sigma_i$.*

There are clearly quantum variants and variants with various oblique planes. Except for some discussion of the face centered cubic lattice at the close of this section we do not make these explicit. Reflections at oblique planes have also been used in [41, 17].

To explain schematically the Peierls-chessboard method, consider a classical spin system and break up the configuration space $K$ into pieces $K_1 \cup \ldots \cup K_m$. (For example, if $K$ is finite, each $K_j$ could be a single point. For the anisotropic classical Heisenberg model, $K$ = unit sphere, and $K_1$ and $K_2$ are the two "polar caps" of the sphere, and $K_3$ is the temperate and tropical regions.) Let $P_\alpha^{(j)}$ be the function which is 1 (resp. 0) if $\sigma_\alpha$ is in $K_j$ (resp. not in $K_j$). Let $\langle A\rangle_{\beta,\Lambda} = \langle Ae^{-\beta H_\Lambda}\rangle_0/\langle e^{-\beta H_\Lambda}\rangle_0$ where $\beta > 0$ and $H_\Lambda$ is the Hamiltonian for the lattice $\Lambda$. Let $\langle\cdot\rangle_{\beta,\infty}$ be some weak-$*$ limit point of $\langle\cdot\rangle_{\beta,\Lambda}$ as $\Lambda\to\mathbb{Z}^\nu$. As we will describe, the Peierls-chessboard method typically allows one to show that for $i \neq j$, $\langle P_\alpha^{(i)}P_\gamma^{(j)}\rangle_{\beta,\Lambda}\to 0$ as $\beta\to\infty$ uniformly in $\Lambda$, $\alpha$, $\gamma$. Suppose that we also know that for $i=1,2$, $\lim_{\beta\to\infty}\langle P_\alpha^{(i)}\rangle_{\beta,\infty} > 0$. Then for large $\beta$, $\langle P_\alpha^{(1)}P_\gamma^{(2)}\rangle_{\beta,\infty} - \langle P_\alpha^{(1)}\rangle_{\beta,\infty}\langle P_\gamma^{(2)}\rangle_{\beta,\infty}$ cannot go to zero in the average, which would be required if $\langle\cdot\rangle_{\beta,\infty}$ were ergodic, so there are two or more phases, and $\langle P^{(2)}\rangle_{\beta,\infty}P^{(1)} - \langle P^{(1)}\rangle_{\beta,\infty}P^{(2)}$ will be a long range order parameter. Actually one can say more; namely if $\lim_{\beta\to\infty}\langle P_\alpha^{(i)}\rangle_{\beta,\infty} > 0$ for $i=1,\ldots,k$ there will be, for $\beta$ large, at least $k$ phases; for, if $\langle\ \rangle_\Lambda$ were a convex combination of $k-1$ or fewer ergodic states, then

$$a_{ij} = \lim_{\Lambda\to\infty} |\Lambda|^{-2} \sum_{\alpha,\beta\in\Lambda} \langle P_\alpha^{(i)}P_\beta^{(j)}\rangle_\infty$$

would exist and would be a matrix of rank at most $k-1$ with $\sum_j a_{ij} = \langle P_\alpha^{(i)}\rangle$. Under the given supposition it has rank at least $k$ for $\beta$ large. See also [9, 5].

How does one show that $\langle P_\alpha^{(i)}P_\gamma^{(j)}\rangle$ is small for $j \neq i$? Let $\Gamma$ be a contour in the elementary Peierls argument (see e.g. [39, 16]) sense. Let $p_i(\Gamma)$ = probability that each spin immediately inside $\Gamma$ is in $K_i$ and each spin outside $\Gamma$ is not in $K_i$. Suppose that $p_i(\Gamma) \leqq e^{-C(\beta)|\Gamma|}$ with $C\to\infty$ as $\beta\to\infty$. Then, by the usual argument for cubes $\Lambda$:

$$\begin{aligned}\langle P_\alpha^{(i)}P_\gamma^{(j)}\rangle &\leqq \sum_{\Gamma\,\text{around}\,\alpha} p_i(\Gamma) + \sum_{\Gamma\,\text{around}\,\beta} p_i(\Gamma) + \sum_{\substack{\Gamma\,\text{wrapped}\\ \text{around}\,|\Lambda|}} p_i(\Gamma)\\ &\leqq \sum_{|\Gamma|=2\nu}^{\infty} (|\Gamma|+1)^N e^{d|\Gamma|}e^{-C(\beta)|\Gamma|}\end{aligned}$$

for suitable $d$ and $N$ independent of $\beta$ (but dependent on $\nu$). Thus to show that $\langle P_\alpha^{(i)}P_\gamma^{(j)}\rangle$ is small uniformly in $\alpha$, $\gamma$, and $\Lambda$ as $\beta\to\infty$, we only need to show that

$$\left\langle \prod_{\alpha\,\text{inside}\,\Gamma} P_\alpha^{(i)} \prod_{\alpha\,\text{outside}\,\Gamma} P_\alpha^{(j_\alpha)}\right\rangle \leqq e^{-C_0(\beta)|\Gamma|} \tag{4.5}$$

for any choice of the $j_\alpha$'s (all distinct from $i$), for then

$$(m-1)^{|\Gamma|}e^{-C_0(\beta)|\Gamma|}\equiv e^{-C(\beta)|\Gamma|}.$$

Finally (4.5) is proven by using chessboard estimates, either directly in the form of Theorem 4.2 or an extended form of Theorem 4.2 which exploits a two site basic element. The net result is that the left side of (4.5) is dominated by the product of $|\Lambda|$ terms (or in the two site picture of $|\Lambda|/2$ terms) most of which are 1. But $0(|\Gamma|)$ of them are of the form $f\equiv\left\langle\prod_{\alpha\in\Lambda}P_\alpha^{(k_\alpha)}\right\rangle^{1/|\Lambda|}$ where $\alpha\to k_\alpha$ is a function that has to be worked out in each case. Typically $f$ can be easily estimated to be small by energetic considerations. See [14, 9, 5, 19] and Paper II for explicit examples.

Of course, that leaves the questions of showing that

$$\lim_{\beta\to\infty}\langle P_\alpha^{(i)}\rangle_{\Lambda=\infty}>0\quad\text{for several } i\text{'s}.$$

We discuss this in detail in Paper II, but note that this often follows from symmetry, or by applying the chessboard estimate to obtain an upper bound on $\left\langle\sum_{k\neq i}P_\alpha^{(k)}\right\rangle_{\Lambda=\infty}$ which is small, see also [9, 5, 19].

Thus far, Peierls-type arguments have not been applicable in cases where a phase transition is accompanied by a spontaneously broken *continuous* symmetry. The only tool available is that invented in [10]: in the notation of Example 1 of § 3, let $\sigma$ be a function on $K$, and let $\sigma_\alpha$ be the function $\sigma$ on the $\alpha$th copy of $K$. For $\Lambda$ a cube, let $p$ be in $\Lambda^*$, the Fourier dual for $\Lambda$ (=1st Brillouin zone; =dual group to $\Lambda$ viewed as a torus) and define

$$\hat{\sigma}_p=\frac{1}{\sqrt{|\Lambda|}}\sum_{\alpha\in\Lambda}e^{ip\alpha}\sigma_\alpha$$

$$g_\Lambda(p)=\langle\hat{\sigma}_p\hat{\sigma}_{-p}\rangle_{\beta,\Lambda}.$$

Suppose that one can prove that for $p\neq 0$:

$$g_\Lambda(p)\leqq 1/2\beta E_p \tag{4.6}$$

for $E_p$ a function satisfying

$$(2\pi)^{-\nu}\int_{\substack{|p_i|\leqq\pi\\ i=1,\ldots,\nu}}E_p^{-1}d^\nu p\equiv C_0<\infty \tag{4.7}$$

and that for $\beta\geqq\beta_0$

$$\langle\sigma_\alpha^2\rangle\geqq D>0. \tag{4.8}$$

Then (following the version of the argument in [4]) for $\beta>\max(\beta_0,\beta_1)$ where $\beta_1=C_0/2D$, we will have (assuming some regularity on $E_p$)

$$\lim_{|\Lambda|\to\infty}[|\Lambda|^{-1}g_\Lambda(p=0)]>0 \tag{4.9}$$

since

$$|\Lambda|^{-1}g_\Lambda(p=0)=|\Lambda|^{-1}\sum_{p\in\Lambda^*}g_\Lambda(p)-|\Lambda|^{-1}\sum_{p\neq 0}g_\Lambda(p)$$

$$\geqq\langle\sigma_\alpha^2\rangle_\Lambda-|\Lambda|^{-1}\sum_{p\neq 0}1/2\beta E_p, \tag{4.10}$$

where the first sum is controlled by a Plancherel formula, and the second by (4.6). With minimal regularity assumptions on $E$,

$$\lim_{\Lambda\to\infty}(4.10)=\langle\sigma_\alpha^2\rangle-(2\beta)^{-1}C_0$$

so (4.9) holds. By an argument of Griffiths (see e.g. [4]), (4.9) implies a first order phase transition with $\sigma_\alpha$ as order parameter.

In certain quantum cases $\left(\text{where } \sum_{\alpha\in\Lambda}\sigma_\alpha \text{ and } H \text{ do not commute}\right)$ and, as we shall see below, for some other than simple, cubic lattices like the face centered cubic lattices, it is necessary or more convenient to rely not on (4.9) but rather on a direct infinite volume argument which is explained in detail in [5, 6, 4].

We note that sometimes (4.8) follows by a symmetry argument (e.g. in the classical Heisenberg model) but that in general one can try to use a chessboard argument to show, e.g. that $\text{Prob}(\sigma_\alpha^2\leqq 2D)\leqq 1/2$ for $\beta\geqq\beta_0$.

The only known way of proving (4.6) is via a "Gaussian domination" or related estimate: Let $K$ be a compact subset of $\mathbb{R}^N$ and let $d\varrho$ be a measure on $K$. Let $\sigma^{(1)},\dots,\sigma^{(N)}$ be the coordinate functions on $K$. Suppose that $H$ has the form

$$H=\frac{1}{2}\sum_{\alpha\neq\gamma}J_{\alpha\gamma}(\sigma_\alpha-\sigma_\gamma)^2\quad\text{(each pair counted once)}$$

and define for $\{h_{\alpha\gamma}\}_{\alpha\neq\gamma}$ real,

$$Z(h_{\alpha\gamma})=\left\langle\exp\left[-\frac{1}{2}\beta\sum_{\alpha\neq\gamma}J_{\alpha\gamma}(\sigma_\alpha-\sigma_\gamma-h_{\alpha\gamma})^2\right]\right\rangle_0,$$

where $\langle\cdot\rangle_0=\int\cdot\pi d\varrho(\sigma_\alpha)$ as usual. We claim that the two conditions: $J_{\alpha\gamma}=J_{\alpha-\gamma}=J_{\gamma-\alpha}$ and

$$Z(h_{\alpha\gamma})\leqq Z(0)\quad\text{(Gaussian domination)}\tag{4.11}$$

imply (4.6) with

$$E_p=\frac{1}{2}\sum_{\alpha\in\Lambda}(1-e^{ip\cdot\alpha})J_{\alpha 0}\,.\tag{4.12}$$

Before proving this, we note that one point of the Definition (3.5) is that it makes $E_p$ independent of $\Lambda$ for $p\in\Lambda^*$.

Since the argument to go from (4.11) to (4.6) is only a mild extension of that in [10], we only sketch the details. By translation invariance, $\frac{d}{d\lambda}Z(\lambda h_{\alpha\gamma})|_{\lambda=0}=0$ so that (4.11) implies that $\frac{d^2}{d\lambda^2}Z(\lambda h_{\alpha\gamma})_{\lambda=0}\leqq 0$. This is equivalent to:

$$\frac{1}{2}\beta^2\left\langle\left|\sum_{\alpha\neq\gamma}J_{\alpha\gamma}(\sigma_\alpha-\sigma_\gamma)\cdot h_{\alpha\gamma}\right|^2\right\rangle\leqq\frac{1}{2}\beta\sum_{\alpha\neq\gamma}|h_{\alpha\gamma}|^2J_{\alpha\gamma}\,.\tag{4.13}$$

(4.13) only holds apriori for real $h_{\alpha\gamma}$ but it extends to complex $h$. Now take $h_{\alpha\gamma}=(e^{ip\cdot\alpha}-e^{ip\cdot\gamma})|\Lambda|^{-1/2}$ and find that (4.13) implies (4.8) with $E_p$ given by (4.12). We summarize:

**Theorem 4.4.** *The Gaussian domination bound* $Z(h_{\alpha\gamma})\leqq Z(0)$ *together with* $J_{\alpha\gamma}=J_{\alpha-\gamma,0}=J_{\gamma-\alpha,0}$ *implies the infrared bound* $g_\Lambda(p)\leqq(2\beta E_p)^{-1}$ *with* $E_p=\frac{1}{2}\sum_{\alpha\in\Lambda}(1-e^{ip\cdot\alpha})J_{\alpha 0}$.

We next turn to a detailed investigation of (4.11).

**Proposition 4.5.** *Suppose that* $J_{\alpha\gamma}\geqq 0$. *Then it suffices to check* (4.11) *for* $h_{\alpha\gamma}$ *of the form* $h_\alpha-h_\gamma$.

*Proof.* Since $J_{\alpha\gamma}\geqq 0$, $Z\to 0$ as any $h_{\alpha\gamma}\to\infty$ and thus $Z$ takes its maximum at some finite point. But $\partial Z/\partial h_{\alpha\gamma}=0$ implies that

$$J_{\alpha\gamma}(h_{\alpha\gamma}-\langle\sigma_\alpha-\sigma_\gamma\rangle)=0$$

for the obvious expectation. Thus, letting $h_\alpha\equiv\langle\sigma_\alpha\rangle$, we see that $h_{\alpha\gamma}=h_\alpha-h_\gamma$ for those $\alpha\gamma$ with $J_{\alpha\gamma}\neq 0$. $Z$ is independent of the other $h_{\alpha\gamma}$ so we can take $h_{\alpha\gamma}=h_\alpha-h_\gamma$ for such $(\alpha\gamma)$ without changing $Z$; i.e. $Z$ takes its maximum value at a point $h_{\alpha\gamma}=h_\alpha-h_\gamma$. □

*Remark.* The proof of Theorem 4.4 only used (4.11) for the special case $h_{\alpha\gamma}=h_\alpha-h_\gamma$ so that Proposition 4.5 is, at this stage, primarily of academic interest. Indeed, there are $J_{\alpha\gamma}$ not all non-negative so that (4.11) holds for $h_{\alpha\gamma}$ of the form $h_\alpha-h_\gamma$, and thus (4.6) holds, even though for such $J$'s, $Z\to\infty$ for a suitable choice of $h_{\alpha\gamma}$ not of the form $h_\alpha-h_\gamma$.

It is an important and interesting open question to characterize the ferromagnetic interactions for which the spin 1/2 Ising model obeys Gaussian domination. We only have partial results on this question relying on reflection positivity. We begin with some examples which delimit the class and, in particular, demonstrate the falseness of the apriori attractive conjecture that Gaussian domination holds for all ferromagnets:

*Example 1.* Consider two spins, one $\sigma_1$, with values $\pm 1$ and the other, $\sigma_2$ with values $\pm 2$, all values having equal apriori weight. Then $\langle e^{-J(\sigma_2-\sigma_1-h)^2}\rangle$ has its maximum near $h=\pm 1$ as $J\to\infty$. This shows that equality of the single spin distributions is essential for Gaussian domination in general (but see Examples 5 and 6).

*Example 2.* Let $\sigma_1=\pm 1$, $\sigma_2=+1$, then

$$Z(h)=\frac{1}{2}[e^{-Jh^2}+e^{-J(2-h)^2}]$$

has its maximum near $h=1$ as $J\to 0$. The given distribution for $\sigma_2$ can be thought of as that of an Ising spin in an intense positive magnetic field. The failure of RP in this case shows that even equality of the magnetic fields at each point is also essential for Gaussian domination.

*Example 3.* (Mean field model.) This is the most involved but also the most significant of the examples we present. $Z(h) \leqq Z(0)$ implies that $M_{\alpha\gamma} = -\left.\frac{\partial^2 \ln Z}{\partial h_\alpha \partial h_\beta}\right|_{h_\alpha = 0}$ is positive semi-definite. For a spin 1/2 model with

$$H = \frac{1}{2} \sum_{\gamma \neq \alpha} J_{\alpha\gamma}(\sigma_\alpha - \sigma_\gamma)^2$$

a simple calculation shows that (set $J_{\alpha\alpha} = 0$)

$$M_{\alpha\gamma} = \delta_{\alpha\gamma}\left(\sum_\delta J_{\alpha\delta}\right) - J_{\alpha\gamma} - \sum_{\delta,\lambda} J_{\alpha\delta} J_{\gamma\lambda} \langle(\sigma_\alpha - \sigma_\delta)(\sigma_\gamma - \sigma_\lambda)\rangle .$$

Take $n+1$ spins, $\sigma_0, \ldots, \sigma_n$ with only $J_{0i} \neq 0$, all equal to $\alpha_n$. Then

$$M_{00} = n\alpha_n - \alpha_n^2[n^2 + n + (n^2 - n)\langle\sigma_i\sigma_j\rangle - 2n^2\langle\sigma_0\sigma_i\rangle] .$$

Take $\alpha_n = 1/\sqrt{n}$, so that $H = -\sigma_0\left(\frac{1}{\sqrt{n}} \sum_1^n \sigma_i\right) + \text{const}$ and thus, as $n \to \infty$, a coupling of a Gaussian and a spin 1/2 spin. Thus as $n \to \infty$

$$\langle\sigma_0\sigma_i\rangle = \frac{1}{\sqrt{n}}\left\langle \sigma_0 \frac{1}{\sqrt{n}} \sum_{i=1}^n \sigma_i \right\rangle \sim c/\sqrt{n}$$

$$\langle\sigma_i\sigma_j\rangle = \frac{1}{n-1}\left[\left\langle\left(\frac{1}{\sqrt{n}} \sum_1^n \sigma_i\right)^2\right\rangle - 1\right] \sim d/n$$

for finite non-zero, $c$ and $d$. Thus

$$M_{00} = -n - 1 + 2c\sqrt{n} + \sqrt{n} + O(1)$$

is negative for $n$ large and therefore $M_{\alpha\gamma}$ is not positive definite for $n$ large.

Our next example, while a trivial extension of RP ideas illustrates that Gaussian domination can hold in some cases where RP fails:

*Example 4.* Let $\langle\cdot\rangle$ be an expectation for a string of $6n$ spins with third neighbor ferromagnetic coupling. Then RP fails both for reflections about the midpoint of bonds and for reflections on sites. Since $Z(h)$ is a product of three nearest neighbor $2n$-point $Z$'s, Gaussian domination for that case yields it for the case at hand.

Our final three examples show that special features of the $J$'s and/or the single spin distributions can allow one to prove Gaussian domination without RP and/or translation invariance. We hasten to add that phase transitions will not occur in Examples 5–7.

*Example 5.* Suppose that $H = \sum J_{\alpha\beta}(\sigma_\alpha - \sigma_\beta)^2$ with $J_{\alpha\beta}$ *arbitrary* positive numbers and that each single spin measures $d\varrho_\alpha(\sigma_\alpha)$ equals $F_\alpha d\sigma_\alpha$ with $F_\alpha$ log concave and even, but not necessarily $\alpha$ independent. Since $e^{-H(\sigma_\alpha - h_\alpha)} \prod_\alpha F_\alpha$ is a log concave function of $\{\sigma_\alpha, h_\alpha\}$, $Z(h_\alpha)$ is log concave in $h_\alpha$ by a general theorem (see e.g. [1]). Since $\partial Z/\partial h_\alpha = 0$, all $h_\alpha = 0$, by symmetry, log concavity implies that $Z(h_\alpha) \leqq Z(0)$.

*Example 6.* Suppose that $H=J_{\alpha\beta}(\sigma_\alpha-\sigma_\beta)^2$ with $J_{\alpha\beta}$ *arbitrary* positive numbers and that each single spin measure $d\varrho_\alpha(\sigma_\alpha)$ equals $F_\alpha d\sigma_\alpha$ with $F_\alpha$ positive definite and real (hence even), but not necessarily $\alpha$ independent. Then

$$Z(h_\alpha)=\int e^{-H(\sigma_\alpha-h_\alpha)}\prod_\alpha[(\int e^{ik_\alpha\sigma_\alpha}d\mu_\alpha(k_\alpha))d\sigma_\alpha]$$

$$=\int\prod_\alpha d\mu(k_\alpha)e^{ik_\alpha h_\alpha}\int\prod_\alpha(d\sigma_\alpha e^{ik_\alpha\sigma_\alpha})e^{-H(\sigma_\alpha)}$$

is positive definite in the $h$'s since the Fourier transform of a Gaussian is a Gaussian. In particular, $Z(h_\alpha)$ takes its maximum value at $h_\alpha=0$ (in essence, the above calculation is proving that the convolution of positive definite functions is positive definite).

*Example 7.* Consider an array of $n$ spin 1/2 Ising spins, $s_1,\dots,s_n$ on a line with arbitrary *positive, nearest neighbor* couplings, $J_{12},J_{23},\dots,J_{n1}$. Let $T(J,h)$ be the two by two matrix

$$\begin{pmatrix} e^{-1/2Jh^2} & e^{-1/2J(2-h)^2} \\ e^{-1/2J(2+h)^2} & e^{-1/2Jh^2} \end{pmatrix}$$

i.e. if we label matrices as $\begin{pmatrix} a_{++} & a_{+-} \\ a_{-+} & a_{--} \end{pmatrix}$, then $T(J,h)_{\sigma_1\sigma_2}=\exp\left(-\frac{1}{2}J(\sigma_1-\sigma_2-h)^2\right)$. We want to note two critical facts about these matrices: first $T(J,0)$ is positive definite and $T(J,0)^{-1/2}T(J,h)T(J,0)^{-1/2}$ is a contraction in the norm $\|(\alpha,\beta)\|=(\alpha^2+\beta^2)^{1/2}$—this is proven in [10]. Secondly, the $T(J,0)$ all commute, are positive definite and when diagonalized simultaneously their largest eigenvalues correspond to a common eigenvector—this follows by noting that $\frac{1}{\sqrt{2}}\begin{pmatrix}1\\1\end{pmatrix}$ and $\frac{1}{\sqrt{2}}\begin{pmatrix}1\\-1\end{pmatrix}$ are the eigenvectors, and the eigenvalue for the first eigenvector is always largest. Since $Z(h_1,\dots,h_n)=\mathrm{Tr}(T(J_{12},h_2-h_1)\dots T(J_{n1},h_1-h_n))$ we can write

$$Z(h_1,\dots,h_n)=\mathrm{Tr}(A_1B_1\dots A_nB_n),$$

where $A_i=T(J_{i-1,i},0)^{1/2}T(J_{i,i+1},0)^{1/2}$ (where $J_{01}\equiv J_{n1}$) and $B_i$ is a contraction, by the first fact noted above. Let $\mu_1(C),\dots,\mu_m(C)$ be the singular values of an $m\times m$ matrix [eigenvalues of $(C^*C)^{1/2}$ ordered so that $\mu_1\geqq\mu_2\geqq--\geqq0$]. An inequality of Horn [21] (see Corollary II.4.1 of [15]) asserts that

$$\sum_{i=1}^m\mu_i(C_1\dots C_k)\leqq\sum_{i=1}^m\mu_i(C_1)\dots\mu_i(C_k).$$

Thus, we have that

$$Z(h_1,\dots,h_n)\leqq\sum_{i=1}^2\mu_i(A_1\dots B_n)$$

$$\leqq\sum_{i=1}^2\mu_i(A_1)\mu_i(A_2)\dots\mu_i(A_n)$$

$$=\sum_{i=1}^2\mu_i(A_1\dots A_n)=Z(0,\dots,0),$$

where we use the fact that $\mu_j(B_i)\leqq 1$, since $B_i$ is a contraction, in the second inequality, and use the second noted fact in the equality that follows.

To illustrate the close connection between chessboard estimates and Gaussian domination, we note:

**Theorem 4.6.** *Let $\Lambda$ be a $2n_1\times\ldots\times 2n_\nu$ rectangle in $\mathbb{Z}^\nu$. Let $J_{\alpha\gamma}$ be a given on $\Lambda$ so that the chessboard estimate (Theorem 4.1) holds for $\langle\cdot\rangle=Z^{-1}\int\cdot e^{-H(\sigma_\alpha)}\prod_{\alpha\in\Lambda}d\varrho(\sigma_\alpha)$ for all $d\varrho$ in $\mathbb{R}^N$ and*

$$H=\frac{1}{2}\sum_{\alpha\neq\gamma}J_{\alpha\gamma}(\sigma_\alpha-\sigma_\gamma)^2\,.$$

*Then the Gaussian domination estimate $Z(h_\alpha)\leqq Z(0)$ holds for arbitrary $d\varrho$ and, in particular, $g_\Lambda(p)\leqq(2\beta E_p)^{-1}$.*

*Proof*. By a limiting argument, we can suppose that $d\varrho(\sigma)=F(\sigma)d^N\sigma$ with $F>0$ on all of $\mathbb{R}^N$. Then, if we define $G_\alpha(\sigma)=F(\sigma+h_\alpha)/F(\sigma)$ we have that

$$\begin{aligned}Z(h_\alpha)&\equiv\int e^{-H(\sigma_\alpha-h_\alpha)}\prod_\alpha d\varrho(\sigma_\alpha)\\&=\int e^{-H(\sigma_\alpha)}\prod_\alpha d\varrho(\sigma_\alpha+h_\alpha)\\&=Z(0)\left\langle\prod_\alpha G_\alpha(\sigma_\alpha)\right\rangle\\&\leqq Z(0)\prod_\alpha\left\langle\prod_\beta G_\alpha(\sigma_\beta)\right\rangle^{1/|\Lambda|}\\&=\prod_\alpha\left[\int e^{-H(\sigma_\gamma)}\prod_\gamma d\varrho(\sigma_\gamma+h_\alpha)\right]^{1/|\Lambda|}=Z(0)\,,\end{aligned}$$

where the inequality is a chessboard estimate and the last equality comes from $H(\sigma_\alpha-h)=H(\sigma_\alpha)$ for constant $h$. □

*Remark.* Using the Dobrushin-Lanford-Ruelle equations one can prove Theorem 4.6 directly in infinite volume for RP Gibbs states.

The above argument has a defect: it does not obviously extend to the quantum case.

Fortunately, one can use a version of the original argument given in [10], based on *Theorem 2.3*: Namely, in the case of $2n$ spins, Theorem 2.3 says that

$$\begin{aligned}|Z(h_{-n+1},\ldots,h_n)|^2\leqq\,&Z(h_{-n+1},\ldots,h_0,h_0,h_{-1},\ldots,h_{-n+1})\\&\cdot Z(h_n,h_{n-1},\ldots,h_1,h_1,\ldots,h_n)\end{aligned}$$

so that translation invariance and the argument in Theorem 4.1 show that $\max|Z(h_i)|$ occurs when all $h$'s are equal. Since $Z(h,\ldots,h)=Z(0)$, the maximum is $Z(0)$. As of now, this is the most widely applicable proof of Gaussian domination we know of.

We remind the reader that in the quantum case there is one additional complication in that Gaussian domination does not lead to a bound on $\langle\hat{\sigma}_p\hat{\sigma}_{-p}\rangle$ but rather on a "Duhamel two point function", $(\hat{\sigma}_p,\hat{\sigma}_{-p})$. This problem and its resolution are discussed in [4], for the case of nearest neighbor interactions. The present generalization is straight forward.

The argument based on Theorem 2.3 has an additional advantage, even in the classical case. Suppose that $\tilde{H}=\frac{1}{2}\sum_{\alpha\neq\gamma} J_{\alpha\gamma}(\sigma_\alpha-\sigma_\gamma)^2+H'$ where $\langle\cdot\rangle_{H'}$ is RP and $J$ obeys (3.3). If $Z(h)=\left\langle \exp-\left(\frac{1}{2}\sum J_{\alpha\gamma}(\sigma_\alpha-\sigma_\gamma-h_\alpha+h_\gamma)^2+H'\right)\right\rangle_0$ then, as above, Theorem 2.3 implies that $Z(h_\alpha)\leqq Z(0)$, and infrared bounds follow. We summarize with

**Theorem 4.7.** *Let $H$ have the form of Theorem 4.6 with $J$ RP. Let $\tilde{H}=H+H'$ with $H'$ RN. Let $Z(h_\alpha)=\langle\exp(H(\sigma_\alpha-h_\alpha)+H')\rangle_0$. Then $Z(h_\alpha)\leqq Z(0)$ and $g_\Lambda(p)\leqq(2\beta E_p)^{-1}$ with $E_p$ depending on $J_{\alpha\gamma}$, as in Theorem 4.4.*

Finally, we want to mention a problem (and its resolution) that occurs for certain special models like the ones on face centered cubic lattices. The infinite volume lattice is reflection invariant about any plane which is the perpendicular bisector of a bond, but any finite volume cutoff will destroy many of these symmetries. The resolution is the following: Let $\langle\cdot\rangle$ denote an infinite volume expectation and, given, $\{h_\alpha\}_{\alpha\in\mathbb{Z}^\nu}$ with only finitely many non-zero $h_\alpha'$'s, let

$$g(h_\alpha)\equiv\left\langle \exp\left(\frac{1}{2}\sum_{\alpha\neq\gamma} J_{\alpha\gamma}[(\sigma_\alpha-\sigma_\gamma)^2-(\sigma_\alpha-\sigma_\gamma-h_\alpha+h_\gamma)^2]\right)\right\rangle.$$

If we can show that $|g(h_\alpha)|\leqq 1$ for all $h_\alpha$, then by following the arguments in [10] one will get infinite volume infrared bounds and therefore long range order. To prove that $|g(h_\alpha)|\leqq 1$, one need only show that $\langle\cdot\rangle$ has a kind of RP about each "bond" plane, i.e. that

$$|g(h_\alpha)|^2\leqq g(h_\alpha')g_\alpha(h_\alpha'') \tag{4.14}$$

where $h_\alpha'$ (resp $h_\alpha''$) is obtained by taking $h_\alpha$ on the left (resp. right) side of the plane and reflecting in the plane. Given (4.14) it is not hard to reduce the proof of $|g(h_\alpha)|\leqq 1$ to showing that $|g(h_\alpha)|^{1/|\Lambda|}\to 1$ for a set of $h_\alpha'$'s constant at $h_0$ on a nice set $\Lambda$. But it is easy to see that $|g(h_\alpha)|\leqq e^{c|\partial\Lambda|}$ for such $h$'s. (Instead one can use Theorem 4.6 in infinite volume; see e.g. [6]).

We can see two ways of proving (4.14). In cases where correlation inequalities are available, one can prove (4.14) for a given plane by taking a suitable sequence of "+ boundary condition states" where the given plane cuts $\Lambda$ exactly in half. Since the limit is independent of the sequence, (4.14) holds for the + boundary condition state. When correlation inequalities are not available, one can at least prove there are multiple phases; for, if not, then all periodic states converge to a *unique* state which would then obey (4.14). If $\langle\sigma_\alpha^2\rangle_{\beta,\infty}$ has a lower bound that is uniform in $\beta$ one would obtain long range order: a contradiction!

## 5. Long Range Models

In [3], Dyson showed that a spin 1/2 Ising model with $J(n)=(1+|n|)^{-\alpha}$ has a phase transition if $1<\alpha<2$ ($\alpha>1$ is needed for sensible thermodynamics), and did not if $2<\alpha$. His method works for any classical model with correlation inequalities such as the plane rotor model [13]. Using similar ideas, Kunz and Pfister [26] treated the two dimensional plane rotor model with $J(n)=(1+|n|)^{-\alpha}$, proving a phase transition if $2<\alpha<4$.

In this section, we illustrate the general methods of this paper by recovering these results (many more examples are presented in [7, 8]) and extending them in several directions: a) cases where correlation inequalities are unknown such as the classical Heisenberg model can be accomodated; b) logarithmic improvements in Dyson's conditions are given; c) certain *quantum* models are accomodated.

We give details in the one dimensional classical case and then treat two dimensions and quantum models in a few remarks. When correlation inequalities of Griffiths type are available, improvements of our results of the following sort are possible: If a phase transition is known for an RP $J_0$ which is also positive, it holds for any larger $J$ *even if the larger J is not RP*. We suppose in all cases that $\sum|J(n)|<\infty$.

We begin our analysis with:

**Theorem 5.1.** *Let K be a compact subset of* $\mathbb{R}^N$ *and let* $d\varrho$ *be a measure different from* $\delta(\sigma)$, *invariant under* $\sigma\to-\sigma$. *Let* $-\beta H=\beta\sum_{i>j}J(i-j)\sigma_i\cdot\sigma_j$ *and let* $E_p=\sum_{n=1}^{\infty}J(n)(1-\cos pn)$. *If* $0\leqq J(n)$ *and J is RP, and if* $g\equiv\int dp/E_p<\infty$, *then there is a first order phase transition with* $\sigma$ *as order parameter, at some sufficiently large, finite* $\beta$.

**Theorem 5.2.** *Let* $J(i-j)$ *be RP. Then the classical isotropic Heisenberg model has a first order phase transition for* $\beta$ *large if and only if* $g\equiv\int dp/E_p<\infty$.

*Proofs.* The *absence* of a first order phase transition (asserted in Theorem 5.2) if $g=\infty$ follows from a slight extension of an argument of Mermin [29], so we concentrate on the existence question. Since $g<\infty$, this follows, according to the strategy of §4, if we show that $\langle\hat{\sigma}_p\hat{\sigma}_{-p}\rangle_{\text{periodic}}\leqq 1/2\beta E_p$ and $\lim_{\beta\to\infty}\langle|\sigma|^2\rangle_{\beta,\infty}>0$. $J$ being RP implies that $\langle\cdot\rangle_{\text{periodic}}$ is RP by Theorems 2.1 and 3.4. The method of §4 then yields the *infrared bounds*. In the case of Theorem 5.2, $\langle|\sigma|^2\rangle_\infty=1$ while in the case of Theorem 5.1, choose $r_0>0$ so that $\int_{|\sigma|>r_0}d\varrho>0$ and use a chessboard estimate to see that $\langle(|\sigma|\leqq r_0)\rangle_{\beta,\infty}\to 0$, as $\beta\to\infty$. The right side of this chessboard estimate is controlled by noting that RP implies that the ground state with the restriction $|\sigma_\alpha|\leqq r_0$ has all spins equal, and then by noting that the energy when all $\sigma_\alpha=r$ is strictly monotone increasing in $|r|$, since $J(n)\geqq 0$. □

These theorems reduce the study of the long range one dimensional case to the study of two questions: 1) When is $J$ RP? 2) When is $\int E_p^{-1}dp<\infty$. In studying the first question the following is useful:

*Definition.* A distribution $F$ on $\mathbb{R}^\nu\backslash\{0\}$ is called *OS positive* (for Osterwalder-Schrader [30]) if and only if $F$ is continuous and

$$\int F(x-y)g(x)\tilde{g}(y)dxdy\geqq 0 \tag{5.1}$$

for all $g\in C_0^\infty(x_1>0)$ where $\tilde{g}(y_1,\ldots,y_\nu)=g(-y_1,y_2,\ldots,y_\nu)$.

**Theorem 5.3.** a) *If $F$ is an OS positive distribution on $R^\nu$ then $J$, defined on $\{(n_1,\dots,n_\nu)|n_1>0\}$ by*

$$J(n)=F(n_1,\dots,n_\nu)\,; \qquad n_1>0,$$

*is RP.*

b) *If $J_1$ and $J_2$ are RP on $\{n_1>0\}\subset\mathbb{Z}^\nu$, then so is $J_1J_2$.*

c) *If $J(n)=\int\limits_0^\infty e^{-ny}d\varrho(y)$, $(n\geqq 1)$, then $J$ is RP on $\mathbb{Z}^1$.*

*Proof.* a) In (5.1), let $g$ approach a sum of delta functions. This shows at once that $J$ is RP.

b) Follows from the fact (Schur's theorem) that if $a_{ij}$ and $b_{ij}$ are positive definite matrices, so is $c_{ij}$ with $c_{ij}=a_{ij}b_{ij}$

c) A restatement of Proposition 3.2; it also follows from a) and well-known structure theorems for OS positive distributions. □

**Proposition 5.4.** *The following functions on $\mathbb{Z}$ are RP in the region $n\geqq 1$:*

a) $J(n)=n^{-\alpha}$, b) $J(n)=(1+n)^{-\alpha}$

*for all $\alpha>0$.*

*Proof.* a) $\int\limits_0^\infty e^{-ny}y^{\alpha-1}dy=\Gamma(\alpha)n^{-\alpha}$ [use Theorem 5.3c];

b) $\int\limits_0^\infty e^{-ny}e^{-y}y^{\alpha-1}dy=\Gamma(\alpha)(n+1)^{-\alpha}$ [use Theorem 5.3c]. □

As for the second question, we note:

**Theorem 5.5.** *Let $E_p=\sum\limits_{n=1}^\infty J(n)(1-\cos pn)$ with $J(n)\geqq 0$. Then*

a) *If $\sum\limits_{n=1}^\infty n^{-3}J(n)^{-1}<\infty$, then $\int dpE_p^{-1}<\infty$.*

b) *If $\limsup\limits_{N\to\infty}(\log N)^{-1}\left[\sum\limits_1^N nJ(n)\right]<\infty$, then $\int dpE_p^{-1}=\infty$.*

*Remarks.* 1. The condition in a) is slightly weaker than the one that Dyson [3] needs for a phase transition. The condition in b) is slightly weaker than the one that Dyson [3] needs to prove that there is no phase transition in the *Ising model*; b) will only imply the absence of continuous symmetry breaking. This is as it must be if the $n^{-2}$ Ising model has a phase transition (as is believed), since $J(n)=n^{-2}$ obeys the conditions of b).

2. b) includes the case $J(n)=n^{-2}$. This case can be done by explicit calculation of $E_p$ (contained in the tables, e.g. (516) of [24]) or by noting that $E_p=f(0)-f(p)$ with $f(p)=\sum\limits_1^\infty n^{-2}\cos pn$ obeying $f''(p)=\pi\delta(p)-\frac{1}{2}$ with periodic boundary conditions at $\pm\pi$. One sees that $E(p)\sim|p|$ in that case.

3. If $J(n)\sim n^{-\alpha}$ at infinity, we are in case (a) if $\alpha<2$ and in case b) if $\alpha\geqq 2$. Actually with regard to a) one cannot improve even logs, since for $J(n)$

$\sim n^{-2}(\log n)\ldots(\log_m n)^{1+\varepsilon}$, then $E_p \sim |p|(\log p)\ldots(\log_m p)^{1+\varepsilon}$. For b), improvements are presumably possible: with little change $(\log N)^{-1}$ can be replaced by $[(\log N)(\log_2)(N)\ldots\log_m(N)]^{-1}$ which allows only $n^{-2}(\log_2 n)\ldots(\log_m n)^{1+\varepsilon}$.

4. If $J(n)=\int_{-1}^{1}\lambda^{|n|-1}d\varrho$, then $E_p \equiv \sum_{n=1}^{\infty} J(n)(1-\cos pn)$ increases when $d\varrho$ increases.

This remark allows one to obtain results for $J$'s which are RP but *not* positive from those in this theorem.

*Proof.* a) We need a lower bound on $E_p = \sum_1^\infty J(n)(1-\cos np)$. For $|x| \leqq \pi$, $(1-\cos x) \geqq \frac{2}{\pi^2}x^2$ so that

$$E_p \geqq \sum_1^{[\pi/|p|]} \frac{2}{\pi^2} p^2 n^2 J(n),$$

where $[x]$ = greatest integer less than $x$. Thus we need only show that

$$\alpha \equiv \int_0^\pi dp \left[\sum_1^{[\pi/|p|]} p^2 n^2 J(n)\right]^{-1} < \infty .$$

By the Schwarz inequality

$$[\pi/p]^2 = \left(\sum_1^{[\pi/p]} 1\right)^2$$

$$\leqq \left[\sum_1^{[\pi/p]} n^2 J(n)\right]\left[\sum_1^{[\pi/p]} (n^2 J(n))^{-1}\right]$$

so that

$$\alpha \leqq \int_0^\pi dp p^{-2}[\pi/p]^{-2} \sum_1^{[\pi/p]} (n^2 J(n))^{-1}$$

$$\leqq \frac{4}{\pi^2}\int_0^\pi dp \sum_1^{[\pi/p]} (n^2 J(n))^{-1}$$

since $n^2[n]^{-2} \leqq ([n]+1)^2[n]^{-2} \leqq 4$ for $n \geqq 1$ and $\pi/p \geqq 1$ for $0 \leqq p \leqq \pi$. Finally we note that

$$\int_0^\pi dp \sum_1^{[\pi/p]} (n^2 J(n))^{-1} = \sum_{n=1}^\infty (n^2 J(n))^{-1} \int_0^{\pi/n} dp$$

$$= \pi \sum_1^\infty n^{-3} J^{-1} .$$

b) We need an upper bound on $E_p$. Since $(1-\cos x) \leqq |x|$ we have that

$$E_p \leqq |p| \sum_1^N n J(n) + 2\sum_N^\infty J(n)$$

for any $N$. To estimate the second term, let $K(j)=\sum_1^j nJ(n)$ so that

$$\sum_N^M J(n)=\sum_N^M \frac{1}{n}[K(n)-K(n-1)]$$
$$=\sum_{n=N}^M \left[\frac{1}{n}K(n)-\frac{1}{n-1}K(n-1)\right]+\frac{1}{n(n-1)}K(n-1)$$
$$=\frac{1}{M}K(M)-\frac{1}{N-1}K(N-1)+\sum_N^M \frac{1}{n(n-1)}K(n-1).$$

Thus, if $\frac{1}{M}K(M)\to 0$ as $M\to\infty$, we have that

$$\sum_N^\infty J(n)\leqq \sum_{N-1}^\infty \frac{1}{n(n+1)}K(n).$$

If $K(n)\leqq C\log n$, we see that

$$E_p\leqq C|p|\log N+\tilde{C}N^{-1}\log N.$$

Choosing $N=[|p|^{-1}]$, we see that $E_p\leqq C|p|C\log(|p|^{-1}))$, so that $\int E_p^{-1}dp=\infty$. □

By combining the previous results of this section we conclude that

**Theorem 5.6.** *If $d\varrho\neq\delta(p)$ is a measure on $\mathbb{R}^N$ symmetric under $\sigma\to-\sigma$ and $J(n)=n^{-\alpha}$, then there is a first order phase transition for the one dimensional spin model when $1<\alpha<2$.*

*Remark.* If $N=1$ (or if $d\varrho$ is anisotropic in a suitable sense) but $d\varrho$ is *not even*, there will be a phase transition in suitable external magnetic field when $1<\alpha<2$; see [10] or [7].

We describe the extensions in a series of remarks:

A) In two dimensions, the functions $p^{\alpha-1}$ have OS positive Fourier transforms for $\alpha>-1$. This follows from

$$\int_0^\infty \frac{dm}{p^2+m^2}m^\alpha=p^{\alpha-1}\int_0^\infty \frac{x^\alpha dx}{x^2+1}$$

and the fact that $(p^2+m^2)^{-1}$ has an OS positive Fourier transform (free Euclidean field [30, 42]). Since $x^{-\beta}$ $(0<\beta<2)$ has a Fourier transform $c_\beta p^{2-\beta}$, we see that $|n|^{-\beta}$ is RP for $0<\beta<2$ by Theorem 5.3a). Then by Theorem 5.3b), we conclude that $|n|^{-\beta}$ is RP for all $\beta>0$. Calculations similar to those above show that in 2 dimensions, $\int dp/E_p<\infty$ if $\sum_{n\neq 0} n^{-6}J(n)^{-1}<\infty$; and for $J(n)=n^{-4}$, an explicit calculation involving periodic Green's functions for $-\Delta$ [and the fact that $\Delta(r^{-2})\sim r^{-4}$ at $\infty$] shows that $E_p\sim p^2\log p+0(p^2)$ at $p=0$, so $\int dp/E_p=\infty$ in that case. We thus obtain:

**Theorem 5.7.** *If $d\varrho\neq\delta(p)$ is a measure on $\mathbb{R}^N$ symmetric under $\sigma\to-\sigma$, and $J(n)=n^{-\alpha}$, then there is a first order phase transition for $2<\alpha<4$, in the two-dimensional spin model.*

This result is of interest only in isotropic cases.

B) It is easy to prove first order phase transitions in suitable quantum systems which are simultaneously real by using the method of [4]. In order for that method to be applicable one must check an algebraic condition; in particular some double commutator should not be large. There are two cases where this condition is easy to verify: in anisotropic models, such as $\sigma_x\sigma_x+\varepsilon\sigma_y\sigma_y$ with $\varepsilon<1$, the double commutator is always small at low temperatures, and in a classical limit, like $S\to\infty$ in Heisenberg models, the double commutator is small, for $S$ sufficiently large, [4]. We conclude:

**Theorem 5.8.** *Fix $J(n)=n^{-\alpha}$ for $1<\alpha<2$. Then the isotropic antiferromagnet with $-H=\sum_{n\neq m}(-1)^{n-m}J(|n-m|)S_n\cdot S_m$ for quantum spins $S_n$ of spin $S$ has a first order phase transition if $S$ is sufficiently large (at some $\beta$ sufficiently large). Moreover, for any $\varepsilon$ with $0<\varepsilon<1$, the spin 1/2 model with $-H=\sum_{n\neq m}J(|n-m|)(S_n^xS_m^x+\varepsilon S_n^yS_m^y)$ has a first order phase transition at some $\beta$ sufficiently large.*

*Acknowledgements.* It is a pleasure to thank F. Dyson, O. Heilmann, L. Rosen, E. Seiler, J. Slawny, and T. Spencer for valuable discussions.

## References


1. Brascamp, H., Lieb, E. H.: Some inequalities for Gaussian measures and the long range order of the one dimensional plasma. In: Functional integration and its applications (ed. A. M. Arthurs), pp. 1—14. Oxford: Clarendon Press 1975
2. Donoghue, W. F., Jr.: Monotone matrix functions and analytic continuation. Berlin-Heidelberg-New York: Springer 1974
3. Dyson, F.: Existence of a phase transition in a one-dimensional ising ferromagnet. Commun. math. Phys. **12**, 91 (1969). Non-existence of spontaneous magnetization in a one-dimensional ising ferromagnet. Commun. math. Phys. **12**, 212 (1969)
4. Dyson, F., Lieb, E. H., Simon, B.: Phase transitions in quantum spin systems with isotropic and nonisotropic interactions. J. Stat. Phys. **18**, 335—383 (1978). See also: Phase transitions in the quantum Heisenberg model. Phys. Rev. Letters **37**, 120—123 (1976)
5. Fröhlich, J.: Phase transitions, goldstone bosons and topological superselection rules. Acta Phys. Austriaca Suppl. **XV**, 133—269 (1976)
6. Fröhlich, J.: The pure phases (harmonic functions) of generalized processes. Or: mathematical physics of phase transitions and symmetry breaking. Invited talk at Jan. 1977 A.M.S., St. Louis meeting. Bull. Am. Math. Soc. (in press)
7. Fröhlich, J., Israel, R., Lieb, E. H., Simon, B.: Phase transitions and reflection positivity. II. Short range lattice models. J. Stat. Phys. (to be submitted)
8. Fröhlich, J., Israel, R., Lieb, E. H., Simon, B.: Phase transitions and reflection positivity. III. Continuous models. Commun. math. Phys. (to be submitted)
9. Fröhlich, J., Lieb, E. H.: Phase transitions in anisotropic lattice spin systems. Commun. math. Phys. **60**, 233—267 (1978)
10. Fröhlich, J., Simon, B., Spencer, T.: Infrared bounds, phase transitions, and continuous symmetry breaking. Commun. math. Phys. **50**, 79 (1976)
11. Gallavotti, G., Miracle-Sole, S.: Equilibrium states of the Ising model in the two-phase region. Phys. Rev. **5**, 2555—2559 (1972)
12. Gel'fand, I. M., Vilinkin, N. Ya.: Generalized functions, Vol. 4. New York: Academic Press 1964
13. Ginibre, J.: General formulation of Griffiths' inequality. Commun. math. Phys. **16**, 310—328 (1970)
14. Glimm, J., Jaffe, A., Spencer, T.: Phase transitions for $\phi_2^4$ quantum fields. Commun. math. Phys. **45**, 203 (1975)
15. Gohberg, I. C., Krein, M. G.: Introduction to the theory of linear non-selfadjoint operators. Providence, RI: American Mathematical Society 1969

16. Griffiths, R.: Phase transitions. In: Statistical mechanics and quantum field theory, Les Houches, 1970, pp. 241—280. New York: Gordon and Breach 1971
17. Hegerfeldt, G. C.: Correlation inequalities for Ising ferromagnets with symmetries. Commun. math. Phys. **57**, 259—266 (1977)
18. Hegerfeldt, G. C., Nappi, C.: Mixing properties in lattice systems. Commun. math. Phys. **53**, 1—7 (1977)
19. Heilmann, O. J., Lieb, E. H.: Lattice models for liquid crystals (in preparation)
20. Holsztynski, W., Slawny, J.: Peierls condition and number of ground states. Commun. math. Phys. **61**, 177—190 (1978)
21. Horn, A.: On the singular values of a product of completely continuous operators. Proc. Nat. Acad. Sci. USA **36**, 374—375 (1950)
22. Israel, R.: Convexity and the theory of lattice gases. Princeton, NJ: Princeton University Press 1978
23. Israel, R.: Phase transitions in one-dimensional lattice systems. Proc. 1977 IUPAP Meeting, Haifa
24. Jolley, C. B. W.: Summation of series. New York: Dover 1961
25. Klein, A.: A characterization of Osterwalder-Schrader path spaces by the associated semigroup. Bull. Am. Math. Soc. **82**, 762—764 (1976)
26. Kunz, H., Pfister, C. E.: First order phase transition in the plane rotor ferromagnetic model in two dimensions. Commun. math. Phys. **46**, 245 (1976)
27. Lieb, E. H.: New proofs of long range order. Proceedings of the International Conference on the Mathematical Problems in Theoretical Physics, Rome, 1977. Lecture notes in physics. Berlin-Heidelberg-New York: Springer (in press)
28. Elliott, R. J.: Phenomenological discussion of magnetic ordering in the rare-earth metals. Phys. Rev. **124**, 346—353 (1961)
29. Mermin, N. D.: Absence of ordering in certain classical systems. J. Math. Phys. **8**, 1061—1064 (1967)
30. Osterwalder, K., Schrader, R.: Axioms for Euclidean green's functions. Commun. math. Phys. **31**, 83 (1973)
31. Osterwalder, K., Seiler, E.: Gauge field theories on the lattice. Ann. Phys. **110**, 440—471 (1978)
32. Kim, D., Thompson, C. J.: A lattice model with an infinite number of phase transitions. J. Phys. A: Math. Gen. **9**, 2097—2103 (1976)
33. Pirogov, S. A., Sinai, Ya. G.: Phase transitions of the first kind for small perturbations of the Ising model. Funct. Anal. Pril. **8**, 25—30 (1974). [Engl. translation: Funct. Anal. Appl. **8**, 21—25 (1974)]
34. Pirogov, S. A., Sinai, Ya. G.: Phase diagrams of classical lattice systems. Teor. Mat. Fiz. **25**, 358—369 (1975). [Engl. translation: Theor. Math. Phys. **25**, 1185—1192 (1975)]
35. Pirogov, S. A., Sinai, Ya. G.: Phase diagrams of classical lattice systems. Continuation. Theor. Mat. Fiz. **26**, 61—76 (1976). [Engl. translation: Theor. Math. Phys. **26**, 39—49 (1971)]
36. Redner, S., Stanley, H. E.: The R-S model for magnetic systems with competing interactions: series expansions and some rigorous results. J. Phys. C. Solid State Phys. **10**, 4765—4784 (1977)
37. Reed, M., Simon, B.: Methods of modern mathematical physics. Vol. II: Fourier analysis, self-adjointness. New York: Academic Press 1975
38. Reed, M., Simon, B.: Methods of modern mathematical physics. Vol. IV: Analysis of operators. New York: Academic Press 1978
39. Ruelle, D.: Statistical mechanics. New York: Benjamin 1969
40. Schoenberg, I. J.: Metric spaces and positive definite functions. Trans. Am. Math. Soc. **44**, 522—536 (1938)
41. Schrader, R.: New Correlation inequalities for the Ising model and $P(\phi)$ theories. Phys. Rev. B **15**, 2798—2803 (1977)
42. Simon, B.: The $P(\phi)_2$ Euclidean (quantum) field theory. Princeton, NJ: Princeton University Press 1974
43. Simon, B.: New rigorous existence theorems for phase transitions in model systems. Proc. 1977 IUPAP Meeting, Haifa

Communicated by J. Glimm

Received April 24, 1978

*Journal of Statistical Physics, Vol. 22, No. 3, 1980*

# Phase Transitions and Reflection Positivity. II. Lattice Systems with Short-Range and Coulomb Interactions

**Jürg Fröhlich,[1] Robert B. Israel,[2] Elliott H. Lieb,[3] and Barry Simon[3]**

*Received June 29, 1979*

We discuss applications of the abstract scheme of part I of this work, in particular of infrared bounds and chessboard estimates, to proving the existence of phase transitions in lattice systems. Included are antiferromagnets in an external field, hard-core exclusion models, classical and quantum Coulomb lattice gases, and six-vertex models.

**KEY WORDS**: Phase transitions; reflection positivity; chessboard estimates; contours.

## 1. INTRODUCTION

This is the second paper in a series describing applications of reflection positivity (RP) to proving the existence of phase transitions in model systems. We exploit: (1) the Peierls chessboard argument first used by Glimm *et al.*[(22)] and further developed by Fröhlich and Lieb[(13)]; (2) the method of infrared bounds, first used by Fröhlich *et al.*[(17)] and extended to quantum systems by Dyson *et al.*[(6)] Reviews of some of these ideas can be found in Refs. 8, 9, 18, 30, 36, and 49. In paper I[(11)] of this series,[4] we presented a general framework, and in a third paper,[(12)] we give applications to quantum field theories. In this paper, we deal with short-range lattice models and Coulomb lattice gases. In I, we discussed long-range lattice models. A further application to a model of a liquid crystal has been found by Heilmann and Lieb[(26)] and one to dipole lattice gases by Fröhlich and Spencer.[(19)]

Research partially supported by Canadian National Research Council Grant A4015 and U.S. National Science Foundation Grants PHY-77-18762, MCS-75-21684-A02, and MCS-78-01885.

[1] IHES, Bures-sur-Yvette, France.

[2] Department of Mathematics, University of British Columbia, Vancouver, Canada.

[3] Departments of Mathematics and Physics, Princeton University, Princeton, New Jersey.

[4] Erratum to paper I[(11)]: The Mermin argument[(40)] uses $J \geqslant 0$, so this hypothesis should be added to Theorem 5.2 of Ref. 11. On the other hand, the *existence* part of the theorem does not require $J \geqslant 0$.

**297**

0022-4715/80/0300-0297$03.00/0 

Typically, the proof of phase transitions by the *method of infrared bounds* involves three steps:

(IR1) Verify reflection positivity for a suitably defined reflection. In many of the models below the reflection is in planes through lattice sites rather than between lattice sites.

(IR2) Choose the quantity, typically the Fourier transform of a two-point function, for which an infrared bound is to be proven, e.g., $\langle \hat{S}(p)\hat{S}(-p)\rangle \leqslant (\beta E_p)^{-1}$, and evaluate $E_p$, the spin wave energy. Check that $\int E_p^{-1}\, dp < \infty$.

(IR3) Complete the argument, typically by proving a sum rule on the Fourier transform. Often symmetry is useful in this last step, but it is not essential; see Ref. 17. For infrared bounds one needs some couplings crossing the plane of reflection.

Typically the proof of phase transitions by the Peierls chessboard method also involves three steps:

(PC1) Identical to (IR1).

(PC2) Choose some decomposition of the configuration space at a single site, or small number of sites, and let $P^{(1)},\ldots, P^{(n)}$ be the corresponding projections. Let $P_\alpha^{(i)}$ be the $P^{(i)}$ associated to site $\alpha$. Use the Peierls chessboard estimate to show that for suitable $i$ and $j$, $\sup_{\alpha,\gamma}\langle P_\gamma^{(i)} P_\alpha^{(j)}\rangle \to 0$ as $\beta$, the inverse temperature, goes to infinity. Here $\langle\cdots\rangle$ is typically an infinite-volume limit of periodic bc equilibrium states. To prove this, one typically must compare the energy per unit volume of a finite number of periodic configurations to the energy per unit volume of some ground state.

(PC3) Completion of the argument: Show that (PC2) implies long-range order. This is discussed below.

We will assume some familiarity with these methods either from paper I or some of the other references given. We expand on step (PC3) in a number of successively more complicated situations.

## 1.1. Two States Related by a Symmetry

This is the situation of the usual spin-$\frac{1}{2}$ Ising model. By symmetry of the interaction and boundary conditions,

$$\langle P_\alpha^{(1)}\rangle = \langle P_\alpha^{(2)}\rangle \tag{1.1}$$

and since only two states are involved, $\langle P^{(1)}\rangle + \langle P^{(2)}\rangle = 1$, so $\langle P_\alpha^{(1)}\rangle = \langle P_\alpha^{(2)}\rangle = 1/2$. Thus, once

$$\sup_{\alpha,\gamma}\langle P_\alpha^{(i)} P_\gamma^{(j)}\rangle \equiv A^{(i,j)} \tag{1.2}$$

is less than 1/4 for $(i,j) = (1,2)$ we have that

$$\lim_{\Lambda\to\infty} |\Lambda|^{-2} \sum_{\alpha,\gamma\in\Lambda} [\langle P_\alpha^{(i)} P_\gamma^{(j)}\rangle - \langle P_\alpha^{(i)}\rangle\langle P_\gamma^{(j)}\rangle] \neq 0 \tag{1.3}$$

implying that $\langle\cdots\rangle$ is not an ergodic state, i.e., multiple phases occur.

## 1.2. Two Important States Related by a Symmetry

This is the situation of the anisotropic classical Heisenberg model[13] or the $\phi_2^4$ field theory.[22] There are now three regions of the single-site configuration space singled out (the two polar regions, call them 1 and 2, and the remainder of the sphere, call it 0). Equation (1.1) still holds and $A^{(ij)}$, given by (1.2), goes to zero for $(ij) = (01), (02), (12)$ as $\beta \to \infty$. Finally, by a chessboard estimate, $\langle P_\alpha^{(0)}\rangle \to 0$ so, by (1.1), $\langle P_\alpha^{(i)}\rangle \to 1/2$ and thus (1.3) still holds.

## 1.3. Two Important States Not Related by a Symmetry

A good example of this is the Pirogov–Sinai triangle model, which we discuss in Section 2. Here there are two important states $P^{(1)}$ and $P^{(2)}$ and a third state $P^{(0)}$ with

$$\langle P_\alpha^{(0)}\rangle_{\beta,\mu} \to 0 \tag{1.4}$$

as $\beta \to \infty$, uniformly as $\mu$, an "external field parameter," varies through a suitable compact set $[a, b]$. Moreover, $A^{(ij)}$, given by (1.2), goes to zero uniformly in $\mu$ for $(ij) = (01), (02), (12)$. However, (1.1) no longer holds. We use a device going back to Ref. 8 (related to an idea in Ref. 17): Suppose that

$$s(\mu) \equiv \langle P_\alpha^{(1)}\rangle_{\beta,\mu} \quad \text{and} \quad t(\mu) \equiv \langle P_\alpha^{(2)}\rangle_{\beta,\mu}$$

obey $s(a) \to 0$, $t(b) \to 0$ as $\beta \to \infty$. Vary $\mu$ between $a$ and $b$ at fixed, large $\beta$. Either $s(\mu)$ varies continuously, in which case there is a range of $\mu_0$ with $s(\mu_0) > 1/3$, $t(\mu_0) > 1/3$ and hence at least two phases, by (1.3), or else $s(\mu)$ varies discontinuously, in which case, picking a $\mu_0$ where $s(\mu)$ is discontinuous, we get at least two phases, by taking (subsequence) limits $\langle\cdots\rangle_{\beta,\mu_0+0}$ and $\langle\cdots\rangle_{\beta,\mu_0-0}$. The main defect in this procedure is that we do not know which possibility occurs (although we note that if there is precisely one $\mu$ with multiple phases, the second is realized) nor do we obtain any kind of continuity of $\mu_0$ in $\beta$ (although typically one can compute $\lim_{\beta\to\infty}\mu_0$).

## 1.4. Three Important States with Two Related by Symmetry

A typical example is the Fisher stabilized antiferromagnet at critical external field: see Section 3. There are four regions $P^{(0)}, \ldots, P^{(3)}$, (1.4) and (1.1) hold, $s(a) \to 0$, and $w(\mu) \equiv \langle P_\alpha^{(3)}\rangle_{\beta,\mu} \to 0$ for $\mu = b$. We now proceed as in Section 1.3 to find either a point with $s(\mu_0) = t(\mu_0) \sim 1/3$ or else two states $\langle\cdots\rangle_{\mu_0-0}$ and $\langle\cdots\rangle_{\mu_0+0}$ with $\langle P^{(3)}\rangle_{\mu_0-0}$ small, $\langle P^{(1)}\rangle_{\mu_0-0} = \langle P^{(2)}\rangle_{\mu_0-0}$ and $\langle P^{(3)}\rangle_{\mu_0+0}$ large. By (1.3), $\langle\cdots\rangle_{\mu_0-0}$ is still not ergodic, so there are at least three states.

## 1.5. Many States Related by a Symmetry

A typical example is the three-state Potts model discussed in Section 3. We cannot just use the lack of ergodicity, so instead we use:

**Theorem 1.1.** Let $\langle\cdots\rangle$ be a translation-invariant state which is a linear combination of $k$ or fewer ergodic states. Let $\{X_\alpha^{(i)}\}_{i=1}^m$ be a set of $m$ distinct observables and let

$$M_{ij} = \lim_{\Lambda\to\infty} |\Lambda|^{-1} \sum_{\alpha,\gamma\in\Lambda} \langle X_\alpha^{(i)} X_\beta^{(j)}\rangle \tag{1.5}$$

Then $M_{ij}$ is a matrix with rank at most $k$.

*Remark.* The limit in (1.5) always exists by the mean ergodic theorem.

*Proof.* Let $\langle\cdots\rangle = \sum_{r=1}^k \lambda_r\langle\cdots\rangle_r$ with $\langle\cdots\rangle_r$ ergodic. Then

$$|\Lambda|^{-2} \sum_{\alpha,\gamma} \langle X_\alpha^{(i)} X_\beta^{(j)}\rangle_r \to \langle X_\alpha^{(i)}\rangle_r \langle X_\alpha^{(j)}\rangle_r$$

so that

$$M_{ij} = \sum_{r=1}^k \lambda_r \langle X_\alpha^{(i)}\rangle_r \langle X_\alpha^{(j)}\rangle_r$$

is rank $k$ or fewer. ∎

If now the $X^{(i)}$ are projections with $\sum P^{(i)} = 1$, then $\sum_j M_{ij} = \langle P_\alpha^{(i)}\rangle$. If $A^{(ij)} \to 0$ (as $\beta\to\infty$) for all $i \neq j$, then $M_{ij}$ goes to zero off-diagonal as $\beta\to\infty$. If we can find an integer $l_0$ such that $l_0$ of the $\langle P_\alpha^{(i)}\rangle$ are all bigger than, say, $(2l_0)^{-1}$, then $M$ will have an $l_0 \times l_0$ block which is very small off-diagonal and $>(2l_0)^{-1}$ on-diagonal; hence $M$ will have rank at least $l_0$, so there will be at least $l_0$ phases by Theorem 1.1. One way of realizing this possibility is to have a symmetry which implies that $l_0$ $P$'s all have the same expectation and a chessboard estimate which shows that the remaining $P$'s are all small.

## 1.6. Many Phases Without Symmetries

As a final level of complication, we want to show how the scheme of Section 1.3 can be extended. We consider a situation with $k + 1$ important states and $k$ external field parameters, say $\lambda_1, \ldots, \lambda_k$, varied in the region $0 \leqslant \lambda_i \leqslant 1$, $\sum_{i=1}^k \lambda_i \leqslant 1$. Let $\lambda_{k+1} = 1 - \sum_1^k \lambda_i$ and suppose that

$$\lim_{\beta\to\infty} \langle P_\alpha^{(i)}\rangle_{\lambda_j} = 0 \quad \text{if} \quad \lambda_i = 0 \quad (\text{uniformly in the other } \lambda\text{'s}) \tag{1.6}$$

We want to show that for $\beta$ large there is some $\lambda^{(0)} = (\lambda_1^{(0)}, \ldots, \lambda_{k+1}^{(0)})$ and a state $\langle\cdots\rangle$ which is a convex linear combination of states which are limits of $\langle\cdots\rangle_{\beta,\lambda}$ as $\lambda \to \lambda^{(0)}$ so that $\langle P^{(i)}\rangle = \langle P^{(1)}\rangle$, $i = 1, \ldots, k+1$. Thus, as in

Section 1.5, there will be at least $k + 1$ phases. Without any change one can accommodate situations with some symmetries and fewer parameters needed. Fix $\beta$ large and let $\Delta_k$ be the simplex

$$\left\{(\lambda_1, \ldots, \lambda_{k+1}): \quad \lambda_i \geqslant 0, \quad \sum_1^{k+1} \lambda_i = 1\right\}$$

Let $f$: $\Delta_k \to \Delta_k$ be given by

$$[f(\lambda^{(0)})]_i = \langle P^{(i)} \rangle_{\beta,\lambda_0} \Big/ \sum_1^{k+1} \langle P^{(j)} \rangle_{\beta,\lambda_0}$$

Let $F(\lambda^{(0)})$ be the convex set in $\Delta_k$ consisting of all limits of convex combinations of limits $\lim_{\lambda_n \to \lambda_0} f(\lambda_n)$. Then $F$ maps $\Delta_k$ into subsets of $\Delta_k$ so that $F(\lambda_0)$ is closed and convex and $F$ is semicontinuous in the sense that if $a_n \to a$ and $a_n \in F(\lambda_n)$ and $\lambda_n \to \lambda$, then $a \in F(\lambda)$. Moreover, by (1.6), $F$ is close to leaving each face of $\Delta_k$ fixed. We now used the following topological results essentially in Brézis.[2]

**Theorem 1.2.** Fix $k$. Let $\Delta_k$ be the $k$-simplex. Then there exists $\epsilon > 0$ so that every map $F$ from $\Delta_k$ into $2^{\Delta_k}$ obeying:

(i) $F(x)$ is convex for all $x$;
(ii) if $a_n \to a$, $x_n \to x$, and $a_n \in F(x_n)$, then $a \in F(x)$;
(iii) for every generalized face $G$ of $\Delta_k$, $x \in G$ and $a \in F(x)$ implies $\operatorname{dist}(a, G) < \epsilon$;

has

$$\left(\frac{1}{k+1}, \ldots, \frac{1}{k+1}\right) \in \bigcup_{x \in \Delta_k} F(x)$$

If one notes that (ii) is equivalent to the statement that the graph $\{(x, a) | a \in F(x)\}$ is closed, then this theorem is equivalent to Theorem A3 proven in the Appendix. This proves the claim above about $k + 1$ phases.

Finally we list the models treated in the remainder of the paper. In Section 2 we "warm up" with the elementary models: the triangle model of Pirogov and Sinai[42] and a model of Ginibre studied by Kim and Thompson.[32] In Section 3 we consider various antiferromagnets: we recover Dobrushin's result[5] on multiple equilibrium states in the Ising antiferromagnet (in the interior of the critical triangle) and extend this result to anisotropic, classical Heisenberg ferromagnets. When next-nearest-neighbor ferromagnetic coupling is added (Fisher stabilization) we obtain three equilibrium states at suitable external field (recovering a result of Pirogov and Sinai) and also multiple states in the classical, *isotropic* antiferromagnet in external field. Furthermore, we discuss a three-state Potts model recently studied by Schick and Griffiths.[46] In Section 4 we treat lattice gases with nearest neighbor exclusion for the square, triangular, and hexagonal lattices, recovering results of Dobrushin,[5]

Heilmann,[25] and Heilmann and Praestgaard.[27] In Section 5 we consider classical and quantum Coulomb monopole gases. Typical is our result that in a lattice where each site can have charge 0, +1, or −1 with Coulomb monopole forces, at low temperature and suitable fugacity there are three phases, one with mainly unoccupied sites and one with the $A$ (resp. $B$) sublattice occupied with positive charges and the $B$ (resp. $A$) sublattice occupied with negative charges. In Section 6 we discuss some six- and eight-vertex models. In Section 7 we describe a useful version of the chessboard Peierls argument which is applicable in particular to the Slawny[50] model of a ferromagnet with an infinity of phases. Finally in Section 8 we describe a rather special model which in $\nu$ dimensions has nearest neighbor ferromagnetic coupling $J$ and next-nearest-neighbor antiferromagnetic coupling $J/2(\nu - 1)$. For this kind of model, discrete symmetries are only presumably broken in three or more dimensions and continuous symmetries only in five or more dimensions. At the end of the paper, we include an Appendix containing a topological result of some significance in applications to complicated models.

We note that our results for the triangular and hexagonal lattice suffer from the same defects noted in Ref. 11 for the face-centered cubic lattice: namely we cannot be certain that the infinite-volume state has reflection positivity unless there is a single state. Thus, typically we can only assert the existence of multiple phases and not of long-range order or number of phases; see Sections 3 and 4.

The methods presented in this series generally yield relatively accurate lower bounds on the transition temperature. If, in addition, some of the tricks of the trade, most especially that of Gallavotti and Miracle-Solé[21] are used, we would expect that, in the simplest models, our bounds are within about 20% of the exact critical temperature. In this paper, we will not push the arguments to the point of explicit lower bounds on transition temperatures, but we will be careful to get pretty good upper bounds on probabilities of contours which then could be used as an element of good bounds on transition temperatures.

## 2. TWO SIMPLE MODELS: THE TRIANGLE ISING MODEL OF PIROGOV–SINAI AND GINIBRE'S MODEL

As a warmup, we begin with two simple models. We are especially interested in illustrating the method for obtaining phase transitions without symmetry.

**Model 2.1.** This is a model of Ginibre, studied in mean field approximation by Kim and Thompson.[32] The basic Hamiltonian $H$ in a finite volume $\Lambda$, a "simple cubic" torus, is

$$H = \sum_{\langle\alpha\beta\rangle} (s_\alpha - s_\beta)^2 + a \sum_\alpha s_\alpha^2 + \mu \sum_\alpha s_\alpha; \qquad a > 0 \tag{2.1}$$

where the first sum is over nearest neighbor pairs. Letting $h = -\frac{1}{2}\mu a^{-1}$, we can add a constant to $H$ and obtain

$$H' = H + \text{const} = \sum_{\langle\alpha\beta\rangle} (s_\alpha - s_\beta)^2 + a \sum (s_\alpha - h)^2 \tag{2.1'}$$

The spins $s_\alpha$ are required to take integer values $0, \pm 1, \ldots$; we allow the possibility of a priori weights $w(n)$ for $\{s | s = n\}$, so that at inverse temperature $\beta$ the partition function is given by

$$Z = \sum_{n_\alpha \in \mathbb{Z}} \left[\prod_{\alpha\in\Lambda} w(s_\alpha)\right] \exp[-\beta H(s_\alpha)]$$

convergent if $w(n) \leqslant \exp(An^B)$, with $B < 2$. Kim and Thompson studied three models: (a) $w(n) = 1$, all $n$; (b) $w(n) = 1$, $|n| \le N$: $w(n) = 0$ for $|n| > N$; and (c) $w(n) = 1$ for $n \geqslant 0$ and $w(n) = 0$ for $n < 0$.

To find the phase structure at low temperatures, it is always useful to look at ground states, i.e., configurations minimizing $H$. From (2.1′) it is obvious that the ground state has $s_\alpha = s^{(0)}$ for all $\alpha$, where $s^{(0)}$ is determined by minimizing $(s^{(0)} - h)^2$. Clearly, if $h \neq \pm\frac{1}{2}, \pm\frac{3}{2}, \ldots$, there is a unique such $s^{(0)}$, but at $h = k + \frac{1}{2} (k \in \mathbb{Z})$ two values (namely $s^{(0)} = k, k + 1$) minimize $H$. Thus at low temperatures, we expect that there are infinitely many $h_k(\beta)$ (near $k + \frac{1}{2}$ as $\beta \to \infty$) with multiple phases. We will prove this assuming $D \geqslant w(n) \geqslant \delta > 0$ for all $n$ [without this assumption one can still show, as below, that if $w(n) \neq 0$ all $n$, then for $\beta > \beta_k$ there are multiple phases at some $h_k(\beta)$ with $|h_k - k - \frac{1}{2}| < \frac{1}{2}$, but $\beta_k$ may go to $\infty$ as $k \to \pm\infty$]. In case $w(n) \neq 0$ for $|n| \leqslant N$ and $w(n) = 0$ for $|n| > N$, we get transitions at $2N$ points. Without loss we look at $k = 0$.

We begin by noting that since $H$ has only nearest neighbor coupling, the model has RP about planes between lattice sites and orthogonal to the coordinate axes. Let $P_\alpha^{\pm}$ be the projection onto $s_\alpha \gtrless \frac{1}{2}$. Let $\Lambda$ have sides $4L \times 4L \times \cdots 4L$ and let $Q$ be the product $\prod_{\alpha\in\Lambda} P_\alpha^{q(\alpha)}$, where $q(\alpha)$ is independent of $\alpha_2 \ldots, \alpha_\nu$ and takes the values $+ + - - + + - - \cdots$ in the $\alpha_1$ direction. Clearly, by taking all spins to be either 0 or 1

$$Z \geqslant \delta^{|\Lambda|} \exp[-\beta c(h)|\Lambda|]$$

where $c(h) = \min(ah^2, a(h - 1)^2)$. On the other hand,

$$Z\langle Q\rangle \leqslant D^{|\Lambda|} \exp[-\tfrac{1}{4}|\Lambda|\beta - \beta d(h, \beta)|\Lambda|]$$

where

$$\exp[-\beta d(h, \beta)] = \sum_n \exp[-\beta a(h - n)^2]$$

It is easy to see that uniformly for $|h - \frac{1}{2}| < \frac{1}{2}$ and $\beta > \beta_0$, $e^{-\beta d(h)} \leqslant A(\beta)e^{-\beta c(h)}$ with $A(\beta_0) \to 1$ as $\beta_0 \to \infty$. Thus, one obtains a uniform bound

$$\langle Q \rangle \leqslant [\epsilon_1(\beta)]^{|\Lambda|} \tag{2.2}$$

with $\epsilon_1 \to 0$ as $\beta \to \infty$ uniformly in $\Lambda$ and $|h - \frac{1}{2}| < \frac{1}{2}$. (In passing, we note that one could be more explicit in the bounds by replacing $P^\pm$ by four $P$'s, namely $P^0$, $P^1$, $P^>$, $P^<$ corresponding to $s_\alpha = 0, 1, \geqslant 2, \leqslant -1$. There would then be six $Q$'s. The one for the pair $P^0$, $P^1$ could be estimated as above. The $Q$ with a $P^<$ or $P^>$ can be estimated by a chessboard estimate and an easy estimate on $\langle \prod_{\alpha\in\Lambda} P^> \rangle$ or $\langle \prod_{\alpha\in\Lambda} P^< \rangle$.) Given (2.2), the chessboard Peierls argument (see Ref. 13 or Ref. 11) immediately implies that

$$\langle P_\alpha{}^+ P_\gamma{}^- \rangle \leqslant \epsilon_2(\beta)$$

uniformly in $\alpha$, $\gamma$, $\Lambda$, and $h$ with $|h - \frac{1}{2}| < \frac{1}{2}$, where $\epsilon_2(\beta) \to 0$ as $\beta \to \infty$.

In case all $w_n = 1$, we are now done, since by symmetry $\langle P_\alpha{}^+ \rangle = \langle P_\alpha{}^- \rangle$ at $h = \frac{1}{2}$. Use the strategy of Section 1.1. In general, we need only note that for $h < \frac{1}{2}$

$$\langle P_\alpha{}^+ \rangle \leqslant \left\langle \prod_{\alpha\in\Lambda} P_\alpha{}^+ \right\rangle^{1/|\Lambda|} \qquad \text{(a chessboard estimate)}$$

$$\leqslant \left\{ \sum_{n\geqslant 1} D \exp[-a\beta(n - h)^2] \right\} / \delta \exp(-a\beta h^2)$$

goes to zero and for $h > \frac{1}{2}$, $\langle P_\alpha{}^- \rangle$ goes to zero. Thus, by the strategy of Section 1.3, there is for $\beta \geqslant \beta_0$ a multiple phase point $h_0(\beta)$ with $h_0(\beta) \to \frac{1}{2}$ as $\beta \to \infty$. Notice that since the estimates on $\beta_0$ only depend on $a$, $\delta$, and $D$ we get a bound uniform in $k$ on $\beta_k$. Thus:

**Theorem 2.1.** Consider the Hamiltonian $H$ of (2.1) with integral spins and a priori weights $w_n$ obeying $0 < \delta \leqslant w_n \leqslant D$. Then there is a $\beta_0$ depending only on $\delta$, $D$, and $a$ so that for $\beta > \beta_0$ there are infinitely many $h_k(\beta)$ where there are multiple phases. Moreover, $|h_k - k - \frac{1}{2}| < \frac{1}{2}$ and

$$\lim_{\beta\to\infty} h_k(\beta) = k + \tfrac{1}{2}$$

uniformly in $k$.

Since the above model is the first one considered, we have given full details. Henceforth, we will be briefer.

**Model 2.2.** (Pirogov–Sinai triangle model). Consider spins $\sigma_\alpha$ taking values $\pm 1$ (with equal a priori weights) on a two-dimensional "simple cubic" torus $\Lambda$. The Hamiltonian is

$$-H = j \sum_{\langle\alpha\beta\rangle} \sigma_\alpha\sigma_\beta - k \sum_{\langle\alpha\beta\gamma\rangle} \sigma_\alpha\sigma_\beta\sigma_\gamma + h \sum_\alpha \sigma_\alpha$$

where

$$j, k, h \geqslant 0 \tag{2.3}$$

the first sum is over all nearest neighbor pairs, and the second sum is over all triples forming an isosceles right triangle with side 1. This model has been considered by Pirogov and Sinai,[42] whose results we recover below. The important fact about this model, both historically and conceptually, is that the usual up–down symmetry of the Ising model is absent. The usual Peierls argument is very difficult, but Pirogov and Sinai succeeded in treating the model by introducing the notion of "contour gas." RP makes this unnecessary.

The periodic state is *not* RP about the "usual" reflections *midway between* lattice planes orthogonal to the coordinate axes. It is, however, RP about reflections in these lattice planes themselves since $-H$ then has the form $\theta A + A$. Notice that this kind of reflection positivity is independent of the signs of $j$, $k$, $h$. However, the equality of the couplings of the four kinds of right triangles is critical (it is irrelevant for the Pirogov–Sinai argument); the equality of the horizontal and vertical $j$'s is not important for RP. The type of chessboard estimate that holds with RP through the sites involves the dual lattice and is given by Theorem 4.3 of Ref. 11; see also the discussion of the AC model in Ref. 13. Given this estimate, we can say two things immediately about the possible ground states:

(i) Among the ground states must be one of the 16 states obtained by picking a configuration of a $2 \times 2$ block and extending it to be periodic with period 2.

(ii) Any other ground state must have the property that every $2 \times 2$ block is among those that yield a minimal energy when extended periodically as in (i).

Because of reflection and rotational symmetries, the 16 states fall into six types given by Table I. The quantity $e(n)$ is the energy per site, i.e., $\langle E \rangle / |\Lambda|$, for the state extended periodically.

From this table and (i) and (ii) above, one sees that as long as

$$2j + |4k - h| > 2k + \tfrac{1}{2}h \tag{2.4}$$

the only candidates for ground states are all plus or all minus and that at $h = 4k$ there is a change over from plus to minus. Thus, we imagine fixing $k, j$ with

$$k < \tfrac{1}{2}j \tag{2.5}$$

**Table I. States for the PS Triangle Model**

| $n$ | Block | $-e(n)$ | $n$ | Block | $-e(n)$ |
|---|---|---|---|---|---|
| 1 | $\begin{smallmatrix}+&+\\+&+\end{smallmatrix}$ | $2j - 4k + h$ | 4 | $\begin{smallmatrix}+&-\\-&+\end{smallmatrix}$ | $-2j$ |
| 2 | $\begin{smallmatrix}-&-\\-&-\end{smallmatrix}$ | $2j + 4k - h$ | 5 | $\begin{smallmatrix}-&-\\+&-\end{smallmatrix}$ | $-2k - \frac{1}{2}h$ |
| 3 | $\begin{smallmatrix}+&-\\+&-\end{smallmatrix}$ | $0$ | 6 | $\begin{smallmatrix}+&+\\-&+\end{smallmatrix}$ | $2k + \frac{1}{2}h$ |

and vary $h$ in a neighborhood of $h = 4k$. By (2.5), condition (2.4) holds for all such $h$. We now want to estimate the probability of a contour separating a set of plus spins from a set of minus spins. A contour of length $l$ runs through precisely $l$ points of the dual lattice. For each such point $i$ we let $F_i$ be the projection onto the state of the four neighboring spins forced by the contour. Using Theorem 4.3 of Ref. 11, we bound the probability of the contour by $l\,\gamma(F_i)$'s and so the statistical weight of any contour of length $l$ is bounded by

$$B^l \equiv [\max(\gamma(F_i))]^l \tag{2.6}$$

Let $-e_\infty = \max(-e(1), -e(2))$. Then, if $i$ is a vertex of a straight portion of contour, $\gamma(F_i)^{|\Lambda|}$ is the expectation of state 3 periodized; hence $\gamma(F_i) \leqslant \exp[-\beta(e(3) - e_\infty)]$. Among the various corners, the worst one is

$$\begin{array}{cc} + & ? \\ \overline{-}\rvert & + \end{array}$$

Replacing ? by $+$ or $-$ in each site and using chessboard again on each site, we see that the quantity $B$ of (2.6) is

$$\begin{aligned} B = \max\{&\exp[-\beta(e(3) - e_\infty)];\ \exp[-\beta(e(6) - e_\infty)] \\ &+ \exp[-\beta(e(4) - e_\infty)]\} \end{aligned}$$

Under (2.5), $B \to 0$ uniformly as $\beta \to \infty$ for $h$ near $4k$. Moreover, for $h > 4k$ (resp. $h < 4k$) $\langle P_- \rangle$ (resp. $\langle P_+ \rangle$) goes to zero as $\beta \to \infty$. By the strategy of Section 1.3, we obtain:

**Theorem 2.2.** Consider the model of (2.3) with $j$, $k$ fixed obeying (2.5). Then for all $\beta$ large, there is an $h(\beta)$ where multiple phases occur. Moreover, $h(\beta) \to 4k$ as $\beta \to \infty$.

When (2.3) fails, one can still analyze the model by the above methods. If $-j > k > 0$, there is a first-order phase transition between states 2 and 4 for $h \simeq 4(k + j)$, and if $j < k < 0$, there is a transition between states 1 and 4 for $h \simeq 4(k - j)$.

One can ask what happens when (2.3) holds but (2.5) fails. We have nothing definite to report, but one can look at the ground state structure to get some idea of what is likely. The structure that results is very similar to that of the hard-core gas with both nearest and second nearest neighbor exclusion (see Section 4). If (2.5) fails, then (2.4) still holds for $|h|$ very large, but for $h \sim 4k$, (2.4) fails. In the region where (2.4) fails, there are four basic blocks: $A = \begin{smallmatrix} + & + \\ - & + \end{smallmatrix}$, $B = \begin{smallmatrix} - & + \\ + & + \end{smallmatrix}$, $C = \begin{smallmatrix} + & - \\ + & + \end{smallmatrix}$, and $D = \begin{smallmatrix} + & + \\ + & - \end{smallmatrix}$; and so by (i) above there are four ground states with period two in both directions. By (ii) above, ground states are precisely configurations obeying:

(iii) In every 2 × 2 block of four spins, one is minus.

(iv) Each minus spin has all its nearest and next nearest neighbors plus.

There are numerous states: Any set of only $A$ and $B$ columns or only $C$ and $D$ columns or any set of only $A$ and $D$ rows or only $B$ and $C$ rows will be a ground state, and these are all the ground states. Since there are no ground states with $A$ and $C$ on the same row, one might think there could be some kind of long-range order; however, there are infinitely many configurations with $A$ and $C$ on the same row, all with energy relative to the ground state uniformly bounded: for example, take a single column of $C$'s imbedded in a "sea" of $D$'s and change a *finite* chain of $D$'s in a single row all to $A$'s. The change in energy needed to insert the $A$'s is bounded uniformly in the length. This suggests that in the region $h \sim 4k$ there is no long-range order. Moreover, at the critical $h$'s where the ground state shifts from type 6 blocks to type 1 blocks or to type 2 blocks, the configurations that arise in a chessboard Peierls argument are ground states. So, *presumably*, there are *no first-order* phase transitions as $h$ is varied when (2.5) fails, but there could well be a higher order phase transition. In that case, the phase diagram for $\beta$ large, $h \sim 4k$, $j \sim 2k$ would involve a shift from a line of first-order phase transitions to a line of higher order transitions.

## 3. ANTIFERROMAGNETS

We begin by considering nearest neighbor antiferromagnets. In zero external field, the classical models are equivalent to the corresponding ferromagnets, so the existence of multiple equilibrium states follows by the arguments of Peierls[41] (for the Ising case) and Malyshev[39] (for the classical Heisenberg case; see also Ref. 13). The quantum Heisenberg models are *not* equivalent, but the quantum antiferromagnet can be treated directly in zero external field; see Ref. 6 for the isotropic case, Ref. 13 for the two-dimensional anisotropic case. The first four models in this section involve antiferromagnets in external field. One no longer has RP under the reflection used in Ref. 6 (see Ex. 3.4 in Ref. 11), i.e., reflection in a plane between lattice sites combined with $\sigma_\alpha \to -\sigma_\alpha$, for the change of sign does not respect the external field. So one reflects in planes containing lattice sites. This introduces two unfortunate limitations. First, quantum spins are no longer allowed, since we have no results on RP for quantum systems when reflections are in planes containing sites. Second, one does not get useful infrared bounds without a next-nearest-neighbor ferromagnetic coupling (see Model 3.4 below).

**Model 3.1.** (Nearest neighbor Ising antiferromagnet in external field) The basic Hamiltonian in $\nu$ dimensions is

$$H = (1/2\nu) \sum_{\langle\alpha\beta\rangle} \sigma_\alpha\sigma_\beta - h \sum_\alpha \sigma_\alpha \tag{3.1}$$

the sum being over nearest neighbors. We will recover the result of Dobrushin[5] that there are multiple phases in the triangle in the $\langle T, h\rangle$ plane with

$$|h| + T/T_0 \leqslant 1 \tag{3.2}$$

(we have made no attempt to compare our $T_0$ with his).

Merely for notational simplicity, we consider the case $\nu = 2$. The periodic states are RP with respect to reflections in lines orthogonal to the coordinate axes and through the sites. Moreover, the formal infinite-volume Hamiltonian has the form $A + \theta A$ with respect to reflections in diagonal planes through the sites. The usual finite-volume cutoffs destroy *this* RP, but we can pick cutoffs which preserve this RP at the cost of losing the other RP; namely one takes a periodic box with sides at 45°. [One can produce a state which is RP in all these planes by the following special argument: first, by changing signs in the even sublattice, transform the model into a ferromagnet in staggered field. By a monotonicity argument[23] and the GKS inequalities, the state in constant field is RP in all planes. By FKG inequalities (see Ref. 16), one can turn on the staggering and by monotonicity again get RP in all planes.]

As in Model 2.2, we must examine six periodic configurations for ground states. The values of $-e(n)$ can be obtained from Table II by setting $\epsilon = 0$ ($\epsilon$ is a parameter which is not relevant here but appears in Model 3.3 as a next-nearest-neighbor ferromagnetic coupling).

Clearly, for $|h| < 1$ there are two ground states obtained by periodizing state 1 and also by translating the resulting state by one unit. Given $\alpha$, let $P_\alpha^{(\pm)}$ denote the projection onto spin $\pm 1$ at $\alpha$. We claim that as long as

$$\beta(1 - |h|) \geqslant T_0^{-1} \tag{3.2$'$}$$

$$\langle P_\alpha(\pm)P_{\alpha+\gamma}(\mp(-1)^{|\gamma|})\rangle < \tfrac{1}{4} \tag{3.3}$$

To prove (3.3) we use a contour argument, but now we draw contours between nearest neighbor spins if they have the *same* sign. We now argue as in Model 2.2. Straight segments of contours contribute $\exp(-\beta/2)$, while the worst corners contribute $\exp[-\frac{1}{2}(\beta - \beta|h|)] + \exp[-(\beta - \beta|h|)]$, so for $|h| > \frac{1}{2}$ the worst contours are ones with only corners (as one should expect!) and these are uniformly small in the region where $\beta(1 - |h|)$ is uniformly large. Thus (3.2′) implies (3.3).

**Table II. States for the Antiferromagnet**

| $n$ | Block | $-e(n)$ | $n$ | Block | $-e(n)$ |
|---|---|---|---|---|---|
| 1 | $\begin{smallmatrix}+&-\\-&+\end{smallmatrix}$ | $\frac{1}{2} + 2\epsilon$ | 4 | $\begin{smallmatrix}+&-\\+&-\end{smallmatrix}$ | $-2\epsilon$ |
| 2 | $\begin{smallmatrix}+&+\\+&+\end{smallmatrix}$ | $-\frac{1}{2} + h + 2\epsilon$ | 5 | $\begin{smallmatrix}+&-\\+&+\end{smallmatrix}$ | $\frac{1}{2}h$ |
| 3 | $\begin{smallmatrix}-&-\\-&-\end{smallmatrix}$ | $-\frac{1}{2} - h + 2\epsilon$ | 6 | $\begin{smallmatrix}-&+\\-&-\end{smallmatrix}$ | $-\frac{1}{2}h$ |

Next, we note that if there is a unique equilibrium state, then it must of necessity be mixing, so that as $\gamma \to \infty$, $\langle P_\alpha(+)P_{\alpha+\gamma}(+)\rangle \to \langle P_\alpha(+)\rangle^2$ and $\langle P_\alpha(-)P_{\alpha+\gamma}(-)\rangle \to \langle P_\alpha(-)\rangle^2$. Since one of the squares is at least 1/4 (since $\langle P_\alpha(+)\rangle$ is monotone in $h$, we know which is larger), (3.3) would fail. That is, (3.3) implies the existence of more than one equilibrium state. Noticing that (3.2 ) is equivalent to (3.2), we have:

**Theorem 3.1.** The Ising model of (3.1) has multiple equilibrium states in the region (3.2) for suitable $T_0$.

Note that our proof has actually shown more, namely nondifferentiability of the pressure in staggered field.

**Model 3.2.** (Nearest neighbor, classical anisotropic Heisenberg antiferromagnet in external field.) One advantage of the chessboard Peierls argument is that it extends so easily to other single-spin distributions. We imagine replacing $\sigma_\alpha$ in Model 3.1 by $\sigma_{\alpha z}$, $\sigma_\alpha\sigma_\beta$ by $\sigma_\alpha \cdot \sigma_\beta$, and $\sigma$ is now a three-vector constrained to lie on the ellipsoid $\sigma_x^2 + \sigma_y^2 + (1-\delta)\sigma_z^2 = 1$ with $\delta > 0$ and the natural induced measure. (This is equivalent to spherical $\sigma$'s with anisotropic coupling). In the same way that Fröhlich and Lieb[13] treat the zero-field model, we introduce a decomposition $P_\alpha(+)$, $P_\alpha(-)$, and $P_\alpha(0)$ corresponding to the two polar caps and the remainder of the ellipsoid. By an elementary chessboard estimate

$$\langle P_\alpha(0)\rangle < \tfrac{1}{3} \quad \text{for} \quad \beta > \beta_0, \quad |h| \leqslant 1 \tag{3.4}$$

[*Warning:* The configuration needed to prove (3.4) is *different* from that in Ref. 13, for the RP used here is different (i.e., reflection through, not between sites); $\langle P_\alpha(0)\rangle_\Lambda \leqslant \langle Q\rangle^{1/|\Lambda|}$, where $Q$ is the projection onto $\prod_\gamma P_\gamma(0)$, the product being over only those $\gamma$ with each component even. In Ref. 13, all $\gamma$ arise. But $\langle Q\rangle^{1/|\Lambda|}$ is small for this $Q$ also.]

Next we claim that for suitable $T_0$

$$\langle P_\alpha(\pm)P_{\alpha+\delta}(\mp(-1)^{|\gamma|})\rangle < \tfrac{1}{9} \tag{3.5}$$

when (3.2′) holds. This is proven as in Model 3.1, but now some extra configurations need to be considered in the contour estimation [put a contour around *all* $P_\alpha(0)$'s and between two neighboring $P_\alpha(+)$'s or $P_\alpha(-)$'s]. Using (3.4), or more properly (3.4) with 1/3 replaced by a much smaller number, each dual lattice site on a contour can be shown to contribute a small amount. Since (3.4) implies that $\langle P_\alpha(+)\rangle > \frac{1}{3}$ or $\langle P_\alpha(-)\rangle > \frac{1}{3}$, (3.5) proves the existence of multiple equilibrium states. Thus:

**Theorem 3.2.** The anisotropic, classical Heisenberg antiferromagnet in external field has multiple phases in the region (3.2), where $T_0$ depends on the amount of anisotropy.

**Model 3.3.** (Fisher stabilized Ising antiferromagnet.) For simplicity we work in two dimensions. The ground state structure of the Hamiltonian $H$ of (3.1) as $h$ is varied is as follows. For $|h| < 1$, there are two ground states and for $|h| > 1$ clearly one. However, for $h = \pm 1$, there are three ground state blocks, and since one of them is given by 5 (or 6) in Table II we can build up infinitely many ground states as we did in Model 2.2 when (2.4) failed. Also, configurations of type 5 destroy any kind of chessboard Peierls argument. Fisher[7] remarked that one can remove the infinite degeneracy by adding a next-nearest-neighbor ferromagnetic coupling, i.e., we take

$$H = \tfrac{1}{4} \sum_{\langle\alpha\beta\rangle} \sigma_\alpha\sigma_\beta - h \sum \sigma_\alpha - \epsilon \sum_{\langle\alpha\beta\rangle'} \sigma_\alpha\sigma_\beta \tag{3.6}$$

where the last sum is over next nearest neighbors and $\epsilon > 0$. Since $\epsilon > 0$, the model still has both RP properties mentioned above (the RP we use in this model is the one in coordinate lines, which holds irrespective of the sign of $\epsilon$, but in Model 3.4 we use the diagonal RP). We will prove the following result, already obtained by Pirogov and Sinai[43,44]:

**Theorem 3.3.** Fix $\epsilon > 0$ in the Hamiltonian (3.6). Then for all $\beta > \beta_0$, there exist $h(\beta)$ so that for $h = h(\beta)$ there are at least three extremal equilibrium states. As $\beta \to \infty$, $h(\beta) \to 1$.

The basic block energy densities are given in Table II (see Ex. 3.1). Near $h = 1$, only blocks 1 and 2 can occur and so there are only three ground states at $h = 1$, in agreement with Fisher's remark. Cover $\Lambda$ with $\tfrac{1}{4}\Lambda$, $2 \times 2$ squares and for each $2 \times 2$ square $a$ let $P_a(i)$, $i = 1, \ldots, 16$, represent the projections onto the 16 possible configurations. Let $i = 1, 2, 3$ correspond to $\begin{smallmatrix}+&-\\-&+\end{smallmatrix}$, $\begin{smallmatrix}-&+\\+&-\end{smallmatrix}$, $\begin{smallmatrix}+&+\\+&+\end{smallmatrix}$, respectively. By an elementary chessboard estimate using Table II

$$\lim_{\beta\to\infty} \langle P_a(i)\rangle = 0 \qquad \text{for} \quad i \geqslant 4$$

uniformly in $\Lambda$ and $h \in [\tfrac{1}{2}, \tfrac{3}{2}]$. Moreover,

$$\lim_{\beta\to\infty} \langle P_a(3)\rangle = 0 \qquad \text{if} \quad h = \tfrac{1}{2}$$

$$\lim_{\beta\to\infty} \langle P_a(1) + P_a(2)\rangle = 0 \qquad \text{if} \quad h = \tfrac{3}{2}$$

and of course by symmetry

$$\langle P_a(1)\rangle = \langle P_a(2)\rangle$$

If we can prove that

$$\lim_{\beta\to\infty} \langle P_a(i)P_b(j)\rangle = 0 \qquad \text{for} \quad i \neq j, \quad i, j = 1, 2, 3 \tag{3.7}$$

uniformly in $a$, $b$, $\Lambda$, $h$, then by the strategy of Section 1.4 the theorem will be proven. (In Section 1 we expressed results in terms of ergodic states and

Cesaro averages; here we need to use extremal Gibbs states and mixing.) We prove (3.7) by drawing contours between $2 \times 2$ blocks with distinct $2 \times 2$ states. The contour surrounding $a$ or $b$ will have state $i(=1, 2,$ or $3)$ on one side and some states (say $k$) on the other. We could attempt a chessboard argument with the full eight-spin configurations, $i$ next to $k$. Instead, we use a device that will be very useful below (the same device is used in Ref. 26): namely, we take a magnifying glass and pick out a convenient part of the eight-spin state.

Specifically, given a contour $\gamma$ of length $L$ with $i$'s inside, we consider the $15^L$ possibilities obtained by specifying the possible states that touch the outside of the contour (actually, if $\gamma$ has obtuse corners, fewer than $15^L$ occur). Each of the $15^L$ possibilities so obtained will have the form

$$\alpha \equiv \left\langle \prod_{I \in \gamma} P_{a_I}(i) P_{b_I}(j(I)) \right\rangle$$

where $I \in \gamma$ labels all dual lattice points in $\gamma$, and $a_I$ and $b_I$ are the $2 \times 2$ blocks inside and outside $\gamma$, respectively. Moreover $j(I) \neq i$. If $j(I) \geqslant 4$, we majorize $P_{a_I}(i)P_{b_I}(j(I))$ by $P_{b_I}(j(I))$. If $j(I) < 4$, we majorize it by $P_c(k)$, where $c$ is the four middle spins of the eight-spin configuration. For example, if $i = {}^{+\,+}_{+\,+}$ and $j(I) = {}^{+\,-}_{-\,+}$ with $a_I$ to the left of $b_I$, then $k = {}^{+\,+}_{+\,-}$. By checking out the possibilities one finds that the $k$'s which arise are always 4 or more. Thus $\alpha$ is dominated by something of the form

$$\left\langle \prod_{J \in Q} P_J(k(J)) \right\rangle$$

where $Q$ contains at least $L/3$ distinct dual lattice sites (each $b$ can be a $b_I$ for three $I$'s!) and $k(J) \geqslant 4$. Thus the probability of a given contour is dominated by $15^L a^L$ where $a^3 = \exp \beta\,[\min_{n \geqslant 3}(-e(n)) - \max_{n=1,2}(-e(n))] \to 0$ as $\beta \to \infty$ uniformly for $h \in [\frac{1}{2}, \frac{3}{2}]$. This proves (3.7) and thus Theorem 3.3.

**Model 3.4.** (Fisher stabilized isotropic classical Heisenberg antiferromagnet in external field.) The basic Hamiltonian is

$$H = (1/2\nu) \sum_{\langle\alpha\beta\rangle} s_\alpha \cdot s_\beta - \epsilon \sum_{\langle\alpha\beta\rangle'} s_\alpha \cdot s_\beta - h \sum_\alpha s_\alpha \cdot \hat{z} \tag{3.8}$$

where $s_\alpha$ is a unit three-vector, the first sum is over nearest neighbors, and the second is over next nearest neighbors. We will only consider $\epsilon \geq 0$. As already noted, one has formal reflection positivity about two kinds of hyperplanes. For this reason, to find the infinite-volume ground-state energy we need only consider configurations with *two* spin values $s$ and $s'$, one at those lattice sites $\alpha$ with $(-1)^\alpha = 1$ and the other at the remaining sites. Thus we want to maximize

$$-e(s, s') = \tfrac{1}{2}h(s + s')\cdot\hat{z} - \tfrac{1}{2}s\cdot s' \tag{3.9}$$

A little vector calculus with Lagrange multipliers shows that the maximum values and configurations are given by [with $s_\perp = (s_x, s_y)$]

$$\text{for } 0 < h < 2 \qquad \max(-e(s, s')) = \tfrac{1}{2} + \tfrac{1}{4}h^2, \quad s' \cdot \hat{z} = s \cdot \hat{z} = \tfrac{1}{2}h, \quad s_\perp = -s_\perp' \tag{3.10}$$

$$\text{for } h \geqslant 2 \qquad \max(-e(s, s')) = -\tfrac{1}{2} + h, \quad s = s' = \hat{z} \tag{3.11}$$

We wish to note two things about (3.10)–(3.11): first, the critical value of $h$ is 2, not 1 as it is in the Ising model. Second, in the case of the plane rotor model, this maximum occurs on a discrete set. One can thus prove by the arguments in Model 3.2 [using $P_\alpha(\pm)$ as the projection onto neighborhoods of $s_{\alpha z} = h/2$, $s_{\alpha x} = \pm[1 - (h/2)^2]^{1/2}$ and $P_\alpha(0)$ the remaining region]:

**Theorem 3.4.** Consider the plane rotor model in two dimensions with Hamiltonian (3.8) and with $\epsilon \geqslant 0$ ($\epsilon = 0$ allowed!). For each $h$, $0 < h < 2$, there exists $\beta_c(h, \epsilon)$ so that for $\beta > \beta_c(h, \epsilon)$ there are multiple equilibrium states.

Returning now to the classical Heisenberg model, the above energetic calculations and chessboard estimates can be used to show that the joint probability distribution of two nearest neighbor spins is more and more concentrated on the maximizing set. In particular for $h$ fixed strictly between 0 and 2 and any $\epsilon \geqslant 0$:

$$\lim_{\beta \to \infty} \langle (s_{\alpha z} - h/2)^2 \rangle = 0 \tag{3.12a}$$

$$\lim_{\beta \to \infty} \langle [|s_{\alpha \perp}| - (1 - \tfrac{1}{4}h^2)^{1/2}]^2 \rangle = 0 \tag{3.12b}$$

$$\lim_{\beta \to \infty} \langle (s_{\alpha \perp} + s_{\gamma \perp})^2 \rangle = 0 \tag{3.12c}$$

for nearest neighbors $\alpha$, $\gamma$; all limits for the torus state, uniform in $\Lambda$. It is now clear that the model looks very much like a plane rotor model, so in two dimensions we expect no long-range order, and suspect that one might be able to prove this by a Mermin[40] argument. In three or more dimensions, we do expect long-range order even when $\epsilon = 0$, but we can only prove it for $\epsilon \neq 0$! The difficulty involves limitations in proving RP and the fact that infrared bounds require couplings that cross the reflecting planes. For this reason, we will require reflection in the diagonal planes, and we therefore pick boundary conditions which will respect RP in diagonals (e.g., periodic with boxes having sides at 45°). Since we lose reflection in site planes, we should remark that (3.12) can be proven using the chessboard estimates that come from diagonal RP. We also note that one can get from any $A$ sublattice site to any other $A$ sublattice site by repeated diagonal reflections.

We consider

$$Z(h_\alpha) = \left\langle \exp\left[ -\frac{\beta}{2\nu} \sum_{\langle\alpha\beta\rangle} s_\alpha \cdot s_\beta + \beta h \sum_\alpha s_\alpha \cdot \hat{z} \right.\right.$$
$$\left.\left. - \frac{\beta\epsilon}{2} \sum_{\langle\alpha\beta\rangle'} (s_\alpha - s_\beta - h_\alpha - h_\beta)^2 \right] \right\rangle$$

Standard theory[17,11] shows that $Z(h_\alpha) \leq Z(0)$ if we use reflection in diagonal planes. From there it is fairly easy, by following the standard theory, to obtain

$$|\langle s_\alpha \cdot s_0 \rangle - 1| \leq \frac{2}{(2\pi)^\nu} \left( \int \frac{d^\nu p}{E_p} \right) \frac{3}{2} (\beta\epsilon)^{-1} \tag{3.13}$$

where $E_p = \nu - \sum \cos p_i$, the integral is over $|p_i| \leqslant \pi$, *and where $\alpha$ is restricted to lie on the sublattice with* $(-1)^\alpha = 1$. Conditions (3.12) and (3.13) imply:

**Theorem 3.5.** Consider the classical Heisenberg Hamiltonian (3.8) with $\nu \geq 3$, $\epsilon > 0$, and $0 < h < 2$. Then for $\beta > \beta_c(\epsilon, \nu, h)$ the periodic state is not mixing and the rotational symmetry about the $z$ axis is spontaneously broken.

We emphasize that our failure to handle the $\epsilon = 0$ case shows once more the limitations of using RP to prove spontaneously broken symmetry!

**Model 3.5.** (The Schick–Griffiths model.) Schick and Griffiths[46] have recently studied a model on a triangular lattice, predicting its phase diagram by a renormalization group calculation; we will prove *part* of their structure with RP methods. We emphasize that a complete analysis of the phase structure in the region where we can only analyze modulo a technical caveat is possible using the Pirogov–Sinai theory.[43,44] Since this is the first model on a triangular lattice we have considered in this series, it is useful to begin with some generalities about such lattices, concentrating particularly on reflections and chessboard estimates. We begin with the infinite lattice. The triangular lattice is the set of vectors in $\mathbb{R}^2$ generated by two unit vectors making an angle of 60°. If bonds are drawn in linking nearest neighbor sites, the plane is divided into equilateral triangles. If one begins by labeling the vertices of one triangle $A$, $B$, $C$, then it is geometrically obvious that one can extend the labeling to the whole lattice in exactly one way so that each triangle has one vertex of each type. This divides the triangular lattice into three sublattices, called respectively the $A$, $B$, $C$ sublattices. If we use $i$ to label the "dual lattice," i.e., the set of triangles, then $A_i$, $B_i$, $C_i$ will denote the $A$, $B$, $C$ vertices of triangle $i$. Clearly the infinite triangular lattice is left invariant by reflections in the three sets of parallel lines obtained by extending lattice bonds. An important geometric fact about these reflections is that *they leave each of the* *basic sublattices invariant*. This has the important consequence that the hexagonal lattice that is obtained by dropping the $B$ sublattice typically has the same RP structure as the triangular lattice, and the analyses of models on the two lattices are closely related (see Remark 2 below).

There is some point in describing the above geometry algebraically, namely to understand the situation in finite volume. If $\mathbf{e}$, $\mathbf{f}$ are the two generating unit vectors, we use $\langle x, y\rangle$ to denote the vector $x\mathbf{e} + y\mathbf{f}$. The six nearest neighbors of $\langle 0, 0\rangle$ are $\langle \pm 1, 0\rangle$, $\langle 0, \pm 1\rangle$, and $\langle \pm 1, \mp 1\rangle$. For $\langle m, n\rangle \in \mathbb{Z}^2$, let $\alpha(\langle m, n\rangle)$ be the congruence class of $m + 2n$ mod 3. Then $\alpha(X) = 1$ or 2 for each neighbor of $\langle 0, 0\rangle$, and thus if the $A$, $B$, $C$ sublattices are defined to be those with $\alpha(X) = 0, 1, 2$, respectively, then neighboring lattice sites are in different sublattices. The invariance of the sublattices is easily checked algebraically. For example, if $r$ is reflection in the plane $y = 0$, then $r\mathbf{e} = \mathbf{e}$ and $r\mathbf{f} = \mathbf{e} - f$, so $r(\langle x, y\rangle) = \langle x + y, -y\rangle$, and thus $\alpha(r\langle m, n\rangle) = m - n \equiv m + 2n$ mod 3.

Now consider imposing periodic boundary conditions on a rhomboidal region like that shown in Fig. 1; so, for example, the spins labeled 1 and 2 are regarded as neighbors in one row. One has the naive perception that the choice of two directions breaks the symmetry of the triangle model under rotations by 60°, and that *no* reflection *symmetry* will be retained since reflections in orthogonal coordinates are involved.

In fact, the naive perception is wrong. If one has an $L \times L$ rhombus, the 60° rotational symmetry is always preserved, and one has reflection invariance in *all three directions* as long as $L$ is even; moreover, one has a breakup into $A$, $B$, $C$ sublattices as long as $L$ is divisible by three. To check these facts, we use the algebraic machinery introduced above. The rhombus should be regarded as equivalence classes in the lattice ($\simeq \mathbb{Z}^2$) modulo some sublattice, namely the sublattice $S_L = \{\langle mL, nL\rangle : m, n \in \mathbb{Z}\}$. To see when the $A$, $B$, $C$

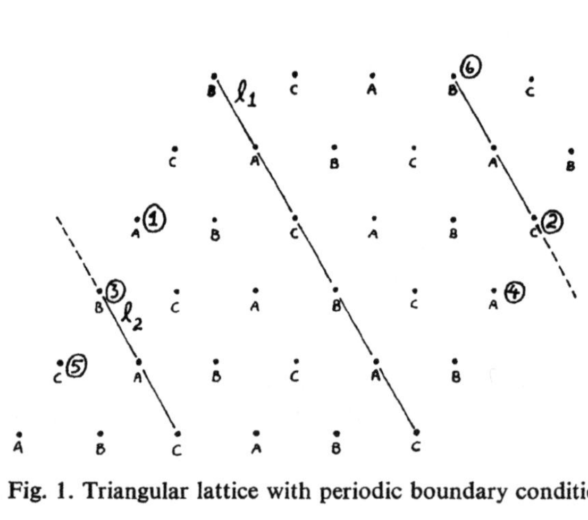

Fig. 1. Triangular lattice with periodic boundary conditions.

substructure is preserved we ask when $\alpha(X + Y) = \alpha(X)$ for all $Y$ in $S_L$, i.e., when $(m + 2n)L$ is always congruent to zero mod 3. This is true if and only if $L$ is divisible by three. In Fig. 1, $L = 6$ and the $A$, $B$, $C$ sublattices have been indicated. The rotation by 60° takes $\mathbf{e}$ to $\mathbf{f}$ and $\mathbf{f}$ to $\mathbf{f} - \mathbf{e}$ and thus $\langle x, y\rangle \to \langle -y, x + y\rangle$. This clearly leaves $S_L$ invariant and thus is definable on the quotient space. The same applies to the reflection $r$ above. This means that if $L$ is even, there will be reflection symmetry in three directions. For example, in Fig. 1 we indicate the two (!) lines $l_1$ and $l_2$ involved in a typical reflection in the nonobvious direction (we say two lines since 2 and 3 are neighboring spins and the two dotted segments are really the same segment). In attempting to obtain reflection positivity along these lines, $\mathfrak{A}_+$ consists of the spins on the lines and in the regions with spins 4 and 5 (which are neighbors!), and $\mathfrak{A}_-$ is the remaining spins and the two lines. If $\theta$ is the reflection in line $l_1$, $\theta$ leaves the spins in $l_1$ fixed but not the spins in $l_2$; for example $\theta(\sigma_3) = \sigma_6$. This means that reflection positivity will fail—it even fails with no coupling between spins, for $\theta(\sigma_3 - \sigma_6) = -(\sigma_3 - \sigma_6)$ and thus $\langle[\theta(\sigma_3 - \sigma_6)][\sigma_3 - \sigma_6]\rangle < 0$.

Let $\langle\cdots\rangle_\infty$ be a limit of periodic states as $L \to \infty$. Since the infinite-volume state for uncoupled spins is RP, we would suppose that $\langle\cdots\rangle_\infty$ is RP if the interaction is of the form $A + \theta A$. We will thus *make a working hypothesis that $\langle\cdots\rangle_\infty$ is RP about all lattice lines.* Below we will see what is implied if this working hypothesis is false.

To understand the chessboard estimates that hold in infinite volume, we begin by abstracting an argument of Seiler and Simon[47]:

**Lemma 3.6.** Let $Y$ be a compact space and let $d\mu$ be a probability measure on $X = \times_{n=-\infty}^{\infty} Y_n$ with each $Y_n$ a copy of $Y$. Suppose that $d\mu$ is invariant under the translation $y_n \to y_{n+2}$, i.e.

$$\int f(y_m, y_{m+1}, \ldots, y_{m+k})\, d\mu = \int f(y_{m+2}, y_{m+3}, \ldots, y_{m+k+2})\, d\mu$$

and reflection positive, i.e., for $f$ real-valued on $Y^k$,

$$\int f(y_1, \ldots, y_k) f(y_0, y_{-1}, \ldots, y_{-k+1})\, d\mu \geqslant 0$$

$$\int f(y_0, \ldots, y_{k-1}) f(y_{-1}, \ldots, y_{-k})\, d\mu \geqslant 0$$

For $g$ a function on $Y$, let

$$\gamma(g) = \overline{\lim} \left[\int \prod_{m=-L+1}^{L} g(y_m)\, d\mu\right]^{1/2L} \tag{3.14}$$

Then for $g_m \geqslant 0$

$$\int \prod_{m=-M+1}^{M} g_m(y_m)\, d\mu \leqslant \prod_{m=M+1}^{M} \gamma(g_m) \tag{3.15}$$

*Proof.* By using the Schwarz inequality about the point $m = \frac{1}{2}$, we can suppose that $g_j = g_{-j+1}$, and by homogeneity that $\|g_j\|_\infty \leqslant 1$. Suppose first that $M = 1$. Then by reflecting successively in $m = \frac{3}{2}, \frac{7}{2}, \ldots, 2^n - \frac{1}{2}$ we see that (3.15) holds. Now suppose that (3.15) is known for $M = K - 1$ and consider $M = K$. Then by repeated use of reflection in $m = -\frac{1}{2}$, we see that

$$\int \prod_{m=-M+1}^{M} g_m(y_m)\, d\mu \leqslant a^{1-1/2^n} b_n$$

where

$$a = \int g_M(y_{-M+1}) \cdots g_2(y_{-1}) g_2(y_0) \cdots g_M(y_{M-2})\, d\mu$$

$$b_n = \left[\int h_M \prod_n g_1(y_j)\, d\mu\right]^{1/2^n}$$

where $h_M$ is a product of $2M - 2$ of the $g$'s and $\prod_n$ is a product of $2^n$ successive sites. By induction, $a \leqslant \prod_{m\neq 0,1} \gamma(g_m)$, and clearly since $\|h_M\|_\infty \leqslant 1$, $b_n \leqslant [\int \prod_n^{\cdot} g_1(y_j)\, d\mu]^{1/2^n}$. Taking $n \to \infty$, (3.15) results. ■

Now use this lemma on $\langle \cdots \rangle_\infty$, supposing that it is RP. Let $F_\alpha$ be a function of the spins around a single triangle $\alpha$. We reflect successively in each of three directions using the lemma. What happens to a single triangle is shown in Fig. 2, where $r$ denotes one of the parallel lines about which one is about to reflect. In the end we get a very large number of factors, say $k$, of $l$th roots of expectations of several parallel arrays of the type shown in Fig. 3. If we started with $L$ of the $F$'s, then we have $lL$ triangles in the $k$ factors. What

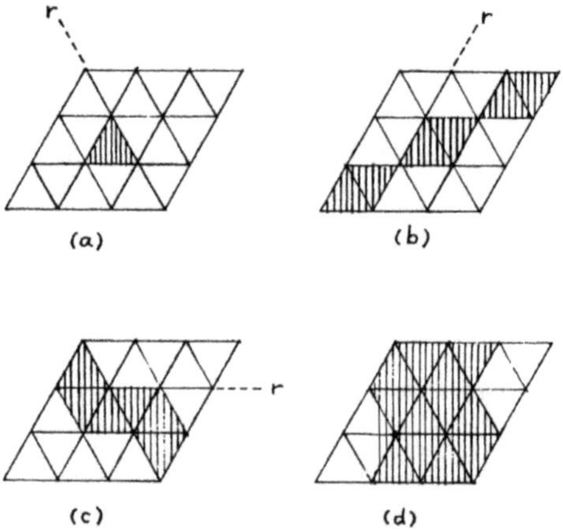


Fig. 2. Arrays produced by repeated reflection.

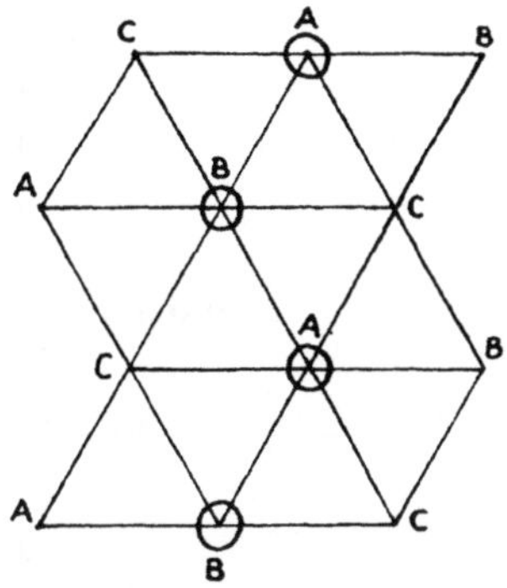


Fig. 3. Blowup of the array of Fig. 2d.

will be critical is that there are the circled vertices interior to the array and that $\frac{1}{2}Ll$ (modulo end effects) triangles have *two interior vertices.*

With these lengthy preliminaries out of the way, we can now describe the Schick–Griffiths model. Each spin takes three possible values, say $a, b, c$. As usual, $P_\alpha(a)$ is the projection onto state $a$ at site $\alpha$. Then the basic Hamiltonian is

$$-H = \sum_{i \in \Lambda^*} (\tfrac{1}{2}MQ_i + \tfrac{1}{2}KR_i) \tag{3.16}$$

where

$$Q_i = \sum_{x=a,b,c} P_{A_i}(x)P_{B_i}(x)P_{C_i}(x)$$

$$R_i = \sum_{x \neq y} P_{A_i}(x)P_{B_i}(x)P_{C_i}(y) + P_{A_i}(x)P_{B_i}(y)P_{C_i}(x) + P_{A_i}(y)P_{B_i}(x)P_{C_i}(x)$$

That is, of the 27 possible configurations for a triangle, the three with all equal spins have energy $-M/2$, the 18 with two equal spins have energy $-K/2$, and the six with three unequal spins have energy 0. It is easy to see that $H$ has the form $A + \theta A$ for any reflection. We claim:

**Theorem 3.7.** Consider the Hamiltonian (3.16). Then *under our working hypothesis* there exists an $N$ so that when $K < -N$ *or* $K - M < N$, the model has multiple equilibrium states (multiple states invariant and ergodic under those translations leaving the three basic sublattices invariant). If $M > 0$, there are at least three, if $M < 0$ at least six, and if $M = 0$ at least nine (see Fig. 4).

We note that as one increases $K$ with $M$ fixed, one has a situation which is very like that in the antiferromagnet as $h$ is increased, so presumably one does not have a first-order phase transition. Our method of proof shows that

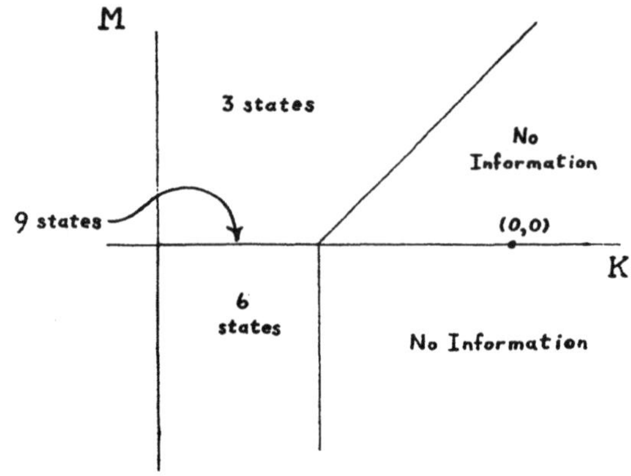


Fig. 4. Phase diagram for the Schick–Griffiths model.

on crossing the line segment $M = 0$, $K < -N$, there is a first-order phase transition: indeed our proof shows that in that region

$$\overline{\lim_{\Lambda\to\infty}} \left\{ \left\langle \left( \frac{1}{2|\Lambda|} \sum_i Q_i \right)^2 \right\rangle_\Lambda - \left\langle \frac{1}{2|\Lambda|} \sum_i Q_i \right\rangle_\Lambda^2 \right\} > 0$$

whence a first-order transition follows by the Griffiths argument (see Ref. 6).

*Proof of Theorem 3.7.* Let us label the 27 states for a triangle, using 1–3 for the ferromagnetic states (all spins equal) and 4–9 for the six antiferromagnetic states (all spins different). Let $P_i(k)$, $i \in \Lambda^*$, $k \in \{1, \ldots, 27\}$ be the obvious projections. We prove below that

$$\langle P_i(k) \rangle \leqslant \exp[\tfrac{1}{4}K - \tfrac{1}{4}\max(M, 0)] \qquad \text{for } k \geqslant 10 \qquad (3.17)$$

By the obvious symmetries of permuting the labels $a$, $b$, $c$ and translation invariance:

$$\langle P_i(1) \rangle = \langle P_i(2) \rangle = \langle P_i(3) \rangle; \qquad \langle P_i(4) \rangle = \cdots = \langle P_i(9) \rangle \qquad (3.18)$$

Moreover, when $M = 0$ there is a special symmetry[46]: namely leave the $A$-lattice spins alone, permute the $B$-lattice spins $a \to b \to c \to a$ and the $C$-lattice spins $a \to c \to b \to a$. Thus

$$\langle P_i(1) \rangle = \cdots = \langle P_i(9) \rangle \qquad \text{if } M = 0$$

so that $\langle P_i(1) \rangle \leqslant \frac{1}{9}$ for $M = 0$. Then, since $\langle \sum_{k=1}^{3} P_i(k) \rangle$ is monotone increasing in $M$ for $K$ fixed,

$$\langle P_i(1) \rangle \leqslant \tfrac{1}{9} \qquad \text{for } M \leqslant 0 \qquad (3.19)$$

Similarly, $\langle \sum_{k=4}^{9} P_i(k) \rangle$ is decreasing in $K$ if $M - K$ is fixed, so

$$\langle P_i(4) \rangle \leqslant \tfrac{1}{9} \qquad \text{for } M \geqslant 0 \tag{3.20}$$

Pick $N$ so that $18 \exp(\frac{1}{4}N) < \frac{1}{6}$. Then by (3.17)–(3.20)

$$\begin{aligned} &\langle P_i(1) \rangle = \ldots = \langle P_i(9) \rangle > 5/54 && \text{if } M = 0, \quad K < -N \\ &\langle P_i(1) \rangle = \langle P_i(2) \rangle = \langle P_i(3) \rangle > 1/18 && \text{if } M \geqslant 0, \quad K < -N + M \\ &\langle P_i(4) \rangle = \cdots = \langle P_i(9) \rangle > 1/12 && \text{if } M \leqslant 0, \quad K < -N \end{aligned}$$

From these facts the theorem is proven if we show that (3.17) holds and for some small (calculable) $\epsilon$

$$\langle P_i(k) P_j(l) \rangle \leqslant \epsilon \qquad \text{all } i, j; \quad k \neq l, \quad k, l \in \{1, \ldots, 9\} \tag{3.21}$$

so long as $N$ is large and $K - \max(M, 0) < N$. Notice that two triangles in different states from 1 to 9 cannot have an edge in common. Thus if $P_i(k) P_j(l) = 1$, there is a chain of triangles in states 10–27 connected vertex-to-vertex and separating $i$ and $j$, as do the shaded triangles in Fig. 5. Such chains (of triangles, rather than line segments) will be considered as "contours." In the usual way,[11,13] (3.21) is proven if we show that the probability of a contour $\gamma$ is dominated by $q^{|\gamma|}$ with $q \to 0$ as $N \to \infty$. We are thus reduced to proving

$$\left\langle \prod_{|\gamma| i\text{'s}} P_i(k(i)) \right\rangle \leqslant \exp\{|\gamma| [\tfrac{1}{4}K - \tfrac{1}{4}\max(M, 0)]\} \tag{3.22}$$

for all $k(i) \geqslant 10$. Notice that (3.22) includes the missing estimate (3.17) as a special case. To prove (3.22), we proceed as indicated in the general discussion above until we reach arrays of the type in Fig. 3. The expectation of such an array is the sum of all configurations consistent with the configurations in the array determined by $\prod_i P_i(k(i))$ divided by the sum of all configurations. For each term in the numerator consider the term in the denominator determined by the following rules: If $M \geqslant 0$, change the spins at the interior $A$ and $B$

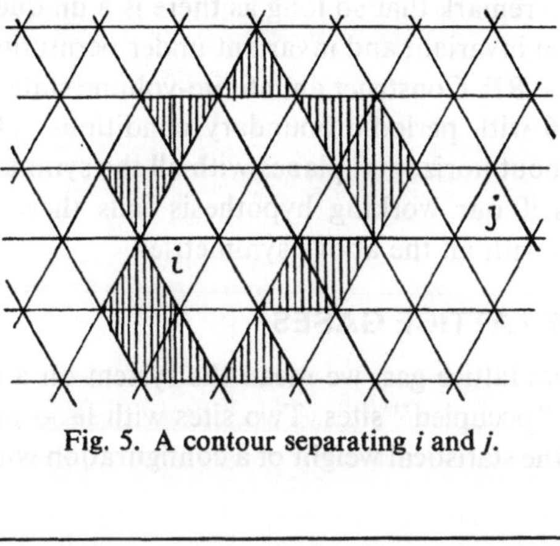


Fig. 5. A contour separating $i$ and $j$.

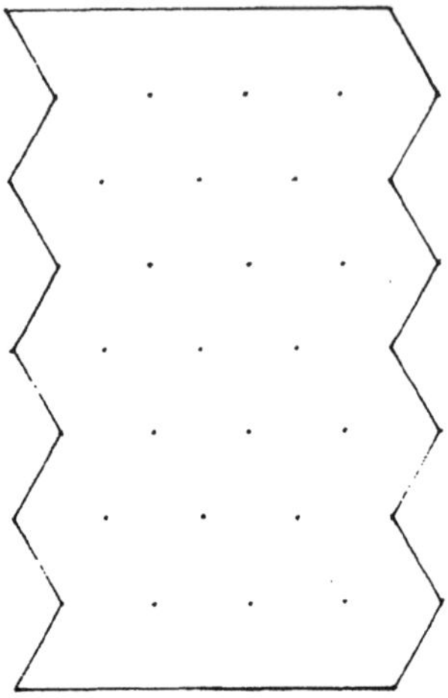

Fig. 6. A special boundary condition.

sites to agree with that on the $C$ sites (which must agree by the method which led to the array). If $M < 0$, change the spins at the interior $A$ and $B$ sites so that the interior $A$, $B$, $C$ triangles have one of each spin. By this procedure the energy is changed by at least $\frac{1}{4}\max(M, 0) - \frac{1}{4}K$ times the number of interior triangles. Since half the triangles are interior, we have proven (3.23). ∎

*Remarks.* 1. The Pirogov–Sinai method[43,44] shows that the theorem holds without any hypothesis on RP for $\langle\cdots\rangle_\infty$.

2. We illustrate our remark on hexagonal lattices in the Schick–Griffiths model. Consider a model on the hexagonal lattice with three states and pair interactions between nearest neighbors with the following energies: $M$ for $bb$; $K$ for $ab$, $bc$, $aa$, $cc$; and 0 for $ac$. This is just the Schick–Griffiths Hamiltonian with an additional interaction forcing the spins at $B$ sites to be in state $b$. The analysis above still works; the net result is an analog of Theorem 3.7 with three, six, or nine states replaced by one, two, or three states.

3. We wish to remark that so long as there is a unique equilibrium state which is translation invariant and invariant under permuting all $a$, $b$, $c$ labels, then $\langle\cdots\rangle_\infty$ will be RP. Construct an infinite-volume state by taking a region like that in Fig. 6 with periodic boundary conditions. The infinite-volume state will be RP about horizontal planes with all the symmetries guaranteeing uniqueness. Thus if our working hypothesis fails there must be multiple equilibrium states with all the above symmetries.

## 4. HARD-CORE LATTICE GASES

By a hard-core lattice gas, we mean the system on a regular array with configurations of "occupied" sites. Two sites with $|\alpha - \beta| \leqslant d$ cannot both be occupied, and the statistical weight of a configuration with $N$ sites occupied is $z^N$. The only parameter of the theory is $z$. The first basic question concerns whether there are multiple equilibrium states for large $z$.

**Model 4.1.** (Nearest-neighbor, hard-core square lattice gas.) We work on a unit square lattice and take $d = 1$. We will recover the result of Dobrushin[5] that there are multiple equilibrium states if $z$ is large.

Let $\sigma_\alpha = +1\ (-1)$ correspond to site $\alpha$ occupied (resp. unoccupied). We first claim that the interaction is RP about any lattice line. Since the statistical weight is just

$$\lim_{J\to\infty} \exp\left[-\sum_{\langle\alpha\beta\rangle} J(\sigma_\alpha + 1)(\sigma_\beta + 1) + \ln z \sum \tfrac{1}{2}(\sigma_\alpha + 1)\right]$$

this follows by our general analysis of interactions which are finite. By Theorem 4.3 of Ref. 11, for nearest neighbors $\alpha, \beta$ in a $2L \times 2L$ lattice

$$\begin{aligned}\langle P_\alpha(-)P_\beta(-)\rangle &\leqslant \sum_{\substack{s=\pm 1\\ t=\pm 1}} \langle P_\alpha(-)P_\beta(-)P_\gamma(s)P_\delta(t)\rangle \\ &\leqslant (1/z^{|\Lambda|/2})^{1/(2|\Lambda|)} + 2(z^{|\Lambda|/4}/z^{|\Lambda|/2})^{1/(2|\Lambda|)} \\ &\leqslant 3\max(z^{-1/4}, z^{-1/8})\end{aligned}$$

where $\alpha, \beta, \gamma, \delta$ form a square.

Since $\langle P_\alpha(+)\rangle = \langle P_\beta(+)\rangle$ and $\langle P_\alpha(+)P_\beta(+)\rangle = 0$, we see that

$$\langle P_\alpha(+)\rangle \geqslant \tfrac{1}{2}(1 - 3z^{-1/8}) \qquad \text{for } z > 1 \tag{4.1}$$

Next we claim that for $\alpha, \gamma$ in *different* sublattices

$$\langle P_\alpha(+)P_\gamma(+)\rangle \leqslant \epsilon(z) \tag{4.2}$$

with $\epsilon \to 0$ as $z \to \infty$. This and (4.1) imply absence of mixing and existence of multiple equilibrium states. To prove (4.2) we let $\beta$ be a neighbor of $\gamma$ and note that $P_\gamma(+) \leqslant P_\beta(-)$. Thus (4.2) follows from

$$\langle P_\alpha(+)P_\beta(-)\rangle \leqslant \epsilon(z) \tag{4.3}$$

for $\alpha, \beta$ in the same sublattice, say $A$. [Notice (4.3) is a direct statement of long-range order]. To prove (4.3), given any configuration we fill in the squares formed by joining together the nearest neighbors of each occupied $A$-lattice site. (These squares have sides which are diagonals of the basic lattice unit square.) The contours are connected components of the boundary of the resulting filled-in region. As usual, (4.3) follows from

$$(\text{probability of contour } \gamma) \leqslant [\delta(z)]^{|\gamma|} \tag{4.4a}$$

with $\delta \to 0$ as $z \to \infty$. We will prove (4.4a) with

$$\delta(z) = z^{-1/4} \tag{4.4b}$$

To prove (4.4), let $l$ be a line in $\gamma$. Since $l$ is on a boundary, the square $\Delta$ containing $l$ has one occupied and three vacant sites. Thus if $P_l$ is the projection onto this configuration

$$(\text{probability of } \gamma) \leqslant \left\langle \prod_{l \in \gamma} P_l \right\rangle \leqslant \langle Q \rangle^{|\gamma|/|\Delta|}$$

where $Q$ is the universal projection obtained by using Theorem 4.3 of Ref. 11, i.e., $Q$ has every lattice site with both coordinates even occupied, and all other sites vacant. Thus

$$\langle Q \rangle \leqslant z^{|\Delta|/4}/z^{|\Delta|/2}$$

proving (4.4).

**Model 4.2.** (Nearest-neighbor, hard-core triangular lattice gas.) We now work on the unit triangular lattice with $d = 1$. As in Model 3.5, we *suppose* that $\langle \cdots \rangle_\infty$ is RP. By the analysis of Model 3.5 and the fact that in an array of Fig. 3 with all sites empty we can gain a factor of $z^{\#\text{triangles}/8}$ by filling in the interior $A$ sites (there is one such site for every eight triangles), we have for $\alpha, \beta, \gamma$ forming a triangle

$$\langle P_\alpha(-) P_\beta(-) P_\gamma(-) \rangle \leqslant z^{-1/8}$$

so that

$$\langle P_\alpha(+) \rangle \geqslant \tfrac{1}{3}(1 - z^{-1/8})$$

As in Model 4.1, we need only show that for $\alpha, \beta$ in the $A$ sublattice

$$\langle P_\alpha(+) P_\beta(-) \rangle \leqslant \epsilon(z)$$

with $\epsilon \to 0$ as $z \to \infty$. Given a configuration, we fill in the hexagons formed by all triangles with an occupied $A$ vertex. Contours are again connected components of boundary. Each edge $J$ of a contour is associated to a unique triangle $i(J)$ with $\sigma_{A_{i(J)}} = -1$. Thus

$$(\text{probability of contour } \gamma) \leqslant \left\langle \prod_{J \in \gamma} P_{A_{i(J)}}(-) P_{B_{i(J)}}(-) P_{C_{i(J)}}(-) \right\rangle$$

$$\leqslant (z^{-1/8})^{|\gamma|}$$

as above. Modulo the hypothesis on $\langle \cdots \rangle_\infty$ being RP, we have proven the existence of at least three equilibrium states, a result of Heilmann.[25] (Without that assumption, we can only conclude that there are at least two states.)

**Model 4.3.** (Nearest-neighbor, hard-core hexagonal lattice gas.) We now work on the unit hexagonal lattice with $d = 1$. Thinking of this as the triangular lattice with $\sigma_\alpha = -1$ for $\alpha$ in the $B$ lattice, we can follow the above analysis directly. Contours are defined as in Model 4.2 and

$$(\text{probability of } \gamma) \leqslant z^{-1/8|\gamma|}$$

still holds. The only difference is that $\langle P_\alpha(+)\rangle \geqslant \frac{1}{2}(1 - z^{-1/8})$ for $\alpha$ in the $A$ or $C$ sublattice. We have at least two states, a result of Heilmann.[25]

We summarize the last three models:

**Theorem 4.1.** The hard-core, nearest-neighbor lattice gas on the square, triangular, and hexagonal lattices have at least two equilibrium states for large $z$, and, for the square lattice, long-range order. If $\langle\cdots\rangle_\infty$ is RP in the triangular and hexagonal lattices, there is long-range order in those cases and at least three states for the triangular lattice.

**Model 4.4.** (Next-nearest-neighbor, hard-core square lattice gas.) This is the model on the square lattice with $d = \sqrt{2}$. It is still RP with respect to reflection in lattice lines, but not with respect to diagonals. The basic chessboard estimate is thus Theorem 4.3 of Ref. 11. One can still attempt to define contours with the basic "blob" being a square with side 2 and sides parallel to the original lattice. However, one can no longer identify the necessary *four-spin* blocks of minus around a contour; indeed, one can see that only *corners* of contours count. This fact suggests that the analog of this model in three or more dimensions has multiple equilibrium states. In two dimensions, one can easily find an infinity of states of finite energy breaking the long-range order. This suggests no multiple states, which is the conventional wisdom.

**Model 4.5.** This is the model on the square lattice which results if one thinks of an occupied site as being the shape shown in Fig. 7. Thus pairs with $d = 1, \sqrt{2}, 2, \sqrt{5}, 3$, and $\sqrt{10}$ are not allowed but $d = 2\sqrt{2}$ is. One can develop a contour analysis about sites $(n, m)$ with $n + m \equiv 0 \bmod 4$ and thereby see that there are four ground states and the Peierls condition of Pirogov and Sinai[42–44] holds. By their theory, one gets four states at large $z$.

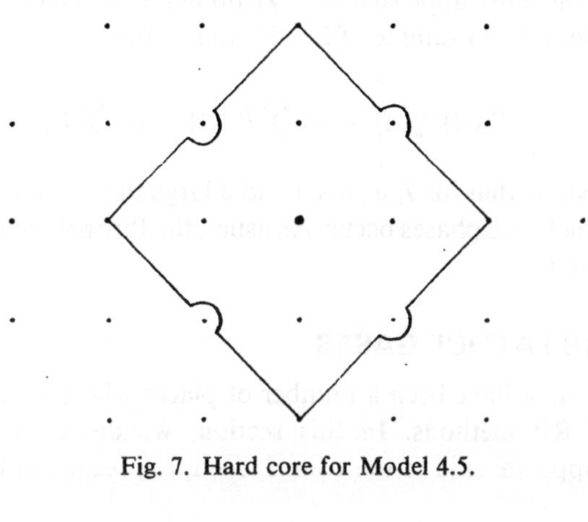

Fig. 7. Hard core for Model 4.5.

Our method yields *no* information since the model has no RP. We see once again the limitations of RP methods!

**Model 4.6.** (Widom–Rowlinson model.) The basic lattice is the square lattice. At each site there are three possible states $A$, $B$, or 0 (=vacant) with the constraint that one cannot have $A$ and $B$ at nearest neighbor sites. The weight of a configuration is $z^m$ with $m$ the number of sites with $A$ or $B$. The applicability of naive contour arguments is due to Gallavotti and Lebowitz,[20] with recent simplifications by Bricmont *et al.*[3] With RP methods it is effortless to prove:

**Theorem 4.2.** The Widom–Rowlinson lattice model has multiple phases for $z$ large, and long-range order with $P_\alpha(A) - P_\alpha(B)$ as order parameter.

*Proof.* RP in lines containing lattice sites and bonds holds as in Model 4.1. By considering the 17 possibilities for filling in a square with one corner vacant, one sees for $z > 1$,

$$\langle P_\alpha(0)\rangle \leqslant 17z^{-1/4}$$

so by symmetry $P_\alpha(A) = P_\alpha(B) \to 1/2$ as $z \to \infty$. Thus we only need to show that $\langle P_\alpha(A)P_\gamma(B)\rangle \to 0$ as $z \to \infty$. Draw a contour of conventional type between dual lattice sites by filling in a square about every vacant site. If $\alpha$ and $\gamma$ are occupied by different species, a contour must separate them. For every dual lattice site on a contour, the corresponding square has at least one vacant and one occupied site. Thus (again counting possibilities)

$$(\text{probability of } \gamma) \leqslant (5z^{-1/4})^{|\gamma|} \quad \blacksquare$$

*Remarks.* 1. The above proof works if there is also next-nearest-neighbor $AB$ exclusion, but longer range exclusion destroys RP!

2. Unlike the naive approach,[3,20] we do not need $A$–$B$ symmetry for the contour argument. For example, if in addition to the $AB$ exclusion one has a Hamiltonian

$$J \sum_{\langle\alpha\beta\rangle} P_\alpha(A)P_\beta(A) - \mu_A \sum P_\alpha(A) - \mu_B \sum P_\alpha(B)$$

one can easily show that for $J$, $\mu_A$ fixed and $\beta$ large there is a $\mu_B(\beta) \to \mu_A + 2J$ as $\beta \to \infty$ at which two phases occur. As usual, the Pirogov–Sinai method also yields these results.

## 5. COULOMB LATTICE GASES

Thus far, there have been a number of places where we emphasized the limitations of RP methods. In this section, we discuss examples which illustrate the opposite side of the coin. Indeed, the results in this section are all new and we do not know how to obtain them with any other methods, for all other methods require contours whose width is the range of the interaction, which is infinite here. In contrast, we will use RP methods and a width-two contour.

In all the models, we will suppose that our lattice is $\mathbb{Z}^\nu$ with $\nu = 3$. It is easy to extend the method to $\nu \geqslant 3$. One can treat the classical case of $\nu = 2$ by replacing the Coulomb potential by a Yukawa potential of mass $\mu$ and then taking $\mu \to 0$. The sole effect is to suppress all nonneutral configurations. In all the models, it is easy to check stability of the energy essentially because the point charges of a lattice Coulomb gas can be replaced by spheres with finite self-energy. For a proof that a sensible thermodynamics exists, see Refs. 14 and 15.

**Model 5.1.** (Coulomb monopole, plus, minus charges.) At each site we have a charge $\sigma_\alpha = \pm 1$ and the Hamiltonian is $H = \sum_{\alpha \neq \beta} \sigma_\alpha \sigma_\beta |\alpha - \beta|^{-1}$. In a periodic boundary condition box, there is some care needed: only neutral configurations have finite energy, and $H^{\text{per}}(\sigma)$ is defined by extending the configuration in $\Lambda$ periodically and letting

$$H^{\text{per}}(\sigma) = \lim_{L \to \infty} \frac{1}{L^3} \sum_{\substack{\alpha, \beta \in L\Lambda \\ \alpha \neq \beta}} \sigma_\alpha \sigma_\beta |\alpha - \beta|^{-1}$$

Alternatively

$$H^{\text{per}}(\sigma) = \sum_{p \in \Lambda^*} \hat{\sigma}_p \hat{\sigma}_{-p} E_p \tag{5.1}$$

where as usual $\hat{\sigma}_p = |\Lambda|^{-1/2} \sum e^{i\alpha \cdot p} \sigma_\alpha$ and $E_p$ is $(2\pi)^{3/2}$ times the Fourier transform of $\sum_{\alpha \neq 0} |\alpha|^{-1} \delta(x - \alpha)$. The $E_p$ can be evaluated by the Poisson summation formula as $\sum_\beta f(p + 2\pi\beta)$, where $f(p) = p^{-2} - \int (p - l)^{-2} \hat{\chi}(l)\, d^3 l$, with $\hat{\chi}$ a function whose Fourier transform vanishes for $|x| > 1$ and $\int \hat{\chi}(l)\, d^3 l = 1$. Notice that $f$ falls off faster than any power, so the sum over $\beta$ is convergent.

We claim next that this model is RP under a $\theta$ obtained by reflecting in a plane and changing the sign of the charge. By the basic machinery of Section 3 of Ref. 11, this follows if we show that for any charge configuration $r$ on the right side of the plane $\alpha_1 = 1/2$ and the configuration $l$ obtained by reflecting and sign shift we have that

$$-\sum_{\substack{\alpha_1 > \frac{1}{2} \\ \beta_1 < \frac{1}{2}}} \sigma_\alpha \sigma_\beta |\alpha - \beta|^{-1} \geqslant 0$$

This, in turn, follows if we prove the stronger result

$$\int_{\substack{x_1 > 0 \\ y_1 < 0}} \rho(x) \rho(y) |x - y|^{-1} \geqslant 0 \tag{5.2}$$

if $\rho(-x_1, x_2, x_3) = \rho(x_1, x_2, x_3)$. The result (5.2) is well known (see, e.g., Section 5 of Ref. 11).

*Remark.* The nearest neighbor antiferromagnet has the same RP, so one can ask if the present model is RP under reflection in lattice planes. The answer is no. This can be seen as follows: if it were, one would have the chessboard estimate, Theorem 4.3 of Ref. 11, so that the probability $p$ of eight neighboring plus charges would be dominated by $q = \langle Q\rangle^{1/|\Lambda|}$, where $Q$ assigns a plus charge to each site. But $q = 0$ by the requirement of overall neutrality, while clearly $p > 0$.

To find the ground states, we need only consider configurations which alternate signs in two directions and are a two-unit reflecting sequence in the third direction, i.e., $xy$ means $\cdots x\, y\, -y\, -x\, x\;\, y\, -y\, -x \cdots$. The two possibilities are given in the first column of Table III with $\mu$, a parameter of Model 5.2, set to zero. In this case $a = 1.74756459$, $b = 1.6170762$, $c = 0.84116805$ ($c$ is irrelevant in this model). $a$ is just the constant computed by Madelung in his famous paper.[38] We have evaluated $b$ and $c$ using Hund's method[29] and the tables in Ref. 45. For additional discussion of Madelung constants, see Refs. 31, 52, and 54–57. For the case at hand, all that matters is that $a > b$. We will show that this implies:

**Theorem 5.1.** The $\sigma_\alpha = \pm 1$ Coulomb monopole gas has at least two equilibrium states at large $\beta$.

*Proof.* As usual let $P_\alpha(\pm)$ be the projection onto $\sigma_\alpha = \pm 1$. Clearly, by symmetry $\langle P_\alpha(+)\rangle = \langle P_\alpha(-)\rangle = 1/2$. Thus we only need to show that

$$\sup_{\alpha,\delta}\langle P_\alpha(+)P_\delta(-(-1)^{|\alpha-\delta|})\rangle \to 0$$

as $\beta \to \infty$, i.e., there is long-range order favoring equal spins on the $A$ sublattice and the opposite spins on the $B$ sublattice. Draw contours by placing square faces between neighboring spins of the same sign. If the contour $\gamma$ has $|\gamma|$ squares, then the probability of $\gamma$ is dominated by a product of $|\gamma|$ projections $P_i$, each one the projection onto two neighboring equal spins. There are three directions in which the axis between the neighbors may point and two possibilities for whether the $A$ site is to the "left" ("above" or $\cdots$) of the $B$ site. By picking the combination that occurs most often, we obtain a bound on the probability of $\gamma$ which has at least $|\gamma|/6$ projections, all of the same type, say $A$ on the left, $B$ on the right. Using a chessboard estimate with

**Table III**

| $n$ | Block | $-e(n)$ | $n$ | Block | $-e(n)$ |
|---|---|---|---|---|---|
| 1 | $+-$ | $a + \mu$ | 3 | $+0$ | $c + \frac{1}{2}\mu$ |
| 2 | $++$ | $b + \mu$ | 4 | $0\,0$ | 0 |

a single unit in two directions and a double unit in the third, we get prob($\gamma$) $\leqslant$ $\alpha^{|\gamma|}$, where $\alpha = \exp[\frac{1}{8}\beta(b - a)]$, which implies the result. ■

**Model 5.2.** (Coulomb monopole gas; +, −, 0 charges.) The model is the same as in Model 5.1, except that now $\sigma_\alpha$ can take the values $\pm 1$ or 0. Now $H$ is $\sum_{\alpha,\beta;\alpha\neq\beta}\sigma_\alpha\sigma_\beta|\alpha - \beta|^{-1} - \mu \sum |\sigma_\alpha|$, so with $z = e^{\beta\mu}$ we have a Coulomb gas with fugacity $z$. Now the relevant energies are given by the four entries in Table III. For $\mu$ large, indeed for $\mu > -a$, we have two phases at low temperatures essentially by the argument in Model 5.1 supplemented by the fact that

$$\langle P_\alpha(0)\rangle \leqslant \exp[-\beta(a + \mu)]$$

by a chessboard estimate. An interesting feature takes place near $\mu \sim -a$. Notice that $2c < a$, so that for $\mu \sim -a$, $c + \frac{1}{2}\mu < a + \mu$. By a contour argument of the type in Model 5.1, we can make

$$\langle P_\alpha(+)[P_\gamma(0) + P_\gamma(-(-1)^{|\alpha-\gamma|})]\rangle$$

small so long as $\exp\{-\beta[e(2) - e(1)]\}$ and $\exp\{-\beta[e(3) - e(1)]\}$ are both small. This will hold uniformly for $\mu \in (-a - \delta, -a + \delta)$ ($\delta$ small) if $\beta$ is large. By the strategy of Section 1.3, we find $\mu(\beta) \to -a$ as $\beta \to \infty$, so that there are three phases at fugacity $z = e^{\beta\mu(\beta)}$. Two phases are like those of Model 5.1 and can be thought of as "crystal" phases. One has mainly empty sites and can be thought of as a "plasma." We summarize with:

**Theorem 5.2.** The Coulomb monopole gas with $\sigma_\alpha = \pm 1, 0$ has at least three phases for $\beta$ large and suitable fugacity $z(\beta)$.

We are indebted to David Brydges for having checked that his proof of Debye screening[4] also applies to the lattice Coulomb gas, provided the activity is sufficiently small and the temperature is sufficiently large. Thus, we conclude that the lattice Coulomb gas has a single-phase region with exponential clustering apart from the two- and three-phase regions exhibited in this paper.

**Model 5.3.** (Monopole gas with a lattice Coulomb potential.) In the last two examples, it appeared that numerical relations among Madelung constants played a significant role. We want to show that this is no accident by considering a more general model. A special case of particular interest of this more general model is the lattice Coulomb potential,

$$V(\alpha - \beta) = \frac{1}{(2\pi)^3}\int_{|p_i|\leqslant\pi} (3 - \cos p_1 - \cos p_2 - \cos p_3)^{-1} e^{ip\cdot(\alpha-\beta)}\, d^3p \tag{5.3}$$

Other examples would include the lattice Yukawa potential [replace 3 in (5.3) by $3 + \mu^2$] and the continuum Yukawa potential restricted to lattice sites. Our Hamiltonian is now

$$H = \sum_{\alpha,\beta} V(\alpha - \beta)\sigma_\alpha\sigma_\beta - \mu \sum_\alpha |\sigma_\alpha| \tag{5.4}$$

where $V$ is assumed RP, i.e.,

$$\sum_{\substack{\alpha_1 \geqslant 1 \\ \beta_1 \geqslant 1}} \bar{z}_\alpha z_\beta V(\alpha_1 + \beta_1 - 1, \alpha_2 - \beta_2, \alpha_3 - \beta_3) \geqslant 0 \tag{5.5}$$

For simplicity, we also suppose that $V(\alpha_1, \alpha_2, \alpha_3)$ is symmetric under permutations of $\alpha_1, \alpha_2, \alpha_3$. As in Model 5.1, we have reflection positivity under $\sigma_\alpha \to -\sigma_{r\alpha}$. By general principles,[11] the periodic state is RP also. Let $\hat{V}$ be the Fourier transform of $V$ normalized by

$$V(\alpha) = \frac{1}{(2\pi)^3} \int_{|p_i| \leqslant \pi} \hat{V}(p) e^{ip\cdot\alpha}\, d^3p \tag{5.6}$$

In finite volume $\Lambda$, one finds that (see Ref. 11)

$$H_\Lambda(\sigma) = \sum_{p \in \Lambda^*} \hat{V}(p)\hat{\sigma}_p\hat{\sigma}_{-p} \tag{5.7}$$

where $\hat{\sigma}_p = |\Lambda|^{-1/2}\sum_{\alpha\in\Lambda} e^{ip\cdot\alpha}\sigma_\alpha$ and $\Lambda^*$ is the dual set of $p$'s to $\alpha \in \Lambda$. If we consider the configuration corresponding to state 1 of Table III, i.e., $\sigma_\alpha = (-1)^{\alpha_1+\alpha_2+\alpha_3}$, we find that $\hat{\sigma}_p = |\Lambda|^{1/2}$ if each $p_i = \pm\pi$ and 0 otherwise. Thus

$$a = -\hat{V}(\pi, \pi, \pi) \tag{5.8a}$$

[In the sum over $\Lambda^*$ only one point of $(\pm\pi, \pm\pi, \pm\pi)$ enters, i.e., these are equivalent $p$'s, and in particular $\hat{V}(\pm\pi, \pm\pi, \pm\pi) = \hat{V}(\pi, \pi, \pi)$]. For the configuration corresponding to state 2, we find that $\hat{\sigma}_p = \frac{1}{2}|\Lambda|^{1/2}(1 \mp i)$ at $p = (\pi, \pm\pi/2, \pi)$ and zero at all nonequivalent points. Thus, noting that $\hat{V}(-\pi/2, \pi, \pi) = \hat{V}(\pi/2, -\pi, -\pi)$ (by reality of $V$) $= \hat{V}(\pi/2, \pi, \pi)$ (since $\pi$ is equivalent to $-\pi$), we see that

$$b = -V(\pi/2, \pi, \pi) \tag{5.8b}$$

For the configuration corresponding to state 3, we have $\hat{\sigma}_p = \frac{1}{4}|\Lambda|^{1/2}(1 \pm i)$ for $p = (\pm\pi/2, \pi, \pi)$, $\frac{1}{2}|\Lambda|^{1/2}$ for $p = (\pi, \pi, \pi)$ and zero for nonequivalent points. Thus

$$c = -\tfrac{1}{4}\hat{V}(\pi, \pi, \pi) - \tfrac{1}{4}\hat{V}(\pi/2, \pi, \pi) \tag{5.8c}$$

We see that $a, b, c$ are not independent, since

$$4c = a + b$$

which can be checked for the Madelung constants! In particular, the two critical inequalities

$$a > b, \qquad a > 2c \tag{5.9}$$

are equivalent to the single inequality

$$\hat{V}(\pi, \pi, \pi) < \hat{V}(\pi/2, \pi, \pi) \tag{5.10}$$

We claim that (5.5) implies (5.10) so long as $V \to 0$ at infinity and $V \not\equiv 0$. For, let $\tilde{V}(n)$ be the Fourier transform of $V(n, m, l)$ in the $m, l$ variables evaluated at $p_2 = p_3 = \pi$. Then (5.5) implies that $\tilde{V}(n)$ is RP as a one-variable object. Thus, by the general theory[11]

$$\tilde{V}(n) = \int_{-1}^{1} d\rho(\lambda)\lambda^{|n|-1}, \qquad |n| \geqslant 1$$

for some positive measure $d\rho$. It follows that (up to a constant which can be absorbed into the definition of $\tilde{V}$)

$$\tilde{V}(p_1, \pi, \pi) = c_0 + \int_{-1}^{1} 2d\rho(\lambda)(\cos p_1 - \lambda)/(1 + \lambda^2 - 2\lambda\cos p_1)$$

so that (5.10) follows from

$$-(\lambda + 1)^{-1} < -\lambda/(\lambda^2 + 1), \qquad \text{for} \quad |\lambda| < 1$$

and the fact that $d\rho$ cannot be concentrated entirely at $\pm 1$, since $\tilde{V}(n) \to 0$ at $\infty$. We have thus proven:

**Theorem 5.3.** The conclusions of Theorems 5.1 and 5.2 remain true if the Coulomb potential $|\alpha - \beta|^{-1}$ is replaced by $V(\alpha - \beta)$ for *any* $V$ obeying (5.5) and not identically zero.

Alternatively, inequalities (5.9) with $>$ replaced by $\geqq$ follow directly from the chessboard estimates (e.g., the probability of configuration 3 of Table III in some finite volume is dominated by the geometric mean of 1 and 4, so $c \geqq \frac{1}{2}a$). In fact, if we check the conditions required for the Cauchy–Schwarz inequalities used to obtain chessboard estimates to be equalities, we find that this would require $V \equiv 0$.

Finally, we should prove that the lattice Coulomb potential obeys (5.5). Letting $\tilde{V}(n, p_2, p_3)$ be the partial Fourier transform of $V$, we see that (5.5) is equivalent to $\tilde{V}(n, p_2, p_3)$ being RP in $n$ for *each* fixed $p_2, p_3$, since

$$\text{LHS of (5.5)} = \sum_{\substack{\alpha_1 \geqslant 1 \\ \beta_1 \geqslant 1}} \int \bar{w}_{\alpha_1 p_2 p_3} w_{\beta_1 p_2 p_3} \tilde{V}(\alpha_1 + \beta_1 - 1, p_2, p_3)\, dp_2\, dp_3$$

with

$$w_{\alpha_1 p_2 p_3} = \text{const} \sum_{\alpha_2 \alpha_3} z_\alpha \exp[i(p_2\alpha_2 + p_3\alpha_3)]$$

Thus, by the one-dimensional theory,[11] (5.5) holds if and only if

$$\tilde{V}(p_1, p_2, p_3) = c(p_2, p_3) + 2\int_{-1}^{1} d\rho(\lambda, p_2, p_3)(\cos p_1 - \lambda)/(1 + \lambda^2 - 2\lambda\cos p_1) \tag{5.11}$$

Letting $\lambda_0(p_2, p_3)$ be defined by $\frac{1}{2}(\lambda_0 + \lambda_0^{-1}) = 2 - \cos p_2 - \cos p_3$ with $0 < \lambda_0 < 1$, and choosing

$$d\rho(\lambda, p_2, p_3) = [2\lambda_0/(1 - \lambda_0^2)]\, \delta(\lambda - \lambda_0), \qquad c = 2/(1 - \lambda_0^2)$$

we find that $\tilde{V}(p_1, p_2, p_3) = (3 - \cos p_1 - \cos p_2 - \cos p_3)^{-1}$ has the form of (5.11), verifying that the lattice Coulomb potential is RP. Our choice of $c$ and $d\rho$ will seem less mysterious if one notes first that

$$2(\cos p_1 - \lambda)(1 + \lambda^2 - 2\lambda\cos p_1)^{-1} = -\lambda^{-1} + (1 + \lambda^2 - 2\lambda\cos p_1)^{-1}(1 - \lambda^2)$$

**Model 5.4.** Consider a Coulomb lattice gas as above, but allow also charges of, say, $\pm 2$, with a different chemical potential ($\mu_1$ for $\pm 1$ charges and $\mu_2$ for $\pm 2$). By the same techniques used above, plus the strategy of Section 1.6, we find that for sufficiently large $\beta$ and some $\mu_1(\beta) \simeq -a$ and $\mu_2(\beta) \simeq -4a$ there are at least five phases: two "crystals" of mainly $\pm 1$ charges, two more "crystals" of mainly $\pm 2$, and a "plasma."

**Model 5.5.** (Dipole gas.) We wish to note that Fröhlich and Spencer[19] have used methods related to those above to prove phase transitions in a model of lattice dipoles where each dipole is required to lie in one of the four positions pointing toward a nearest neighbor.

**Model 5.6.** (Quantum monopole gas; following Fröhlich.)[10] We consider a quantum model where one can place either of two fermions at each lattice site, one with charge $+1$ and one with charge $-1$. We will show that at suitable temperatures and "fugacities," crystals form, i.e., there is long-range correlation for having opposite charges on the even and odd sublattices. For simplicity, we work in $\nu \geqslant 3$ dimensions and exploit infrared bounds. In two dimensions with a Yukawa potential, one can prove similar results using the method of "exponential localization."[13] With minor changes, one can accommodate general RP potentials and/or bosons with hard cores (i.e., no more than one particle allowed at each site) in place of fermions; indeed, many of the complications in the fermion case are absent for bosons.

We begin by recalling a few facts about fermion systems with $m < \infty$ degrees of freedom (in our case $m$ will be $2|\Lambda|$). There are $m$ operators $\psi_1, \ldots, \psi_m$ on $\mathbb{C}^M$ ($M = 2^m$) obeying

$$\{\psi_i, \psi_j^*\} = \delta_{ij} \tag{5.12a}$$

$$\{\psi_i, \psi_j\} = 0 \tag{5.12b}$$

where $\{A, B\} = AB + BA$ is the anticommutator. All sets on $\mathbb{C}^M$ obeying (5.12) are unitarily equivalent. The existence and uniqueness questions for the relations (5.12) are standard; see Ref. 48 or the remarks below. Define

$$n_i = \psi_i^* \psi_i \tag{5.13}$$

Then (5.12) implies that

$$[n_i, \psi_j] = 0 \quad (i \neq j); \qquad n_i \psi_i = (n_i - 1)\psi_i^* = 0$$

Since (5.12a) implies that Ran $\psi_i$ + Ran $\psi_i^* = \mathbb{C}^M$, we conclude that $n_i$ has eigenvalues 0 and 1 only and that Ran $\psi_i \perp$ Ran $\psi_i^*$. Two families of unitaries are important: first let

$$u_i = (-1)^{n_i} \tag{5.14a}$$

Then

$$u_i \psi_j u_i^{-1} = \psi_j \quad (j \neq i); \qquad u_i \psi_i u_i^{-1} = -\psi_i \tag{5.14b}$$

and

$$u_i^2 = 1; \qquad u_i u_j = u_j u_i; \qquad u_i^* = u_i \tag{5.14c}$$

Second, let

$$v_i = \left(\prod_{j \neq 1} u_j\right)\left((\psi_i + \psi_i^*)\right) \tag{5.15a}$$

Then

$$v_i \psi_j v_i^{-1} = \psi_j \quad (j \neq i); \qquad v_i \psi_i v_i^{-1} = \psi_i^* \tag{5.15b}$$

and

$$v_i^2 = 1; \qquad v_i v_j = -v_j v_i \quad (i \neq j); \qquad v_i^* = v_i \tag{5.15c}$$

We will need two additional properties of the $\psi_i$:

**Proposition 5.4.** (a) The $\psi_i$ have a simultaneous real matrix representation in which $n_i$, $u_i$, $v_i$ are also real.

(b) The $2^{2m}$ operators $\psi_1^{a_1}(\psi_1^*)^{b_1} \cdots \psi_m^{a_m}(\psi_m^*)^{b_m}$, with each $a_i$ and $b_i$ zero or one, are linearly independent and span Hom($\mathbb{C}^M$). Moreover,

$$\mathrm{Tr}(\psi_1^{a_1}(\psi_1^*)^{b_1} \cdots) = \prod_{i=1}^{m} [(1 - \tfrac{1}{2}a_i)\, \delta_{a_i b_i}] \tag{5.16}$$

*Remark.* Part (b) implies in particular that any map between two families obeying the CAR (5.12) extends to an automorphism of the corresponding algebras and thus is unitarily implemented since every automorphism of Hom($\mathbb{C}^M$) is so implemented. This yields a proof of uniqueness for the CAR; existence is proven in (a).

*Proof.* (a) Write $\mathbb{C}^M$ as the $m$-fold tensor product $\mathbb{C}^2 \otimes \cdots \otimes \mathbb{C}^2$. Let $n_i = 1 \otimes \cdots \otimes n \otimes \cdots \otimes 1$ ($n$ is the $i$th factor) with $n = \begin{pmatrix} 1 & 0 \\ 0 & 0 \end{pmatrix}$, so that

$u_i = (-1)^{n_i} = 1 \otimes \cdots \otimes u \otimes \cdots \otimes 1$ with $u = \left(\begin{smallmatrix}-1 & 0\\ 0 & 1\end{smallmatrix}\right)$. Let $p_i = 1 \otimes \cdots \otimes p \otimes \cdots \otimes 1$ with $p = \left(\begin{smallmatrix}0 & 0\\ 1 & 0\end{smallmatrix}\right)$. Then $n_i = p_i{}^*p_i$, $\{p_i, p_i{}^*\} = 1$, and $[p_i, p_j] = [p_i, p_j{}^*] = 0$ $(i \neq j)$. To get the full CAR, we employ a Jordan–Wigner transformation and let

$$\psi_i = \Big(\prod_{j<i} u_j\Big) p_i$$

(b) Let $C = \psi_1^{a_1}(\psi_1{}^*)^{b_1} \cdots \psi_m^{a_m}(\psi_m{}^*)^{b_m}$. If $a_k \neq b_k$, then $u_k C u_k^{-1} = -C$, so $\mathrm{Tr}(C) = \mathrm{Tr}(-C) = 0$. Since the $n_i$ commute and $v_j n_i v_j^{-1} = n_i$ $(i \neq j)$, $v_j n_j v_j^{-1} = 1 - n_j$, we see that any product of $(n_i - \frac{1}{2})$'s has zero trace for the same reason, so that (5.16) holds. Now let $A_j = \psi_j + \psi_j{}^*$, $B_j = i^{-1}(\psi_j - \psi_j{}^*)$. Relation (5.16) and the commutation relations among the $A$'s and $B$'s imply that the operators $A_1^{a_1}B_1^{b_1}\cdots$ are orthonormal in the trace inner product $(C, D) = \mathrm{Tr}(C^*D)$. They are thus independent, so that the $\psi_1^{a_1}\ldots$; which have the same span and as many elements, are also independent. That the span is $\mathrm{Hom}(\mathbb{C}^M)$ follows from counting dimensions. ∎

With these preliminaries out of the way, we can describe the model and analyze it. In finite volume $\Lambda \subset \mathbb{Z}^\nu$, we take $m = 2|\Lambda|$ and label the $\psi$'s as $\{\psi_\alpha{}^\pm\}_{\alpha\in\Lambda}$. We let

$$q_\alpha = n_\alpha^{(+)} - n_\alpha^{(-)} \tag{5.17a}$$

$$T = -\sum_{\langle\alpha,\beta\rangle\subset\Lambda} (\psi_\alpha{}^*\psi_\beta + \psi_\beta{}^*\psi_\alpha) \tag{5.17b}$$

the sum being over nearest neighbor pairs. We will take in a torus $\Lambda$

$$H_\Lambda = T + \sum_{\alpha,\beta\in\Lambda} V^{(\Lambda)}(\alpha - \beta) q_\alpha q_\beta - \mu \sum_{\alpha\in\Lambda} q_\alpha{}^2 \tag{5.17c}$$

where $V$ is the periodic Coulomb potential and $T$ has periodic boundary conditions. The $T$ in (5.17b) is just the second quantization of $-2\nu I - \Delta$, where $\Delta$ is the lattice Laplacian. It describes "hopping" between neighboring sites. Often one describes these kinds of models with the sign of $T$ reversed. Since $T$ and $-T$ are unitarily equivalent under $U = \prod_{\alpha\in\Lambda_{\text{even}}} u_\alpha{}^+ u_\alpha{}^-$ with $Uq_\alpha U^{-1} = q_\alpha$, the models are equivalent. Our choice of the sign is convenient since it makes $T$ look "ferromagnetic." We aim to prove:

**Theorem 5.5.** For some $\beta_c, \mu_c$ the model (5.17) has staggered long-range order for $\mu \geqslant \mu_c$, $\beta \geqslant \beta_c$ in the sense that for such $\mu, \beta$ and $\langle\cdots\rangle$ an infinite-volume state

$$\langle P(q_\alpha = 1) P(q_\beta = (-1)^{|\alpha|+|\beta|})\rangle > \tfrac{1}{4}$$

uniformly in $\alpha, \beta$.

The first step of the proof is familiar from other models with staggered long-range order, such as the antiferromagnet; namely we flip spins on the odd sublattice. Thus let $v$ be given by (5.15) and define

$$U = \prod_{\alpha \in \Lambda_{\text{odd}}} v_\alpha{}^+ v_\alpha{}^- \tag{5.18a}$$

$$\tilde{H} = UHU^{-1} \tag{5.18b}$$

Then, since $Uq_\alpha U^{-1} = (-1)^{|\alpha|} q_\alpha$,

$$\tilde{H} = \tilde{T} + \sum_{\alpha,\beta\in\Lambda} W(\alpha - \beta) q_\alpha q_\beta - \mu \sum_{\alpha\in\Lambda} q_\alpha{}^2 \tag{5.18c}$$

where (for each nearest neighbor pair $\langle\alpha, \beta\rangle$, $\alpha$ is the even one)

$$\tilde{T} = - \sum_{\langle\alpha,\beta\rangle\subset\Lambda} (\psi_\alpha{}^* \psi_\beta{}^* + \psi_\beta \psi_\alpha) \tag{5.18d}$$

and

$$W(\alpha - \beta) = (-1)^{|\alpha-\beta|} V(\alpha - \beta) \tag{5.19}$$

We are thus reduced to showing that the $\tilde{H}$ model has long-range order in the unstaggered sense. The kinetic energy term in $\tilde{H}$ is already in a "ferromagnetic" form. To see that the $W$ term is also, we note:

**Proposition 5.7.** Suppose that $V$ is a function on $\mathbb{Z}^\nu$ which is RP with respect to reflections between lattice sites. Then $W(\alpha) = (-1)^{|\alpha|} V(\alpha)$ is RN.

*Proof.* Let $r(\alpha_1, \ldots, \alpha_\nu) = (1 - \alpha_1, \alpha_2, \ldots, \alpha_\nu)$ for $\alpha_1 \geqslant 1$. Then for any $c_\alpha$'s

$$\sum_{\substack{\alpha_1 \geqslant 1 \\ \beta_1 \geqslant 1}} V(\alpha - r\beta) \bar{c}_\alpha c_\beta \geqslant 0 \tag{5.20}$$

Let $d_\alpha = (-1)^{|\alpha|} c_\alpha$, so that

$$\sum W(\alpha - r\beta) \bar{d}_\alpha d_\beta = -\sum V(\alpha - r\beta) \bar{c}_\alpha c_\beta$$

since $|\alpha - r\beta| \equiv |\alpha| + |\beta| + 1 \pmod 2$. Thus (5.20) implies that $W$ is RN. ■

The general framework of Ref. 11 for quantum systems required distinguished algebras $\mathfrak{A}_+$ and $\mathfrak{A}_-$ and a morphism $\theta: \mathfrak{A}_+ \to \mathfrak{A}_-$. The machinery worked smoothly if $\mathfrak{A}_+$ and $\mathfrak{A}_-$ commute. For this reason, we make a preliminary Jordan–Wigner (=Klein) transformation to make the left and right commute. Thus we suppose that $\Lambda$ consists of sites $\alpha$ with $\alpha_1 = -L + 1, \ldots, L$. For $\alpha_1 \geqslant 1$, let $\phi_\alpha{}^\pm = \psi_\alpha{}^\pm$ and for $\alpha_1 \leqslant 0$, let $\phi_\alpha{}^\pm = (-1)^{N_R} \psi_\alpha{}^\pm$, where

$$N_R = \sum_{\alpha_1 \geqslant 1} q_\alpha$$

So

$$(-1)^{N_R} = \prod_{\alpha_1 \geqslant 1} u_\alpha{}^+ u_\alpha{}^-$$

Notice that $n_\alpha{}^\pm = (\phi_\alpha{}^\pm)^* \phi_\alpha{}^\pm$ and that

$$\tilde{H} = \tilde{T}_L + \tilde{T}_R + \tilde{T}_{LR} + \sum W(\alpha - \beta) q_\alpha q_\beta - \mu \sum q_\alpha{}^2$$

where (with $\alpha$ the even member of the pair $\langle\alpha\beta\rangle$)

$$\tilde{T}_L = - \sum_{\substack{\langle\alpha,\beta\rangle \\ \alpha_1, \beta_1 \leqslant 0}} (\phi_\alpha{}^* \phi_\beta{}^* + \phi_\beta \phi_\alpha)$$

$$\tilde{T}_R = - \sum_{\substack{\langle\alpha,\beta\rangle \\ \alpha_1, \beta_1 \geqslant 1}} (\phi_\alpha{}^* \phi_\beta{}^* + \phi_\beta \phi_\alpha)$$

$$\tilde{T}_{LR} = - \sum_{\substack{\langle\alpha,\beta\rangle \\ \alpha_1 \geqslant 1, \beta_1 \leqslant 0 \text{ or } \alpha_1 \leqslant 0, \beta_1 \geqslant 1}} [\phi_\alpha{}^* (-1)^{N_R} \phi_\beta{}^* + \phi_\beta (-1)^{N_R} \phi_\alpha]$$

Let $\mathfrak{A}_L$ (resp. $\mathfrak{A}_R$) denote the real algebra generated by $\{\phi_\alpha, \phi_\alpha{}^*\}$ with $\alpha_1 \leqslant 0$ (resp. $\alpha_1 \geqslant 1$). Motivated by the formula for $\tilde{T}_{LR}$, we define $\theta\colon \mathfrak{A}_L \to \mathfrak{A}_R$ by

$$\theta(\phi_\alpha) = \phi_{r\alpha}(-1)^{N_R} \quad \text{if} \quad (-1)^{|\alpha|} = 1 \tag{5.21a}$$

$$\theta(\phi_\alpha) = (-1)^{N_R} \phi_{r\alpha} \quad \text{if} \quad (-1)^{|\alpha|} = -1 \tag{5.21b}$$

using Proposition 5.4 to extend $\theta$ to a *-automorphism (to be able to do this one needs to check that the image under $\theta$ of the $\phi$'s obeys the CAR). The asymmetry in the definition of $\theta$ comes from the fact that we have taken $\phi_\beta \leftrightarrow \phi_\beta{}^*$ for $(-1)^{|\beta|} = -1$.

With this definition of $\theta$, $\theta(n_\alpha{}^\pm) = n_{r\alpha}^\pm$ and by (5.16), for any $A \in \mathfrak{A}_L$

$$\operatorname{Tr}(\theta(A)) = \operatorname{Tr} A \tag{5.22}$$

By Proposition 5.4(a), Tr is real on $\mathfrak{A}$, the real algebra generated by $\mathfrak{A}_R$ and $\mathfrak{A}_L$. Finally by (5.21), $-\tilde{H}$ has the form $A + \theta A + \sum B_l \theta(B_l)$, so by the general theory[6,11] $\langle \cdots \rangle$ is RP and moreover

$$Z(h_\alpha) \leqslant Z(h_\alpha \equiv 0) \tag{5.23}$$

where

$$Z(h_\alpha) = \operatorname{Tr} \exp\left[-\tilde{T} + \sum W(\alpha - \beta)(q_\alpha - h_\alpha - q_\beta + h_\beta)^2 + \mu' \sum q_\alpha{}^2\right]$$

with $\mu'$ defined so that when $h_\alpha \equiv 0$, $Z(h_\alpha)$ is the partition function for $\tilde{H}$. In the usual way[6,11] (5.23) leads to an inequality in the Duhamel two-point function (for $k \neq 0$)

$$(\hat{q}(k), \hat{q}(-k)) \leqslant -[2\beta \hat{W}(k)]^{-1} \tag{5.24}$$

The proof of Theorem 5.5 is completed as follows: If long-range order is not present, then by (5.24)

$$(q_0, q_0) \leqslant \tfrac{1}{2}$$

for $\beta$ large. A contradiction is obtained by using the Falk–Bruch inequality if we show that

$$\langle q_0{}^2\rangle \to 1 \qquad (\beta, \mu \text{ large}) \tag{5.25}$$

and

$$\langle [q_0, [q_0, \tilde{H}]]\rangle \to 0 \qquad (\beta, \mu \text{ large}) \tag{5.26}$$

By a direct calculation $[q_0, [q_0, \tilde{H}]]$ is a multiple of

$$\sum_{\substack{\langle\alpha,\beta\rangle\\ \alpha=0 \text{ or } \beta=0}} \psi_\alpha{}^*\psi_\beta{}^* + \psi_\beta\psi_\alpha$$

and this has small expectation between all states with $\langle q_0{}^2\rangle$ and $\langle q_\alpha{}^2\rangle$ ($|\alpha| = 1$) close to 1. Thus (5.25) implies (5.26).

Condition (5.25) can be proven either by thermodynamic considerations or by a chessboard estimate. Here is the thermodynamic proof. As above, write

$$-\tilde{H} = -\tilde{T} + \sum W(\alpha - \beta)(q_\alpha - q_\beta)^2 + \mu' \sum q_\alpha{}^2$$

with $\mu' = \mu$ − const. Since $W$ is RN, the maximum value of $-\sum W(\alpha - \beta) \times (q_\alpha - q_\beta)^2$ occurs with all $q_\alpha$ equal, i.e., it is zero. Thus

$$\langle -\tilde{H}/|\Lambda|\rangle \leqslant \mu'\langle q_0{}^2\rangle + 2\nu \tag{5.27}$$

On the other hand, using

$$\mathrm{Tr}(1) = 4^{|\Lambda|}$$

and

$$\mathrm{Tr}\{\exp[-\beta H(\mu')]\} \geqslant \exp(-\beta\langle a|\tilde{H}|a\rangle)$$

(where $a$ is the state with all $q_\alpha = +1$) and convexity of $\ln \mathrm{Tr}\, e^{-B}$, we see that

$$\langle -\tilde{H}/|\Lambda|\rangle \geqslant \mu' - \beta^{-1}\ln 4 \tag{5.28}$$

Then (5.27) and (5.28) imply that

$$\langle q_0{}^2\rangle \geqslant 1 - (\mu')^{-1}\beta^{-1}\ln 4 - 2(\mu')^{-1}\nu$$

proving (5.25). This completes the proof of Theorem 5.5. ∎

## 6. SIX- AND EIGHT-VERTEX MODELS

In this section we will discuss several aspects of planar six- and eight-vertex models in zero external field. We remind the reader that for these models, the free energy is known exactly(33–35,51,1) (for a comprehensive review of the six-vertex models see Ref. 37). We will leave extensions to 16-vertex models and to higher dimensions to the reader's imagination!

The "spins" in this model are assignments of a direction (or arrow) to each bond of a square lattice. Of the 16 possible configurations at each lattice site only the first six (resp. all eight) of those pictured in Fig. 8 are allowed. The basic Hamiltonian is

$$H = \sum_{i=1}^{6(8)} N_i \epsilon_i \tag{6.1}$$

where $N_i$ is the number of vertices of type $i$.

**Model 6.1.** (Six-vertex $F$-model; $\epsilon_1 = \epsilon_2 = \epsilon_3 = \epsilon_4 > \epsilon_5 = \epsilon_6$.) Consider a vertical line midway between lattice sites. We let $\mathfrak{A}_0$ denote functions of the arrows that intersect this line and $\mathfrak{A}_r$ (resp. $\mathfrak{A}_l$) functions of the arrows on the right (resp. left). We define $\theta: \mathfrak{A}_+ = [\mathfrak{A}_r \cup \mathfrak{A}_0] \to \mathfrak{A}_- = [\mathfrak{A}_r \cup \mathfrak{A}_0]$ as follows (here $[\mathfrak{A}]$ is the algebra generated by $\mathfrak{A}$): If $\sigma_i$ ($i$ a *bond*) is $\pm 1$, depending on whether the arrow points right (resp. up) or left (resp. down) for $i$ horizontal (resp. vertical), then $\theta(\sigma_i) = \sigma_{\theta i}$ for horizontal bonds and $\theta(\sigma_i) = -\sigma_{\theta i}$ for vertical bonds. By the general theory(11) the uncoupled state is RP. Moreover, the induced map from configurations at site $\alpha$ to those at site $\theta(\alpha)$ is $\theta(1) = (3)$, $\theta(2) = (4)$, $\theta(3) = (1)$, $\theta(4) = (2)$, $\theta(5) = (6)$, $\theta(6) = (5)$. Since

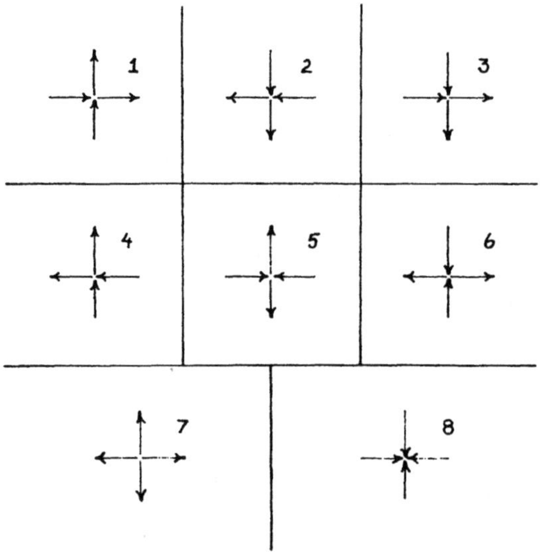


Fig. 8. Configurations of the six- and eight-vertex models.

the basic interactions only involve spins in $\mathfrak{A}_+$ or $\mathfrak{A}_-$, $H$ is of the form $A + \theta A$ so long as $\epsilon_1 = \epsilon_3$, $\epsilon_2 = \epsilon_4$, $\epsilon_5 = \epsilon_6$. Again by the general theory[(11)] the interacting state will be RP. If

$$\epsilon_1 = \epsilon_2 = \epsilon_3 = \epsilon_4; \qquad \epsilon_5 = \epsilon_6 \tag{6.2}$$

we will have RP in both horizontal and vertical planes. In that case, by mimicking the proof of Theorem 4.3 of Ref. 11, we will obtain the following chessboard estimate:

**Theorem 6.1.** For any function $F$ of the arrows coming into a single site $\mu$, let $\gamma(F) = \langle\prod_{\alpha\in\Lambda}\tau_\alpha(F)\rangle^{1/|\Lambda|}$, where $\tau_\alpha(F)$ is a translate of $F$ if $\alpha_1 - \mu_1 \equiv \alpha_2 - \mu_2 \equiv 0 \pmod 2$, a reflected (by $\theta$ as above) translate if $\alpha_1 - \mu_1 + 1 \equiv \alpha_2 - \mu_2 \equiv 0 \pmod 2$, etc. Then whenever (6.2) holds, we have in a $2L$ by $2L$ torus

$$\left\langle \prod_{\alpha\in\Lambda} F_\alpha \right\rangle \leqslant \prod_{\alpha\in\Lambda} \gamma(F_\alpha) \tag{6.3}$$

where $F_\alpha$ is a function of the arrows at site $\alpha$.

Now suppose that (6.2) holds and $\epsilon_5 < \epsilon_1$. Then there are two ground states, as is easy to see from RP: one obtained from state (5) and reflection, and the other from state (6). They are interchanged by translations of one unit or by flipping all arrows. We can organize contours as follows: let $s_i = +1$ if the arrow on bond $i$ is in the direction it would be in one ground state, $-1$ if it is in the other direction. We now tilt our heads by 45° and think of the system as an Ising model on a square lattice with spacing $\frac{1}{2}\sqrt{2}$ obtained by putting dots on the midpoints of bonds. Place contours in the standard way for an Ising model, so in terms of the original lattice a contour $\gamma$ is a collection of $|\gamma|$ diagonal segments of length $\frac{1}{2}\sqrt{2}$. The six-vertex condition implies that contours cannot turn at vertices in the original lattice but only at points in the dual lattice of the original lattice (see Fig. 9 for a typical contour). This reduces the number of contours to be counted when one pushes through to an estimate on transition temperatures. Clearly $\gamma$ passes through $\frac{1}{2}|\gamma|$ original lattice sites, and at each site the configuration is forced to be one of

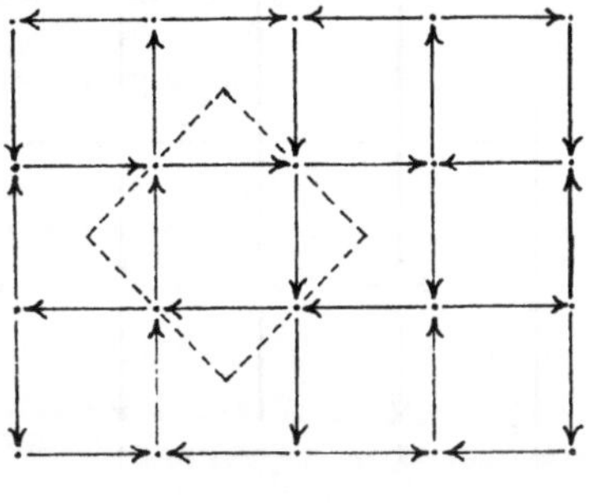

Fig. 9. A typical contour.

1–4. Moreover, under repeated reflection the projection onto any of the configurations 1–4 leads to an array like that shown in Fig. 10 with energy per site $\frac{1}{4}(\epsilon_1 + \epsilon_2 + \epsilon_3 + \epsilon_4)$. Thus, by Theorem 6.1,

$$\mathrm{Prob}(\gamma) \leqslant \exp[-\tfrac{1}{2}\beta|\gamma|(\epsilon_1 - \epsilon_5)]$$

in the usual way. Thus, since $\langle\sigma_i\rangle = 0$ in the periodic state, we obtain:

**Theorem 6.2.** The Hamiltonian (6.1) with (6.2) and $\epsilon_1 > \epsilon_5$ has at least two equilibrium states for $\beta$ large, with $\langle\sigma_i\rangle \neq 0$ and long-range order.

If $\epsilon_1 \leqslant \epsilon_5$, we obtain a structure very similar to that in Model 2.2 when (2.5) fails. Thus one expects no long-range order; indeed this follows from the known analyticity of the pressure in that case.[37]

**Model 6.2.** (Eight-vertex model.) We have RP in the same planes as in Model 6.1 so long as $\epsilon_7 = \epsilon_8$ in addition to (6.2). If $\epsilon_5 < \min(\epsilon_1, \epsilon_7)$, one obtains phases with the same structure as in the earlier model. Indeed, the only change in the analysis is that contours can now have corners at original lattice sites. If $\epsilon_7 < \min(\epsilon_1, \epsilon_5)$, there are again two ground states and by mimicking the argument in the last model, one obtains multiple equilibrium states. If $\epsilon_5 = \epsilon_7$ (i.e., $\epsilon_1 = \epsilon_2 = \epsilon_3 = \epsilon_4 = \epsilon$ and $\epsilon_5 = \epsilon_6 = \epsilon_7 = \epsilon_8 = -\epsilon$), then the model is invariant under reversal of all the vertical arrows in any column or all the horizontal arrows in any row, which suggests the absence of any long-range order. This is supported by the following exact calculation of the partition function with free boundary conditions. One can make an arbitrary choice of all vertical arrows and of one horizontal arrow in each row; once this choice has been made, the eight-vertex condition determines all the other horizontal arrows. Moreover, the equality of energies in two sets implies that the energy of a configuration is completely determined by the vertical arrows. Therefore if there are $M$ rows and $L$ columns, we have $Z = 2^M Q^L$, where $Q$ is the partition function for a single column. For each column, the position of the bottom arrow is irrelevant and then we pick up a

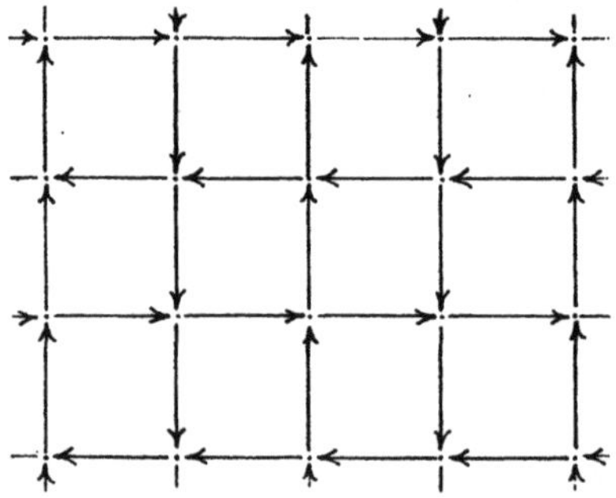

Fig. 10. A universal configuration.

**Phase Transitions and Reflection Positivity. II.** 339

factor of $\pm\epsilon$, depending on whether each successive arrow is parallel or antiparallel to the preceding one. Thus $Q = 2^{M+1}(\cosh \beta\epsilon)^M$, so

$$Z = 2^{M+L}(2 \cosh \beta\epsilon)^{LM}$$

**Model 6.3.** (Diagonal RP in the generalized $F$ model.) As an exercise in RP, we want to examine when Model 6.1 is RP in a diagonal line through sites. Thus, imagine a diagonal line going from the lower left to the upper right. The geometric $\theta'$ on bonds takes vertical bonds into horizontal bonds. We will define $\theta'$ on the bond arrows by $\theta'(\uparrow) = \leftarrow$, $\theta'(\downarrow) = \rightarrow$, $\theta'(\rightarrow) = \downarrow$, $\theta'(\leftarrow) = \uparrow$, i.e., geometric reflection followed by flip. Then $\theta'$ will leave states 3–6 unchanged and will interchange states 1 and 2. Thus if $\epsilon_1 = \epsilon_2$, the part of the Hamiltonian that does not involve sites on the line will have the form $A + \theta A$. To see whether the remaining piece of $-H$ has the form $\sum B_i \theta B_i$ we need to look at whether the matrix of statistical weights for a state $\mu$ above and $\nu$ below is positive definite. There are four possibilities for $\mu$: $a = \leftarrow^{\downarrow}$, $b = \rightarrow^{\uparrow}$, $c = \rightarrow^{\downarrow}$, $d = \leftarrow^{\uparrow}$. Thus $\theta a = {}_{\uparrow}\!\rightarrow$, $\theta b = {}_{\downarrow}\!\leftarrow$, $\theta c = {}_{\downarrow}\!\rightarrow$, $\theta d = {}_{\uparrow}\!\leftarrow$, and the matrix $A_{\mu\nu} = \exp[-\beta \text{ energy of } (\mu\theta\nu)]$ has the form

$$\begin{pmatrix} e^{-\beta\epsilon_6} & e^{-\beta\epsilon_2} & 0 & 0 \\ e^{-\beta\epsilon_1} & e^{-\beta\epsilon_5} & 0 & 0 \\ 0 & 0 & e^{-\beta\epsilon_3} & 0 \\ 0 & 0 & 0 & e^{-\beta\epsilon_4} \end{pmatrix}$$

So, with the condition $\epsilon_1 = \epsilon_2$ already required, positive-definiteness is equivalent to $\exp[-\beta(\epsilon_5 + \epsilon_6)] - \exp[-\beta(\epsilon_1 + \epsilon_2)] \geqslant 0$, i.e., $\epsilon_5 + \epsilon_6 \leqslant \epsilon_1 + \epsilon_2$. As the general theory (Schoenberg's Theorem) predicts,[11] the condition is independent of $\beta$. Thus:

**Theorem 6.3.** The six-vertex model is RP with respect to reflections in diagonals as described above if and only if $\epsilon_1 = \epsilon_2$ and $\epsilon_5 + \epsilon_6 \leqslant \epsilon_1 + \epsilon_2$ (the conditions $\epsilon_3 = \epsilon_4$ and $\epsilon_5 = \epsilon_6$ are not required).

By using this theorem and reflecting in two diagonal directions, one can extend Theorems 6.1 and 6.2 to the case where $\epsilon_1 = \epsilon_2$, $\epsilon_3 = \epsilon_4$, $\epsilon_5 + \epsilon_6 < 2 \min(\epsilon_1, \epsilon_3)$.

## 7. THE PROJECTED PEIERLS ARGUMENT

In this section we want to describe a method which allows one to study a $\nu$-dimensional model by considering only contours in a two-dimensional plane. This argument applies to some situations where a naive $\nu$-dimensional contour argument is not applicable. We illustrate the method in a simple case:

**Model 7.1.** ($\nu$-dimensional Ising model.) Consider the nearest neighbor, spin-$\frac{1}{2}$ Ising model in $\nu$ ($>2$) dimensions. As usual, let $P_\alpha^\pm$ be the projection onto $s_\alpha = \pm 1$. Suppose we show that $\langle P_\alpha^+ P_\beta^- \rangle \leqslant \epsilon$ for all $\alpha, \beta$ in a common two-dimensional plane of the lattice [there are $\binom{\nu}{2}$ types of such planes]. Then for *any* $\alpha, \gamma$ we can find $\beta_1, \ldots, \beta_k$ with $k = [\frac{1}{2}(\nu - 1)]$ and the pairs $(\alpha, \beta_1)$, $(\beta_1, \beta_2), \ldots, (\beta_k, \gamma)$ in common two-dimensional planes. But since (with $\beta_0 = \alpha$, $\beta_{k+1} = \gamma$)

$$\langle P_\alpha^+ P_\gamma^- \rangle \leqslant \sum_{j=0}^{k} \langle P_{\beta_j}^+ P_{\beta_{j+1}}^- \rangle$$

we see that $\langle P_\alpha^+ P_\gamma^- \rangle \leqslant (k+1)\epsilon$. Thus a proof of two-dimensional "long-range order" implies full long-range order.

Now let $\alpha, \beta$ lie in a common plane. By considering configurations of the spins in that plane, we can dominate $\langle P_\alpha^+ P_\beta^- \rangle$ by a sum, over contours $\Gamma$ in that plane, of the probability $P(\Gamma)$ that in that plane all the spins inside $\Gamma$ are plus and those outside $\Gamma$ are minus. But by chessboard estimates, $P(\Gamma) \leqslant \exp(-a|\Gamma|)$ with $a \to \infty$ as $T \to 0$. This proves long-range order.

**Model 7.2.** (The Slawny model.) Take $\nu = 3$ for simplicity and spin-$\frac{1}{2}$ spins. For each $a$, a square face of some cube in the basic lattice $\mathbb{Z}^3$, let $t_a$ be the product of the spins at the corners of $a$. The Slawny Hamiltonian[50] is

$$H = \sum_{a\Lambda} t_a$$

The reason for the interest in this Hamiltonian is that is possesses an infinite local symmetry group; namely, flipping all spins in any family of parallel planes leaves $H$ invariant. This implies that there will be infinitely many phases once some kind of long-range order is shown. We remark that, while there are some similarities between the Slawny model and the currently fashionable $\mathbb{Z}_2$-gauge model, they are very different. The latter model has symmetries that affect only a finite number of spins, and no long-range order.

Two proofs of such long-range order have been given: in Ref. 50, Slawny used correlation inequalities to compare the model to the two-dimensional Ising model, and in Ref. 28, Holsztyński and Slawny introduced a sophisticated contour argument applicable to this model. Here, we will prove a kind of long-range order. Going from this to the infinite number of states will be left to the reader. Note first that one has RP in planes between lattice sites. Fix a horizontal plane, and for each (original lattice) site $\alpha$ in the plane let $z_\alpha$ be the product of the spin at $\alpha$ and the spin directly above $\alpha$. Let $P_\alpha^\pm$ be the projection onto $z_\alpha = \pm 1$. We will show that $\langle P_\alpha^+ P_\gamma^- \rangle$ is small for $\beta$ sufficiently large, by a two-dimensional contour analysis. In the usual way, the probability of a contour $\Gamma$ will be dominated by $a^{|\Gamma|/4}$, where $a = \langle P_{\text{univ}} \rangle_\Lambda^{4/|\Lambda|}$.

Here $P_{\text{univ}}$ results from taking $P_\alpha{}^+ P_{\alpha+\delta}^-$ [$\delta = (1, 0, 0)$] and making repeated reflections, with a two-element period in the vertical and $\delta$ directions and a one-element period in the third direction. Since in the resulting configuration $1/12$ of the squares have $t_\alpha = -1$, we obtain $a \to 0$ as $\beta \to \infty$. Finally, we note that $\langle P_\alpha{}^+ \rangle = \langle P_\alpha{}^- \rangle = \frac{1}{2}$ by the previously mentioned symmetry. This implies long-range order.

**Model 7.3.** Let us contrive a model where the method of this section works, but where no other method we know will work. Take the Slawny model and make a small (proportional to $\epsilon$) perturbation by adding a nearest neighbor antiferromagnetic coupling and a small external field. This perturbation destroys the ferromagnetic properties that make the correlation inequalities work, and the decomposition property of Ref. 28 fails. One can use the Pirogov–Sinai method, but the resulting bound on the transition temperature will go to zero as $\epsilon$ does. By modifying the method of Model 7.2 we obtain a lower bound going to the bound of that model as $\epsilon \downarrow 0$.

## 8. THE BALANCED MODEL

On the lattice $\mathbb{Z}^\nu$ the basic Hamiltonian is

$$-\beta H = \beta \sum_{\langle \alpha\gamma \rangle} \sigma_\alpha \sigma_\gamma - \frac{\beta}{2(\nu - 1)} \sum_{\langle \alpha\gamma \rangle}{}' \sigma_\alpha \sigma_\gamma \tag{8.1}$$

where the first sum is over all nearest-neighbor pairs and the second sum is over all next-nearest-neighbor pairs. This model entered naturally in our analysis in Ref. 11. If $-1/[2(\nu - 1)]$ is replaced by $\alpha$ in (8.1), then one has RP about planes between sites if and only if $2|\alpha|(\nu - 1) \leqslant 1$, i.e., (8.1) is at the borderline of RP. Moreover, it is at a point of balance between ferromagnetism and antiferromagnetism. The interaction of a single spin with a neighboring hyperplane of all plus spins is zero due to a precise cancellation of the ferromagnetic and antiferromagnetic interactions. For this reason, we call the model the "balanced model."

**Model 8.1.** (Balanced Ising model; breaking of discrete symmetry.) Here we have nothing definite to report; we will describe what we believe happens. First, there should not be a first-order phase transition in two dimensions: for, if one uses the naive Peierls argument, one finds that only corners of contours are "bad" in that they make positive contributions to the energy. Thus, there are infinitely many contours (namely those with four corners) with the same energy shift, and one expects entropy to overwhelm energy and prevent the occurrence of long-range order.

We believe that in $\nu \geqslant 3$ dimensions, there are at least two phases (indeed, as we discuss below, probably exactly two). Various methods have not yet yielded a proof:

(i) The basic chessboard Peierls argument fails for an interesting reason. In finite volume, one must consider three kinds of contours: ones that surround $\alpha$, ones that surround $\gamma$, and ones that wrap around the torus. The first two classes typically have negligible probability as $|\Lambda| \to \infty$. However, in this case there are contours of zero energy wrapping around the torus, so that the chessboard Peierls argument fails.

(ii) The naive Peierls argument comes very close to working: use plus boundary conditions and put in conventional contours $\Gamma$ (which are two-dimensional if $\nu = 3$). The set of "edges" of $\Gamma$ is what counts. Given a set $e$ of edges, of length $|e|$, the number of $\Gamma$ with those edges is easily bounded by $A^{|e|}$. Moreover, the set of *connected* edge graphs that surround a vertex $\alpha$ is easily bounded by $BC^{|e|}$. The problem is that the set of edges need not be connected. Indeed, if $\Gamma$ is a large cube of side $l$ with $l$ unit cubes removed from the surface, $|e| \leqslant 12l + 12l$, but the number of ways of removing the cubes goes like

$$\binom{l^2}{l} \simeq e^{al \ln l}$$

which overwhelms the energy factor.

(iii) The analysis in (ii) suggests that we try a procedure of removing a single connected piece of edge contour. This is essentially the method that Holsztyński and Slawny[28] use. The problem is that in their language the decomposition property fails, and it is far from clear how to make this procedure work.

(iv) One can try the projected Peierls argument of Section 7. As in the analysis of the Slawny model (Model 7.2), one easily shows that with $P^{\pm}$ as in that model

$$\langle P_\alpha{}^+ P_\gamma{}^- \rangle \to 0 \qquad \text{as} \quad \beta \to \infty \tag{8.2}$$

The difference from the Slawny model is the following: in that model, $\langle P_\alpha{}^+\rangle = \langle P_\alpha{}^-\rangle = \frac{1}{2}$ by symmetry, so (8.2) implies that long-range order occurs. In the case at hand, it seems quite likely that $\langle P_\alpha{}^-\rangle \to 0$ as $\beta \to \infty$, preventing long-range order of the $t_\alpha = \sigma_\alpha \sigma_{\alpha+\delta}$. Of course, $\langle P_\alpha{}^-\rangle$ small means that neighboring spins tend to have the same sign, which strongly suggests that there is long-range order of the $\sigma_\alpha$'s.

(v) The following chessboard Peierls argument works, modulo a geometrical lemma which we have not been able to prove. Consider four sites $\alpha_1, \ldots, \alpha_4$ forming the corners of a rectangle $R$ with sides parallel to the $x$, $y$ axes. We wish to show that $\langle \prod \sigma_{\alpha_i} \rangle > \frac{1}{2}$ for large $\beta$, uniformly in $R$. This will imply long-range order of some sort, although not necessarily a spontaneous magnetization. The geometrical result is the following:

**Conjecture.** In any configuration with $\sigma_{\alpha_1}\sigma_{\alpha_2}\sigma_{\alpha_3}\sigma_{\alpha_4} = -1$ there is some $L$ and some connected set of contour edges of length $L$ within a distance $2L$ of one of the $\alpha_i$.

Given this conjecture, the inequality follows by chessboard estimates and easy bounds on the number of connected sets of edges of length $L$ in a given volume. The conjecture would follow in turn from a second conjecture:

**Conjecture.** In any configuration as above, there is a connected component $E$ of contour edges such that either $E$ is unbounded and intersects $R$, or $E$ is bounded and the smallest box with sides parallel to the axes containing $E$ also contains one of the $\alpha_i$.

While these conjectures seem reasonable, we have not been able to prove them.

(vi) One can try infrared bounds. As we shall see, these only work if $\nu \geqslant 5$, where indeed one can prove that some long-range order occurs.

The Hamiltonian (8.1) has an infinity of ground states; indeed, any array of hyperplanes with constant $\sigma_\alpha$ in each hyperplane is a ground state. On this basis, one might expect an infinity of phases, but we believe this is unlikely, for the ground states are *not* related by a symmetry. For this reason, one should expect the need for small external fields to manifest the instability associated with the ground state. Indeed, adding a term $\beta C \sum_\alpha t_\alpha$ to the $H$ of (8.1), one can still prove (8.2) uniformly in $C$ small. On the other hand, it is easy to see that for suitable $C_0$ and $\beta_0$

$$\langle P_\alpha{}^{\pm} \rangle < \tfrac{1}{2}$$

for $\mp C \geqslant C_0\beta^{-1}$, $\beta > \beta_0$. This means that there will be multiple phases at a point $C(\beta)$ with $|C(\beta)| \leqslant C_0\beta^{-1}$. More generally one expects, given $n$, to be able to add a suitable small external field and obtain at least $n$ phases.

**Model 8.2.** (Balanced classical Heisenberg model; breaking of continuous symmetry.) The spin wave energy in (8.1) is

$$E_p = \sum_{i=1}^{\nu} \cos p_i - \frac{1}{2(\nu - 1)} \sum_{i=1}^{\nu} \sum_{j>i} [\cos(p_i + p_j) + \cos(p_i - p_j)] \tag{8.3}$$

There are several important features of (8.3):

(a) $E_p \geqslant 0$; $E_p = 0$ if and only if at least $\nu - 1$ of the $p_i$ vanish.

(b) $E_p \sim p^4$ for $p$ near 0; $E_p \sim (p - p_0)^2$ for $p_0 \neq 0$, but $\nu - 1$ components of $p_0$ zero.

For $\nu \geqslant 5$, $\int E_p^{-1}\, d^\nu p < \infty$, so that the FSS method[17] of infrared bounds implies that the two-point function is a measure with some concentration on the "manifold" of zeros of $E_p$. This in turn implies the existence of multiple phases. This argument also applies to the classical Heisenberg model

[$\sigma_\alpha\sigma_\gamma$ in (8.1) replaced by $\boldsymbol{\sigma}_\alpha\cdot\boldsymbol{\sigma}_\gamma$, $\boldsymbol{\sigma}$ a unit vector]. If our picture (in the discrete case) of instability but no infinite number of qualitatively different phases is correct here too, then one expects the singularity in the two-point function to occur only at $p = 0$.

## APPENDIX. A TOPOLOGICAL THEOREM

In our discussion of multiphase results we required a theorem (Theorem A3 below) which is a special case of Theorem 13 of Brezis.[2] Since this reference is of limited availability, we provide here for the reader's convenience a proof patterned after ideas from [2] and also some ideas contained in a proof given us by J. Mather at a time when we were unaware of Brezis' paper.

**Lemma A1.** (Variational inequality of Hartman and Stampacchia.)[24] Let $A$ be a continuous map from $C$, a compact convex subset of $\mathbb{R}^n$, to $\mathbb{R}^n$. Then there exists $u \in C$ so that $(A(u), u - v) \leqslant 0$ for all $v \in C$.

*Proof* (Brezis,[2] Lemma 4). Since $C$ is compact and convex, for any $z \in \mathbb{R}^n$ there exists a unique point $P(z) \in C$ with

$$\|P(z) - z\| = \inf_{x\in C}\|x - z\|$$

Notice that $P(z) = u$ if and only if for all $v \in C$, $(d/d\lambda)\|u + \lambda(v - u) - z\||_{\lambda=0} \geqslant 0$, which holds if and only if $(u - z, v - u) \leqslant 0$ for all $v \in C$. Now let $T: C \to C$ be defined by

$$T(u) = P(u - A(u))$$

By the Brouwer fixed-point theorem,[53] there is some $u$ with $T(u) = u$, i.e., with $(u - (u - A(u)), v - u) \leqslant 0$ for all $v \in C$. ∎

**Theorem A2.** Let $\Delta_n$ be the canonical $n$-simplex $\{(t_0, \ldots, t_n) \in \mathbb{R}^{n+1}: t_i \geqslant 0, \sum t_i = 1\}$ and let $\{E_j\}_{j=1}^{J_n}$ ($J_n = 2^{n+1} - 2$) be the family of generalized edges (subsets where some nonempty subset $\{t_i\}_{i\in I_j}$ of the $t$'s are all zero and the other $t$'s are all nonzero). Let $F: \Delta_n \to \Delta_n$ be a continuous map with $\mathrm{dist}(F(u), E_j) < (n + 1)^{-1}$ for all $u \in E_j$ and all $j$. Then there exists some $u \in \Delta_n$ with

$$F(u) = x_0 \equiv \left(\frac{1}{n + 1}, \ldots, \frac{1}{n + 1}\right)$$

*Proof.* Let $A(u) = F(u) - x_0$. By Lemma $A1$, there exists $u_0 \in \Delta_n$ with $(F(u_0) - x_0, u_0 - v) \leqslant 0$ for all $v \in \Delta_n$. If $u_0$ is an interior point of $\Delta_n$, then $F(u_0) - x_0$ must be zero, so the theorem is proven. Since $\partial\Delta_n = \cup_j E_j$, we will be done if we can use the condition $\mathrm{dist}(F(u), E_j) < (n + 1)^{-1}$ to prove that $u_0$ cannot lie in any $E_j$. Consider first an $E_j$ with a single zero coordinate.

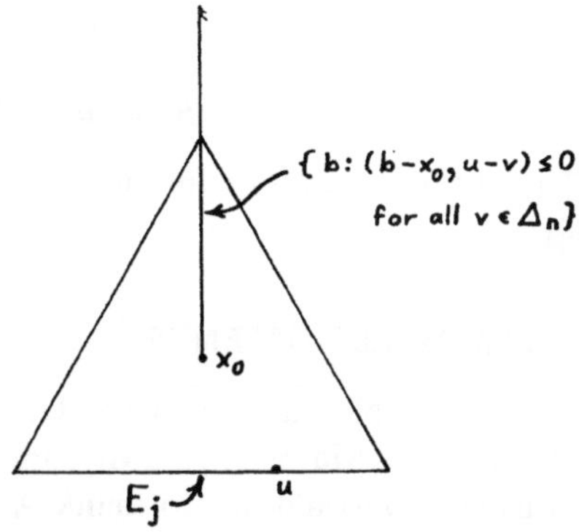


Fig. 11

Fix $u \in E_j$. Then $\{a: \sum a_i = 0,\ (a, u - v) \leqslant 0$ for all $v \in \Delta_n\}$ is a half-line perpendicular to the face $E_j$. Thus $(F(u) - x_0, u - v) \leqslant 0$ implies that $F(u)$ must lie in a half-line moving up from $x_0$ (see Fig. 11), and this is not allowed by the condition $\text{dist}(F(u), E_j) < (n + 1)^{-1}$. A similar geometric argument works for each face. ■

**Theorem A3.** (A rather special case of Theorem 13 of Ref. 2). Let $F$ be a map of $\Delta_n$ into convex subsets of $\Delta_n$ so that $\{(x, u): x \in \Delta_n, u \in F(x)\}$ is closed. Suppose that for some $A < (n + 1)^{-1}$ and each generalized edge $E_j$ we have that $x \in E_j$ and $u \in F(x)$ implies that $\text{dist}(u, E_j) \leqslant A$. Then there exists $u \in \Delta_n$ so that

$$x_0 = \left(\frac{1}{n+1}, \ldots, \frac{1}{n+1}\right) \in F(u)$$

*Proof.* Suppose for each $\delta$ we find $F_\delta: \Delta_n \to \Delta_n$, so that for $\text{dist}(u, \partial\Delta_n) > \delta$, $\text{dist}(F_\delta(u), F(u)) < \delta$. Then by Theorem A2 we can find, for $\delta$ small, $u_\delta$ such that $\text{dist}(u_\delta, \partial\Delta_n) > \delta$ and $\text{dist}(x_0, F(u_\delta)) < \delta$. Thus by the compactness of $\Delta_n$ there is $u$ such that $x_0 \in F(u)$. Let $G(u)$ be the barycenter of $F(u)$ for $u \in \Delta_n$, and otherwise $G(u) = P(u)$, the nearest point to $u$ in $\Delta_n$. $G$ is not continuous, but if $j_\epsilon$ is a smooth function supported in $\{x: \text{dist}(x, 0) < \epsilon\}$ and $\int j_\epsilon d^{n+1}x = 1$, then $G_\epsilon = G * j_\epsilon$ is smooth and $\text{dist}(G_\epsilon(u), F(u))$ is small by the hypotheses on $F$. Choosing $F_\delta = G_{\epsilon(\delta)}$ for suitable $\epsilon(\delta)$, the result follows. ■

*Remark.* Here is an alternate proof of Theorem A2 which depends on the following lemma of Sperner[(58)]:

**Lemma A4.** Let $\Delta_n$ be an $n$-simplex with $(n - 1)$-dimensional faces $G_1, \ldots, G_{n+1}$. Let $B_1, \ldots, B_{n+1}$ be closed subsets of $\Delta_n$ with $B_i \cap G_i = \varnothing$. Then either $\cap_j B_j \neq \varnothing$ or $\cap_j(\Delta_n \backslash B_j) \neq \varnothing$. In particular, if $\cup_j B_j = \Delta_n$, then $\cap_j B_j \neq \varnothing$.

*Alternate proof of Theorem A2.* Let $F_i(u)$ denote the coordinates of $F(u)$. Let

$$B_j = \{u: \quad F_j(u) \geqslant 1/(n+1)\}$$

Then $\cup_j B_j = \Delta_n$. By hypothesis, $B_j \cap G_j = \varnothing$, so by the lemma $\cap_j B_j \neq \varnothing$. ■

## ACKNOWLEDGMENTS

It is a pleasure to thank H. Brézis, J. Bricmont, D. Brydges, O. J. Heilmann, J. Mather, J. Slawny, and T. Spencer for valuable discussions. One of us (R. I.) would like to thank A. Jaffe for the hospitality of the Harvard Physics Department, where some of this work was done.

## REFERENCES

1. R. J. Baxter, *Phys. Rev. Lett.* **26**:832 (1971).
2. H. Brezis, *Séminaire Choquet* **6** (18) (1966/67).
3. J. Bricmont, J. L. Lebowitz, and C. E. Pfister, in preparation.
4. D. Brydges, *Comm. Math. Phys.* **58**:313 (1978).
5. R. L. Dobrushin, *Func. Anal. Appl.* **2**: 302 (1968).
6. F. J. Dyson, E. H. Lieb, and B. Simon, *J. Stat. Phys.* **18**:335 (1978).
7. M. E. Fisher, unpublished.
8. J. Fröhlich, *Acta Phys. Aust. Suppl.* **XV**:133 (1976).
9. J. Fröhlich, *Bull. Am. Math. Soc.* **84**:165 (1978).
10. J. Fröhlich, Princeton University Lectures (Spring 1977).
11. J. Fröhlich, R. B. Israel, E. H. Lieb, and B. Simon, *Comm. Math. Phys.* **62**:1 (1978).
12. J. Fröhlich, R. B. Israel, E. H. Lieb, and B. Simon, in preparation.
13. J. Fröhlich and E. H. Lieb, *Comm. Math. Phys.* **60**:233 (1978).
14. J. Fröhlich and Y. M. Park, *Comm. Math. Phys.* **59**:235 (1978).
15. J. Fröhlich and Y. M. Park, in preparation.
16. J. Fröhlich and B. Simon, *Ann. Math.* **105**:493 (1977).
17. J. Fröhlich, B. Simon, and T. Spencer, *Comm. Math. Phys.* **50**:79 (1976).
18. J. Fröhlich and T. Spencer, *Cargèse Lectures* (1976).
19. J. Fröhlich and T. Spencer, in preparation.
20. G. Gallavotti and J. L. Lebowitz, *J. Math. Phys.* **12**:1129 (1971).
21. G. Gallavotti and S. Miracle-Solé, *Phys. Rev. B* **5**:2555 (1972).
22. J. Glimm, A. Jaffe, and T. Spencer, *Comm. Math. Phys.* **45**:203 (1975).
23. F. Guerra, L. Rosen, and B. Simon, *Ann. Math.* **101**:111 (1975).
24. P. Hartman and G. Stampacchia, *Acta Math.* **115**:271 (1966).
25. O. J. Heilmann, *Comm. Math. Phys.* **36**:91 (1974); *Lett. Nuovo Cimento* **3**:95 (1972).
26. O. J. Heilmann and E. H. Lieb, *J. Stat. Phys.*, to appear.
27. O. J. Heilmann and E. Praestgaard, *J. Stat. Phys.* **9**:23 (1973); *J. Phys. A* **7**:1913 (1974).
28. W. Holsztyński and J. Slawny, unpublished (1976), and in preparation.
29. F. Hund, *Z. Physik* **94**:11 (1935).
30. R. B. Israel, in *Statistical Physics—'Statphys 13,'* D. Cabib, C. G. Kuper, and I. Riess, eds., *Ann. Israel Phys. Soc.* **2**:528 (1978).

31. Q. C. Johnson and D. H. Templeton, *J. Chem. Phys.* **34**:2004 (1961).
32. D. Kim and C. J. Thompson, *J. Phys. A* **9**:2097 (1976).
33. E. H. Lieb, *Phys. Rev.* **18**:692 (1967).
34. E. H. Lieb, *Phys. Rev.* **18**:108 (1967).
35. E. H. Lieb, *Phys. Rev.* **19**:103 (1967).
36. E. H. Lieb, in *Springer Lecture Notes in Physics*, No. 80 (1978), pp. 59–67.
37. E. H. Lieb and F. Y. Wu, in *Phase Transitions and Critical Phenomena*, C. Domb and M. S. Green, eds. (Academic Press, 1972), Vol. 1, p. 331.
38. E. Madelung, *Phys. Z.* **19**:524 (1978).
39. V. A. Malyshev, *Comm. Math. Phys.* **40**:75 (1975).
40. N. D. Mermin, *J. Math. Phys.* **8**:1061 (1967).
41. R. Peierls, *Proc. Camb. Phil. Soc.* **32**:477 (1936).
42. S. A. Pirogov and Ja. G. Sinai, *Func. Anal. Appl.* **8**:21 (1974).
43. S. A. Pirogov and Ja. G. Sinai, *Theor. Math. Phys.* **25**:1185 (1975).
44. S. A. Pirogov and Ja. G. Sinai, *Theor. Math. Phys.* **26**:39 (1976).
45. Y. Sakamoto and U. Takahasi, *J. Chem. Phys.* **30**:337 (1959).
46. M. Schick and R. B. Griffiths, *J. Phys. A* **10**:2123 (1977).
47. E. Seiler and B. Simon, *Ann. Phys.* **97**:470 (1976).
48. B. Simon, in *Mathematics of Contemporary Physics*, R. Streater, ed. (Academic Press, 1972), pp. 18–76.
49. B. Simon, in *Statistical Physics—'Statphys 13,'* D. Cabib, C. G. Kuper, and I. Riess, eds., *Ann. Israel Phys. Soc.* **2**:287 (1978).
50. J. Slawny, *Comm. Math. Phys.* **35**:297 (1974).
51. B. Sutherland, *Phys. Rev. Lett.* **19**:103 (1967).
52. M. P. Tosi, *Solid State Phys.* **16**:1 (1970).
53. J. W. Vick, *Homology Theory; An Introduction to Algebraic Topology* (Academic Press, 1973).
54. P. P. Ewald, *Ann. Phys.* **64**:253 (1921).
55. H. M. Evjen, *Phys. Rev.* **39**:675 (1932).
56. F. C. Frank, *Phil. Mag.* **41**:1287 (1950).
57. M. H. Cohen and F. Keffer, *Phys. Rev.* **99**:1128 (1955).
58. E. Sperner, *Abh. Math. Sem. Univ. Hamburg* **6**:265 (1928).

*Journal of Statistical Physics, Vol. 20, No. 6, 1979*

# Lattice Models for Liquid Crystals

**Ole J. Heilmann[1] and Elliott H. Lieb[2]**

*Received November 28, 1978*

A problem in the theory of liquid crystals is to construct a model system which at low temperatures displays long-range orientational order, but not translational order in all directions. We present five lattice models (two two-dimensional and three three-dimensional) of hard-core particles with attractive interactions and prove (using reflection positivity and the Peierls argument) that they have orientational order at low temperatures; the two-dimensional models have no such ordering if the attractive interaction is not present. We cannot prove that these models do not have complete translational order, but their zero-temperature states are such that we are led to conjecture that complete translational order is always absent.

**KEY WORDS:** Lattice models; liquid crystals; phase transitions.

## 1. INTRODUCTION

Since the work of Onsager,(1) hard-rod models have been widely used to explain the existence of liquid crystals, especially the transition from an isotropic liquid phase to a liquid crystal. DiMarzio(2) applied the Bethe–Guggenheim approximation to the model of hard rods of finite length on a cubic lattice, and showed that in this approximation one would obtain a phase with orientational ordering at sufficiently high density. It is doubtful, however, whether hard rods on a cubic lattice without any additional interaction do indeed undergo a phase transition (see de Gennes,(3) in particular Fig. 2.6). The only rigorously known result is the absence of phase transitions for dimers (Heilmann and Lieb(4)).

It is possible to apply Peierls' argument to prove that hard rods on lattices

Work of EHL supported by U.S. National Science Foundation Grant MCS 75-21684 A02. Financial assistance from the Danish Natural Science Research Council is also gratefully acknowledged.

[1] Department of Chemistry, H. C. Ørsted Institute, University of Copenhagen, Copenhagen, Denmark.

[2] Departments of Mathematics and Physics, Princeton University, Princeton, New Jersey.



0022-4715/79/0600-0679$03.00/0 

undergo phase transitions when subjected to more artificial restrictions (Lebowitz and Gallavotti,[5] Heilmann[6]) or to appropriate interactions between the rods (Runnels and Freasier,[7] Heilmann and Præstgaard[8]). A characteristic of all these models, however, is that their ordered states have the configuration at every vertex completely specified. Thus, these states also have complete crystalline (i.e., translational) ordering and are not, therefore, like liquid crystals.

The purpose of this paper is to present five lattice models (defined in Section 2) which we believe have a phase transition from a liquid to a liquid crystal. (These results were announced in Ref. 17.) We shall prove that these models do have a phase transition from an isotropic to an anisotropic phase. At present we are not able to demonstrate rigorously that the anisotropic phase is not crystalline, but, in contrast to the above-mentioned cases, the proof of orientational ordering that we give does not imply that the anisotropic phase is crystalline. In the discussion at the end of this paper we shall give some additional arguments for the conjecture that the anisotropic phase of these models does not possess *complete translational ordering*. (Complete translational ordering means that translational invariance is broken in all directions, as in a crystal.) For model I it is possible to go a bit further. By combining the results of this paper with a Kirkwood–Salzburg equation analysis, it can be shown that the conjecture is true at sufficiently low temperature. A paper by Heilmann and Kjær[9] will present the result.

In three of the models, dimers are placed on either a quadratic or a cubic lattice, while in the two other models the molecules are quadratic fourmers placed on a cubic lattice. In all five cases we include attractive forces which stabilize the anisotropic states without forcing the formation of a crystal. This combination of repulsive forces (i.e., steric hindrance) and attractive forces may be important for the formation of real liquid crystals.

In Section 3 we prove that these models possess a *reflection positivity*, a concept which goes back to the work of Osterwalder and Schrader[10] in field theory and which has recently been used in lattice statistical mechanical models by Frohlich *et al.*[11] and Dyson *et al.*[12] For the general theory of reflection positivity in statistical mechanics, see Ref. 18.

In Section 4 we show how reflection positivity can be used to prove that empty vertices are unlikely at high enough dimer fugacity and at low enough temperature. In Section 5 we define the contours which are needed for a Peierls argument[13] and extend the argument of Section 4 to prove that contours are also unlikely. Finally, in Section 6 the results of Sections 4 and 5 are combined with Peierls' argument to give a simple proof of the existence of an anisotropic phase. The combination of Peierls' argument and reflection positivity was first used in field theory by Glimm *et al.*[14] and in statistical mechanics by Frohlich and Lieb.[15] Unfortunately, reflection positivity cannot be extended (at least not at present) to systems in which the molecules are longer than dimers. Therefore the proof given here cannot be generalized to larger molecules, although we believe, of course, that the phase transition persists if the molecule length is increased from two to any larger value. All these models have the property of no ordering of any kind at high temperatures. This can be seen by standard high-temperature expansion methods. Therefore, all these models have a phase transition.

## 2. THE MODELS

In models I and II we place hard dimers on a (two-dimensional) quadratic lattice; i.e., a vertex of the lattice is either empty or covered by at most one dimer. A dimer covers two vertices that are connected by an edge. In model I there is a contribution $-a$ to the energy for each pair of neighboring, collinear dimers (see Fig. 1). In model II there is a contribution $-b$ to the energy for each edge of the lattice that has both its endpoints covered by dimers perpendicular to the edge (see Fig. 2).

In model III we place quadratic fourmers on a (three-dimensional) cubic lattice, again with the rule that a vertex is either empty or covered by at most one fourmer; a fourmer covers four vertices, which together constitute a minimal square of the cubic lattice, so that there are three orientations of fourmers altogether. An attractive energy $-a$ is included between coplanar fourmers that occupy two pairs of neighboring vertices in the cubic lattice (see Fig. 3). Model IV is identical to model III except that the energy is $-b$ for each edge that connects vertices covered by (distinct) coplanar fourmers (see Fig. 4).

Finally, model V is the three-dimensional version of model II, i.e., we

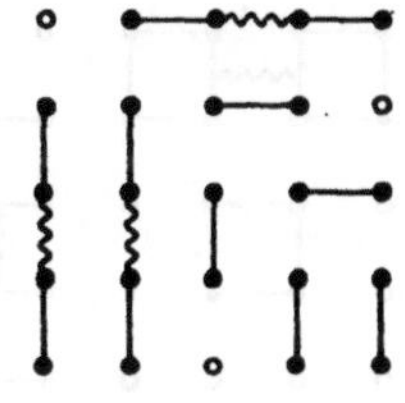

**Fig. 1. A dimer arrangement on a subset of the square lattice. Uncovered vertices are shown as open circles. The attractive interactions corresponding to model I are shown as wavy lines.**

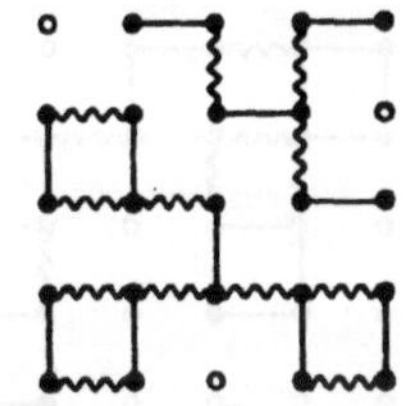

**Fig. 2. The same as Fig. 1 but with the attractive interactions corresponding to model II shown as wavy lines.**

place hard dimers on a cubic lattice and add $-b$ to the energy for each edge that connects vertices covered by parallel dimers perpendicular to the edge. Models II and V have been treated in the DiMarzio approximation by Cotter and Martire.[16]

In addition to the above-mentioned energies we also add a chemical potential term to the Hamiltonian (i.e., we use the grand canonical ensemble in which the particle number is not fixed):

$$-\mu N_d$$

where $N_d$ is the number of particles (dimers or fourmers.)

For simplicity of exposition we confine our attention in the following sections to the two-dimensional systems; the extension to three dimensions is obvious and the relevant results will be stated at the end of the relevant sections.

In order to have a convenient notation we introduce a coordinate system in the plane of the quadratic lattice such that the vertices of the lattice are the points with integer coordinates ($\mathbb{Z}^2$). The reflection positivity only holds for finite systems if we have periodic boundary conditions and an even number of vertices in each direction; consequently we take as our domain $\Lambda$ a rectangular subset of size $2N \times 2M$:

$$\Lambda = \{(x, y):\quad x = 0, 1, \ldots, 2N - 1, \quad y = 0, 1, \ldots, 2M - 1\}$$

and compute $x$-coordinates modulo $2N$ onto $0 \leqslant x < 2N$ and $y$-coordinates modulo $2M$ onto $0 \leqslant y < 2M$ whenever necessary, i.e., $\Lambda$ is wrapped on a torus. (In three dimensions we take $0 \leqslant z < 2L$.)

A dimer placed on the quadratic lattice is identified by the position, in Cartesian coordinates, of its midpoint, i.e., $(x + 1/2, y)$ for the dimer that

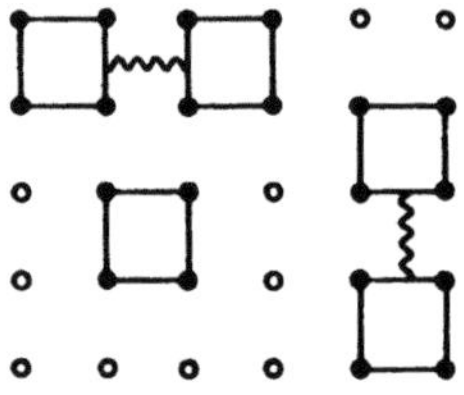

Fig. 3. An arrangement of coplanar fourmers. Uncovered vertices are shown as open circles. The attractive interactions corresponding to model III are shown as wavy lines.

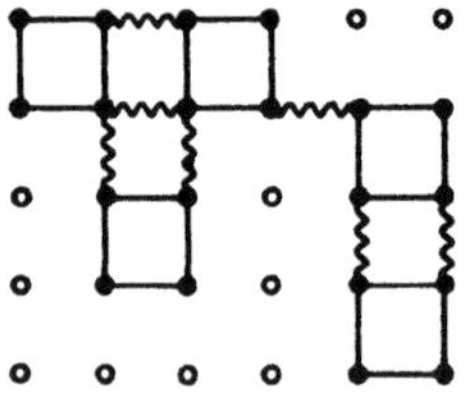

Fig. 4. The same as Fig. 3, but with the attractive interactions corresponding to model IV shown as wavy lines.

covers the vertices $(x, y)$ and $(x + 1, y)$ and $(x, y + 1/2)$ for the dimer that covers $(x, y)$ and $(x, y + 1)$. The former are called *horizontal dimers* and the latter *vertical dimers* in order to distinguish between the two possible orientations. The set of all possible dimer positions is denoted $\mathscr{B}$. Sometimes we will call $\mathscr{B}$ the "edges of $\Lambda$."

A dimer arrangement on $\Lambda$ is an allowed configuration of dimers on the edges of $\Lambda$ (i.e., without multiple occupancy of a vertex). The set of all possible dimer arrangements on $\Lambda$ (including the empty one) will be denoted $\mathscr{D}$. A convenient way to describe $\mathscr{D}$ is the following: To each of the $8NM$ possible dimer positions in $\mathscr{B}$ one attaches a copy of the two point space $\{0, 1\}$ (i.e., a variable which is either zero or one). Then $\mathscr{D}$ can be identified in a natural way with a subset of $\mathscr{C} = \{0, 1\}^{8NM}$, letting 1 correspond to the presence and 0 to the absence of a dimer. We call $\mathscr{C}$ the *phase space* and introduce the characteristic function $\chi$ defined for all $\xi \in \mathscr{C}$ by

$$\chi(\xi) = \begin{cases} 1 & \text{if} \quad \xi \in \mathscr{D} \\ 0 & \text{if} \quad \xi \notin \mathscr{D} \end{cases} \tag{1}$$

with the above-mentioned identification between $\mathscr{C}$ and dimer configurations being implicitly understood in (1).

## 3. REFLECTION POSITIVITY

We first introduce the reflection lines: Let $j$ be an integer satisfying $0 \leqslant j \leqslant N - 1$ and consider the pair of "vertical" lines in the plane

$$\begin{aligned} L_j^- &= \{(j + \tfrac{1}{2}, y): \ y \in \mathbb{R}\} \\ L_j^+ &= \{(j + N + \tfrac{1}{2}, y): \ y \in \mathbb{R}\} \end{aligned} \tag{2}$$

Let $\mathscr{B}_j^0$ be the (horizontal) dimer positions that are in $L_j^- \cup L_j^+$ (i.e., $\mathscr{B}_j^0 = \{(x, y): x = j + 1/2 \text{ or } j + N + 1/2,\ y = 0, 1,..., 2M - 1\}$) and let $\mathscr{B}_j^+$ (resp. $\mathscr{B}_j^-$) be the dimer positions to the right of $L_j^-$ and to the left of $L_j^+$ (resp. to the right of $L_j^+$ or the left of $L_j^-$). The dimer positions $\mathscr{B}$ are thus partioned into three sets, $\mathscr{B}_j^0$, $\mathscr{B}_j^+$, and $\mathscr{B}_j^-$, where the first set contains $4M$ points and the two latter contain $2M(2N - 1)$ points each. The phase space $\mathscr{C}$ is then given by

$$\mathscr{C} = \mathscr{C}_j^0 \times \mathscr{C}_j^+ \times \mathscr{C}_j^-$$

where $\mathscr{C}_j^0 = \{0, 1\}^{4M}$ corresponds to the dimer configurations on $\mathscr{B}_j^0$, etc.

There is a natural involution $\theta_j$ of $\Lambda$ onto $\Lambda$ obtained by reflection

$$\theta_j: \ (x, y) \to (2j + 1 - x, y) \tag{3}$$

The reflection $\theta_j$ maps $\mathscr{B}_j^+$ onto $\mathscr{B}_j^-$ and $\mathscr{B}_j^-$ onto $\mathscr{B}_j^+$, while it is the identity on $\mathscr{B}_j^0$. If lifts to an involution of $\mathscr{C}$ onto $\mathscr{C}$ mapping $\mathscr{C}_j^+$ onto $\mathscr{C}_j^-$ and $\mathscr{C}_j^-$ onto $\mathscr{C}_j^+$ and leaving $\mathscr{C}_j^0$ invariant. We shall use $\theta_j$ as a symbol for all these maps.

We also consider the natural action of $\theta_j$ as an involution on (complex-valued) functions on $\mathscr{C}$. If $f$ is such a function, then $\theta_j f$ is the function given by

$$(\theta_j f)(\xi) = f(\theta_j \xi), \qquad \xi \in \mathscr{C} \tag{4}$$

By $\mathscr{F}_j^+$ we denote the functions that depend only on the $\mathscr{C}_j^0$ and $\mathscr{C}_j^+$ coordinate, i.e., a point $\xi \in \mathscr{C}$ is a triplet $\xi = (\xi^-, \xi^0, \xi^+)$ with $\xi_j^i \in \mathscr{C}^i$, and $f \in \mathscr{F}_j^+$ if $f(\xi^-, \xi^0, \xi^+)$ is independent of $\xi^-$. Similarly $\mathscr{F}_j^-$ denotes the functions that are independent of $\mathscr{C}_j^+$. Clearly $\theta_j \mathscr{F}_j^+ = \mathscr{F}_j^-$. A function that depends only on $\mathscr{C}_j^0$ is in both $\mathscr{F}_j^+$ and $\mathscr{F}_j^-$.

**Lemma 1.** Let $f \in \mathscr{F}_j^+$. Then

$$S = \sum_{\xi \in \mathscr{C}} \overline{f(\xi)}(\theta_j f)(\xi) \geqslant 0$$

*Proof.* Write $\xi \in \mathscr{C}$ as $\xi = (\xi^-, \xi^0, \xi^+)$. Then

$$f(\xi^-, \xi^0, \xi^+) = g(\xi^0, \xi^+)$$

and

$$S = \sum_{\xi^0 \in \mathscr{C}_j^0} \sum_{\xi^- \in \mathscr{C}_j^-} \sum_{\xi^+ \in \mathscr{C}_j^+} \bar{g}(\xi^0, \xi^+) g(\xi^0, \xi^-)$$

For each $\xi^0$

$$\sum_{\xi^+ \in \mathscr{C}_j^+} g(\xi^0, \xi^+) = \sum_{\xi^- \in \mathscr{C}_j^-} g(\xi^0, \xi^-)$$

so $S \geqslant 0$. ∎

We now turn our attention to the dimer arrangement space $\mathscr{D} \subset \mathscr{C}$ and to the Hamiltonian $H$ of the dimer system. If $f$ is any (complex-valued) function on $\mathscr{D}$, it can be extended to a function on $\mathscr{C}$ in many ways. However, $\chi f$, where $\chi$ is given by (1), has a natural extension, i.e.,

$$(\chi f)(\xi) = \begin{cases} f(\xi), & \xi \in \mathscr{D} \\ 0, & \xi \notin \mathscr{D} \end{cases} \tag{5}$$

$\chi$ itself can be written as

$$\chi = \chi_j^- \chi_j^+$$

with $\chi_j^+ \in \mathscr{F}_j^+$, $\chi_j^- = \theta_j \chi_j^+$, and $\chi^\pm$ take on the values 1 or 0.

The Hamiltonians discussed in Section 2 can all be written in the following canonical way: For $\xi = (\xi^-, \xi^0, \xi^+) \in \mathscr{D}$, $\xi^i \in \mathscr{C}_j^i$,

$$H(\xi) = k(\xi^0) + h(\xi^0, \xi^+) + (\theta_j h)(\xi^0, \xi^-) - \sum_i g_i(\xi^0, \xi^+)(\theta_j g_i)(\xi^0, \xi^-) \tag{6}$$

$h(\xi^0, \xi^+)$ [resp. $(\theta_j h)(\xi^0, \xi^-)$] is the interaction energy (including the chemical potential) of all the dimers on $\mathscr{B}_j^+ \cup \mathscr{B}_j^0$ [resp. $\mathscr{B}_j^- \cup \mathscr{B}_j^0$]. Here $-k(\xi^0)$ is the interaction energy (if any) of the dimers on $\mathscr{B}_j^0$ together with their chemical potential. The term $k$ is included in $H$ in order that the energy of $\mathscr{B}_j^0$ not be counted twice. The only relevant fact about these functions is that they are real. The sum in (6) is the interaction of the dimers on $\mathscr{B}_j^+$ with those on $\mathscr{B}_j^-$. The functions $g_i$ are real and nonnegative; the minus sign in front of the summation expresses the fact that the interaction is *attractive*. The minus sign is crucial. The sum on $i$ will have $4M$ terms in the simple nearest neighbor case, but for the purpose of the following Theorem 2 it could have more terms if more than nearest neighbor term were present.

The partition function of the system of interacting dimers at the temperature $T = \beta^{-1}$ is given by

$$Z = \sum_{\xi \in \mathscr{D}} \exp[-\beta H(\xi)] \tag{7}$$

If $f$ is any (complex-valued) function on $\mathscr{D}$, then its expectation value is

$$\langle f \rangle = Z^{-1} \sum_{\xi \in \mathscr{D}} f(\xi) \exp[-\beta H(\xi)] \tag{8}$$

**Theorem 2.** Let $f$ be a (complex-valued) function on $\mathscr{D}$ and $f \in \mathscr{F}_j^+$. Then

$$\langle \bar{f}(\theta_j f) \rangle \geq 0$$

*Proof.* For $\xi \in \mathscr{D}$ let $G(\xi) = \exp[-\beta H(\xi)]$. Consider $\tilde{G} = \chi G$ as a function on $\mathscr{C}$ [see (5)]. Clearly $\tilde{G} = G_j^+ G_j^- G_j'$, where

$$G_j^+(\xi^-, \xi^0, \xi^+) = \chi_j^+(\xi^0, \xi^+) \exp\{-\beta[\tfrac{1}{2}k(\xi^0) + h(\xi^0, \xi^+)]\}$$
$$G_j^- = \theta_j G_j^+$$
$$G_j' = \chi(\xi) \exp\left[\beta \sum_i g_i(\xi^0, \xi^+)(\theta_j g_i)(\xi^0, \xi^-)\right]$$

and $G_j^+ \in \mathscr{F}_j^+$. We can make a power series expansion of the exponential in $G_j'$ and thereby obtain for $\tilde{G}$ a sum of terms, each with a positive coefficient. Each term is clearly of the form

$$\chi(\xi)\mu(\xi^0, \xi^+)(\theta_j \mu)(\xi^0, \xi^-) = \rho(\xi^0, \xi^+)(\theta_j \rho)(\xi^0, \xi^-)$$

where $\rho = \chi_j^+\mu$. Thus, $\tilde{G}$ is a sum (with positive coefficients) of terms of the form $\phi(\xi^0, \xi^+)(\theta_j\phi)(\xi^0, \xi^-)$ with $\phi \in \mathscr{F}_j^+$ and $\phi$ real. Theorem 2 now follows from Lemma 1. ∎

**Corollary 3.** Let $f$ and $g$ be (complex-valued) functions on $\mathscr{D}$ with $f \in \mathscr{F}_j^+$ and $g \in \mathscr{F}_j^-$. Then

$$|\langle fg\rangle|^2 \leqslant \langle \bar{f}(\theta_j f)\rangle\langle \bar{g}(\theta_j g)\rangle$$

*Proof.* Standard Schwarz inequality. ∎

Corollary 3 is the desired goal of this section. For Models I and II the decomposition of $H$ in the form (6) holds for every $j = 0, 1,..., N-1$ and *also* for every pair of *horizontal* lines through the center of the vertical edges and separated by a distance $M$.

For the three three-dimensional models on a lattice $2N \times 2M \times 2L$ we have, by a similar argument, reflection positivity (i.e., the analog of Corollary 3) for reflection through pairs of *planes* separated by $N$ or $M$ or $L$.

## 4. A LOWER BOUND ON THE DIMER DENSITY

As the first application of reflection positivity we shall show how a useful upper bound on the density of empty vertices can be obtained. In the two-dimensional case let $(x, y) \in \Lambda$ be a vertex and let $P_{(x,y)}$ be the projection onto the configurations in which there is no dimer on $(x, y)$, i.e., for $\xi \in \mathscr{D}$

$$P_{(x,y)} = \begin{cases} 0 & \text{if there is a dimer on } (x, y) \text{ in } \xi \\ 1 & \text{if there is no dimer on } (x, y) \text{ in } \xi \end{cases} \tag{9}$$

In conformity with the convention in Section 3, $P_{(x,y)}$ should really be thought of as a function on the bonds incident on $(x, y)$. By translational invariance

$$1 - 2\rho = \langle P_{(x,y)}\rangle$$

is independent of $(x, y)$ ($\rho$ is the dimer density).

If, in Corollary 3, we take $f = P_{(x,y)}$ and $g = 1$ (assuming $j + 1 \leqslant x \leqslant j + N$} we obtain ($w = 2j + 1 - x + 2N$)

$$\langle P_{(x,y)}\rangle^2 \leqslant \langle P_{(x,y)}P_{(w,y)}\rangle\langle 1\rangle$$

so that

$$1 - 2\rho = \langle P_{(x,y)}\rangle \leqslant \langle P_{(x,y)}P_{(w,y)}\rangle^{1/2}$$

Our goal in the next theorem is to extend this argument to every vertex in $\Lambda$, not just two. The advantage of this is clear: instead of having to estimate

$\langle P_{(x,y)}\rangle$, i.e., the expectation value of a local quantity, we need only estimate the expectation value of a global quantity, which is much easier.

**Theorem 4.** Let $T$ be the set of projections of the form $P = P_{z1}P_{z2}\cdots P_{zk}$, where $z1, z2,..., zk$ are distinct vertices in $\Lambda$ and $1 \leqslant k \leqslant 4NM$; $|P| = k$. Then

$$\max_{P\in T} \langle P\rangle^{1/P} = \langle \hat{P}\rangle^{1/4MN}$$

where $\hat{P} = \prod_{z\in\Lambda} P_z$. In particular

$$1 - 2\rho \leqslant \langle \hat{P}\rangle^{1/4MN} \tag{10}$$

*Proof.* Define $f(P) = \langle P\rangle^{1/|P|}$ for $P \in T$ and $f(I) = 1$. (Here, $I$ is the unit operator, i.e., the unit function on configurations, $|I| = 0$.) Let $t \subset T$ be the set of $P$'s that maximize $f(P)$. If $\theta$ is a reflection operator and $P \in T$, then we can write $P = QR$, where $Q \in \mathscr{F}^-$ and $R \in \mathscr{F}^+$. Now, $Q$ or $R$ might be $I$, but not both; $Q$ and $R$ belong to $T$ except if they are $I$. Let $\tilde{Q} = \theta Q$, $\tilde{R} = \theta R$; then $|\tilde{Q}| = |Q|$, $|\tilde{R}| = |R|$, and $|Q| + |R| = |P|$. By Corollary 3

$$\langle P\rangle^2 \leqslant \langle Q\tilde{Q}\rangle\langle R\tilde{R}\rangle$$

so

$$f(P) \leqslant f(Q\tilde{Q})^{|Q|/|P|} f(R\tilde{R})^{|R|/|P|} \tag{11}$$

This holds even if $R = I$. From (11) we see that if $R \neq I$, $P = QR$, and $P \in t$, then $R\tilde{R} \in t$ also. Let $P$ be any element in $t$ ($t$ cannot be empty since $T$ is finite) and suppose that $P_{(x,y)}$ is a factor in $P$. Reflect in the vertical line through $(x + 1/2, y)$, then in the line through $(x + 3/2, y)$, then in the line through $(x + 5/2, y)$, etc. In this way we eventually obtain a $P' \in t$ containing a horizontal chain of $P_z$'s of length $2N$ (namely $P_{(i,y)}$, $i = 0, 1,..., 2N - 1$). If we then reflect $P'$ in the horizontal lines with ordinates $y + 1/2$, $y + 3/2$, $y + 5/2$, etc., we eventually obtain $\hat{P} \in t$. ■

Next we use Theorem 4 to obtain an upper bound on $1 - 2\rho$. Let $E_0$ be the energy of a close-packed configuration with the lowest possible energy and $\epsilon_0$ be the corresponding energy per vertex (i.e., $\epsilon_0 = E_0/4NM$ in two dimensions and $\epsilon_0 = E_0/8NML$ in three dimensions); then we have the following result:

**Theorem 5**

$$1 - 2\rho \leqslant \exp(\beta\epsilon_0)$$

*Proof.* By considering only a term with energy $E_0$ in $Z$ [Eq. (7)], we find $Z \geqslant \exp(-\beta E_0)$. If $\hat{P}$ is as in Theorem 4, then the numerator in (8) for $\langle\hat{P}\rangle$

contains only the term with no dimers; this term is one. Thus $\langle P \rangle = Z^{-1}$ and Theorem 5 follows from Theorem 4. ■

The following are the values of $\epsilon_0$ in the five models mentioned in Section 2:

$$\begin{aligned} \epsilon_0(\mathrm{I}) &= -\mu/2 - a/2, & \epsilon_0(\mathrm{II}) &= -\mu/2 - b \\ \epsilon_0(\mathrm{III}) &= -\mu/4 - a/2, & \epsilon_0(\mathrm{IV}) &= -\mu/4 - b \\ \epsilon_0(\mathrm{V}) &= -\mu/2 - 2b \end{aligned} \tag{12}$$

Clearly, if $\mu$ and $a$ (resp. $b$) are chosen such that $\epsilon_0 < 0$, then $1 - 2\rho$ goes to zero as $T \to 0$, i.e., all vertices become occupied by dimers.

## 5. ESTIMATES FOR THE PROBABILITY OF BAD SQUARES

In the previous section reflection positivity was used to provide an upper bound for the probability of empty vertices. Here we extend the argument to somewhat more complex events called "bad squares." These estimates will be used in the next section, where a Peierls-type argument will be given to show that there is long-range order at low temperatures.

In two dimensions, let $S_z$ denote an elementary square of the lattice centered at $z = (l + 1/2, m + 1/2)$, $0 \leqslant l < 2L$, $0 \leqslant m < 2M$, and consisting of the four vertices $(l, m)$, $(l + 1, m)$, $(l, m + 1)$, and $(l + 1, m + 1)$. Henceforth, for simplicity, we will assume $L$ and $M$ are even. There are four classes of squares, labeled $\sigma_{ee}$, $\sigma_{eo}$, $\sigma_{oe}$, $\sigma_{oo}$, according as $l$ (resp. $m$) is even or odd, and there are $ML$ squares in each class.

The extensions to three dimension is obvious. $S_z$ is then a "cube" of six vertices centered at $(l + 1/2, m + 1/2, n + 1/2)$ with $0 \leqslant l < 2L$, $0 \leqslant m < 2M$, $0 \leqslant n < 2N$. The $L$, $M$, and $N$ are assumed to be even. There are eight equivalence classes: $\sigma_{eee}, \ldots, \sigma_{ooo}$ with $LMN$ cubes in each.

Fix $z$ and consider a configuration $\xi$. We say that $S_z$ is "bad" if and only if $\xi$ restricted to $S_z$ is not compatible with a ground state. Specifically this means: $S_z$ is bad if and only if there is one or more empty vertice on $S_z$ or else the dimers (resp. fourmers) on the vertices of $S_z$ are not all of the same orientation. [Note: The dimers (fourmers) on $S_z$ refer to all the dimers (fourmers) incident on the vertices of $S_z$, including those that connect vertices of $S_z$ to some other vertex.]

Let $Q_z$ be the projection onto configurations in which $S_z$ is "bad." Thinking of $Q_z$ as a function on dimer configurations (in two dimensions), the support of $Q_z$ is the 12 bonds connected to $S_z$. The three-dimensional analogue is obvious.

If $A$ denotes a nonempty subset of square (cube) midpoints, and $|A|$ the number of its elements, let

$$g = \max_{A \neq \varnothing} \left\langle \prod_{z \in A} Q_z \right\rangle^{1/|A|} \tag{13}$$

Our aim is to show that $g \to 0$ as $\beta \to \infty$ (exponentially fast, in fact), uniformly in $L$ and $M$. We will show:

**Lemma.** If $Q_z$ is the projection onto a bad square (resp. cube) centered at $z$ and if $g$ is given by (13), then

$$g \leqslant ce^{-\beta\alpha}$$

with $c = \{4$ (in 2 dim.), 8 (in 3 dim.)$\}$ and $\alpha > 0$. Here $\alpha$ depends on the model and its parameters and is given by the following (note $\alpha > 0$ as long as $\epsilon_0 < 0$):

| | |
|---|---|
| model 1 | $\alpha = \frac{1}{4}[\mu + a - \max(0, \mu)]$ |
| model 2 | $\alpha = \frac{1}{4}[\mu + 2b - \max(0, \mu + b)]$ |
| model 3 | $\alpha = \frac{1}{8}[\mu + 2a - 1/4 \max(0, \mu) - 1/4 \max(0, 3\mu + 4a)]$ |
| model 4 | $\alpha = \frac{1}{8}[\mu + 4b - 1/4 \max(0, \mu) - 1/4 \max(0, 3\mu + 8b)]$ |
| model 5 | $\alpha = \frac{1}{8}[2\mu + 8b - \frac{1}{2} \max(0, 3\mu + 10b) - \frac{1}{2} \max(0, \mu + 2b)]$ |

*Proof.* As a preliminary step, let $Q(A) = \prod_{z \in A} Q_z$. Write $Q(A) = Q_{ee}(A)Q_{eo}(A)Q_{oe}(A)Q_{oo}(A)$ [in three dimensions $Q(A) = Q_{eee}(A) \cdots Q_{ooo}(A)$]. Here

$$Q_{ee}(A) = \prod_{z \in A \cap \sigma_{ee}} Q_z$$

and so forth. Let $j$ denote the subscript $(eo)$, etc. Since $Q_j(A) \leqslant 1$, $\langle Q(A) \rangle \leqslant \langle Q_j(A) \rangle$ for any $j$. There is at least one $j$ such that $|A_j| \geqslant |A|/4$ ($|A|/8$ in three dimensions). Thus,

$$g \leqslant \max\{\langle Q(A) \rangle^{1/4|A|} : \quad \varnothing \neq A \subseteq \sigma_{ee}\} \tag{14}$$

(resp. $8|A|$ in three dimensions). In other words, (14) says that at the expense of a factor of 4 we can restrict attention to $A$'s such that all squares are in the same equivalence class. The arbitrary choice of $\sigma_{ee}$ is based on the symmetry with respect to the class.

There are $LM$ (resp. $LMN$) possible pairs of reflection lines (planes). Of these there are $LM/4$ (resp. $LMN/8$) which carry an $S_z$ into an $S_{z'}$ (and $Q_z$ into $Q_{z'}$) provided $z \in \sigma_{ee}$. In two dimensions these are the lines $\{(l + 1/2, y)$: $l$ odd, $y \in \mathbb{R}\}$ and $\{(x, m + 1/2)$: $m$ odd, $x \in \mathbb{R}\}$. Thus, by the same reasoning as for the empty vertex estimate (Section 4), a maximizing $A$ for (14) is $A = \sigma_{ee}$, i.e., $Q(A)$ is the projection onto configurations in which *every* $(ee)$

square is bad. Then $|A| = LM$. (In three dimensions $A = \sigma_{eee}$, $|A| = LMN$). Let $E$ be the minimum energy under the condition that every $(ee)$ square is bad. Then

$$g \leqslant (\# e^{-\beta E})^{1/4LM}/e^{-\beta\epsilon_0}$$

(resp. $8LMN$ in three dimensions), since $Z \geqslant \exp(-4\beta LM\epsilon_0)$ (resp. $8LMN$).

Here # is the total number of possible configurations. In two dimensions $\# \leqslant 2^{8LM}$ since every edge has a dimer, or it does not. ($\# \leqslant 2^{24LMN}$ in three dimensions).

The lemma is proved if we show $E/4LM$ (resp. $E/8LMN$) $- \epsilon_0 \equiv \alpha > 0$ Specifically,

$$\begin{aligned}
E &\geqslant -[\mu + a + \max(0,\mu)]LM && \text{model 1}\\
&\geqslant -[\mu + 2b + \max(0,\mu + b)]LM && \text{model 2}\\
&\geqslant -[\mu + 2a + 1/4\max(0,\mu) + 1/4\max(0,3\mu + 4a)]LMN && \text{model 3}\\
&\geqslant -[\mu + 4b + 1/4\max(0,\mu) + 1/4\max(0,3\mu + 8b)]LMN && \text{model 4}\\
&\geqslant -[2\mu + 8b + \tfrac{1}{2}\max(0,3\mu + 10b) + \tfrac{1}{2}\max(0,\mu + 2b)]LMN && \text{model 5}
\end{aligned}$$

We give the proof for model 1. The reader can easily do the other cases. The energy for a configuration $\xi$ can be thought of as a sum of vertex energies, namely 0 (for an empty vertex), $-\frac{1}{2}\mu$ for a dimer with an unsaturated end (absence of a wiggly line in the figures), and $-\mu/2 - a/2$ for a dimer with a saturated end. Now *every* vertex in $\Lambda$ belongs to exactly one square in $\sigma_{ee}$. The total vertex energy of a bad square is not less than 0 (all empty), $-\mu/2$ (three empty), $-\mu - a$ (two empty), $-3\mu/2 - a$ (one empty), $-2\mu - a$ (none empty). Thus, this square energy is not less than $-\mu - a - \max(0, \mu)$, and this leads to the $E$ given above. ■

## 6. THE PEIERLS ARGUMENT FOR LONG-RANGE ORDER

In Section 5 we showed that the probability of finding $k$ bad squares at arbitrary locations is not greater than $g^k$, with $g = c\exp(-\beta\alpha)$. Both $\alpha$ and $c$ depend on the model, but $\alpha < 0$ (since $\epsilon_0 < 0$ by hypothesis). Thus $g \to 0$ as $\beta \to \infty$. We will use this information in a Peierls-type argument to prove that if a dimer (fourmer) is placed at a fixed vertex (say $z = 0$), then the probability of finding a dimer of the *same* orientation at a vertex $z \neq 0$ is greater than 1/2 for large $\beta$, uniformly in $z$.

We will assume that the reader is familiar with the Peierls argument and will explain only the novel features for our liquid crystal models. In particular, we assume familiarity with the nonessential technical problems arising from the use of periodic boundary instead of free boundary conditions (contours that run around the torus), and that as $L$, $M$, $N \to \infty$ one needs a condition such as $\min(L, M, N)/\log \max(L, M, N) \to \infty$. It is also well known, and easy to prove, that our definition of long-range order implies other definitions of phase transitions, such as the existence of at least two Gibbs states or the existence of a spontaneous polarization or the discontinuity of the derivative free energy/unit volume with respect to an external polarizing field at zero field. References 15 and 18 can be consulted about some of these points.

With a dimer (fourmer) fixed at 0, we will show that the probability of *not* having an identically oriented object at $z \neq 0$ is less than 1/2 for large $\beta$.

Call the object (dimer or fourmer) at 0 an $h$-object (e.g., a horizontal dimer in models 1 and 2). Let $\xi$ be a configuration with an $h$-object at 0 and no $h$-object at $z$. A *plus vertex* is defined as one which (a) has an $h$-object and (b) belongs to at least one good (i.e., not bad) square (cube). Otherwise, the vertex is said to be *minus*.

Let $\Lambda^*$ be the lattice of midpoints of squares in $\Lambda$ together with the usual notion of nearest neighbor points. Two points in $\Lambda^*$ are said to be connected if and only if there is a connected path from one to the other. Squares in $\Lambda$ are said to be connected if and only if their midpoints are connected in $\Lambda^*$.

Suppose that in $\xi$ there are two nearest neighbor vertices, one of which is plus and the other minus. They have two squares (resp. four cubes) in common, *all of which must be bad*. The midpoints of these two squares (resp. four cubes) can be connected by a line (resp. square) which we call *a piece of contour*.

Now we refer to the usual Peierls argument for the Ising model. Since the origin is plus and $z$ is minus, there is a closed contour $\gamma$ in $\Lambda^*$ separating the two points. On one side of the contour there are plus vertices and on the other side there are minus vertices.

As we saw above, every vertex of the contour (in $\Lambda^*$) is the midpoint of a bad square. Hence

$$\text{Prob}(\gamma) < g^{|\gamma|}$$

with $|\gamma|$ being the length (area) of $\gamma$.

We sketch the remainder of the Peierls argument: There must be a closed contour surrounding 0 or surrounding $z$. If $P$ is the probability that there is no $h$-object at $z$, we then have

$$P \leqslant 2 \sum_{|\gamma| \geqslant 4} g^{|\gamma|} 3^{|\gamma|} (|\gamma|/4)^2$$

in two dimensions, and with a similar expression in three dimensions. $P \to 0$ as $\beta \to \infty$. Here $3^{|\gamma|}$ is an upper bound for the number of contours of length $|\gamma|$ and $(|\gamma|/4)^2$ is an upper bound to their area, i.e., to the number of ways they can be placed to enclose a given point.

## 7. CONCLUSION

We have demonstrated that at low enough temperatures (and not too negative a chemical potential) each of the five models presented in Section 2 will exist in a phase where one of the two (three) possible orientations is preferred over the other. The question of a possible positional ordering remains open. We conjecture that long-range positional ordering does not occur, for the following reasons (as exemplified for model I): Presumably, if there is no positional ordering for large $\beta$, there is none for any $\beta$. In the ground state all dimers are of one orientation, say horizontal. Thus, for large $\beta$, the system is like a product of uncorrelated one-dimensional systems because there is no interaction among the rows unless defects are present. These we have shown to be rare for large $\beta$. Along a column there is not likely to be long-range ordering because in one dimension any defects, however rare, will destroy ordering. However, this argument leads us to expect that the (finite) correlation lengths will be very different in the direction of the dimers and in the orthogonal direction, the former tending to infinity as $\beta \to \infty$.

A similar argument applies to the two fourmer models when one considers the ordering of fourmer positions among planes parallel to the preferred orientation of the planes of the fourmers. However, the situation is quite different if one considers possible ordering within a plane. The attractive interaction in model III clearly favors a complete ordering within a plane and it is easy to prove by Peierls' argument that the two-dimensional version (where one only has fourmers of one orientation) does exist in a completely ordered phase at low enough temperatures. This effect would be expected to carry over to the three-dimensional model, in which case it raises the intriguing question of whether the model can exist in six different phases.

In model IV the attractive interaction is designed to allow "sliding" of a row of fourmers relative to a neighboring row in the same plane and one would not expect the two-dimensional version to exhibit a phase transition (Nisbet and Farquhar[19]). This two-dimensional problem has not yet been resolved, however.

## REFERENCES


1. L. Onsager, *Ann. N.Y. Acad. Sci.* **51**:627 (1949).
2. E. A. Di Marzio, *J. Chem. Phys.* **35**:658 (1961).
3. P. G. de Gennes, *The Physics of Liquid Crystals* (Clarendon Press, Oxford, 1974).
4. O. J. Heilmann and E. H. Lieb, *Phys. Rev. Lett.* **24**:1412 (1970); *Comm. Math. Phys.* **25**:190 (1972).
5. J. L. Lebowitz and G. Galavotti, *J. Math. Phys.* **12**:1129 (1971).
6. O. J. Heilmann, *Lett. Nuovo Cim.* **3**:95 (1972).
7. L. K. Runnels and B. C. Freasier, *Comm. Math. Phys.* **32**, 191 (1973).

8. O. J. Heilmann and E. Præstgaard, *Chem. Phys.* **24**:119 (1976).
9. O. J. Heilmann and K. H. Kjær, in preparation.
10. Osterwalder and Schrader, *Helv. Phys. Acta* **46**:277 (1973); *Comm. Math. Phys.* **31**:83 (1973).
11. J. Frohlich, B. Simon, and T. Spencer, *Comm. Math. Phys.* **50**:79 (1976).
12. F. Dyson, E. H. Lieb, and B. Simon, *Phys. Rev. Lett.* **37**:120 (1976); *J. Stat. Phys.* **18**:335 (1978).
13. R. Peierls, *Proc. Camb. Phil. Soc.* **32**:477 (1936).
14. J. Glimm, A. Jaffe, and T. Spencer, *Comm. Math. Phys.* **45**:203 (1975).
15. J. Frohlich and E. H. Lieb, *Phys. Rev. Lett.* **38**:440 (1977); *Comm. Math. Phys.* **60**:233 (1978).
16. M. A. Cotter and D. E. Martire, *Mol. Cryst. Liq. Cryst.* **7**:295 (1969).
17. E. H. Lieb, in *Proceedings of the International Conference on the Mathematical Problems in Theoretical Physics*, Springer Lecture Notes in Physics, **80** (1978).
18. J. Frohlich, R. Israel, E. H. Lieb, and B. Simon, *Comm. Math. Phys.* **62**:1 (1978).
19. R. M. Nisbet and I. E. Farquhar, *Physica* **73**, 351 (1974).

With T. Kennedy and S. Shastry in J. Statist. Phys. *53*, 1019–1030 (1988)

*Journal of Statistical Physics, Vol. 53, Nos. 5/6, 1988*

# Existence of Néel Order in Some Spin-1/2 Heisenberg Antiferromagnets

**Tom Kennedy,[1] Elliott H. Lieb,[1] and B. Sriram Shastry[1,2]**

*Received May 31, 1988; revision received June 23, 1988*

The methods of Dyson, Lieb, and Simon are extended to prove the existence of Néel order in the ground state of the 3D spin-1/2 Heisenberg antiferromagnet on the cubic lattice. We also consider the spin-1/2 antiferromagnet on the cubic lattice with the coupling in one of the three lattice directions taken to be $r$ times its value in the other two lattice directions. We prove the existence of Néel order for $0.16 \leqslant r \leqslant 1$. For the 2D spin-1/2 model we give a series of inequalities which involve the two-point function only at short distances and each of which would by itself imply Néel order.

**KEY WORDS**: Néel order; spin-1/2 antiferromagnets; infrared bounds; Gaussian domination.

The existence or absence of Néel order in various spin-1/2 Heisenberg antiferromagnets is still unresolved after 50 years of study. Interest in this question has been revived recently in the context of certain models proposed for high-$T_c$ superconductors. In particular Anderson[(2)] suggested that the "resonating valence bond" (RVB) state, a state with strong antiferromagnetic correlations but without Néel order, is relevant in this context. There has been considerable subsequent debate on whether the 2D spin-1/2 Heisenberg antiferromagnet on the square lattice possesses Néel order. Recent numerical work of Liang *et al.*[(11)] shows that the energy of the RVB state is very close to the energy of Néel-like states, but other recent numerical work of Reger and Young[(13)] and Gross *et al.*[(7)] suggests the existence of Néel order. In view of the delicate nature of the question, it is important to review the rigorous results on the existence of Néel order

[1] Department of Physics, Princeton University, Princeton, New Jersey 08544.
[2] Permanent address: Tata Institute of Fundamental Research, Bombay, India.



0022-4715/88/1200-1019$06.00/0 

for quantum antiferromagnets and to extend them to the case of spin 1/2 in two and three dimensions.

The rigorous results on the existence of Néel order in quantum Heisenberg antiferromagnets are all based on the work of Dyson *et al.*,[3] who proved that there is Néel order at low temperatures if the spin is at least 1 and if the dimension is three or more, and for spin 1/2 if the dimension is sufficiently large. This work extended the results of Fröhlich *et al.*[5] for the classical Heisenberg model. In two dimensions the Mermin–Wagner–Hohenberg theorem states that there is no Néel order at low but nonzero temperatures, but the question of whether or not there is Néel order in the ground state is nontrivial. Jordaõ-Neves and Fernando-Perez[10] observed that the methods of Dyson *et al.* can be applied to the ground state in two dimensions. These methods show that the ground state of the two-dimensional Heisenberg antiferromagnet has Néel order if the spin is at least 1. (Because of a numerical error, they asserted the result only for spin greater than 1; this numerical oversight was corrected in ref. 1.) Two of the most interesting cases from the physical point of view, spin 1/2 in two and three dimensions, have thus far eluded rigorous results.

In this paper we show that a simple extension of the above methods proves the existence of Néel order for the case of spin 1/2 in three dimensions. We also consider a spin-1/2 model which interpolates between two and three dimensions. This model is the three-dimensional cubic lattice with the coupling constant in two of the three lattice directions taken to be 1, but in the third lattice direction it is taken to be $r$. When $r=0$ we recover the case of two dimensions, while $r=1$ yields three dimensions. For this model we prove that there is Néel order if $1 \geqslant r \geqslant 0.16$. (Although we only consider the ground states of these models, the techniques we use may be combined with the techniques of Dyson *et al.* for nonzero temperatures to prove the existence of a phase transition for $1 \geqslant r \geqslant 0.16$.) The case of spin 1/2 in two dimensions remains an open problem. Recently Gross *et al.*[7] have performed a quantum Monte Carlo simulation that shows that there is Néel order in the two-dimensional model. Of course one can argue that the two-point function does in fact decay to zero, but this decay is too slow to be seen in their simulation. We give a series of inequalities which involve the two-point function only at short distances, each of which by itself would imply Néel order.

This paper consists of two parts. In the first part we recall the basic inequality and strategy from ref. 3. We then show how to extend these methods to the case of spin 1/2 in three dimensions and to the model that interpolates between two and three dimensions. The second part of the paper provides a simple proof—directly adapted to the ground state—of the key inequality from ref. 3. This proof is included for the convenience of the reader. It can be skipped on a first reading of the paper.

To review the methods of Dyson *et al.* we consider a finite lattice $\Lambda$ with an even number of sites in every direction and periodic boundary conditions. We define the Fourier transform of $S_x^3$,

$$S_q = |\Lambda|^{-1/2} \sum_{x \in \Lambda} e^{-iq \cdot x} S_x^3$$

Here $q$ is in the reciprocal lattice. The Fourier transform of the two-point function in the ground state is then

$$g_q = \langle S_{-q} S_q \rangle \geqslant 0$$

where $\langle \cdot \rangle$ denotes expectation in the ground state. Note that the ground state is unique for the antiferromagnet.[12] Dyson *et al.* proved a pointwise upper bound on the analog of this function for nonzero temperatures. Their bound holds for all values of $q$ except $Q$. By $Q$ we denote either $(\pi, \pi)$ or $(\pi, \pi, \pi)$, depending on whether we are considering two or three dimensions. The zero-temperature limit of their inequality (as derived in ref. 10) in dimension $d$ is

$$g_q \leqslant f_q, \qquad q \neq Q \tag{1}$$

where $f_q = (e_0 E_q / 6dE_{q-Q})^{1/2}$, $E_q = \sum_{i=1}^d (1 - \cos q_i)$, and $-e_0$ is the ground-state energy per site. Note that $f_q$ depends on the spin $S$ only through the dependence of $e_0$ on $S$. Inequality (1) is called an infrared bound, and a direct proof of it will be given later in this paper.

For an antiferromagnet the existence of Néel order corresponds to $g_q$ containing a $\delta$ function at $Q$ in the infinite-volume limit. Let $m^2$ be the coefficient of this delta function. If we integrate $g_q$ (in the infinite-volume limit) over the Brillouin zone, we obtain the value of the two-point function at zero separation. The bound (1) then implies

$$m^2 + \int d^d q \, f_q \geqslant \int d^d q \, g_q = S(S+1)/3 \tag{2}$$

where

$$\int d^d q = (2\pi)^{-d} \int_0^{2\pi} dq_1 \cdots \int_0^{2\pi} dq_d$$

A simple argument by Anderson (reproduced in Appendix C of ref. 3) shows that $e_0 \leqslant S(dS + \frac{1}{2})$. Thus, inequality (2) forces $m^2$ to be nonzero if $S$

is large enough. Numerical evaluation of the resulting integrals shows that in both two and three dimensions $S=1$ is large enough.

To extend these methods, we make use of another piece of information about $g_q$, namely

$$\int d^dq\, g_q \cos q_i = \langle S_0^3 S_{\delta_i}^3 \rangle = -e_0/3d, \qquad i=1,2,3 \tag{3}$$

Here $\delta_i$ is the unit vector in the $i$ direction. (The factor of $1/3d$ appears because there are $d$ bonds per site in $d$ dimensions and the expectation of $S_x^3 S_y^3$ is one-third of the expectation of $\mathbf{S}_x \cdot \mathbf{S}_y$.) If there is no Néel order, i.e., there is no $\delta$-function at $q=Q$, then the upper bound (1) implies

$$e_0/3d \leqslant (e_0/6d)^{1/2} \int d^dq\, (E_q/E_{q-Q})^{1/2}\, d^{-1} \left( - \sum_{i=1}^{d} \cos q_i \right)_+ \tag{4}$$

where the positive part $F_+$ of a function $F$ equals $F$ when $F$ is positive and is zero otherwise.

Turning now to $d=3$, we can evaluate the integral in (4) numerically and we find that the right side equals $0.0824(e_0)^{1/2}$. Thus, (4) implies that $e_0 \leqslant 0.550$. However, taking the Néel state as a variational state shows that $-e_0$ is less than $-3/4$ for $d=3$. This contradiction shows there must be Néel order when $d=3$ and $S=1/2$.

In two dimensions the above argument only shows that $e_0 \leqslant 1.064$. The numerical estimates[7–9,11,13] of $e_0$ are all around 0.67, so we cannot conclude from inequality (4) that there is Néel order when $S=1/2$.

The model that interpolates between two and three dimensions is obtained by considering a three-dimensional lattice with the Hamiltonian

$$H = \sum_{\{xy\}} J_{xy} \mathbf{S}_x \cdot \mathbf{S}_y \tag{5}$$

where the coupling constant $J_{xy}$ equals 1 for bonds $\{xy\}$ in one of the first two coordinate directions and equals $r$ for bonds in the third coordinate direction. We will prove that this model has Néel order if $1 \geqslant r \geqslant 0.16$ and $S=1/2$.

Letting $g_q^r$ denote the Fourier transform of the two-point function for this model, we will show

$$0 \leqslant g_q^r \leqslant f_q^r, \qquad q \neq Q \tag{6}$$

where

$$\begin{aligned} f_q^r &= (e_0^r E_q^r / 12 E_{q-Q}^r)^{1/2} \\ E_q^r &= 2 - \cos q_1 - \cos q_2 + r(1 - \cos q_3) \end{aligned} \tag{7}$$

where $-e_0^r$ is the ground-state energy per site. We now consider the following mathematical problem. Assuming $m^2=0$, maximize $I=\int d^3q\, g_q$ over all functions $g_q$ subject to both inequality (6) and to

$$\int d^3q\, g_q(\cos q_1+\cos q_2+r\cos q_3)=-e_0^r/3 \tag{8}$$

which is the analogue of (3) when $r\neq 1$. If the maximum of $I$ is less than 1/4, then we have a contradiction, so $m^2$ must be nonzero, i.e., there must be Néel order.

We claim that the maximum of $I$ is attained either by

$$g_q=f_q^r\chi(\cos q_1+\cos q_2+r\cos q_3<\alpha) \tag{9a}$$

or

$$g_q=f_q^r\chi(\cos q_1+\cos q_2+r\cos q_3>-\alpha) \tag{9b}$$

for some $\alpha\geqslant 0$. The characteristic function $\chi(\cdot)$ equals 1 if the expression inside $(\cdot)$ is true and 0 otherwise. Which of the two cases (9a) and (9b) we must choose and the value of the parameter $\alpha$ are determined by the constraint (8). For a given value of $r$, we determine $\alpha$ by numerically computing the integral in (8) using (9). With this $\alpha$ we then compute $I$. We do not know the exact value of $e_0^r$, so we carry out this calculation for several values of $e_0$ ranging from the Néel bound of $(2+r)/4$ to the Anderson bound of $(3+r)/4$. The critical value of $r$ ranges from 0.16 when the Néel bound is used to 0.14 when the Anderson bound is used.

To show that the maximum is attained by (9), let $g_q$ be a function which satisfies (6) and (8). Consider the two regions $R_+$ and $R_-$ defined as follows:

$$R_\pm=\{q\colon \pm(\cos q_1+\cos q_2+r\cos q_3)>0 \text{ and } g_q<f_q\}$$

Suppose *both* these regions have nonzero measure. Then we can increase $g_q$ slightly in *both* regions in such a way that conditions (6) and (8) still hold. The new function has a larger $I$. Thus, we need only consider $g_q$ such that *only one* of $R_+$ and $R_-$ has positive measure.

Let us assume that $R_+$ has nonzero measure and $R_-$ has zero measure. We shall show that this leads to (9a). [The other case, leading to (9b), is similar.] Suppose $g_q$ is not given by (9a) for any $\alpha$. Then there exists an $\alpha>0$ such that both of the following sets have positive measure.

$$R_>=\{q\colon (\cos q_1+\cos q_2+r\cos q_3)>\alpha \text{ and } g_q>0\}$$
$$R_<=\{q\colon (\cos q_1+\cos q_2+r\cos q_3)<\alpha \text{ and } g_q<f_q\}$$

We can then decrease $g_q$ slightly in $R_>$ and increase it slightly in $R_<$ in such a way that (6) and (8) still hold. Since $\cos q_1 + \cos q_2 + r\cos q_3$ is greater on $R_>$ than on $R_<$, we must increase $g_q$ more than we decrease it, i.e., $\int d^3q\, g_q$ must get larger. Thus, the maximizing $g_q$ is given by (9a).

Although we cannot prove the existence of Néel order for spin 1/2 in two dimensions, we will show how to obtain sufficient conditions for the existence of Néel order which only involve the two-point function at relatively short distances. Define $\bar{g}(n)$ as follows:

$$\bar{g}(n) = \frac{1}{n+1} \sum_{m=0}^{n} (-1)^m \langle S_0^3 S_{m\delta_i}^3 \rangle \tag{10}$$

for $i = 1$ or 2. (Recall that $\delta_i$ is the unit vector in the $i$ direction.) The two cases of $i = 1$ or 2 give the same result because of the invariance under rotations of the lattice by $\pi/2$. If there is no Néel order, then the infrared bound implies

$$\begin{aligned} \bar{g}(n) &= \int dq \frac{1}{2(n+1)} \sum_{m=0}^{n} (-1)^m [\cos(mq_1) + \cos(mq_2)]\, g_q \\ &\leqslant \int dq \left\{ \frac{1}{2(n+1)} \sum_{m=0}^{n} (-1)^m [\cos(mq_1) + \cos(mq_2)] \right\}_+ f_q \end{aligned} \tag{11}$$

The last integral is then computed numerically. Table I shows the resulting upper bound on $\bar{g}(n)$.

**Table I. The Upper Bound on $\bar{g}(n)$ [See Eqs. (10) and (11)] Which Follows from the Infrared Bound (1) and the Assumption That There Is No Néel Order in the 2D, $S = 1/2$ Model, and the Value of $\bar{g}(n)$ Obtained Using the Numerical Results of Ref. 7[a]**

| $n$ | Bound on $\bar{g}(n)$ | $\bar{g}(n)$ using results of ref. 7 |
|---|---|---|
| 1 | 0.228 | 0.184 |
| 2 | 0.170 | 0.145 |
| 3 | 0.137 | 0.124 |
| 4 | 0.115 | 0.110 |
| 5 | 0.099 | 0.100 |
| 6 | 0.088 | 0.093 |
| 7 | 0.079 | 0.087 |
| 8 | 0.072 | 0.082 |

[a] Their numerical results were obtained using a Monte Carlo simulation on a $24 \times 24$ lattice and may require sizable corrections arising from the lack of manifest rotation invariance in the simulations. A contradiction for any one value of $n$ implies that there must be Néel order.

Recently Gross *et al.*[7] have numerically computed the two-point function along a coordinate direction for distances up to 11. The numbers in the second column of Table I are the "raw" data from ref. 7 and should in principle be corrected for by extrapolating to zero temperature and infinite size, and for spin space isotropy. (We have simply assumed that the quoted $\langle S^z S^z \rangle$ correlations at distance $n$ are estimates for one-third of the infinite-volume $\langle \mathbf{S} \cdot \mathbf{S} \rangle$ correlations. The authors of ref. 7. have alerted us to the possibility of sizable corrections originating from the lack of manifest rotation invariance in the simulations.)

We now give the proof of inequalities (1) and (6). Inequality (1) may be obtained by taking the zero-temperature limit of Dyson *et al.*'s bound on $g_q$. It is possible to prove inequality (1) directly in the ground state. Such a direct proof has not appeared in the literature as far as we know, so we provide it here. We start with the model defined on a finite lattice with periodic boundary conditions.

It is convenient to introduce the (positive) spectral weight function

$$R(\omega) = \frac{1}{2} \sum_n \left[ |(\phi_n, S_q \phi_0)|^2 + |(\phi_n, S_{-q} \phi_0)|^2 \right] \delta(\omega - e_n + e_0)$$

where $\phi_n$ are the energy eigenstates, $e_n$ are the corresponding eigenvalues, and $\phi_0$ is the unique ground state. Then

$$g_q = \int_0^\infty d\omega \, R(\omega)$$

The susceptibility is

$$\chi_q = \int_0^\infty d\omega \, R(\omega) \, \omega^{-1}$$

By the Cauchy–Schwarz inequality,

$$\left[ \int_0^\infty d\omega \, R(\omega) \right]^2 \leqslant \int_0^\infty d\omega \, R(\omega) \, \omega^{-1} \int_0^\infty d\omega \, R(\omega) \omega \tag{12}$$

The last integral is

$$\int_0^\infty d\omega \, R(\omega) \omega = \tfrac{1}{2} \langle [[S_q, H], S_{-q}] \rangle = 2 e_0 E_q / 3d \tag{13}$$

for the usual Heisenberg model after some computation.[3] We will show later that

$$\chi_q \leqslant \frac{1}{4} \frac{1}{E_{q-Q}}, \qquad q \neq Q \tag{14}$$

Combining (12)–(14) yields (1). Note that in this argument the Cauchy–Schwarz inequality plays the role played by the Falk–Bruch inequality[(4)] in the nonzero-temperature case.

For the model (5) which interpolates between two and three dimensions, the bound (14) holds with $E_{q-Q}$ replaced by $E^r_{q-Q}$. The double commutator (13) equals

$$\tfrac{2}{3}[(2-\cos q_1-\cos q_2)\,\rho_1+r(1-\cos q_3)\,\rho_3]$$

where $\rho_1$ and $\rho_3$ are the expectations of $\mathbf{S}_x\cdot\mathbf{S}_y$ for bonds $\{xy\}$ in the first and third lattice directions, respectively. We will show that $0\leqslant\rho_3\leqslant\rho_1$. This fact, together with $e^r_0=2\rho_1+r\rho_3$, implies that the double commutator is bounded above by $e^r_0E^r_q/3$. This, in turn, implies (6). First of all, $\rho_3\geqslant 0$, for if $\rho_3<0$, we would have that $-e^r_0>2\rho_1=\langle H^1+H^2\rangle$, where $H^1$ (resp. $H^2$) is the interaction in the 1 (resp. 2) direction in the lattice. We could then lower the energy, $-e^r_0$, by replacing the ground state $\phi_0$ by the ground state for $H^1+H^2$. [This ground state is the product of the unique two-dimensional ground state for each (12) plane and has the property that $\rho_3=0$.] Now suppose that $\rho_3>\rho_1$ and assume that the lattice is cubic (i.e., the number of sites in each direction is the same). Then we can simply rotate the lattice about the 1 axis so that directions 2 and 3 are interchanged. The energy would then be $-\rho_1-r\rho_1-\rho_3$, which cannot be less than $-e^r_0=-2\rho_1-r\rho_3$. Thus, $\rho_3\leqslant\rho_1$.

We now turn to the bound on the susceptibility, inequality (14). The Heisenberg antiferromagnetic Hamiltonian is unitarily equivalent to the Hamiltonian

$$\sum_{\{xy\}}(-S^1_xS^1_y+S^2_xS^2_y-S^3_xS^3_y) \tag{15}$$

The unitary transformation is rotation by $\pi$ about the 2 axis in the spin space at site $x$ for all sites $x$ with odd $|x|$. We work in the usual basis in which the matrices of $S^1$ and $S^3$ have only real entries, while $S^2$ has only purely imaginary entries. Define $T^1=S^1$, $T^2=iS^2$, $T^3=S^3$. Then the matrices $T^i$ all have only real entries and the above Hamiltonian is

$$-\sum_{\langle xy\rangle}(T^1_xT^1_y+T^2_xT^2_y+T^3_xT^3_y) \tag{16}$$

Let $h=h_x$ be a real-valued function on the sites. We define an $h$-dependent Hamiltonian as follows:

$$H(h)=\frac{1}{2}\sum_{\{xy\}}[(T^1_x-T^1_y)^2+(T^2_x-T^2_y)^2+(T^3_x-T^3_y-h_x+h_y)^2] \tag{17}$$

When $h=0$ this agrees with the above Hamiltonian except for a constant term. Let $E(h)$ be the ground-state energy of $H(h)$. We will show later that

$$E(h) \geqslant E(0), \qquad \forall h \tag{18}$$

Hence

$$\frac{d^2}{d\lambda^2} E(\lambda h)\bigg|_{\lambda=0} \geqslant 0 \tag{19}$$

We use perturbation theory to compute this derivative and take $h_x$ to be $\cos q \cdot x$. The unitary transformation of rotation by $\pi$ about the 2 axis changes $S_x^3$ to $(-1)^{|x|} S_x^3$, and so changes $S_q$ to $S_{q-Q}$. The bound (14) then follows from (19).

Notice that (18) is equivalent to proving that $E(h)$ attains its minimum when $h_x$ is a constant. Suppose $E(h)$ attains its minimum at a function $\bar{h}$, and there is a bond $\{x_0 y_0\}$ with $\bar{h}_{x0} \neq \bar{h}_{y0}$. The energy $E(h)$ may attain its minimum at more than one configuration, in which case we choose a minimizing configuration $\bar{h}$ with the least number of bonds $\{xy\}$ with $\bar{h}_x \neq \bar{h}_y$. We will then construct another function $h'$ which is also a minimizer for $E$, but has fewer bonds with $h'_x \neq h'_y$. This contradiction will imply that $E(h)$ must attain its minimum at $h_x = \text{const}$.

We draw a plane through the midpoint of the bond $x_0 y_0$ and perpendicular to the bond. We also draw a second plane parallel to the first but shifted by $L/2$. ($L$ is the number of sites in a single lattice direction, which we assume to be even.) These two planes, which will be denoted collectively by $P$, divide the lattice into two halves which we refer to as the right and left halves. (Remember that we are using periodic boundary conditions.)

We work in the usual real, orthonormal basis of $S^3$ eigenstates. Let $\psi_\alpha^L$, $\psi_\beta^R$ denote the basis vectors associated with the left and right half Hilbert spaces, respectively, so $\psi_\alpha^L \otimes \psi_\beta^R$ is a basis for the full Hilbert space. It is crucial to note that the Hamiltonian $H(\bar{h})$ has *real* matrix elements in this basis, so the ground state $\psi$ can be written as

$$\psi = \sum_{\alpha,\beta} c_{\alpha\beta} \psi_\alpha^L \otimes \psi_\beta^R$$

for *real* numbers $c_{\alpha\beta}$. We will think of $c_{\alpha\beta}$ as the elements of a matrix which we denote by $c$.

There are three types of bonds. Bonds with both endpoints in the left half will be referred to as "left" bonds. Bonds with one endpoint in the left half and one in the right half will be referred to as "crossing." The "right" bonds are defined in the obvious way.

We denote the bonds crossing $P$ by $\{x_i y_i\}$ with $x_i$ in the left half and $y_i$ in the right half. For these bonds we write

$$\begin{aligned}(T_x^1-T_y^1)^2&+(T_x^2-T_y^2)^2+(T_x^3-T_y^3-\hbar_x+\hbar_y)^2\\&=(T_x^1)^2+(T_y^1)^2-2T_x^1T_y^1+(T_x^2)^2+(T_y^2)^2-2T_x^2T_y^2\\&\quad+(T_x^3-\hbar_x)^2+(T_y^3-\hbar_y)^2-2(T_x^3-\hbar_x)(T_y^3-\hbar_y)\end{aligned}\tag{20}$$

Define $H^L$ to be the sum of all the terms in $H(\hbar)$ labeled by left bonds plus the terms $(T_x^1)^2+(T_x^1)^2+(T_x^3-\hbar_x)^2$ from the crossing bonds. $H^R$ is defined analogously. Then

$$H=H^L+H^R-2\sum_i\,[T_{x_i}^1T_{y_i}^1+T_{x_i}^2T_{y_i}^2+(T_{x_i}^3-\hbar_{x_i})(T_{y_i}^3-\hbar_{y_i})]\tag{21}$$

Let

$$H_{\alpha\gamma}^L=(\psi_\alpha^L,\,H^L\psi_\gamma^L)$$

$$X_{\alpha\gamma}^{L,i}=(\psi_\alpha^L,\,T_{x_i}^1\psi_\gamma^L),\qquad Y_{\alpha\gamma}^{L,i}=(\psi_\alpha^L,\,T_{x_i}^2\psi_\gamma^L),\qquad Z_{\alpha\gamma}^{L,i}=(\psi_\alpha^L,\,(T_{x_i}^3-\hbar_{x_i})\,\psi_\gamma^L)$$

and similarly for $H_{\alpha\gamma}^R$, $X_{\alpha\gamma}^{R,i}$, $Y_{\alpha\gamma}^{R,i}$, $Z_{\alpha\gamma}^{R,i}$ with $x_i$ replaced by $y_i$. It is important to note that all these matrix elements are real. (It is here that the reality of all the vectors and operators is used.) We let $X^{L,i}$ denote the matrix whose $(\alpha,\gamma)$ element is $X_{\alpha\gamma}^{L,i}$. Then the transpose of $X^{L,i}$ is the same as the adjoint of $X^{L,i}$. The latter is denoted by $(X^{L,i})^\dagger$. The same notation and remark apply to the other quantities $Y$, $Z$, $H$.

Remembering that the $c_{\alpha\beta}$ are real, we obtain

$$\begin{aligned}E(\hbar)&=(\psi,\,H(\hbar)\psi)\\&=\sum_{\alpha\beta\gamma}c_{\alpha\beta}c_{\gamma\beta}H_{\alpha\gamma}^L+\sum_{\alpha\beta\gamma}c_{\alpha\beta}c_{\alpha\gamma}H_{\beta\gamma}^R\\&\quad-2\sum_i\sum_{\alpha\beta\gamma\delta}c_{\alpha\beta}c_{\gamma\delta}(X_{\alpha\gamma}^{L,i}X_{\beta\delta}^{R,i}+Y_{\alpha\gamma}^{L,i}Y_{\beta\delta}^{R,i}+Z_{\alpha\gamma}^{L,i}Z_{\beta\delta}^{R,i})\\&=\operatorname{Tr}cc^\dagger H^L+\operatorname{Tr}c^\dagger cH^R-2\sum_i\operatorname{Tr}[c^\dagger X^{L,i}c(X^{R,i})^\dagger\\&\quad+c^\dagger Y^{L,i}c(Y^{R,i})^\dagger+c^\dagger Z^{L,i}c(Z^{R,i})^\dagger]\end{aligned}\tag{22}$$

The next step is to prove a trace inequality. Let $c$, $M$, $N$ be matrices, not necessarily real. Then we shall prove that

$$|\operatorname{Tr}c^\dagger McN^\dagger|^2\leqslant\operatorname{Tr}c_LM^\dagger c_LM\operatorname{Tr}c_RNc_RN^\dagger\tag{23}$$

where $c_L=(cc^\dagger)^{1/2}$ and $c_R=(c^\dagger c)^{1/2}$. Both $c_L$ and $c_R$ are Hermitian positive semidefinite and, by the polar decomposition theorem, there is a unitary matrix $U$ such that $c=Uc_R$ and $c^\dagger=c_RU^\dagger$. Thus, by the cyclicity of the trace

$$\operatorname{Tr}c^\dagger McN^\dagger=\operatorname{Tr}AB$$

with

$$A = c_R^{1/2} U^\dagger M U c_R^{1/2}, \qquad B = c_R^{1/2} N^\dagger c_R^{1/2}$$

By the Schwarz inequality for traces

$$|\mathrm{Tr}\, c^\dagger M c N^\dagger|^2 \leqslant \mathrm{Tr}\, A^\dagger A \,\mathrm{Tr}\, B^\dagger B = \mathrm{Tr}\, \alpha M^\dagger \alpha M \,\mathrm{Tr}\, c_R N c_R N^\dagger$$

with $\alpha = U c_R U^\dagger$. Since $\alpha^2 = U c_R^2 U^\dagger = cc^\dagger$, and since $cc^\dagger$ has a unique square root $c_L$, we have that $\alpha = c_L$. This proves (23).

To apply (23) to our case, consider $M = X^{L,i}$, $N = X^{R,i}$. Since $2ab \leqslant a^2 + b^2$, we have

$$2\,\mathrm{Tr}\, c^\dagger X^{L,i} c (X^{R,i})^\dagger \leqslant \mathrm{Tr}\, c_L X^{L,i} c_L (X^{L,i})^\dagger + c_R X^{R,i} c_R (X^{R,i})^\dagger \qquad (24)$$

A similar inequality obviously holds with the $X$ matrices replaced by $Y$ or $Z$ matrices.

The definitions of $c_L$ and $c_R$ and (24) imply

$$\begin{aligned} E(\bar{h}) \geqslant\ & \mathrm{Tr}\, c_L^2 H^L + \mathrm{Tr}\, c_R^2 H^{R'} \\ & - \sum_i \mathrm{Tr}[c_L X^{L,i} c_L (X^{L,i})^\dagger + c_L Y^{L,i} c_L (Y^{L,i})^\dagger + c_L Z^{L,i} c_L (Z^{L,i})^\dagger] \\ & - \sum_i \mathrm{Tr}[c_R X^{R,i} c_R (X^{R,i})^\dagger + c_R Y^{R,i} c_R (Y^{R,i})^\dagger + c_R Z^{R,i} c_R (Z^{R,i})^\dagger] \end{aligned} \qquad (25)$$

Let $h_x^R$ denote the function which agrees with $\bar{h}_x$ on the right sites and on the left sites equals the reflection of $\bar{h}_x$ in the planes $P$. The $h^L$ is defined analogously. Recall that $\bar{h}_x \neq \bar{h}_y$ for at least one crossing bond. Hence, at least one choice, $h^R$ or $h^L$, has the property that it has strictly fewer bonds with $h_x \neq h_y$ than does the original $\bar{h}$.

Let

$$\psi^L = \sum_{\alpha,\beta} (c_L)_{\alpha\beta}\, \psi_\alpha^L \otimes \psi_\beta^R$$

with $\psi^R$ defined analogously using $(c_R)_{\alpha\beta}$. (Note that $\|\psi^L\| = \|\psi^R\| = \|\psi\|$.) Then the right side of (25) equals

$$\tfrac{1}{2}(\psi^R, H(h^R)\,\psi^R) + (\psi^L, H(h^L)\,\psi^L) \geqslant \tfrac{1}{2}E(h^R) + \tfrac{1}{2}E(h^L)$$

We have chosen $\bar{h}$ so that $E(\bar{h})$ is a minimum, so this inequality implies that $E$ also attains its minimum at both $h^R$ and $h^L$. This contradicts the minimality of the number of bonds such that $\bar{h}_x \neq \bar{h}_y$, so (18) is proven. This concludes the proof of (14).

We have demonstrated the existence of Néel order in the ground state of the 3D spin-1/2 Heisenberg antiferromagnet. Although our methods do not work for the 2D model, they work surprisingly well for the model which interpolates between $d=2$ and $d=3$. The bound (1) on $g_q$ and the value of the nearest neighbor correlation (3) are not by themselves sufficient to show the existence of Néel order in $d=2$ with $S=1/2$, so the obvious question is what additional information might show the existence of Néel order in this case. Rigorous lower bounds on the short-range correlations could prove the existence of Néel order (see Table I). Another possibility is to improve the bound on $g_q$. The two places where one could hope to improve this bound are to improve the bound on the susceptibility $\chi_q$ (14) or to improve inequality (12) resulting from the use of the Cauchy–Schwarz inequality. For the 1D spin-1/2 Heisenberg antiferromagnet the exact value[6,14] of $\chi_0$ is $1/\pi^2$ while the bound (14) equals 1/8 at $q=0$. Thus, one cannot hope to improve the bound (14) by, for example, an overall factor of 2. If the Cauchy–Schwarz inequality (12) is close to being an equality, then $R(\omega)$ must be close to a $\delta$-function at a single value of $\omega$. Although this appears unlikely, we have not been able to improve inequality (12).

## ACKNOWLEDGMENTS

We thank the authors of ref. 7 for correspondence concerning their work. T. K. is a National Science Foundation Postdoctoral Fellow. The work of E. H. L. was supported in part by National Science Foundation grant PHY-85-152-88-A02. The work of B. S. S. was supported in part by National Science Foundation grant DMR 8518163.

## REFERENCES

1. I. Affleck, T. Kennedy, E. H. Lieb, and H. Tasaki, *Commun. Math. Phys.* **115**:477–528 (1988).
2. P. W. Anderson, *Science* **235**:1196–1198 (1987).
3. F. J. Dyson, E. H. Lieb, and B. Simon, *J. Stat. Phys.* **18**:335–383 (1978).
4. H. Falk and L. W. Bruch, *Phys. Rev.* **180**:442–444 (1969).
5. J. Fröhlich, B. Simon, and T. Spencer, *Commun. Math. Phys.* **50**:79–95 (1976).
6. R. B. Griffiths, *Phys. Rev. A* **133**:768–775 (1964).
7. M. Gross, E. Sánchez-Velasco, and E. Siggia, Ground state properties of the 2-dimensional antiferromagnetic Heisenberg model, Cornell University preprint.
8. D. A. Huse, *Phys. Rev. B* **37**:2380–2382 (1988).
9. D. A. Huse and V. Elser, *Phys. Rev. Lett.* **60**:2531 (1988).
10. E. Jordão Neves and J. Fernando Perez, *Phys. Lett.* **114A**:331–333 (1986).
11. S. Liang, B. Doucot, and P. W. Anderson, *Phys. Rev. Lett.* **61**:365 (1988).
12. E. H. Lieb and D. C. Mattis, *J. Math. Phys.* **3**:749–751 (1962).
13. J. D. Reger and A. P. Young, *Phys. Rev. B* **37**:5978–5981 (1988).
14. C. N. Yang and C. P. Yang, *Phys. Rev.* **151**:258–264 (1966).

VOLUME 61, NUMBER 22 PHYSICAL REVIEW LETTERS 28 NOVEMBER 1988

# The *XY* Model Has Long-Range Order for All Spins and All Dimensions Greater than One

Tom Kennedy,[a] Elliott H. Lieb, and B. Sriram Shastry[b]

*Department of Physics, Princeton University, P.O. Box 708, Princeton, New Jersey 08544*

(Received 13 July 1988)

The quantum *XY* model of interacting spins on a hypercubic lattice has long-range order in the ground state for all values of the spin and all dimensions greater than one. We also show that in the limit of high dimension the spontaneous magnetization converges to the spontaneous magnetization of the Néel state.

PACS numbers: 64.60.Cn, 05.30.−d, 75.10.Jm

We consider the question of the existence of long-range order (LRO) in the ground state of the quantum *XY* model on the hypercubic lattice in two and higher dimensions. (We note that there cannot be any long-range order in two dimensions at positive temperature $T$, as the Mermin-Wagner-Hohenberg theorem shows.) The Hamiltonian is

$$H=-\sum_{\langle x,y\rangle}(S_x^1S_y^1+S_x^2S_y^2), \tag{1}$$

with $\langle x,y\rangle$ denoting nearest-neighbor pairs. Dyson, Lieb, and Simon (DLS)[1] showed the existence of LRO for spin $S=\frac{1}{2}$ in 3D for positive temperature $T$. Neves and Perez[2] noticed that one could take the $T\to 0$ limit of the DLS formalism and thereby prove LRO for the ground state of the *XXX* Heisenberg model in two dimensions for $S\geq 1$. Kubo[3] used the methods in Refs. 1 and 2 to show the existence of LRO for $S\geq 1$ in 3D and for $S\geq\frac{3}{2}$ in 2D for the *XY* model in (1).

In a recent paper,[4] we were able to improve on the DLS method and applied this improvement to the *XXX* Heisenberg antiferromagnet to show ground-state LRO for $S=\frac{1}{2}$ in 3D. Here we show that this method is also capable of proving ground-state LRO for the *XY* model (1) for $S\geq\frac{1}{2}$ and all dimensions greater than one. The case $S=\frac{1}{2}$ corresponds to bosons with hard-core repulsion (at half-filling) via the lattice-gas analogy of Matsubara and Matsuda[5] and proves that Bose-Einstein condensation (or off-diagonal long-range order) occurs in this system despite hard-core repulsion in real space. *In fact, these $S=\frac{1}{2}$ systems provide the only examples known to us of interacting particles in which Bose-Einstein condensation has been proved to occur.*

For a finite $\nu$-dimensional hypercubic lattice $\Lambda$ the ground state is unique.[6] The object of principal interest is the Fourier transform of the two-point function. That is, for $\alpha=1$, 2, or 3 we set

$$\hat{S}_p^\alpha=|\Lambda|^{-1/2}\sum_{x\in\Lambda}S_x^\alpha\exp(ip\cdot x),$$

and then set

$$g_p^\alpha=\langle\hat{S}_p^\alpha\hat{S}_{-p}^\alpha\rangle,$$

with the angular brackets denoting ground-state expectation value. By Parseval's identity

$$|\Lambda|^{-1}\textstyle\sum_p g_p^\alpha=\langle(S_0^\alpha)^2\rangle,$$

and by taking the usual infinite-volume limit and passing from sums to integrals, we obtain

$$(2\pi)^{-\nu}\int d^\nu p\, g_p^1=\langle(S_0^1)^2\rangle, \tag{2}$$

where the integral is over the cube $|p_i|\leq\pi$ for $i=1,\dots,\nu$. Note that in (2), and henceforth, $g_p^\alpha$ denotes the infinite-volume limit of $g_p^\alpha$ for finite volume.

Equation (2) will not be used here but instead we follow Ref. 4 and introduce the additional sum rule which is derived in the same way as (2) for $i=1,\dots,\nu$,

$$(2\pi)^{-\nu}\int d^\nu p\, g_p^1\cos p_i=\langle S_0^1S_{\delta_i}^1\rangle\equiv e_1, \tag{3}$$

where $\delta_i$ is the nearest neighbor to 0 in the direction $i$. Clearly $\langle S_0^1S_{\delta_i}^1\rangle$ is independent of $i$ in the infinite-volume limit. By symmetry $e_1$ is minus half the ground-state energy per bond. (Note that we are treating the ferromagnetic *XY* model, so that $e_1>0$. The antiferromagnetic and ferromagnetic *XY* models are isomorphic.[1]) It might be thought that (3) demands more information (i.e., $e_1$) than does (2), but this is misleading because our bound on $g_p^1$ below also requires knowledge of $e_1$.

The bound on $g_p^1$ for $p\neq 0$ is

$$0\leq g_p^1\leq\tfrac{1}{2}\left[\sum_{i=1}^{\nu}(e_1-e_3\cos p_i)\Big/\sum_{i=1}^{\nu}(1-\cos p_i)\right]^{1/2}, \tag{4}$$

where $e_3=\langle S_0^3S_{\delta_i}^3\rangle$. This bound is true for every finite $\Lambda$ and hence remains true in the limit $\Lambda\to\infty$. It appears several times in the literature and so we shall not attempt to outline its lengthy derivation. It can be derived as a $T\to 0$ limit of the bound in DLS[1] as done (in the *XXX* context) in Ref. 2. It can also be derived directly in the ground state, as done in the *XXX* context in Ref. 4. As Kubo points out, $|e_3|\leq e_1$ (for otherwise the energy could be lowered by interchanging the 1 and 3 spin

VOLUME 61, NUMBER 22 PHYSICAL REVIEW LETTERS 28 NOVEMBER 1988

directions), so that

$$0 \le g_p^1 \le \tfrac{1}{2}(e_1)^{1/2}\left[\left(v+\left|\sum_{i=1}^{v}\cos p_i\right|\right)\Big/\sum_{i=1}^{v}(1-\cos p_i)\right]^{1/2} \quad (5)$$

The sum rule (3) says

$$e_1 = (2\pi)^{-v}\int d^v p\, g_1^p v^{-1}\sum_{i=1}^{v}\cos p_i . \quad (6)$$

Let $m^2 = \lim_{|x|\to\infty}\langle S_0^1 S_x^1\rangle$ be the square of the spontaneous magnetization in the 1 direction. Our goal, LRO, is equivalent to $m \ne 0$. Since the right-hand side of (5) is integrable, $m \ne 0$ if and only if $g_p^1$ (in the infinite-volume limit) has a delta function at $p=0$. In fact, $(2\pi)^v m^2$ is the coefficient of this delta function. The bounds (5) and the sum rule (6) imply

$$e_1 \le \tfrac{1}{2}(e_1)^{1/2} I(v) + m^2, \quad (7)$$

where

$$I(v) = (2\pi)^{-v}\int d^v p\left[\sum_{i=1}^{v}(1+\cos p_i)\Big/\sum_{i=1}^{v}(1-\cos p_i)\right]^{1/2}\left\{v^{-1}\sum_{i=1}^{v}\cos p_i\right\}_+ , \quad (8)$$

and $\{a\}_+ = \max\{a,0\}$. Numerically we find $I(2)=0.65$ and $I(3)=0.35$. A simple variational argument (using the wave function with all spins aligned in the 1 direction) shows $e_1 \ge \frac{1}{2}S^2$ in all dimensions. Thus for any spin $S \ge \frac{1}{2}$ and $v=2$ or $v=3$ we conclude from (7) that $m \ne 0$, thereby implying there must be LRO in the *XY* model.

To prove LRO for $v>3$ we must bound $I(v)$. We shall now prove that $I(v) \le I(2)$ for $v \ge 3$, and thereby establish [from (7)] LRO for all $v>3$. The proof is as follows. Let $F(x) = x[(1+x)/(1-x)]^{1/2}$ for $0 \le x \le 1$. One checks that $F$ is monotone increasing and convex. For $i,j=1,\dots,v$ let $Y_{ij}(p) = \frac{1}{2}(\cos p_i + \cos p_j)$ and let $Y(p) = v^{-1}\sum_{i=1}^{v}\cos p_i$. The integral in (8) is $F(\{Y(p)\}_+)$. Now

$$Y(p) = (v^2-v)^{-1}\sum_{i\ne j} Y_{ij}(p).$$

Since $\{a+b\}_+ \le \{a\}_+ + \{b\}_+$, we have

$$\{Y(p)\}_+ \le (v^2-v)^{-1}\sum_{i\ne j}\{Y_{ij}(p)\}_+ .$$

Since $F$ is monotone,

$$F(\{Y(p)\}_+) \le F\left[(v^2-v)^{-1}\sum_{i\ne j}\{Y(p)\}_+\right].$$

Since $F$ is convex,

$$F(\{Y(p)\}_+) \le (v^2-v)^{-1}\sum_{i\ne j}F(\{Y_{ij}(p)\}_+).$$

But

$$(2\pi)^{-v}\int F(\{Y_{ij}(p)\}_+)d^v p = I(2).$$

This proves $I(v) \le I(2)$. By the same analysis one can also prove that $I(v) \le I(\mu)$ whenever $v > \mu$.

Having achieved our goal of proving LRO, we now turn to an additional fact about the spontaneous magnetization in the limit of high dimension. We shall show that $m^2$ converges to the classical value $S^2/2$ as $v\to\infty$. First, we prove that $I(v)\to 0$ as $v\to\infty$. Using $\{Y(p)\}_+ \le |Y(p)| \le 1$ and the Schwarz inequality, we have

$$I(v)^2 \le 2(2\pi)^{-2v}\int d^v p[1-Y(p)]^{-1}\int d^v p\, Y(p)^2 .$$

The last integral is $(2\pi)^v/2v$ (since $\int \cos p_i \cos p_j = 0$). To bound the first integral we note that $1/(1-x)$ is convex and so, imitating the above analysis for $v \ge 3$,

$$(2\pi)^{-v}\int d^v p[1-Y(p)]^{-1} \le (2\pi)^{-3}\int d^3 p[1-\tfrac{1}{3}(\cos p_1+\cos p_2+\cos p_3)]^{-1},$$

which is finite. This proves our assertion.

The importance of the assertion that $\lim_{v\to\infty} I(v) = 0$ is that inequality (7) and the inequality $m^2 \le e_1$ (which follows from reflection positivity; see Ref. 4) then imply that as $v\to\infty$ the spontaneous magnetization approaches the energy $e_1$. As we will show below the energy $e_1$ converges to $\frac{1}{2}S^2$ as $v\to\infty$. Thus as $v\to\infty$ the spontaneous magnetization converges to the classical value. The same proof and conclusion applies to the Heisenberg antiferromagnet considered in Ref. 4. This result validates the folklore that as $v\to\infty$ the ground-state correlations converge to the classical values.

To show that $\lim_{v\to\infty} e_1 = \frac{1}{2}S^2$, we first note (as above) that $e_1 \ge \frac{1}{2}S^2$. To obtain an upper bound to $e_1$ we write the Hamiltonian (1) as the average of $\sum_{\langle xy\rangle}(-S_x^1S_y^1 - S_x^2S_y^2 + S_x^3S_y^3)$ and $\sum_{\langle xy\rangle}(-S_x^1S_y^1 - S_x^2S_y^2 - S_x^3S_y^3)$. The first Hamiltonian is unitarily equivalent to the antiferromagnetic Heisenberg Hamiltonian, and so Anderson's bound[1] says that its ground-state energy per bond is $\ge -(S^2+S/2v)$. The second Hamiltonian is the Heisenberg ferromagnet, and so its ground-state energy per bond is $-S^2$. Thus the ground-state energy per bond of Hamiltonian (1) is not less than $-(S^2+S/4v)$, and so $e_1 \le \frac{1}{2}S^2 + S/8v$. By combining the two bounds it follows that $\lim_{v\to\infty} e_1 = \frac{1}{2}S^2$.

One can also consider admixing some 3-component, i.e., the two-site interaction is changed to $-S_i^1S_j^1 - S_i^2S_j^2 + \Delta S_i^3S_j^3$. Then our method will show LRO for small $\Delta$ in all the above cases. It does not show LRO for

VOLUME 61, NUMBER 22 PHYSICAL REVIEW LETTERS 28 NOVEMBER 1988

$\Delta = 1$ with $\nu = 2$, but it does for $\nu = 3$, as shown in Ref. 4.

T.K. acknowledges support from the NSF. E.H.L. and B.S.S. were supported in part by NSF Grants No. PHY-85-15288-A02 and No. DMR 8518163, respectively.

[(a)]Permanent address: Mathematics Department, University of Arizona, Tucson, AZ 85721.

[(b)]On leave from Tata Institute of Fundamental Research, Bombay, India; now at AT&T Bell Laboratories, Murray Hill, NJ 07974.

[1]F. Dyson, E. H. Lieb, and B. Simon, J. Stat. Phys. **18**, 335 (1978).

[2]E. Jordao Neves and J. Fernando Perez, Phys. Lett. **114A**, 331 (1986).

[3]K. Kubo, Phys. Rev. Lett. **61**, 110 (1988).

[4]T. Kennedy, E. H. Lieb, and B. S. Shastry, J. Stat. Phys. (to be published).

[5]T. Matsubara and H. Matsuda, Prog. Theor. Phys. **16**, 569 (1956).

[6]E. Lieb and D. C. Mattis, J. Math. Phys. 3, 749 (1962).

# Part V

# Classical Thermodynamics

*Journal of Statistical Physics, Vol. 24, No. 1, 1981*

# The Third Law of Thermodynamics and the Degeneracy of the Ground State for Lattice Systems

**Michael Aizenman[1] and Elliott H. Lieb[1,2]**

*Received February 11, 1980*

The third law of thermodynamics, in the sense that the entropy per unit volume goes to zero as the temperature goes to zero, is investigated within the framework of statistical mechanics for quantum and classical lattice models. We present two main results: (i) For all models the question of whether the third law is satisfied can be decided completely in terms of ground-state degeneracies alone, provided these are computed for all possible "boundary conditions." In principle, there is no need to investigate possible entropy contributions from low-lying excited states. (ii) The third law is shown to hold for ferromagnetic models by an analysis of the ground states.

**KEY WORDS:** Third law; entropy; thermodynamics; lattice systems; statistical mechanics.

## 1. INTRODUCTION

### 1.1. Questions Raised in the Past

The third law of thermodynamics, in Planck's form, is that the entropy density $S$ for a bulk system at the temperature $T$ approaches zero as $T \to 0$. The discussion of that rule from the vantage point of statistical mechanics has centered both on the question of its validity (it is known to have some notable exceptions, some of which we shall mention) and on its relation to the nondegeneracy of the system's Hamiltonian at its lowest, or ground state, energy.

Dedicated to Pierre Résibois. Work supported in part by NSF grant PHY-7825390 A01.

[1] Department of Physics, Princeton University, Princeton, New Jersey.

[2] Also at the Department of Mathematics, Princeton University, Princeton, New Jersey.

While the latter aspect of the third law is often put forward in textbooks, under closer scrutiny it was seriously questioned by Griffiths, among others (see Ref. 3). He pointed out[1,2] that:

1. For any finite system the entropy as $T \to 0$ is determined by the ground state degeneracy. However, for bulk systems it would be necessary to achieve a very low $T$ in order to be sure that the system is effectively in its ground state. This $T$ is usually unattainably small. In other words, what is effectively done in the laboratory corresponds to taking the thermodynamic limit first and then the limit $T \to 0$. Hence the interchange of the limits could lead to misleading conclusions about real situations.

2. There are lattice models which in finite volumes have nondegenerate ground states, but for which, nevertheless, the third law is not satisfied.

3. For (infinitely) large systems, the ground state contribution to the partition function is negligible at any nonzero temperature. Thus $S_0$, the limiting entropy density at $T = 0$, may depend crucially also on the distribution of low-energy excitations.

The discussion of the third law in Ref. 2 concludes with: "My good wishes to anyone who wants to embark on this quest, but let him remember that he must do *more* than examine the ground state!."

After the above warning, rigorous proofs of the third law for ferromagnetic lattice systems[4,5] avoided references to ground states and relied on less direct arguments, such as correlation inequalities or Lee–Yang techniques.

Our purpose in this paper is to show, for quantum and classical lattice systems, that the entropy density at $T = 0$ is indeed directly related to the degeneracy of the ground state when the latter is suitably interpreted. We will establish various forms of this relationship. These have value both conceptually and as a simpler tool for proving the third law for certain lattice systems (and computing the nonzero entropy for others).

We find it very useful to study the problem using the infinite system formalism. With regard to calculations of the zero-temperature entropy which are based on finite-volume ensembles, with the correct order of limits and some specified boundary conditions, we reach the following conclusions. For simplicity, these are stated for classical systems with finite-range interactions.

1. Griffiths is correct in asserting the need to take into account low-lying excitations. As he showed by an example (discussed here as Example 7) these can have bulk contribution to the entropy at $T = 0$. However, as we shall point out next, there is another way of finding all the contributions of these excitations.

2. If, for a sequence of finite regular domains, the number of excitations with energy of the order $o(V)$ grows exponentially with the volume $V$,

then, *necessarily*, for some proper boundary conditions the system has a highly degenerate ground state. The multitude of low-lying excitations can always be viewed as a result of the imposition of degeneracy-breaking boundary conditions. In the example referred to above, the degenerate ground states are clearly visible.

3. The contribution to the entropy, at $T = 0$, of the above-mentioned low-lying excitations is completely accounted for by counting the entropy of the ground states which correspond to boundary conditions with the highest degeneracy. Furthermore, there is no other subtle source of bulk entropy.

In Section 6 we use the relation which we have established between $S_0$ and the ground state degeneracy to give a simple proof of the third law for certain, classical and quantum, ferromagnetic models.

We have been recently informed that some of our results were also derived by Slawny.[14]

## 1.2. Definition of $S_0$

Entropy is conveniently viewed as a function of energy, since as such it is convex. The thermodynamic relation we shall use to define it in a finite volume $\Lambda$ is

$$F_\Lambda = E - \beta^{-1} S_\Lambda(E) \tag{1.1}$$

where $\beta = 1/kT$ and the free energy of a system in the region $\Lambda$, $F_\Lambda = F_\Lambda(\beta)$, is obtained from its partition function $Z_\Lambda(\beta)$:

$$F_\Lambda(\beta) = -(1/\beta) P_\Lambda(\beta); \qquad P_\Lambda(\beta) = |\Lambda|^{-1} \ln Z_\Lambda(\beta) \tag{1.2}$$

with $\beta$ such that

$$E = -(\partial/\partial\beta) P_\Lambda(\beta) \tag{1.3}$$

The existence of the thermodynamic limit for $F_\Lambda(\beta)$ is well understood, at least for a large class of lattice systems.[6] By a standard application of the convexity in $\beta$ of $P_\Lambda(\beta)$, this implies that:

1. There is a thermodynamic limit $e_0$ for the ground state energy density and for the maximal energy density $e_{max}$.

2. In the thermodynamic limit, $S_\Lambda(E)$ converges pointwise, in the interval $(e_0, e_{max})$, to a function which we denote by $S(E)$.

3. The limit of $S_\Lambda(E)$ is independent of boundary conditions, for $E \in (e_0, e_{max})$, and the function $S(\cdot)$ is convex (and, thus, continuous).

However, the above arguments do not yield convergence at the boundary point $e_0$. In fact, the limit of $S_\Lambda(e_0(\Lambda))$ may depend on the boundary conditions.

The main subject of our discussion is

$$S_0 = \lim_{E \downarrow e_0} S(E) \equiv \lim_{E \downarrow e_0} \lim_{\Lambda \uparrow \infty} S_\Lambda(E) \tag{1.4}$$

which *defines* the entropy density at $T = 0$ in the thermodynamic limit.

Thus we follow the canonical ensemble formalism, whose advantage is the convexity of $S_\Lambda(\cdot)$. With the physical restriction of $T \geqslant 0$, the above procedure does not define $S(E)$ on the full range of possible values of the energy density. This definition can be accomplished by considering all $\beta \in (-\infty, \infty)$. Alternatively, there is a variational characterization of $S(\cdot)$ on its full domain, to be mentioned in the next section, which avoids any reference to the temperature. Other definitions of $S(E)$ exist, corresponding to the various other ensembles, but they are known to agree with the above one in the thermodynamic limit.[1,6]

## 2. THE INFINITE-SYSTEM FORMULATION OF THE THERMODYNAMIC LIMIT

The analysis of the thermodynamic limit is facilitated by the infinite-system formalism,[6] which is very useful for this purpose. Our discussion will center on classical and quantum lattice systems.

With each lattice site $i \in \mathsf{L} = \mathsf{Z}^d$ ($d$ is the dimension of the space) there is associated either a finite discrete space $\mathcal{H}_i^{(c)}$ or a finite-dimensional Hilbert space $\mathcal{H}_i^{(q)}$. For any finite $\Lambda \subset \mathsf{L}$ the observables measurable in $\Lambda$ form an algebra $\mathcal{Q}_\Lambda$, which corresponds either to functions on $\Omega_\Lambda = \times_{i \in \Lambda} \mathcal{H}_i^{(c)}$ or operators (the *full* matrix algebra) on $\mathcal{H}_\Lambda = \times_{i \in \Lambda} \mathcal{H}_i^{(q)}$.

In both cases states of the system are represented by expectation value functionals (i.e., positive and normalized) $\rho$ on the algebra of (quasilocal) observables $\mathcal{Q} = \overline{V \mathcal{Q}_\Lambda}$. In the case of classical systems, states correspond to probability measures on the "phase space" of the spin configurations of the infinite system, $\Omega = \times_{i \in \mathsf{L}} \mathcal{H}_i^{(c)}$.

For any $\Lambda \subset \mathsf{L}$, $\pi_\Lambda$ will stand for the projection, or restriction, of the corresponding object to $\Lambda$. Thus $\pi_\Lambda \rho$ is either the probability distribution on $\Omega_\Lambda$ or, if applicable, the density operator on $\mathcal{H}_\Lambda$ which gives the restriction of $\rho$ to $\mathcal{Q}_\Lambda$.

We denote the set of translation-invariant states by $\mathcal{I}$. The (information-theoretic) *entropy of a state* $\rho \in \mathcal{I}$ is defined via

$$s_\Lambda(\rho) = -|\Lambda|^{-1} \operatorname{tr}_\Lambda(\pi_\Lambda \rho) \ln(\pi_\Lambda \rho) \tag{2.1}$$

as the limit[7]

$$s(\rho) = \lim_{\Lambda \uparrow \mathsf{L}} s_\Lambda(\rho) \tag{2.2}$$

Here $\Lambda \uparrow \mathsf{L}$ means convergence over any sequence of finite domains in $\mathsf{L}$

which tend to $\mathbf{L}$ in the van Hove sense,[6] and the limit in (2.2) is independent of the sequence. In the classical case $\mathrm{tr}_\Lambda$ represents summation over $\Omega_\Lambda$, which is a commonly used a-priori measure.

The Hamiltonian of the system is formally

$$H = \sum_{B \subset \mathbf{L}} \Phi_B \tag{2.3}$$

with $\Phi_B \in \mathcal{Q}_B$. We shall assume translation invariance of both $\{\mathcal{H}_i\}$ and $\{\Phi_B\}$. In that case the Hamiltonian density is given by

$$h = \sum_{B \ni 0} \frac{1}{|B|} \Phi_B \tag{2.4}$$

The partition function mentioned in (1.2) is

$$Z_\Lambda = \mathrm{tr}_\Lambda \exp\Big( - \sum_{B \in \Lambda} \Phi_B \Big) \tag{2.5}$$

and a sufficient condition for the convergence mentioned there, with

$$e_0 = \inf_{(\min)} \{ \rho(h) \mid \rho \in \mathcal{I} \} \tag{2.6}$$

is that

$$\|\Phi\| \equiv \sum_{B \ni 0} \frac{1}{|B|} \|\Phi_B\| < \infty \tag{2.7}$$

$\|\Phi_B\|$ being the corresponding "sup" norm.

We will always assume $\|\Phi\| < \infty$.

The two entropy functions to which we have referred are related by the following variational principle:

$$S(E) = \sup\{ s(\rho) \mid \rho \in \mathcal{I}, \rho(h) = E \} \tag{2.8}$$

for any $E \in (e_0, e_{\max})$.

Equation (2.8) follows from the variational principle for $P(\beta)$, found in Refs. 6 and 8, by the usual Legendre-transform technique, which is applicable because $P(\cdot)$ and $S(\cdot)$ are convex.

The supremum in (2.8) is always attainable. The translation-invariant states for which $s(\rho) = S(E)$, with $E = \rho(h)$, are all the equilibrium states, i.e., the translation-invariant Gibbs states,[6,8,9] or possibly [if $E(T)$ is discontinuous] convex combinations of such states.

## 3. A VARIATIONAL PRINCIPLE FOR $S_0$

At the end of the previous section we saw that the thermodynamic entropy density $S(E)$ is the entropy density $s(\rho)$ for certain (entropy-maximizing) states appropriate to the energy $E$. We shall now extend this result to $S_0$ [which corresponds to the boundary of the domain of definition of $S(\cdot)$]. We have $S_0 = s(\rho)$ for an appropriate, most degenerate, class of (ground) states.

**Proposition 1.** If $\|\Phi\| < \infty$, then

$$S_0 = \max\{s(\rho) \mid \rho \in \mathcal{I}, \quad \rho(h) = e_0\} \tag{3.1}$$

*Proof.* (i) Assume the existence of $\rho \in \mathcal{I}$ such that

$$\rho(h) = e_0, \qquad s(\rho) > S_0 \tag{3.2}$$

Let $\rho' \in \mathcal{I}$ be a state with $\rho'(h) > e_0$ and consider the states $\rho_\lambda = (1-\lambda)\rho + \lambda\rho'$, $\lambda \in [0, 1]$. Both $\rho_\lambda(h)$ and $s(\rho_\lambda)$ are continuous (in fact affine[(6)]) functions of $\lambda$. {See Fig. 1, where the point $A$ (resp. $B$) gives $s(\rho)$ [resp. $s(\rho')$] and the dotted line is $s(\rho_\lambda)$ } Since $S(\cdot)$ is continuous, (3.2) implies that for small enough $\lambda \in (0, 1]$

$$s(\rho_\lambda) > s(\rho_\lambda(h)) \tag{3.3}$$

That would contradict (2.6). Thus

$$S_0 \geqslant \sup\{s(\rho) \mid \rho \in \mathcal{I}, \quad \rho(h) = e_0\} \tag{3.4}$$

(ii) To conclude the proof, choose a weakly convergent sequence $\rho_n \xrightarrow{w} \bar{\rho}$ for which

$$\rho_n(h) \to e_0, \qquad s(\rho_n) \geqslant S(\rho_n(h)) - 2^{-n} \tag{3.5}$$

The existence of such a sequence follows from (2.8). Since $s(\rho)$ is an upper semicontinuous function on $\mathcal{I}$,[(6,7)] we get

$$s(\bar{\rho}) \geqslant \lim_{n\to\infty} \sup s(\rho_n) \geqslant \lim_{n\to\infty} S(\rho_n(h)) = S_0 \tag{3.6}$$

while

$$\bar{\rho}(h) = \lim_{n\to\infty} \rho_n(h) = e_0$$

Combining (3.4) with (3.6), we see that the supremum in (3.4) is attained and that (3.1) holds. □

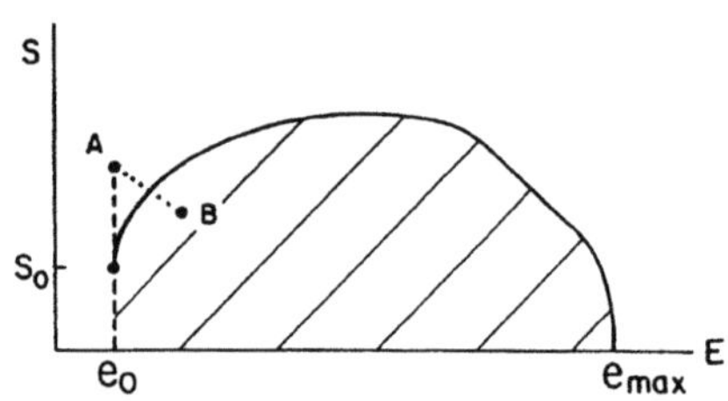


Fig. 1. The general form of $S(E)$. If $S_0 \neq 0$, then the system may have various translation-invariant, energy-minimizing states with different entropies (see Section 4). The dotted line is to illustrate an argument made in the proof of Proposition 1. The shaded area is the range of possible values of $(\rho(h), s(\rho))$ for $\rho \in \mathcal{I}$.

**Corollary 1.** If $\rho \in \mathcal{G}$ is a weak limit of Gibbs states $\rho_n \in \mathcal{G}$ at temperatures $T_n$, with $T_n \to 0$, then

$$s(\rho) = S_0 \tag{3.7}$$

*Proof.* For Gibbs states $s(\rho_n) = S(\rho_n(h))$. Furthermore, $T_n \to 0$ implies that $\rho_n(h) \to e_0$. Thus the arguments in the second part of the above proof apply to $\rho_n$, and prove (3.7). □

*Remark.* Corollary 1 provides us with a method of computing $S_0$. For example, it is known (e.g., by Peierls' argument) that the $T = 0$ limit of the "+ states" of the Ising model in $d \geqslant 2$, with $h \geqslant 0$, is concentrated on the single configuration: $\sigma_i = +1$, $\forall i \in \mathsf{L}$. Thus for this model $S_0 = 0$. The Peierls argument was mentioned merely to illustrate the use of Corollary 1. A better proof of the third law for ferromagnets will proceed directly from Theorem I and will be given in Section 6.

## 4. $S_0$ AND THE DEGENERACY OF GROUND STATES FOR CLASSICAL LATTICE SYSTEMS

Proposition 1 provides a characterization of $S_0$ as the maximal entropy density of a translation-invariant ground state. For classical systems we can also relate it to the number of ground state configurations. Let us first clarify this concept.

**Definition 1.** Let $\Phi$ be an interaction of a classical system with $\sum_{B \ni 0} \|\Phi_B\| < \infty$. A spin configuration $\sigma \in \Omega$ is a *ground state configuration for* $\Lambda \subset \mathsf{L}$ if for any $\sigma' \in \Omega$ that differs from $\sigma$ only in $\Lambda$, i.e., $\sigma_{\Lambda^c} = \sigma'_{\Lambda^c}$,

$$\sum_{\substack{B \subset \mathsf{L} \\ B \cap \Lambda \neq 0}} \left[\Phi_B(\sigma') - \Phi_B(\sigma)\right] \geqslant 0 \tag{4.1}$$

We denote the collection of such configurations by $G_\Lambda$ and call the elements of

$$G = \bigcap_{\substack{\Lambda \subset \mathsf{L} \\ |\Lambda| < \infty}} G_\Lambda$$

*ground state configurations.*

The local energy minimization condition (4.1) gives rise to the set $\pi_\Lambda G_\Lambda \in \Omega_\Lambda$. Notice that while

$$\pi_\Lambda G \subset \pi_\Lambda G_\Lambda \tag{4.2}$$

equality in (4.2) is not generally true. There may be configurations in $\Lambda$ which for certain boundary conditions (i.e., extensions to the whole of $\mathsf{L}$) minimize the energy with respect to variations in $\Lambda$, but which nevertheless cannot be completed to ground state configurations on the whole lattice $\mathsf{L}$.

*Example 1.* An example of the above situation is provided by the nearest neighbor, ferromagnetic Ising spin configuration in a $4L \times L$ box ($d = 2$), for which $\sigma \equiv -1$ on a $2L \times L$ column, symmetrically placed between two $L \times L$ columns on which $\sigma \equiv +1$. This configuration can be extended to L by making the $2L \times L$ column into a $2L \times \infty$ column of minus spins, all the others being plus. This extended configuration cannot be changed inside $\Lambda$ without raising the energy. However, there is no extension with the property that the energy cannot be lowered by changing the spins in some finite box.

It should not be assumed that the presence of both negative and positive spins in $\sigma$ is the reason that $\sigma$ is not a ground state configuration. The configurations in which $\sigma_i = -1$ if $i_x \leqslant x_0$ $[i = (i_x, i_y)]$ and $\sigma_i = +1$ if $i_x > x_0$ are in $G$.

*Example 2.* For an interaction of finite range one can ask whether all the ground states can be characterized by a local condition, i.e., whether $G$ is identical with the set

$$\{\sigma \in \Omega \,|\, \pi_\Lambda \sigma \in \pi_\Lambda G, \quad \forall \Lambda \subset \mathsf{L} \quad \text{with diam} \quad \Lambda < R\}$$

for some $R < \infty$. That is not the case. For example, for the nearest neighbor ferromagnetic Ising system ($d = 2$), and any $R$, the configuration

$$\sigma = \begin{cases} +1 & |x|, |y| \leqslant R/2 \\ -1 & \text{otherwise} \end{cases}$$

is in the above set, but not in $G$.

**Definition 2.** The quantities

$$D_\Lambda = \ln \operatorname{card}(\pi_\Lambda G), \qquad D_\Lambda^* = \ln \operatorname{card}(\pi_\Lambda G_\Lambda)$$

where card $(A)$ is number of elements in a set $A$, are called *ground state configurational entropies in the domain* $\Lambda$.

The quantities $D_\Lambda$ and $D_\Lambda^*$ have some properties similar to those of entropy of states. It is easy to see that $D_\Lambda$ and $D_\Lambda^*$ are subadditive, i.e., if $\Lambda = \Lambda_1 \cup \Lambda_2$ with $\Lambda_1, \Lambda_2$ disjoint, then

$$D^{(*)}_\Lambda \leqslant D^{(*)}_{\Lambda_1} + D^{(*)}_{\Lambda_2} \tag{4.3}$$

Furthermore, by (4.2)

$$D_\Lambda \leqslant D_\Lambda^* \tag{4.4}$$

**Remark.** Entropy is not only subadditive but also strongly subadditive, i.e.,

$$S_{\Lambda_1 \cup \Lambda_2 \cup \Lambda_3} + S_{\Lambda_3} \leqslant S_{\Lambda_1 \cup \Lambda_3} + S_{\Lambda_2 \cup \Lambda_3} \tag{4.5}$$

for three disjoint domains. We do not know under which conditions (4.5) holds for $D^*_\Lambda$.

By standard arguments,[6] (4.5) implies the following result:

**Proposition 2.** The following two limits exist:

$$d^{(*)} = \lim_{\Lambda \uparrow \mathsf{L}} |\Lambda|^{-1} D^{(*)}_\Lambda \tag{4.6}$$

for any sequence of rectangles increasing to L, or for any regular sequence in the sense of Ref. 10. The limit is independent of the sequence.

Now we come to the main facts about $D^{(*)}_\Lambda$ and $S_0$.

**Proposition 3.**

$$d = d^* = S_0 \tag{4.7}$$

*Proof.* In view of (4.4), it is enough to show that

$$d \geqslant S_0 \geqslant d^* \tag{4.8}$$

(i) Let $\rho \in \mathcal{G}$, $\rho(h) = e_0$. We claim this implies that

$$\rho(G) = 1 \tag{4.9}$$

(Here $\rho$ is being thought of as a probability measure on the set of configurations.) For any finite cubic region $\Lambda$ there is a transformation $T_\Lambda : \Omega \to G_\Lambda$ which modifies any configuration $\sigma$ only in $\Lambda$, mapping it on one which minimizes the energy subject to the boundary conditions $\sigma_\Lambda$. If $\rho(G) < 1$, then for some $\Lambda$ this map lowers the energy with positive probability, i.e.,

$$\rho\Big(\sum_{B \cap \Lambda \neq \emptyset} [\Phi_B(T_\Lambda \cdot) - \Phi_B(\cdot)]\Big) = -\Delta E < 0 \tag{4.10}$$

Now choose $k$ large enough so that

$$\epsilon(k) \equiv \sum_{\substack{B \ni 0 \\ \operatorname{diam} B > k}} \|\Phi_B\| < \frac{\Delta E}{10|\Lambda|} \tag{4.11}$$

and let $\{\Lambda_n\}_{n=1,2,\ldots}$ be the collection, ordered in some way, of the translates of $\Lambda$ by vectors in the sublattice $(k + \operatorname{diam}\Lambda)\mathsf{L}$. In order to produce a state with a lower energy density than $\rho$, we construct

$$\rho_n(\cdot) = T_{\Lambda_n} \cdot T_{\Lambda_{n-1}} \cdots T_{\Lambda_1}\rho(\cdot)$$

While the energy decreases produced by subsequent applications of $T_{\Lambda_n}$ may be smaller than $\Delta E$, we still have (since $\{\Lambda_n\}$ are at least the distance $k$ apart)

$$\begin{aligned}\rho_n\Big(\sum_{B\cap\Lambda_n\neq\emptyset}\Phi_B(\cdot)\Big)-\rho_{n-1}\Big(\sum_{B\cap\Lambda_n\neq\emptyset}\Phi_B(\cdot)\Big)\\ \leqslant\rho\Big(\sum_{B\cap\Lambda_n\neq\emptyset}\big[\Phi_B(T_{\Lambda_n}\cdot)-\Phi_B(\cdot)\big]\Big)+\epsilon(k)|\Lambda|\leqslant-\tfrac{4}{5}\Delta E\end{aligned}\tag{4.12}$$

Thus the total gain in energy is proportional to $n$:

$$\begin{aligned}\rho_n\Big(\sum_{B\cap(\cup_1^n\Lambda_m)\neq\emptyset}\Phi_B(\cdot)\Big)-\rho\Big(\sum_{B\cap(\cup_1^n\Lambda_m)\neq\emptyset}\Phi_B(\cdot)\Big)\\ =\sum_{m=1}^{n}\rho_n\Big(\sum_{B\cap\Lambda_n\neq\emptyset}\Phi_B(\cdot)\Big)-\rho_{n-1}\Big(\sum_{B\cap\Lambda_n\neq\emptyset}\Phi_B(\cdot)\Big)\leqslant-\frac{4}{5}\Delta E\cdot n\end{aligned}\tag{4.13}$$

In the limit $n\to\infty$ the states $\rho_n$ converge locally (i.e., weakly) to some state $\rho_\infty$. Averaging $\rho_\infty$ over translations, we obtain a state $\bar{\rho}\in\mathfrak{T}$, for which [using (4.13)]

$$\bar{\rho}(h)\leqslant e_0-(k+\operatorname{diam}\Lambda)^{-d}\tfrac{4}{5}\Delta E<e_0\tag{4.14}$$

(4.14) is a contradiction, which proves (4.9).

The measure $\pi_\Lambda\rho$ is thus concentrated on $\pi_\Lambda G$. By a well-known upper bound on the entropy of a probability distribution over a finite set of $N$ points, $S\leqslant\ln N$, this implies that

$$s_\Lambda(\rho)\leqslant(1/|\Lambda|)D_\Lambda\tag{4.15}$$

(4.15), combined with Proposition 1, proves the first inequality in (4.8).

(ii) We shall prove the second inequality by a variational argument. First, notice the following bound on the interaction across the boundary of any domain $\Lambda\subset\mathsf{L}$:

$$\sum_{\substack{B\cap\Lambda\neq\emptyset\\ B\cap\Lambda^c\neq\emptyset}}\|\Phi_B\|=\sum_{X\in\Lambda}\frac{1}{B\cap\Lambda}\|\Phi_B\|\leqslant|\Lambda|b_\Lambda\tag{4.16}$$

with

$$f_\Lambda(k)=|\{x\in\Lambda\,|\,\operatorname{dist}(x,\Lambda^c)=k\}|/|\Lambda|$$

$$b_\Lambda=\sum_{k=0}^{\infty}f_\Lambda(k)\epsilon(k)$$

and the $\epsilon(k)$ defined in (4.11).

Equation (4.16) implies for any ground state configuration in $\Lambda$, $\sigma\in G_\Lambda$, that for all $\hat{\sigma}\in\Omega$

$$\frac{1}{|\Lambda|}\sum_{B\subset\Lambda}\Phi_B(\sigma)\leqslant\frac{1}{|\Lambda|}\sum_{B\subset\Lambda}\Phi_B(\hat{\sigma})+2b_\Lambda\tag{4.17}$$

With $\sigma$ fixed, take expectation values of both sides of (4.17) in any translation-invariant ground state $\hat{\rho}$ for $\hat{\sigma}$. Using (4.16) once again, we find that for any $\sigma \in G_\Lambda$

$$\frac{1}{|\Lambda|} \sum_{B \subset \Lambda} \Phi_B(\sigma) \leqslant e_0 - \hat{\rho}\left[ \sum_{\substack{B \cap \Lambda \neq \emptyset \\ B \cap \Lambda^c \neq \emptyset}} \frac{|B \cap \Lambda|}{|B|} \Phi_B(\hat{\sigma}) \right] + 2b_\Lambda \leqslant e_0 + 3b_\Lambda \tag{4.18}$$

Consider now a rectangular partition of $\mathsf{L}$, $\Lambda$ being the basic cube, and let $\rho^{(\Lambda)}$ be the (product) state for which the spins in each cube are equally distributed (with equal weights) on all the configurations in $\pi_\Lambda G_\Lambda$, independently of the spins in the other cubes. If $\hat{\rho}^{(\Lambda)} \in \mathcal{G}$ is the state obtained from $\rho^{(\Lambda)}$ by averaging over the translations then, since $\hat{\rho}^{(\Lambda)}$ and $\rho^{(\Lambda)}$ have the same entropies for a corresponding subgroup of translations (by the argument in Ref. 6, Proposition 7.2.3),

$$s(\hat{\rho}^{(\Lambda)}) = (1/|\Lambda|)D^*_\Lambda \tag{4.19}$$

and, using (4.18) and (4.16),

$$\hat{\rho}^{(\Lambda)}(h) \leqslant e_0 + 4b_\Lambda \tag{4.20}$$

By the variational principle for $S(\cdot)$, (2.8) and (3.1), this implies

$$\sup\{ S(e) \mid e \in [e_0, e_0 + 4b_\Lambda]\} \geqslant \frac{1}{|\Lambda|} \quad \ln \operatorname{card}(\pi_\Lambda G_\Lambda) \tag{4.21}$$

Letting $\Lambda \uparrow \mathsf{L}$ (along a sequence of cubes), we have

$$\lim_{\Lambda \uparrow \mathsf{L}} b_\Lambda = \lim_{\Lambda \uparrow \mathsf{L}} \sum_{k=0}^{\infty} f_\Lambda(k)\epsilon(k) = 0 \tag{4.22}$$

since $f_\Lambda(k) \to 0$ for any $k$ and $\epsilon(k) \to 0$ as $k \to \infty$, while $\sum_{k=0}^{\infty} f_\Lambda(k) = 1$. This leads to the conclusion that $S_0 \geqslant d^*$. □

In step (i) of the above proof it was shown that $\rho(G) = 1$ for any state $\rho$ such that (1) $\rho \in \mathcal{G}$ and (2) $\rho(h) = e_0$. In many situations (see Examples 3 and 4) there is a smaller set of ground state configurations $\hat{G} \subset G$ which supports the translation-invariant ground states, i.e., such that $\rho(\hat{G}) = 1$ for any $\rho$ with the above properties 1 and 2. We shall generically denote such a set by $\hat{G}$, although we have not defined it uniquely for the general case. A very useful (see Section 6) strengthening of Proposition 2 is its following corollary, which holds for any version of a set $\hat{G}$ with the above property.

**Corollary 2.**

$$\lim_{\Lambda \uparrow \mathsf{L}} \frac{1}{|\Lambda|} \quad \ln \operatorname{card} \pi_\Lambda \hat{G} = S_0 \tag{4.23}$$

for any regular sequence of domains.

*Proof.* The conclusion of step (i) of the proof of Proposition 2 [which implies (4.22) for $G$ rather than $\hat{G}$] holds for $\hat{G}$ by its defining characterization. The only other property of $G$ on which the proof relied is (4.2), and it is also satisfied by $\hat{G}$. □

*Example 3.* For the two-dimensional Ising model (with $h = -\sum_{|i|=1}\frac{1}{2}\sigma_0\sigma_i$), $e_0 = -2$. Thus $\rho \in \mathcal{I}$, $\rho(h) = e_0$ implies that $\sigma_i\sigma_j = +1$ with $\rho$ probability 1, for any $i, j \in \mathbf{Z}^2$, $|i-j| = 1$. It follows that one may choose $\hat{G}$ to consist of the two configurations $\sigma_i \equiv +1$ and $\sigma_i \equiv -1$. On the other hand, $G$ for this system is an infinite set, as we saw at the end of the discussion of Example 2.

As we shall see now, for finite-range interactions we can restrict $\hat{G}$ by a rigidness condition, whenever $S_0 = 0$.

**Definition 3.** A ground state configuration $\sigma \in G$ is *rigid* if, for any finite $\Lambda \subset \mathbf{L}$, $\sigma_\Lambda$ is the unique energy-minimizing configuration with the boundary condition $\sigma_{\Lambda^c}$.

*Example 4.* The Ising model, $d = 2$, configuration, for which $\sigma_i = +1$ if $i_x, i_y \geqslant 0$ and $\sigma_i = -1$ otherwise, is a ground state configuration which is not rigid. The same is true for any configuration with a single contour which has a step (whose position cannot be fixed by a boundary condition). However, the configuration mentioned at the end of Example 2, with a single straight contour line, is rigid.

**Proposition 4.** If $S_0 = 0$ for a system with finite-range interaction, then any translation-invariant ground state, i.e., $\rho \in \mathcal{I}$ with $\rho(h) = e$, is supported on the set of rigid configurations in $G$.

*Proof.* Let $\rho$ be a state for which the above assumptions are satisfied, $\Lambda$ a finite domain in $\mathbf{L}$, and $R$ the interaction range. Viewing $\rho$ as a measure on the space of the configurations $\sigma$, let $\epsilon$ be the probability that the boundary condition $\sigma_{\Lambda^c}$ does not have a unique ground state in $\Lambda$. We shall now show that

$$S_0 \geqslant \epsilon(R + \operatorname{diam}\Lambda)^{-d}\tfrac{1}{2}\ln 2 \tag{4.24}$$

(4.2) would clearly imply the stated conclusion.

Consider the collection $\{\Lambda_n\}$ of translates of $\Lambda$ by vectors in the sublattice $(R + \operatorname{diam}\Lambda)\mathbf{L}$. Let $g_k(\sigma)$ be the function which for each configuration gives the fraction of boxes, among $\{\Lambda_1, \ldots, \Lambda_k\}$, in which $\sigma_{\Lambda_i^c}$ does not have a unique ground state. By the translation invariance of $\rho$, the expectation value of $g_k(\sigma)$ is $\epsilon$. Since $0 \leqslant g_k(\cdot) \leqslant 1$

$$\epsilon = \rho(g_k(\cdot)) \leqslant \epsilon/2 + \rho(\{\sigma \in \Omega \,|\, g_k(\sigma) \geqslant \epsilon/2\}) \tag{4.25}$$

and we see that

$$\rho(\{\sigma \in \Omega \,|\, g_k(\sigma) \geqslant \epsilon/2\}) \geqslant \epsilon/2 \tag{4.26}$$

We now define $\rho'$ as the state obtained from $\rho$ by redistributing $\sigma$ in $\bigcup_1^\infty \Lambda_n$, for any given values of $\sigma$ in

$$A = \left( \bigcup_1^\infty \Lambda_n \right)^c,$$

equally among all the ground states in that set. Since the $\{\Lambda_n\}$ are at least a distance $R$ apart, this conditional distribution has the product structure. It follows therefore, using (4.26), that the entropy density of $\rho'$ with respect to the sublattice $(R + \operatorname{diam}\Lambda)\mathbb{L}$, is at least $(R + \operatorname{diam}\Lambda)^{-d}\epsilon/2\ln 2$. By averaging $\rho'$ over translations we obtain a translation-invariant ground state $\bar{\rho}$ with the same entropy (by a standard argument to which we referred in the proof of Proposition 2).

Now (4.24) follows by the variational principle of Proposition 1, applied to $\bar{\rho}$. □

For systems with finite-range interactions we also have:

**Corollary 3.** Let $N_\Lambda(b_\Lambda)$ denote the degeneracy of the ground state in $\Lambda$ with the boundary condition $b_\Lambda \in \Omega_{\Lambda^c}$. If $\Phi$ is an interaction of finite range, the following limit exists and

$$\lim_{\Lambda \uparrow \infty} \sup_{b_\Lambda \in \Omega_{\Lambda^c}} \frac{1}{|\Lambda|} \ln N_\Lambda(b_\Lambda) = S_0 \tag{4.27}$$

*Proof.* Let the interaction range be $R$. The number of inequivalent boundary conditions for $\Lambda$ is bounded by $\exp(\alpha C_d R|\partial\Lambda|)$, where $\exp(\alpha)$ is the number of points in $\mathcal{H}_0^{(c)}$ and $C_d$ is a dimension-dependent constant. Therefore

$$N_\Lambda(b_\Lambda) \leqslant \operatorname{card} \pi_\Lambda G_\Lambda \leqslant N_\Lambda(b_\Lambda) \exp(\alpha C_d R|\partial\Lambda|) \tag{4.28}$$

Substituting (4.28) in (4.3), we get (4.27). □

Thus we have found an answer to Griffith's puzzle. While the third law is not implied by the nondegeneracy of the finite-volume systems with certain boundary conditions, $S_0$ still counts the maximal degeneracy. What for a given boundary condition may seem as a bulk contribution to the entropy due to low-lying excitations, may also be accounted for as a boundary effect!

Systems for which $S_0 \neq 0$ may, nevertheless, have ground states with zero entropy which satisfy the $T = 0$ version of the Dobrushin–Lanford–Ruelle equilibrium condition.[8] An example is the $d = 2$ antiferromagnetic triangular lattice Ising model discussed in Ref. 9, in which it is possible to suppress the ground state degeneracy by fixing the spins in one row to be all $+$. By Proposition 1 such states cannot be attained as limits of equilibrium states with $T \to 0$. We can now see the reason for it. $S_0 \neq 0$ implies that the system has more degenerate boundary conditions. If at any fixed $T \neq 0$ one tries to impose the nondegenerate boundary, as soon as $\Lambda$ is large enough free energy considerations favor spontaneous change of the spins in a boundary layer.

## 5. QUANTUM LATTICE SYSTEMS

We shall now introduce the notion of ground states for infinite quantum lattice systems. This will enable us to extend to such systems the basic relation between $S_0$ and the degeneracy in finite volume. However, there will not be a simple analog of Corollary 3.

A basic property of ground states, and of the ground state configurations discussed in Section 4, is the minimization of the energy with respect to local perturbations. The formalization of the notion of local perturbation of a quantum system leads to an interesting observation. A natural definition, which draws on the properties of the mapping induced on $\mathcal{Q}$ by the mapping $T_\Lambda : \Omega \to \Omega$ discussed in Section 4, is:

**Definition 4.** Let $\mathcal{Q}$ be the algebra of observables of an infinite quantum lattice system. A local perturbation in $\Lambda \subset \mathsf{L}$ is a linear mapping $T$: $\mathcal{Q} \to \mathcal{Q}$ such that:

(a) $T(A) \geqslant 0$ for any $A \geqslant 0$.

(b) $T(A) = A$ for any $A \in \mathcal{Q}_{\Lambda^c}$ [in particular $T(1) = 1$].

(In our shorthand notation $A = 1_\Lambda \otimes A \in \mathcal{Q}$, for $A \in \mathcal{Q}_{\Lambda^c}$.)

It turns out[12] that quite generally the above properties imply also:

(c) $T(X) \in \mathcal{Q}_\Lambda$ for any $X \in \mathcal{Q}_\Lambda$.

(d) $T(XY) = T(X)Y$ for any $X \in \mathcal{Q}_\Lambda$, $Y \in \mathcal{Q}_{\Lambda^c}$.

(e) $T$ is completely positive.

It is gratifying to have (e), which is sometimes imposed on physical grounds.[13] However, it may seem surprising that (c) is true, because it is not implied in the classical case. In Section 4 we used local perturbations, which also satisfy (a) and (b), for which the change produced in $\Lambda$ could also depend on the configuration in $\Lambda^c$ (without modifying it). As we see, there are no such perturbations, in the sense of action on the observables in a state-independent way, for quantum systems. In view of the fact that a classical system (e.g., Ising model) can be thought of as a special quantum system, it may seem that there is a paradox here. There is none. The reason that (c) is true quantum mechanically is that we insisted upon $\mathcal{Q}_\Lambda$ being the full matrix algebra. Noncommutativity plays a role in (c).

If it is desired to have consistency between the quantum and classical definitions of local perturbations and of the ground states (to be defined shortly), one could follow either of the following paths:

1. Impose (c) as a restriction in the definition of local perturbations in Section 4.

2. Define local perturbations directly on states. Or, equivalently, use a state-dependent definition, replacing (b) by

(b′) $\rho(T(A)) = \rho(A)$ for any $A \in \mathfrak{A}_{\Lambda^c}$.

We shall not bother to do either. Let us, however, remark that, had we followed path 1, the set of classical ground states $G$ would not have been changed, although $G_\Lambda$ would have been.

**Definition 5.** Let $\Phi$ be an interaction of a quantum system with $\sum_{B \ni 0} \|\Phi_B\| < \infty$. A state $\rho$ is a ground state in $\Lambda$ if for any local perturbation $T$ in $\Lambda$

$$\sum_{B \cap \Lambda \neq \varnothing} \left[\rho(T\Phi_B) - \rho(\Phi)\right] \geqslant 0 \tag{5.1}$$

We denote the collection of ground states in $\Lambda$ by $\mathcal{G}_\Lambda$, and call the elements of

$$\mathcal{G} = \bigcap_{\substack{\Lambda \subset L \\ |\Lambda| < \infty}} \mathcal{G}_\Lambda$$

the ground states of the systems.

For quantum systems we define $Q$ and $Q^*$, in analogy to $D^{(*)}$, by

$$Q^{(*)} = \sup\{s_\Lambda(\rho) \,|\, \rho \in \mathcal{G}_{(\Lambda)}\} \tag{5.2}$$

For classical systems, when viewed as quantum, $Q$ and $D$ are the same (but $Q^*$ and $D^*$ are not).

With the corresponding substitutions, the proof of Proposition 2 carries through, almost verbatim, to the quantum case. In particular, (4.3) corresponds to a well-known subadditivity of entropy. Thus we have:

**Proposition 5.** The following limits, through sequences of regular domains, exist and

$$\lim_{\Lambda \uparrow L} \frac{1}{|\Lambda|} Q_\Lambda = \lim_{\Lambda \uparrow L} \frac{1}{|\Lambda|} Q_\Lambda^* = S_0 \tag{5.3}$$

## 6. APPLICATIONS OF THE THEORY TO SPECIFIC MODELS

### 6.1. The Third Law for Ferromagnetic Models

We shall now demonstrate how to use the relation of $S_0$ to the ground state degeneracy, or entropy, to prove the third law for certain systems. A common simplifying feature in Examples 5 and 6 is that all the terms of the interaction $\{\Phi_B\}$ can attain their minima simultaneously. Consequently, $e_0$ can be computed, and $\mathcal{G}$ consists of states which minimize each $\Phi_B$. This does not apply to the third example (and to that referred to in Ref. 9), where nevertheless one may still find $e_0$ by grouping $\{\Phi_B\}$ into larger units.

*Example 5.* Ferromagnetic Ising systems. These are classical systems with the spin variables $\sigma_i$ having values in $\mathcal{K}_i = \{-\iota, -\iota + 1, \ldots, \iota\}$ and the Hamiltonian

$$H = -\sum J_B \sigma_B \tag{6.1}$$

where $\sigma_B = \prod_{i \in B} \sigma_i$ and

$$J_B \geqslant 0 \qquad \forall B \subset \mathsf{L} \tag{6.2}$$

**Proposition 6.** $S_0 = 0$ for any ferromagnetic Ising system with $H \neq 0$.

*Proof.* The interactions $\Phi_B = J_B \sigma_B$ attain their minima on the configuration $\sigma \equiv +1$ (among others, possibly). Thus $e_0 = \sum_{B \ni 0} (1/|B|) J_B$, and $\rho(h) = e_0$ if and only if

$$\rho(\sigma_B) = \iota^{|B|} \tag{6.3}$$

for any $B \subset \mathsf{L}$ such that $J_B \neq 0$.

Since $\iota^{|B|}$ is the maximal value that $\sigma_B$ takes, (6.3) implies that $\hat{G}$, the support of the translation-invariant, energy-minimizing states (cf. Section 4), consists of configurations for which

$$\sigma_B = 1 \tag{6.4}$$

whenever $J_B \neq 0$.

We now claim that for any cube $\Lambda$

$$\operatorname{card} \pi_\Lambda \hat{G} \leqslant 2^{C|\partial \Lambda|} \tag{6.5}$$

with some fixed $C = C(J) < \infty$. To see this choose $B_0 \subset \mathsf{L}$ such that $J_{B_0} \neq 0$. Then (6.5) follows from the observation that, using (6.4) over all the translates of $B_0$, the values of $\sigma$ in $\Lambda$, for $\sigma \in \hat{G}$, can be uniquely reconstructed from the values in the shell of width $\operatorname{diam} B_0$ surrounding $\Lambda$.

(6.5) implies that for any $\rho$ as above

$$s_\Lambda(\rho) \leqslant C(|\partial\Lambda|/|\Lambda|) \ln 2 \to 0 \tag{6.6}$$

as $\Lambda \uparrow \mathsf{L}$. By Proposition 1 this proves that $S_0 = 0$. □

The above result is not new, having been proven in Refs. 4 and 5. However, as we have already mentioned, the previous proofs relied on less direct arguments, such as correlation inequalities or Lee–Yang methods.

*Example 6.* Quantum Heisenberg ferromagnet. The system consists of the usual quantum spin-1/2 variables $\boldsymbol{\sigma}_i = (\sigma_i^{(x)}, \sigma_i^{(y)}, \sigma_i^{(z)})$, interacting via

$$H = -\sum J_{ij}\boldsymbol{\sigma}_i \cdot \boldsymbol{\sigma}_j \tag{6.7}$$

with translation-invariant $J_{ij} \geqslant 0$. In the most common model $J$ couples nearest neighboring spins, but we shall not make this assumption.

**Proposition 7.** $S_0 = 0$ for any ferromagnetic Heisenberg model with nonzero interaction.

*Proof.* By well-known properties of spins,

$$\boldsymbol{\sigma}_i \cdot \boldsymbol{\sigma}_j = \tfrac{1}{2}(\boldsymbol{\sigma}_i + \boldsymbol{\sigma}_j)^2 - \tfrac{1}{2}(\boldsymbol{\sigma}_i^2 + \boldsymbol{\sigma}_j^2) = P_{ij} - \tfrac{3}{4} \tag{6.8}$$

where $P_{ij}$ is the projection on the symmetric subspace for the $(i, j)$ permutation. Thus, while the $\{\Phi_B\}$ do not commute, any finite number of them can attain their minima simultaneously (on the corresponding totally symmetric, or maximal angular momentum space). Therefore we can compute $e_0$ and, what is more important, conclude that $\rho \in \mathcal{G}$ and $\rho(h) = e_0$ imply

$$\rho(\boldsymbol{\sigma}_i \cdot \boldsymbol{\sigma}_j) = \tfrac{1}{4}, \qquad \rho(P_{ij}) = 1 \tag{6.9}$$

for each $(i, j)$ with $J_{ij} \neq 0$.

For the sake of clarity, we shall first consider the usual nearest neighbor model. Generally $\rho(P_{ij}) = \rho(P_{jk}) = 1$ implies $\rho(P_{ik}) = 1$. Thus the transitivity of $J$ and (6.9) imply that for any $\rho$ that satisfies the above conditions, (6.9) holds for any $(i, j)$. Consequently, for any $\Lambda \subset \mathbb{L}$, $\pi_\Lambda \rho$ is concentrated on the $(|\Lambda| + 1)$-dimensional (Hilbert space) subspace of completely symmetric functions, corresponding to

$$\Big(\sum_{i \in \mathfrak{L}} \boldsymbol{\sigma}_i\Big)^2 = \tfrac{1}{2}|\Lambda|\big(\tfrac{1}{2}|\Lambda| + 1\big)$$

By a well-known upper bound on the entropy of a density operator of finite rank, this leads to

$$s(\rho)_\Lambda \leqslant (1/|\Lambda|)\ln(|\Lambda| + 1) \tag{6.10}$$

Invoking Proposition 1, we conclude that $S_0 = 0$.

To conclude the proof, we note that in the general case any $\Lambda$ is decomposed into a finite number of connected components, by $J_{ij}$. As in Example 1, this number is bounded by const $\cdot$ (diam $B_0$) $\cdot$ $|\partial\Lambda|$. In each component the total angular momentum must be maximal. Therefore $s_\Lambda(\rho)$ is bounded by the sum of the right sides of (6.6) and (6.10). Again, $S_0 = 0$. □

The above argument does not apply to antiferromagnets, since there the $\{\Phi_B\}$ do not attain their minima simultaneously. The proof that $S_0 = 0$ for such systems is a very intriguing *open problem*.

## 6.2. Examples of Systems Which Violate the Third Law

*Example 7.* The following system, with $S_0 \neq 0$, has been considered by Griffiths,[1,2] who used it to demonstrate the need to consider low-lying excitations. The system, in $d = 3$, consists of spin variables which take the values $\sigma_i = 0, \pm 1$, and interact via the Hamiltonian

$$H = 3\sum_i \sigma_i^2 - \frac{1}{2} \sum_{|i-j|=1} \sigma_i^2 \sigma_j^2 \tag{6.11}$$

To facilitate the analysis of the system in finite domains $\Lambda$, let us rewrite the Hamiltonian as

$$H_\Lambda = \frac{1}{2} \sum_{\substack{i,j \in \Lambda \\ |i-j|=1}} \left(\sigma_i^2 - \sigma_j^2\right)^2 + \frac{1}{2} \sum_{i \in \Lambda} \sigma_i^2 \operatorname{card}\{ j \notin \Lambda \,|\, |j-i| = 1\} \tag{6.12}$$

The first term in (6.11) attains its absolute minimum on any configuration with $\sigma_i^2 = \text{const}$. Equation (6.12) clearly shows that for the "free" boundary conditions in (6.11) the only ground state is $\sigma_i \equiv 0$. As Griffiths pointed out, from this point of view the system has at least $2^{|\Lambda|}$ excitations with the low energy $\frac{1}{2}|\partial\Lambda|$, corresponding to configurations on which $\sigma_i = \pm 1$, but not 0. Thus, he concluded, $S_0 \geqslant \ln 2$, despite the nondegeneracy in finite volumes (for the above "free" boundary conditions).

Our approach permits us to discard the second term in (6.12) [which has no effect on $S(E)$ in the thermodynamic limit]. . We then have a case where all the pair energies can be minimized simultaneously. Following, mutatis mutandis, the analysis of Example 5, we conclude that the restricted set of ground state configurations $\hat{G}$ consists of all those for which $\sigma_i = \pm 1$, but not 0, and the one where $\sigma_i \equiv 0$. Thus $S_0 = \ln 2$, i.e., the lower bound described by Griffiths gives the full value of $S_0$.

There are other spectacular examples of lattice models which have $S_0 \neq 0$. One is the $d = 2$, triangular Ising antiferromagnet. Wannier[11] studied it in detail and calculated $S_0$ by explicitly calculating the limit in (1.4). By counting some of the ground state configurations (using an argument which he attributes to Anderson), Wannier found an entropy lower bound which was lower than $S_0$, as it should be. The point of our paper is that $S_0$ can really be computed by this direct route, even though the calculation is not an elementary matter.

Other examples for which $S_0 \neq 0$ can be calculated are the dimer systems[17,18] and the ice models.[19] The latter, having been proposed by Pauling[16] to account for the observed residual entropy of ice, must be considered to be one of the more successful applications of statistical mechanics to the real world.[15] Our point, once again, is that calculating the ground state configurational degeneracy is legitimate.

All these models also illustrate another important point. By choosing particular boundary conditions one can actually construct states for which the entropy is less than $S_0$, in fact zero. At first sight this might be considered conceptually puzzling, because if the ground state degeneracy is the only important quantity for $S_0$, then which one should be used? Answer: the maximal degeneracy.

As a final remark, let us remind the reader that in many models the set of parameters (of the Hamiltonian) on which the third law is violated forms a submanifold of lower dimension. A good general result in this direction still remains to be proven. A somewhat easier open problem is to decide whether the set of interactions of any given finite range for which $S_0 = 0$ is open.

## ACKNOWLEDGMENT

We thank W. Thirring for stimulating discussions.

## REFERENCES


1. R. B. Griffiths, *J. Math. Phys.* **6**:1447 (1965).
2. R. B. Griffiths, in *A Critical Review of Thermodynamics*, E. B. Stuart, B. Gal-Or, and A. J. Brainard, eds. (Mono Book Corp., Baltimore, 1970).
3. H. B. G. Casimir, *Z. Physik* **171**:246 (1963); M. J. Klein, in *Termodinamica Dei Processi Irreversibili*, S. R. De Groot, ed. (N. Zamichelli, Bologna, 1960), p. 1.
4. H. S. Leff, *Phys. Rev. A* **2**:2368 (1970).
5. J. Slawny, *Comm. Math. Phys.* **34**:271 (1973); **46**:75 (1976).
6. D. Ruelle, *Statistical Mechanics* (W. A. Benjamin, New York, 1969).
7. A. Wehrl, *Rev. Mod. Phys.* **50**:221 (1978).
8. R. Israel, *Convexity in the Theory of Lattice Gases* (Princeton Univ. Press, 1979).
9. R. L. Dobrushin, *Theor. Prob. Appl.* **13**:197 (1968); O. E. Lanford III and D. Ruelle, *Comm. Math. Phys.* **13**:194 (1969).
10. E. H. Lieb and J. L. Lebowitz, *Adv. Math.* **9**:316 (1972), p. 351.
11. G. H. Wannier, *Phys. Rev.* **79**:357 (1950).
12. M. Aizenman, E. B. Davies, and E. H. Lieb, *Adv. Math.* **28**:84 (1978).
13. E. B. Davies, *Quantum Theory of Open Systems* (Academic Press, New York, 1976).
14. J. Slawny, private communication.
15. W. F. Giauque and M. A. Ashley, *Phys. Rev.* **43**:81 (1933); W. F. Giauque and J. W. Stout, *J. Am. Chem. Soc.*, **58**:1144 (1936).
16. L. Pauling, *J. Am. Chem. Soc.* **57**:2680 (1935).
17. P. W. Kasteleyn, *Physica* **27**:1209 (1962).
18. H. N. V. Temperley and M. E. Fisher, *Phil. Mag.* **6**:1061 (1961).
19. E. H. Lieb, *Phys. Rev. Lett.* **18**:692 (1967); *Phys. Rev.* **162**:162 (1967).

With J. Yngvason in Notices of the AMS *45*, 571–581 (1998)

# A Guide to Entropy and the Second Law of Thermodynamics

*Elliott H. Lieb and Jakob Yngvason*

This article is intended for readers who, like us, were told that the second law of thermodynamics is one of the major achievements of the nineteenth century—that it is a logical, perfect, and unbreakable law—but who were unsatisfied with the "derivations" of the entropy principle as found in textbooks and in popular writings.

A glance at the books will inform the reader that the law has "various formulations" (which is a bit odd, as if to say the Ten Commandments have various formulations), but they all lead to the existence of an entropy function whose reason for existence is to tell us which processes can occur and which cannot. We shall abuse language (or reformulate it) by referring to the existence of entropy as *the* second law. This, at least, is unambiguous. The entropy we are talking about is that defined by thermodynamics (and *not* some analytic quantity, usually involving expressions such as $-p \ln p$, that appears in information theory, probability theory, and statistical mechanical models).

There are three laws of thermodynamics (plus one more, due to Nernst, which is mainly used in low-temperature physics and is not immutable—as are the others). In brief, these are:

*The Zeroth Law*, which expresses the transitivity of thermal equilibrium and which is often said to imply the existence of temperature as a parametrization of equilibrium states. We use it below but formulate it without mentioning temperature. In fact, temperature makes no appearance here until almost the very end.

*The First Law*, which is conservation of energy. It is a concept from mechanics and provides the connection between mechanics (and things like falling weights) and thermodynamics. We discuss this later on when we introduce simple systems; the crucial usage of this law is that it allows energy to be used as one of the parameters describing the states of a simple system.

*The Second Law.* Three popular formulations of this law are:

> *Clausius:* No process is possible, the sole result of which is that heat is transferred from a body to a hotter one.
>
> *Kelvin (and Planck):* No process is possible, the sole result of which is that a body is cooled and work is done.
>
> *Carathéodory:* In any neighborhood of any state there are states that cannot be reached from it by an adiabatic process.

All three formulations are supposed to lead to the entropy principle (defined below). These steps can be found in many books and will not be trod-

*Elliott H. Lieb is professor of mathematics and physics at Princeton University. His e-mail address is* `lieb@math.princeton.edu`. *Work partially supported by U.S. National Science Foundation grant PHY95-13072A01.*

*Jakob Yngvason is professor of theoretical physics at Vienna University. His e-mail address is* `yngvason@thor.thp.univie.ac.at`. *Work partially supported by the Adalsteinn Kristjansson Foundation, University of Iceland.*

den again here. Let us note in passing, however, that the first two use concepts such as hot, cold, heat, cool that are intuitive but have to be made precise before the statements are truly meaningful. No one has seen "heat", for example. The last (which uses the term "adiabatic process", to be defined below) presupposes some kind of parametrization of states by points in $\mathbf{R}^n$, and the usual derivation of entropy from it assumes some sort of differentiability; such assumptions are beside the point as far as understanding the meaning of entropy goes.

Why, one might ask, should a mathematician be interested in this matter, which historically had something to do with attempts to understand and improve the efficiency of steam engines? The answer, as we perceive it, is that the law is really an interesting mathematical theorem about an ordering on a set, with profound physical implications. The axioms that constitute this ordering are somewhat peculiar from the mathematical point of view and might not arise in the ordinary ruminations of abstract thought. They are special but important, and they are driven by considerations about the world, which is what makes them so interesting. Maybe an ingenious reader will find an application of this same logical structure to another field of science.

The basic input in our analysis is a certain kind of ordering on a set, and denoted by

$$\prec$$

(pronounced "precedes"). It is transitive and reflexive, as in A1, A2 below, but $X \prec Y$ and $Y \prec X$ do not imply $X = Y$, so it is a "preorder". The big question is whether $\prec$ can be encoded in an ordinary, real-valued function, denoted by $S$, on the set, such that if $X$ and $Y$ are related by $\prec$, then $S(X) \leq S(Y)$ if and only if $X \prec Y$. The function $S$ is also required to be additive and extensive in a sense that will soon be made precise.

A helpful analogy is the question: When can a vector-field, $V(x)$, on $\mathbf{R}^3$ be encoded in an ordinary function, $f(x)$, whose gradient is $V$? The well-known answer is that a necessary and sufficient condition is that $\operatorname{curl} V = 0$. Once $V$ is observed to have this property, one thing becomes evident and important: it is necessary to measure the integral of $V$ only along some curves—not all curves—in order to deduce the integral along *all* curves. The encoding then has enormous predictive power about the nature of future measurements of $V$. In the same way, knowledge of the function $S$ has enormous predictive power in the hands of chemists, engineers, and others concerned with the ways of the physical world.

Our concern will be the existence and properties of $S$, starting from certain natural axioms about the relation $\prec$. We present our results without proofs, but full details and a discussion of related previous work on the foundations of classical thermodynamics are given in [7]. The literature on this subject is extensive, and it is not possible to give even a brief account of it here, except for mentioning that the previous work closest to ours is that of [6] and [2] (see also [4], [5], and [9]). These other approaches are also based on an investigation of the relation $\prec$, but the overlap with our work is only partial. In fact, a major part of our work is the derivation of a certain property (the "comparison hypothesis" below), which is taken as an axiom in the other approaches. It was a remarkable and largely unsung achievement of Giles [6] to realize the full power of this property.

Let us begin the story with some basic concepts.

1. *Thermodynamic system:* Physically this consists of certain specified amounts of certain kinds of matter, e.g., a gram of hydrogen in a container with a piston, or a gram of hydrogen and a gram of oxygen in two separate containers, or a gram of hydrogen and two grams of hydrogen in separate containers. The system can be in various states which, physically, are *equilibrium states*. The space of states of the system is usually denoted by a symbol such as $\Gamma$ and states in $\Gamma$ by $X, Y, Z$, etc.

Physical motivation aside, a state-space, mathematically, is just a set to begin with; later on we will be interested in embedding state-spaces in some convex subset of some $\mathbf{R}^{n+1}$; i.e., we will introduce coordinates. As we said earlier, however, the entropy principle is quite independent of coordinatization, Carathéodory's principle notwithstanding.

2. *Composition and scaling of states:* The notion of Cartesian product, $\Gamma_1 \times \Gamma_2$, corresponds simply to the two (or more) systems being side by side on the laboratory table; mathematically it is just another system (called a *compound system*), and we regard the state-space $\Gamma_1 \times \Gamma_2$ as being the same as $\Gamma_2 \times \Gamma_1$. Points in $\Gamma_1 \times \Gamma_2$ are denoted by pairs $(X, Y)$, as usual. The subsystems comprising a compound system are physically independent systems, but they are allowed to interact with each other for a period of time and thereby to alter each other's state.

   The concept of scaling is crucial. It is this concept that makes our thermodynamics inappropriate for microscopic objects like atoms or cosmic objects like stars. For each state-space $\Gamma$ and number $\lambda > 0$ there is another state-space, denoted by $\Gamma^{(\lambda)}$, with points denoted by $\lambda X$. This space is called a *scaled copy* of $\Gamma$. Of course we identify $\Gamma^{(1)} = \Gamma$ and $1X = X$. We also require $(\Gamma^{(\lambda)})^{(\mu)} = \Gamma^{(\lambda\mu)}$ and $\mu(\lambda X) = (\mu\lambda)X$. The physical interpretation of $\Gamma^{(\lambda)}$ when $\Gamma$ is the space of one gram of hydrogen is simply the state-space of $\lambda$ grams of hydrogen. The state $\lambda X$ is the state of $\lambda$ grams of hydrogen with the same "intensive" properties as $X$, e.g., pressure, while "extensive" properties like energy, volume, etc., are scaled by a factor $\lambda$ (by definition).

For any given $\Gamma$ we can form Cartesian product state-spaces of the type $\Gamma^{(\lambda_1)} \times \Gamma^{(\lambda_2)} \times \cdots \times \Gamma^{(\lambda_N)}$. These will be called *multiple-scaled copies* of $\Gamma$.

The notation $\Gamma^{(\lambda)}$ should be regarded as merely a mnemonic at this point, but later on, with the embedding of $\Gamma$ into $\mathbb{R}^{n+1}$, it will literally be $\lambda\Gamma = \{\lambda X : X \in \Gamma\}$ in the usual sense.

3. *Adiabatic accessibility:* Now we come to the ordering. We say $X \prec Y$ (with $X$ and $Y$ possibly in *different* state-spaces) if there is an *adiabatic process* that transforms $X$ into $Y$.

What does this mean? Mathematically, we are just given a list of pairs $X \prec Y$. There is nothing more to be said, except that later on we will assume that this list has certain properties that will lead to interesting theorems about this list and will lead, in turn, to the existence of an *entropy function, S*, characterizing the list.

The physical interpretation is quite another matter. In textbooks a process is usually called adiabatic if it takes place in "thermal isolation", which in turn means that "no heat is exchanged with the surroundings". Such statements appear neither sufficiently general nor precise to us, and we prefer the following version (which is in the spirit of Planck's formulation of the second law [8]). It has the great virtue (as discovered by Planck) that it avoids having to distinguish between work and heat—or even having to define the concept of heat. We emphasize, however, that the theorems do not require agreement with our physical definition of adiabatic process; other definitions are conceivably possible.

*A state Y is adiabatically accessible from a state X, in symbols $X \prec Y$, if it is possible to change the state from X to Y by means of an interaction with some device consisting of some auxiliary system and a weight in such a way that the auxiliary system returns to its initial state at the end of the process, whereas the weight may have risen or fallen.*

The role of the "weight" in this definition is merely to provide a particularly simple source (or sink) of mechanical energy. Note that an adiabatic process, physically, does not have to be gentle, or "static" or anything of the kind. It can be arbitrarily violent! (See Figure 1.)

An example might be useful here. Take a pound of hydrogen in a container with a piston. The states are describable by two numbers, energy and volume, the latter being determined by the position of the piston. Starting from some state $X$, we can take our hand off the piston and let the volume increase explosively to a larger one. After things have calmed down, call the new equilibrium state $Y$. Then $X \prec Y$. Question: Is $Y \prec X$ true? Answer: No. To get from $Y$ to $X$ we would have to use some machinery and a weight, with the machinery returning to its initial state, and there is no way this can be done. Using a weight, we can indeed recompress the gas to its original volume, but we will find that the energy is then larger than its original value.

On the other hand, we could let the piston expand very, very slowly by letting it raise a carefully calibrated weight. No other machinery is involved. In this case, we can reverse the process (to within an arbitrarily good accuracy) by adding a tiny bit to the weight, which will then slowly push the piston back. Thus, we could have (in principle, at least) both $X \prec Y$ and $Y \prec X$, and we would call such a process a *reversible adiabatic process.*

Let us write

$$X \prec\prec Y \quad \text{if} \quad X \prec Y$$

but not

$$Y \prec X \quad (\text{written } Y \not\prec X).$$

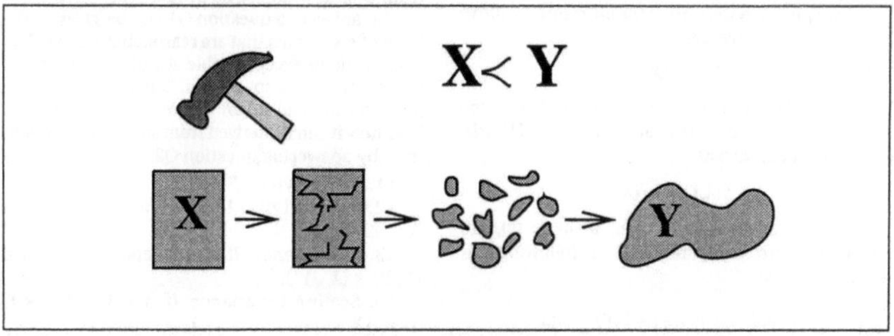


**Figure 1. A violent adiabatic process connecting equilibrium states *X* and *Y*.**

In this case we say that we can go from $X$ to $Y$ by an *irreversible adiabatic process*. If $X \prec Y$ and $Y \prec X$ (i.e., $X$ and $Y$ are connected by a reversible adiabatic process), we say that $X$ and $Y$ are *adiabatically equivalent* and write

$$X \stackrel{A}{\sim} Y.$$

Equivalence classes under $\stackrel{A}{\sim}$ are called *adiabats*.

4. *Comparability*: Given two states $X$ and $Y$ in two (same or different) state-spaces, we say that they are comparable if $X \prec Y$ or $Y \prec X$ (or both). This turns out to be a crucial notion. Two states are not always comparable; a necessary condition is that they have the same material composition in terms of the chemical elements. Example: Since water is $H_2O$ and the atomic weights of hydrogen and oxygen are 1 and 16 respectively, the states in the compound system of 2 grams of hydrogen and 16 grams of oxygen are comparable with states in a system consisting of 18 grams of water (but not with 11 grams of water or 18 grams of oxygen).

Actually, the classification of states into various state-spaces is done mainly for conceptual convenience. The second law deals only with states, and the only thing we really have to know about any two of them is whether or not they are comparable. Given the relation $\prec$ for all possible states of all possible systems, we can ask whether this relation can be encoded in an entropy function according to the following:

**Entropy principle:** *There is a real-valued function on all states of all systems (including compound systems) called* **entropy**, *denoted by $S$, such that*

*a)* Monotonicity: *When $X$ and $Y$ are comparable states, then*

$$X \prec Y \quad \text{if and only if} \quad S(X) \le S(Y)\,. \tag{1}$$

*b)* Additivity and extensivity: *If $X$ and $Y$ are states of some (possibly different) systems and if $(X, Y)$ denotes the corresponding state in the compound system, then the entropy is additive for these states; i.e.,*

$$S(X, Y) = S(X) + S(Y)\,. \tag{2}$$

*$S$ is also extensive; i.e., for each $\lambda > 0$ and each state $X$ and its scaled copy $\lambda X \in \Gamma^{(\lambda)}$ (defined in 2, above)*

$$S(\lambda X) = \lambda S(X)\,. \tag{3}$$

A formulation logically equivalent to (a), not using the word "comparable", is the following pair of statements:

$$X \stackrel{A}{\sim} Y \Longrightarrow S(X) = S(Y) \quad \text{and} \\ X \prec\prec Y \Longrightarrow S(X) < S(Y). \tag{4}$$

The last line is especially noteworthy. It says that entropy must increase in an irreversible adiabatic process.

The additivity of entropy in compound systems is often just taken for granted, but it is one of the startling conclusions of thermodynamics. First of all, the content of additivity, (2), is considerably more far-reaching than one might think from the simplicity of the notation. Consider four states, $X, X', Y, Y'$, and suppose that $X \prec Y$ and $X' \prec Y'$. One of our axioms, A3, will be that then $(X, X') \prec (Y, Y')$, and (2) contains nothing new or exciting. On the other hand, the compound system can well have an adiabatic process in which $(X, X') \prec (Y, Y')$ but $X \not\prec Y$. In this case, (2) conveys much information. Indeed, by monotonicity there will be many cases of this kind, because the inequality $S(X) + S(X') \le S(Y) + S(Y')$ certainly does not imply that $S(X) \le S(Y)$. The fact that the inequality $S(X) + S(X') \le S(Y) + S(Y')$ tells us *exactly* which adiabatic processes are allowed in the compound system (among comparable states), independent of any detailed knowledge of the manner in which the two systems interact, is astonishing and is at the *heart of thermodynamics*. The second reason that (2) is startling is this: From (1) alone, restricted to one system, the function $S$ can be replaced by $29S$ and still do its job, i.e., satisfy (1). However, (2) says that it is possible to calibrate the entropies of all systems (i.e., simultaneously adjust all the undetermined multiplicative constants) so that the entropy $S_{1,2}$ for a compound $\Gamma_1 \times \Gamma_2$ is $S_{1,2}(X, Y) = S_1(X) + S_2(Y)$, even though systems 1 and 2 are totally unrelated!

We are now ready to ask some basic questions.

Q1: Which properties of the relation $\prec$ ensure existence and (essential) uniqueness of $S$?

Q2: Can these properties be derived from simple physical premises?

Q3: Which convexity and smoothness properties of $S$ follow from the premises?

Q4: Can temperature (and hence an ordering of states by "hotness" and "coldness") be defined from $S$, and what are its properties?

The answer to question Q1 can be given in the form of six axioms that are reasonable, simple, "obvious", and unexceptionable. An additional, crucial assumption is also needed, but we call it a hypothesis instead of an axiom because we show later how it can be derived from some other axioms, thereby answering question Q2.

**A1. Reflexivity.** $X \stackrel{A}{\sim} X$.

**A2. Transitivity.** If $X \prec Y$ and $Y \prec Z$, then $X \prec Z$.

**A3. Consistency.** If $X \prec X'$ and $Y \prec Y'$, then $(X, Y) \prec (X', Y')$.

**A4. Scaling Invariance.** If $\lambda > 0$ and $X \prec Y$, then $\lambda X \prec \lambda Y$.

**A5. Splitting and Recombination.** $X \stackrel{A}{\sim}$

$((1-\lambda)X, \lambda X)$ for all $0 < \lambda < 1$. Note that the state-spaces are not the same on both sides. If $X \in \Gamma$, then the state-space on the right side is $\Gamma^{(1-\lambda)} \times \Gamma^{(\lambda)}$.

**A6. Stability.** If $(X, \varepsilon Z_0) \prec (Y, \varepsilon Z_1)$ for some $Z_0$, $Z_1$, and a sequence of $\varepsilon$'s tending to zero, then $X \prec Y$. This axiom is a substitute for continuity, which we cannot assume because there is no topology yet. It says that "a grain of dust cannot influence the set of adiabatic processes".

An important lemma is that (A1)–(A6) imply the *cancellation law*, which is used in many proofs. It says that for any three states $X, Y, Z$

(5) $$(X,Z) \prec (Y,Z) \implies X \prec Y\,.$$

The next concept plays a key role in our treatment.

**CH. Definition: We say that the *Comparison Hypothesis* (CH) holds for a state-space $\Gamma$ if all pairs of states in $\Gamma$ are comparable.**

Note that A3, A4, and A5 automatically extend comparability from a space $\Gamma$ to certain other cases; e.g., $X \prec ((1-\lambda)Y, \lambda Z)$ for all $0 \le \lambda \le 1$ if $X \prec Y$ and $X \prec Z$. On the other hand, comparability on $\Gamma$ alone does not allow us to conclude that $X$ is comparable to $((1-\lambda)Y, \lambda Z)$ if $X \prec Y$ but $Z \prec X$. For this, one needs CH on the product space $\Gamma^{(1-\lambda)} \times \Gamma^{(\lambda)}$, which is not implied by CH on $\Gamma$.

The significance of A1–A6 and CH is borne out by the following theorem:

***Theorem 1 (Equivalence of entropy and A1–A6, given CH).*** *The following are equivalent for a state-space $\Gamma$:*

*i) The relation $\prec$ between states in (possibly different) multiple-scaled copies of $\Gamma$, e.g., $\Gamma^{(\lambda_1)} \times \Gamma^{(\lambda_2)} \times \cdots \times \Gamma^{(\lambda_N)}$, is characterized by an entropy function, $S$, on $\Gamma$ in the sense that*

(6) $$(\lambda_1 X_1, \lambda_2 X_2, \ldots) \prec (\lambda_1' X_1', \lambda_2' X_2', \ldots)$$

*is equivalent to the condition that*

(7) $$\sum_i \lambda_i S(X_i) \le \sum_j \lambda_j' S(X_j')$$

*whenever*

(8) $$\sum_i \lambda_i = \sum_j \lambda_j'.$$

*ii) The relation $\prec$ satisfies conditions (A1)–(A6), and (CH) holds for every multiple-scaled copy of $\Gamma$.*

*This entropy function on $\Gamma$ is unique up to affine equivalence; i.e., $S(X) \to aS(X) + B$, with $a > 0$.*

That (i) $\implies$ (ii) is obvious. The proof of (ii) $\implies$ (i) is carried out by an explicit construction of the entropy function on $\Gamma$, reminiscent of an old definition of heat by Laplace and Lavoisier in terms of the amount of ice that a body can melt.

**Basic Construction of $S$** (Figure 2): Pick two reference points $X_0$ and $X_1$ in $\Gamma$ with $X_0 \prec\prec X_1$. (If such points do not exist, then $S$ is the constant function.) Then define for $X \in \Gamma$

(9) $$S(X) := \sup\{\lambda : ((1-\lambda)X_0, \lambda X_1) \prec X\}.$$

*Remarks:* As in axiom A5, two state-spaces are involved in (9). By axiom A5, $X \stackrel{A}{\sim} ((1-\lambda)X, \lambda X)$, and hence, by CH in the space $\Gamma^{(1-\lambda)} \times \Gamma^{(\lambda)}$, $X$ is comparable to $((1-\lambda)X_0, \lambda X_1)$. In (9) we allow $\lambda \le 0$ and $\lambda \ge 1$ by using the convention that $(X, -Y) \prec Z$ means that $X \prec (Y, Z)$ and $(X, 0Y) = X$. For (9) we need to know only that CH holds in twofold scaled products of $\Gamma$ with itself. CH will then automatically be true for all products. In (9) the reference points $X_0, X_1$ are fixed and the supremum is over $\lambda$. One can ask how $S$ changes if we change the two points $X_0, X_1$. The answer is that the change is affine; i.e., $S(X) \to aS(X) + B$, with $a > 0$.

Theorem 1 extends to products of multiple-scaled copies of different systems, i.e., to general *compound* systems. This extension is an immediate consequence of the following theorem, which is proved by applying Theorem 1 to the product of the system under consideration with some standard reference system.

***Theorem 2 (Consistent entropy scales).*** *Assume that CH holds for all compound systems. For each system $\Gamma$ let $S_\Gamma$ be some definite entropy function on $\Gamma$ in the sense of Theorem 1. Then there are constants $a_\Gamma$ and $B(\Gamma)$ such that the function $S$, defined for all states of all systems by*

(10) $$S(X) = a_\Gamma S_\Gamma(X) + B(\Gamma)$$

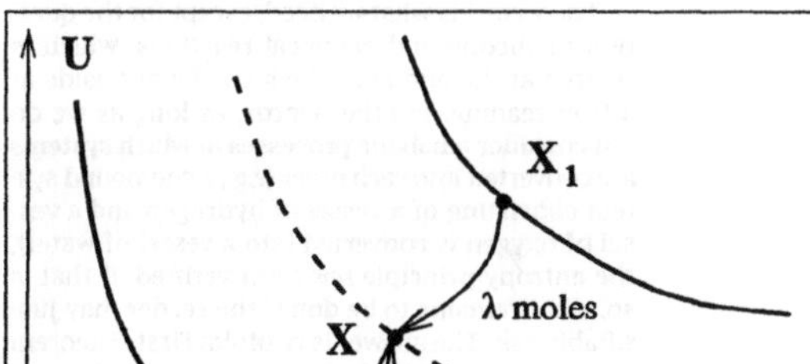


**Figure 2. The entropy of $X$ is determined by the largest amount of $X_1$ that can be transformed adiabatically into $X$, with the help of $X_0$.**

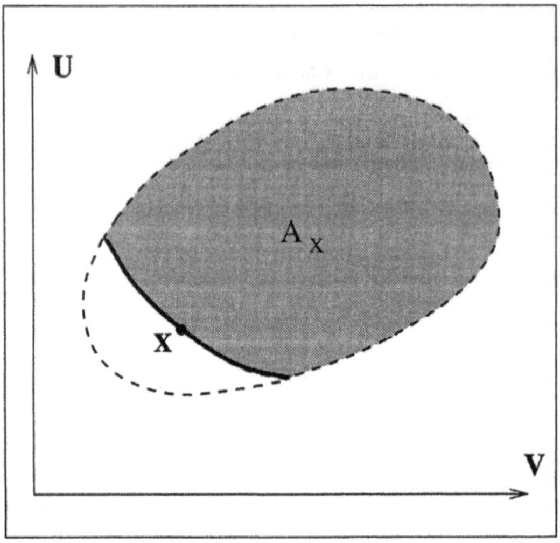


**Figure 3. The coordinates $U$ and $V$ of a simple system. The state-space (bounded by dashed line) and the forward sector $A_X$ (shaded) of a state $X$ are convex, by axiom A7. The boundary of $A_X$ (full line) is an adiabat.**

*for $X \in \Gamma$, satisfies additivity (2), extensivity (3), and monotonicity (1) in the sense that whenever X and Y are in the same state-space, then*

$$X \prec Y \quad \text{if and only if} \quad S(X) \leq S(Y). \tag{11}$$

Theorem 2 is what we need, except for the question of mixing and chemical reactions, which is treated at the end and which can be put aside at a first reading. In other words, as long as we do not consider adiabatic processes in which systems are converted into each other (e.g., a compound system consisting of a vessel of hydrogen and a vessel of oxygen is converted into a vessel of water), the entropy principle has been verified. If that is so, what remains to be done? the reader may justifiably ask. The answer is twofold: First, Theorem 2 requires that CH hold for *all* systems, and we are not content to take this as an axiom. Second, important notions of thermodynamics such as "thermal equilibrium" (which will eventually lead to a precise definition of temperature) have not appeared so far. We shall see that these two points (i.e., thermal equilibrium and CH) are not unrelated.

As for CH, other authors—[6], [2], [4], and [9]—essentially *postulate* that it holds for all systems by making it axiomatic that comparable states fall into equivalence classes. (This means that the conditions $X \prec Z$ and $Y \prec Z$ always imply that $X$ and $Y$ are comparable; likewise, they must be comparable if $Z \prec X$ and $Z \prec Y$). By identifying a state-space with an equivalence class, the comparison hypothesis then holds in these other approaches *by assumption* for all state-spaces. We, in contrast, would like to derive CH from something that we consider more basic. Two ingredients will be needed: the analysis of certain special but commonplace systems called "simple systems" and some assumptions about thermal contact (the "zeroth law") that will act as a kind of glue holding the parts of a compound system in harmony with each other. The simple systems are the building blocks of thermodynamics; all systems we consider are compounds of them.

## Simple Systems

A **Simple System** is one whose state-space can be identified with some open convex subset of some $\mathbf{R}^{n+1}$ with a distinguished coordinate denoted by $U$, called the *energy*, and additional coordinates $V \in \mathbf{R}^n$, called *work coordinates*. The energy coordinate is the way in which thermodynamics makes contact with mechanics, where the concept of energy arises and is precisely defined. The fact that the amount of energy in a state is independent of the manner in which the state was arrived at is, in reality, the first law of thermodynamics. A typical (and often the only) work coordinate is the volume of a fluid or gas (controlled by a piston); other examples are deformation coordinates of a solid or magnetization of a paramagnetic substance.

Our goal is to show, with the addition of a few more axioms, that CH holds for simple systems and their scaled products. In the process we will introduce more structure, which will capture the intuitive notions of thermodynamics; thermal equilibrium is one.

First, there is an axiom about convexity:

**A7. Convex combination.** If $X$ and $Y$ are states of a simple system and $t \in [0, 1]$, then

$$(tX, (1-t)Y) \prec tX + (1-t)Y,$$

in the sense of ordinary convex addition of points in $\mathbf{R}^{n+1}$. A straightforward consequence of this axiom (and A5) is that the **forward sectors** (Figure 3)

$$A_X := \{Y \in \Gamma : X \prec Y\} \tag{12}$$

of states $X$ in a simple system $\Gamma$ are *convex* sets.

Another consequence is a connection between the existence of irreversible processes and Carathéodory's principle [3, 1] mentioned above.

***Lemma 1.*** *Assume (A1)-(A7) for $\Gamma \subset R^{n+1}$ and consider the following statements:*

*a) Existence of irreversible processes: For every $X \in \Gamma$ there is a $Y \in \Gamma$ with $X \prec\prec Y$.*

*b) Carathéodory's principle: In every neighborhood of every $X \in \Gamma$ there is a $Z \in \Gamma$ with $X \not\prec Z$.*

*Then (a) $\Longrightarrow$ (b) always. If the forward sectors in $\Gamma$ have interior points, then (b) $\Longrightarrow$ (a).*

We need three more axioms for simple systems, which will take us into an analytic detour. The first of these establishes (a) above.

**A8. Irreversibility.** For each $X \in \Gamma$ there is a point $Y \in \Gamma$ such that $X \prec\prec Y$. (This axiom is implied by A14, below, but is stated here separately because important conclusions can be drawn from it alone.)

**A9. Lipschitz tangent planes.** For each $X \in \Gamma$ the *forward sector* $A_X = \{Y \in \Gamma : X \prec Y\}$ has a *unique* support plane at $X$ (i.e., $A_X$ has a *tangent plane* at $X$). The tangent plane is assumed to be a *locally Lipschitz continuous* function of $X$, in the sense explained below.

**A10. Connectedness of the boundary.** The boundary $\partial A_X$ (relative to the open set $\Gamma$) of every forward sector $A_X \subset \Gamma$ is connected. (This is technical and conceivably can be replaced by something else.)

Axiom A8 plus Lemma 1 asserts that every $X$ lies on the boundary $\partial A_X$ of its forward sector. Although axiom A9 asserts that the convex set $A_X$ has a true tangent at $X$ only, it is an easy consequence of axiom A2 that $A_X$ has a true tangent everywhere on its boundary. To say that this tangent plane is locally Lipschitz continuous means that if $X = (U^0, V^0)$, then this plane is given by

$$(13) \qquad U - U^0 + \sum_1^n P_i(X)(V_i - V_i^0) = 0$$

with locally Lipschitz continuous functions $P_i$. The function $P_i$ is called the generalized *pressure* conjugate to the work coordinate $V_i$. (When $V_i$ is the volume, $P_i$ is the ordinary pressure.)

Lipschitz continuity and connectedness are well known to guarantee that the coupled differential equations

$$(14) \qquad \frac{\partial U}{\partial V_j}(V) = -P_j(U(V), V) \quad \text{for } j = 1, \dots, n$$

not only have a solution (since we know that the surface $\partial A_X$ exists) but this solution must be unique. Thus, if $Y \in \partial A_X$, then $X \in \partial A_Y$. In short, the surfaces $\partial A_X$ foliate the state-space $\Gamma$. What is less obvious but very important because it instantly gives us the comparison hypothesis for $\Gamma$ is the following.

***Theorem 3 (Forward sectors are nested).*** *If $A_X$ and $A_Y$ are two forward sectors in the state-space $\Gamma$ of a simple system, then exactly one of the following holds.*

a) $A_X = A_Y$; i.e., $X \overset{A}{\sim} Y$.

b) $A_X \subset$ Interior $(A_Y)$; i.e., $Y \prec\prec X$.

c) $A_Y \subset$ Interior $(A_X)$; i.e., $X \prec\prec Y$.

It can also be shown from our axioms that the orientation of forward sectors with respect to the energy axis is the same for *all* simple systems. By convention we choose the direction of the energy axis so that the energy always *increases* in adiabatic processes at fixed work coordinates. When temperature is defined later, this will imply that temperature is always positive.

Theorem 3 implies that $Y$ is on the boundary of $A_X$ if and only if $X$ is on the boundary of $A_Y$. Thus the adiabats, i.e., the $\overset{A}{\sim}$ equivalence classes, consist of these boundaries.

Before leaving the subject of simple systems let us remark on the connection with Carathéodory's development. The point of contact is the fact that $X \in \partial A_X$. We assume that $A_X$ is convex and use transitivity and Lipschitz continuity to arrive even-

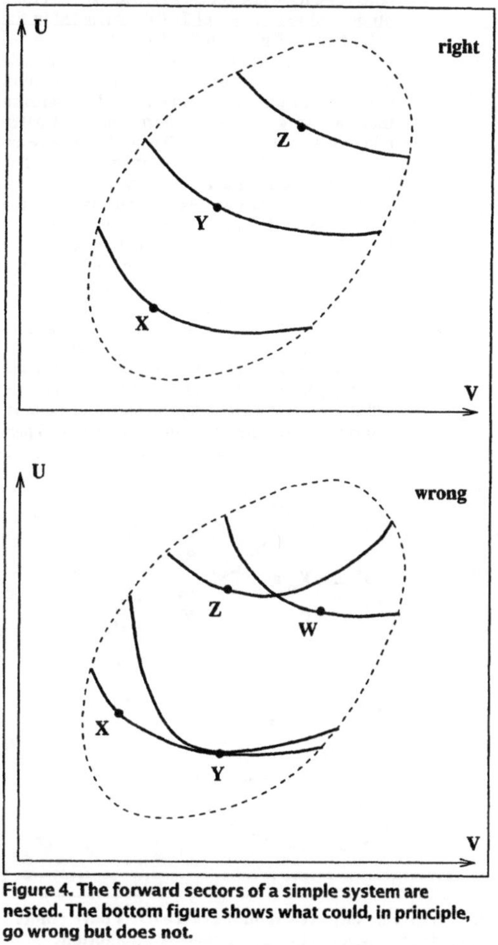


**Figure 4. The forward sectors of a simple system are nested. The bottom figure shows what could, in principle, go wrong but does not.**

tually at Theorem 3. Carathéodory uses Frobenius's theorem plus assumptions about differentiability to conclude the existence locally of a surface containing $X$. Important *global* information, such as Theorem 3, is then not easy to obtain without further assumptions, as discussed, e.g., in [1].

### Thermal Contact

Thermal contact and the zeroth law entail the very special assumptions about $\prec$ that we mentioned earlier. It will enable us to establish CH for products of several systems and thereby show, via Theorem 2, that entropy exists and is additive. Although we have established CH for a simple system, $\Gamma$, we have not yet established CH even for a product of two copies of $\Gamma$. This is needed in the definition of $S$ given in (9). The $S$ in (9) is determined up to an affine shift, and we want to be able to calibrate the entropies (i.e., adjust the multiplicative and additive constants) of all systems so that they work together to form a global $S$ satisfying the entropy principle. We need five more axioms. They might look a bit abstract, so a few words of introduction might be helpful.

In order to relate systems to each other in the hope of establishing CH for compounds and thereby an additive entropy function, some way must be found to put them into contact with each other. Heuristically we imagine two simple systems (the same or different) side by side and fix the work coordinates (e.g., the volume) of each. Connect them with a "copper thread", and wait for equilibrium to be established. The total energy $U$ will not change, but the individual energies $U_1$ and $U_2$ will adjust to values that depend on $U$ and the work coordinates. This new system (with the thread permanently connected) then behaves like a simple system (with one energy coordinate) but with several work coordinates (the union of the two work coordinates). Thus, if we start initially with $X_1 = (U_1, V_1)$ for system 1 and $X_2 = (U_2, V_2)$ for system 2 and if we end up with $X = (U, V_1, V_2)$ for the new system, we can say that $(X_1, X_2) \prec X$. This holds for every choice of $U_1$ and $U_2$ whose sum is $U$. Moreover, after thermal equilibrium is reached, the two systems can be disconnected, if we wish, to once more form a compound system, whose component parts we say are in thermal equilibrium. That this is transitive is the zeroth law.

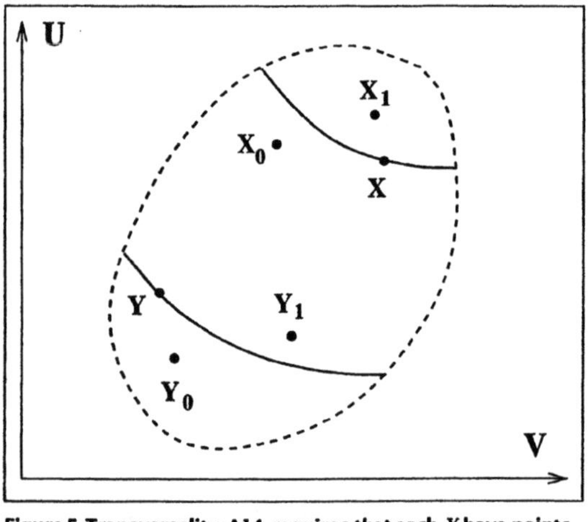


**Figure 5. Transversality, A14, requires that each $X$ have points on each side of its adiabat that are in thermal equilibrium.**

Thus, we cannot only make compound systems consisting of independent subsystems (which can interact, but separate again), we can also make a new simple system out of two simple systems To do this an energy coordinate has to disappear, and thermal contact does this for us. All of this is formalized in the following three axioms.

**A11. Thermal contact.** For any two simple systems with state-spaces $\Gamma_1$ and $\Gamma_2$ there is another *simple* system, called the *thermal join* of $\Gamma_1$ and $\Gamma_2$, with state-space

$$\Delta_{12} = \{(U, V_1, V_2) : U = U_1 + U_2 \text{ with } (U_1, V_1) \in \Gamma_1, (U_2, V_2) \in \Gamma_2\}. \tag{15}$$

Moreover,

$$\Gamma_1 \times \Gamma_2 \ni ((U_1, V_1), (U_2, V_2)) \prec (U_1 + U_2, V_1, V_2) \in \Delta_{12}. \tag{16}$$

**A12. Thermal splitting.** For any point $(U, V_1, V_2) \in \Delta_{12}$ there is at least one pair of states, $(U_1, V_1) \in \Gamma_1$, $(U_2, V_2)) \in \Gamma_2$, with $U = U_1 + U_2$, such that

$$(U, V_1, V_2) \stackrel{A}{\sim} ((U_1, V_1), (U_2, V_2)). \tag{17}$$

If $(U, V_1, V_2) \stackrel{A}{\sim} ((U_1, V_1), (U_2, V_2))$, we say that the states $X = (U_1, V_1)$ and $Y = (U_2, V_2))$ are in *thermal equilibrium* and write

$$X \stackrel{T}{\sim} Y.$$

**A13. Zeroth law of thermodynamics.** If $X \stackrel{T}{\sim} Y$ and if $Y \stackrel{T}{\sim} Z$, then $X \stackrel{T}{\sim} Z$.

A11 and A12 together say that for each choice of the individual work coordinates there is a way to divide up the energy $U$ between the two systems in a stable manner. A12 is the stability statement, for it says that joining is reversible; i.e., once the equilibrium has been established, one can cut the copper thread and retrieve the two systems back again, but with a special partition of the energies.

This reversibility allows us to think of the thermal join, which is a simple system in its own right, as a special subset of the product system $\Gamma_1 \times \Gamma_2$, which we call the *thermal diagonal*. In particular, A12 allows us to prove easily that $X \stackrel{T}{\sim} \lambda X$ for all $X$ and all $\lambda > 0$.

A13 is the famous zeroth law, which says that the thermal equilibrium is transitive and hence an equivalence relation. Often this law is taken to mean that the equivalence classes can be labeled by an "empirical" temperature, but we do not want to mention temperature at all at this point. It will appear later.

Two more axioms are needed.

A14 requires that for every adiabat (i.e., an equivalence class w.r.t. $\overset{A}{\sim}$) there exists at least one isotherm (i.e., an equivalence class w.r.t. $\overset{T}{\sim}$) containing points on both sides of the adiabat. Note that, for each given $X$, only two points in the entire state-space $\Gamma$ are required to have the stated property. This assumption essentially prevents a state-space from breaking up into two pieces that do not communicate with each other. Without it, counterexamples to CH for compound systems can be constructed. A14 implies A8, but we listed A8 separately in order not to confuse the discussion of simple systems with thermal equilibrium.

A15 is technical and perhaps can be eliminated. Its physical motivation is that a sufficiently large copy of a system can act as a heat bath for other systems. When temperature is introduced later, A15 will have the meaning that all systems have the same temperature range. This postulate is needed if we want to be able to bring every system into thermal equilibrium with every other system.

**A14. Transversality.** If $\Gamma$ is the state-space of a simple system and if $X \in \Gamma$, then there exist states $X_0 \overset{T}{\sim} X_1$ with $X_0 \prec\prec X \prec\prec X_1$.

**A15. Universal temperature range.** If $\Gamma_1$ and $\Gamma_2$ are state-spaces of simple systems, then, for every $X \in \Gamma_1$ and every $V$ belonging to the projection of $\Gamma_2$ onto the space of its work coordinates, there is a $Y \in \Gamma_2$ with work coordinates $V$ such that $X \overset{T}{\sim} Y$.

The reader should note that the concept "thermal contact" has appeared, but not temperature or hot and cold or anything resembling the Clausius or Kelvin-Planck formulations of the second law. Nevertheless, we come to the main achievement of our approach: *With these axioms we can establish CH for products of simple systems* (each of which satisfies CH, as we already know). First, the thermal join establishes CH for the (scaled) product of a simple system with itself. The basic idea here is that the points in the product that lie on the thermal diagonal are comparable, since points in a simple system are comparable. In particular, with $X, X_0, X_1$ as in A14, the states $((1-\lambda)X_0, \lambda X_1)$ and $((1-\lambda)X, \lambda X)$ can be regarded as states of the *same* simple system and are therefore comparable. *This is the key point needed for the construction of S, according to (9).* The importance of transversality is thus brought into focus.

With some more work we can establish CH for multiple-scaled copies of a simple system. Thus, we have established $S$ within the context of one system and copies of the system, i.e., condition (ii) of Theorem 1. As long as we stay within such a group of systems there is no way to determine the unknown multiplicative or additive entropy constants. The next task is to show that the multiplicative constants can be adjusted to give a universal entropy valid for copies of *different* systems, i.e., to establish the hypothesis of Theorem 2. This is based on the following.

***Lemma 2 (Existence of calibrators).*** *If $\Gamma_1$ and $\Gamma_2$ are simple systems, then there exist states $X_0, X_1 \in \Gamma_1$ and $Y_0, Y_1 \in \Gamma_2$ such that*

$$X_0 \prec\prec X_1 \quad \text{and} \quad Y_0 \prec\prec Y_1$$

*and*

$$(X_0, Y_1) \overset{A}{\sim} (X_1, Y_0).$$

The significance of Lemma 2 is that it allows us to fix the *multiplicative* constants by the condition

$$S_1(X_0) + S_2(Y_1) = S_1(X_1) + S_2(Y_0). \tag{18}$$

The proof of Lemma 2 is complicated and really uses all the axioms A1 to A14. With its aid we arrive at our chief goal, which is CH for compound systems.

***Theorem 4 (Entropy principle in products of simple systems).*** *The comparison hypothesis CH is valid in arbitrary scaled products of simple systems. Hence, by Theorem 2, the relation $\prec$ among states in such state-spaces is characterized by an entropy function S. The entropy function is unique, up to an overall multiplicative constant and one additive constant for each simple system under consideration.*

At last we are ready to define *temperature.* Concavity of $S$ (implied by A7), Lipschitz continuity of the pressure, and the transversality condition, together with some real analysis, play key roles in the following, which answers questions Q3 and Q4 posed at the beginning.

***Theorem 5 (Entropy defines temperature).*** *The entropy S is a concave and continuously differentiable function on the state-space of a simple system. If the function T is defined by*

$$\frac{1}{T} := \left(\frac{\partial S}{\partial U}\right)_V, \tag{19}$$

*then $T > 0$ and T characterizes the relation $\overset{T}{\sim}$ in the sense that $X \overset{T}{\sim} Y$ if and only if $T(X) = T(Y)$. Moreover, if two systems are brought into thermal contact with fixed work coordinates, then, since the total entropy cannot decrease, the energy flows from the system with the higher T to the system with the lower T.*

The temperature need not be a strictly monotone function of $U$; indeed, it is not so in a "multiphase region". It follows that $T$ is not always capable of specifying a state, and this fact can cause some pain in traditional discussions of the second law if it is recognized, which usually it is not.

### Mixing and Chemical Reactions

The core results of our analysis have now been presented, and readers satisfied with the entropy principle in the form of Theorem 4 may wish to stop at this point. Nevertheless, a nagging doubt will occur to some, because there are important adiabatic processes in which systems are not conserved, and these processes are not yet covered in the theory. A critical study of the usual textbook treatments should convince the reader that this subject is not easy, but in view of the manifold applications of thermodynamics to chemistry and biology it is important to tell the whole story and not ignore such processes.

One can formulate the problem as the determination of the additive constants $B(\Gamma)$ of Theorem 2. As long as we consider only adiabatic processes that preserve the amount of each simple system (i.e., such that Eqs. (6) and (8) hold), these constants are indeterminate. This is no longer the case, however, if we consider mixing processes and chemical reactions (which are not really different, as far as thermodynamics is concerned). It then becomes a nontrivial question whether the additive constants can be chosen in such a way that the entropy principle holds. Oddly, this determination turns out to be far more complex mathematically and physically than the determination of the multiplicative constants (Theorem 2). In traditional treatments one usually resorts to *gedanken* experiments involving strange, nonexistent objects called "semipermeable membranes" and "van t'Hofft boxes". We present here a general and rigorous approach which avoids all this.

What we already know is that every system has a well-defined entropy function—e.g., for each $\Gamma$ there is $S_\Gamma$—and we know from Theorem 2 that the multiplicative constants $a_\Gamma$ can be determined in such a way that the sum of the entropies increases in any adiabatic process in any compound space $\Gamma_1 \times \Gamma_2 \times \ldots$. Thus, if $X_i \in \Gamma_i$ and $Y_i \in \Gamma_i$, then

$$(20)\qquad (X_1, X_2, \ldots) \prec (Y_1, Y_2, \ldots) \quad \text{if and only if} \quad \sum_i S_i(X_i) \le \sum_j S_j(Y_j),$$

where we have denoted $S_{\Gamma_i}$ by $S_i$ for short. The additive entropy constants do not matter here, since each function $S_i$ appears on both sides of this inequality. It is important to note that this applies even to processes that, in intermediate steps, take one system into another, provided the total compound system is the same at the beginning and at the end of the process.

The task is to find constants $B(\Gamma)$, one for each state-space $\Gamma$, in such a way that the entropy defined by

$$(21)\qquad S(X) := S_\Gamma(X) + B(\Gamma) \quad \text{for} \quad X \in \Gamma$$

satisfies

$$(22)\qquad S(X) \le S(Y)$$

whenever

$$X \prec Y \quad \text{with} \quad X \in \Gamma,\ Y \in \Gamma'.$$

Moreover, we require that the newly defined entropy satisfy scaling and additivity under composition. Since the initial entropies $S_\Gamma(X)$ already satisfy them, these requirements become conditions on the additive constants $B(\Gamma)$:

$$(23)\qquad B(\Gamma_1^{(\lambda_1)} \times \Gamma_2^{(\lambda_2)}) = \lambda_1 B(\Gamma_1) + \lambda_2 B(\Gamma_2)$$

for all state-spaces $\Gamma_1, \Gamma_2$ under consideration and $\lambda_1, \lambda_2 > 0$. Some reflection shows us that consistency in the definition of the entropy constants $B(\Gamma)$ requires us to consider all possible chains of adiabatic processes leading from one space to another via intermediate steps. Moreover, the additivity requirement leads us to allow the use of a "catalyst" in these processes, i.e., an auxiliary system that is recovered at the end, although a state change *within* this system might take place. With this in mind we define quantities $F(\Gamma, \Gamma')$ that incorporate the entropy differences in all such chains leading from $\Gamma$ to $\Gamma'$. These are built up from simpler quantities $D(\Gamma, \Gamma')$, which measure the entropy differences in one-step processes, and $E(\Gamma, \Gamma')$, where the catalyst is absent. The precise definitions are as follows. First,

$$(24)\qquad D(\Gamma, \Gamma') := \inf\{S_{\Gamma'}(Y) - S_\Gamma(X) \ :\ X \in \Gamma,\ Y \in \Gamma',\ X \prec Y\}.$$

If there is no adiabatic process leading from $\Gamma$ to $\Gamma'$, we put $D(\Gamma, \Gamma') = \infty$. Next, for any given $\Gamma$ and $\Gamma'$, we consider all finite chains of state-spaces $\Gamma = \Gamma_1, \Gamma_2, \ldots, \Gamma_N = \Gamma'$ such that $D(\Gamma_i, \Gamma_{i+1}) < \infty$ for all i, and we define

$$(25)\qquad E(\Gamma, \Gamma') := \inf\{D(\Gamma_1, \Gamma_2) + \cdots + D(\Gamma_{N-1}, \Gamma_N)\},$$

where the infimum is taken over all such chains linking $\Gamma$ with $\Gamma'$. Finally we define

$$(26)\qquad F(\Gamma, \Gamma') := \inf\{E(\Gamma \times \Gamma_0, \Gamma' \times \Gamma_0)\},$$

where the infimum is taken over all state-spaces $\Gamma_0$. (These are the catalysts.)

The importance of the $F$'s for the determination of the additive constants is made clear in the following theorem:

***Theorem 6 (Constant entropy differences).*** *If $\Gamma$ and $\Gamma'$ are two state-spaces, then for any two states*

$X \in \Gamma$ *and* $Y \in \Gamma'$

$$X \prec Y \quad \text{if and only if} \quad S_\Gamma(X) + F(\Gamma,\Gamma') \le S_{\Gamma'}(Y). \tag{27}$$

An essential ingredient for the proof of this theorem is Eq. (20).

According to Theorem 6 the determination of the entropy constants $B(\Gamma)$ amounts to satisfying the inequalities

$$-F(\Gamma',\Gamma) \le B(\Gamma) - B(\Gamma') \le F(\Gamma,\Gamma') \tag{28}$$

together with the linearity condition (23). It is clear that (28) can only be satisfied with finite constants $B(\Gamma)$ and $B(\Gamma')$ if $F(\Gamma,\Gamma') > -\infty$. To exclude the pathological case $F(\Gamma,\Gamma') = -\infty$, we introduce our last axiom, A16, whose statement requires the following definition.

***Definition.*** A state-space $\Gamma$ is said to be *connected* to another state-space $\Gamma'$ if there are states $X \in \Gamma$ and $Y \in \Gamma'$, and state-spaces $\Gamma_1,\dots,\Gamma_N$ with states $X_i, Y_i \in \Gamma_i$, $i = 1,\dots,N$, and a state-space $\Gamma_0$ with states $X_0, Y_0 \in \Gamma_0$, such that

$$(X, X_0) \prec Y_1, \quad X_i \prec Y_{i+1}, \ i = 1,\dots,N-1, \quad X_N \prec (Y, Y_0).$$

**A16. Absence of sinks:** If $\Gamma$ is connected to $\Gamma'$, then $\Gamma'$ is connected to $\Gamma$.

This axiom excludes $F(\Gamma,\Gamma') = -\infty$ because, on general grounds, one always has

$$-F(\Gamma',\Gamma) \le F(\Gamma,\Gamma'). \tag{29}$$

Hence $F(\Gamma,\Gamma') = -\infty$ (which means, in particular, that $\Gamma$ is connected to $\Gamma'$) would imply $F(\Gamma',\Gamma) = \infty$, i.e., that there is no way back from $\Gamma'$ to $\Gamma$. This is excluded by axiom 16.

The quantities $F(\Gamma,\Gamma')$ have simple subadditivity properties that allow us to use the Hahn-Banach theorem to satisfy the inequalities (28), with constants $B(\Gamma)$ that depend linearly on $\Gamma$, in the sense of Eq. (23). Hence we arrive at

***Theorem 7 (Universal entropy).*** *The additive entropy constants of all systems can be calibrated in such a way that the entropy is additive and extensive and $X \prec Y$ implies $S(X) \le S(Y)$, even when $X$ and $Y$ do not belong to the same state-space.*

Our final remark concerns the remaining nonuniqueness of the constants $B(\Gamma)$. This indeterminacy can be traced back to the nonuniqueness of a linear functional lying between $-F(\Gamma',\Gamma)$ and $F(\Gamma,\Gamma')$ and has two possible sources: one is that some pairs of state-spaces $\Gamma$ and $\Gamma'$ may not be connected; i.e., $F(\Gamma,\Gamma')$ may be infinite (in which case $F(\Gamma',\Gamma)$ is also infinite by axiom A16). The other is that there might be a true gap; i.e.,

$$-F(\Gamma',\Gamma) < F(\Gamma,\Gamma') \tag{30}$$

might hold for some state-spaces, even if both sides are finite.

In nature only states containing the same amount of the chemical elements can be transformed into each other. Hence $F(\Gamma,\Gamma') = +\infty$ for many pairs of state-spaces, in particular, for those that contain different amounts of some chemical element. The constants $B(\Gamma)$ are, therefore, never unique: For each equivalence class of state-spaces (with respect to the relation of connectedness) one can define a constant that is arbitrary except for the proviso that the constants should be additive and extensive under composition and scaling of systems. In our world there are 92 chemical elements (or, strictly speaking, a somewhat larger number, $N$, since one should count different isotopes as different elements), and this leaves us with at least 92 free constants that specify the entropy of one gram of each of the chemical elements in some specific state.

The other possible source of nonuniqueness, a nontrivial gap (30) for systems with the same composition in terms of the chemical elements, is, as far as we know, not realized in nature. (Note that this assertion can be tested experimentally without invoking semipermeable membranes.) Hence, once the entropy constants for the chemical elements have been fixed and a temperature unit has been chosen (to fix the multiplicative constants), the universal entropy is completely fixed.

We are indebted to many people for helpful discussions, including Fred Almgren, Thor Bak, Bernard Baumgartner, Pierluigi Contucci, Roy Jackson, Anthony Knapp, Martin Kruskal, Mary Beth Ruskai, and Jan Philip Solovej.

### References

[1] J. B. Boyling, *An axiomatic approach to classical thermodynamics*, Proc. Roy. Soc. London **A329** (1972), 35–70.

[2] H. A. Buchdahl, *The concepts of classical thermodynamics*, Cambridge Univ. Press, Cambridge, 1966.

[3] C. Carathéodory, *Untersuchung über die Grundlagen der Thermodynamik*, Math. Ann. **67** (1909), 355–386.

[4] J. L. B. Cooper, *The foundations of thermodynamics*, J. Math. Anal. Appl. **17** (1967), 172–193.

[5] J. J. Duistermaat, *Energy and entropy as real morphisms for addition and order*, Synthese **18** (1968), 327-393.

[6] R. Giles, *Mathematical foundations of thermodynamics*, Pergamon, Oxford, 1964.

[7] E. H. Lieb and J. Yngvason, *The physics and mathematics of the second law of thermodynamics*, preprint, 1997; Phys. Rep. (to appear); Austin Math. Phys. arch. 97-457; Los Alamos arch. cond-mat/9708200.

[8] M. Planck, *Über die Begrundung des zweiten Hauptsatzes der Thermodynamik*, Sitzungsber. Preuss. Akad. Wiss. Phys. Math. Kl. (1926), 453–463.

[9] F. S. Roberts and R. D. Luce, *Axiomatic thermodynamics and extensive measurement*, Synthese **18** (1968), 311–326.

# A Fresh Look at Entropy and the Second Law of Thermodynamics

**The existence of entropy, and its increase, can be understood without reference to either statistical mechanics or heat engines.**

Elliott H. Lieb and Jakob Yngvason

In days long gone, the second law of thermodynamics (which predated the first law) was regarded as perhaps the most perfect and unassailable law in physics. It was even supposed to have philosophical import: It has been hailed for providing a proof of the existence of God (who started the universe off in a state of low entropy, from which it is constantly degenerating); conversely, it has been rejected as being incompatible with dialectical materialism and the perfectibility of the human condition.

Alas, physicists themselves eventually demoted the second law to a lesser position in the pantheon—because (or so it was declared) it is "merely" statistics applied to the mechanics of large numbers of atoms. Willard Gibbs wrote: "The laws of thermodynamics may easily be obtained from the principles of statistical mechanics, of which they are the incomplete expression"[1]—and Ludwig Boltzmann expressed similar sentiments.

Is that really so? Is it really true that the second law is merely an "expression" of microscopic models, or could it exist in a world that was featureless at the $10^{-8}$ cm level? We know that statistical mechanics is a powerful tool for understanding physical phenomena and calculating many quantities, especially in systems at or near equilibrium. We use it to calculate entropy, specific and latent heats, phase transition properties, transport coefficients, and so on, often with good accuracy. Important examples abound, such as Max Planck's realization that by staring into a furnace he could find Avogadro's number, and Linus Pauling's highly accurate back-of-the-envelope calculation of the residual entropy of ice. But is statistical mechanics essential for the second law?

In any event, it is still beyond anyone's computational ability (except in idealized situations) to account for a very precise, essentially infinitely accurate law of physics from statistical mechanical principles. No exception to the second law of thermodynamics has ever been found—not even a tiny one. Like conservation of energy (the "first" law), the existence of a law so precise and so independent of details of models must have a logical foundation that is independent of the fact that matter is composed of interacting particles. Our aim here is to explore that foundation. The full details can be found in reference 2.

As Albert Einstein put it,[3] "A theory is the more impressive the greater the simplicity of its premises, the more different kinds of things it relates, and the more extended its area of applicability. Therefore the deep impression which classical thermodynamics made upon me. It is the only physical theory of universal content concerning which I am convinced that, within the framework of the applicability of its basic concepts, it will never be overthrown."[3]

In an attempt to reaffirm the second law as a pillar of physics in its own right, we have returned to a little-noticed movement that began in the 1950s with the work of Peter Landsberg,[4] Hans Buchdahl,[5] Gottfried Falk, Herbert Jung,[6] and others[2] and culminated in the book of Robin Giles,[7] which must be counted one of the truly great, but unsung works in theoretical physics. It is in these works that the concept of "comparison" (explained below) emerges as one of the key underpinnings of the second law. The approach of these authors is quite different from lines of thought in the tradition of Sadi Carnot, which base thermodynamics on the efficiency of heat engines. (See reference 8, for example, for modern expositions of the latter approach.)

## The basic question

The paradigmatic event that the second law deals with can be described as follows. Take a macroscopic system in an equilibrium state $X$ and place it in a room along with a gorilla equipped with arbitrarily complicated machinery (a metaphor for the rest of the universe), and a weight—and close the door. As in the old advertisement for indestructible luggage, the gorilla can do anything to the system—including tearing it apart. At the end of the day, however, when the door is opened, the system is found to be in some other equilibrium state, $Y$, the gorilla and machinery are found in their original state, and the only other thing that has possibly changed is that the weight has been raised or lowered. Let us emphasize that although our focus is on equilibrium states, the processes that take one such state into another can be arbitrarily violent. The gorilla knows no limits. (See figure 1.)

The question that the second law answers is this: What distinguishes those states $Y$ that can be reached from $X$ in this manner from those that cannot? The

ELLIOTT LIEB *is a Higgins Professor of Physics and a professor of mathematics at Princeton University in Princeton, New Jersey.* JAKOB YNGVASON *is a professor of theoretical physics at the University of Vienna and president of the Erwin Schrödinger Institute for Mathematical Physics in Vienna, Austria.*

FIGURE 1. THE SECOND LAW OF THERMODYNAMICS says that increased entropy characterizes those final states of a macroscopic system that can be reached from a given initial state without leaving an imprint on the rest of the universe, apart from the displacement of a weight. The scenario shown here illustrates that the process can be quite violent. (a) A system in an equilibrium state $X$ (blue) is placed in a room with a gorilla, some intricate machinery (green), and a weight. (b) The gorilla, machinery, and system interact and the system undergoes a violent transition. (c) The system is found in a new equilibrium state $Y$ (red), the gorilla and machinery are found in their original state, while the weight may have been displaced. The role of the weight is to supply energy (via the machinery) both for the actions of the gorilla and for bringing the machinery and gorilla back to their initial states. The recovery process may involve additional interactions between machinery, system, and gorilla—interactions besides those indicated in (b).

answer: There is a function of the equilibrium states, called entropy and denoted by $S$, that characterizes the possible pairs of equilibrium states $X$ and $Y$ by the inequality $S(X) \leq S(Y)$. The function can be chosen so as to be additive (in a sense explained below), and with this requirement it is unique, up to a change of scale. Our main point is that the existence of entropy relies on only a few basic principles, independent of any statistical model—or even of atoms.

What is exciting about this seemingly innocuous statement is the uniqueness of entropy, for it means that all the different methods for measuring or computing entropy must give the same answer. The usual textbook derivation of entropy as a state function, starting with some version of "the second law," proceeds by considering certain slow, almost reversible processes (along adiabats and isotherms). It is not at all evident that a function obtained in this way can contain any information about processes that are far from being slow or reversible. The clever physicist might think that with the aid of modern computers, sophisticated feedback mechanisms, unlimited amounts of mechanical energy (represented by the weight) and lots of plain common sense and funding, the system could be made to go from an equilibrium state $X$ to a state $Y$ that could not be reached by the primitive quasistatic processes used to define entropy in the first place. This cannot happen, however, no matter how clever the experimenter or how far from equilibrium one travels!

What logic lies behind this law? Why can't one gorilla undo what another one has wrought? The atomistic foundation of the logic is not as simple as is often suggested. It concerns not only such matters as the enormous number of atoms involved ($10^{23}$), but also other aspects of statistical mechanics that are beyond our present mathematical abilities. In particular, the interaction of a system with the external world (represented by the gorilla and machinery) cannot be described in any obvious way by Hamiltonian mechanics. Although irreversibility is an important open problem in statistical mechanics, it is fortunate that the logic of thermodynamics itself is independent of atoms and can be understood without knowing the source of irreversibility.

The founders of thermodynamics—Rudolf Clausius, Lord Kelvin, Planck, Constantin Carathéodory, and so on—clearly had transitions between equilibrium states in mind when they stated the law in sentences such as "No process is possible, the sole result of which is that a body is cooled and work is done" (Kelvin). Later it became tacitly understood that the law implies a continuous increase in some property called entropy, which was supposedly defined for systems out of equilibrium. The ongoing, unsatisfactory debates (see reference 9, for example) about the definition of this nonequilibrium entropy and whether it increases shows, in fact, that what is supposedly "easily" understood needs clarification. Once again, it is a good idea to try to understand first the meaning of entropy for equilibrium states—the quantity that our textbooks talk about when they draw Carnot cycles. In this article we restrict our attention to just those states; by "state" we always mean "equilibrium state." Entropy, as the founders of thermodynamics understood the quantity, is subtle enough, and it is worthwhile to understand the "second law" in this restricted context. To do so it is

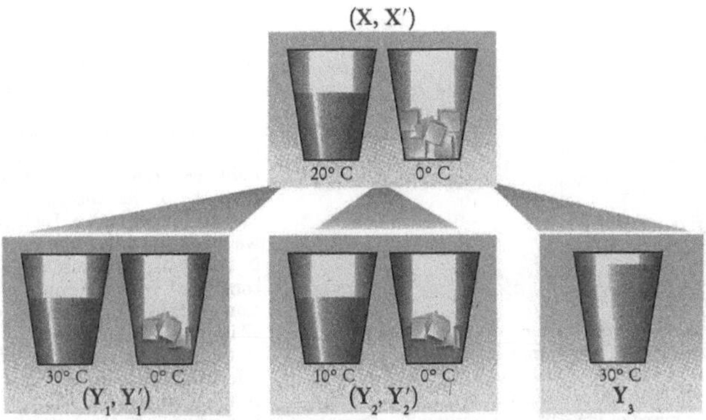


FIGURE 2. ADIABATIC STATE CHANGES for a compound system consisting of a glass of whiskey and a glass of ice. The states $(Y_1,Y_1')$, $(Y_2,Y_2')$, and $Y_3$ are all adiabatically accessible from $(X,X')$. $(Y_1,Y_1')$ can be reached by weight-powered stirrers (not shown) acting on each glass. $(Y_2,Y_2')$ is obtained by bringing the two subsystems temporarily into thermal contact. $Y_3$ is obtained by pouring the whiskey on the ice and stirring; this is a mixing process and changes the system. The original state $(X,X')$ is *not* adiabatically accessible from any of the three states $(Y_1,Y_1')$, $(Y_2,Y_2')$, or $Y_3$.

not necessary to decide whether Boltzmann or Gibbs had the right view of irreversibility. (Their views are described in Joel L. Lebowitz's article, "Boltzmann's Entropy and Time's Arrow," PHYSICS TODAY, September 1993, page 32.)

## The basic concepts

To begin at the beginning, we suppose we know what is meant by a thermodynamic system and equilibrium states of such a system. Admittedly, these are not always easy to define, and there are certainly systems, such as a mixture of hydrogen and oxygen or an interstellar ionized gas, capable of behaving as though they were in equilibrium even if they are not truly so. The prototypical system is a so-called "simple system," consisting of a substance in a container with a piston. But a simple system can be much more complicated than that. Besides its volume, it can have other coordinates, which can be changed by mechanical or electrical means—shear in a solid or magnetization, for example. In any event, a state of a simple system is described by a special coordinate $U$, which is its energy, and one or more other coordinates (such as the volume $V$) called work coordinates. An essential point is that the concept of energy, which we know about from moving weights and Newtonian mechanics, can be defined for thermodynamic systems. This fact is the content of the first law of thermodynamics.

Another type of system is a "compound system," which consists of several different or identical independent, simple systems. By means of mixing or chemical reactions, systems can be created or destroyed.

Let us briefly discuss some concepts that are relevant for systems and their states, which are denoted by capital letters such as $X, X', Y, \ldots$. Operationally, the composition, denoted $(X,X')$, of two states $X$ and $X'$ is obtained simply by putting one system in a state $X$ and one in a state $X'$ side by side on the experimental table and regarding them jointly as a state of a new, compound system. For instance, $X$ could be a glass containing 100 g of whiskey at standard pressure and 20 °C, and $X'$ a glass containing 50 g of ice at standard pressure and 0 °C. To picture $(X,X')$, one should think of the two glasses standing on a table without touching each other. (See figure 2.)

Another operation is the "scaling" of a state $X$ by a factor $\lambda > 0$, leading to a state denoted $\lambda X$. Extensive properties such as mass, energy, and volume are multiplied by $\lambda$, while intensive properties such as pressure stay intact. For the states $X$ and $X'$ as in the example above, $\frac{1}{2}X$ is 50 g of whiskey at standard pressure and 20 °C, and $\frac{1}{5}X'$ is 10 g of ice at standard pressure and 0 °C. Compound systems scale in the same way: $\frac{1}{5}(X,X')$ is 20 g of whiskey and 10 g of ice in separate glasses with pressure and temperatures as before.

A central notion is adiabatic accessibility. If our gorilla can take a system from $X$ to $Y$ as described above—that is, if the only net effect of the action, besides the state change of the system, is that a weight has possibly been raised or lowered, we say that $Y$ is adiabatically accessible from $X$ and write $X \prec Y$ (the symbol $\prec$ is pronounced "precedes"). It has to be emphasized that for macroscopic systems the relation is an absolute one: If a transition from $X$ to $Y$ is possible at one time, then it is *always* possible (that is, it is reproducible), and if it is impossible at one time, then it *never* happens. This absolutism is guaranteed by the large powers of 10 involved—the impossibility of a chair's spontaneously jumping up from the floor is an example.

## The role of entropy

Now imagine that we are given a list of all possible pairs of states $X,Y$ such that $X \prec Y$. The foundation on which thermodynamics rests, and the essence of the second law, is that this list can be simply encoded in an entropy function $S$ on the set of all states of all systems (including compound systems), so that when $X$ and $Y$ are related at all, then

$$X \prec Y \text{ if and only if } S(X) \leq S(Y).$$

Moreover, the entropy function can be chosen in such a way that if $X$ and $X'$ are states of two (different or identical) systems, then the entropy of the compound system in this pair of states is given by

$$S(X,X') = S(X) + S(X').$$

This additivity of entropy is a highly nontrivial assertion. Indeed, it is one of the most far-reaching properties of the second law. In compound systems such as the whiskey/ice example above, all states $(Y,Y')$ such that $X \prec Y$ and $X' \prec Y'$ are adiabatically accessible from $(X,X')$. For instance, by letting a falling weight run an electric generator one can stir the whiskey and also melt some ice. But it is important to note that $(Y,Y')$ can be adiabatically accessible from $(X,X')$ without $Y$ being adiabatically accessible from $X$. Bringing the two glasses into contact and separating them again is adiabatic for the compound system,

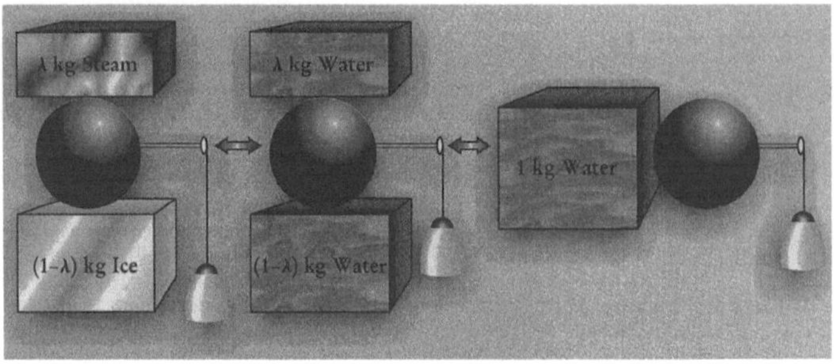


FIGURE 3. DEFINITION OF ENTROPY. One can define the entropy of 1 kg of water in a given state (represented by the orange color) by obtaining the state from a fraction λ kg of steam in a fixed, standard state (red) and a fraction 1 − λ kg of ice in a fixed, standard state (blue), with the aid of a device (green) and a weight (yellow). The device returns to its initial state at the end of the process, but the weight may end up raised or lowered. The entropy $S_{water}$, measured in units of $S_{steam}$, is the maximum fraction $\lambda = \lambda_{max}$ for which the transformation to 1 kg of water in the given (orange) state is possible. The system of steam and ice is used here only for illustration. The definition of entropy need not involve phase changes.

but the resulting cooling of the whiskey is not adiabatic for the whiskey alone. The fact that the inequality $S(X) + S(X') \leq S(Y) + S(Y')$ *exactly* characterizes the possible adiabatic transitions for the compound system, even when $S(X) \geq S(Y)$, is quite remarkable. It means that it is sufficient to know the entropy of each part of a compound system to decide which transitions due to interactions between the parts (brought about by the gorilla) are possible.

Closely related to additivity is extensivity, or scaling of entropy,

$$S(\lambda X) = \lambda S(X),$$

which means that the entropy of an arbitrary mass of a substance is determined by the entropy of some standard reference mass, such as 1 kg of the substance. Without this scaling property, engineers would have to use different steam tables each time they designed a new engine.

In traditional presentations of thermodynamics, based for example on Kelvin's principle given above, entropy is arrived at in a rather roundabout way that tends to obscure its connection with the relation $\prec$. The basic message we wish to convey is that the existence and uniqueness of entropy are equivalent to certain simple properties of the relation $\prec$. This equivalence is the concern of reference 2.

An analogy leaps to mind: When can a vector field $\mathbf{E}(x)$ be encoded in an ordinary function (potential) $\phi(x)$ whose gradient is $\mathbf{E}$? The well-known answer is that a necessary and sufficient condition is that curl $\mathbf{E} = 0$. The importance of this encoding does not have to be emphasized to physicists; entropy's role is similar to the potential's role, and the existence and meaning of entropy are not based on any formula such as $S = -\Sigma_i p_i \ln p_i$, involving probabilities $p_i$ of "microstates." Entropy is derived (uniquely, we hope) from the list of pairs $X \prec Y$; our aim is to figure out what properties of this list (analogous to the curl-free condition) will allow it to be described by an entropy. That entropy will then be endowed with an unambiguous physical meaning independent of anyone's assumptions about "the arrow of time," "coarse graining," and so on. Only the list, which is given by physics, is important for us now.

The required properties of $\prec$ do *not* involve concepts like "heat" or "reversible engines"; not even "hot" and "cold" are needed. Besides the "obvious" conditions "$X \prec X$ for all $X$" (reflexivity) and "$X \prec Y$ and $Y \prec Z$ implies $X \prec Z$" (transitivity), one needs to know that the relation behaves reasonably with respect to the composition and scaling of states. By this we mean the following:

▷ Adiabatic accessibility is consistent with the composition of states: $X \prec Y$ and $Z \prec W$ implies $(X,Z) \prec (Y,W)$.

▷ Scaling of states does not affect adiabatic accessibility: If $X \prec Y$, then $\lambda X \prec \lambda Y$.

▷ Systems can be cut adiabatically into two parts: If $0 < \lambda < 1$, then $X \prec ([1 - \lambda]X, \lambda X)$, and the recombination of the parts is also adiabatic: $([1 - \lambda]X, \lambda X) \prec X$.

▷ Adiabatic accessibility is stable with respect to small perturbations: If $(X, \varepsilon Z) \prec (Y, \varepsilon W)$ for arbitrarily small $\varepsilon > 0$, then $X \prec Y$.

These requirements are all very natural. In fact, in traditional approaches they are usually taken for granted, without mention. They are not quite sufficient, however, to define entropy. A crucial additional ingredient is the comparison hypothesis for the relation $\prec$. In essence, this is the hypothesis that equilibrium states, whether simple or compound, can be grouped into classes such that if $X$ and $Y$ are in the same class, then either $X \prec Y$ or $Y \prec X$. In nature, a class consists of all states with the same mass and chemical composition—that is, with the same amount of each of the chemical elements. If chemical reactions and mixing processes are excluded, the classes are smaller and may be identified with the "systems" in the usual parlance. But it should be noted that systems may be compound, or consist of two or more vessels of different substances. In any case, the role of the comparison hypothesis is to ensure that the list of pairs $X \prec Y$ is sufficiently long. Indeed, we shall give an example later of a system whose pairs satisfy all the other axioms, but that is *not* describable by an entropy function.

## Construction of entropy

Our main conclusion (which we do not claim is obvious, but whose proof can be found in reference 2) is that the existence and uniqueness of entropy is a consequence of the comparison hypothesis and the assumptions about adiabatic accessibility stated above. In fact, if $X_0$, $X$, and $X_1$ are three states of a system and $\lambda$ is any scaling factor between 0 and 1, then either $X \prec ([1 - \lambda]X_0, \lambda X_1)$ or $([1 - \lambda] X_0, \lambda X_1) \prec X$, by the comparison hypothesis. If *both* alternatives hold, then the properties of entropy demand that

$$S(X) = (1 - \lambda)S(X_0) + \lambda S(X_1).$$

If $S(X_0) \neq S(X_1)$, then this equality can hold for at most one $\lambda$. With $X_0$ and $X_1$ as reference states, the entropy is therefore fixed, apart from two free constants, namely the values $S(X_0)$ and $S(X_1)$.

From the properties of the relation $\prec$ listed above, one can show that there is, indeed, always a $0 \leq \lambda \leq 1$ with the required properties, provided that $X_0 \prec X \prec X_1$. It is equal to the *largest* $\lambda$, denoted $\lambda_{max}$, such that $([1-\lambda]X_0, \lambda X_1) \prec X$. Defining the entropies of the reference states arbitrarily as $S(X_0) = 0$ and $S(X_1) = 1$ unit, we obtain the following simple formula for entropy:

$$S(X) = \lambda_{max} \text{ units.}$$

The scaling factors $(1-\lambda)$ and $\lambda$ measure the amount of substance in the states $X_0$ and $X_1$, respectively. The formula for entropy can therefore be stated in the following words: $S(X)$ is the maximal fraction of substance in the state $X_1$ that can be transformed adiabatically (that is, in the sense of $\prec$) into the state $X$ with the aid of a complementary fraction of substance in the state $X_0$. This way of measuring $S$ in terms of substance is reminiscent of an old idea, suggested by Pierre Laplace and Antoine Lavoisier, that heat be measured in terms of the amount of ice melted in a process. As a concrete example, let us assume that $X$ is a state of liquid water, $X_0$ of ice and $X_1$ of vapor. Then $S(X)$ for a kilogram of liquid, measured with the entropy of a kilogram of water vapor as a unit, is the maximal fraction of a kilogram of vapor that can be transformed adiabatically into liquid in state $X$ with the aid of a complementary fraction of a kilogram of ice. (See figure 3.)

In this example the maximal fraction $\lambda_{max}$ cannot be achieved by simply exposing the ice to the vapor, causing the former to melt and the latter to condense. That would be an irreversible process—that is, it would not be possible to reproduce the initial amounts of vapor and ice adiabatically (in the sense of the definition given earlier) from the liquid. By contrast, $\lambda_{max}$ is uniquely determined by the requirement that one can pass adiabatically from $X$ to $([1-\lambda_{max}]X_0, \lambda_{max}X_1)$ *and* vice versa. For this transformation it is necessary to extract or add energy in the form of work—for example by running a little reversible Carnot machine that transfers energy between the high-temperature and low-temperature parts of the system (see figure 3). We stress, however, that neither the concept of a "reversible Carnot machine" nor that of "temperature" is needed for the logic behind the formula for entropy given above. We mention these concepts only to relate our definition of entropy to concepts for which the reader may have an intuitive feeling.

By interchanging the roles of the three states, the definition of entropy is easily extended to situations where $X \prec X_0$ or $X_1 \prec X$. Moreover, the reference points $X_0$ and $X_1$, where the entropy is defined to be 0 and 1 unit respectively, can be picked consistently for different systems such that the formula for entropy will satisfy the crucial additivity and extensivity conditions

$$S(X,X') = S(X) + S(X') \quad \text{and} \quad S(\lambda X) = \lambda S(X).$$

It is important to understand that once the existence and uniqueness of entropy have been established, one need not rely on the $\lambda_{max}$ formula displayed above to determine it in practice. There are various experimental means to determine entropy that are usually much more practical. The standard method consists of measuring pressures, volumes, and temperatures (on some empirical scale), as well as specific and latent heats. The empirical temperatures are converted into absolute temperatures $T$ (by means of formulas that follow from the mere existence of entropy but do not involve $S$ directly), and the entropy is computed by means of formulas like $\Delta S = \int (dU + PdV)/T$, with $P$ the pressure. The existence and uniqueness of entropy implies that this formula is independent of the path of integration.

## Comparability of states

The possibility of defining entropy entirely in terms of the relation $\prec$ was first clearly stated by Giles.[7] (Giles's definition is different from ours, albeit similar in spirit.) The importance of the comparison hypothesis had been realized earlier, however.[4-6] All the authors take the comparison hypothesis as a postulate—that is, they do not attempt to justify it from other, simpler premises. However, it is in fact possible to *derive* comparability for any pair of states of the same system from some natural and directly accessible properties of the relation $\prec$.[2] The derivation uses the customary parameterization of states in terms of energy and work coordinates. But such parameterizations are irrelevant, and therefore not used, for our definition of entropy—once the comparison hypothesis has been established.

To appreciate the significance of the comparison hypothesis, it may be helpful to consider the following example. Imagine a world whose thermodynamical systems consist exclusively of incompressible solid bodies. Moreover, all adiabatic state changes in this world are supposed to be obtained by means of the following elementary operations:

▷ Mechanical rubbing of the individual systems, increasing their energy.

▷ Thermal equilibration in the conventional sense (by bringing the systems into contact).

The state space of the compound system consisting of two identical bodies, 1 and 2, can be parameterized by their energies, $U_1$ and $U_2$. Figure 4 shows two states, $X$ and $Y$, of this compound system, and the states that are adiabatically accessible from each of these states. It is evident from the picture that neither $X \prec Y$ nor $Y \prec X$ holds. The comparison hypothesis is therefore violated in this hypothetical example, and so it is not possible to characterize adiabatic accessibility by means of an additive entropy function. A major part of our work consists of understanding why such situations do not happen—why the comparison hypothesis appears to hold true in the real world.

The derivation of the comparison hypothesis is based on an analysis of simple systems, which are the building blocks of thermodynamics. As we already mentioned, the states of such systems are described by an energy coordinate $U$ and at least one work coordinate, such as the volume $V$. The following concepts play a key role in this analysis:

▷ The possibility of forming "convex combinations" of states with respect to the energy $U$ and volume $V$ (or other work coordinates). This means that given any two states $X$ and $Z$ of one kilogram of our system, we can pick any state $Y$ on the line between them in $U,V$ space and, by taking appropriate fractions $\lambda$ and $1-\lambda$ in states $X$ and $Z$, respectively, there will be an adiabatic process taking this pair of states into state $Y$. This process is usually quite elementary. For example, for gases and liquids one need only remove the barrier that separates the two fractions of the system. The fundamental property of entropy increase will then tell us that $S(Y) \geq \lambda S(X) + (1-\lambda)S(Z)$. As Gibbs emphasized, this "concavity" is the basis for thermodynamic stability—namely positivity of specific heats and compressibilities.

▷ The existence of at least one irreversible adiabatic state change, starting from any given state. In conjunction with the concavity of $S$, this seemingly weak requirement excludes the possibility that the entropy is constant in a whole neighborhood of some state. The classical formulations of the second law follow from this.

▷ The concept of thermal equilibrium between simple systems, which means, operationally, that no state changes take place when the systems are allowed to

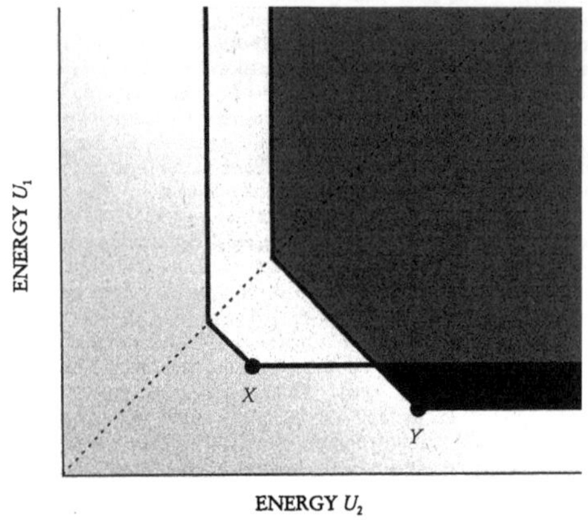


FIGURE 4. HYPOTHETICALLY NONCOMPARABLE STATES. The graph shows the state space of a pair of identical, incompressible solids with the energies $U_1$ and $U_2$ as the only coordinates of the compound system. The states adiabatically accessible from $X$ (yellow/orange) and $Y$ (red/orange) are shown under the assumption that the only adiabatic changes consist in combinations of rubbing (increasing $U_1$ or $U_2$) and thermal equilibration (moving to the diagonal $U_1 = U_2$). In this example, adiabatic accessibility *cannot* be characterized by an entropy function, because neither a transformation from $X$ to $Y$ nor from $Y$ to $X$ is possible. The comparison hypothesis does not hold here. In the real world, however, it *always* holds.

exchange energy with each other at fixed work coordinates. The zeroth law of thermodynamic says that if two systems are in thermal equilibrium with a third, then they are in thermal equilibrium with one another. This property is essential for the additivity of entropy, because it allows a consistent adjustment of the entropy unit for different systems. The zeroth law leads to a definition of temperature by the usual formula $1/T = (\partial S/\partial U)_V$.

Using these notions (and a few others of a more technical nature), the comparison hypothesis can be established for all simple systems and their compounds.

It is more difficult to justify the comparability of states if mixing processes or chemical reactions are taken into account. In fact, although a mixture of whiskey and water at 0 °C is obviously adiabatically accessible from separate whiskey and ice by pouring whiskey from one glass onto the rocks in the other glass, it is not possible to reverse this process adiabatically. Hence it is not clear that a block of a frozen whiskey/water mixture at −10 °C, say, is at all related in the sense of $\prec$ to a state in which whiskey and water are in separate glasses. Textbooks usually appeal here to *gedanken* experiments with "semipermeable membranes" that let only water molecules through and hold back the whiskey molecules, but such membranes really exist only in the mind.[10] However, without invoking any such device, it turns out to be possible to shift the entropy scales of the various substances in such a way that $X \prec Y$ always implies $S(X) \leq S(Y)$. The converse assertion, namely, $S(X) \leq S(Y)$ implies $X \prec Y$ provided that $X$ and $Y$ have the same chemical composition, cannot be guaranteed *a priori* for mixing and chemical reactions, but it is empirically testable and appears to be true in the real world. This aspect of the second law, comparability, is not usually stressed, but it is important; it is challenging to figure out how to turn the frozen whiskey/water block into a glass of whiskey and a glass of water without otherwise changing the universe, except for moving a weight, but such an adiabatic process is possible.

## What has been gained?

The line of thought that started more than 40 years ago has led to an axiomatic foundation for thermodynamics. It is appropriate to ask what if anything has been gained in comparison to the usual approaches involving quasi-static processes and Carnot machines on the one hand and statistical mechanics on the other hand. There are several points. One is the elimination of intuitive but hard-to-define concepts such as "hot," "cold," and "heat" from the foundations of thermodynamics. Another is the recognition of entropy as a codification of possible state changes, $X \prec Y$, that can be accomplished without changing the rest of the universe in any way except for moving a weight. Temperature is eliminated as an *a priori* concept and appears in its natural place—as a quantity derived from entropy and whose consistent definition really depends on the existence of entropy, rather than the other way around. To define entropy, there is no need for special machines and processes on the empirical side, and there is no need for assumptions about models on the statistical mechanical side. Just as energy conservation was eventually seen to be a consequence of time translation invariance, in like manner entropy can be seen to be a consequence of some simple properties of the list of state pairs related by adiabatic accessibility.

If the second law can be demystified, so much the better. If it can be seen to be a consequence of simple, plausible notions, then, as Einstein said, it cannot be overthrown.

*We are grateful to Shivaji Sondhi and Roderich Moessner for helpful suggestions. Lieb's work was supported by the National Science Foundation. Yngvason's work was supported by the Adalsteinn Kristjansson Foundation and the University of Iceland.*

## References

1. C. Kittel, H. Kroemer, *Thermal Physics*, Freeman, New York (1980), p. 57.
2. E. H. Lieb, J. Yngvason, Phys. Rep. **310**, 1 (1999); erratum **314**, 669 (1999). For a summary, see Notices Amer. Math. Soc. **45**, 571 (1998).
3. A. Einstein, autobiographical notes in *Albert Einstein: Philosopher-Scientist*, P. A. Schilpp, ed., Library of Living Philosophers, Cambridge U.P., London (1970), vol. VII, p. 33.
4. P. T. Landsberg, Rev. Mod. Phys. **28**, 363 (1956).
5. H. A. Buchdahl, *The Concepts of Classical Thermodynamics*, Cambridge U.P., London (1966).
6. G. Falk, H. Jung, *Handbuch der Physik* **III/2**, S. Flügge, ed., Springer, Berlin (1959), p. 199.
7. R. Giles, *Mathematical Foundations of Thermodynamics*, Pergamon, Oxford, England (1964).
8. D. R. Owen, *A First Course in the Mathematical Foundations of Thermodynamics*, Springer, Heidelberg, Germany (1984). J. Serrin, Arch. Rat. Mech. Anal. **70**, 355 (1979). M. Silhavý, *The Mechanics and Thermodynamics of Continuous Media*, Springer, Heidelberg, Germany (1997). C. A. Truesdell, S. Bharata, *The Concepts and Logic of Thermodynamics as a Theory of Heat Engines*, Springer, Heidelberg, Germany (1977).
9. J. L. Lebowitz, I. Prigogine, D. Ruelle, Physica A **263**, 516, 528, 540 (1999).
10. E. Fermi, *Thermodynamics*, Dover, New York, (1956), p. 101. ■

Part VI

# Lattice Systems

JOURNAL OF MATHEMATICAL PHYSICS VOLUME 8, NUMBER 5 MAY 1967

# Properties of a Harmonic Crystal in a Stationary Nonequilibrium State*

Z. RIEDER, J. L. LEBOWITZ, AND E. LIEB†
*Belfer Graduate School of Science, Yeshiva University, New York, New York*

(Received 14 July 1966)

The stationary nonequilibrium Gibbsian ensemble representing a harmonic crystal in contact with several idealized heat reservoirs at different temperatures is shown to have a Gaussian Γ space distribution for the case where the stochastic interaction between the system and heat reservoirs may be represented by Fokker–Planck-type operators. The covariance matrix of this Gaussian is found explicitly for a linear chain with nearest-neighbor forces in contact at its ends with heat reservoirs at temperatures $T_1$ and $T_N$, $N$ being the number of oscillators. We also find explicitly the covariance matrix, but not the distribution, for the case where the interaction between the system and the reservoirs is represented by very "hard" collisions. This matrix differs from that for the previous case only by a trivial factor. The heat flux in the stationary state is found, as expected, to be proportional to the temperature difference $(T_1 - T_N)$ rather than to the temperature gradient $(T_1 - T_N)/N$. The kinetic temperature of the $j$th oscillator $T(j)$ behaves, however, in an unexpected fashion. $T(j)$ is essentially constant in the interior of the chain *decreasing* exponentially in the direction of the *hotter* reservoir rising only at the end oscillator in contact with that reservoir (with corresponding behavior at the other end of the chain). No explanation is offered for this paradoxical result.

## 1. INTRODUCTION

In a series of papers[1–3] Lebowitz and Bergmann developed a general formalism for describing the time evolution of a Gibbs ensemble representing a system in contact with one or more idealized heat reservoirs (temperature baths). They imagine the reservoirs made up of an infinite number of identical noninteracting components each of which interacts with the system at most once. This interaction is impulsive and it is assumed that prior to this interaction the components of each reservoir have an equilibrium distribution with some specified temperature $T_\alpha$, where $\alpha = 1, \cdots, n$, specifies the different reservoirs. Under these conditions the Γ space ensemble density of the system $\mu(x, t)$ satisfies the generalized Liouville equation

$$\frac{\partial\mu(x, t)}{\partial t} + (\mu, H) = \sum_{\alpha=1}^{n} \int [K_\alpha(x, x')\mu(x', t) - K_\alpha(x', x)\mu(x, t)]\, dx'. \quad (1.1)$$

Here $x = (\mathbf{q}_1, \cdots, \mathbf{q}_N, \mathbf{p}_1, \cdots, \mathbf{p}_N)$ is a point in the phase space of the system, $H(x)$ is the Hamiltonian of the system, $(\mu, H)$ is the Poisson bracket between $\mu$ and $H$, and the right side of (1.1) represents the effect of collisions with reservoir components on the evolution of $\mu$. $K(x, x')\, dx\, dt$ is the conditional probability that when the system is at the point $x'$ in its Γ space it will suffer a collision in the time interval $dt$ as a result of which it will jump to the region $(x, x + dx)$.

Under very general conditions $\mu(x, t)$ approaches, as $t \to \infty$, a stationary distribution $\mu_s(x)$. This stationary distribution will correspond to the system being in equilibrium if the temperature of all the reservoirs is the same; otherwise $\mu_s(x)$ will represent a stationary nonequilibrium state in which there are heat currents flowing through the system. (More general nonequilibrium situations may also be represented in this manner.[1,4]) It is to be expected for a *physical* system of macroscopic size, whose interaction with the heat reservoirs is confined to specified "surface regions," that its bulk properties in the stationary state will depend only on the temperature of the reservoirs and not on the details of the interaction (this, of course, is expected to be true when the reservoirs all have the same temperature); e.g., the properties of a "long" metal bar should not depend on whether its ends are in contact with water or with wine "heat reservoirs" at temperature $T_1$ and $T_2$. (We are assuming here "good" heat contact between reservoirs and system so that regions of the system in direct contact with a given reservoir are essentially at the "temperature" of that reservoir.) This belief justifies the idealization of the reservoirs already made in deriving (1.1) and the further drastic simplification made below, and thus, we expect, for realistic systems,

* Based in part on a Ph.D thesis submitted by Z. Rieder to Yeshiva University.

† Present address: Physics Department, Northeastern University, Boston, Massachusetts.

[1] P. G. Bergmann and J. L. Lebowitz, Phys. Rev. 99, 578 (1955); J. L. Lebowitz and P. G. Bergmann, Ann. Phys. (N.Y.) 1, 1 (1957).

[2] J. L. Lebowitz, Phys. Rev. 114, 1192 (1959).

[3] J. L. Lebowitz, Rend. Sculola Intern. Fis. XIV Corso Bologna, Italy (1961).

[4] J. L. Lebowitz and A. Shimony, Phys. Rev. 128, 1945 (1962); E. P. Gross and J. L. Lebowitz, *ibid.* 104, 1528 (1956).

that the stationary state found from our model will correctly represent, in the Gibbs ensemble sense, such a physical system in a steady nonequilibrium state.

To obtain an explicit simple form for the right side of (1.1) we imagine the system to contain at its surface $n$ pistons of mass $M_\alpha$. The $\alpha$th reservoir will consist of point particles of mass $m_\alpha$ at uniform densities $\rho_\alpha$ *always* having a Maxwellian velocity distribution at temperature $T_\alpha$ prior to a collision with the $\alpha$th piston. During such an elastic collision there will be an exchange of momentum in some specified direction. Under these conditions the kernel $K_\alpha(x, x')$ may be specified explicitly [cf. Eq. (2.3), Ref. 2]. Still further simplification is achieved when $m_\alpha \ll M_\alpha$ so that the piston velocity is changed very little during a collision. The effect of the collisions with the reservoirs on the time evolution of $\mu(x, t)$ may then be represented by a Fokker–Planck-type term,[2,5] and (1.1) assumes the form

$$\frac{\partial\mu(x,t)}{\partial t} + (\mu, H) = \sum_{\alpha=1}^{n}\lambda_\alpha\frac{\partial}{\partial P_\alpha}\left[P_\alpha\mu + kT_\alpha M_\alpha\frac{\partial}{\partial P_\alpha}\mu\right]. \tag{1.2}$$

Here $(Q_\alpha, P_\alpha)$ are the coordinates and momentum of the $\alpha$th piston (the pistons being part of the system) and $\lambda_\alpha$ is the "friction constant" of the $\alpha$th piston given by [Eq. (3.3), Ref. 2]

$$\lambda_\alpha = \rho_\alpha A_\alpha(8m_\alpha kT_\alpha/\pi M_\alpha^2)^{\frac{1}{2}}, \tag{1.3}$$

where $A_\alpha$ is the collision cross section (or area) of the $\alpha$th piston. It is easy to show[2] that $\mu(x, t)$ satisfying (1.2) will in general approach, as $t \to \infty$, a stationary value $\mu_s(x)$.

Up to now we have not specified the nature of our system which determines $H(x)$. We now consider the case where our system is a harmonic crystal and the pistons are just some of the particles of the system. (Their location need not be specified at the moment.) It is then shown in Sec. 2 that the stationary solution of (1.2), which $\mu(x, t)$ will approach as $t \to \infty$, is a Gaussian in the coordinates and momenta of the system (corresponding to the canonical distribution when the temperatures of all the reservoirs are equal; $T_\alpha = T$). The explicit form of the stationary distribution, i.e., the covariance matrix of the Gaussian, is found in Sec. 3 for the special case of a one-dimensional crystal of $N$ particles with nearest-neighbor interactions in which the first particle is in contact with a reservoir at temperature $T_1$ and the last with a reservoir at temperature $T_N$ and $\lambda_1 = \lambda_N$. The general form of the distribution for this system, including its time dependence, has also been discussed independently by Bils.[6] Since we are interested in this system solely as a model, we do not worry about the drastic simplifications made in the right side of (1.1) to arrive at (1.2) or the further one, $\lambda_1 = \lambda_N$, required to obtain an explicit stationary nonequilibrium ensemble in $\Gamma$ space. What is unfortunate, however, is that the harmonic crystal is *not* a realistic physical system. As is well known,[7] the harmonic crystal has an "infinite" heat conductivity; i.e., the heat flux is not proportional to the temperature gradient when one considers the relaxation of this system from some initial nonequilibrium state. This is reflected in the true stationary state considered here by the fact that the heat flux is proportional (when $N \gg 1$, or strictly speaking in the limit $N \to \infty$), to the temperature difference between the ends of the system, $(T_1 - T_N)$, rather than to the temperature gradient $(T_1 - T_N)/N$, which would be the case if there was any anharmonic coupling. This is also reflected in the form of $T(j)$, the kinetic temperature of the $j$th harmonic oscillator which is uniform throughout the linear chain, being equal to $\frac{1}{2}(T_1 + T_N)$, except near the edges where it varies exponentially in a backward way; i.e., with $T_1 > T_N$, $T(j)$ will *decrease* from its mean value as $j \to 1$ jumping to a higher value, close to $T_1$, for $j = 1$. Also the heat flux, $J(\lambda)$, will vary with $\lambda$, the strength of the coupling to the reservoirs, in an unphysical way, reaching a maximum at $\lambda = \frac{1}{2}\sqrt{3}\,\omega$, (where $m\omega^2$ is the force constant between the oscillators), and vanishing as $\omega^2/\lambda$ for $\lambda \to \infty$. This may perhaps be understood as a mismatching between the frequencies of the reservoirs and the oscillators. We have no explanation for the abnormal behavior of $T(j)$.

An alternate idealization of the stochastic interaction between the reservoirs and the system is to imagine that after *each* collision with a component of the $\alpha$th reservoir the momentum $P_\alpha$ will have a Maxwellian distribution at the temperature $T_\alpha$,

$$h_\alpha(P) = (M_\alpha kT_\alpha/2\pi)^{\frac{1}{2}} \exp\left[-P^2/2M_\alpha kT_\alpha\right]. \tag{1.4}$$

This is an opposite extreme of the small momentum transfer considered before and corresponds to the pistons and reservoir components having the same mass.[1,2] We simplify this further by assuming that

[5] J. L. Lebowitz and P. Resibois, Phys. Rev. **139**, A1101 (1965).

[6] O. Bils, "On the Non-Stationary Equilibrium of a Finite Chain of Coupled Oscillators" (to be published).

[7] G. Klein and I. Prigogine, Physica **19**, 1053 (1953); P. C. Hemmer, Kgl. Norske Videnskab. Selskab Fork. **33**, 101 (1960); E. I. Takizawa and K. Kobayasi, Chinese J. Phys. **1**, 59 (1963); E. Teramoto, Progr. Theoret. Phys. (Kyoto) **28**, 1059 (1962); R. Rubin, Phys. Rev. **131**, 964 (1963).

the probability of a collision with a reservoir component in a time interval $dt$ is given by $\lambda'_\alpha\, dt$, independent of the state of the system. These assumptions lead to a modified Krook type of collision kernel[4] and Eq. (1.2) assumes the form

$$\frac{\partial\mu(x,t)}{\partial t}+(\mu,H)=\sum_\alpha\lambda'_\alpha\left\{h_\alpha(P_\alpha)\int\mu(x,t)\,dP_\alpha-\mu(x,t)\right\}. \quad (1.5)$$

While the stationary solution of (1.5) for a harmonic crystal is no longer a Gaussian the stationary covariance matrix of the linear chain and hence the kinetic temperature and heat flux, is of the same form as before. The only change is that the system acts as if the temperature difference $(T_1 - T_N)$ was reduced by the factor $[1+(\omega^2/\lambda^2)\varphi_1]$, where $\varphi_1$ depends on $\omega$ and $\lambda$. These assertions about the covariance matrix are proved in Sec. 3.

## 2. STATIONARY STATE OF A HARMONIC CRYSTAL

The Hamiltonian of a harmonic crystal containing $\mathcal{N}$ particles, each being $s$ dimensional, may be written in the general form[8]

$$H=\tfrac{1}{2}\sum_{i=N}^{2N}x_i^2+\tfrac{1}{2}\sum_{i,j=1}^{N}\Phi_{ij}x_ix_j;\quad N=s\mathcal{N}. \quad (2.1)$$

Here the $x_i$, $i=1,\cdots,N$, are the Cartesian coordinates of the particles, (relative to their equilibrium positions), while $x_j$, $j=i+N$, is the momentum conjugate to $x_i$ (we have set the mass of the particles equal to unity). The generalized Liouville equation (1.2) now has the form

$$\frac{\partial\mu(x,t)}{\partial t}=\sum_{i=1}^{2N}\frac{\partial}{\partial x_i}(\xi_i\mu)+\tfrac{1}{2}\sum_{i,j=1}^{2N}\frac{\partial^2}{\partial x_i\partial x_j}(d_{ij}\mu), \quad (2.2)$$

where

$$\xi_i=\sum_{j=1}^{2N}a_{ij}x_j \quad (2.3)$$

and $a_{ij}$ and $d_{ij}$ are elements of $2N$ by $2N$ matrices $\mathbf{a}$ and $\mathbf{d}$ which we write in the partitioned form

$$\mathbf{a}=\begin{pmatrix}\mathbf{0}&-\mathbf{I}\\ \mathbf{\Phi}&\mathcal{R}\end{pmatrix},\quad \mathbf{d}=\begin{pmatrix}\mathbf{0}&\mathbf{0}\\ \mathbf{0}&\boldsymbol{\epsilon}\end{pmatrix}. \quad (2.4)$$

Here $\mathbf{0}$ and $\mathbf{I}$ are the null and unit $N$ by $N$ matrices, $\Phi_{ij}$ is defined in (2.1), $\mathcal{R}_{ij}=\lambda_\alpha\delta_{\alpha i}\delta_{ij}$ [$\lambda_\alpha$ given in (1.3) with $M_\alpha=1$] and $\epsilon_{ij}=2kT_i\mathcal{R}_{ij}$. The general time-dependent solution of (2.2) may be found[6] by diagonalizing the right side of (2.2) as was done by Wang and Uhlenbeck[9] for fluctuations in electrical circuits. (Wang and Uhlenbeck consider only the case corresponding to all the $T_\alpha$ being the same.) It is clear, however, from an inspection of (2.2) that its stationary solution $\mu_s$ [corresponding to setting $\partial\mu/\partial t=0$ in (2.2)], which is all that is of interest to us in this problem, has the general form

$$\mu_s(x)=(2\pi)^{-N}\operatorname{Det}[\mathbf{b}^{-\frac{1}{2}}]\exp\left[-\tfrac{1}{2}\sum_{i,j=1}^{2N}b_{ij}^{-1}x_ix_j\right]. \quad (2.5)$$

The matrix $\mathbf{b}$ is the positive definite covariance matrix, and is related to expectation values in the stationary state by

$$b_{ij}=\langle x_ix_j\rangle=\int\mu_s(x)x_ix_j\,dx \quad (2.6)$$

and we have

$$A_i=\langle x_i\rangle=\int\mu_s(x)x_i\,dx=0. \quad (2.7)$$

Substituting (2.5) into (2.2) and equating terms yields the basic, necessary, and sufficient equation

$$\mathbf{a}\cdot\mathbf{b}+\mathbf{b}\cdot\mathbf{a}^\dagger=\mathbf{d}, \quad (2.8)$$

where $\mathbf{a}^\dagger$ is the transpose of $\mathbf{a}$. Once $\mathbf{b}$ is known all the properties of the stationary state, e.g., heat flux, local kinetic temperature, etc., are readily available. [It is clear that when all the $T_\alpha=T$ then $\boldsymbol{\epsilon}=2kT\mathcal{R}$ and $\mu_s(x)\sim e^{-\beta H(x)}$, $\beta=(kT)^{-1}$; i.e, the stationary state is the equilibrium state at temperature $T$.]

The uniqueness of the stationary solution $\mu_s(x)$ for the case where the coupling with the reservoirs does not vanish and the phase space of the crystal is not divided into different isolated parts (i.e., the representative phase point of the system can move between any two regions via a combination of its natural motion and collision with the reservoirs) follows from the general results of Ref. 1, explicitly verifiable here, that an *arbitrary* initial distribution will approach a unique $\mu_s(x)$ as $t\to\infty$. For the harmonic crystal in which there are no "torn bonds" isolating some parts this condition of ergodicity is clearly satisfied. The uniqueness of $\mu_s$ for the linear chain is shown explicitly in the next section.

Equations (2.7) and (2.8) are consequences of the general equations satisfied by the time-dependent expectation values $A_i(t)$ and $b_{ij}(t)$ defined with $\mu_s(x)\to\mu(x,t)$ in (2.6)–(2.7). We then have from (1.2)

$$(d/dt)\mathbf{A}(t)=-\mathbf{a}\cdot\mathbf{A}(t) \quad (2.9)$$

[8] Cf., for example, A. A. Maradudin, E. W. Montroll, and G. H. Weiss, *Theory of Lattice Dynamics in the Harmonic Approximation* (Academic Press Inc., New York, 1963); E. W. Montroll, Third Berkeley Symp. Math. Stat. and Prob. 3, 209 (1957).

[9] M. C. Wang and G. E. Uhlenbeck, Rev. Mod. Phys. 17, 323 (1945).

and

$$(d/dt)\mathbf{b}(t) = \mathbf{d} - \mathbf{a}\cdot\mathbf{b}(t) - \mathbf{b}(t)\cdot\mathbf{a}^\dagger. \tag{2.10}$$

For the case where $\mu(x, t)$ satisfies Eq. (1.5) the expectation values of the coordinates and momenta $\mathbf{A}'(t)$ again satisfy (2.9) (with $\lambda_\alpha$ replaced by $\lambda'_\alpha$) while the covariance matrix $\mathbf{b}'(t)$ now satisfies the equation

$$(d/dt)\mathbf{b}'(t) = \tfrac{1}{2}\mathbf{d} - \mathbf{a}\cdot\mathbf{b}'(t) - \mathbf{b}'(t)\cdot\mathbf{a}^\dagger + \mathbf{r}\cdot\mathbf{b}'(t)\cdot\mathbf{r} \tag{2.11}$$

with

$$\mathbf{r} = \begin{pmatrix} 0 & 0 \\ 0 & \mathfrak{R}^{\frac{1}{2}} \end{pmatrix}. \tag{2.12}$$

In the stationary state $\mathbf{A}'$ and $\mathbf{A}$ again vanish while $\mathbf{b}'$ or $\mathbf{b}$ satisfy (2.11) or (2.10) with the left sides set equal to zero.

### 3. EXPLICIT SOLUTION FOR A LINEAR CHAIN

We consider now a one-dimensional harmonic crystal (chain of pistons) with nearest-neighbor interactions, whose ends are rigidly fixed.[8] The interaction with the reservoirs takes place at the first and last piston, $\alpha = 1, N$ and we set

$$\lambda_1 = \lambda_N = \lambda, \quad T_1 = T(1+\eta), \quad T_N = T(1-\eta); \quad |\eta| \le 1. \tag{3.1}$$

The $N$ by $N$ matrices $\mathbf{\Phi}$, $\mathfrak{R}$, and $\boldsymbol{\epsilon}$ now have the form

$$\mathbf{\Phi} = \omega^2\mathbf{G}; \quad G_{ij} = 2\delta_{ij} - \delta_{i+1,j} - \delta_{i,j+1} \quad (\text{for}, j = 1 \text{ through } N-1), \tag{3.2}$$

$$G_{Nj} = G_{jN} = \begin{cases} 0, & j < N-1, \\ -1, & j = N-1, \\ 2, & j = N, \end{cases}$$

$$\mathfrak{R} = \lambda\mathbf{R}; \quad \mathbf{R}_{ij} = \delta_{ij}(\delta_{i1} + \delta_{iN}), \tag{3.3}$$

$$\boldsymbol{\epsilon} = 2kT\lambda(\mathbf{R} + \eta\mathbf{E}); \quad E_{ij} = \delta_{ij}(\delta_{i1} - \delta_{iN}). \tag{3.4}$$

We now write the $2N$ by $2N$ covariance matrix $\mathbf{b}$ in the partitioned form

$$\mathbf{b} = \begin{pmatrix} \mathbf{x} & \mathbf{z} \\ \mathbf{z}^\dagger & \mathbf{y} \end{pmatrix}. \tag{3.5}$$

The $N$ by $N$ matrices $\mathbf{x}$, $\mathbf{y}$, and $\mathbf{z}$ give, respectively, the correlations in the stationary state, among the coordinates, momenta and between the coordinates and momenta

$$x_{ij} = \langle q_i q_j \rangle, \quad y_{ij} = \langle p_i p_j \rangle, \quad z_{ij} = \langle q_i p_j \rangle. \tag{3.6}$$

To obtain the deviation of these correlations from their equilibrium values at uniform temperature $T$, corresponding to $\eta = 0$, we write

$$\mathbf{x} = (kT/\omega^2)[\mathbf{G}^{-1} + \eta\mathbf{X}], \tag{3.7}$$

$$\mathbf{y} = kT[\mathbf{I} + \eta\mathbf{Y}], \tag{3.8}$$

$$\mathbf{z} = \lambda^{-1}kT\eta\mathbf{Z}. \tag{3.9}$$

Using now (2.8) we find the following equations for $\mathbf{X}$, $\mathbf{Y}$, $\mathbf{Z}$:

$$\mathbf{Z} = -\mathbf{Z}^\dagger, \tag{3.10}$$

$$\mathbf{Y} = \mathbf{XG} + \mathbf{ZR}, \tag{3.11}$$

$$2\mathbf{E} - \mathbf{YR} - \mathbf{RY} = \nu[\mathbf{GZ} - \mathbf{ZG}]. \tag{3.12}$$

In addition, $\mathbf{X}$ and $\mathbf{Y}$ are required to be symmetric

$$\mathbf{X} = \mathbf{X}^\dagger, \quad \mathbf{Y} = \mathbf{Y}^\dagger, \tag{3.13}$$

while $\mathbf{b}$ is required to be positive definite. The quantity $\nu$ in (3.12) is $\nu = \omega^2/\lambda^2$, and is the only dimensionless parameter to remain in the problem.

To obtain an explicit solution of (3.10)–(3.13) we first note that the left side of (3.12), $2\mathbf{E} - \mathbf{YR} - \mathbf{RY}$, is a bordered matrix (it has nonvanishing elements only in the first and last rows and columns). Hence $\mathbf{GZ} - \mathbf{ZG}$ must also be bordered. Using the explicit form of $G$, (3.2), together with the antisymmetry requirement (3.10), it is easy to show that $\mathbf{Z}$ is necessarily a skew-symmetric Toeplitz matrix when $\mathbf{GZ} - \mathbf{ZG}$ is a bordered matrix, and $\mathbf{Z}$ may therefore be written in the form

$$\mathbf{Z} = \begin{pmatrix} 0 & \varphi_1 & \varphi_2 & \cdots & \varphi_{N-2} & \varphi_{N-1} \\ -\varphi_1 & \ddots & \ddots & \ddots & & \varphi_{N-2} \\ -\varphi_2 & \ddots & \ddots & \ddots & \ddots & \vdots \\ \vdots & \ddots & \ddots & \ddots & \ddots & \varphi_2 \\ & & \ddots & \ddots & \ddots & \varphi_1 \\ -\varphi_{N-1} & \cdots & & -\varphi_2 & -\varphi_1 & 0 \end{pmatrix}. \tag{3.14}$$

The quantities $\varphi_1, \cdots, \varphi_N$ are simply related to the entries in the bordered matrix in the left side of (3.12) and turn out to be

$$\nu\varphi_j = \delta_{j1} - Y_{1j} = \delta_{j1} + Y_{N,N-j+1}, \tag{3.15}$$

where $\varphi_N \equiv 0$ by definition. Equation (3.15) implies certain obvious restrictions on $\mathbf{Y}$ in order that (3.12) has a solution.

Next, Eq. (3.11), together with the fact that $\mathbf{Y}$ is symmetric implies that

$$\mathbf{XG} - \mathbf{GX} = -(\mathbf{RZ} + \mathbf{ZR}). \tag{3.16}$$

Once again, the right side of (3.16) is a bordered matrix, which is known in terms of the $\varphi$'s. Unlike $\mathbf{Z}$, however, **X** is required to be symmetric and we find that one solution is a Hankel matrix:

$$\mathbf{X} = \begin{pmatrix} \varphi_1 & \varphi_2 & \varphi_3 & \cdots & \cdots & \varphi_{N-2} & \varphi_{N-1} & 0 \\ & & & & & & & -\varphi_{N-1} \\ & & & & & & & \cdot \\ & & & & & & & \cdot \\ & & & & & & & \cdot \\ & & & & & & & -\varphi_2 \\ & & & & & & & -\varphi_1 \end{pmatrix}. \tag{3.17}$$

The solution to (3.16) is, however, not unique. We could add to (3.17) any symmetric matrix that commutes with **G**. Because all the eigenvalues of **G** are nondegenerate, such a matrix must be of the form $\tilde{\mathbf{X}} = P(\mathbf{G})$, where $P$ is any polynomial. Nevertheless, $\tilde{\mathbf{X}}$ is required to vanish as a consequence of our last condition, (3.11).

Substitution of (3.14) and (3.17) into the right side of (3.11) gives an expression for **Y** in terms of the $\varphi$'s. This, when combined with (3.15) yields an equation for the $\varphi$'s, namely,

$$\sum_{j=1}^{N-1} K_{ij}\varphi_j = \delta_{i1}, \tag{3.18}$$

where **K** is the $(N-1)$-square matrix

$$\mathbf{K} = \begin{pmatrix} \nu+2 & -1 & & & \\ -1 & \nu+2 & -1 & & \\ & -1 & \ddots & \ddots & \\ & & \ddots & \ddots & -1 \\ & & -1 & \nu+2 & -1 \\ & & & -1 & \nu+2 \end{pmatrix}. \tag{3.19}$$

The matrix **Y** is then the Hankel matrix

$$\mathbf{Y} = \mathbf{E} - \nu\mathbf{X}. \tag{3.20}$$

It is to be noted that both **X** and **Y** are antisymmetric about the "cross" diagonal [i.e., the $(1, N)$–$(N, 1)$ diagonal], a state of affairs that reflects the fact that changing the sign of $\eta$ corresponds to interchanging the reservoirs at the ends of the chain. Were we to add a matrix $\tilde{\mathbf{X}} = P(\mathbf{G})$ to **X**, as discussed above, then **X** would acquire a symmetric component about the "cross" diagonal. By (3.11) **Y**, too, would acquire such a component. But (3.15) precludes this possibility.

We conclude then, that the solution to (3.10)–(3.13) is given *uniquely* by (3.14), (3.17), and (3.20), assuming, of course that (3.18) has a unique solution. That this is so follows from the fact that **K** is positive definite for all $\nu \geq 0$.

To find $\varphi$ we use Cramer's rule. Consider the general $M$-square version of $K$ and let $D_M$ denote its determinant. Then

$$\varphi_j = (D_{N-1})^{-1} \times (\text{cofactor of } K_{1j} \text{ in } D_{N-1}). \tag{3.21}$$

But, it is easily seen that the cofactor of $K_{1j}$ is simply $D_{N-1-j}$. The computation of $D_M$ is simple. We observe that if we consider $D_M$ to be a function of the parameter $x = 1 + \frac{1}{2}\nu$, then $D_M(x) = 2xD_{M-1}(x) - D_{M-2}(x)$, while $D_0(x) = 1$ and $D_1(x) = 2x$. This is just the recursion relation for $U_M(x)$, the Chebyshev polynomial of the second kind. Using the fact that $U_M(\cosh\theta) = \sinh(M+1)\theta/\sinh\theta$, and defining $\alpha$ by

$$\cosh\alpha = 1 + \tfrac{1}{2}\nu \tag{3.22}$$

we obtain

$$\varphi_j = \sinh(N-j)\alpha/\sinh N\alpha. \tag{3.23}$$

For large $N$ and fixed $j$ we have the asymptotic formula:

$$\varphi_j = e^{-j\alpha} = (\varphi_1)^j, \tag{3.24}$$

and

$$\varphi_1 = e^{-\alpha} = 1 + \tfrac{1}{2}\nu - \tfrac{1}{2}(4\nu + \nu^2)^{\frac{1}{2}}. \tag{3.25}$$

For $\nu \to 0$ or $\infty$ we have

$$\varphi_1(\nu) \to \begin{cases} 1 - \nu^{\frac{1}{2}}, & \text{as } \nu \to 0, \\ \nu^{-1}, & \text{as } \nu \to \infty. \end{cases} \tag{3.26}$$

Finally, we note that **b** is necessarily positive definite, a property that can be proved directly from (2.8) using the fact that $\mathbf{\Phi}$ is positive definite.

For the case of hard collisions, $\mu(x, t)$ satisfying (1.5) and $\mathbf{b}'(t)$ satisfying (2.11), the corresponding matrices $\mathbf{X}'$, $\mathbf{Y}'$, and $\mathbf{Z}'$ will also satisfy Eq. (3.10) and (3.11) while (3.12) is now replaced by

$$\mathbf{E} - \mathbf{Y}'\mathbf{R} - \mathbf{R}\mathbf{Y}' + \mathbf{R}\mathbf{Y}'\mathbf{R} = \nu[\mathbf{G}\mathbf{Z}' - \mathbf{Z}'\mathbf{G}] \tag{3.27}$$

with $\nu$ now given by $\lambda'^2/\omega^2$. These equations may be brought into the same form as (3.10)–(3.12) by the replacements

$$\mathbf{X}' = (1 + \nu\varphi_1)^{-1}\mathbf{X}, \quad \mathbf{Y}' = (1 + \nu\varphi_1)^{-1}\mathbf{Y}, \quad \mathbf{Z}' = (1 + \nu\varphi_1)^{-1}\mathbf{Z}, \tag{3.28}$$

which corresponds simply to the replacement of the temperature difference $\eta$ by $\eta^*$, where

$$\eta^* = \frac{\eta}{(1 + \nu\varphi_1)} \to \begin{cases} \eta & \text{as } \nu \to 0, \\ \frac{1}{2}\eta & \text{as } \nu \to \infty. \end{cases} \tag{3.29}$$

## 4. PROPERTIES OF THE STATIONARY STATE

### Kinetic Temperature

The kinetic temperature of the $j$th particle is given by

$$kT(j, \nu; N) = \langle p_j^2 \rangle = kT[1 + \eta Y_{jj}] = \langle x_i\, \partial H/\partial x_i \rangle \tag{4.1}$$

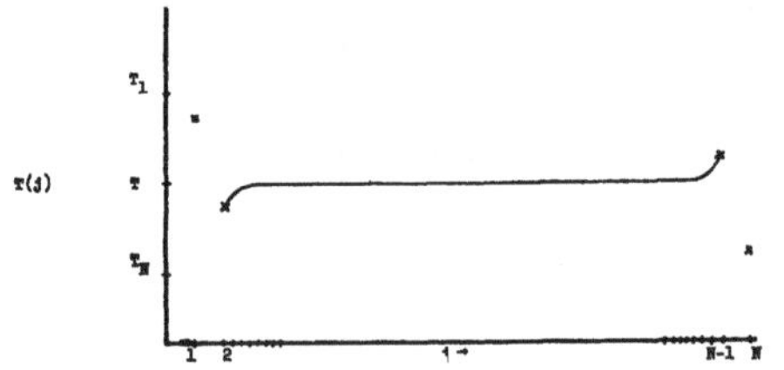


FIG. 1. Typical behavior of $T(j)$.

the last equality holding for all $i = 1, \cdots, 2N$ indicating some kind of equipartition for the stationary nonequilibrium state. In the limit of $N \to \infty$ we have

$$T(j, \nu) = \begin{cases} T[1 - \eta\nu(\varphi_1)^{2j-1}], & 1 < j < \frac{1}{2}N, \\ T[1 + \eta\nu(\varphi_1)^{2j'-1}], & 1 < j' = N - j < \frac{1}{2}N, \end{cases} \tag{4.2}$$

and

$$\begin{aligned} T(1, \nu) &= T_1 - \nu\varphi_1 T\eta, \\ T(N, \nu) &= T_N + \nu\varphi_1 T\eta, \end{aligned} \tag{4.3}$$

with $\varphi_1$ given by (3.25) and $T_1 = T(1 + \eta)$, $T_N = T(1 - \eta)$. The temperature of the linear chain thus deviates from its average value $T = \frac{1}{2}(T_1 + T_N)$ only at its edges where it changes exponentially over a length $\ell(\nu)$;

$$\ell(\nu) \to \begin{cases} \frac{1}{2}\nu^{-\frac{1}{2}}, & \nu \to 0, \\ \frac{1}{2}(\ln \nu)^{-1}, & \nu \to \infty. \end{cases}$$

It is a totally unexpected result of this model that the kinetic temperature *drops* below the average value at the second particle and then *increases* exponentially as we move *away* from the hot reservoir (cf. Fig. 1).

For the case of hard collisions (4.2) is unchanged except for the replacement of $\eta$ by $\eta^*$, while (4.3) now has the form

$$\begin{aligned} T'(1, \nu) &= T_1 - 2\nu\varphi_1 T\eta^*, \\ T'(N, \nu) &= T_N - 2\nu\varphi_1 T\eta^*. \end{aligned} \tag{4.5}$$

### Heat Flux

It is easy to show that the energy flux across a plane separating the $(i - 1)$th particle from the $i$th particle is given by[7]

$$j_{i-1,i} = \omega^2\langle q_{i-1}p_i\rangle = \omega^2 Z_{12} \equiv J, \quad i = 2, \cdots, N - 1.$$

The equalities hold in the stationary state where the flux is constant throughout the system and coincides with the energy flux $j_1 = -j_N$, coming from the reservoir at the left and going into the reservoir at the right, which is given by[2] $\lambda k[T_1 - T(1, \nu)]$. We then have using (3.25)

$$\begin{aligned} J(\lambda, \omega) &= (\omega^2/\lambda)kT\varphi_1\eta \xrightarrow[N\to\infty]{} \tfrac{1}{2}(\omega^2/\lambda)kT[1 + (\nu/2) \\ &\qquad - \tfrac{1}{2}\nu(1 + 4/\nu)^{\frac{1}{2}}](T_1 - T_N) \\ &= \tfrac{1}{2}(\omega^2/\lambda)k(T_1 - T_N) \quad \text{for } \lambda \gg \omega \\ &= \tfrac{1}{2}\lambda k(T_1 - T_N) \quad \text{for } \lambda \ll \omega. \end{aligned} \tag{4.6}$$

As expected the heat flux for the harmonic crystal is proportional to $(T_1 - T_N)$ rather than to the temperature gradient $(T_1 - T_N)/N$, i.e., the "heat" conductivity[7] is proportional to the size of the system.

The behavior of $J(\lambda, \omega)$ for fixed $\omega$ is very peculiar. For small $\lambda$, $\lambda \ll \omega$, $J$ is proportional to $\lambda$ as it should be, and is independent of $\omega$, the whole chain behaving as if it were just one tight piston. As $\lambda$ increases $J$ reaches a maximum at $\lambda = \frac{1}{2}\sqrt{3}\,\omega$, $J_{\max} = \frac{1}{6}k(T_1 - T_N)\omega$, and then decreases, vanishing as $\lambda^{-1}$ when $\lambda \to \infty$; the system now behaves, in the stationary state, as if the oscillators at the two ends are at the temperatures of the corresponding reservoirs, while the remainder are in equilibrium at temperature $T$. This latter behavior is quite unexpected. For a physical system (anharmonic coupling) we would expect $J(\lambda)$ to reach a limiting value proportional to the heat conductivity of the system times the temperature gradient.

For the case of hard collisions

$$J'(\lambda, \omega) = (1 + \nu\varphi_1)^{-1}J(\lambda, \omega) \to \begin{cases} J(\lambda, \omega), & \nu \to 0, \\ \frac{1}{2}J(\lambda, \omega), & \nu \to \infty. \end{cases}$$


## ACKNOWLEDGMENTS

One of us (J. L. L.) would like to thank Dr. O. Bils for stimulating correspondence.

This work was supported by the U.S. Air Force Office of Scientific Research under Grant No. 508-66.

With H.J. Brascamp and J.L. Lebowitz in Bull. Int. Statist. Inst. *46*, 393–404 (1975)

# THE STATISTICAL MECHANICS OF ANHARMONIC LATTICES

*Herm Jan Brascamp, Elliot H. Lieb, Joel L. Lebowitz* — **USA**

## I. INTRODUCTION

The equilibrium and non-equilibrium properties of regular harmonic crystals are well understood (Montroll, 1956, Lanford and Lebowitz, 1974). For them it is possible to define an infinite volume equilibrium state for dimensionality three and higher. This state is a Gaussian measure. For one (resp. two) dimensions it is not possible to define such a state because the mean square displacement of any particle diverges as $|\Lambda|$ (resp. $ln|\Lambda|$) as $|\Lambda|\to\infty$, where $|\Lambda|$ is the volume of the system. In all dimensions, however, it is possible to define a state on the algebra generated by the difference variables (Lanford and Lebowitz, 1974).

It is natural to ask, as we do here, whether this dependence on dimensionality, and essential independence of the detailed properties of the harmonic force matrix, remains true for anharmonic forces. In one dimension, the only case for which the answer can be computed explicitly, anharmonic and harmonic yield the same result: there is no long-range order — by which we mean that the expected value of the square displacement of a particle at the center diverges as $|\Lambda|\to\infty$. It seems likely that this analogy also holds in two and three dimensions, and in this paper we present some evidence that it does.

To simplify the situation as much as possible we restrict ourselves to nearest neighbor interactions on a simple cubic lattice and we assume the displacement of each particle to be one-dimensional; really it should be a vector. Most of our results can be readily generalized to more complicated situations.

The setting is the lattice $Z^\nu$, $\nu=1$, 2 or 3. Associated with each point $n\in Z^\nu$ is a variable $x_n\in R$ which we call the particle coordinate at $n$. Let $\Lambda$ be the cube in $Z^\nu$, centered at the origin 0, of side $2L+1$ and volume $|\Lambda|=(2L+1)^\nu$. Finally, let $v{:}R\to R$ be a function such that

$$I(\alpha)=\int e^{-\alpha v}<\infty,\forall\alpha>0,$$

and $v(x)=v(-x)$. We set $X=\{x_i\}_{i\in\Lambda}$.

393

The potential energy of the particles in $\Lambda$ is

$$H(X) = \sum_{<i,j>} v(x_i - x_j) \tag{1}$$

with the summation being over all distinct pairs of points $<i,j>$, $i \in \Lambda$, $j \in Z^\nu$, i and $j$ nearest neighbors in $Z^\nu$, i.e. $|i-j| = 1$. If $j \notin \Lambda$ we set $x_j = 0$ in (1). Thus, the particles at the boundary of $\Lambda$ are "tied down".

With $dX = \prod_{i \in \Lambda} dx_i$, let

$$Z_\Lambda = \int dX \exp[-H(X)] \tag{2}$$

be the partition function. We are interested in knowing whether the marginal distribution of $x_0$ has a limit as $\Lambda \to \infty$. E.g. does

$$<x_0^2>_\Lambda = \int dX x_0^2 \exp\ [-H(X)]/Z_\Lambda \tag{3}$$

have a finite limit? When $\nu = 1$ the situation is clear because the increments are essentially independent. Asymptotically, $<x_0^2>_\Lambda \sim |\Lambda|$.

Consider the **harmonic** crystal wherein $v(x) = ax^2$, $\alpha > 0$. In this case all the integrals are Gaussians and can be calculated exactly in terms of normal modes (Montroll, 1956). As $\Lambda \to \infty$

$$\frac{1}{|\Lambda|} lnZ_\Lambda \to -\frac{1}{2}(2\pi)^{-\nu} \int d^\nu k ln\left[\nu - \sum_{j=1}^{\nu} \cos k_j\right] - \frac{1}{2} ln(2a/\pi) \tag{4}$$

where the integration is over the cube $[-\pi, \pi]^\nu$. Similarly,

$$<x_0^2>_\Lambda \to (4\alpha)^{-1}(2\pi)^{-\nu} \int d^\nu k \left[\nu - \sum_{j=1}^{\nu} \cos k_j\right]^{-1} \tag{5}$$

which diverges for $\nu = 1$ and 2 but converges for $\nu = 3$. A more careful estimate shows that the divergence is proportional to $|\Lambda|$ (resp. $ln|\Lambda|$) for $\nu = 1$ (resp. 2).

In Section V we shall show that for $\nu = 2$, $<x_0^2>_\Lambda$ goes to infinity at least as fast as $ln|\Lambda|$ for potentials that satisfy the hypothesis of Theorem 4. This is a large class, including non-convex potentials, but it does not include potentials with hard walls, i.e. $v(x) = \infty$, $|x| > M$. For $\nu = 3$ our results, in Section VI, are more meager and are confined to convex potentials which increase at least as fast as $x^2$ as $|x| \to \infty$. However, if the potential is too flat near $x = 0$, it must increase precisely as $x^2$ for large $x$. For-a decorated lattice, and convex $v(x)$, we can prove that $<x_0^2>_\Lambda$ is bounded in three dimensions.

Our theorems do not prove the **conjecture** that $<x_0^2>_\Lambda$ **always** diverges when $\nu = 2$ and **always** stays bounded when $\nu = 3$, but they make it plausible. We have been unable to find a counterexample.

A noteworthy point is that the integral in (5) occurs in the theory of the random walk and is related to the reciprocal of the probability of not returning to the origin. This leads us to suspect that for general $v(x)$ there is also some connection between the random walk problem and $<x_0^2>_\Lambda$, but we do not know what it is. We shall, however, have a little to say about this in Section VII.

## II. THE THERMODYNAMIC LIMIT OF THE FREE ENERGY

One question that is easy to dispose of is that if $g_\Lambda \equiv |\Lambda|^{-1} ln Z_\Lambda$ then

$$\lim_{\Lambda \to \infty} g_\Lambda = g \tag{6}$$

exists. We shall discuss the two-dimensional case, but the argument is general. An upper bound to $Z_\Lambda$ is obtained by writing $\exp(-H) = \exp(-H_1)\exp(-H_2)$ where $H_1$(resp. $H_2$) contains all the "horizontal" (resp. "vertical") terms in (1). By Schwarz's inequality $Z_\Lambda^2 \leqslant \int \exp(-2H_1) \int \exp(-2H_2)$. Using the same inequality, $\int \exp(-2H_1) = \int \exp(-2H_2) \leqslant I(2)^{2L(2L+1)} I(4)^{2L+1}$. Thus $g_\Lambda <$ const. $< \infty$.

Now consider a sequence of domains $\Lambda_j$ with $L_j = 2^j, j = 1, 2, \ldots$, and define $g_j = |\Lambda_j| g_{\Lambda j} 2^{-2j-2}$. In the integral for $Z_j$, do the integral over all $x_n$ except when $n = (0, m)$. With $Y = \{x_{o,m} - 2^j \leqslant m \leqslant 2^j\}$, $Z_j$ has the form $Z_j = \int W(Y)^2 \exp[-H(Y)]\, dY$ and $H$ is the energy of the middle column. Clearly, $\int W(Y) \exp[-H(Y)] \times \exp[-K(Y)] = \hat{Z}_j$, where $\hat{Z}_j$ is the partition function of a $(2^{j+1}+1) \times (2^j+1)$ rectangle and $K(Y) = \sum_{m\Lambda}(x_{0,m})$. Using Schwarz's inequality, $\hat{Z}_j^2 \leqslant Z_j \int \exp[-2K(Y) - H(Y)]\, dY \leqslant Z_j I(2)^{2^j} I(4)^{2^{j+1}}$. Now splitting $\hat{Z}_j$ again into two pieces, we finally get

$$g_j \geqslant g_{j-1} - R_j \tag{7}$$

where $|R_j| <$ (const.) $2^{-j}$. Since $g_j$ is bounded above, (7) implies that $g_j$ has a limit. Since $|\Lambda_j| 2^{-2j-2} \to 1$, $g_{\Lambda j}$ has the same limit.

## III. A COMPARISON THEOREM

We return to the problem of evaluating $<x_0^2>_\Lambda$, and ask if there is any way to relate it to the calculation (5) for the pure harmonic case. A useful theorem is the following (Brascamp and Lieb, 1974, 1975).

T h e o r e m 1: *Let* $G = exp[-(x, Bx)]$, $B > 0$, *be a Gaussian on* $R^n$ *with covariance matrix* $1/2\ B^{-1}$. *If* $V(x)$ *is convex (resp. concave) and if* $M$ *is the covariance matrix of* $G$ *exp* $(-V)$, *then* $M \leqslant 1/2 B^{-1}$ *(resp.* $M \geqslant 1/2 B^{-1}$*).*

This theorem can settle the question for $v = 3$ when $v''(x) \geqslant 2\alpha > 0, \forall x$. In this case $<x_0^2>_\Lambda \leqslant$ eqn. (5). See Section VI.

The following two theorems will also be useful in dealing with convex potentials. A function $F(x)$ is said to be log concave if $F(x) = \exp[-f(x)]$, with $f(x)$ convex.

Theorem 2: *(Prékopa, 1971, 1973, Leindler, 1972, Rinott, 1973, Brascamp and Lieb, 1974, 1975): Let $F(x, y)$ be log concave in $(x, y) \in R^m \times R^n$. Then*

$$G(x) \equiv \int F(x, y)dy$$

*is log concave in $x \in R^m$.*

Theorem 2 is the basic for proving Theorem 1 (Brascamp and Lieb, 1974).

An important consequence of Theorem 2 is that when $v(x)$ is convex, the distribution of $x_0$ is log concave and, of course, even and monotone non-increasing on $(0, \infty)$. In this case, it is easy to see that if $\langle x_0^2 \rangle_\Lambda$ stays bounded, then all moments, $\langle x_0^p \rangle_\Lambda$, stay bounded.

A sharpened version of Theorem 2 is the following (Brascamp and Lieb, 1975).

Theorem 3: *Let $f(x, y)$ be convex in $(x, y) \in R \times R$, and let $f \in C^2(R^2)$. Define $F(x, y) = exp[-f(x, y)]$,* and

$$exp[-g(x)] = \int F(x, y)dy.$$

*The last integral and the integrals*

$$\int f_{xx} F\, dy, \quad \int (f_x)^2 F\, dy$$

*are assumed to converge uniformly in a neighborhood of a given point $x_0$. Then $g(x)$ is twice continuously differentiable near $x_0$. Its second derivative at $x_0$ satisfies*

$$g''(x_0) \geqslant \langle f_{xx} - (f_{xy})^2/f_{yy} \rangle \geqslant 0; \qquad (8)$$

*the average is taken with the normalized weight proportional to $F(x_0, y)$.*

Remarks: 1. Theorem 3 generalizes to $(x, y) \in R^m \times R^n$; for the purpose of this paper, however, we can restrict ourselves to the given case.

2. Since $f$ is convex,

$$f_{yy} \geqslant 0 \text{ and } (f_{xy})^2 \leqslant f_{xx} f_{yy}.$$

Thus, if we set

$$[f_{xx} - (f_{xy})^2/f_{yy}]\,(x_0, y) = 0$$

when $f_{yy}(x_0, y) = 0$, the inequality (8) is true.

## IV. STIFFENING THE SPRINGS DOES NOT NECESSARILY DECREASE $\langle x_0^2 \rangle$

It would be very helpful, if true, to know that increasing some of the terms in (1) decreases $\langle x_0^2 \rangle$. Consider the harmonic case and allow each term in (1) to be different, i.e.

$$H = \sum_{\langle i,j \rangle} \alpha_{ij}(x_i - x_j)^2, \quad \alpha_{ij} \geqslant 0.$$

It is elementary to see that increasing **any** $\alpha_{ij}$ does not increase $<x_0^2>$, because the total covariance matrix is decreased in the sense of forms. (As the following example shows, however, it is possible that $<x_0^2>$ is independent of some $\alpha_{ij}$.)

Now let us do the same thing for a general $v(x)$, i.e.

$$H \to \sum_{<i,j>} \alpha_{ij} v(x_i - x_j)$$

and we can even assume that $v$ is convex. We shall give a simple counter-example to the proposal that increasing any $\alpha_{ij}$ does not increase $<x_0^2>$.

Consider the following case with three particles, i.e. $H = v(x)+v(x-y)+v(y-z)+v(z)+\alpha v(x-z)$ and $x_0 = y$. Let $v(x) = x^2+\varepsilon x^4$, $\varepsilon > 0$. We want to show that **increasing** $\alpha$ from 0 can **decrease** $<x_0^2>$. Let $g_\infty$ (resp. $g_0$) be $<x_0^2>$ for $\alpha = \infty$ (resp. $\alpha = 0$). Then $g_\infty = 2 \int y^2 G(y)/\int G(y)$ and $g_0 = \int y^2 F(y)/\int F(y)$, with $G(y) = \exp(-2v(y))$ and $F(y) = R(y)^2$ where $R = \exp(-v) * \exp(-v)$. A simple calculation shows that for the pure harmonic case ($\varepsilon = 0$), $g_0 = g_\infty = 1/2$. When $\varepsilon > 0$ it is impossible to calculate the integrals, but it is possible to calculate $g_i = dg_i/d\varepsilon|_{\varepsilon=0}$. One finds that $g_\infty = -3/4$ and $g_0 = -9/8$. Thus, for small, positive $\varepsilon$, $g_\infty > g_0$, which is the contradiction we wished to demonstrate.

## V. TWO DIMENSIONS

We shall show that, for a large class of potentials $v$, $<x_0^2>_\Lambda$ increases at least as fast as $\ln|\Lambda|$ as $|\Lambda| \to \infty$. The method given here follows the argument by Hohenberg (1967), Mermin and Wagner (1966), Mermin (1967, 1968). We thank Dr. B. Halperin for showing us how the ideas in these references apply to the present problem.

Let $\varphi_1$, $\varphi_2$ be vectors in $R^n$ with $\|\varphi_1\| = 1$, and define

$$y_i = (\varphi_i, X); \quad \partial H/\partial y_i = (\varphi_i, \nabla H). \tag{9}$$

Let $T$ be a linear orthogonal transformation, $T$: $x_i \to \mathfrak{x}_i$, such that $\mathfrak{x}_1 = y_1$. Then, an integration by parts given that

$$\int dX y_2 (\partial H/\partial y_1) \exp(-H(X)) = \int dX \exp[-H(X)] \partial y_2/\partial y_1. \tag{10}$$

In this section we shall assume $v \in C^2$ and, to justify eliminating the boundary terms when integrating by parts, we assume that

$$[|x| + |v'(x)|] \exp[-v(x)] \to 0 \text{ as } |x| \to \infty. \tag{11}$$

Now $\partial y_2/\partial y_1 = (\varphi_1, \varphi_2)$. Application of Schwarz's inequality to the left side of (10) gives

$$<y_2^2> \geqslant (\varphi_1, \varphi_2)^2 / < (\partial H/\partial y_1) >^2. \tag{12}$$

Consider the matrix

$$(M)_{ij} = < (\partial H/\partial x_i)(\partial H/\partial x_j) > = <\partial^2 H/\partial x_i \partial x_j>, \tag{13}$$

where the last equality follows from an integration by parts and (11). Obviously, $M$ is a real, positive definite matrix. In terms of $M$, (12) reads

$$<y_2^2> \geqslant (\varphi_1, \varphi_2)^2/(\varphi_1, M\varphi_1) \qquad (12a)$$

If we let $\varphi = \varphi_2$, $y = (\varphi, X)$ and $\varphi_1 = M^{-1}\varphi/\|M^{-1}\varphi\|$, then we obtain

$$<y^2>_\Lambda \geqslant (\varphi, M^{-1}\varphi). \qquad (14)$$

Let us investigate the matrix $M$. By (13),

$$M_{ii} = \sum_{j;\, <i,j>} <v''(x_i - x_j)>, \text{ if } i \in \Lambda \qquad (15)$$

$$M_{ij} = -<v''(x_i - x_j)>, \text{ if } i, j \text{ are nearest neighbors and } i, j \in \Lambda,$$

$$M_{ij} = 0, \quad \text{otherwise.}$$

The sum in (15) is over the 4 nearest neighbors of $i$; if $i \in \partial\Lambda$, it should be understood that $x_j = 0$ if $j \notin \Lambda$. Note that

$$(\varphi, M\varphi) = \sum_{<i,j>} (\varphi_i - \varphi_j)^2 <v''(x_i - x_j)>_\Lambda.$$

Let us assume, for the moment, that there is a positive constant $A$, independent of $i$, $j$ and $\Lambda$, such that

$$<v''(x_i - x_j)>_\Lambda \leqslant A < \infty \qquad (16)$$

Then we have the matrix inequality

$$M \leqslant -A\Delta \qquad (17)$$

where $\Delta \leqslant 0$ is defined by

$$\Delta_{ii} = -4, i \in \Lambda,$$

$$\Delta_{ij} = 1, \text{ if } i,j \text{ are nearest neighbors and } i \text{ and } j \in \Lambda,$$

$$\Delta_{ij} = 0, \text{ otherwise.}$$

The matrix $\Delta$ arises precisely in the case of pure harmonic forces,

$$H(X) = \sum_{<i,j>} (x_i - x_j)^2 = -(X, \Delta X),$$

so that for harmonic forces

$$<y^2>_\Lambda = -(\varphi, \Delta^{-1}\varphi)/2. \qquad (18)$$

By (14, 17, 18), and with $y = x_0$, $<x_0^2>_\Lambda(v) \geqslant (2/A) <x_0^2>_\Lambda(v = x^2)$. Since the latter behaves as $ln|\Lambda|$ as $\Lambda \to \infty$, our statement is proved under the assumptions (11, 16). Let us give some simple conditions on $v$ that are sufficient for (16).

Obviously, it suffices that

$$v''(x) \leqslant A, \forall\, x. \qquad (19)$$

On the other hand, since

$$<v''(x_i-x_j)> = <v'(x_i-x_j)(\partial H/\partial x_i)>,$$

(16) is certainly satisfied if, for all $i$, $j$,

$$<[v'(x_i-x_j)]^2> \leqslant A/4. \tag{20}$$

Thus another sufficient condition is that

$$|v'(x)| \leqslant B < \infty, \forall x. \tag{21}$$

We finally consider convex potentials $v(x)$. Write

$$\int dX \exp[-H(X)]\delta(x_i-x_j-x) = \exp[-v(x)]W(x),$$

which defines $W(x)$. By Theorem 2, $W(x)$ is log concave if $v$ is assumed to be convex. We always assume $v(x)$ to be even, which implies that $W(x)$ is even. Hence, $W(x)$ is decreasing for $x>0$. Note further that $v'(x)$ is positive and increasing for $x>0$, since $v(x)$ is even and convex. Altogether,

$$\begin{aligned} <[v'(x_i-x_j)]^2> &= \int_0^\infty dx[v'(x)]^2\, W(x)\exp[-v(x)]/\int_0^\infty dx\, W(x)\exp[-v(x)] \\ &\leqslant \int_0^\infty dx[v'(x)]\exp[-v(x)]/\int_0^\infty dx\exp[-v(x)]. \end{aligned} \tag{22}$$

The last inequality follows from the fact that $[v'(x)]^2$ is increasing and $W(x)$ is decreasing. Thus, in the case of convex potentials, it suffices that the last member of (22) is finite.

We summarize the results of this section:

Theorem 4: *In two dimensions,* $<x_0^2>_\Lambda$ *increases at least as* $ln\,|\Lambda|$ *as* $\Lambda\to\infty$ *if* $v(x)$ *satisfies*

(*i*) $[|x|+|v'(x)|]exp[-v(x)]\to 0$ *as* $|x|\to\infty$

(*ii*) *One of the following three conditions:*

(*a*) $v''(x) \leqslant A < \infty, \forall x$

(*b*) $|v'(x)| \leqslant B < \infty, \forall x$

(*c*) $v(x)$ *is convex and*

$$\int dx[v'(x)]^2 exp[-v(x)] < \infty.$$

By taking suitable limits this class includes such diverse potentials as $|x|^\gamma$, $1 \leqslant \gamma < \infty$, $x^2-|x|$ (which has a double minimum) and max $(|x|^\gamma, 1)$, $\forall \gamma > 0$ (which has a flat bottom). It does not include $|x|^\gamma$, $0<\gamma<1$ or $|x|^\gamma$, $\gamma=\infty$. By the last expression we mean the hammock potential, i.e.

$$v(x) = \begin{cases} 0, & |x| \leqslant 1 \\ \infty, & |x| > 1 \end{cases}$$

A final remark is that when $v(x)$ satisfies the hypothesis of Theorem 6 then, by the methods of Section VI, $<x_0^2>_\Lambda$ does not diverge faster than $ln\,|\Lambda|$.

## VI. THREE DIMENSIONS

As already remarked after Theorem 1, $<x_0>_\Lambda$ is bounded in three dimensions if $v(x)$ is convex in the following strict sense:

$$v''(x) > 2a \geqslant 0, \forall x.$$

We exploit this idea a bit further for functions $v \in C^2$.

Let us split the lattice points $n = (n_1, n_2, n_3) \in Z^3$ according to whether $|n| \equiv n_1 + n_2 + n_3$ is even or odd. To emphasize the distinction, let us write $x_n = y_n$, if $|n|$ is even; $x_n = z_n$, if $|n|$ is odd. Then we can write

$$H(X) = \sum_{|m| \text{odd}} \sum_{n:<m,n>} v(y_n - z_m),$$

so that

$$\int \exp[-H(X)]\, dZ = \prod_{|m| \text{odd}} w(Y_m), \tag{23}$$

where $Y_m$ stands for the 6 nearest neighbors of $m$, and

$$w(Y_m) = \int dz_m \exp[- \sum_{n:<n,m>} v(y_n - z_m)]. \tag{24}$$

Notice that the $y'$ s occupy a face-centered cubic lattice, consisting of the points $n$ with $n_1$, $n_2$, $n_3$ even, together with the centres of the faces of the resulting $2 \times 2 \times 2$ cubes.

We shall show (Theorem 5) that, under certain conditions on $v(x)$,

$$w(Y_m) = \exp[-\frac{1}{2} a \sum_{n:<n,m>} \sum_{k:<m,k>} (y_n - y_k)^2 - f(Y_m)], \tag{25}$$

where $f$ is a convex function (of its 6 variables jointly) and $a$ is a suitable positive constant.

Let us further split the variables $\{Y_m\}$ into $\{u_i\}$ and $\{v_j\}$, with the $u_i$ corresponding to the corners of the $2 \times 2 \times 2$ cubes and the $v_i$ to the centres of the faces. Then it follows from (23, 24, 25) and Theorem 2 that

$$\int \exp[-H(X)]\, dZ\, dV = \exp[-a \sum_{<i,j>} (u_i - u_j)^2 - g(U)]. \tag{26}$$

where $g$ is a convex function of the $U$-variables and where the summation is over all pairs of nearest neighbors on the lattice, $(2Z)^3$, of $2 \times 2 \times 2$ cubes.

Since the required $<x_0^2>$ is obviously equal to $<u_0^2>$ with the weight (26), Theorem 1 implies that $<x_0^2>$ is bounded above by (5) and is thus bounded.

Let us now give a sufficient condition on $v(x)$ so that (25) is satisfied.

T h e o r e m 5: *Let* $v(x)$ *be convex, and let*

(*i*) $0 < A \leqslant v''(x) \leqslant B < \infty$, *if* $|x| \geqslant M$

(*ii*) $|v(x) - Cx^2| \leqslant D < \infty$, $\forall\, x$

*with strictly positive constants A,B,C and M. Define*

400

$$exp[-g(y_1, ..., y_k)] = \int dz\, exp[-\sum_{j=1}^{k} v(y_j - z)].$$

*Then*

$$g(y_1, ..., y_k) = \frac{1}{2} a \sum_{i,j=1}^{k} (y_i - y_j)^2 + h(y_1, ..., y_k),$$

*with a constant $a > 0$ and with the function h convex.*

Proof: Apply Theorem 3 to the second derivative of $g$ in a direction $(\lambda_1, ..., \lambda_k)$ at a given point. Thus

$$\sum_{i,j} \lambda_i \lambda_j \partial^2 g / \partial y_i \partial y_j \geqslant < \sum_i \lambda_i^2 v''(y_i - z) - \frac{\sum_{i,j} \lambda_i \lambda_j v''(y_i - z) v''(y_j - z)}{\sum_i v''(y_i - z)} > =$$

$$= < \frac{\sum_{i,j} (\lambda_i - \lambda_j)^2 v''(y_i - z) v''(y_j - z)}{2 \sum_i v''(y_i - z)} >.$$

By condition (i), this exceeds

$$\frac{A^2}{2kB} \sum_{i,j} (\lambda_i - \lambda_j)^2 \int_{|z - y_i| > M} dz \exp[-\sum_i v(y_i - z)] / \int dz \exp[-\sum_i v(y_i - z)].$$

Condition (ii) in turn implies that this is not smaller than

$$\frac{A^2}{2kB} \exp(-2kD) \sum_{i,j} (\lambda_i - \lambda_j)^2 \int_{|z| > kM} dz \exp(-kCz^2) / \int dz \exp(-kCz^2) \equiv \frac{1}{2} a \sum_{i,j} (\lambda_i - \lambda_j)^2.$$

This proves Theorem 5. Q.E.D.

Remarks: 1. Theorem 5 obviously fails if $v(x)$ increases slower than quadratically as $|x| \to \infty$, because then also $g'' \to 0$ as $|y_i - y_j| \to \infty$.

2. It is less obvious, but true, that Theorem 5 also fails if $v(x)$ increases faster than quadratically, and if $v'' = 0$ somewhere.
Take, for example, $v(x) = x^4$, and let

$$\exp[-g(x,y)] = \int dz \exp[-z^4 - (x-z)^4 - (y-z)^4].$$

By Theorem 2, $g_{xx} \geqslant 0$. Also, simple differentiation gives that

$$g_{xx} \leqslant 12 < (x-z)^2 >.$$

In particular, for $y = 2x$

$$g_{xx}(x, 2x) < 12 \int dz\, z^2 \exp[-3z^4 - 12z^2x^2] / \int dz \exp[-3z^4 - 12z^2x^2].$$

Hence, $g_{xx}(x, 2x) \to 0$ as $|x| \to \infty$. Note however, that $g_{xx}(x, y) \to \infty$ as $(x, y) \to \infty$ in any other direction than $y = 2x$. The situation is a bit worse when there are 6 neighbors, but in any case Theorem 5 only just barely fails for $v(x) = x^4$. This supports the conjecture that $< x_0^2 >$ is bounded for the $x^4$ interaction. More evidence in this direction is supplied by the fact that $< x_0^2 >$ can be proved to be bounded if some interactions are removed.

Namely, remove all the lattice points that do not lie on a corner or on an edge of some $2\times2\times2$ cube. Then we have what is called a decorated lattice and the points in the middle of an edge have only two nearest neighbors. One can use the fact that

$$\int dz \exp[-v(x-z)-v(y-z)] = \exp[-a(x-y)^2-f(x-y)] \qquad (27)$$

with $f$ convex, for a much wider class of potentials than ones in Theorem 5. For example, (27) holds true $v(x) = |x|^\gamma$, $\gamma > 2$, including the hammock potential ($\gamma = \infty$).

Theorem 6: *In three dimensions,* $\langle x_0^2\rangle_\Lambda$ *is bounded in* $\Lambda$ *if* $v(x)$ *satisfies one of the following conditions*

(a) $v(x) = ax^2+f(x)$, *$f$ convex,* $a>0$

(b) *conditions (i) and (ii) of Theorem* 5.

As we said above, $x^4$ just fails. However, $v(x) = x^4, |x| \leqslant 1$; $2x^2-1, |x| \geqslant 1$ does satisfy the hypothèses of Theorem 6 and therefore, if $x^4$ fails in reality, it cannot be due to small amplitude fluctuations alone.

## VII. A TENUOUS CONNECTION WITH THE RANDOM WALK PROBLEM

As we said in the introduction, the various tricks we have employed to show that $\langle x_0^2\rangle_\Lambda$ goes to infinity in one and two dimensions and stays bounded in three dimensions do not really go to the heart of the problem. Somehow there must be a more immediate connection with some property of the lattice more directly related to its geometry.

To this end, let $\Delta$ be the second difference operator on $Z^\nu$ with zero boundary conditions on $\Lambda$. I.e., for $n\in\Lambda$,

$$(\Delta f)(n) = -2\nu f(n)+\sum_j f(j)\chi(n,j) \qquad (28)$$

where $f(j) \equiv 0$ if $j\notin\Lambda_n$ and $\chi(n,j) = 1$ if $n$ and $j$ are nearest neighbors and zero otherwise. Consider the Green function for $n$, $m\in\Lambda$

$$-\Delta_n G(n,m) = \delta_{n,m}$$

with $\delta$ being the Kronecker delta. If a random walker starting at $n$ is observed for a very long time, then $G(n,m)$ is proportional to the number of times, on the average, that the walker visits the site $m$. Clearly $G(n,m) = G(m,n)$ and as $\Lambda\to\infty$, $G(n,n)$ remains finite for $\nu = 3$ and goes to infinity for $\nu = 1,2$.

To use $G$, we write

$$x_0 = \int_{j\in\Lambda} x_j\delta_{j,0} = -\sum_{j\in\Lambda} x_j(\Delta G(\,\cdot\,,0))(j) \qquad (29)$$

$$= (1/2)\sum_j\sum_n (\nabla x)(j,n)\,(\nabla G(\,\cdot\,,0))(j,n) \qquad (30)$$

$$= -\sum_{j\in\Lambda}(\Delta x)(j)G(j,0) \tag{31}$$

where $(\nabla f)(j,n) \equiv [f(j)-f(n)]\chi(j,n)$, and $f(k) \equiv 0$ if $k \notin \Lambda$.

In one dimension, with $\Lambda = [-L,L]$, $G(n,0) = L+1-|n|$, and (30) reads

$$x_0 = 1/2\{(x_0-x_1)+(x_1-x_2)+\ldots+(x_L)\}+1/2\{(x_0-x_{-1})+(x_{-1}-x_{-2})+ +\ldots+(x_{-L})\}.$$

We now square the right side of (30) and insert it into (3), the expression for $<x_0^2>_\Lambda$. Suppose we consider only the "diagonal terms", i.e. when $j'=j$ and $n'=n$. Assuming that $<(x_j-x_n)^2>\chi(j,n)\leqslant A<\infty$, these terms would contribute

$$(A/4)\sum_j\sum_n \chi(j,n)(\nabla G(\cdot,0))(j,n)^2,$$

and this is finite in three dimensions because $G(j,0)\sim|j|^{-1}$ as $j\to\infty$ and $G(j,0)$ is bounded. For the same reason, using Cauchy's inequality, we could handle any finite number of off-diagonal terms, i.e. those for which $|j-j'|<B$ and $B$ is fixed. We cannot handle all the terms this way because we would end up with $\left(\sum_{n:|n|\geqslant 1}|n|^{-2}\right)^2$ and this diverges at infinity. However, the method would work (by Young's inequality) if we knew that the correlations

$$<(x_j-x_n)(x_k-x_l)>\chi(j,n)\chi(k,l)$$

vanished faster than $|j-k|^{-2-\varepsilon}(\varepsilon>0)$ as $|j-k|\to\infty$.

We almost have here a proof by contradiction for $\nu = 3$. Loosely speaking, if the correlations do not decay, then we have long-range order. But then one would expect that $<x_0^2>_\Lambda$ remains bounded, because $<x_0^2>_\Lambda\to\infty$ means there is no long-range order, i.e. the center of the crystal does not feel the effect of the tied down boundaries.

When $\nu = 2$, the use of (30) almost shows that $<x_0^2>_\Lambda$ diverges, because $G(j,0)\sim -ln|j|$ for small $j$ and large $\Lambda$. This argument is insufficient, however, because there might be a cancellation of alternating terms.

## ACKNOWLEDGEMENTS

This work originated from discussions with Dr. O. Lanford III, to whom we are grateful for valuable comments. We are indebted to Dr. B. Halperin for showing us how to apply the method by Hohenberg, Mermin and Wagner, and Mermin to the two-dimensional case.

## REFERENCES

H.J. Brascamp and E.H. Lieb (1974), *Some inequalities for Gaussian Measures, International Conference on Functional Integration and its Applications*, Cumberland Lodge, England, April 1—4, 1974.

H.J. Brascamp and E.H. Lieb (1975), *On extensions of the Brunn-Minkowski and*

*Prékopa-Leindler theorems, including inequalities for log concave functions, and with an application to the diffusion equation, preprint.*
P.C. Hohenberg (1967), *Phys. Rev.* 158, 383.
O.E. Lanford III and J.L. Lebowitz (1974), *Time Evolution and Ergodic Properties of Harmonic System, 1974 Battelle Rencontres*, Springer-Verlag, New York–Heidelberg–Berlin (in press).
L. Leindler (1972), *Acta Sci. Math. Szeged* 33, 217.
N.D. Mermin (1967), *J. Math. Phys.* 8, 1061.
N.D. Mermin (1968), *Phys. Rev.* 176, 250.
N.D. Mermin and H. Wagner (1966), *Phys. Rev. Letters* 17, 1133.
E.W. Montroll (1956), in: *Proceedings of the Third Berkeley Symposium on Mathematical Statistics and Probability*, Vol. III, Univ. of Calif. Press, Berkeley, page 209.
A. Prékopa (1971), *Acta Sci. Math. Szeged* 32, 301.
A. Prékopa (1973), *Acta Sci. Math. Szeged* 34, 335.
Y. Rinott (1973), *Thesis*, Weizmann Institute, Rehovot, Israel.

## SUMMARY

It is well known that for a lattice of oscillators coupled harmonically there is no long-range order in one and two dimensions, but there is in three dimensions. Long-range order means that the mean square displacement of an oscillator remains finite as the size of the lattice increases. This fact is related to the probability of return to the origin of a random walk. The question to be discussed here is whether the above facts are a consequence of geometry or whether they depend on the harmonic (i.e. $x^2$) nature of the potential. In the anharmonic case no explicit solution exists. We show that for a large class of potentials the existence or non-existence of order depends only on dimensionality.

## RÉSUMÉ

On sait depuis longtemps que, pour un système d'oscillateurs en interaction harmonique sur un réseau, il existe un ordre macroscopique dans le cas de trois dimensions, mais non dans le cas d'une ou de deux dimensions.

L'existence d'un ordre macroscopique signifie que la moyenne quadratique de l'élongation d'un oscillateur reste finie quand la taille du réseau augmente indéfiniment. Cela est lié à la probabilité pour une marche au hasard de retourner à son point de départ.

La question que nous discutons ici est la suivante: les résultats précédents sont-ils une conséquence de la géométrie, ou dépendent-ils de la nature harmonique (i.e. $\sim x^2$) du potentiel?

Dans le cas d'un potentiel anharmonique, il n'existe pas de solution explicite. Nous montrons que, pour une vaste classe de potentiels, l'existence ou la non-existence d'un ordre macroscopique ne dépend que du nombre de dimensions.

*Journal of Statistical Physics, Vol. 16, No. 6, 1977*

# Time Evolution of Infinite Anharmonic Systems

**Oscar E. Lanford III,[1] Joel L. Lebowitz,[2,3] and Elliott H. Lieb[4]**

*Received December 28, 1976*

We prove the existence of a time evolution for infinite anharmonic crystals for a large class of initial configurations. When there are strong forces tying particles to their equilibrium positions then the class of permissible initial conditions can be specified explicitly; otherwise it can only be shown to have full measure with respect to the appropriate Gibbs state. Uniqueness of the time evolution is also proven under suitable assumptions on the solutions of the equations of motion.

**KEY WORDS:** Existence of time evolution; uniqueness; strong restoring forces.

## 1. INTRODUCTION

The time evolution of classical (or quantum) Hamiltonian dynamical systems containing an infinite number of particles is of great interest in statistical mechanics, being an essential ingredient in the study of nonequilibrium phenomena in macroscopic systems. There are many difficulties, however, in dealing with the dynamics of infinite systems and the available results on the existence of the time evolution of such systems are not entirely satisfactory. It is only for one-dimensional classical systems[(1)] or harmonic crystals[(2)] that we have a strong evolution theorem, i.e., we can specify explicitly a class of initial conditions for which a time evolution exists. This set of initial conditions is furthermore appropriate for a large class of interactions between the particles and has full equilibrium measure for all these interactions. In contrast, all that has been proven so far for higher dimensions[(3–5)] is the

Supported in part by NSF Grants #MCS 75-05576-A01 (to O.E.L.), #MPS 75-20638 (to J.L.L.), and #MCS 75-21684 (to E.H.L.).

[1] University of California at Berkeley, Berkeley, California.
[2] Institut des Hautes Etudes Scientifiques, Bures-sur-Yvette, France.
[3] Permanent address: Belfer Graduate School of Science, Yeshiva University, New York, New York.
[4] Princeton University, Princeton, New Jersey.

existence of a time evolution for a given interaction on some unspecified set of initial configurations which has full measure with respect to the equilibrium state for that interaction. It is the purpose of this paper to prove the existence of a strong time evolution for a certain class (where condition $A_3$ of Section 2 holds) of anharmonic crystals[6] in arbitrary dimensions and a weaker time evolution for very general anharmonic systems (Section 4).

## 2. EXISTENCE OF TIME EVOLUTION

The setting is the lattice $\mathbb{Z}^\nu$. At each point $i \in \mathbb{Z}^\nu$ we have an oscillator with coordinate $q_i \in \mathbb{R}$ and momentum $p_i \in \mathbb{R}$. Really, we should take $q_i$ and $p_i$ in $\mathbb{R}^k$ for some $k$; with $k = \nu$ this would represent, physically, the fact that each point of $\mathbb{Z}^\nu$ is the equilibrium position of a particle. To avoid complicating the notation, we take $k = 1$, but our results obviously go through for general $k$. By $\mathbf{q}$ (resp. $\mathbf{p}$) we denote the collection of oscillator coordinates (resp. momenta).

The oscillator variables are regarded as functions of time $t$, $\{q_i(t), p_i(t)\}$, and are represented collectively by $\mathbf{q}(t)$ and $\mathbf{p}(t)$. They satisfy the following infinite set of coupled differential equations:

$$dq_i(t)/dt = p_i(t) \tag{1a}$$

$$dp_i(t)/dt = F_i = -\partial U_i(q_i(t))/\partial q_i + R_i(\mathbf{q}(t)) \tag{1b}$$

In Eq. (1b) we wrote $F_i$, the force acting on the $i$th particle, as a sum of two terms: a gradient of a "self-energy" term $U_i(q_i)$ and a force $R_i$, which we shall take (but need not have) to be the gradient of some interaction energy

$$R_i = -\sum_j \partial V_j(\mathbf{q})/\partial q_i \tag{1c}$$

Our basic assumption in this part is that the self-energy $U_i(q_i)$ is such a steeply increasing function of $q_i$ that it "dominates" the motion of the particles when they are far from their equilibrium positions. We also assume that the interactions have a finite range $D$ (this is convenient but not essential). Stated precisely, we assume:

$A_1$. $V_i(q)$ depends only on those $q_j$ for which the Euclidean distance $|i - j| \leqslant D$.

$A_2$. Each $U_j(q_j)$ and $V_j(\mathbf{q})$ is a twice continuously differentiable function of its arguments.

$A_3$. $|q_j| \leqslant C_1 U_j(q_j) + C_2$, $C_1$ and $C_2$ nonnegative constants.

$A_4$. There exist nonnegative bounded constants $A_{ij} < C$, $A_{ij} = 0$ for $|i - j| > D$, such that

$$|p_i R_i(\mathbf{q})| \leqslant \sum_j A_{ij} \mathcal{L}_j \tag{2a}$$

where

$$\mathscr{L}_i(p_i, q_i) = \tfrac{1}{2}p_i^2 + U_i(q_i) + K \geqslant 0, \qquad K \text{ a constant} \tag{2b}$$

**Example.** Conditions A will be satisfied if $U_i(q_i)$ is a polynomial of degree $2n$ whose leading coefficient $\lambda_i$ is strictly positive, $\lambda_i \geqslant \lambda > 0$, and $R_i(q)$ is a multinomial of degree at most $n$.

The nonnegative functions $\mathscr{L}_i$ will play an important role in establishing the existence and uniqueness of solutions to the equations of motion (1). They are similar to self-energy or Lyapunov functions.

The problem posed by Eqs. (1a)–(1c) is the following: Given suitable initial data $\mathbf{q}(0)$, $\mathbf{p}(0)$, find $\mathbf{q}(t)$, $\mathbf{p}(t)$ that agree with the initial data at $t = 0$ and satisfy (1a)–(1c). This problem is equivalent to another one: Find $\mathbf{q}(t)$ such that

$$q_i(t) = q_i(0) + p_i(0)t + \int_0^t (t-s)F_i(\mathbf{q}(s))\, ds \tag{3}$$

Any solution to (3) will satisfy the initial condition and will be differentiable. One can then define $p_i(t) = dq_i(t)/dt$ and (1a)–(1c) will be satisfied. Conversely, any solution to (1a)–(1c) satisfies (3).

**Definition.** We denote by $B_r$ the real Banach space of sequences $\boldsymbol{\xi} = \{\xi_j\}$, $j \in \mathbb{Z}^\nu$, such that the norm

$$\|\xi\|_r = \sup_{j\in\mathbb{Z}^\nu} \{[\exp(-|j|r)]|\xi_j|\} \tag{4}$$

is finite.

**Lemma 1.** Let $\mathbf{q}(t)$, $\mathbf{p}(t)$ be solutions of (1a)–(1c) defined for $0 \leqslant t \leqslant T$, with initial data $\mathbf{q}(0)$ such that $\mathscr{L}(0) = \{\mathscr{L}_i(0)\} \in B_r$, where we have written $\mathscr{L}_i(t)$ for $\mathscr{L}_i(p_i(t), q_i(t))$. Then there is a constant $a$, independent of the initial condition but depending on $r$, such that

$$\|\mathscr{L}(t)\|_r \leqslant [\exp(at)]\|\mathscr{L}(0)\|_r \tag{5}$$

*Proof.* Using the equations of motion (1a)–(1c), we have

$$d\mathscr{L}_i(t)/dt = p_i(t)R_i(\mathbf{q}(t)) \tag{6}$$

By conditions $A_3$ and $A_4$

$$(d/dt)\mathscr{L}_i(t) \leqslant |d\mathscr{L}_i(t)/dt| \leqslant \sum_j A_{ij}\mathscr{L}_j(t) \tag{7}$$

where the $A_{ij}$ are constants, independent of $t$; $0 \leqslant A_{ij} \leqslant C$; and $A_{ij} = 0$ for $|i-j| > D$, the range of the potential. If $\mathbf{A}$ denotes the matrix with elements $A_{ij}$, then[(2)] $\boldsymbol{\Psi}(t) = [\exp(\mathbf{A}t)]|\mathscr{L}(0)|$ is a solution of the equations

$$d\Psi_i/dt = \sum A_{ij}\Psi_j(t), \qquad \Psi_j(0) = |\mathscr{L}_j(0)| \tag{8}$$

Standard arguments show that $|\mathscr{L}_i(t)| \leqslant \Psi_i(t)$, so Eq. (5) follows from (7) with $a$ equal to the $r$-norm of the bounded operator $\mathbf{A}$ on $B_r$.

**Theorem 1.** Let $\mathbf{q}(0)$, $\mathbf{p}(0)$ be such that $\mathscr{L}(0)$ (defined in Lemma 1) belongs to $B_r$. There then exists a $\in B_r$ solution of Eqs. (1a)–(1c) defined for all $t$.

*Proof.* We shall first consider the case of a finite system in a bounded region $\Lambda_\alpha \subset \mathbb{Z}^\nu$. Let $q_i^\alpha(t)$, $p_i^\alpha(t)$ be the solutions of the equations

$$\left.\begin{aligned} dq_i^\alpha/dt &= p_i^\alpha(t) \\ dp_i^\alpha/dt &= F_i(\mathbf{q}^\alpha(t)) \end{aligned}\right\} \quad \text{for} \quad i \in \Lambda_\alpha \tag{9a, 9b}$$

$$dq_i^\alpha/dt = dp_i^\alpha/dt = 0 \qquad \text{for} \quad i \notin \Lambda_\alpha \tag{9c}$$

with the initial conditions $q_i^\alpha(0) = q_i(0)$, $p_i^\alpha(0) = p_i(0)$, i.e., Eqs. (9a)–(9c) are time evolution equations for $\mathbf{q}^\alpha(t)$, $\mathbf{p}^\alpha(t)$ with *all* the particles outside $\Lambda_\alpha$ "tied down" to their initial positions.[2,3] Solutions of (9a)–(9c) are prevented from going to infinity in finite time by Lemma 1; they therefore exist for all time. The time evolution mappings $T_t^\alpha$ generated by (9a)–(9c) leave invariant the energy in $\Lambda_\alpha$,

$$H_\alpha(\mathbf{q}, \mathbf{p}) = \sum_{i\in\Lambda_\alpha} [\tfrac{1}{2}p_i^2 + U_i(q_i)] + \sum_j{}' V_j(\mathbf{q})$$

where $\sum'$ is the sum over all $j$ such that $\operatorname{dist}(j, \Lambda_\alpha) \leqslant D$. The solutions of (9a)–(9c) will satisfy the equations

$$\left.\begin{aligned} q_i^\alpha(t) &= q_i^\alpha(0) + p_i^\alpha(0)t + \int_0^t (t-s)F_i(\mathbf{q}^\alpha(s))\,ds \\ p_i^\alpha(t) &= p_i^\alpha(0) + \int_0^t F_i(\mathbf{q}^\alpha(s))\,ds \end{aligned}\right\} \quad \text{for} \quad i \in \Lambda_\alpha \tag{10a, 10b}$$

$$q_i^\alpha(t) = q_i^\alpha(0), \qquad p_i^\alpha(t) = p_i^\alpha(0) \qquad \text{for} \quad i \notin \Lambda_\alpha \tag{10c}$$

Using now the bound (5) for the time evolution $T_t^\alpha$, we have, by condition $A_3$, that $|F_i(\mathbf{q}^\alpha(t))| < K_i$ for $t \in [0, T]$ with $K_i < \infty$ independent of $\Lambda_\alpha$. Hence by the Arzela–Ascoli theorem we can choose sequences $\Lambda_\alpha \to \mathbb{Z}^\nu$ such that $q_i^\alpha(t)$, $p_i^\alpha(t) \to q_i(t)$, $p_i(t)$ uniformly on $[0, T]$. This is true for all finite $T$, so the sequence can be further refined to get uniform convergence on *every* bounded interval. The $q_i(t)$ will satisfy Eq. (3), so the $(\mathbf{q}(t), \mathbf{p}(t))$ satisfy (1a)–(1c), the equations of motion for the infinite system, with the initial conditions $(\mathbf{q}(0), \mathbf{p}(0))$.

By our assumption, $\mathscr{L}(0) \in B_r$. Hence, by (5), we also have an estimate of the form

$$\mathscr{L}_j(t) = \tfrac{1}{2}p_i^2(t) + U_j(q_j(t)) + K < K' \exp(r|j|), \qquad |t| \leqslant T$$

for each $T$ (but where $K'$ grows with $T$). By $A_3$ this implies

$$|q_j(t)| < C' \exp(r|j|), \qquad |p_j(t)| < C'' \exp(\tfrac{1}{2}r|j|) \tag{11}$$

This gives us rather good control over the time evolution, e.g., if the initial values are bounded, $|q_i(0)| < C$ and $|p_i(0)| < C$, then $q_i(t)$ and $p_i(t)$ will also be bounded for all finite $t$.

## 3. UNIQUENESS OF TIME EVOLUTION

Having established the existence of solutions of Eqs. (1a)–(1c) for a large class of initial conditions, we now consider their uniqueness. As is generally the case, e.g., for harmonic systems[2] we can obtain uniqueness only if we impose some conditions on how the solution $\{q_j(t), p_j(t)\}$ grows with $|j|$.

**Definition.** For any family $\mathbf{B} = \{B_i\}$ of positive constants, define $\Delta(\mathbf{B}) = \{\mathbf{q}: |q_i| \leqslant B_i \text{ for all } i\}$ and define $\bar{B}_k = \sup\{B_i: |i| \leqslant k\}$, $k = 1, 2, \ldots$. We will say $\mathbf{B}$ is a *sequence of uniqueness* if (a) the following holds:

$$\limsup_{k\to\infty} \bar{B}_k^{1/k} < \infty \tag{12a}$$

and (b) there exists a constant $c$ such that

$$\sup_{\mathbf{q}\in\Delta(\mathbf{B})} \sum_j |\partial F_i(\mathbf{q})/\partial q_j| \leqslant ci^2 \qquad \text{for all } i \tag{12b}$$

**Theorem 2.** Let $\mathbf{B}$ be a sequence of uniqueness. Then two solutions $\mathbf{q}^{(1)}(t)$ and $\mathbf{q}^{(2)}(t)$ of (3), both defined on $[0, T]$ and both taking values in $\Delta(\mathbf{B})$, are identical on $[0, T]$.

***Proof.*** Assume the contrary. Then we can assume that there are arbitrarily small, positive $t$'s for which $\mathbf{q}^{(1)}(t) \neq \mathbf{q}^{(2)}(t)$. We will show that this leads to a contradiction. Writing out (3) for $\{q^{(1)}(t)\}$ and $\{q^{(2)}(t)\}$ and subtracting the two gives

$$q_i^{(1)}(t) - q_i^{(2)}(t) = \int_0^t dt_1\,(t - t_1)[F_i(\mathbf{q}^{(1)}(t_1)) - F_i(\mathbf{q}^{(2)}(t_1))]$$

Let

$$\delta_n(t) = \sup\{|q_i^{(1)}(t) - q_i^{(2)}(t)|: \quad |i| \leqslant nD\}$$

where $D$ is the range of the potential, as defined in $A_1$. We then get, using (12b),

$$\delta_n(t) \leqslant \left[\int_0^t dt_1\,(t - t_1)\delta_{n+1}(t_1)\right] cn^2$$

Iterating this $k$ times, then using the bound $\delta_{n+k}(t) \leqslant 2\bar{B}_{(n+k)D}$, we obtain

$$\delta_n(t) \leqslant \left[\int_0^t dt\,(t-t_1)^{2k-1}\delta_{n+k}(t)\right]c^k[n(n+1)\cdots(n+k-1)]^2$$

$$\leqslant \frac{t^{2k}}{(2k)!}(2\bar{B}_{(n+k)D})c^k[n(n+1)\cdots(n+k-1)]^2$$

Thus, letting $k \to \infty$, we find that $\delta_n(t) = 0$ for

$$0 < t < \left\{\limsup_{k\to\infty}\left[\frac{2\bar{B}_{(n+k)D}c^k[n(n+1)\cdots(n+k-1)]^2}{(2k)!}\right]^{1/2k}\right\}^{-1}$$

$$= 2(\tfrac{1}{2}\sqrt{c}\limsup_{k\to\infty}\bar{B}_k^{D/2k})^{-1}$$

This is true for all $n$, so

$$q_i^{(1)}(t) = q_i^{(2)}(t)$$

for all $i$, provided

$$t < 4[c\limsup_{k\to\infty}(B_k^{D/k})]^{-1/2}$$

which proves the theorem.

**Example.** If (as in the example of Section 2) there exists a constant $c_1$ such that

$$|\partial F_i/\partial q_j| \leqslant c_1(\sup\{|q_j|:\ |i-j| \leqslant D\})^{2n-2} \tag{13}$$

then any sequence of the form $B_j = b|j|^{1/(n-1)}$ is a sequence of uniqueness if $n \geqslant 2$.

This means that we have uniqueness in the class of solutions such that

$$\sup_{|t|\leq\tau}\sup\{|q_j(t)|/(|j|^{1/(n-1)}+1)\} < \infty \qquad \text{for all } \tau$$

Arguments similar to those leading to Eq. (11) show that if $\mathscr{L}_j(0)$ grows no faster than $|j|^{1/(n-1)}$, then there does exist a solution in this class.

In the harmonic case ($n = 1$), condition (12b) is vacuous and any sequence $(B_j)$ such that

$$\sup_k\{\bar{B}_k^{1/k}\} < \infty$$

is a sequence of uniqueness (compare Ref. 2).

## 4. WEAK TIME EVOLUTION FOR GENERAL INTERACTIONS

In this section we sketch a proof that, under very general assumptions, solutions to the equations of motion exist for almost all initial conditions with respect to any Gibbs state. We do not assume here that conditions $A_3$ and $A_4$ hold. The proof is very simple and almost nothing needs to be assumed about the interaction, but it should be noted that very reasonable interactions —such as the one-dimensional harmonic chain with formal interaction energy $\frac{1}{2}\sum_i (q_{i+1} - q_i)^2$—do not have any Gibbs states at all.[2] About such interactions our theorem evidently says nothing. We refer the reader to recent work for an analysis of Gibbs states for the kind of system considered here.[7-9]

We will assume as before that our interaction is of Hamiltonian form with range $D$, i.e., we assume that $A_1$ and $A_2$ hold. In addition, we assume that:

**B₁**. For each finite subset $\Lambda_\alpha$ of $\mathbb{Z}^\nu$, the equations of motion (9a)–(9c) admit solutions for all time for all initial points.

**B₂**. For each $\Lambda_\alpha$, each $\beta > 0$, and each specification of the $q_i$ for $i \notin \Lambda_\alpha$, the measure

$$\exp[-\beta H_\alpha(\mathbf{q}, \mathbf{p})] \prod_{j \in \Lambda_\alpha} dq_j \, dp_j \tag{14}$$

with $H_\alpha$ given in (10) is finite (normalizable) on $(R \times R)^{\Lambda_\alpha}$.

Condition $\mathbf{B}_2$ makes it possible to define Gibbs states by an obvious adaptation of the definitions used in other cases, but it does not imply the *existence* of nontrivial Gibbs states.

We note that (i) by conservation of energy and Liouville's theorem, any Gibbs state is invariant under $T_t^\alpha$ for all $\alpha$, $t$; (ii) with respect to any Gibbs state, the $p_i$ are independent, identically distributed, Gaussian random variables of mean zero.

**Theorem 3.** Let $\mu$ be a Gibbs state for the interaction under consideration. For $\mu$-almost all initial points $\{q_i, p_i\}$, there exists a solution $\{q_i(t)\}$ of Eq. (3) defined for all $t$ and satisfying

$$\sup_t \sup_i \frac{|q_i(t) - q_i|}{(1 + t^2)[\log_+(i)]^{1/2}} < \infty \tag{15}$$

where $\log_+(j) = \sup[\log|j|, 1]$

***Proof.*** (The argument here is similar to that used in Ref. 3.) For any $x = \{q_i, p_i\}$ define

$$B(x) = \sup_i \frac{|p_i|}{[\log_+(q_i)]^{1/2}}$$

$$\bar{B}_\alpha(x) = \int_{-\infty}^{\infty} \frac{dt}{1 + t^2} B(T_t^\alpha x), \qquad \bar{B}_\infty(x) = \liminf_{\alpha \to \infty} \bar{B}_\alpha(x)$$

It follows from (ii) that $\int B\,d\mu < \infty$; hence, from (i) and Fubini's theorem, $\int \bar{B}_\alpha\,d\mu$ is finite and independent of $\alpha$. By Fatou's lemma, $\int \bar{B}_\alpha\,d\mu < \infty$. We will show: If $\bar{B}_\infty(x) < \infty$, then there exists a solution to (3) satisfying (15).

To see this, note first that there must then exist a sequence $\alpha_n \to \infty$ and a constant $C$ such that

$$\bar{B}_{\alpha_n}(x) \leqslant C \qquad \text{for all } n$$

Hence,

$$\begin{aligned} |q_i^{\alpha_n}(t) - q_i| &\leqslant \left| \int_0^t dt_1\, p_i^{\alpha_n}(t_1) \right| \leqslant (1 + t^2) \int_{-\infty}^{\infty} \frac{dt_1}{1 + t_1^2} |p_i^{\alpha_n}(t_1)| \\ &\leqslant (1 + t^2)[\log_+(i)]^{1/2} \bar{B}_{\alpha_n}(x) \\ &\leqslant (1 + t^2)[\log_+(i)]^{1/2} C \qquad \text{for all } i, n, t \end{aligned} \tag{16}$$

Since $F_i$ depends only on a finite number of the $q_j$, and since each $V_j$ is continuously differentiable, this bound implies a family of bounds of the form

$$|dp_i^{\alpha_n}(t)/dt| \leqslant K_i(|t|)$$

where each $K_i$ is a nondecreasing function of $t$ (which does not depend on $n$). The proof of the existence of solutions is now completed in the same way as in Theorem 1; (15) follows from (16) by passage to the limit.

There remains the question of uniqueness. Suppose that (13) holds with some $n > 1$. If $\{q_i, p_i\}$ satisfies

$$\sup_{i \in \mathbb{Z}^\nu} (|q_i|/|i|^{1/(n-1)}) < \infty \tag{17}$$

and if there exists a solution to (3) satisfying (15), then

$$\sup_{|t| \leq \tau} \sup_i [|q_i(t)|/|i|^{1/(n-1)}] \quad \text{is finite for all } \tau$$

Theorem 2 asserts that the solution is unique in this class. We would therefore like to know whether the condition (17) holds $\mu$ almost everywhere. A sufficient condition is given by the following:

**Proposition.** If there exists $\gamma > \nu(n - 1)$ and $C$ such that

$$\int |q_i|^\gamma\, d\mu < C \qquad \text{for all } i \tag{18}$$

then

$$\sup_{i \in \mathbb{Z}^\nu} (|q_i|/|i|^{1/(n-1)}) < \infty \qquad \mu \text{ almost everywhere}$$

*Proof.*

$$\mu\{|q_i| > |i|^{1/(n-1)}\} = \mu\{|q_i|^\gamma > |i|^{\gamma/(n-1)}\}$$
$$\leqslant \left(\int |q_i|^\gamma \, d\mu\right) / i^{\gamma/(n-1)}$$

Since $\gamma(n-1) > \nu$,

$$\sum_{i \in \mathbb{Z}^\nu} \mu\{|q_i| > |i|^{1/(n-1)}\} < \infty$$

so by the Borel–Cantelli lemma[10]

$$\limsup_i (|q_i|/|i|^{1/(n-1)}) \leqslant 1 \qquad \mu \text{ almost everywhere}$$

Collecting the above results, we have the following:

**Theorem 4.** Let the interactions satisfy conditions $A_1$, $A_2$, $B_1$, and $B_2$ and also (13) with $n > 1$. Let $\mu$ be a Gibbs state for this interaction such that, for some $\gamma > \nu(n-1)$, (18) is satisfied. Then for $\mu$-almost all $\{q_i, p_i\}$ there exists a solution to (3) such that

$$\sup_{|t| \leq \tau} \sup_{i \in \mathbb{Z}^\nu} [|q_i(t)|/|i|^{1/(n-1)}] < \infty \qquad \text{for all } \tau$$

and this solution is unique.

## REFERENCES


1. O. E. Lanford III, *Commun. Math. Phys.* **9**:169 (1969); **11**:257 (1969).
2. O. E. Lanford III and J. L. Lebowitz, in *Lecture Notes in Physics*, No. 38, Springer-Verlag (1975), p. 144; J. L. van Hemmen, Thesis, University of Groningen (1976).
3. O. E. Lanford III, in *Lecture Notes in Physics*, No. 38, Springer-Verlag (1975), p. 1.
4. Ya. G. Sinai, *Vestnik Markov. Univ. Ser. I, Math. Meh.* **1974**:152.
5. C. Marchioro, A. Pellegrinotti, and E. Presutti, *Commun. Math. Phys.* **40**:175 (1975).
6. N. W. Ashcroft and N. D. Mermin, *Solid State Physics*, Holt, Rinehart and Winston (1976).
7. H. J. Brascamp, E. H. Lieb, and J. L. Lebowitz, The Statistical Mechanics of Anharmonic Lattices, in *Proceedings of the 40th Session of the International Statistics Institute*, Warsaw (1975).
8. D. Ruelle, *Commun. Math. Phys.* **50**:189 (1976).
9. J. L. Lebowitz and E. Presutti, *Commun. Math. Phys.* **50**:195 (1976).
10. L. Breiman, *Probability*, Addison-Wesley, Section 3.14.

Commun. Math. Phys. 78, 545–566 (1981)

Communications in Mathematical Physics



# Lattice Systems with a Continuous Symmetry

## III. Low Temperature Asymptotic Expansion for the Plane Rotator Model

Jean Bricmont[1,*,++], Jean-Raymond Fontaine[2,**,++], Joel L. Lebowitz[2,**,†], Elliott H. Lieb[3,+], and Thomas Spencer[4,***]

[1] Department of Mathematics, Princeton University, Princeton, NJ 08540, USA
[2] Department of Mathematics, Rutgers University, New Brunswick, NJ 08903, USA
[3] Department of Mathematics and Physics, Princeton University, Princeton, NJ 08540, USA
[4] Department of Mathematics, Rutgers University, New Brunswick, NJ 08903, USA

**Abstract.** We prove that the expansion in powers of the temperature $T$ of the correlation functions and the free energy of the plane rotator model on a $d$-dimensional lattice is asymptotic to all orders in $T$. The leading term in the expansion is the spin wave approximation and the higher powers are obtained by the usual perturbation series. We also prove the inverse power decay of the pair correlation at low temperatures for $d=3$.

## I. Introduction

We investigate the low temperature properties of the classical plane rotator model described by the Hamiltonian:

$$-\beta H=\beta\sum_{\langle ij\rangle}\cos(\phi_i-\phi_j),\qquad \phi_i\in[-\pi,\pi],\tag{1}$$

$\beta$ is the inverse temperature $T$ and $\langle i,j\rangle$ are nearest neighbor pairs of sites on the $d$-dimensional simple cubic lattice $Z^d$.

It has been known for a long time that the SO(2) symmetry of this model is only broken in $d\geqq 3$ where there is a spontaneous magnetization at low temperatures [8, 13]. These results provide a qualitative justification of the spin wave picture. In this paper we prove that in any dimension $d$, the free energy and the correlations have a low temperature expansion about the spin wave approximation valid to all orders in $T$. In particular we show how to get higher order correction in $T$ to the spontaneous magnetization ($d\geqq 3$). The zeroth order value for the spontaneous magnetization was obtained in [13].

* Supported by NSF Grant No. MCS 78-01885
** Supported by NSF Grant No. PHY 78-15920
*** Supported by NSF Grant No. DMR 73-04355
+ Supported by NSF Grant No. PHY-7825390 A01
++ On leave from: Institut de Physique Théorique, Université de Louvain, Belgium
† Also: Department of Physics

0010-3616/81/0078/0545/$04.40

The formal expansion in $T$ is obtained [6, 11] by making a change of variables, $\phi_i = \sqrt{T}\phi'_i$, and then expanding the cosine into a power series so that up to a constant,

$$-\beta H = -\frac{1}{2}\sum_{\langle ij\rangle}(\phi'_i - \phi'_j)^2 + \frac{1}{4!}T\sum_{\langle i,j\rangle}(\phi'_i - \phi'_j)^4 + \dots, \qquad \phi'_i \in [-\pi\sqrt{\beta}, \pi\sqrt{\beta}].$$

We see that there are two perturbations of the massless gaussian field (the spin wave approximation): The first one is the power series in $T$ for $-\beta H$ and the other one is the restriction of $\phi'$ to the interval $|\phi'_i| \leqq \pi\sqrt{\beta}$, i.e. the Gibbs factor $\exp[-\beta H]$ has to be multiplied by a product of the characteristic functions $\chi(|\phi'_i| \leqq \pi\sqrt{\beta})$. The first perturbation, at least when the series is truncated at a given order, can be treated using methods developed in [3] (Part I of this series). The aim of this paper is to get rid of the second one (and of the truncation of the power series). This is done using infrared bounds [13]. We prove that the contribution of this second perturbation is exponentially small in $T$ (when $T \to 0$) and therefore does not appear in the power series expansion. In two dimensions the formal perturbation theory is not defined for functions like the spontaneous magnetization which vanishes. However we can compute asymptotics for all nonvanishing correlations.

As in [3], our method does not give directly any results about the decay of the correlations nor about analyticity in $T$ for $T \neq 0$. However, using ideas from [4] and an improvement of the results in [25] we show, for instance, that the two point function $\langle \sin\phi_0 \sin\phi_x \rangle$ behaves exactly like $|x|^{-1}$ for $d=3$ whenever there is a spontaneous magnetization.

The outline of the paper is as follows: In Sect. II we describe the model and some of its known properties that are used later. In Sect. III we state and prove the main result. In Sect. IV we give an alternative proof of a part of the theorem. In Sect. V we study the decay of the two point correlation function.

## II. Definition of the Model

Let $H_\Lambda$ be the Hamiltonian defined in (1) with periodic boundary conditions on a parallelepiped $\Lambda \subset Z^d$, centered on the origin. We also consider the Hamiltonian:

$$\beta H_{\Lambda,h} = \beta H_\Lambda - h\sum_{i\in\Lambda}\cos\phi_i. \tag{2}$$

A probability measure on $[-\pi, \pi]^{|\Lambda|}$ is defined ($|\Lambda|$ = number of sites in $\Lambda$) by:

$$d\mu_{\Lambda,h} = Z_{\Lambda,h}^{-1}\exp\left\{\beta\sum_{\langle i,j\rangle\subset\Lambda}\cos(\phi_i - \phi_j) + h\sum_{i\in\Lambda}\cos\phi_i\right\}. \tag{3}$$

We shall consider the correlations functions:

$$\langle\cos m\phi\rangle_{\Lambda,h} = \int\cos m\phi\, d\mu_{\Lambda,h}(\phi), \tag{4}$$

where $m\phi = \sum_i m(i)\phi_i$, and $m: Z^d \to Z$ is a function of compact support, and the pressure $P(T)$ defined by:

$$P(T) = \lim_{\Lambda\uparrow Z^d}|\Lambda|^{-1}\log Z_{\Lambda,h=0}.$$

It is easy to show that $P(T)$ exists. Using the Lee-Yang theorem [20, 9], we can also show that the thermodynamic limit of the correlation functions exists for $h \neq 0$, and that the state obtained is clustering [12, 18, 23]. The "+" state is defined by:

$$\lim_{h \downarrow 0} \langle \cos m\phi \rangle_h \equiv \langle \cos m\phi \rangle_+ .$$

The limit exists by Ginibre's inequalities. By symmetry, $\langle \sin m\phi \rangle_+ \equiv 0$.

In $d=2$, the uniqueness of the translation invariant equilibrium state [2] implies:

$$\langle \cos m\phi \rangle_+ = \lim_{\Lambda \to Z^d} \langle \cos m\phi \rangle_{\Lambda, h=0} = 0 .$$

In $d=3$, it has been proven [13] that $\langle \cos \phi \rangle_+ \equiv \bar{m}(\beta) > 0$ for $\beta$ large. This is a consequence of the infrared bounds which we now recall.

*Notation.* We let

$$\sigma_i = (\sigma_i^0, \sigma_i^1) = (\cos \phi_i, \sin \phi_i) \in R^2 .$$

As in [3], the unit vectors along the coordinate axis are denoted by $e_\alpha$, $\alpha = 1, \dots, d$. Given a function $f: Z^d \to R^2$ of compact support, we write

$$V_i^\alpha f = f(i) - f(i + e_\alpha) = (V^\alpha f)(i) ,$$

$$\sum_{e_\alpha, \alpha = 1, \dots, d} \equiv \sum_\xi , \qquad \sigma(g) = \sum_i g(i) \cdot \sigma_i .$$

Lattice sites will sometimes also be denoted by $x$, $y$ or $z$.

*Infrared Bounds* [13]

IR 1. Let $g^\alpha : Z^d \to R^2$, $\alpha = 1, \dots, d$ be functions of compact support. Then

$$\left\langle \exp \sigma \left( \sum_{\alpha=1}^{d} V^\alpha g^\alpha \right) \right\rangle_{\Lambda, h} \leqq \exp \left[ \sum_{\alpha, i} (g^\alpha(i))^2 \right] \Big/ 2\beta .$$

When $f \in R^2$ is such that $\sum_i f(i) = 0$, then $g^\alpha(i) = -\Delta^{-1} V_i^\alpha f$ is well defined. Applying IR 1 to this particular $g^\alpha(i)$ we get [13]:

$$\langle \exp \sigma(f) \rangle_{\Lambda, h} \leqq \exp[(f, (-\Delta^{-1}) f)/2\beta] .$$

Using the fact that $\langle \ \rangle_h$ is ergodic, IR 1 has the following consequence:

IR 2 [13]. For $d \geqq 3$, let $\hat{\sigma}_i = \sigma_i - \langle \sigma_i \rangle_h$, then

$$\langle \exp \hat{\sigma}(g) \rangle_h \leqq \exp[(g, (-\Delta)^{-1} g)/2\beta]$$

for any $g$ of compact support.

From IR 2, one obtains the bound,

$$[\bar{m}(\beta)]^2 = \langle \cos \phi \rangle_+^2 \geqq 1 - I(d) T , \tag{5}$$

where

$$I(d) \equiv (2\pi)^{-d} \int_{|k_\xi| \leqq \pi} d^d k \left[ \sum_\xi (1 - \cos k_\xi) \right]^{-1} .$$

As we shall see later, the r.h.s. of (5) are the first terms in the asymptotic expansion of $\bar{m}^2$.

## III. The Main Result

**Theorem 1.** *For any d, the perturbation expansion in T for the "excess free energy" $Q(T) \equiv P(T) - d\beta - \frac{1}{2}\ln T$ and the correlation functions $\langle \cos m\phi \rangle_+$ is asymptotic to all orders in T.*

### *1. Strategy of the Proof*

The proof will be divided into two parts; we shall first prove the result for $Q(T)$ and for correlation functions of the type

a) $\langle \cos m\phi \rangle_+$ with $\sum_i m(i) \equiv \underline{m} = 0$,

and then for correlation functions of the type

b) $\langle \cos m\phi \rangle_+$ with $\underline{m} \neq 0$.

In some way, Parts a) and b) correspond to the cases $\langle (\nabla_0 \phi)^2 \rangle$ and $\langle \phi_0^2 \rangle$ of [3]. For technical reasons, we are not able to generalize the proof given in [3] for $\langle \phi_0^2 \rangle$ to Case b). Using however the existence of an asymptotic expansion for $\langle \cos(\phi_0 - \phi_x) \rangle_+$ and the decay of the truncated two point function, $\langle \cos\phi_0 \cos\phi_x \rangle_+ - \langle \cos\phi_0 \rangle_0^2 < c|x|^{-1}$, $d \geqq 3$, which we get from IR bounds [4, 25], we are able to generate an asymptotic expansion for $\langle \cos\phi \rangle_+$. Using reflection positivity, we generalize our argument to any $\langle \cos m\phi \rangle_+$ with $\underline{m} \neq 0$. For $d<3$, Part b) is trivial since by the Dobrushin-Schlosman theorem [8], $\langle \cos m\phi \rangle_+ \equiv 0$ for $m \neq 0$.

To keep matters simple, we shall first consider in a) $d \geqq 3$ and then indicate the changes required for $d=2$. In Sect. 4 we sketch an alternative proof of a) which should also work for non nearest neighbor interactions. This proof uses correlation inequalities.

### *2. Asymptotic Expansion for $\langle \cos m\phi \rangle$ with $\underline{m}=0$, $d \geqq 3$*

Before giving the proof, we shall derive two technical lemmas from the infrared bound IR 2. Those are the key ingredients which will be used to deal with the second perturbation of the free massless Gaussian field described in the introduction, namely $\chi(|\phi_j| \leqq \pi\sqrt{\beta})$.

We introduce the periodic function $a(\phi): \mathbb{R} \to [-\pi, \pi]$, s.t. $a(\phi + 2\pi n) = \phi, n \in N$.

**Lemma 1.** *For $d \geqq 3$ there exists positive constants $c_1, c_2$ independent of T and h, such that*

$$\langle \exp(\sqrt{\beta}|a(\phi)|/c_1) \rangle_h \leqq c_2 < \infty .$$

*Proof.* We first prove $\langle \exp(2\sqrt{\beta}a^2(\phi)/\pi^2) \rangle < \infty$. IR 2 with $g = (-\sqrt{\beta}\delta_{i0}, 0)$ implies

$$\langle \exp[-\sqrt{\beta}(\cos\phi - \langle \cos\phi \rangle_h)] \rangle_h \leqq \exp[C_{00}/2] \leqq c_2' \tag{6}$$

with $C_{ij} = (-\Delta)^{-1}(i,j)$ the covariance of the Gaussian measure corresponding to the spin wave approximation. Taking the square root of (5) we obtain,

$$\langle \cos\phi \rangle_h \geqq 1 - \tfrac{1}{2} T I(d) \varepsilon(T) \tag{7}$$

with $\varepsilon(T)\to 1$ when $T\to 0$. The inequalities (6), (7) imply:

$$\langle \exp(\sqrt{\beta}[(1-1/2I(d)T\varepsilon(T)-\cos\phi])\rangle_h \leqq c_2',$$

$1-\cos\phi \geqq 2/\pi^2\phi^2$ for $\phi\in[-\pi,\pi]$, so that

$$\langle \exp(2\sqrt{\beta}a^2(\phi)/\pi^2)\rangle_h \leqq c_2' \exp[1/2I(d)\sqrt{T}\varepsilon(T)] \leqq c_2''. \tag{8}$$

We now use again IR 2 with $\hat{\sigma}(g)=\sqrt{\beta}\sin\phi$. Let us first remark that for any $k<\pi$ $\exists c>0$ such that $|x|\leqq c|\sin x|$ for $|x|\leqq k$. This implies

$$\exp(\sqrt{\beta}|x|/c) \leqq \exp[\sqrt{\beta}\sin x] + \exp[-\sqrt{\beta}\sin x] \quad \text{for} \quad |x|<k. \tag{9}$$

Defining $\mu_h(\phi)$ by $\langle F(\phi)\rangle_h = \int F(\phi)d\mu_h(\phi)$,

$$\int_{-\pi}^{\pi} \exp(\sqrt{\beta}|\phi|/c)d\mu_h(\phi) = \int_{-\pi/2}^{\pi/2} \exp(\sqrt{\beta}|\phi|/c)d\mu_h(\phi) + \int_{|\phi|>\pi/2} \exp(\sqrt{\beta}|\phi|/c)d\mu_h(\phi).$$

To estimate the first integral we use (9):

$$\begin{aligned}\int_{-\pi/2}^{\pi/2} \exp(\sqrt{\beta}|\phi|/c)d\mu_h(\phi) &\leqq \int_{-\pi/2}^{\pi/2} (\exp\sqrt{\beta}\sin\phi + \exp-\sqrt{\beta}\sin\phi)d\mu_h(\phi)\\ &\leqq \int_{-\pi}^{\pi} (\exp\sqrt{\beta}\sin\phi + \exp-\sqrt{\beta}\sin\phi)d\mu_h(\phi)\\ &\leqq c_3 \text{ uniformly in } T \text{ by IR 2}.\end{aligned}$$

To estimate the second integral, we use the Chebyshev's inequalities and (8):

$$\begin{aligned}\int_{|\phi|>\pi/2} \exp(\sqrt{\beta}|\phi|/c)d\mu_h(\phi) &\leqq (\exp\sqrt{\beta}\pi/c)\mu_h\{\phi||\phi|\geqq\pi/2\}\\ &\leqq (\exp\sqrt{\beta}\pi/c)\exp(-2\sqrt{\beta}\pi^2/\pi^2 4)\\ &= \exp[-\sqrt{\beta}(1/2-\pi/c)] \leqq c_4\end{aligned}$$

uniformly in $T$ for $c>2\pi$.

*Remark.* Setting $\sqrt{\beta}\phi=\phi'$ Lemma 1 implies:

$$\int \exp(|\phi'|/c)d\mu'(\phi') \leqq \text{const}$$

when $d\mu'$ is the measure obtained by the change of variables. But,

$$|\phi'|^n/c^n \leqq n!\exp(|\phi'|/c)$$

hence,

$$\int |\phi'|^n d\mu'(\phi') \leqq \text{const}\, n!c^n, \tag{10}$$

where the const is $T$ independent.

**Lemma 2.** *For* $d\geqq 3$, $\langle\delta(\phi\pm\pi)\rangle_+ \leqq c_1\exp[-c_2\beta^{1/2}]$, *where* $c_1$ *and* $c_2$ *are positive constants independent of* $T$ *and* $\delta$ *is considered as a periodic* $\delta$*-function.*

*Proof.* The proof will be based on Lemma 1 and the DLR equations [7, 17]. If $\mu$ is the equilibrium measure corresponding to the "+" state, the DLR equations [7, 17] read:

$$\langle \delta(\phi-\pi)\rangle_+ = \int d\mu(\phi_1 \dots \phi_{2d}) F(\phi_1 \dots \phi_{2d}),$$

where

$$F(\phi_1 \dots \phi_{2d}) = \left[\int d\phi \exp\left(\beta \sum_{i=1}^{2d} \cos(\phi-\phi_i)\right)\right]^{-1} \cdot \int \delta(\phi-\pi) \exp\left(\beta \sum_{i=1}^{2d} \cos(\phi-\phi_i)\right) d\phi,$$

and $\{\phi_i\}_{i=1}^{2d}$ are the nearest neighbors of $\phi$. Let

$$A = \{\phi_1 \dots \phi_{2d} \| \phi_1| < a, \dots, |\phi_{2d}| < a\},$$

($a=\pi/32$ for instance) and let us estimate the integral

$$\int_A F(\phi_1 \dots \phi_{2d}) d\mu(\phi_1 \dots \phi_{2d})$$

by finding a bound for $\sup_A |F(\phi_1 \dots \phi_{2d})|$:

$$\begin{aligned} F(\phi_1 \dots \phi_{2d}) &= \left[\int_{-\pi}^{\pi} \exp\left(\beta \sum_{i=1}^{2d} [\cos(\phi-\phi_i) - \cos(\pi-\phi_i)]\right) d\phi\right]^{-1} \\ &\leqq \left[\int_{-a}^{a} \exp\left(\beta \sum_{i=1}^{2d} [\cos(\phi/2-\phi_i)\cos(\phi/2)]\right) d\phi\right]^{-1} \\ &\leqq [2a \exp 2\beta c']^{-1}, \end{aligned}$$

where $c' = 2d\cos(3a/2)\cos(a/2)$. To prove the result, we still have to estimate

$$\int_{A^c} F(\phi_1 \dots \phi_{2d}) d\mu(\phi_1 \dots \phi_{2d}) \leqq \mu(A^c) \sup_{A^c} F(\phi_1 \dots \phi_{2d}).$$

By Chebyshev and Lemma 1,

$$\mu(A^c) \leqq \text{const} \exp[-c\beta^{1/2}], \quad (c>0).$$

Hence

$$\begin{aligned} \sup_{A^c} F(\phi_1 \dots \phi_{2d}) &\leqq \left[\int_{-\pi}^{\pi} d\phi \exp(-2d\beta |\cos(\phi/2)|)\right]^{-1} \\ &\leqq \left[\int_{-1/\beta}^{0} d\phi \exp(-2d\beta |\sin(\phi/2)|)\right]^{-1} \\ &\leqq \text{const}\, \beta. \end{aligned}$$

*Proof of Case a) for $d=3$.* We now prove that the expansion for $\langle \cos \nabla_0^e \phi \rangle_+$ is asymptotic up to second order. The proof can be easily generalized to all orders and to all correlation functions with $\underline{m}=0$.

Making the change of variables $\phi_i' = \sqrt{\beta}\phi_i$,

$$\langle \cos \nabla_0^e \phi \rangle_{\Lambda,h} = \frac{\int_{-\pi\sqrt{\beta}}^{\pi\sqrt{\beta}} \prod_{i\in\Lambda} d\phi_i \cos(\sqrt{T}\nabla_0^e\phi)\exp\left[\beta \sum_{i\in\Lambda,\xi} (\cos\sqrt{T}\nabla_i^\xi\phi - 1) + h\sum_{i\in\Lambda}\cos\sqrt{T}\phi_i\right]}{\int_{-\pi\sqrt{\beta}}^{\pi\sqrt{\beta}} \prod_{i\in\Lambda} d\phi_i \exp\left[\beta \sum_{i\in\Lambda,\xi} (\cos\sqrt{T}\nabla_i^\xi\phi - 1) + h\sum_{i\in\Lambda}\cos\sqrt{T}\phi_i\right]}.$$

*N.B.* We shall generally not indicate explicitly whether we are using the original angle variable $\phi$ or the scaled $\phi'$ as this should be clear from the context.

We now expand $\cos(\sqrt{T}\nabla_0^e\phi)$ up to second order

$$\cos(\sqrt{T}\nabla_0^e\phi) = 1 - T\frac{(\nabla_0^e\phi)^2}{2} + T^2\frac{(\nabla_0^e\phi)^4}{4!} - T^3\frac{(\nabla_0^e)^6}{6!}\cos\sigma,$$

where $\sigma$ is fixed by Taylor's theorem. Clearly we have to expand $\langle(\nabla_0^e\phi)^2\rangle$ up to order 1 and $\langle(\nabla_0^e\phi)^4\rangle$ up to order 0. The third term

$$\frac{T^3}{6!}|\langle(\nabla_0^e\phi)^6\cos\sigma\rangle| \leq \frac{T^3}{6!}\langle(\nabla_0^e\phi)^6\rangle \leq \text{const}\, T^3$$

because of Lemma 1 (see Remark).

As noted in the introduction there are two perturbations of the massless Gaussian lattice field: a power series in $T$ for

$$\beta\cos(\sqrt{T}\nabla_i^\xi\phi) + \tfrac{1}{2}(\nabla_i^\xi\phi)^2 - 1,$$

and a characteristic function $\chi(|\phi_i| < \pi\sqrt{\beta})$. The perturbation $h\sum_{i\in\Lambda}\cos\sqrt{T}\phi$ is irrelevant because $h$ will be set equal to zero.

As in [3] the expansion for $\langle(\nabla_0^e\phi)^2\rangle$ and $\langle(\nabla_0^e\phi)^4\rangle$ will be generated by a regularized form of the integration by parts (I.P.) formula that we now recall: if $F$ is a function of the Gaussian variables $\phi_l$, $l = 1, \ldots, n$

$$\begin{aligned}\int \phi_0 F(\{\phi_l\}_{l=1}^n) d\mu_{0\Lambda} &= \sum_{i\in\Lambda} C_{0i}^{\Lambda,m} \int \frac{d}{d\phi_i} F(\{\phi_l\}_{l=1}^n) d\mu_{0\Lambda} \\ &\quad + m^2 \sum_{i\in\Lambda} C_{0i}^{\Lambda,m} \int \phi_i F(\{\phi_l\}_{l=1}^n) d\mu_{0\Lambda},\end{aligned} \tag{13}$$

where $d\mu_{0\Lambda}$ is the Gaussian measure of the massless field with periodic boundary conditions in a box $\Lambda$, and $C_{0i}^{\Lambda,m}$ is the covariance of the massive Gaussian field with periodic boundary conditions in the box $\Lambda$.

As in [3] $m$ will be $T$-dependent. Let us apply (13) to $\langle(\nabla_0^e\phi)^2\rangle$. After regrouping some terms, see [3, Eq. (9)], we obtain

$$\begin{aligned}\langle(\nabla_0^e\phi)^2\rangle_{\Lambda,h} &= \nabla_0^e\nabla_0^e C_{00}^{\Lambda,m} - \beta\sum_{i,\xi}\nabla_0^e\nabla_i^\xi C_{0i}^{\Lambda,m}\langle\nabla_0^e\phi[-\sqrt{T}\sin\sqrt{T}\nabla_i^\xi\phi + T(\nabla_i^\xi\phi)\rangle_{\Lambda,h} \\ &\quad + -\sqrt{T}h\sum_{i\in\Lambda}\nabla_0^e C_{0i}^{\Lambda,m}\langle\nabla_0^e\phi\sin\sqrt{T}\phi_i\rangle_{\Lambda,h} \\ &\quad + \sum_i m^2\nabla_0^e C_{0i}^{\Lambda,m}\langle\nabla_0^e\phi\phi_i\rangle_{\Lambda,h} \\ &\quad + \sum_i \nabla_0^e C_{0i}^{\Lambda,m}\langle\nabla_0^e\phi[\delta(\phi + \pi\sqrt{\beta}) + \delta(\phi - \pi\sqrt{\beta})]\rangle_{\Lambda,h}.\end{aligned} \tag{14}$$

Taking $\lim_{h\to 0}\lim_{\Lambda\to\infty}$ in (14) we obtain the same equation in the limiting state $\langle\ \rangle$ except that the term proportional to $h$ has disappeared, because $\langle|\nabla_0^e\phi|\rangle_{\Lambda,h}$ is uniformly bounded (see Remark after Lemma 1), thus

$$\begin{aligned}\langle(\nabla_0^e\phi)^2\rangle = \nabla_0^e\nabla_0^e C_{00}^m - \beta\sum_{i,\xi}\nabla_0^e\nabla_i^\xi C_{0i}^m \cdot \langle\nabla_0^e\phi[-\sqrt{T}\sin\sqrt{T}\nabla_i^\xi\phi + T(\nabla_i^\xi\phi)]\rangle \\ + \sum_i m^2\nabla_0^e C_{0i}^m\langle\nabla_0\phi\phi_i\rangle \\ + \sum_i \nabla_0^e C_{0i}^m\langle\nabla_0^e\phi[\delta(\phi+\pi\sqrt{\beta})+\delta(\phi-\pi\sqrt{\beta})]\rangle. \qquad (15)\end{aligned}$$

This equation is similar to Eq. (10) of [3] except for the presence of the last term of the r.h.s. This term is the contribution to the expansion of the characteristic functions $\chi(|\phi_i|\leqq\pi\sqrt{\beta})$. We shall use Lemmas 1 and 2 to prove that it is exponentially small in $T$ (as $T\to 0$).

Choosing $m=m(T)=\exp[-(\ln T)^2]$, the first term on the r.s. of (15) gives the 0-th order of the expansion since

$$|\nabla_0^e\nabla_0^e C_{00}^m - \nabla_0^e\nabla_0^e C_{00}^{m=0}| \leqq \text{const}\, m$$

(see [3, Appendix B]). The mass terms

$$m^2\sum_i \nabla_0^e C_{0i}^m\langle\nabla_0^e\phi\phi_i\rangle \leqq \text{const}\, m^2 m^{-1}|\langle\nabla_0^e\phi\phi_i\rangle|$$

because $\sum_i|\nabla_0^e C_{0i}^m|\leqq Cm^{-1}$, see [3, Appendix A].

By Lemma 1 $|\langle\nabla_0^e\phi\phi_i\rangle|\leqq$const uniformly in $T$. So the mass term is bounded by const$m$. By our choice of $m$ it is exponentially small with $T$. The 4-th term of the r.h.s. of (15) is bounded by

$$\text{const}\, m^{-1}|\langle\nabla_0^e\phi\delta(\phi\pm\pi\sqrt{\beta})\rangle|$$

but

$$\langle\nabla_0^e\phi\delta(\phi\pm\pi\sqrt{\beta})\rangle \leqq c_1\exp(-c_2\beta^{1/2}) \quad \text{by Lemma 2}.$$

Therefore this term is still exponentially small in $T$ (as $T\to 0$). Applying Taylor's theorem to $\sin(\sqrt{T}(\nabla_i^\xi\phi))$ and using $\sum_i|\nabla_0^e\nabla_i^\xi C_{0i}^m|<\text{const}\ln m$, see [3, Appendix A], the second term (the temperature term) is bounded by $CT\ln m|\langle\nabla_0^e\phi(\nabla_i^\xi\phi)^3\rangle|$ and hence by const $T\ln m$.

In general when we apply the integration by parts formula there appear 4 terms as in (15), the last two are exponentially small and they disappear from the expansion. The second term called, the temperature term or II-term is small compared to the first term called the I-term.

To get the first order of the expansion of $\langle(\nabla_0^e\phi)^2\rangle$, we have to apply I.P. once more to the II-term in (15). This yields after having applied Taylor's theorem to the sin:

$$\begin{aligned}\mathrm{II} = \frac{\beta}{3!}\sum_{i,\xi}\nabla_0^e\nabla_i^\xi C_{0i}^m\langle\nabla_0^e\phi T^2(\nabla_i^\xi\phi)^3\rangle \\ + \frac{\beta}{5!}\sum_{i,\xi}\nabla_0^e\nabla_i^\xi C_{0i}^m\langle\nabla_0^e\phi T^3(\nabla_i^\xi\phi)^5\cos\sigma\rangle.\end{aligned}$$

The second term above is bounded by const $T^3 \ln m$ and is small compared to $T$.

We now apply I.P. to $\langle \nabla_0^e\phi(\nabla_i^\xi\phi)^3\rangle$ and then we apply repeatedly I.P. to the I-term produced by the preceding I.P. until we have a purely Gaussian expectation value. By what we have explained before all the remainder terms are small with respect to the purely Gaussian expectation value. We apply the same procedure to $\langle(\nabla_0^e\phi)^4\rangle$ to get the 0-th order. So

$$\langle\cos\nabla_0^e\phi\rangle_+ = 1 - \frac{T}{2}\langle(\nabla_0^e\phi)^2\rangle_G^m - \frac{T^2}{4!}\sum_{i,\xi}\langle(\nabla_0^e\psi)^2(\nabla_i^\xi\phi)^4\rangle_G^m$$
$$+\frac{T^2}{4!}\langle(\nabla_0^e\phi)^4\rangle_G^m + O(T^{2+\varepsilon}),$$

where $\langle\ \rangle_G^m$ is the expectation value in the massive Gaussian field. This yields

$$\langle\cos\nabla_0^e\phi\rangle_+ = 1 - T[C_{00}^m - C_{0e}^m] - \tfrac{3}{2}T^2[C_{00}^m - C_{0e}^m]^2 + O(T^{2+\varepsilon})$$

As already proved in [3, Appendix B], if

$$\langle\cos\nabla_0^e\phi\rangle = 1 + a_1(m(T))T + a_2(m(T))T^2 + O(T^{2+\varepsilon})$$

then

$$|a_i(m(T)) - a_i(0)| \leqq m(T) = e^{-(\ln T)^2};$$

so

$$\langle\cos\nabla_0^e\phi\rangle_+ = 1 - T[C_{00} - C_{0e}] - \tfrac{3}{2}T^2[C_{00} - C_{0e}]^2 + O(T^{2+\varepsilon}).$$

But $C_{00} - C_{0e} = \frac{1}{2d}$. So finally,

$$\langle\cos\nabla_0^e\phi\rangle_+ = 1 - T/2d - 3T^2/8d^2 + O(T^{2+\varepsilon}).$$

### 3. *Asymptotic Expansion of* $\langle\cos m\phi\rangle$ *with* $\sum_i m(i) = 0$ *in* $d=2$

In two dimensions there is no breakdown of the SO(2) symmetry so the measure $d\mu(\phi)$ is not concentrated around $\phi = 0$ and we do not expect Lemmas 1 and 2 to be true. We shall however prove similar results for the difference variables $(\phi_x - \phi_y)$ even when $|x-y| \sim \exp[\beta^{1/4}]$. In $d=2$, $\langle\ \rangle$ will denote $\lim_{\Lambda\to\infty}\langle\ \rangle_{\Lambda,h=0}$.

**Lemma 3.** *For* $d=2$ *let* $x, y \in Z^2$ *and* $|x-y| = O(\exp\beta^{1/4})$ *then*

$$\langle\exp(\beta^{1/4}|a(\phi_x - \phi_y)|/\pi\rangle \leqq c < \infty,$$

*where c is T independent.*

*Proof*. From IR 1, we have,

$$\langle\exp[\pm\beta^{1/4}(\cos\phi_x - \cos\phi_y)]\rangle \leqq \exp[\beta^{-1/2}(C_{00} - C_{yx})]$$
$$\leqq \exp[c_1\beta^{-1/2}\ln|x-y|] \leqq c_2, \tag{17}$$

where $c_1$ and $c_2$ are positive, $T$-independent constants, and we have used $|x-y| = O(\exp\beta^{1/4})$. Similarly,

$$\langle\exp(\pm\beta^{1/4}(\sin\phi_x - \sin\phi_y))\rangle \leqq c_3. \tag{18}$$

The bounds (17) and (18) imply

$$\langle \exp\{\pm 2\beta^{1/4}\sin[(\phi_x+\phi_y)/2\rangle \sin[(\phi_x-\phi_y)/2]\} + \exp\{\pm 2\beta^{1/4}\cos[(\phi_x+\phi_y)/2]\sin[(\phi_x-\phi_y)/2]\} \leqq c_4,$$

which clearly gives:

$$\langle \exp[\beta^{1/4}|\sin(\phi_x-\phi_y)/2|]\rangle \leqq c_4. \tag{19}$$

The lemma then follows by noting that $|a(\phi)|/\pi \leqq \sin\phi/2$.

*Remark.* When $|x-y|=O(1)$, we obviously have

$$\langle \exp\beta^{1/2}|a(\phi_x-\phi_y)|\rangle \leqq c < \infty \tag{20}$$

($c$ is $T$ independent).

**Lemma 4.** *For $d=2$, let $|x-x_0|=c\exp[\beta^{1/4}]$. Then*

$$\langle \delta(\phi_{x_0}-\phi_x\pm\pi)\rangle \leqq c_1\exp(-c_2\beta^{1/4}),$$

*where $c_1$ and $c_2$ are positive, T-independent, constants.*

*Proof*. The proof is similar to the one of Lemma 2 given Lemma 1. By the D.L.R. equations [7, 17], if $\mu$ is the equilibrium measure corresponding to $\langle\ \rangle$ then:

$$\langle \delta(\phi_{x_0}-\phi_x-\pi)\rangle = \int d\mu(\phi_1,\ldots,\phi_{2d},\psi_1,\ldots,\psi_{2d})F(\phi_1,\ldots,\phi_{2d},\psi_1,\ldots,\psi_{2d}),$$

where

$$F=\left[\int\delta(\phi_{x_0}-\phi_x-\pi)\exp\left(\beta\sum_{i=1}^{2d}[\cos(\phi_{x_0}-\phi_i)+\cos(\phi_x-\psi_i)]\right)d\phi_{x_0}d\phi_x\right] \cdot\left[\int d\phi_{x_0}d\phi_x\exp\left(\beta\sum_{i=1}^{2d}[\cos(\phi_{x_0}-\phi_i)+\cos(\phi_x-\psi_i)]\right)\right]^{-1}, \tag{21}$$

where $\{\phi_i\}_{i=1}^{2d}$ and $\{\psi_i\}_{i=1}^{2d}$, are respectively the nearest neighbor variables to $\phi_{x_0}$ and $\phi_x$.

Let $A=\{\phi_i,\psi_j\,|\,|\phi_i-\psi_i|<a,\ |\psi_j-\psi_{j'}|<a,\ |\phi_i-\phi_{i'}|<a \text{ and } i,j=1,\ldots,2d\}$. ($a$ is for instance $\pi/32$.)

Our proof is in 2 steps.

1) Estimate of $\int_A F d\mu$.

$$\int_A d\mu(\phi_1,\ldots,\phi_{2d},\psi_1,\ldots,\psi_{2d})F \leqq \sup_A F(\phi_1,\ldots,\phi_{2d},\psi_1,\ldots,\psi_{2d}).$$

Using the double angle formula, the numerator of (21) is bounded by $\exp[2\beta\cos(\pi/2-a/2)]$. The denominator $D$ of (21) obeys:

$$D \geqq \int_{|\phi_{x_0}-\phi_{i_0}|<a} d\phi_{x_0}\exp\left(\beta\sum_i\cos(\phi_0-\phi_i)\right)\cdot\int_{|\phi_x-\psi_{i_0}|<a} d\phi_x\exp\left(\sum_i\cos(\phi_x-\psi_i)\right),$$

where $i_0\in\{1,\ldots,2d\}$. $\forall i$, $|\phi_i-\phi_{i_0}|<a$ because $\phi_i,\phi_{i_0}\in A$ and therefore $|\phi_0-\phi_i|<2a$. So

$$D\geqq a^2\exp(\beta 2d 2\cos 2a)\geqq a^2\exp[2d\beta].$$

Finally

$$\sup_A F \leqq a^{-2}\exp(-2\beta(d-\cos(\pi/2-a/2))) \leqq a^{-2}\exp(-c\beta). \tag{22}$$

2) Estimate of $\int_{A^c} F d\mu$.

$$\int_{A^c} F d\mu < |F|_\infty \mu(A^c).$$

By Chebyshev inequality and Lemma 3, $\mu(A^c) \leqq \exp(-\beta^{1/4}a/\pi)$. To estimate $|F|_\infty$ we note that

$$\begin{aligned}|F| &= \left[\int \exp\left(\beta\sum_i [\cos(\phi_{x_0}-\phi_i)+\cos(\phi_{x_0}+\pi-\psi_i)]\right) d\phi_{x_0}\right]\\ &\quad\cdot\left[\int \exp\left(\beta\sum_i [\cos(\phi_{x_0}-\phi_i)+\cos(\phi_x-\psi_i)]\right) d\phi_{x_0} d\phi_x\right]^{-1}\\ &\leqq \sup_{\phi_{x_0}}\Big[\exp(\beta[\cos(\phi_{x_0}-\phi_i)+\cos(\phi_{x_0}+\pi-\psi_i)])\\ &\quad\cdot\left(\int \exp\left(\beta\sum_i [\cos(\phi_{x_0}-\phi_i)+\cos(\phi_x-\psi_i)]\right) d\phi_x\right)^{-1}\Big] \equiv S.\end{aligned}$$

By compactness, $\exists\tilde{\phi}_{x_0} \in [-\pi, \pi]$ such that

$$\begin{aligned}S &= \exp\left(\beta\sum_i [\cos(\tilde{\phi}_{x_0}-\phi_i)+\cos(\tilde{\phi}_{x_0}+\pi-\psi_i)]\right)\\ &\quad\cdot\left(\int \exp\left(\beta\sum_i [\cos(\tilde{\phi}_{x_0}-\phi_i)+\cos(\phi_x-\psi_i)]\right) d\phi_x\right)^{-1}\\ &= \left\{\int \exp\left(2\beta\sum_i \sin[(\phi_x-\tilde{\phi}_{x_0})/2+\pi/2-\psi_i]\sin[(\phi_x-\tilde{\phi}_{x_0})/2-\pi/2]\right) d\phi_x\right\}^{-1}\\ &\leqq \left[\int_{|\phi_x|<\beta^{-1}} \exp(-2\beta|\phi_x|) d\phi_x\right]^{-1} < \text{const}\,\beta,\end{aligned}$$

where the const is $T$-independent.

Given Lemmas 4 and 5, we shall perform the expansion in a way which is similar to the case $d=2$ of [3]. We still need an additional lemma:

**Lemma 5.** *Let $F(\phi_0, \ldots, \phi_N)$ be a periodic function of $\phi_0, \ldots, \phi_N$ with period $2\pi$ and such that*

$$F(\phi_0+c, \ldots, \phi_N+c) = F(\phi_0, \ldots, \phi_N). \tag{23}$$

*Then*

$$\frac{1}{(2\pi)^N}\int F(\phi_0 \ldots \phi_N) d\phi_0 \ldots d\phi_N = \frac{1}{(2\pi)^{N-1}}\int F(0, \phi_1 \ldots \phi_N) d\phi_1 \ldots d\phi_N.$$

*Proof.*

$$+\frac{1}{2\pi}\int F(\phi_0 \ldots \phi_N) d\phi_0 \ldots d\phi_N = \int d\phi_0 \left[\int F(\phi_0, \phi_1+\phi_0 \ldots \phi_N+\phi_0) d\phi_1 \ldots d\phi_N\right]$$

by periodicity. The result then follows from (23). As a direct consequence of Lemma 5 we have:

**Corollary 1.** *If* $\sum_i m(i)=0$, *then*

$$\langle \cos m\phi \rangle_{\Lambda, h=0} = \langle \cos m\phi \rangle'_{\Lambda, h=0},$$

*where the* $\langle\ \rangle'$ *state is the state* $\langle\ \rangle$ *with the restriction that* $\phi_{x_0}=0$. *The point* $x_0$ *can be chosen arbitrarily.*

*Proof of Case a) for* $d=2$. We expand around a massive Gaussian field with mass $m(T)=e^{-(\ln T)^2}$ in a finite periodic cubical box $\Lambda_0 \subset \Lambda$ with sides $R(T)=\exp[\beta^{1/4}]$. In expanding $\langle \cos m\phi\rangle'$ defined in Corollary 1 we shall choose $x_0$ just outside of $\Lambda_0$, but within $\Lambda$ with $d(x_0, \Lambda_0)=2$. By Lemmas 4 and 5, $\forall x \subset \Lambda_0$, the "spin" at the point $x$ in our system in $\Lambda$ will be with high probability in the same direction as the one at $x_0$. Writing $H_\Lambda = H^G_{\Lambda_0} + H'_\Lambda$ with

$$H^G_{\Lambda_0} = \tfrac{1}{2} \sum_{\langle ij\rangle \subset \Lambda_0} (\phi_i - \phi_j)^2 + m^2(T) \sum_{i\in\Lambda_0} \phi_i^2 + \tfrac{1}{2} \sum_{i\in\partial\Lambda_0} (\phi_i - \phi_{\bar{i}})^2,$$

$$H'_\Lambda = \tfrac{1}{2} \sum_{\substack{\langle i,j\rangle \subset \Lambda \\ i\in\Lambda_0 j\in\Lambda_0}} (\phi_i - \phi_j)^2 + \tfrac{1}{2} \sum_{\langle i,j\rangle \subset \Lambda - \Lambda_0} (\phi_i - \phi_j)^2 - m^2(T) \sum_{i\in\Lambda_0} \phi_i^2$$

$$- \tfrac{1}{2} \sum_{i\in\partial\Lambda_0} (\phi_i - \phi_{\bar{i}})^2 + \sum_{\langle i,j\rangle\subset\Lambda} \beta\left(\cos \sqrt{T}(\phi_i - \phi_j) - 1 + \frac{T}{2}(\phi_i - \phi_j)^2\right),$$

where $\bar{i}$ and $i$ are nearest neighbor in $\Lambda_0$ with periodic boundary conditions.

To expand $\langle \cos \nabla^e_0 \phi\rangle = \langle \cos \nabla^e_0 \phi\rangle'$, we make as before the change of variable $\phi' = \sqrt{T}\phi$. We have to perform an expansion for $\langle (\nabla^e_0\phi)^2\rangle'$. Let us first consider the zeroth order of $\langle (\nabla^e_0\phi)^2\rangle'$ by doing an integration by parts with respect to $H^G_{\Lambda_0}$. This yields for $\Lambda \uparrow Z^2$,

$$\begin{aligned}
\langle (\nabla^e_0\phi)^2\rangle' &= \nabla^e_0 \nabla^e_0 C^{\Lambda_0, m}_{00} && \text{(a)}\\
&+ \sum_{\substack{i\in\Lambda_0\\ \xi}} \nabla^e_0 \nabla^\xi_i C^{\Lambda_0, m}_{0i} \langle \nabla^e_0\phi(-\beta\sqrt{T} \sin \sqrt{T}\nabla^\xi_i\phi + \nabla^\xi_i\phi)\rangle' && \text{(b)}\\
&+ \sum_{\substack{i\in\partial\Lambda_0\\ |i-i'|=1, i'\in\Lambda\setminus\Lambda_0}} \nabla^e_0 C^{\Lambda_0, m}_{0i} \langle \nabla^e_0\phi(-\beta\sqrt{T} \sin \sqrt{T}(\phi_i - \phi_{i'}))\rangle' && \text{(c)}\\
&+ \sum_{i\in\partial\Lambda_0} \nabla^e_0 C^{\Lambda_0, m}_{0i} \langle \nabla^e_0\phi(\phi_i - \phi_{\bar{i}})\rangle' && \text{(d)}\\
&+ m^2_{0i} \sum_{i\in\Lambda_0} \nabla_0 C^{\Lambda_0, m}_{0i} \langle \nabla^e_0 \phi\phi_i\rangle' && \text{(e)}\\
&+ \sum_{i\in\Lambda_0} \nabla_0 C^{\Lambda_0, m}_{0i} \langle \nabla^e_0\phi[\delta(\phi_i + \pi) + \delta(\phi_i - \pi)]\rangle'. && \text{(f)} \qquad (24)
\end{aligned}$$

Before considering each term in (24), let us express Lemmas 3 and 4 in the state $\langle\ \rangle'$ using Lemma 5. They are

$$\langle \exp(\beta^{1/4}|a(\phi_i - \phi_{x_0})|)\rangle = \langle \exp(\beta^{1/4}|a(\phi_i)|)\rangle' \leq c \leq \infty, \quad \forall i\in\Lambda_0, \tag{25a}$$

$$\forall i\in\Lambda_0 \langle \delta(\phi_i - \phi_{x_0} \pm \pi)\rangle = \langle \delta(\phi_i \pm \pi)\rangle' \leq c_1 \exp(-c_2\beta^{1/4}), \quad \forall i\in\Lambda_0. \tag{25b}$$

Using now Appendices A and B in [3] we show that the r.s. of (24) has the following properties: (a) is the zeroth order term, (b) is estimated as in the case $d=3$ using (20), (c)–(f) are negligible, with our choice of $m(T)$ and $R(T)$ to all orders in $T$, e.g.

$$(\mathrm{e}) \leqq m\beta \sum_{i\in\partial\Lambda_0} |\nabla_0^e C_{0i}^{\Lambda_0,m}| \leqq \mathrm{const}\, R(T)\exp(-m(T)R(T)),$$

$$(\mathrm{f}) \leqq \mathrm{const}\, m^{-1}\exp(-c_1\beta^{1/4}).$$

To finish case a we still have to prove the result for the "free energy" $Q(T)=P(T)-\frac{d}{T}-\frac{1}{2}\ln T$. This follows from the existence of an asymptotic expansion for $\langle\cos\nabla_0^e\phi\rangle$ and the formula

$$Q_\Lambda(T)-Q_\Lambda(a)=\int_a^T -\frac{d}{\mu^2}\left[\langle\cos\nabla\phi\rangle_\Lambda(\mu)-1+\frac{\mu}{2d}\right]d\mu$$

when we take in both sides $a\to 0$.

### *4) Asymptotic Expansion for $\langle\cos m\phi\rangle$ with $\sum_i m(i)\neq 0$, $d>3$*

The method used here is somewhat indirect and consists of two steps:

– find a sequence of functions depending on $\{\nabla_x^e\phi\}$ whose expectations values approximate $\langle\cos m\phi\rangle$, or a product of such expectations; this is basically a clustering property of the state; for example, $\langle\cos(\phi_0-\phi_x)\rangle$ converges to $\langle\cos\phi_0\rangle^2$ as $|x|\to\infty$ (see below).

– Define a new function of the temperature, by letting $x$ depend on $\beta$ above such that the deviation from the desired function, for example $\langle\cos(\phi_0-\phi_x)\rangle-\langle\cos\phi_0\rangle^2$, is of order $T^n$ and perform an asymptotic expression for this new function up to order $n$.

We first carry out this proof for the magnetization, (a) and (b) below, and then show that reflection positivity allows us to make an inductive argument for all other functions $\langle\cos m\phi\rangle$, with $\underline{m}\neq 0$, (c) below.

(a) We write

$$\langle\cos\phi_0\rangle^2=\langle\cos(\phi_0-\phi_x)\rangle-(\langle\cos(\phi_0-\phi_x)\rangle-\langle\cos\phi_0\rangle^2). \tag{26}$$

Using reflecting positivity, and the infrared bounds one shows [4] that $0\leqq\langle\cos(\phi_0-\phi_x)\rangle-\langle\cos\phi_0\rangle^2\leqq|x|^{-1}\ln|x|$. Using correlation inequalities one may improve this to a $|x|^{-1}$ bound [25]. (See also Sect. V for further discussion.) If we want an asymptotic expansion up to order $n$ we choose $|x|$ to be of order $\beta^{n+1}$ so that the second term in (26) is negligible and we have only to do the expansion of $\langle\cos(\phi_0-\phi_x)\rangle$ which now depends on $\beta$ not only because of the measure, but also via $x$.

(b) As before we expand $\cos\sqrt{T}(\phi_0-\phi_x)$ into a power series and generate the expansion of the terms in this series, e.g. $\langle(\phi_0-\phi_x)^{2k}\rangle T^k$, by integration by parts, but with the rule that $\langle(\phi_0-\phi_x)^{2k}\rangle$ is expanded until order $2k(n+1/2)+n+1$. One estimates the remainder as before: when $(\phi_0-\phi_x)$ is integrated by parts it produces $C_{0i}-C_{0x}$ (or $\nabla_i^\xi C_{0i}-\nabla_i^\xi C_{ix}$) which one estimates by $|x||\nabla_0^e C_{0i}|$ (or $|x||\nabla_0^e\nabla_i^\xi C_{0i}|$). So at the end the remainder is multiplied by $|x|^{2k}=\beta^{2k(n+1)}$. The terms which were exponentially small are not affected by this factor. The only terms we have to worry about are the temperature terms. Since $\langle(\phi_0-\phi_x)^{2k}\rangle$ is expanded until order $2k(n+1/2)+n+1$, these temperature terms have a factor

$$T^{2k(n+1/2)}T^k T^{n+1}\beta^{2k(n+1)}=T^{n+1}$$

and we are in the same situation as before: $T^{n+1}(\ln m)^{2n(n+1/2)}$ is negligible with respect to $T^n$.

All that is left to consider now are the Gaussian terms produced in the I.P. which depend on $\beta$ via $x$. It is easy to see that each of these terms can be written as a sum of Gaussian expectation involving only $\phi_0$ or only $\phi_x$ (and the latter do not depend on $x$ by translation invariance) and of terms mixing $\phi_0$ and $\phi_x$ as for example

$$\sum_{\substack{x_1\dots x_l\\ \xi_1\dots\xi_l}} \nabla_{x_1}^{\xi_1} C_{0x_1} \prod_{(ij)} \nabla_{x_i}^{\xi_i}\nabla_{x_j}^{\xi_j} C_{x_ix_j} \nabla_{x_l}^{\xi} C_{x_l\cdot x} \tag{27}$$

or products of such terms.

For the terms involving only $\phi_0$ or $\phi_x$, we use the estimates of Appendix B in [3] to show that the difference between these terms and those with a massless covariance is of order $\exp(-(\ln T)^2)$. We now show that terms like (27) are small compared to $T^n$:

Call the term (27) $F(x)$ and $G(x)=xF(x)$. Then

$$\sup_x |x||F(x)| \leq \int_{-\pi}^{\pi} |\tilde{G}(p)| d^d p = \int_{-\pi}^{\pi} \sum_{i=1}^{d} \left|\frac{d}{dp_i}\tilde{F}(p)\right| d^d p. \tag{28}$$

Now, by explicit computation $\tilde{F}(p)$ is of the form

$$\prod_e (\exp i p_e - 1)^{n_e} \Big(\sum_\xi (1-\cos p_\xi) + m^2(T)\Big)^{-(l+1)}, \quad \text{with} \quad \sum n_e = 2l.$$

(28) is therefore bounded by $\ln m(T) \sim (\ln T)^2$ which implies that

$$|F(x)| \leq |x|^{-1}(\ln T)^2 = T^{n+1}(\ln T)^2$$

which is small compared to $T^n$.

This finishes the proof for the spontaneous magnetization. Its expansion (or rather the square of it) will be given in terms of these graphs mentioned above which involve only $\phi_0$ or only $\phi_x$ (with the massless covariance).

(c) Now we give the general inductive argument which allows us to prove the asymptotic expansion for all correlation functions $\langle \cos \underline{m}\phi \rangle$, $\underline{m} \equiv \sum_i m(i) \neq 0$. This uses heavily the known decay in $|x|^{-1}$ of $\langle \cos\phi_0 \cos\phi_x\rangle - \langle \cos\phi_0\rangle^2$ and of $\langle \sin\phi_0 \sin\phi_0\rangle$ [4, 25] and also Theorem 3 of [4] which gives a kind of "domination by the two point function" based on reflection positivity.

1. By symmetry we have only to consider $\underline{m} \geq 0$. We start with the case $\underline{m}=1$, and write

$$\begin{aligned}\langle \cos m\phi \rangle = [&\langle \cos(m\phi - \phi_x)\rangle - (\langle \cos(m\phi-\phi_x)\rangle \\ &- \langle \cos m\phi\rangle\langle\cos\phi_0\rangle)]/\langle\cos\phi_0\rangle\end{aligned}$$

noting that by translation invariance $\langle\cos\phi_0\rangle = \langle\cos\phi_x\rangle$.

We already have the asymptotic expansion for $\langle\cos\phi_0\rangle$ [it starts with $(1+O(T)]$ and can therefore be inverted); $\langle\cos(m\phi-\phi_x)\rangle$ is a function of the differences and can be treated as $\langle\cos(\phi_0-\phi_x)\rangle$ above. So, all we have to show is that $\cos(m\phi-\phi_x)-\langle\cos m\phi\rangle\langle\cos\phi_0\rangle$ is of order $T^n$ for $|x|$ increasing like some power of $\beta$ (actually here we take $|x|=\beta^{2n}$). We write $\cos(m\phi-\phi_x)=\cos m\phi\cos\phi_x+\sin m\phi\sin\phi_x$; and suppose for simplicity that $x$ is along the $e_1$ axis, that $i_1<0$ whenever $m(i)\neq 0$. Using reflection positivity with respect to the plane perpendicular to the $e_1$ axis at zero, $x\to\bar{x}$, then gives

$$|\langle\sin m\phi\sin\phi_x\rangle|\leqq c_m\langle\sin\phi_x\sin\phi_{\bar{x}}\rangle^{1/2},$$

where $c_m$ depends only on $m$. (This is similar to Theorem 3 in [4].) Since the r.s. decays like $|x|^{-1}$ [25], we get, with $|x|\sim\beta^{2n}$,

$$|\langle\sin m\phi\sin\phi_x\rangle|\leqq c_m T^n.$$

Similarly we write,

$$\begin{aligned}\langle\cos m\phi\cos\phi_x\rangle-\langle\cos m\phi\rangle\langle\cos\phi_0\rangle&=\langle\cos m\phi(\cos\phi_x-\langle\cos\phi_0\rangle)\rangle\\&\leqq c_m\langle(\cos\phi_x-\langle\cos\phi_0\rangle)(\cos\phi_{\bar{x}}-\langle\cos\phi_0\rangle)\rangle^{1/2}\\&=c_m(\langle\cos\phi_x\cos\phi_{\bar{x}}\rangle-\langle\cos\phi_0\rangle^2)^{1/2}.\end{aligned}$$

2. For general $m$, let us assume that we have done the proof of the expansion for $\underline{m}\leqq k$ and let $\underline{m}=k+1$ and $\underline{n}=k$. Let $n_x$ denote $n$ translated by an amount $x$ in the positive $e_1$ direction, and $y$ a point in the positive $e_2$ direction. $m$ is assumed to be such that $i_1<0$, $i_2<0$ if $m(i)\neq 0$. We shall prove that $\langle\cos(m\phi-n_x\phi-\phi_y)\rangle$ is close to $\langle\cos m\phi\rangle\langle\cos n\phi\rangle\langle\cos\phi_0\rangle$ with an error of order $T^n$ if $|x|$ and $|y|$ are of order $\beta^{2n}$. Then, since we know the expansion of $\langle\cos n\phi\rangle$, $\langle\cos\phi_0\rangle$ by our recursive hypothesis, and can invert them, we obtain the expansion of $\langle\cos m\phi\rangle$ by the same arguments as above.

To prove the cluster property, we write

$$\cos(m\phi-n_x\phi-\phi_y)=\cos(m\phi-\phi_y)\cos n_x\phi+\sin(m\phi-\phi_y)\sin n_x\phi.$$

Now, by reflection positivity,

$$|\langle\sin(m\phi-\phi_y)\sin n_x\phi\rangle|\leqq c_m(\langle\sin n_x\phi\sin n_{\bar{x}}\phi\rangle)^{1/2}.$$

To bound $\langle\sin n_x\phi\sin n_{\bar{x}}\phi\rangle$ we use the fact that, by reflection positivity,

$$\langle\cos n_x\phi\cos n_x\phi\rangle-\langle\cos n\phi\rangle^2=1/2\langle(\cos n_x\phi-\langle\cos n_x\phi\rangle)(\cos n_{\bar{x}}\phi-\langle\cos n_x\phi\rangle)\rangle$$

is positive and therefore

$$\langle\sin n_x\phi\sin n_{\bar{x}}\phi\rangle\leqq\langle\cos(n_x\phi-n_{\bar{x}}\phi)\rangle-\langle\cos n\phi\rangle^2.$$

In the right hand side of this inequality, the first term can be expanded in powers of $T$ because it is a function of the differences and the second by assumption because $\underline{n}=k$. But the coefficients in this expansion cancel up to order $n$ except for an error of order $T^{2n}\ln T$ if $|x|\sim\beta^{2n}$.

We apply the same argument to control the difference between $\langle\cos(m\phi-\phi_y)\cos n_x\phi\rangle$ and $\langle\cos(m\phi-\phi_y)\rangle\langle\cos n_x\phi\rangle$ and then between $\langle\cos(m\phi-\phi_y)\rangle$ and $\langle\cos m\phi\rangle\langle\cos\phi_0\rangle$.

## IV. An Alternative Proof of Part a) of Theorem 1

In this section, we carry out a proof of Part a) of the Theorem which does not use the infrared bounds. It proceeds in three steps:

1. It follows from Ginibre's inequalities [14] that for any

$$\Lambda, \Lambda' \subset Z^d, \quad \langle \cos m\phi \rangle^0_{\Lambda'} \leqq \langle \cos m\phi \rangle \leqq \langle \cos m\phi \rangle^+_\Lambda, \tag{30}$$

where $\langle\ \rangle^0_{\Lambda'}$ is the Gibbs measure with free b.c. (i.e. no coupling with $\Lambda'^c$) on $\Lambda'$, and $\langle\ \rangle^+_\Lambda$ is the Gibbs measure with 0 boundary condition ($\phi_i = 0 \forall i \in \Lambda^c$). We carry out the expansion of the right and left side of (30), where we let $\Lambda$ and $\Lambda'$ depend on the temperature in such a way, [e.g. $|\Lambda|, |\Lambda'| \sim \exp(\sqrt{\beta})$] that the coefficients of the asymptotic expansion of both sides of (30) are the same, giving the expansion of $\langle \cos m\phi \rangle$.

2. We show (Proposition 2) that, if $|\Lambda|, |\Lambda'|$ grow like $\exp(\sqrt{\beta})$ then the left and right hand side of (30) are exponentially close as $\beta \to \infty$ to the corresponding expectation values conditioned on the event: $|\phi_i - \phi_j| \leqq \eta$, $\eta > 0$ for all nearest neighbor pairs $\langle ij \rangle$, with $\langle ij \rangle \cap \Lambda \neq \emptyset$ or $\langle ij \rangle \subset \Lambda'$.

3. We show that, for $\eta < \pi/2$, these conditioned expectation values are equal to the ones where we let all $\phi_i$ run from $-\infty$ to $+\infty$ (Proposition 2 and Corollary 2). Once this is done, we integrate by parts with respect to a massive Gaussian as before.

We define $\langle\ \rangle^+_{\Lambda,\eta}$ and $\langle\ \rangle^0_{\Lambda,\eta}$ as the expectation value $\langle\ \rangle^+_\Lambda$ or $\langle\ \rangle^0_\Lambda$ conditioned on the event that for all n.n. pairs, $|\phi_i - \phi_j| \leqq \eta \bmod 2\pi$, that is

$$d\mu^+_{\Lambda,\eta} = (Z_{\Lambda,\eta})^{-1} \exp(-\beta H_{\Lambda,+}) \prod_{\langle i,j \rangle \cap \Lambda \neq \emptyset} \chi(|\phi_i - \phi_j| \leqq \eta \bmod 2\pi) \cdot \prod_{i \in \Lambda} d\phi_i$$

with $\phi_j = 0$ if $j \notin \Lambda$ and similarly for $d\mu^0_{\Lambda,\eta}$.

Then we have:

**Proposition 1.** *There exists constants $c_1, c_2$ such that, for all $\beta, \Lambda, \eta$ and for $\gamma = +$ or $0$*

$$|\langle \cos m\phi \rangle^\gamma_\Lambda - \langle \cos m\phi \rangle^\gamma_{\Lambda,\eta}| \leqq c_2 [|\Lambda| \exp(-c_1 \beta \eta^2)] \exp(c_2 |\Lambda| \exp(-c_1 \beta \eta^2)).$$

*Proof*. The proof is the same for $+$ or $0$; we write

$$\begin{aligned}
\langle \cos m\phi \rangle^+_\Lambda &= \sum_X \langle \cos m\phi \rangle^+_{\Lambda,X} P(X) \\
&= \langle \cos m\phi \rangle^+_{\Lambda, X=\emptyset} P(X=\emptyset) + \sum_{X \neq \emptyset} \langle \cos m\phi \rangle^+_{\Lambda,X} P(X) \\
&\langle \cos m\phi \rangle^+_{\Lambda,\eta} + \sum_{X \neq \emptyset} (\langle \cos m\phi \rangle^+_X - \langle \cos m\phi \rangle^+_{\Lambda,\eta}) P(X),
\end{aligned}$$

where the sum over $X$ runs over all subsets of $\{\langle ij \rangle \cap \Lambda \neq \emptyset\}$ (or $\{\langle ij \rangle \subset \Lambda\}$ for free b.c.). $\langle\ \rangle_{\Lambda,X}$ is the expectation value conditioned on the event that $|\phi_i - \phi_j| \geqq \eta \bmod 2\pi$ for $\langle ij \rangle \in X$ and $|\phi_i - \phi_j| \leqq \eta \bmod 2\pi$ for $\langle ij \rangle \notin X$; $P(X)$ is the probability of that event. Since $\chi \leqq 1$ we estimate

$$\begin{aligned}
P(X) &\leqq \Big\langle \prod_{\langle ij \rangle \in X} \chi(|\phi_i - \phi_j| < \eta \bmod 2\pi) \Big\rangle \\
&\leqq \exp(-c_1 \beta \eta^2 |X|)
\end{aligned}$$

by the contour estimates of [1]. Therefore

$$\begin{aligned}|\langle\cos m\phi\rangle_\Lambda^+ - \langle\cos m\phi\rangle_{\Lambda,\eta}^+| &\leqq 2\sum_{X\neq\emptyset}\exp(-c_1\beta\eta^2|X|)\\ &\leqq 2[(1+\exp(-c_1\beta\eta^2))^{2d|\Lambda|}-1]\\ &\leqq c_2(|\Lambda|\exp(-c_1\beta\eta^2))\exp(c_2|\Lambda|\exp(-c_1\beta\eta^2)).\end{aligned}$$

In the second step we use the fact that the sum runs over all subsets of $\{\langle ij\rangle\cap\Lambda\neq\emptyset\}$ whose number is less than $2d|\Lambda|$. In the last step we use

$$\begin{aligned}(1+x)^n-1 &= \exp(n\log(1+x))-1\\ &\leqq \exp(nx)-1 \leqq nx\exp(nx), \qquad (x>0).\end{aligned}$$

**Proposition 2.** *Let $\Lambda$ be a hypercube in $Z^d$ and let $k\in Z^d$ be, either in $\Lambda$, or nearest neighbor to some point in $\Lambda$. $B(\{\phi_i\},\phi_k)$, where $\{\phi_i\}$ is the set $\{\phi_i\}_{i\in\Lambda'}$ is any periodic function (in $L^1([-\pi,\pi]^{|\Lambda|})$) of period $2\pi$ in each of its variables. Then for any $\eta<\pi/2$ and any $\phi_k\in[-\pi,\pi]$,*

$$\begin{aligned}L &\equiv \int_{-\pi}^{\pi} B(\{\phi_i\},\phi_k)\prod_{\langle ij\rangle\subset\Lambda\cup\{k\}}\chi(|\phi_i-\phi_j|\leqq\eta \bmod 2\pi)\prod_{\substack{i\in\Lambda\\ i\neq k}}d\phi_i\\ &= \int_{-\infty}^{+\infty} B(\{\phi_i\},\phi_k)\prod_{\langle ij\rangle\subset\Lambda\cup\{k\}}\chi(|\phi_i-\phi_j|\leqq\eta)\prod_{\substack{i\in\Lambda\\ i\neq k}}d\phi_i\equiv R.\end{aligned} \tag{31}$$

*Proof*. In $L$ we write

$$\chi(|\phi_i-\phi_j|\leqq\eta \bmod 2\pi)=\sum_{n_{ij}\in Z}\chi(|\phi_i-\phi_j+2\pi n_{ij}|\leqq\eta)$$

and expand the product over $\langle ij\rangle$. We always write the subscripts $ij$ in lexicagraphic order corresponding to a directed bond. This gives

$$L=\sum_{\substack{\{n_{ij}\}\\ n_{ij}\in Z}}\int_{-\pi}^{+\pi} B(\{\phi_i\},\phi_k)\prod_{\langle ij\rangle\subset\Lambda\cup\{k\}}\chi(|\phi_i-\phi_j+2\pi n_{ij}|\leqq\eta)\prod_{i\in\Lambda}d\phi_i. \tag{32}$$

In $R$ we write for each integral

$$\int_{-\infty}^{+\infty} = \ldots = \sum_{m\in Z}\int_{2\pi m-\pi}^{2\pi m+\pi}\ldots.$$

After changing variables $\phi_i\to\phi_i-2\pi m_i$ (which leaves $B$ unchanged because of periodicity) we get:

$$R=\sum_{\substack{\{m_i\}\\ m_i\in Z}}\int_{-\pi}^{+\pi} B(\{\phi_i\},\phi_k)\cdot\prod_{\langle ij\rangle\subset\Lambda\cup\{k\}}\chi(|\phi_i-\phi_j+2\pi(m_i-m_j)|\leqq\eta)\prod_{i\in\Lambda}d\phi_i \tag{33}$$

($m_k=0$ in the above expression). All we have to show in order to prove the equality of (32) and (33) is that each set $\{n_{ij}\}$ for which the integral does not vanish is such that $n_{ij}$ can be written as $n_{ij}=m_i-m_j$ for all oriented pairs $\langle ij\rangle$. Given such a set

$\{n_{ij}\}$ we define for all $m$, $n$, $i$, $j$, $n_{ji}=n_{ij}$. Since we may take $m_k=0$ and since for any $j\in\Lambda$ there exists always a path $(ki_1),\dots,(i_n,j)$ of n.n. connecting $k$ and $j$ we may define $-m_j=n_{ki_1}+n_{i_1i_2}+\dots+n_{i_n,j}$. Then, of course $n_{ij}=m_i-m_j$ but the trouble might be that two different paths may lead to two different definitions of $m_j$. This does not occur however if for each closed loop on the lattice $\sum_{\text{loop}} n_{ij}=0$.

For this to hold it is enough that $\sum_{\text{Sq.}} n_{ij}=0$ for each elementary square with 4 n.n. pairs in the lattice. But $\sum_{\text{Sq.}}(\phi_i-\phi_j)=0$ and we may write

$$2\pi\left|\sum_{\text{Sq.}} n_{ij}\right| = \left|\sum_{\text{Sq.}}(2\pi n_{ij}+\phi_i-\phi_j)\right|$$

$$\leqq \sum_{\text{Sq.}} |2\pi n_{ij}+\phi_i-\phi_j| \leqq 4\eta < 2\pi$$

by our condition on $\eta$ and the fact that the integral in (32) vanishes unless $|2\pi n_{ij}+\phi_i-\phi_j|\leqq\eta$. Since $|\sum n_{ij}|$ is a positive integer strictly less than one, it is zero.

We apply now Proposition 2 to the measures $\mu^{\gamma}_{\Lambda,\eta}$. Let $\Lambda$ be a cube in $Z^d$: $\Lambda=[-L,L]^d$, $l=(L,0,\dots,0)$, $k=(L+1,0,\dots,0)$, $\Lambda'=\Lambda\cup\{k\}$. We define two measures on $R^{||\Lambda|}$:

$$d\mu^{(1)}_{\Lambda}=(Z^{(1)}_{\Lambda})^{-1}\prod_{\langle ij\rangle\subset\Lambda}\exp(\beta\cos(\phi_i-\phi_j))\chi(|\phi_i-\phi_j|\leqq\eta)$$

$$\cdot\exp\beta\cos\phi_k)\chi(|\phi_k|\leqq\eta)\prod_{i\in\Lambda}d\phi_i, \tag{34}$$

$$d\mu^{(2)}_{\Lambda}=(Z^{(2)})^{-1}\prod_{\langle ij\rangle\subset\Lambda}\exp(\beta\cos(\phi_i-\phi_j))$$

$$\cdot\chi(|\phi_i-\phi_j|\leqq\eta)\prod_{i\in\partial\Lambda}\exp(\beta\cos\phi_i)\chi(|\phi_i|\leqq\eta)\prod_{i\in\Lambda}d\phi_i, \tag{35}$$

where in $Z^{(1)}_{\Lambda}, Z^{(2)}_{\Lambda}$ the integration runs from $-\infty$ to $+\infty$.

**Corollary 2.** *For $\eta<\pi/2$ and any $m$ with* $(\operatorname{supp} m)\subset\Lambda$ *and with $\underline{m}=\sum m(i)=0$,*

$$\langle\cos m\phi\rangle^{0}_{\Lambda',\eta}=\langle\cos m\phi\rangle^{(1)}_{\Lambda}, \tag{36}$$

$$\langle\cos m\phi\rangle^{+}_{\Lambda,\eta}=\langle\cos m\phi\rangle^{(2)}_{\Lambda}. \tag{37}$$

*Proof*. For the free b.c. we first remark that, since we integrate in the numerator and the denominator of the l.h.s. of (36) functions which are invariant under rotations of all the spins in $\Lambda'$, we may as well set $\phi_k=0$ since that integration is redundant. Then we apply Proposition 2 to the N. and D. with $\phi_k=0$ being the spin fixed outside $\Lambda$ and

$$B=\exp\left(\beta\left(\sum_{\langle ij\rangle\subset\Lambda}\cos(\phi_i-\phi_j)\right)+\beta\cos\phi_k\right)$$

in the D. and the same thing multiplied by $\cos m\phi$ in the N.

For the +b.c. we apply also the identity (31) to the N. and D. including in $B$ a factor

$$\prod_{\substack{j\notin\Lambda\\ j\neq k}} \prod_{\substack{i\in\Lambda\\ |i-j|=1}} \exp(\beta\cos\phi_i)\chi(|\phi_i|\leqq\eta \bmod 2\pi) \tag{38}$$

(which is periodic). We have to do this because Proposition 2 only allows one spins fixed outside of $\Lambda$ (at site $k$). But now we have a periodic characteristic function which is left. In order to get (37) (no periodic $\chi$ in $\mu_\Lambda^{(2)}$) we write

$$\chi(|\phi_i|\leqq\eta \bmod 2\pi) = \sum_{n_i\in Z} \chi(|\phi_i - 2\pi n_i|\leqq\eta)$$

and expand the product over $i$ in (36) both in the N. and D. We shall see that only the term where all $n_i=0$ contributes and this will prove (37). Let us go around $\partial\Lambda$ starting from $l$ and let $j$ be the first point where $n_j\neq 0$. Then there is a $j'\in\partial\Lambda$ (just before $j$) with $|j-j'|=1$, $n_{j'}=0$. Since $\langle jj'\rangle\subset\Lambda$, $|\phi_j-\phi_{j'}|\leqq\eta$ and since $n_{j'}=0$, $|\phi_{j'}|\leqq\eta$ (for the integrand not be zero). So $|\phi_j|\leqq 2\eta$ and $2\pi|n_j|\leqq|2\pi n_j-\phi_j|+|\phi_j|\leqq 3\eta<2\pi$ and so $n_j=0$.

Now it is easy to check that the measures (34) and (35) are log concave perturbation of Gaussian ones: for $|\phi_i-\phi_j|<\pi/2$ one can find a $\tau>0$ such that $\cos(\phi_i-\phi_j)+\tau(\phi_i-\phi_j)^2$ is concave [5].

Now we generate the asymptotic expansion of $\langle\cos m\phi\rangle_\Lambda^{(1)}$, $\langle\cos m\phi\rangle_\Lambda^{(2)}$ by integrating by parts as before. We let the radius of the box $\Lambda, R(T)=\exp(\sqrt{\beta})$, which grows sufficiently fast to show that the boundary terms are negligible $[m(T)R(T)\sim\exp-(\log T)^2\exp(\sqrt{\beta})\to\infty$ as $T\to 0]$. Since our measures are log concave perturbations of Gaussian ones, we can control the $(\nabla\phi')^{2n}$, $\phi'=\sqrt{\beta}\phi$ and the $\delta$ functions $\delta(|\nabla\phi|=\eta)$ by the Brascamp-Lieb inequalities [5]. We also control the expectation values appearing in the mass term with these inequalities, because $\langle\phi^2\rangle_\Lambda^{(i)}$, $i=1,2$ is bounded by the corresponding Gaussian expectation which is finite for $d\geqq 3$ and diverge like $\log\Lambda\sim\sqrt{\beta}$ for $d=2$ [for $d=1$ one would have to choose $R(T)=\exp(-3/2(\ln T)^2)$] (see [3]).

*Remark.* 1. Since we do not use reflection positivity, the above proof works, in principle, for ferromagnetic finite range interactions instead of nearest-neighbor ones. One would have to check the Gaussian estimates of Appendices A and B of [3] for non-nearest-neighbor Gaussian measures. The proof however only works for $\langle\cos m\phi\rangle$ with $\underline{m}=0$.

2. One could, with this method, perform directly the expansion for $\langle\cos m\phi\rangle$ with $\sum m_i\neq 0$ and, in particular show that $\langle\cos\phi_0\rangle\neq 0$ for $T$ small enough if one could prove one of the following statements, which are presumably true:

1. The difference $|\langle\cos m\phi\rangle_\Lambda-\langle\cos m\phi\rangle_{\Lambda,\eta}|$ between the correlation functions and the one conditioned on $|\phi_i-\phi_j|\leqq\eta$ for all pairs $\langle i,j\rangle$ is exponentially small as $\beta\to\infty$ uniformly in $\Lambda$.

2. The expectation value $\langle\cos m\phi\rangle_\Lambda^+$ converge sufficiently fast to their infinite volume limit. A convergence of order $(\log|\Lambda|)^{-1}$ would suffice to prove the phase transition.

3. The limit of $\langle\cos m\phi\rangle_h$ as $h\downarrow 0$ is fast enough; again $(\log h)^{-1}$ would be enough to prove the phase transition.

Any of these statements would provide a proof of phase transition for three or more dimensional rotator models without using reflection positivity.

## V. On the Decay of the Two-Point Function

We show here that whenever there is a spontaneous magnetization, the transverse two-point function of the plane rotator, $\langle \sin\phi_0 \sin\phi_x \rangle$, behaves exactly like $\beta^{-1}|x|^{-(d-2)}$ for large $|x|$. This is what the Gaussian (spin-wave) approximation would predict. The upper bound was proven in [25]. We prove here the lower bound, which improves Goldstone theorem type of result (e.g. [19]) showing that $\sum_x \langle \sin\phi_0 \sin\phi_x \rangle^2$ diverges. The proof relies on the infrared bound, the Mermin-Wagner argument and the correlation inequalities of [22, 24, 15].

**Theorem.** *For any $d \geqq 3$, there exists constants $c_1, c_2$ such that for all $x \in Z^d$*

a) $$\frac{c_2 \bar{m}^2}{\beta |x|^{d-2}} \leqq \langle \sin\phi_0 \sin\phi_x \rangle \leqq \frac{c_1}{\beta |x|^{d-2}},$$

b) $$\frac{c_2 \bar{m}^2}{\beta |x|^{2(d-2)}} \leqq \langle \cos\phi_0 \cos\phi_x \rangle - \bar{m}^2 \leqq \frac{c_1}{\beta |x|^{d-2}},$$

*where $\bar{m} = \langle \cos\phi_0 \rangle$ is the spontaneous magnetization and $|x| = \sum_{e=1}^{d} |x_e|$.*

*Proof.* The upper bounds are proven in [25]. b) Follows from a), the Dunlop-Newman inequality [9] and the inequality $\langle \cos\phi_0 \cos\phi_x \rangle - \bar{m}^2 \leqq \langle \sin\phi_0 \sin\phi_x \rangle$ proven in [10, 16]. Let us prove a) for $d=3$ (the general case is similar). By the infrared bounds and the Mermin-Wagner argument, (see e.g. [21]) $S(p)$, the Fourier transform of $\langle \sin\phi_0 \sin\phi_x \rangle$ satisfies the bounds,

$$\frac{c_2' \bar{m}^2}{\beta |p|^2} \leqq S(p) \leqq \frac{c_i'}{\beta |p|^2}.$$

Therefore, if $f_L(x)$ is the characteristic function of a cube $\Lambda \leqq Z^3$ centered at the origin of volume $L^3$:

$$c_2'' \beta^{-1} L^5 \leqq \sum_{x,y} \langle \sin\phi_x \sin\phi_y \rangle f_L(x) f_L(y) \leqq c_1'' \beta^{-2} L^5$$

$\left(\text{one has only to show that } \int_{-\pi}^{\pi} p^{-2} |\hat{f}_L^2(p)|^2 d^3p \sim L^5, \text{ see e.g. [4]}\right)$.

By positivity of $\langle \sin\phi_x \sin\phi_y \rangle$ and translation invariance, we have:

$$\left(\frac{L}{2}\right)^3 \sum_{x \in 1/2\Lambda} \langle \sin\phi_0 \sin\phi_x \rangle \leqq \sum_{x,y} \langle \sin\phi_x \sin\phi_y \rangle f_L(x) f_L(y)$$
$$\leqq c_1'' \beta^{-1} L^5 \qquad (40)$$

and

$$(L)^3 \sum_{x \in \Lambda} \langle \sin\phi_0 \sin\phi_x \rangle \geqq \sum_{x,y} \langle \sin\phi_x \sin\phi_y \rangle f_L(x) f_L y)$$
$$\geqq c_2'' \beta^{-1} L^5, \qquad (41)$$

where $k\Lambda$ is a cube of size $(kL)^3$ centered at the origin.

Now we use the fact that, by correlation inequalities [22, 24, 15] $\langle\sin\phi_0 \sin\phi_x\rangle$ reaches its maximum for $x$ inside $\Lambda$ at the corners of $\Lambda$ and its maximum outside $\Lambda$ for $x$ along the axis $|x|=\left[\frac{L}{2}\right]+1$. (The minimum property is used in [25] for the upper bound.) To get the lower bound, let

$$x=\left(\left[\frac{L}{2}\right]+1,0,\ldots,0\right).$$

Then

$$\begin{aligned}
&\langle\sin\phi_0\sin\phi_x\rangle\\
&\quad\geqq((kL^3)-(L)^3)^{-1}\sum_{\substack{x\in k\Lambda\\ x\notin\Lambda}}\langle\sin\phi_0\sin\phi_x\rangle\\
&\quad=((kL)^3-(L)^3)^{-1}\left(\sum_{x\in k\Lambda}\langle\sin\phi_0\sin\phi_x\rangle-\sum_{x\in\Lambda}\langle\sin\phi_0\sin\phi_x\rangle\right)\\
&\quad\geqq((kL)^3-(L)^3)^{-1}\frac{c_2''(kL)^2}{\beta}-\frac{2^5c_1''L^2}{\beta}
\end{aligned}$$

using inequalities (40) and (41). This proves the lower bound when $x$ is along a coordinate axis, if we choose $k$ such that $c_2''k^2\geqq 2^5c_1''+1$ and $c_2=2(k^3-1)^{-1}$.

When $x$ is not along a coordinate axis we use the result of [15] which says that for given $|x|=\sum_{e=1}^{d}|x_e|$, $\langle\sin\phi_0\sin\phi_x\rangle$ reaches its maximum along the coordinate axis. This is proven in [15] for Ising models, but the same inequalities hold for the plane rotator model: one has to use the fact that the state $\langle\ \rangle$ can be obtained as a limit of Gibbs states in $\Lambda_n$ for any increasing sequence $\Lambda_n\uparrow Z^d$, [2, 23], so that we can use all the inequalities of [22, 24, 15] at once.

## References


1. Bricmont, J., Fontaine, J.R.: Correlation inequalities and contour estimates (in preparation)
2. Bricmont, J., Fontaine, J.R., Landau, L.J.: Commun. Math. Phys. **56**, 281 (1977)
3. Bricmont, J., Fontaine, J.R., Lebowitz, J.L., Spencer, T.: Lattice systems with a continuous symmetry. I. Commun. Math. Phys. **78**, 281–302 (1980)
4. Bricmont, J., Fontaine, J.R., Lebowitz, J.L., Spencer, T.: Lattice systems with a continuous symmetry. II. Commun. Math. Phys. **78**, 363–371 (1981)
5. Brascamp, H.J., Lieb, E.H.: J. Funct. Anal. **22**, 366 (1976)
6. Brezin, E.: Proceedings of the 13th IUPAP Conference on Statistical Physics [(ed.) Cabib, D., Kuper, C., and Reiss]. Haifa 1977
7. Dobrushin, R.L.: Funct. Anal. and Appl. **2**, 292 (1968)
8. Dobrushin, R.L., Shlosman, S.B.: Commun. Math. Phys. **42**, 31 (1975)
9. Dunlop, F., Newman, C.: Commun. Math. Phys. **44**, 223 (1975)
10. Dunlop, F.: Commun. Math. Phys. **49**, 247 (1976)
11. Elitzur, S.: The applicability of perturbation expansion to two-dimensional Goldstone systems. I.A.S. Princeton preprint 1019
12. Fröhlich, J., Guerra, F., Robinson, D., Stora, R. (eds.): Marseille Conference C.N.R.S. 1975
13. Fröhlich, J., Simon, B., Spencer, T.: Commun. Math. Phys. **50**, 79 (1976)
14. Ginibre, J.: Commun. Math. Phys. **16**, 310 (1970)
15. Hegerfeldt, G.: Commun. Math. Phys. **57**, 259 (1977)

16. Kunz, H., Pfister, C.E., Vuillermot, J.: Phys. A. Math. Gen. **9**, 1673 (1976)
17. Lanford, O., Ruelle, D.: Commun. Math. Phys. **13**, 194 (1969)
18. Lebowitz, J., Penrose, O.: Commun. Math. Phys. **11**, 99 (1968)
19. Lebowitz, J., Penrose, O.: Phys. Rev. Lett. **35**, 549 (1975)
20. Lee, T.D., Yang, C.N.: Phys. Rev. **87**, 410 (1952)
21. Mermin, D.: J. Math. Phys. **6**, 1061 (1967)
22. Messager, A., Miracle-Sole, S.: J. Stat. Phys. **17**, 245 (1977)
23. Messager, A., Miracle-Sole, S., Pfister, C.E.: Commun. Math. Phys. **58**, 19 (1978)
24. Schrader, R.: Phys. Rev. B **15**, 2798 (1977)
25. Sokal, A.: In preparation

Communicated by A. Jaffe

Received August 12, 1980

# Part VII

# Miscellaneous

Commun. math. Phys. 28, 251—257 (1972)


# The Finite Group Velocity of Quantum Spin Systems

Elliott H. Lieb*

Dept. of Mathematics, Massachusetts Institute of Technology
Cambridge, Massachusetts, USA

Derek W. Robinson**

Dept. of Physics, Univ. Aix-Marseille II, Marseille-Luminy, France

Received May 15, 1972

**Abstract.** It is shown that if $\Phi$ is a finite range interaction of a quantum spin system, $\tau_t^\Phi$ the associated group of time translations, $\tau_x$ the group of space translations, and $A, B$ local observables, then

$$\lim_{\substack{|t| \to \infty \\ |x| > v|t|}} \| [\tau_t^\Phi \tau_x(A), B] \| \, e^{\mu(v)t} = 0$$

whenever $v$ is sufficiently large ($v > V_\Phi$) where $\mu(v) > 0$. The physical content of the statement is that information can propagate in the system only with a finite group velocity.

## 1. Introduction

In [2] it was demonstrated that for a large class of translationally invariant interactions, time translations of quantum spin systems can be defined as automorphisms of a $C^*$-algebra, $\mathscr{A}$, of quasi-local observables, i.e. the abstract algebra generated by the spin operators. This should allow one to discuss features of the dynamical propagation of physical effects in an algebraic manner independent of the state of the system, i.e. independent of the kinematical data. It is expected that this propagation has many features in common with the propagation of waves in continuous matter and the point of this paper is to demonstrate such a feature, namely a finite bound for the group velocity of a system with finite range interaction. This result is obtained by a simple estimation derived from the equations of motion and it is possible that more detailed estimations would give more precise information of the form of spin-wave propagation. We briefly discuss this possibility at the end of Section 3.

* Work supported by National Science Foundation Grant N°: GP–31674 X.
** Work supported by National Science Foundation Grants N°: GP–31239 X and GP–30819 X.

## 2. Basic Notation

We use the formalism introduced in [1] and [2]. For completeness we recall the basic definitions which will be used in the sequel.

The kinematics of a quantum spin system constrained to a $\nu$-dimensional cubic lattice, $\mathbb{Z}^\nu$, are introduced by associating with each point $x \in \mathbb{Z}^\nu$ an $N$-dimensional vector space $\mathscr{H}_x$ and with each finite set $\Lambda \subset \mathbb{Z}^\nu$ the direct product space

$$\mathscr{H}_\Lambda = \prod_{x \in \Lambda}^{\otimes} \mathscr{H}_x .$$

The algebra of strictly local observables, $\mathscr{A}_\Lambda$, of the subsystem $\Lambda$, is defined to be the algebra of all matrices acting on $\mathscr{H}_\Lambda$. If $\Lambda_1 \subset \Lambda_2$, the algebra $\mathscr{A}_{\Lambda_1}$ acting on $\mathscr{H}_{\Lambda_1}$ can be identified with the algebra $\mathscr{A}_{\Lambda_1} \otimes \mathbb{1}_{\Lambda_2 \setminus \Lambda_1}$ acting on $\mathscr{H}_{\Lambda_2}$ ($\mathbb{1}_{\Lambda_2 \setminus \Lambda_1}$ is the identity operator on $\mathscr{H}_{\Lambda_2 \setminus \Lambda_1}$) and with this identification $\mathscr{A}_{\Lambda_1} \subset \mathscr{A}_{\Lambda_2}$. Due to this isotony relation the set theoretic union of all $\mathscr{A}_\Lambda$, with $\Lambda \subset \mathbb{Z}^\nu$ finite is a normed *-algebra and we define the completion of this algebra to be the $C^*$-algebra $\mathscr{A}$ of (quasi-)local kinematical observables of the spin system. The group $\mathbb{Z}^\nu$ of space translations is a subgroup of the automorphism group of $\mathscr{A}$, and we denote the action of this group by $A \in \mathscr{A}_\Lambda \to \tau_x A \in \mathscr{A}_{\Lambda+x}$ for $x \in \mathbb{Z}^\nu$.

To define the dynamics of our system we introduce an interaction $\Phi$ as a function from the finite sets $X \subset \mathbb{Z}^\nu$ to elements $\Phi(X) \subset \mathscr{A}_X$. In contrast to [2] we will only consider finite range interactions in the sequel. Thus we demand that $\Phi$ satisfies:

1. $\Phi(X)$ is Hermitian for $X \subset \mathbb{Z}^\nu$.
2. $\Phi(X+a) = \tau_a \Phi(X)$ for $X \subset \mathbb{Z}^\nu$ and $a \in \mathbb{Z}^\nu$.
3. The union $R_\Phi$ of all $X$ such that $X \ni 0$ and $\Phi(X) \neq 0$ is a finite subset of $\mathbb{Z}^\nu$.

[Physically only particles situated at the points $x \in R_\Phi$ have a nonzero sinteraction with a particle at the origin.]

The Hamiltonian of a finite system $\Lambda$ with interaction $\Phi$ is defined by:

$$H_\Phi(\Lambda) = \sum_{X \subset \Lambda} \Phi(X) .$$

In [2] it was established that each interaction $\Phi$ defines a strongly continuous, one-parameter group of automorphisms $\tau_t^\Phi$ of $\mathscr{A}$. Explicitly we have for each $A \in \mathscr{A}$ and $t \in R$ an element $\tau_t^\Phi(A) \in \mathscr{A}$ such that:

$$\lim_{\Lambda \to \infty} \| \tau_t^\Phi(A) - e^{itH_\Phi(\Lambda)} A e^{-itH_\Phi(\Lambda)} \| = 0$$

$$\lim_{t \to 0} \| \tau_t^\Phi(A) - A \| = 0$$

$$\tau_t^\Phi(A B) = \tau_t^\Phi(A)\, \tau_t^\Phi(B) , \qquad \text{etc.} \ldots .$$

In fact $t \to \tau_t^\Phi(A)$ with $A \in \mathscr{A}_\Lambda$ is analytic in a strip $|\mathrm{Imt}| < a_\Phi$ with $a_\Phi > 0$ and further:

$$\lim_{\Lambda \to \infty} \left\| \frac{d}{dt} \tau_t^\Phi(A) - \frac{d}{dt} e^{itH_\Phi(\Lambda)} A e^{-itH_\Phi(\Lambda)} \right\| = 0 .$$

For details see [2].

## 3. Local Commutativity

Our aim is to discuss the behaviour of commutators

$$C_{A,B}(x,t) = [\tau_t^\Phi \tau_x(A), B]$$

for $A$ and $B$ strictly local, i.e. contained in some $\mathscr{A}_\Lambda$. We wish to examine the magnitude of these commutators for large $x$ and $t$ and to show that information propagates with a finite group velocity $V_\Phi$. More precisely, we have the following.

**Theorem.** *For each finite range interaction $\Phi$ there exists a finite group velocity $V_\Phi$ and a strictly positive increasing function $\mu$ such that for $v > V_\Phi$*

$$\lim_{\substack{|t| \to \infty \\ |x| > v|t|}} e^{\mu(v)|t|} \, \| [\tau_t^\Phi \tau_x(A), B] \| = 0$$

*for all strictly local $A$ and $B$.*

*Proof.* First we note that it is sufficient to prove the theorem for $A, B \in \mathscr{A}_{\{0\}}$. This is because each strictly local $A, B \in \mathscr{A}_\Lambda$ can be written as a polynomial in elements of $\mathscr{A}_{\{x\}}$ with $x \in \Lambda$. Hence the norm of the general commutator can be bounded above by a finite sum of norms of similar commutators but with $A \in \mathscr{A}_{\{x\}}$, $B \in \mathscr{A}_{\{y\}}$ and $x, y \in \Lambda$. Using translation invariance each of these commutators can be reduced to a commutator with $A, B \in \mathscr{A}_{\{0\}}$.

As $\mathscr{H}_0$ is finite dimensional we can choose a finite basis $a_1, \ldots, a_{N^2}$ of $\mathscr{A}_{\{0\}}$ closed under multiplication with $\|a_i\| = 1$ and such that every $A \in \mathscr{A}_{\{0\}}$ has a unique decomposition of the form

$$A = \sum_{i=1}^{N^2} C_i(A) \, a_i , \qquad C_i(A) \in \mathbb{C} .$$

Further, if $A \in \mathscr{A}_{\{x_1, \ldots, xn\}}$, then $A$ has a unique decomposition as a polynomial

$$A = \sum_{i_1=1}^{N^2} \cdots \sum_{in=1}^{N^2} e(i_1, \ldots, i_n; A) \prod_{j=1}^{n} \tau_{x_j}(a_{i_j}) , \qquad e \in \mathbb{C} .$$

Next, with $B \in \mathscr{A}_{\{0\}}$ fixed, consider

$$C_i(x,t) = [\tau_t^{\Phi} \tau_x(a_i), B]$$

and

$$F_i(x,t) = \| C_i(x,t) \| .$$

From the definition of the time translation automorphisms and their properties cited in the previous section, one obtains the differential equations

$$\frac{d}{dt} C_i(x,t) = i \sum_{X \ni 0} [\tau_t^{\Phi} \tau_x([\Phi(X), a_i]), B] .$$

For each $X$ in this sum, the corresponding $\Phi(X)$ can be written as a polynomial in the set of elements $\tau_y(a_j)$, $y \in X$, $j = 1, 2, \ldots, N^2$. The commutator $D_i(X) = [\Phi(X), a_i]$ is then a polynomial of the same kind. Each monomial, $M$, in $D_i(X)$ is of the form

$$M = \prod_{y \in X} \tau_y(a_{j(y)}) .$$

Hence

$$\tau_t^{\Phi} \tau_x(M) = \prod_{y \in X} \tau_t^{\Phi} \tau_{x+y}(a_{j(y)}) .$$

The commutator $[\tau_t^{\Phi} \tau_x(M), B]$ will have $N(X)$ terms (the number of points in $X$), each obtained by taking the commutator $[\tau_t^{\Phi} \tau_{x+y}(a_{i(y)}), B]$ and leaving the other elements in $\tau_t^{\Phi} \tau_x(M)$ as coefficients. Each of these coefficients has norm one (the $a_j$ have norm one and automorphisms preserve the norm). Hence

$$\| [\tau_t^{\Phi} \tau_x(M), B] \| \leqq \sum_{y \in X} \| [\tau_t^{\Phi} \tau_{x+y}(a_{j(y)}), B] \|$$

and

$$\begin{aligned} \left\| \frac{dC_i(x,t)}{dt} \right\| &\leqq \sum_{y \in R^{\Phi}} \sum_{j=1}^{N^2} d_{ji}(\Phi; y) F_j(x+y,t) \\ &\leqq \sum_{y \in R^{\Phi}} \sum_{j=1}^{N^2} d_{ji}(\Phi) F_j(x+y,t) \\ &= (\mathscr{L} F)_i (x,t) \end{aligned}$$

where $d_{ji}(\Phi; y)$ is a non-negative coefficient that depends on the interaction $\Phi$ and the point $y$, whilst $d_{ji}(\Phi) = \max_{y \in R_{\Phi}} d_{ji}(\Phi; y)$. Using the triangle inequality it is easy to verify that

$$\left\| \frac{dC_i(x,t)}{dt} \right\| \geqq \mathscr{D}_t F_i(x,t) ,$$

where $\mathscr{D}_t$ denotes the upper derivative, i.e.

$$(\mathscr{D}_t f)(t) = \limsup_{\varepsilon \to 0} \frac{f(t+\varepsilon) - f(t)}{\varepsilon} .$$

Thus $F_i$ satisfies the differential inequality

$$(\mathscr{D}_t F)_i(x,t) \leqq (\mathscr{L} F)_i(x,t)$$

and the initial data

$$F_i(x,0) = \delta_{x,0}\,\omega_i$$

where

$$\omega_i = \| [a_i, B] \| .$$

[For the sequel note that we could have used the analyticity of $\tau_t^{\Phi}$, etc. to deduce the partial difference inequality

$$\frac{F_i(x,t+h) - F_i(x,t)}{h} \leqq (\mathscr{L} F)_i(x,t) + hK$$

for $h$ small, where $K$ is a positive constant. This would have avoided the introduction of the upper derivative to circumvent the possible non-differentiability of $F_i$.]

Next, consider the Green's function $G_i^k(x,t)$ defined by the differential equations

$$\frac{\partial G_i^k(x,t)}{\partial t} = (\bar{\mathscr{L}} G^k)_i(x,t)$$

and

$$G_j^k(x,0) = \delta_{x,0}\,\delta_{j,k}$$

($\bar{\mathscr{L}}$ is the adjoint of $\mathscr{L}$).

Subsequently we shall use the non-negativity of the coefficients $d_{ij}(\Phi)$ to conclude that the $G_i^k$ are non-negative, for all $t > 0$, $x$, $i$ and $k$, that they are real analytic in $t$, and that there exists a constant $V_\Phi > 0$, and an increasing function $\mu(v) > 0$ such that

$$\lim_{\substack{t\to\infty \\ |x| > vt}} e^{\mu(v)t}\, G_i^k(x,t) = 0\,, \qquad v > V_\Phi\,.$$

For the moment we accept these properties and reconsider the differential inequalities satisfied by the $F_i$. We deduce from the non-negativity of the $G_i^k$ that

$$\sum_{i=1}^{N^2} \sum_{x\in\mathbf{Z}^\nu} \int_0^{t'} dt\, G_i^k(y-x, t'-t)\, (\mathscr{D}_t F)_i(x,t)$$
$$\leqq \sum_{i=1}^{N^2} \sum_{x\in\mathbf{Z}^\nu} \int_0^{t'} dt\, G_i^k(y-x, t'-t)\, (\mathscr{L} F)_i(x,t)\,.$$

Upon integrating by parts on the left hand side, a process which is legitimized by the real analyticity of $G_i^k$ in $t$ and the interpretation of the differential inequality as a partial difference inequality, we find that

$$0 \leqq F_k(x,t) \leqq \sum_i G_i^k(x,t)\,\omega_i$$

which concludes the proof for $t\to\infty$. The limit $t\to-\infty$ is easily handled by noting that $\tau^{\Phi}_{-t}=\tau_t^{-\Phi}$ and the differential inequality that we have derived is invariant under interchange of $\Phi$ by $-\Phi$.

The task remains of corroborating our assertions concerning the Greens functions. These functions can be explicitly calculated in matrix form and one finds:

$$G(x,t)=\frac{1}{(2\pi)^\nu}\int_0^{2\pi} d\theta_1 \ldots \int_0^{2\pi} d\theta_\nu e^{i\theta\cdot x}\exp\{t\omega(\theta)\,D\}$$

where $D$ and $G$ are matrices whose $(i,j)$ elements are respectively $d_{ij}$ and $G^i_j(x,t)$ and

$$\omega(\theta)=\sum_{y\in R_\Phi} e^{+i\theta\cdot y}.$$

The analyticity of $G$ follows from this formula. The non-negativity is best seen from the differential equation which gives directly

$$G^k_i(x,0)\geqq 0\,,\quad \frac{dG^k_i}{dt}(x,0)\geqq 0,\ldots,\frac{d^n G^k_i}{dt^n}(x,0)\geqq 0\,,$$

Next, note that for all $y\in R^\nu$

$$|\omega(\theta+i\gamma)|\leqq\omega(i\gamma)=\sum_{y\in R_\Phi} e^{-\gamma\cdot y},$$

and that $\omega$ has period $2\pi$ in each of the coordinates $\theta_i$. Thus by contour integration we have

$$G(vt,t)=\frac{1}{(2\pi)^\nu}\int_0^{2\pi} d\theta_1 \ldots \int_0^{2\pi} d\theta_\nu e^{i(\theta+i\gamma)\cdot vt}\exp\{t\omega(\theta+i\gamma)\,D\}$$

and

$$\|G(vt,t)\|\leqq e^{-\gamma\cdot vt}\exp\{t\omega(i\gamma)\,\|D\|\}\,.$$

If $|v|$ is sufficiently large there is certainly a $\gamma$ such that

$$\gamma\,.\,v\geqq\|D\|\,\omega(i\gamma)$$

and the asymptotic properties of $G$ follow immediately.

*Remark 1.* The conclusions of the above theorem can be established, by a simple refinement of the above argument, for a class of infinite range interactions with exponential decrease. An example of such an interaction would be a two-body interaction (i.e. $\Phi(X)=0$ unless $N(X)=2$) with the property that

$$\|\Phi(\{0,x\})\|\leqq e^{-a|x|}\,,\qquad a>0\,.$$

It is also clear that the argument can be refined to apply to interactions which are more slowly decreasing, but in this case the commutators will not tend to zero exponentially. Instead their decrease will be linked to that of the interaction.

*Remark 2.* Our method of estimating the group velocity $V_\Phi$ is crude, but nevertheless yields an upper bound that is not difficult to calculate in specific cases, e.g. the nearest neighbour Heisenberg model. One could systematically improve our method of calculating $V_\Phi$. To do so one would prescribe a hierarchy of commutators; the first category would include all two point commutators, the second all three point commutators, etc. In this way one obtains a set of coupled diffusion equations with the property that truncation at any stage, and estimation of the above type, gives an upper bound for $V_\Phi$. It is interesting to ask whether these successive approximations actually converge to the true $V_\Phi$.

Finally for each finite range interaction $\Phi$, $\mathscr{A}$ is asymptotically abelian for the group of space-time automorphisms $(x, t) \in \mathbb{Z}^\nu x R \to \tau_x \tau_t^\Phi$ in the cone $C_\Phi = \{(x, t); |x| > V_\Phi |t|\}$, i.e.

$$\lim_{\substack{|t| \to \infty \\ |x| > V_\Phi |t|}} \| [\tau_x \tau_t^\Phi(A), B] \| = 0$$

for all $A, B \in \mathscr{A}$. This follows because the $A$ and $B$ can be approximated in norm by strictly local elements $A_L$ and $B_L$ of $\mathscr{A}$ and further

$$\| \tau_x \tau_t^\Phi(A - A_L) \| = \| A - A_L \| , \quad \text{etc.}$$

for all $(x, t) \in \mathbb{Z}^\nu x R$.

*Acknowledgments.* This work was undertaken whilst one of us (D. W. Robinson) was a guest of the Department of Physics, Harvard University. The hospitality of Profs. S. Coleman, S. Glashow and A. Jaffe during this period is gratefully acknowledged.

## References

1. Robinson, D. W.: Commun. math. Phys. **6**, 151 (1967).
2. Robinson, D. W.: Commun. math. Phys. **7**, 337 (1968). — See also; Streater, R. F.: Commun. math. Phys. **7**, 93 (1968). — Ruskai, M. B.: Commun. math. Phys. **20**, 193 (1971).

D. W. Robinson
Centre de Physique Théorique
C.N.R.S.
31, chemin J. Aiguier
F-13 Marseille 9°, France

Commun. math. Phys. 31, 327—340 (1973)


# The Classical Limit of Quantum Spin Systems

Elliott H. Lieb*

Institut des Hautes Etudes Scientifiques, Bures-sur-Yvette, France

Received February 28, 1973

**Abstract.** We derive a classical integral representation for the partition function, $Z^Q$, of a quantum spin system. With it we can obtain upper and lower bounds to the quantum free energy (or ground state energy) in terms of two classical free energies (or ground state energies). These bounds permit us to prove that when the spin angular momentum $J \to \infty$ (but after the thermodynamic limit) the quantum free energy (or ground state energy) is equal to the classical value. In normal cases, our inequality is $Z^C(J) \leqq Z^Q(J) \leqq Z^C(J+1)$.

## I. Introduction

It is generally believed in statistical mechanics that if one takes a quantum spin system of $N$ spins, each having angular momentum $J$, normalizes the spin operators by dividing by $J$, and takes the limit $J \to \infty$, then one obtains the corresponding classical spin system wherein the spin variables are replaced by classical vectors and the trace is replaced by an integration over the unit sphere. Indeed, Millard and Leff [1] have shown this to be true for the Heisenberg model *when N is held fixed.* Their proof is quite complicated and it is therefore not surprising that this goal was not achieved before 1971. Despite that success, however, the problem is not finished. One wants to show that one can interchange the limit $N \to \infty$ with the limit $J \to \infty$, i.e. is the classical system obtained if we first let $N \to \infty$ and then let $J \to \infty$? In the Millard-Leff proof the control over the $N$ dependence of the error is not good enough to achieve this desideratum.

A more useful result, and one which would include the above, would be to obtain, for each $J$, upper and lower bounds to the quantum free energy in terms of the free energies of two classical systems such that those two bounds have a common classical limit as $J \to \infty$. In this paper we do just that, and the result is surprisingly simple: In most cases of interest (including the Heisenberg model), the classical upper bound is

* On leave from the Department of Mathematics, M.I.T., Cambridge, Mass. 02139, USA. Work partially supported by National Science Foundation Grant GP-31674X and by a Guggenheim Memorial Foundation Fellowship.

Commun. Math. Phys. *31*, 327–340 (1973)



obtained by replacing the quantum spin by $(J+1)$ times the classical unit vector, while the lower bound is obtained by using $J$ instead of $(J+1)$. Symbolically,

$$Z^C(J) \leqq Z^Q(J) \leqq Z^C(J+1). \tag{1.1}$$

In other cases the result is a little more complicated to state, but it is of the same nature. With an upper and lower bound in hand, it is then possible to derive rigorous bounds on expectation values, as we shall describe in Sections V and VI.

The main tool in our derivation will be what has been termed by Arrechi *et al.* [2] the Bloch coherent state representation. These states and some of their properties were obtained earlier [3, 4], but the most complete account is in Ref. [2]. Our lower bound is obtained by a variational calculation, while the upper bound is obtained from a representation of the quantum partition function that bears some similarity to the Wiener (or path) integral. Apart from its use in deriving the upper bound, the representation may be of theoretical value in proving other properties of quantum spin systems. In particular, it provides a sensible definition of the quantum partition function for all complex $J$, not just when $J$ is half an integer, and one may discuss the existence or non-existence of a phase transition as a function of the continuous parameter $J$.

In a forthcoming paper [7] it will be shown how to apply the methods and bounds developed herein (using not only the Bloch states but the Glauber coherent photon states as well) to certain models of the interaction of atoms with a quantized radiation field, for example the Dicke Maser model.

## II. Bloch Coherent States

In this section we recapitulate results derived in Refs. [2] and [3]. We consider a single quantum spin of fixed total angular-momentum and shall denote by $S \equiv (S_x, S_y, S_z)$ the usual angular momentum operators:

$$[S_x, S_y] = iS_z, \quad \text{and cyclically,} \quad S_\pm = S_x \pm iS_y. \tag{2.1}$$

We denote by $J$ the total angular momentum, i.e.

$$S^2 = S_x{}^2 + S_y{}^2 + S_z{}^2 = J(J+1). \tag{2.2}$$

The Hilbert space on which these operators act has dimension $2J+1$, i.e. it is $\mathbb{C}^{2J+1}$.

On the classical side, we denote by $\mathscr{S}$ the unit sphere in three dimensions:

$$\mathscr{S} = \{(x, y, z) \mid x^2 + y^2 + z^2 = 1\}, \tag{2.3}$$

and by $L^2(\mathscr{S})$ the space of square integrable functions on $\mathscr{S}$ with the usual measure

$$\Omega = (\theta, \varphi), \ 0 \leqq \theta \leqq \pi, \ 0 \leqq \varphi < 2\pi, \tag{2.4}$$

$$d\Omega = \sin\theta \, d\theta \, d\varphi, \tag{2.5}$$

$$x = \sin\theta \cos\varphi, \ y = \sin\theta \sin\varphi, \ z = \cos\theta. \tag{2.6}$$

(Note: In Ref. [2], but not Ref. [3] the "south pole", instead of the customary "north pole" corresponds to $\theta = 0$. Hence our formulas will differ from Ref. [2] by the replacement $\theta \to \pi - \theta$).

With $|J\rangle \in \mathbb{C}^{2J+1}$ being a normalized "spin up" state, $S_z|J\rangle = J|J\rangle$, one defines the Bloch state $|\Omega\rangle \in \mathbb{C}^{2J+1}$ by

$$\begin{aligned} |\Omega\rangle &= \exp\{\tfrac{1}{2}\theta[S_- e^{i\varphi} - S_+ e^{-i\varphi}]\} \, |J\rangle \\ &= [\cos\tfrac{1}{2}\theta]^{2J} \exp\{(\tan\tfrac{1}{2}\theta)\, e^{i\varphi} S_-\} \, |J\rangle \\ &= \sum_{M=-J}^{J} \binom{2J}{M+J}^{1/2} (\cos\tfrac{1}{2}\theta)^{J+M} (\sin\tfrac{1}{2}\theta)^{J-M} \exp[i(J-M)\varphi] \, |M\rangle \end{aligned} \tag{2.7}$$

where $|M\rangle$ is the normalized state

$$|M\rangle = \binom{2J}{M+J}^{-1/2} [(J-M)!]^{-1} (S_-)^{J-M} |J\rangle \tag{2.8}$$

such that

$$S_z|M\rangle = M|M\rangle. \tag{2.9}$$

It is clear from (2.7) that the set of states $|\Omega\rangle$ are complete in $\mathbb{C}^{2J+1}$. Their overlap is given by

$$\begin{aligned} K_J(\Omega', \Omega) &\equiv \langle \Omega' | \Omega \rangle \\ &= \{\cos\tfrac{1}{2}\theta \cos\tfrac{1}{2}\theta' + e^{i(\varphi-\varphi')} \sin\tfrac{1}{2}\theta \sin\tfrac{1}{2}\theta'\}^{2J} \end{aligned} \tag{2.10}$$

so that if we think of $K_J(\Omega', \Omega)$ as the kernel of a linear transformation on $L^2(\mathscr{S})$ it is selfadjoint and compact. In fact, it is positive semidefinite. We also have

$$|K_J(\Omega', \Omega)|^2 = [\cos\tfrac{1}{2}\Theta]^{4J}, \tag{2.11}$$

where

$$\cos\Theta = \cos\theta \cos\theta' + \sin\theta \sin\theta' \cos(\varphi - \varphi') \tag{2.12}$$

is the cosine of the angle between $\Omega$ and $\Omega'$. In particular $|\Omega\rangle$ is normalized since $K_J(\Omega, \Omega) = 1$.

Now let $\mathcal{M}^{2J+1}$ be the set of linear transformations on $\mathbb{C}^{2J+1}$ (i.e. operators on the spin space) and, for a given $G \in L^1(\mathcal{S})$, define $A_G \in \mathcal{M}^{2J+1}$ by

$$A_G = \frac{2J+1}{4\pi} \int d\Omega\, G(\Omega)\, |\Omega\rangle \langle \Omega| . \tag{2.13}$$

$\left(\text{Note: } \int d\Omega \text{ always means } \int_{\mathcal{S}} d\Omega\right)$. Since the Hilbert space is finite dimensional there is no problem in giving a meaning to (2.13). It is a remarkable fact that every operator in $\mathcal{M}^{2J+1}$ can be written in the form (2.13). In particular,

$$1 = \frac{2J+1}{4\pi} \int d\Omega\, |\Omega\rangle \langle \Omega| . \tag{2.14}$$

Thus, to every operator $A \in \mathcal{M}^{2J+1}$ there correspond two functions:

$$g(\Omega) = \langle \Omega| A |\Omega\rangle , \tag{2.15}$$

and the $G(\Omega)$ of (2.13). The former is, of course, unique, but the latter is not. However, it is always possible to choose $G(\Omega)$ to be infinitely differentiable. In Table 1 we list some function pairs for operators of common interest and useful formulas for calculation are given in Appendix A.

Table 1. Expectation values, $g(\Omega)$, and operator kernels, $G(\Omega)$, [cf. (2.13), (2.15)] for various operators commonly appearing in quantum spin Hamiltonians

| Operator | $g(\Omega)$, (2.15) | $G(\Omega)$, (2.13) |
|---|---|---|
| $S_z$ | $J \cos\theta$ | $(J+1)\cos\theta$ |
| $S_x$ | $J \sin\theta \cos\varphi$ | $(J+1)\sin\theta\cos\varphi$ |
| $S_y$ | $J \sin\theta \sin\varphi$ | $(J+1)\sin\theta\sin\varphi$ |
| $S_z^2$ | $J(J-\frac{1}{2})(\cos\theta)^2 + J/2$ | $(J+1)(J+3/2)(\cos\theta)^2 - \frac{1}{2}(J+1)$ |
| $S_x^2$ | $J(J-\frac{1}{2})(\sin\theta\cos\varphi)^2 + J/2$ | $(J+1)(J+3/2)(\sin\theta\cos\varphi)^2 - \frac{1}{2}(J+1)$ |
| $S_y^2$ | $J(J-\frac{1}{2})(\sin\theta\cos\varphi)^2 + J/2$ | $(J+1)(J+3/2)(\sin\theta\cos\varphi)^2 - \frac{1}{2}(J+1)$ |

We need three final remarks. The first is that if we consider $|\Omega\rangle\langle\Omega'| \in \mathcal{M}^{2J+1}$ then

$$\operatorname{Tr} |\Omega\rangle \langle \Omega'| = K_J(\Omega', \Omega) \tag{2.16}$$

(where Tr means Trace) as may be seen from (2.7). Hence, from (2.13)

$$\operatorname{Tr} A_G = \frac{2J+1}{4\pi} \int d\Omega\, G(\Omega)\,. \tag{2.17}$$

The second is that

$$\frac{2J+1}{4\pi} \int d\Omega\, K_J(\Omega', \Omega)\, K_J(\Omega, \Omega'') = K_J(\Omega', \Omega'')\,, \tag{2.18}$$

as may be seen from (2.14). Thus, $K_J$ reproduces itself under convolution.

The third remark is that for any $A \in \mathscr{M}^{2J+1}$ we can use (2.14) to obtain

$$\begin{aligned} \operatorname{Tr} A &= \frac{2J+1}{4\pi} \int d\Omega \operatorname{Tr} |\Omega\rangle \langle\Omega| A \\ &= \frac{2J+1}{4\pi} \int d\Omega \sum_{M=-J}^{J} \langle M|\Omega\rangle \langle\Omega| A |M\rangle \\ &= \frac{2J+1}{4\pi} \int d\Omega\, \langle\Omega| A |\Omega\rangle\,. \end{aligned} \tag{2.19}$$

## III. Lower Bound to the Quantum Partition Function

We consider a system of $N$ quantum spins and shall label the operators and the angular momenta (which need not all be the same) by a superscript $i$, $i = 1, \ldots, N$. The Hamiltonian, $H$, can be completely general but, in any event, it can always be written as a polynomial in the $3N$ spin operators. The partition function is

$$Z^Q = \alpha_N \operatorname{Tr} \exp(-\beta H)\,, \tag{3.1}$$

where

$$\alpha_N = \prod_{i=1}^{N} (2J^i + 1)^{-1}\,. \tag{3.2}$$

[The normalization factor $\alpha_N$ is inessential; it is chosen to agree with the classical partition function when $\beta = 0$]. The Hilbert space is

$$\mathscr{H}_N = \bigotimes_{i=1}^{N} \mathscr{H}^i = \bigotimes_{i=1}^{N} \mathbb{C}^{2J^i+1}\,. \tag{3.3}$$

We denote by $|\Omega_N\rangle$ the complete, normalized set of states on $\mathscr{H}_N$ defined by

$$|\Omega_N\rangle = \bigotimes_{i=1}^{N} |\Omega^i\rangle\,, \tag{3.4}$$

Commun. Math. Phys. *31*, 327–340 (1973)



by $\mathcal{S}_N$ the Cartesian product of $N$ copies of the unit sphere, and by $d\Omega_N$ the product measure (2.4), (2.5) and (2.6) on $\mathcal{S}_N$. Using (2.19),

$$Z^Q = (4\pi)^{-N} \int d\Omega_N \langle \Omega_N | e^{-\beta H} | \Omega_N \rangle . \tag{3.5}$$

By the Peierls-Bogoliubov inequality, $\langle \psi | e^X | \psi \rangle \geqq \exp \langle \psi | X | \psi \rangle$ for any normalized $\psi \in \mathcal{H}_N$ and $X$ selfadjoint. Thus,

$$Z^Q \geqq (4\pi)^{-N} \int d\Omega_N \exp\{-\beta \langle \Omega_N | H | \Omega_N \rangle\} . \tag{3.6}$$

Suppose, at first, that the polynomial, $H$, is linear in the operators $\boldsymbol{S}^i$ of each spin. That is we allow multiple site interactions of arbitrary complexity such as $S_x^{\,1} S_y^{\,2} S_y^{\,3} S_z^{\,4}$, but do not allow monomials such as $(S_x^{\,1})^2$ or $S_x^{\,1} S_y^{\,1}$. In this case, which we shall refer to as *the normal case*, we see from (2.15) and Table 1 that the right side of (3.6) is precisely the classical partition function in which each $\boldsymbol{S}^i$ is replaced by $J^i$ times a vector in $\mathcal{S}$. I.e.

$$\boldsymbol{S}^i \to J^i (\sin\theta^i \cos\varphi^i, \sin\theta^i \sin\varphi^i, \cos\theta^i) . \tag{3.7}$$

Thus, in the normal case,

$$Z^Q \geqq Z^C(J^1, \ldots, J^N) , \tag{3.8}$$

where $Z^C$ means the classical partition function (with the normalization $(4\pi)^{-N}$).

In more complicated cases, (3.7) is not correct and $S_z^{\,1}$, for example, has to be replaced by $J^1 \cos\theta^1$ if it appears linearly in $H$, $(S_z^{\,1})^2$ has to be replaced by $[J^1 \cos\theta^1]^2 + J^1 (\sin\theta^1)^2/2$ and so forth (see Table 1). However, to leading order in $J^i$, (3.7) is correct.

We note in passing that it is not necessary to use the Peierls-Bogoliubov inequality for all operators appearing in $H$. Thus, suppose the whole Hilbert space is $\mathcal{H}' = \mathcal{H} \otimes \mathcal{H}_N$ where $\mathcal{H}$ is the Hilbert space of some additional degrees of freedom (which may or may not themselves be spins) and $H$ is selfadjoint on $\mathcal{H}'$. Then (by a generalized Peierls-Bogoliubov inequality)

$$\begin{aligned} Z^Q &= \alpha_N \operatorname{Tr}_{\mathcal{H}'} \operatorname{Tr}_{\mathcal{H}} \exp(-\beta H) \\ &\geqq \operatorname{Tr}_{\mathcal{H}} (4\pi)^{-N} \int d\Omega_N \exp\{-\beta \langle \Omega_N | H | \Omega_N \rangle\} \end{aligned} \tag{3.9}$$

where $\langle \Omega_N | H | \Omega_N \rangle$ is a partial expectation value and defines a selfadjoint operator on $\mathcal{H}$. We shall give an example of (3.9) in Appendix B. It is clear that if $\mathcal{H}$ is itself a spin space, then (3.9) gives a better bound than (3.6) applied to the full space $\mathcal{H}'$.

## IV. Upper Bound to the Quantum Partition Function

Returning to the definitions (3.1) and (3.3) we note that

$$Z^Q = \lim_{n\to\infty} Z(n)\,, \tag{4.1}$$

where

$$Z(n) = \alpha_N \operatorname{Tr}(1 - \beta n^{-1} H)^n\,. \tag{4.2}$$

Now, let $H$ be represented by some $G(\Omega_N)$ as in (2.13), whence $1 - \beta n^{-1} H$ is represented by

$$F_n(\Omega_N) = 1 - \beta n^{-1} G(\Omega_N)\,. \tag{4.3}$$

Using (2.10), (2.13) and (2.16), we can represent $Z_n$ as an $nN$ fold integral:

$$Z(n) = \alpha_N \int d\Omega_N^{\,1} \cdots \int d\Omega_N^{\,n} \prod_{j=1}^{n} F_n(\Omega_N^{\,j})\, L_J(\Omega_N^{\,j}, \Omega_N^{\,j+1}) \tag{4.4}$$

with $n+1 \equiv 1$ in the last factor, and where

$$L_J(\Omega_N', \Omega_N) \equiv (4\pi)^{-N} \alpha_N^{-1} \prod_{i=1}^{N} K_{J^i}(\Omega'^{\,i}, \Omega^i)\,. \tag{4.5}$$

Thus

$$L_J(\Omega_N, \Omega_N) = (4\pi)^{-N} a_N^{-1}\,. \tag{4.6}$$

$$\int d\Omega_N\, L_J(\Omega_N', \Omega_N)\, L_J(\Omega_N, \Omega_N'') = L_J(\Omega_N', \Omega_N'')\,. \tag{4.7}$$

Equations (4.1) and (4.4) are our desired integral representation for $Z^Q$. To use them to obtain a bound, we think of $F_n$ as a multiplication operator and of $L_J$ as the kernel of a compact, selfadjoint operator on $L^2(\mathscr{S}_N)$. If $B(\Omega_N', \Omega_N)$ is such a kernel, then

$$\operatorname{Tr} B = \int d\Omega_N\, B(\Omega_N, \Omega_N) \tag{4.8}$$

is the trace on $L^2(\mathscr{S}_N)$. Thus,

$$Z(n) = \alpha_N \operatorname{Tr}(F_n L_J)^n\,. \tag{4.9}$$

In general, if $m = 2^j$, $j = 0, 1, 2, 3, \ldots$,

$$|\operatorname{Tr}(AB)^{2m}| \leq \operatorname{Tr}(A^2 B^2)^m \leq \operatorname{Tr} A^{2m} B^{2m}\,, \tag{4.10}$$

whenever $A$ and $B$ are selfadjoint. This follows from the Schwarz inequality (see Ref. [5] for details). Hence, if we take a sequence $n = 2^j$, $j = 1, 2, \ldots$ in (4.2) and use (4.7) $n$ times and (4.6), we obtain, in the limit

Commun. Math. Phys. 31, 327–340 (1973)



$n \to \infty$,

$$Z^Q \leqq (4\pi)^{-N} \int d\Omega_N \exp[-\beta G(\Omega_N)]. \tag{4.11}$$

(4.11) is our desired classical upper bound. It is just like (3.6). In the normal case we see from Table 1 that $\boldsymbol{S}^i$ is replaced by $(J^i+1)$ times a classical unit vector. In other cases, $G(\Omega_N)$ is a bit more complicated, but the same remarks as in Section III apply. Thus, in the normal case

$$Z^C(J^1, \ldots, J^N) \leqq Z^Q \leqq Z^C(J^1+1, \ldots, J^N+1). \tag{4.12}$$

This inequality says that as $J$ increases the quantum and classical free energies form two decreasing, *interlacing* sequences.

As in Section III, if $\mathscr{H}' = \mathscr{H} \otimes \mathscr{H}_N$ an inequality similar to (4.11) can be shown to hold, i.e.

$$Z^Q \leqq \mathrm{Tr}_{\mathscr{H}} (4\pi)^{-N} \int d\Omega_N \exp[-\beta H(\cdot, \Omega_N)], \tag{4.13}$$

where $H(\cdot, \Omega_N)$ is a selfadjoint operator on $\mathscr{H}$ obtained by replacing each monomial in the spin operators in $H$ by the appropriate $G(\Omega_N)$ function found in Table 1. We shall illustrate (4.13) in Appendix B. If $\mathscr{H}$ is a spin space then (4.13) gives a better bound than (4.11) applied to the full $\mathscr{H}'$.

## V. Bounds on Expectation Values and the Ground State Energy

The expectation value of a quantum operator (observable), $A$, is

$$\langle A \rangle^Q = \mathrm{Tr}\, A \exp(-\beta H)/\mathrm{Tr} \exp(-\beta H). \tag{5.1}$$

We can always assume $A$ is selfadjoint (otherwise consider $A + A^\dagger$ and $iA - iA^\dagger$), in which case the Peierls-Bogoliubov inequality reads, for $\lambda$ real,

$$\lambda \langle A \rangle^Q \geqq f(\lambda) - f(0), \tag{5.2}$$

where

$$f(\lambda) = -\beta^{-1} \ln \mathrm{Tr} \exp[-\beta(H + \lambda A)], \tag{5.3}$$

is a free energy. Hence, with $\lambda > 0$,

$$[f(0) - f(-\lambda)]/\lambda \geqq \langle A \rangle^Q \geqq [f(\lambda) - f(0)]/\lambda. \tag{5.4}$$

The upper and lower bounds to $f(\lambda)$ derived in the preceding two sections can be used to advantage in (5.4). In particular, we use (5.4) in the next section to derive $J \to \infty$ limits of quantum expectation values.

If we take the limit $\beta \to \infty$ in (3.1) we obtain bounds on the quantum ground state energy:

$$E_-^C \leqq E^Q \leqq E_+^C, \tag{5.5}$$

where $E^C$ is the classical ground state energy (i.e. the minimum of the classical Hamiltonian over $\mathscr{S}_N$) and the + (resp. −) refers to the substitution of the appropriate $G(\Omega_N)$ (resp. $g(\Omega_N)$) functions from Table 1. In the normal case

$$E^C(J^1, \ldots, J^N) \geqq E^Q \geqq E^C(J^1+1, \ldots, J^N+1). \tag{5.6}$$

As ground state expectation values obey an inequality similar to (5.2), with $f$ replaced by $E$, a bound similar to (5.4) holds for $E$. This is merely the variational principle.

The upper bound in (5.6) is easy to obtain directly by a variational calculation, but the lower bound is not. It is not easy to find a direct proof of it in a system consisting of three spins antiferromagnetically coupled to each other.

## VI. The Thermodynamic Limit

### A. The Free Energy

We shall, for simplicity, consider only the normal case here. The general case can be handled in a similar manner.

Let $H_N$ be a Hamiltonian (polynomial) of $N$ spins in which each spin has angular momentum one. Replace each spin operator $\boldsymbol{S}^i$ by $(J)^{-1}\boldsymbol{S}^i$ and let $\boldsymbol{S}^i$ now have angular momentum $J$. We shall denote this symbolically by $H_N^Q(J)$ and the partition function, (3.1), by $Z_N^Q(J)$. [It would equally be possible to allow different $J$ values for different spins, but that is a needless complication. Also, the factor $J^{-1}$ is not crucial. One could as well use $J^{-1/2}(J+1)^{-1/2}$]. Denoting the free energy per spin by $f_N(J) = -(N\beta)^{-1} \ln Z_N(J)$, the theorem to be proved is that

$$\lim_{J\to\infty} \lim_{N\to\infty} f_N^Q(J) = f^C \equiv \lim_{N\to\infty} f_N^C, \tag{6.1}$$

where $f_N^C$ is the free energy per spin of the classical partition function in which each $\boldsymbol{S}^i$ is replaced by a classical unit vector. It is assumed that $H_N$ is known to have a thermodynamic limit for the free energy per spin. We also want to prove an analogous formula for the ground state energy per spin. Our bounds are

$$f_N^C \geqq f_N^C(J) \geqq f_N^C(\delta_J), \tag{6.2}$$

Commun. Math. Phys. 31, 327–340 (1973)



where the right side is the classical free energy per spin in which each vector is multiplied by $\delta_J \equiv (J+1)/J$.

If we think of $\delta_J$ as a variable, $\delta$, then $H_N^C(\delta)$, the classical Hamiltonian as a function of $\delta$, is continuous in $\delta$. Moreover, $N^{-1} H_N^C(\delta)$ is equicontinuous in $N$, i.e. given any $\varepsilon > 0$ it is possible to find a $\gamma > 0$ such that $\|N^{-1}[H_N^C(\delta+x) - H_N^C(\delta)]\| \leqq \varepsilon$ for $|x| < \gamma$, independent of $N$, where $\| \ \|$ means the uniform on $\mathscr{S}_N$. Hence, the limit function

$$f^C(\delta) \equiv \lim_{N\to\infty} f_N^C(\delta) \tag{6.3}$$

is continuous in $\delta$. This, together with (6.2), proves (6.1).

The same equicontinuity holds for the classical ground state energy. Thus, the analogue of (6.1) is also true for the ground state energy per spin:

$$\lim_{J\to\infty} \lim_{N\to\infty} N^{-1} E_N^Q(J) = \lim_{N\to\infty} E_N^C . \tag{6.4}$$

*B. Expectation Values*

We consider expectation values of intensive observables $N^{-1} A_N$. For example, $A_N$ might be the Hamiltonian itself, in which case $\langle N^{-1} A_N \rangle$ is the energy per spin. Alternatively, $A_N$ could be $\sum_{i=1}^{N} S_z^i$ so that $\langle N^{-1} A_N \rangle$ is the magnetization per spin. As before, we replace each $\boldsymbol{S}^i$ by $(J)^{-1}$ times a quantum spin of angular momentum $J$, *both* in the Hamiltonian and in $A_N$. Then, using inequality (5.4) and the bounds (6.2) we have, for each positive $\lambda$, fixed $N$ and fixed $J$,

$$\begin{aligned} \lambda^{-1}[f_N^C(0;1) - f_N^C(-\lambda;\delta_J)] &\geqq N^{-1}\langle A_N \rangle^Q \\ &\geqq \lambda^{-1}[f_N^C(\lambda;\delta_J) - f_N^C(0;1)] , \end{aligned} \tag{6.5}$$

where $f_N^C(\lambda;\delta)$ is the classical free energy per spin when the Hamiltonian is $H_N^C + \lambda A_N^C$ and where each classical spin unit vector in $H_N^C$ and $A_N^C$ is multiplied by $\delta$. We are interested in $\delta_J = (J+1)/J$.

Now take the limit $N\to\infty$ and then the limit $J\to\infty$ in (6.5). By the same equicontinuity remark as in Section VI.A, for each $\lambda > 0$,

$$\begin{aligned} &\limsup_{J\to\infty} \limsup_{N\to\infty} N^{-1}\langle A_N \rangle^Q \leqq \lambda^{-1}[f^C(0) - f^C(-\lambda)] , \\ &\liminf_{J\to\infty} \liminf_{N\to\infty} N^{-1}\langle A_N \rangle^Q \geqq \lambda^{-1}[f^C(\lambda) - f^C(0)] . \end{aligned} \tag{6.6}$$

In (6.5), $f^C(\lambda)$ is the limiting classical free energy per spin for the Hamiltonian $H_N^C + \lambda A_N^C$ (with $\delta = 1$). It is easy to see that $f^C(\lambda)$ is concave in $\lambda$

and hence $\lim_{\lambda\downarrow 0} \lambda^{-1}[f^C(\lambda)-f^C(0)] \equiv G^+$ and $\lim_{\lambda\downarrow 0} \lambda^{-1}[f^C(0)-f^C(-\lambda)] \equiv G^-$ exist everywhere. If $G^+ = G^-$ (i.e. the right derivative equals the left derivative) then by a theorem of Griffiths (6)

$$\lim_{N\to\infty} \frac{d}{d\lambda} f_N^C(\lambda) = \frac{d}{d\lambda} f^C(\lambda). \tag{6.7}$$

This is the case in which the classical expectation value $N^{-1}\langle A_N\rangle^C$ has a well defined limit. Call it $\alpha$. Then

$$\lim_{J\to\infty} \lim_{N\to\infty} N^{-1}\langle A_N\rangle^Q = \alpha, \tag{6.8}$$

as one sees by taking the limit $\lambda\to 0$ in (6.6). In other words, we have proved that *for intensive observables, as defined above, the quantum expectation value equals the classical expectation value after first taking the thermodynamic limit and then taking the classical limit $J\to\infty$*. If one takes the limits in the opposite order the theorem is trivially true and uninteresting. Note that we have not proved that the quantum thermodynamic limit, $\lim_{N\to\infty} N^{-1}\langle A_N\rangle^Q$ exists. It may not.

The same proof obviously goes through for ground state expectation values, as in Section VI.A, because the ground state energy is also concave in $\lambda$.

*Acknowledgements.* The author thanks the Institut des Hautes Etudes Scientifiques for its hospitality, as well as the Chemistry Laboratory III, University of Copenhagen where part of this work was done. The financial assistance of the Guggenheim Memorial Foundation is gratefully acknowleged. The author also acknowledges his gratitude to Dr. N. W. Dalton who suggested the problem to him in 1967.

## Appendix A: Some Useful Formulas

The algebra $\mathscr{M}^{2J+1}$ has $S_+$, $S_-$ and $S_z$ as generators. Hence, the following generating function permits, by differentiation, easy calculation of $g(\Omega)$ in (2.15) or Table 1 for any operator. It is to be found, with appropriate modifications, in Ref. [2].

$$\begin{aligned}&\langle\Omega|\exp(\gamma S_+)\exp(\beta S_z)\exp(\alpha S_-)|\Omega\rangle \\ &= \{[e^{-i\varphi}\sin\tfrac{1}{2}\theta + \gamma\cos\tfrac{1}{2}\theta][e^{i\varphi}\sin\tfrac{1}{2}\theta + \alpha\cos\tfrac{1}{2}\theta]\,e^{-\beta/2} + e^{\beta/2}[\cos\tfrac{1}{2}\theta]^2\}^{2J}.\end{aligned} \tag{A.1}$$

Turning to (2.13), we calculate $A_G$ for a sufficiently large class of functions $G(\Omega)$. Let

$$G(\Omega) = e^{im\varphi}(\cos\tfrac{1}{2}\theta)^p(\sin\tfrac{1}{2}\theta)^q \tag{A.2}$$

Commun. Math. Phys. 31, 327–340 (1973)



where $m$ is an integer and $p$ and $q$ are complex numbers. Defining $A(m, p, q) \equiv A_G$, the matrix elements of this operator can be calculated using (2.7) to be

$$A(m, p, q; M, M') = \delta(M - M' - m)\,\Gamma(J+\alpha+1+p/2)\,\Gamma(J-\alpha+1+q/2)$$
$$\cdot\,[(J+\alpha+m/2)!\,(J+\alpha-m/2)!\,(J-\alpha-m/2)!\,(J-\alpha+m/2)!]^{-1/2} \quad \text{(A.3)}$$
$$\cdot\,(2J+1)!/\Gamma(2J+2+p/2+q/2),$$

where $\delta$ is the Kroenecker delta function, $\Gamma$ is the gamma function and $\alpha = (M + M')/2$. This formula has been used to calculate Table 1.

## Appendix B: Application to the One Dimensional Heisenberg Chain

To illustrate the methods of this paper, we derive bounds for the free energy of a Heisenberg chain whose Hamiltonian is

$$H = -\sum_{i=1}^{N-1} \boldsymbol{S}^i \cdot \boldsymbol{S}^{i+1}. \quad \text{(B.1)}$$

Each spin is assumed to have angular momentum $J$. We have chosen the isotropic case for simplicity, but one could equally well handle the anisotropic Hamiltonian with a magnetic field. Note that $\beta > 0$ is the ferromagnetic case while $\beta < 0$ is the antiferromagnetic case.

The classical partition function is

$$Z_N^C(\beta, x) = (4\pi)^{-N} \int d\Omega_N \exp\left\{\beta x^2 \sum_{i=1}^{N-1} \boldsymbol{\Omega}^i \cdot \boldsymbol{\Omega}^{i+1}\right\} \quad \text{(B.2)}$$

with free energy per spin

$$f^C(\beta, x) = -\lim_{N\to\infty} (N|\beta|)^{-1} \ln Z_N^C(\beta, x). \quad \text{(B.3)}$$

Our bounds are that

$$f^C(\beta, J) \geqq f^Q(\beta, J) \geqq f^C(\beta, J+1). \quad \text{(B.4)}$$

It is easy to evaluate (B.2) by the transfer matrix method. The normalized eigenfunction (of $\Omega$) giving the largest eigenvalue is obviously the constant function $(4\pi)^{-1/2}$. Thus,

$$f^C(\beta, x) = -|\beta|^{-1} \ln A(\beta, x), \quad \text{(B.5)}$$

where

$$\begin{aligned} A(\beta, x) &= (4\pi)^{-1} \int d\Omega \exp\{\beta x^2 \boldsymbol{\Omega} \cdot \boldsymbol{\Omega}'\} \\ &= (\beta x^2)^{-1} \sinh(\beta x^2), \end{aligned} \tag{B.6}$$

and $A(\beta, x)$ is independent of $\Omega'$ as it should be. In this approximation, (B.4), one cannot distinguish between the ferro- and antiferromagnetic cases as far as the free energy is concerned.

To illustrate the idea mentioned at the ends of Sections III and IV, we suppose that the chain has $2N+1$ spins and we let $\mathscr{H}_N$ (resp. $\mathscr{H}$) be the Hilbert space for the odd (resp. even) numbered spins. $\mathscr{H}' = \mathscr{H} \otimes \mathscr{H}_N$ is the whole space. Our bounds are

$$g(\beta, J) \geqq f^Q(\beta, J) \geqq g(\beta, J+1), \tag{B.7}$$

where

$$g(\beta, x) = -\lim_{N\to\infty} (2N|\beta|)^{-1} \ln\{(2J+1)^{-N} \tilde{Z}_N(\beta, x)\}, \tag{B.8}$$

$$\tilde{Z}_N(\beta, x) = (4\pi)^{-N} \int d\Omega_N \operatorname{Tr} \exp\left\{\beta x \sum_{i=1}^{N} \boldsymbol{S}^{2i} \cdot (\boldsymbol{\Omega}^{2i-1} + \boldsymbol{\Omega}^{2i+1})\right\} \tag{B.9}$$

and where $d\Omega_N = d\Omega^1\, d\Omega^3 \ldots d\Omega^{2N+1}$ and the trace is over the Hilbert space of $\boldsymbol{S}^2, \boldsymbol{S}^4, \ldots, \boldsymbol{S}^{2N}$.

Since the remaining spin operators no longer interact, it is easy to calculate the trace. For a single spin:

$$\operatorname{Tr} \exp[b\boldsymbol{S} \cdot \boldsymbol{v}] = \sum_{M=-J}^{J} \exp[bMv] \tag{B.10}$$

where $b$ is a constant and $\boldsymbol{v}$ is a vector of length $v$. Now we can do the integration over $\mathscr{S}_N$ by the transfer matrix method (with the same eigenvector $(4\pi)^{-1/2}$) and obtain

$$g(\beta, x) = -\tfrac{1}{2}|\beta|^{-1} \ln[A(\beta, x)/(2J+1)], \tag{B.11}$$

where

$$\begin{aligned} A(\beta, x) &= (4\pi)^{-1} \int d\Omega \sum_{M=-J}^{J} \exp\{\beta x M |\boldsymbol{\Omega} + \boldsymbol{\Omega}'|\} \\ &= 2\int_0^1 y\, dy \sinh[(2J+1)\beta x y]/\sinh[\beta x y]. \end{aligned} \tag{B.12}$$

Again, no distinction between the ferro- and antiferromagnetic cases appears.

Commun. Math. Phys. *31*, 327–340 (1973)



## References

1. Millard, K., Leff, H.: J. Math. Phys. **12**, 1000—1005 (1971).
2. Arecchi, F. T., Courtens, E., Gilmore, R., Thomas, H.: Phys. Rev. A**6**, 2211—2237 (1972).
3. Radcliffe, J. M.: J. Phys. A**4**, 313—323 (1971).
4. Kutzner, J.: Phys. Lett. A**41**, 475—476 (1972).
   Atkins, P. W., Dobson, J. C.: Proc. Roy. Soc. (London) A, A**321**, 321—340 (1971).
5. Golden, S.: Phys. Rev. B**137**, 1127—1128 (1965).
6. Griffiths, R. B.: J. Math. Phys. **5**, 1215—1222 (1964).
7. Hepp, K., Lieb, E. H.: The equilibrium statistical mechanics of matter interacting with the quantized radiation field. Preprint.

E. H. Lieb
I.H.E.S.
F-91440 Bures-sur-Yvette, France

Commun. Math. Phys. 77, 127–135 (1980)

Communications in Mathematical Physics



# A Refinement of Simon's Correlation Inequality*

Elliott H. Lieb

Departments of Mathematics and Physics, Princeton University, Princeton, NJ 08544, USA

**Abstract.** A general formulation is given of Simon's Ising model inequality: $\langle\sigma_\alpha\sigma_\gamma\rangle \leqq \sum_{b\in B} \langle\sigma_\alpha\sigma_b\rangle\langle\sigma_b\sigma_\gamma\rangle$ where $B$ is any set of spins separating $\alpha$ from $\gamma$. We show that $\langle\sigma_\alpha\sigma_b\rangle$ can be replaced by $\langle\sigma_\alpha\sigma_b\rangle_A$ where $A$ is the spin system "inside" $B$ containing $\alpha$. An advantage of this is that a finite algorithm can be given to compute the transition temperature to any desired accuracy. The analogous inequality for plane rotors is shown to hold if a certain conjecture can be proved. This conjecture is indeed verified in the simplest case, and leads to an upper bound on the critical temperature. (The conjecture has been proved in general by Rivasseau. See notes added in proof.)

In an accompanying paper [1] in this volume Simon proves a correlation inequality with important consequences. For a finite range pairwise interacting (generalized) Ising ferromagnet (the spins take on values $2M, 2M-2, \ldots, -2M$), Simon shows that

$$\langle\sigma_\alpha\sigma_\gamma\rangle \leqq \sum_{b\in B} \langle\sigma_\alpha\sigma_b\rangle\langle\sigma_b\sigma_\gamma\rangle, \tag{1}$$

where $B$ is any set of spins separating $\alpha$ from $\gamma$ (i.e. any path from $\alpha$ to $\gamma$ must run through $B$). Aizenman and Simon [2] have proved a related inequality for $N$-component spins. In this paper we shall generalize (1) in the following way: $\langle\sigma_\alpha\sigma_b\rangle$ can be replaced by $\langle\sigma_\alpha\sigma_b\rangle_A$, where $A$ is the connected component of the lattice containing $\alpha$ and $B$ and $\langle\cdot\rangle_A$ denotes expectation values in the $A$ system alone. The possibility of extending this inequality to plane rotors is also discussed, but the proof is carried to completion only in a special case. (See notes added in proof.)

In [1] Simon discusses the consequences of (1) and our generalization. We shall not repeat them, except to note that the most interesting consequence of the extension is that for the first time one has an algorithm for computing the transition temperature, $T_c$ (in the sense that above, but not below $T_c$ there is

* Work partially supported by U.S. National Science Foundation grant PHY-7825390 A01

0010-3616/80/0077/0127/$01.80

exponential decay of the two point function $\langle\sigma_0\sigma_x\rangle$), to arbitrary accuracy. Take $\alpha=0$ and let $B$ be the spins on the boundary of a square of side $L$ centered at 0. By boundary we mean all points within a distance $R$ of the geometric boundary, where $R$ is not less than the range of the interaction. The $A$ system is the inside of the square alone. $\langle\sigma_0\sigma_b\rangle_A$ can be computed explicitly, and if

$$\sum_{b\in B}\langle\sigma_0\sigma_b\rangle<1 \tag{2}$$

for some $T$, then there is exponential decay for that $T$. This sets an upper bound to $T_c$. It is easy to see [1], however, that as $L\to\infty$, $T_L$ [the $T$ for which equality holds in (2)] approaches $T_c$. While the convergence of $T_L$ to $T_c$ is expected to be extremely slow, the mere existence of the algorithm is an interesting matter of principle. It is not known if $T_L$ is necessarily monotone decreasing in $L$; this is an open question.

A consequence of our generalization is the continuity of the mass gap as function of the interaction, for nearest neighbor ferromagnetic interactions, proven in [1]. A more general stability of the mass gap, $m$, under perturbations was pointed out by Aizenman (private communication). It is expressed by the *lower semicontinuity of $m$, as function of the interaction, in the cone of pairwise ferromagnetic interactions of any fixed finite range.* This is proven in the following way. Suppose the (finite range) Hamiltonian $H$ is given and $T$ is such that for the infinite system

$$\langle\sigma_0\sigma_x\rangle<C_\varepsilon\exp[(-m+\varepsilon)|x|]$$

for all $x$, all $\varepsilon>0$, but not $\varepsilon<0$. $m$ is then the mass gap and it will be assumed that $m>0$. Given $\varepsilon>0$, it is easy to see that for any $R$ there must be a finite box such that

$$\sum_{b\in B}\langle\sigma_0\sigma_b\rangle\exp[\mu|b|]<1 \tag{3}$$

for $\mu=m-\varepsilon$. Conversely, our generalization of (1) shows that if (3) holds with some $\mu$ for some box, then the mass gap is not less than $\mu$.

Since condition (3) (with $\mu=m-\varepsilon$) refers to a finite system, by continuity it continues to hold (with $\mu=m-2\varepsilon$) when the Hamiltonian is changed from $H$ to $H+K$ and $\|K\|<\delta_0$, for some $\delta_0>0$ and independent of $K$. If we also require that $H+K$ is pairwise ferromagnetic and has range $\leqq R$, then (3) (with $\mu=m-2\varepsilon$) and our generalization of (1) imply that the new mass gap is not less than $m-2\varepsilon$.

Simon's proof of (1) uses a graphical expansion. The analysis presented here will not use this explicitly, but instead will use certain "gaussian correlation inequalities" of Newman [3]. While it is true that Newman's inequalities are themselves proved by graphical means, it is hoped that the decomposition of the problem into the two steps given here will be useful.

Let us begin with some definitions. The system under consideration is viewed as the union of two subsystems of spins $A$ and $C$.

$$A\cap C=B$$

is the set of spins common to both. To say that the $B$ spins separate $A$ from $C$ means that

$$H_{A+C}=H_A+H_C, \tag{4}$$

where the $H$'s are Hamiltonians. The symbols $Z$ denote partition functions, $\langle\cdot\rangle$ denote expectation values and $(\cdot)=Z\langle\cdot\rangle$ denote unnormalized expectation values – all at reciprocal temperature $\beta$. Thus, for example,

$$\begin{aligned}&(\sigma_A)_A=\mathrm{Tr}\,\sigma_A\exp(-\beta H_A)\\&Z_A=(1)_A=\mathrm{Tr}\exp(-\beta H_A)\\&\langle\sigma_A\rangle_A=(\sigma_A)_A/Z_A.\end{aligned} \tag{5}$$

Here $\sigma_A$ is some observable in the $A$ system. It may, of course, depend on the $B$ spins since they are in $A$.

The spins that are mostly relevant to our analysis are the $B$ spins. The word "spin" is to some extent a misnomer, for the only hypothesis is that at each point $b\in B$ there is an independent a-priori probability measure $d\mu_b$ on some measure space $\Omega_b$. For simplicity we take these to be independent of $b$. Let $\{\phi^n\}$ be a *complete* orthonormal family of functions in $L^2(d\mu)$. The choice of the $\{\phi^n\}$ is important because the hypotheses made later can be expected to hold, if at all, only for special choices. With $\mathbf{n}=(n_1,n_2,\ldots)$ a multi-index on $B=(b_1,b_2,\ldots)$, we denote the following orthonormal functions on $\prod_{b\in B}\Omega_b$:

$$\phi_B^{\mathbf{n}}=\phi_{b_1}^{n_1}\phi_{b_2}^{n_2}\ldots. \tag{6}$$

*Example 2 (Spin $\frac{1}{2}$ Ising Model).* Here $\Omega=\{-1,1\}$, $\mu$ gives weight $\frac{1}{2}$ to each point and $\{\phi^n\}=\{\phi^0,\phi^1\}$ with $\phi^0(\sigma)=1$, $\phi^1(\sigma)=\sigma$.

*Example 2 (Plane Rotor).* $\Omega$ is the unit circle $0\leqq\theta<2\pi$, $d\mu(\theta)=d\theta/2\pi$ is the uniform measure, and $\phi^n(\theta)=\exp(in\theta)$ with $n=0,\pm1,\pm2,\ldots$.

The constitution of the remainder of the $A$ and $C$ systems is irrelevant to the general formalism we present. It can be composed of quarks, for example. $\sigma_A$ (resp. $\sigma_C$)will denote observables in the $A$ (resp. $C$) systems and they can both depend on the $B$ spins. Note that the functions $\phi_B^{\mathbf{n}}$ can be regarded either as $A$ or as $C$ observables.

A formula connecting $A$, $C$ and $A+C$ expectations is required. In other words, we have to "glue" the $A$ and $C$ systems together to form the $A+C$ system.

**Lemma 1.**

$$(\sigma_A\sigma_C)_{A+C}=\sum_{\mathbf{n}}(\sigma_A\phi_B^{\mathbf{n}})_A(\sigma_C\bar{\phi}_B^{\mathbf{n}})_C. \tag{7}$$

*In particular,*

$$Z_{A+C}=\sum_{\mathbf{n}}(\phi_B^{\mathbf{n}})_A(\bar{\phi}_B^{\mathbf{n}})_C. \tag{8}$$

*Proof.* In a schematic notation, let $x,y$, and $z$ respectively stand for the $B$ variables, the $A$ variables other than $B$, and the $C$ variables other than $B$. The

Boltzmann factor is $M(x,y)N(x,z)$ where $M(x,y)=\exp[-\beta H_A(x,y)]$ and $N(x,z)=\exp[-\beta H_C(x,z)]$. Let the a-priori measure be $d\mu_\beta(x)\,d\mu_\alpha(y)\,d\mu_\gamma(z)$ and let $F(x)=\int d\mu_\alpha(y)\,\sigma_A(x,y)\,N(x,y)$, $G(x)=\int d\mu_\gamma(z)\sigma_C(x,z)M(x,z)$. Then, by Parseval's theorem,

$$(\sigma_A\sigma_C)_{A+C}=\int d\mu_\beta(x)F(x)G(x)=\sum_{\mathbf{n}}D_{\mathbf{n}}E_{\mathbf{n}}$$

with $D_{\mathbf{n}}=\int d\mu_\beta(x)\phi_B^{\mathbf{n}}(x)F(x)$ and $E_{\mathbf{n}}=\int d\mu_\beta(x)\bar{\phi}_B^{\mathbf{n}}(x)G(x)$. But this sum on $\mathbf{n}$ is precisely the right side of (7). □

Henceforth we fix the observables $\sigma_A$ and $\sigma_C$, the Hamiltonians $H_A$ and $H_C$, and make the following hypotheses (with respect to $\sigma_A$ and $\sigma_C$) about the $A$ and $C$ systems.

*H.C1 (Positivity).* $\langle\sigma_C\bar{\phi}_B^{\mathbf{n}}\rangle_C\geqq 0$ for all $\mathbf{n}$. (9)

*H.A1 (The Gaussian-Type Inequality* [3]*).* There exists a function $F(\mathbf{n})$, not necessarily nonnegative, of the multi-index such that

$$\langle\sigma_A\phi_B^{\mathbf{m}}\rangle_A\leqq\sum_{\mathbf{n}}F(\mathbf{n})\langle\sigma_A\phi_B^{\mathbf{n}}\rangle_A\langle\bar{\phi}_B^{\mathbf{n}}\phi_B^{\mathbf{m}}\rangle_A \tag{10}$$

for all $\mathbf{m}$ such that $\langle\sigma_C\bar{\phi}_B^{\mathbf{m}}\rangle_C>0$.

The meaning of H.A1 will become clear later when we consider the Ising and plane rotor models as examples. For now we note that comparatively little is required of system $C$. The main theorem is the following:

**Theorem 1.** *Under hypotheses H.A1 and H.C1*

$$\langle\sigma_A\sigma_C\rangle_{A+C}\leqq\sum_{\mathbf{n}}F(\mathbf{n})\langle\sigma_A\phi_B^{\mathbf{n}}\rangle_A\langle\bar{\phi}_B^{\mathbf{n}}\sigma_C\rangle_{A+C}\,. \tag{11}$$

*Proof.* Multiply (11) by $Z_AZ_{A+C}$ and use Lemma 1. We require that

$$Z_A\sum_{\mathbf{m}}(\sigma_A\phi_B^{\mathbf{m}})_A(\bar{\phi}_B^{\mathbf{m}}\sigma_C)_C$$
$$\leqq\sum_{\mathbf{m}}\left\{\sum_{\mathbf{n}}F(\mathbf{n})(\sigma_A\phi_B^{\mathbf{n}})_A(\bar{\phi}_B^{\mathbf{n}}\phi_B^{\mathbf{m}})_A\right\}(\bar{\phi}_B^{\mathbf{m}}\sigma_C)_C\,. \tag{12}$$

Here, $\phi_B^{\mathbf{n}}$ has been regarded as an $A$ observable. In view of H.C1 it suffices to prove (12) for each $\mathbf{m}$ but, if we divide by $Z_A^2$, this is seen to be H.A1. □

The analogue of Simon's inequality [1] would have $\langle\ \rangle_{A+C}$ instead of $\langle\ \rangle_A$ on the right side of (11). There are then two natural questions: When does the Simon type of inequality hold and when is it weaker than Theorem 1, as it is for the Ising model? The following hypotheses help to answer this.

*H.C2.* $\langle\bar{\phi}_B^{\mathbf{n}}\rangle_C\geqq 0$, all $\mathbf{n}$. (13)

*H.A2 (inequality of the second Griffiths type).*

$$\langle\sigma_A\phi_B^{\mathbf{n}}\rangle_A\langle\phi_B^{\mathbf{m}}\rangle_A\leqq\langle\sigma_A\phi_B^{\mathbf{n}}\phi_B^{\mathbf{m}}\rangle_A,\quad\text{all }\mathbf{n} \tag{14}$$

whenever $\langle\bar{\phi}_B^{\mathbf{m}}\rangle_C>0$.

**Theorem 2.** *Suppose H.A2 and H.C2 hold. Then*

$$\langle \sigma_A \phi_B^{\mathbf{n}} \rangle_A \leqq \langle \sigma_A \phi_B^{\mathbf{n}} \rangle_{A+C}, \quad \text{all } \mathbf{n}. \tag{15}$$

*Proof.* (15) is equivalent to

$$\sum_{\mathbf{mn}} (\phi_B^{\mathbf{m}})_A (\bar{\phi}_B^{\mathbf{m}})_C (\sigma_A \phi_B^{\mathbf{n}})_A \leqq Z_A \sum_{\mathbf{m},\mathbf{n}} (\sigma_A \varphi_B^{\mathbf{n}} \phi_B^{\mathbf{m}})_A (\bar{\phi}_B^{\mathbf{m}})_C,$$

but this is implied by (13), (14). □

**Corollary 1.** *Suppose* (11) *and* (15) *hold and* $F(\mathbf{n}) \geqq 0$. *Then*

$$\langle \sigma_A \sigma_C \rangle_{A+C} \leqq \sum_{\mathbf{n}} F(\mathbf{n}) \langle \sigma_A \phi_B^{\mathbf{n}} \rangle_{A+C} \langle \bar{\phi}_B^{\mathbf{n}} \sigma_C \rangle_{A+C}. \tag{16}$$

Moreover, the right side of (16) is not less than the right side of (11).

If $F(\mathbf{n})$ is not nonnegative, (16) can still be proved under a further hypothesis:

*H.A3.*

$$\langle \sigma_A \phi_B^{\mathbf{m}} \rangle_A \langle \phi_B^{\mathbf{k}} \rangle_A \leqq \sum_{\mathbf{n}} F(\mathbf{n}) \langle \sigma_A \phi_B^{\mathbf{n}} \phi_B^{\mathbf{k}} \rangle_A \langle \bar{\phi}_B^{\mathbf{n}} \phi_B^{\mathbf{m}} \rangle_A. \tag{17}$$

whenever both $\langle \sigma_C \bar{\phi}_B^{\mathbf{m}} \rangle_C > 0$ and $\langle \bar{\phi}_B^{\mathbf{k}} \rangle_C > 0$.

**Theorem 3.** (16) *holds under hypothese H.C1, H.C2, and HA3.*

The proof of Theorem 3 is an imitation of the proofs of Theorems 1 and 2. Note that under these hypothese one cannot say that (16) is weaker than (11).

The following is a trivial consequence of the definitions

**Lemma 2.** *If* $F(\mathbf{n}) \geqq 0$ *then H.A1 and H.A2 imply H.A3.*

## The Ising Model as an Example

*Spin 1/2 Ising Models*

The $\phi^n$ are given in Example 1. We take $\sigma_A$ and $\sigma_C$ each to be products of an odd number of spins. H.C1, H.C2, and H.A2 are Griffiths' inequalities. Newman's inequality [3] states, in particular, that if $F$ is a family of partitions of $K = \{1, \dots, k\}$ into two disjoint subsets then (with $\sigma_D = \sigma_a \sigma_b \dots \sigma_d$ when $D = \{a, b, \dots, d\}$)

$$\langle \sigma_K \rangle \leqq \sum_{f \in F} \langle \sigma_{f_1} \rangle \langle \sigma_{f_2} \rangle, \tag{18}$$

whenever $|K| = 2L$ is even and *every partition of* $K$ into $L$ pairs is a refinement of some $f \in F$. Sylvester [14] also gives a proof of (18).

Let the spins in $B$ be labeled $\sigma_1, \dots, \sigma_M$. In (10), $\mathbf{m}$ can be thought of as a subset of $\{1, \dots, M\}$. Clearly, $\langle \sigma_A \phi_B^{\mathbf{m}} \rangle_A > 0$ implies that $|\mathbf{m}|$ is odd.

Assume that $\sigma_A$ is just one spin, $\sigma_\alpha$, and, without loss, that $\alpha \notin B$. Taking $K = \{\alpha\} \cup \mathbf{m}$, and all $f_1$ of the form $\{\alpha, i\}$ with $i \in \mathbf{m}$, (18) implies (10) with

$$F(\mathbf{n}) = 1 \quad \text{if} \quad |\mathbf{n}| = 1$$
$$= 0 \quad \text{otherwise}. \tag{19}$$

Commun. Math. Phys. 77, 127–135 (1980)



[Note: There are more terms on the right side of (10) than the right side of (18). The excess terms are nonnegative by Griffiths first inequality.] In this case we conclude that

$$\langle \sigma_a \sigma_C \rangle_{A+C} \leqq \sum_{b \in B} \langle \sigma_a \sigma_b \rangle_A \langle \sigma_b \sigma_C \rangle_{A+C} \tag{20}$$

as stated in the introduction. It was not assumed that $|C|=1$.

If $\sigma_A$ is a product of $N$(odd) spins then (18) implies (10) with

$$\begin{aligned} F(\mathbf{n}) &= 1 \quad \text{if} \quad |\mathbf{n}| = 1, 3, \ldots, N \\ &= 0 \quad \text{otherwise}. \end{aligned} \tag{21}$$

Then (20) changes to

$$\langle \sigma_A \sigma_C \rangle_{A+C} \leqq \sum_{\substack{b \subset B \\ |b| \leqq |A|}} \langle \sigma_A \sigma_b \rangle_A \langle \sigma_b \sigma_C \rangle_{A+C} . \tag{22}$$

*Other Ising Models*

One generalization is to spin $M > \frac{1}{2}$ with $\sigma = 2M, \ldots, -2M$. A way to proceed would be to use an appropriate orthonormal basis $\{\phi^n\}$ of dimensions $2M+1$. We have not pursed this possibility. A second method is to use Griffiths' trick [5] of writing a spin $M$ as $M$ ferromagnetically coupled spin $\frac{1}{2}$ spins. H.C1, H.C2, and H.A2 follow from this, as does (20) and (21) by summing over the "component" spins. Much is lost this way, however.

Another generalization, which we shall not explicate, is to allow multi-spin interactions.

## The Plane Rotor Model

We consider pairwise ferromagnetic interactions; the interaction between two spins $\vec{\sigma}_a$ and $\vec{\sigma}_b$ is $-J_{ab}\vec{\sigma}_a \cdot \vec{\sigma}_b = -J_{ab}\cos(\theta_a - \theta_b)$, with $J_{ab} \geqq 0$. The basis $\{\phi^n\}$ is given in Example 2.

There is some reason to believe that the analogue of (20) holds in the following sense:

$$\langle \vec{\sigma}_a \cdot \vec{\sigma}_c \rangle_{A+C} \leqq \sum_{b \in B} \langle \vec{\sigma}_a \cdot \vec{\sigma}_b \rangle_A \langle \vec{\sigma}_b \cdot \vec{\sigma}_c \rangle_{A+C} . \tag{23}$$

when $\vec{\sigma}_a$ and $\vec{\sigma}_c$ are single spins. In terms of the $\phi^n$ we have

$$2\vec{\sigma}_a \cdot \vec{\sigma}_c = \phi_a^1 \phi_c^{-1} + \phi_a^{-1} \phi_c^1 .$$

Therefore we require that (9) and (10) hold when $\sigma_C$ (resp. $\sigma_A$) is $\phi_c^1$ or $\phi_c^{-1}$ (resp. $\phi_a^1$ or $\phi_a^{-1}$) and $F$ is given by (19). With these choices for $\sigma_A$ and $\sigma_C$, (9), and also (14) hold [6]. The difficulty lies with (10).

We do not have a proof of (10), but believe it to be true. A possibility would be to try to imitate the graphical proof [3,4] that is successful in the Ising case. It would then be necessary to deal with directed graphs (digraphs). The following, if it were true would immediately yield a proof of (10):

*Conjecture.* Let $G$ be a finite direted graph (possibly with several edges between two vertices) and let the valence at vertex $i$ (the number of arrows in minus the number out of $i$) be $M_i$. (Clearly, $\Sigma M_i = 0$.) Suppose $M_1 = +1$, and $M_2, M_3, \ldots, M_k < 0$, and $M_i \geqq 0$ otherwise. Let $N$ be the number of subgraphs (subsets of edges) of G, including the empty graph, having valence 0 at each vertex. Let $K$ be the number of subgraphs of $G$ with the following property: vertex 1 has valence $+1$, some vertex $j$, with $2 \leqq j \leqq k$, has valence $-1$, and all other valences are 0. Then $N \leqq K$.

There is one special but important case in which the conjecture, and hence (23) holds. Suppose $\vec{\sigma}_a$ is connected to $n$ nearest neighbors, which we take to be $B$. It is immaterial whether the $B$ spins are connected together, for any such interaction can be regarded as part of $H_C$. In a $v$ dimensional cubic lattice $n = 2v$.

The graph $G$ in the conjecture then has the following structure: it has $n+1$ vertices and is star-like with edges only between the central vertex $V_1$ (which is really the original vertex marked $a$) and the other $n$. Suppose $M_2 < 0$. Then it is easy to see that the conjecture is verified if the following is true: Let $\tilde{G}$ be the subgraph of $G$ consisting of vertices 1 and 2 and all the edges between them. Let $\tilde{N}$ be the number of valence 0 subgraphs of $\tilde{G}$, and let $\tilde{K}$ be the number of subgraphs of $\tilde{G}$ with $M_1 = +1$, $M_2 = -1$. Then $\tilde{N} \leqq \tilde{K}$. The easy proof of this is left to the reader.

This simple case can also be conveniently viewed in terms of (10) directly. We require that

$$\langle \phi_a^1 \phi_B^{\mathbf{m}} \rangle \leqq \sum_{b=1}^{n} \langle \phi_a^1 \phi_b^{-1} \rangle \langle \phi_b^1 \phi_B^{\mathbf{m}} \rangle . \tag{24}$$

Both sides of (24) vanish unless $\sum_1^n m_i = -1$. Let $P(m)$ be the Fourier transform of $\exp[\beta \cos\theta]$, namely

$$P(m) = I_m(\beta) > 0, m \in \mathbb{Z}, \tag{25}$$

where $I_m$ is the modified Bessel function. Then (24) reads

$$\{P(0)^n\} \left\{ \prod_{i=1}^{n} P(m_i) \right\} \leqq \sum_{i=1}^{n} \{P(0)^{n-1} P(1)\} \left\{ P(m_i + 1) \prod_{j \neq i} P(m_j) \right\} . \tag{26}$$

Suppose $m_1 < 0$, say. It is sufficient to have

$$P(0) P(m_1) \leqq P(1) P(m_1 + 1) . \tag{27}$$

If both sides of (27) are expanded in a power series in $\beta$, (27) is true term by term. This is just the graphical exercise mentioned above. However, the following stronger result, which implies (27), holds.

**Lemma 2.** *Fix $\beta \geqq 0$. The function $m \to I_m(\beta)$ is log concave on the integers, i.e.*

$$I_m(\beta)^2 \geqq L_{m+1}(\beta) I_{m-1}(\beta), \quad m \in \mathbb{Z} . \tag{28}$$

Commun. Math. Phys. 77, 127–135 (1980)



*Proof.* $I_m(\beta)=I_{-m}(\beta)$. (28) is trivial for $m=0$, so it is sufficient to consider $m\geqq 1$. If both sides of (28) are expanded in a power series in $\beta$, we claim (28) holds termwise. Use

$$I_m(\beta)=\sum_{j=0}^{\infty}(\beta/2)^{2j+m}[j!(j+m)!]^{-1}.$$

Thus for the coefficient of $\beta^{2m+2t}$, $t\geqq 0$, we require

$$\sum_{j=0}^{t} g_0(j-t/2)g_m(j-t/2)\geqq \sum_{j=0}^{t} g_0(j-t/2)g_m(j+1-t/2), \tag{29}$$

where $g_p(x)=[\Gamma(t/2+1+p+x)\Gamma(t/2+1+p-x)]^{-1}\theta(x)$ and $\theta(x)=1$ if $|x|\leqq t/2+p$ and $=0$ otherwise. Now $\Gamma(x)$ is log convex for $x>0$ and hence $g_p(x)$ is even and log concave. Thus, for $j$ an integer in $[0,t]$, $g_0(j-t/2)$ is a positive sum of functions of the form $\mu_a(j)=1$ if $a\leqq j\leqq t-a$ and $=0$ otherwise, for $a=0,1,\ldots[t/2]$. Hence, it suffices to have $\sum_{j=a}^{t-a} g_m(j-t/2)-g_m(j+1-t/2)\geqq 0$. But this is true because $g_m(j-t/2)$ is also a positive sum of the $\mu_b(j)$ functions (with $-m\leqq b\leqq[t/2]$) and $\sum_{j=a}^{t-a}\mu_b(j+r)$ is decreasing for $r\geqq 0$, $r\in\mathbb{Z}$. □

Since (23) holds when $\vec{\sigma}_a$ is a single spin and $B$ are its neighbors, we can conclude using (2) that

**Theorem 4.** *For the plane rotor model on a ν-dimensional hypercubic lattice there is exponential fall-off of the two-point function if* $\langle\vec{\sigma}_a\cdot\vec{\sigma}_b\rangle<1/2\nu$ *for the two-spin system consisting of a and b alone. This is equivalent to* $I_1(\beta)/I_0(\beta)<1/2\nu$ *if* $J_{ab}=1$ *is assumed for the nearest neighbor coupling constant. In particular*

$$\beta_c\geqq 0.52\quad(\nu=2);\qquad \beta_c\geqq 0.34\quad(\nu=3).$$

For $\nu=2$, Fröhlich [7] has shown that $\beta_c\geqq 0.64$, and Aizenman and Simon [8] have shown that $\beta_c\geqq 0.88$.

*Acknowledgements.* I should like to thank M. Aizenman, J. Bricmont, and B. Simon for helpful conversations.

## References


1. Simon, B.: Commun. Math. Phys. **77**, 111–126 (1980). See also Phys. Rev. Lett. **44**, 547–549 (1980)
2. Aizenman, M., Simon, B.: Commun. Math. Phys. **77**, 137–143 (1980)
3. Newman, C.M.: Wahrsch. **33**, 75–93 (1975)
4. Sylvester, G.S.: Commun. Math. Phys. **42**, 209–220 (1975)
5. Griffiths, R.B.: J. Math. Phys. (N.Y.) **10**, 1559–1565 (1969)
6. Ginibre, J.: Commun. Math. Phys. **16**, 310–328 (1970)
7. Fröhlich, J.: Private communication
8. Aizenman, M., Simon, B.: A comparison of plane rotor and ising models. Phys. Lett. A (submitted)


Communicated by A. Jaffe

Received March 21, 1980

**Notes added in proof.** (1) Simon's inequality (1) for the Ising model is a special case of a class of inequalities and identities discussed by Boel and Kasteleyn [9, 10]. They found necessary and sufficient conditions for such inequalities to hold; therefore (1) can be proved by their methods.

(2) The conjecture in this paper has been proved by Rivasseau [11]. Thus inequality (23) for rotors holds for all subsystems $A$, not merely for the case of the star graph proved here.

## References

9. Boel, R.J., Kasteleyn, P.W.: Commun. Math. Phys. **61**, 191 (1978); Commun. Math. Phys. **66**, 167 (1979); Physica **93** A, 503 (1978)
10. Kasteleyn, P.W., Boel, R.J.: Phys. Lett. **70** A, 220 (1979)
11. Rivasseau, V.: Lieb's correlation inequality for plane rotors, Commun. Math. Phys. **77**, 145–147 (1980)

Vol. 71, No. 2 DUKE MATHEMATICAL JOURNAL © August 1993

# FLUXES, LAPLACIANS, AND KASTELEYN'S THEOREM

ELLIOTT H. LIEB AND MICHAEL LOSS

**1. Introduction.** The genesis of this paper was an attempt to understand a problem in condensed matter physics related to questions about electron correlations, superconductivity, and electron-magnetic field interactions. The basic idea, which was proposed a few years ago, is that a magnetic field can lower the energy of electrons when the electron density is not small. Certain very specific and very interesting mathematical conjectures about eigenvalues of the Laplacian were made, and the present paper contains a proof of some of them. Furthermore, those conjectures lead to additional natural conjectures about determinants of Laplacians which we both present and prove here. It is not clear whether these determinantal theorems have physical applications but they might, conceivably in the context of quantum field theory. Some, but not all, of the results given here were announced earlier in [LE].

The setting is quantum mechanics on a graph or lattice. (All our terminology will be precisely defined in the sequel.) Physically, the vertices of our graph $\Lambda$ can be thought of either as a discretization of space (i.e., replace the Laplacian by a finite difference operator), or they can be seen as locations of atoms in a solid. There are $|\Lambda|$ vertices. In the atomic interpretation the edges become electron bonds joining the atoms, and the model is known as the tight-binding model or Hückel model. The natural Laplacian $\mathscr{L}$ associated with $\Lambda$ is a $|\Lambda| \times |\Lambda|$ matrix indexed by the vertices of $\Lambda$ and whose diagonal elements satisfy $-\mathscr{L}_{xx} =$ number of attached edges (or valency) of vertex $x$. The other elements are $\mathscr{L}_{xy} = 1$ if $x$ and $y$ are connected by an edge, and zero otherwise.

For us it is more convenient to consider the matrix $\hat{\mathscr{L}}$ which is the Laplacian without the diagonal term, i.e., $\mathscr{L}_{xx}$ is replaced by zero. In the context of graph theory $\hat{\mathscr{L}}$ is also known as the adjacency matrix. There are three excuses for this: (i) in the solid state context, $\hat{\mathscr{L}}$ is the natural object because atoms do not bond to themselves; (ii) most of the graphs that are considered in the physics literature have constant valency, and so $\hat{\mathscr{L}}$ and $\mathscr{L}$ have the same spectrum modulo a constant which is equal to this valency; (iii) mathematically, $\hat{\mathscr{L}}$ seems to be the more natural object—from our point of view, at least—because its spectrum on a bipartite graph is always a union of pairs $\lambda$ and $-\lambda$ (when $\lambda \neq 0$), as explained in Section 2. The spectrum of $\mathscr{L}$ generally does not have any such symmetry.

Received 2 November 1992.
Lieb's work partially supported by U.S. National Science Foundation grant PHY90-19433A01
Loss's work partially supported by U.S. National Science Foundation grant DMS92-07703

We label the eigenvalues of $\hat{\mathscr{L}}$ by $\lambda_1(\hat{\mathscr{L}}) \geqslant \lambda_2(\hat{\mathscr{L}}) \geqslant \cdots$. The Hamiltonian for a single electron is $-\mathscr{L}$ or $-\hat{\mathscr{L}}$, and we take it to be $-\hat{\mathscr{L}}$ here. If our system has $M$ *free* electrons, the rule of quantum mechanics is that the eigenvalues of our system are all the possible sums of $M$ of the $(-\lambda_i)$'s in which each $-\lambda_i$ is allowed to appear at most twice in the sum. There are $\binom{2|\Lambda|}{M}$ eigenvalues. In particular, if $M = 2N$, the smallest eigenvalue is

$$E_0^{(N)} = -2 \sum_{j=1}^{N} \lambda_j(\hat{\mathscr{L}}). \tag{1.1}$$

A (spatially varying) magnetic field is now added to the system in the following way. $\hat{\mathscr{L}}_{xy}$ is replaced by $T_{xy} = \hat{\mathscr{L}}_{xy} \exp[i\theta(x, y)]$, with $\theta$ real and with $\theta(x, y) = -\theta(y, x)$ so that $T$ is Hermitian. The function $\theta(x, y)$ is interpreted physically as the integral of a magnetic vector potential from the point $x$ to the point $y$. This $T$ is the discrete analogue of replacing the Laplacian on $\mathbb{R}^n$ by $(\nabla - iA(x))^2$ (with $\nabla$ = gradient), which is the Laplacian on a $U(1)$ bundle.

The central question that we address is this: *What choice of $\theta$ minimizes $E_0^{(N)}$ for a given $N$?*

In order to appreciate this question, consider the $N = 1$ case. Then $\theta \equiv 0$ is an answer because (with $\phi$ being the normalized largest eigenvector of $T$) $\lambda_1(T) = \sum \bar{\phi}_x \phi_y \hat{\mathscr{L}}_{xy} \exp[i\theta(x, y)] \leqslant \sum |\phi_x| |\phi_y| \hat{\mathscr{L}}_{xy} \leqslant \lambda_1(\hat{\mathscr{L}})$. This proof that $\theta \equiv 0$ is optimum also works in a more general setting, namely for the lowest eigenvalue of the "Schrödinger operator" $-T + V$, where $V$ is any real diagonal matrix. Again, $T = \hat{\mathscr{L}}$, or $\theta \equiv 0$, minimizes $-\lambda_1(T - V)$. The same is true in $\mathbb{R}^n$ for $-(\nabla - iA(x))^2 + V(x)$; the minimum occurs when $A(x) \equiv 0$. This conclusion is known as the *diamagnetic inequality* and states, physically, that "a magnetic field raises the energy".

It was discovered by [AM] and [KG] that the situation can be quite different when $N$ is close to $(1/2)|\Lambda|$. (When $N = |\Lambda|$, $E_0^{(N)} = \mathrm{Tr}\, T = 0$ for all $\theta$; hence $N = (1/2)|\Lambda|$ is the most extreme case.) Since then, the problem has been investigated for various lattices and $N$'s by several authors such as [BR], [BBR], [HLRW], [RD], [WP], and [WWZ], some of whom consider it to be important in the theory of high temperature superconductivity. [HLRW], for example, start with the square lattice $\mathbf{Z}^2$, take $\Lambda$ to be a large rectangular subset of $\mathbf{Z}^2$, and then let $|\Lambda| \to \infty$ and $N \to \infty$ with $N/|\Lambda|$ fixed. They also take the magnetic flux (which is the sum of the $\theta$'s around the edges of a face, and which is defined in Section 2) to have the *same value* in each square box of $\mathbf{Z}^2$. On the basis of their numerical evidence they proposed that flux/box $= 2\pi N/|\Lambda|$ is the optimal choice. In [AM] the term "flux phase" was introduced to describe this state in which the presence of a magnetic field lowers the energy.

It should be pointed out that the spectrum of $T$ for $\mathbf{Z}^2$ as a function of constant flux/box was discussed by many authors for many years; it was Hofstadter [HD] who grasped the full beauty of this object—which is anything but a continuous function of the flux and which is full of gaps—and called it a "butterfly". The spectrum can be found by solving a one-dimensional difference equation, due to Harper [HP], which is a discrete analogue of, but more complicated than, Mathieu's equation. The spectrum is such a complicated function of the flux that it is difficult to decide on the optimum flux for a given $N$.

The most striking case is $N/|\Lambda| = 1/2$, or $M = |\Lambda|$, which is called the **half-filled band**. The optimal flux is supposed to be $\pi$, which is the maximum possible flux since flux is determined only modulo $2\pi$ and since flux and $-$flux yield identical spectra. It is this case that we investigate in this paper in an attempt to verify the rule just stated and which appears in [AM], [HLRW], and [RD]. We are completely successful only in some special cases, but we have been able to generalize the problem in several interesting directions. For example, one of our main results is Theorem 3.1. It completely solves the problem for determinants (i.e., for products of eigenvalues instead of sums of eigenvalues) on bipartite planar graphs.

Our determinant theorem turns out to be closely related to Kasteleyn's famous 1961 theorem about planar graphs, which allowed him to solve (in principle) the dimer problem and Ising model for all planar graphs. Our route, via fluxes, gives an alternative proof of Kasteleyn's theorem and, we believe, a more transparent one. This is presented in the appendix.

The setting we adopt is a general graph $\Lambda$, with no particular symmetry such as $\mathbf{Z}^2$ enjoys, and an *arbitrary*, but *fixed* amplitude $|t_{xy}| > 0$ given on each edge ($|t_{xy}| = 1$ in the case of $\hat{\mathscr{L}}$). The problem is to determine $\theta$ and $T := \{t_{xy}\}_{x,y\in\Lambda}$ with $t_{xy} = |t_{xy}| \exp[i\theta(x, y)]$ so as to minimize the (absolute) **ground state energy**

$$E_0(T) := -\mathrm{Tr}\,|T| - \mathrm{Tr}\,T = -\mathrm{Tr}\,|T|, \tag{1.2}$$

with Tr = Trace. The right side of (1.2) is twice the sum of the negative eigenvalues of $-T$. For a bipartite graph this is the sum of the $|\Lambda|/2$ or $(|\Lambda| - 1)/2$ lowest eigenvalues of $-T$.

A word has to be said here about different definitions of ground state energy. Electrons have two spin states available to them and the Pauli exclusion principle states that each eigenstate of $-T$ can be occupied by at most one electron of each kind. Thus, each eigenstate can be occupied by 0 or 1 (twice) or 2 electrons. That explains the factor of 2 in (1.1): there the lowest $N$ eigenstates of $-T$ are each occupied by two electrons. Our definition of $E_0(T)$ in (1.2) is the absolutely lowest ground state energy and corresponds to the electron number being twice the number of negative eigenvalues. On the other hand, the half-filled band would have the electron number equal to $|\Lambda|$ by definition. If $|\Lambda| = 2N$ is even, then the half-filled band ground state energy is given by (1.1) with $N = |\Lambda|/2$. If $|\Lambda| = 2N + 1$, the half-filled band ground state energy is $-2\sum_{j=1}^{N} \lambda_j - \lambda_{N+1}$. It is this half-filled band energy that is mostly considered in the physics literature. However, we regard our definition (1.2) as mathematically more natural and physically as interesting as the strict half-filled band definition. *For bipartite graphs* (defined at the end of Section 2) *$E_0(T)$ and $E_0^{(N)}$ with $N = |\Lambda|/2$ or $(|\Lambda| - 1)/2$ agree with each other.* (Note that if $\Lambda$ is bipartite and $|\Lambda| = 2N + 1$, then $\lambda_{N+1} = 0$.)

The two definitions can produce strikingly different conclusions, however, in special cases. In [RD] the ground state energy $E_0(T)$ of $N$ electrons (including spin) hopping on a ring of $N$ sites is considered. By Theorem 4.1 we know that, for a ring with $N$ odd (and which is therefore not bipartite), the expression (1.2) is minimized by flux $\pi$ and flux 0. However, it has been shown in some cases (see [RD]) that the half-filled band energy for such a ring is minimized by the flux $\pi/2$ (which, incidentally, we call the canonical flux in this paper).

There is an important difference between our minimization problem and the one in [HLRW] and some other papers in the physics literature. For a regular structure like $Z^2$ we allow *different fluxes in different boxes*. In the physics literature the problem is sometimes stated with *constant fluxes* or with periodic fluxes. We find our formulation (with arbitrary fluxes) to be more natural mathematically, and we believe it to be more natural in those physical problems where this theory might be applicable.

Besides the ground state energy, we consider other functions of $T$, such as $\ln|\det T| = \operatorname{Tr} \ln|T|$. A particularly important one, physically, is $\ln \Xi$ where $\Xi$ is the **grand canonical partition function** with chemical potential $\mu$ and inverse temperature $\beta$, given by

$$\Xi = \sum_{m_1} \cdots \sum_{m_{|\Lambda|}} \sum_{n_1} \cdots \sum_{n_{|\Lambda|}} \exp\left[\beta \sum_{j=1}^{|\Lambda|} (\lambda_j + \mu)(n_j + m_j) + \beta\mu\right] = \prod_{j=1}^{|\Lambda|} \{1 + \exp[\beta(\lambda_j + \mu)]\}^2, \tag{1.3}$$

where the sum on each $n_i$ and $m_i$ is over the set $\{0, 1\}$. The physical **free energy** is defined by $\mathscr{F} = -\beta^{-1} \ln \Xi$. We consider only $\mu = 0$ here because that corresponds to a half-filled band in the bipartite case (see (4.5) and footnote).

Another important quantity is the **gap**, $G(T)$, which is *not* defined by a trace. We define it to be

$$G^{(N)}(T) := -\lambda_{N+1} + \lambda_N. \tag{1.4}$$

When $N$ is the number of negative eigenvalues of $-T$, we denote $G^{(N)}(T)$ by $G(T)$. Clearly, $G(T)$ is the minimum energy to excite the system from the $2N$-electron ground state. For the half-filled band on a bipartite graph with $|\Lambda| = 2N$, $G(T) = 2\lambda_N$. This, however, may not be mathematically interesting because $\lambda_N$ may be automatically zero for dimensional reasons. That is, if $A$ and $B$ are the two subsets of vertices of $\Lambda$ that define the bipartite structure, then $T$ always has at least $||B| - |A||$ zero eigenvalues. For this reason we define $\tilde{G}(T)$ for a *bipartite* $\Lambda$ (with $|\Lambda|$ odd or even) to be

$$\tilde{G}(T) := \lambda_{|A|} - \lambda_{|B|+1} = 2\lambda_{|A|}, \tag{1.5}$$

assuming $|B| \geqslant |A|$. We can then ask the question: *Which flux maximizes $G(T)$, or $\tilde{G}(T)$ in the bipartite case?*

So far we have discussed free—or noninteracting—electrons. The same questions can be asked for interacting electrons and very much less is known in that case. In Section 8, however, we are able to carry over our techniques to one example—the Falicov-Kimball model.

Many of these results were announced in [LE]. We thank P. Wiegmann for bringing this problem to our attention and, along with I. Affleck, D. Arovas, J. Bellissard, and J. Conway, for helpful discussions.

**2. Definitions and properties of fluxes.** A **graph** $\Lambda$ is a finite set of **vertices** (or **sites**), usually denoted by lower case roman letters $x, y, z$, etc., together with **edges** (or **bonds**), which are certain unordered pairs of *distinct* sites and are denoted by $(x, y)$, equivalently $(y, x)$. Thus there will be at most one edge between two vertices. The set of sites or vertices will be denoted by $V$ or $V(\Lambda)$ and the number of them by $|\Lambda|$. The set of edges will be denoted by $E$ or $E(\Lambda)$. If $(x, y) \in E$, the sites $x$ and $y$ are said to be **end points** of the edge $(x, y)$.

A graph $\Lambda$ is connected if for every pair of sites $x$ and $y$ there is a **path** $P$ in $\Lambda$ connecting $x$ and $y$, i.e., there is a sequence of points $x = x_0, x_1, x_2, \dots, x_n = y$ such that $(x_i, x_{i+1})$ is an edge for every $0 \leqslant i < n$. Although $\Lambda$ is not just the set of vertices, but also contains the edges, we shall nevertheless sometimes write $x \in \Lambda$ where $x$ is a site in $\Lambda$.

A **hopping matrix** $T$ associated with a graph $\Lambda$ is a Hermitian $|\Lambda| \times |\Lambda|$ matrix indexed by the sites of $\Lambda$, with elements denoted by $t_{xy} = \overline{t_{yx}}$ for $x, y \in \Lambda$, and with the important property that $t_{xy} \neq 0$ only if $(x, y) \in E$, i.e., if $x$ and $y$ are connected by an edge. In particular, $t_{xx} = 0$ for all $x \in V$. The $T$ matrix is the important object here. For that reason, if $t_{xy} = 0$ for any edge $(x, y)$, we might as well delete this edge from the graph $\Lambda$. Thus, without loss of generality, we can assume that $t_{xy}$ is nonzero and that the corresponding graph $\Lambda$ is connected. If it is not connected, $T$ breaks up into blocks which can be considered separately. We call $|t_{xy}|$ the **hopping amplitudes**. No other assumption is made about $t_{xy}$ unless explicitly stated otherwise. The eigenvalues of $T$ are usually denoted by $\lambda$, and sometimes by $\lambda(T)$ to be more specific.

A **circuit** $C$ of length $\ell$ in $\Lambda$ is an ordered sequence of *distinct* sites $x_1, \dots, x_\ell$ with the property that $(x_i, x_{i+1})$ is an edge for $i = 1, \dots, \ell$ with $x_{\ell+1} \equiv x_1$. We explicitly *include* $\ell = 2$. Note that $x_2, \dots, x_\ell, x_1$ is the same circuit as $C$, but $x_\ell, x_{\ell-1}, \dots, x_1$ is different.

If $C$ is a circuit, then we can define the **flux** $\Phi_C$ **of** $T$ **through** $C$, which is a number in $[0, 2\pi)$, as

$$\Phi_C := \arg\left(\prod_{i=1}^{n} t_{x_i, x_{i+1}}\right) =: \arg\left(\prod_C T\right). \tag{2.1}$$

The symbol $\prod_C T$ has an evident meaning.

A **gauge transformation** is a diagonal unitary transformation $U$ with elements, $u_{xy} = \exp[i\phi_x]\delta_{xy}$, where $\phi_x \colon V \to \mathbb{R}$ is a function on the sites. Obviously, a gauge transformation $T \to U^*TU$ leaves the spectrum of $T$ and all the fluxes unchanged.

2.1. LEMMA (Fluxes determine the spectrum). *Let $T$ and $T'$ be two hopping matrices which have the same hopping amplitudes and the same flux through each circuit $C$ of the graph $\Lambda$. Then there is a gauge transformation $U$ such that $T' = U^*TU$.*

*Proof.* By our convention, the $t_{xy}$ and $t'_{xy}$ are never zero. Thus $W_{xy} \equiv t_{xy}/t'_{xy}$ satisfies $|W_{xy}| = 1$ for all edges $(x, y)$, and the flux of $W$ through each circuit of $\Lambda$ is zero. Let $x_0$ be an arbitrary, but henceforth fixed, site in $\Lambda$. For any $x$ we can pick a path $P$ connecting $x_0$ and $x$ and define $\phi_x = \arg(\prod_P W)$. The value of $\phi_x$ does not depend on the choice of the path because, if $P'$ is another path connecting $x_0$ and $x$, we have that $\arg(\prod_P W) = \arg(\prod_{P'} W)$ since the flux through the loop given by connecting $x_0$ to $x$ along $P$ and then connecting $x$ to $x_0$ along $P'$ is zero. If $x$ and $y$ are arbitrary sites in $\Lambda$ with $(x, y) \in E$ and if we take $P_x$ to be a path connecting $x_0$ to $x$ and $P_y$ a path connecting $x_0$ to $y$, we observe that $1 = \prod_{P_x} W\, W_{xy} \prod_{P_y} \overline{W}$ since $P_x$ followed by $(x, y)$ is a path from $x_0$ to $y$. But this equals $\exp[i(\phi_x - \phi_y)]W_{xy}$, and hence $W_{xy} = \exp[-i(\phi_x - \phi_y)]$. Thus $t'_{xy} = \exp[i\phi_x]t_{xy}\exp[-i\phi_y]$, which proves the lemma. ■

Up to now, a graph has been regarded as an abstract object consisting of vertices and edges. Now we wish to regard graphs as embedded either in $\mathbb{R}^2$ or $\mathbb{R}^3$. This means that the sites of $\Lambda$ can be regarded as distinct fixed points in $\mathbb{R}^3$ and each edge $(x, y)$ will be identified with exactly one piecewise linear curve between $x$ and $y \in \mathbb{R}^3$. It is convenient to *exclude* the end points $x$ and $y$ in the definition of an edge. We require that any one edge does not intersect the other edges or the sites. Circuits are then identified with simple, oriented closed curves.

Obviously, any graph can be embedded in $\mathbb{R}^3$ but only some graphs, called **planar graphs**, can be embedded in $\mathbb{R}^2$. It is these graphs that will mostly concern us in this paper.

The set of edges $E$ and sites $V$ of a planar graph, regarded as a set in $\mathbb{R}^2$, is closed. Its complement is therefore open, and this complement has a finite number of connected components which we label $F_0, F_1, F_2, \ldots$. We define $F_0$ to be the open set that contains the point at infinity (i.e., the exterior of the graph). The others we call the **faces** of the graph $\Lambda$. Each face has a boundary and this boundary is composed of a subset of the edges and sites of $\Lambda$. For later purposes we call **elementary circuits** those circuits which are entirely contained in the boundary of a single face.

If the graph is planar, a circuit $C$ of length greater than 2 will have an inside and an outside. The interior, which is an open set, is then the union of a certain number of faces, edges, and vertices called **interior faces, interior edges**, and **interior vertices**. We denote their numbers by $f$, $e$, and $v$. We can speak of the **orientation** of $C$ as being either positive (anticlockwise) or negative (clockwise) according as the winding number with respect to a point in its interior is either $+1$ or $-1$.

In general, an arbitrary specification of fluxes through the circuits of $\Lambda$ may be inconsistent in the sense that there may not exist a choice of $T$ with the prescribed fluxes. Some kind of divergence—or closedness condition is needed. In two dimensions, however, the following lemma shows how fluxes can be specified in a consistent way.

2.2. LEMMA (Construction of phases from fluxes). *Let $\Lambda$ be a planar graph and let $F_1, F_2, \ldots, F_f$ be its faces. Let $\Phi^1, \Phi^2, \ldots, \Phi^f$ be any given numbers in $[0, 2\pi)$. (We call $\Phi^j$ the* **flux through** *$F_j$.) Then there is a function $\theta(x, y)$: $E(\Lambda) \to [0, 2\pi)$ so that the fluxes defined by (2.1) with $t_{xy} := \exp[i\theta(x, y)]$ satisfy $\Phi_C = \sum_{\text{interior faces of } C} \Phi^j$ for every positively oriented circuit $C$ on $\Lambda$.*

*Proof.* Pick a point $z_j = (z_j^1, z_j^2) \in \mathbb{R}^2$ in each face $F_j$ and consider the vector (really, a one-form) field on $\mathbb{R}^2$, $\vec{A}(x) = \vec{A}(x^1, x^2)$, with singularities at the $z_j$'s, given by

$$\vec{A}(x) = \sum_{\text{all faces } F_j \text{ of } \Lambda} \Phi^j \vec{a}(x - z_j)$$

with $\vec{a}(x) = (2\pi)^{-1}[(x^1)^2 + (x^2)^2]^{-1}(-x^2, x^1)$. For each pair $x, y \in V(\Lambda)$ with $(x, y) \in E(\Lambda)$ define $\theta(x, y) = \int_x^y \vec{A}$, i.e., the integral from $x$ to $y$ along the curve representing the edge $(x, y)$. The flux $\Phi_C$ through any circuit $C$ on $\Lambda$ is given by $\oint_C A$. By Cauchy's integral formula (or Stokes's theorem), this integral equals $\Phi_C$ given above. ■

*Remark.* A more practical and direct way to construct phases $\theta(x, y)$ satisfying the conclusion of Lemma 2.2 is to concentrate the vector field $\vec{a}(x - z_j)$ along a line. More precisely, let $L_j$ denote some semi-infinite line starting from $z_j$ and extending to infinity, but which does not intersect any of the sites of $\Lambda$ and whose intersection with an edge is always transverse, i.e., nontangent. We orient $L_j$ from $z_j$ outwards. For each *ordered* pair of sites $(x, y)$, with $(x, y) \in E$, first orient the curve representing $(x, y)$ in the direction from $x \in \mathbb{R}^2$ to $y \in \mathbb{R}^2$. Call this oriented curve $\varepsilon$. Then let $\mu_j(x, y) \equiv \sum_{L_j \cap \varepsilon} (\pm 1)$, where the sum is over all the points of intersection of $L_j$ with this curve $\varepsilon$ and where the $+$ sign (resp. $-$ sign) is taken if the (counterclockwise) angle from $L_j$ to $\varepsilon$ is less than (resp. more than) $\pi$. Finally, we set $\theta(x, y) = \sum_{j=1}^f \Phi^j \mu_j(x, y)$.

It is a fact that every planar graph $\Lambda$ can be **triangulated**, i.e., that there is a planar graph $\Lambda'$ with precisely the *same* vertices as $\Lambda$ and whose edges $E'$ (which are now sets of curves) contain $E$, the edge-set of $\Lambda$, and with the property that $\Lambda'$ is **triangular**. The concept of a planar graph $\Lambda'$ being triangular means that every one of the faces of $\Lambda'$ has as its boundary the union of three edges and three vertices. We say that $\Lambda'$ is a **triangulation** of $\Lambda$. It is easy to check that the three edges must always form a circuit. Note that, in general, a graph $\Lambda$ can be triangulated in several ways.

2.3. LEMMA (Number of triangles in a circuit). *Let $\Lambda$ be a triangular planar graph and let $C$ be a circuit in $\Lambda$ of length $\ell \geqslant 2$. Let $f$ and $v$ denote the number of interior faces and interior vertices of $C$. Then*

$$\ell - f + 2v = 2. \tag{2.2}$$

For $\ell = 2$, (2.2) is clearly true (with $f = v = 0$). Therefore, we need consider only $\ell > 2$.

*First proof.* Let $\beta$ be one of the $\ell$ edges in $C$. This edge must be part of the boundary of exactly one of the inner triangles, which we call $\tau$. The boundary of $\tau$ contains 3 edges, $\beta_1$, $\beta_2$, and $\beta_3$. There are two cases.

(a) $\beta_1 = \beta$ and $\beta_2$, $\beta_3$ are interior to $C$.
(b) $\beta_1 = \beta$, $\beta_2$ is an edge of $C$, and $\beta_3$ is interior to $C$.

In case (a) we consider the circuit $C'$ whose edges are the same as those of $C$ except that $\beta$ is replaced by the two edges $\beta_2$ and $\beta_3$. In case (b) we remove $\beta_1$ and $\beta_2$ from $C$ and replace them with $\beta_3$. It is easy to check that $\ell' - f' + 2v' = \ell - f + 2v$ and that $f' = f - 1$. By successively removing triangles in this way, we eventually have only one triangle left, in which case $\ell = 3$, $f = 1$, $v = 0$.

*Second proof.* Euler's formula says that (total number of vertices) + (total number of faces) − (total number of edges) = 1. Since $\ell$ also equals the number of vertices in $C$, we have that $1 = (v + \ell) + (f) - (\ell + e)$, where $e$ is the number of interior edges. Each edge in $C$ lies in the boundary of precisely one interior triangle, while each interior edge lies in the boundary of two such triangles. Since each triangle has three edges in its boundary and since $C$ has $\ell$ edges, we have $3f = 2e + \ell$. Therefore, $e = 3f/2 - \ell/2$ and $1 = v + f - e = v - f/2 + \ell/2$. ■

2.4. Corollary ($f$ is independent of triangulation). *Let $\Lambda$ be an arbitrary planar graph and let $\Lambda'$ denote a triangulation of $\Lambda$. For each circuit $C$ in $\Lambda$ the number of triangular faces of $\Lambda'$ that are interior to $C$ is independent of the triangulation $\Lambda'$.*

*Proof.* The result follows from (2.2) since $\ell$ and $v$ do not depend on the chosen triangulation $\Lambda'$. ■

With the aid of triangulation we can describe the **canonical flux distribution** for *any* planar graph $\Lambda$. Choose any triangulation $\Lambda'$ and place flux $\pi/2$ in every triangular face. By Lemma 2.2, this defines phases $\theta(x, y)$ on $E(\Lambda')$ and hence on $E(\Lambda)$. A priori, these phases might depend on the triangulation, but by Corollary 2.4, all triangulations give rise to the same set of fluxes through the circuits of $\Lambda$. By Lemma 2.1 the $\theta(x, y)$'s are uniquely defined up to a gauge transformation, i.e., $\theta(x, y) \to \theta(x, y) + \phi_x - \phi_y$ with the function $\phi_x$ being the only quantity that might depend on the triangulation. Since the flux distribution is invariant under gauge transformations, the canonical flux distribution is well defined!

Of special interest to us are bipartite planar graphs. In general, a **bipartite graph** is a graph $\Lambda$ whose vertex set $V$ is the union of two disjoint sets $A$ and $B$ with the property that $(x, y)$ is never an edge of $\Lambda$ if $x \in A$ and $y \in A$ or $x \in B$ and $y \in B$. We shall assume $|B| \geqslant |A|$. If $\Lambda$ is a planar bipartite graph, the canonical flux will always be $\pi$ through every elementary square, zero through every elementary hexagon, etc. However, one has to be cautious about this because one could have, for instance, a square with vertices $a$, $b$, $c$, and $d$ and a fifth vertex $g$ inside the square connected by an edge only to $a$. In this case our rule says that the canonical flux through the circuit $a$, $b$, $c$, and $d$ is zero and not $\pi$.

A special feature of bipartite graphs, planar or otherwise, is that the nonzero eigenvalues of any hopping matrix $T$ come in opposite pairs; i.e., if $\lambda$ is an eigenvalue of $T$, then so is $-\lambda$. This follows from $T = -U^*TU$ where $U$ is the diagonal unitary matrix with $+1$ on the $A$-sites and $-1$ on the $B$-vertices. $T$ itself can be written in the form $\begin{pmatrix} 0 & M \\ M^* & 0 \end{pmatrix}$, where $M$ contains the matrix elements between $A$-and $B$-sites.

**3. Determinants for planar graphs.** One of the main theorems of this paper is Theorem 3.1 about determinants of bipartite graphs, and one of the concepts needed there is that of the **dimer partition function** $D(T)$ of the graph $\Lambda$ with hopping matrix $T$. A **dimer covering** (or matching) of $\Lambda$ is a subset $\{e_1, e_2, \ldots, e_n\}$ of $E$ such that every site in $\Lambda$ is an end point of precisely one of the $e_i$'s. In general, $\Lambda$ has many dimer coverings, but it may have none at all. In particular, if $|\Lambda|$ is odd or if $\Lambda$ is bipartite and $|A| \neq |B|$, then there are no dimer coverings.

We define the **dimer partition function** to be

$$D(T) = \sum_{\text{dimer coverings}} \prod_i |t_{x_i,y_i}|, \tag{3.1}$$

where the product is over all the edges $e_i = (x_i, y_i)$ that consitute a particular dimer covering. If $|t_{xy}| = 1$, then $D(T)$ is just the number of dimer coverings of $\Lambda$. Note that $D(T)$ depends only on the $|t_{xy}|$'s and is therefore independent of the fluxes. In particular, $D(T)$ is determined by the upper triangular array $\{t_{xy}\}_{x \leqslant y}$. (See the appendix.)

3.1. THEOREM (Canonical flux counts dimers and maximizes bipartite graph determinants). *Let $\Lambda$ be a planar graph and let $|t_{xy}|$ be given positive numbers for all edges $(x, y)$ in $\Lambda$. For the canonical flux distribution*

$$\det T = (-1)^{|\Lambda|/2} D(T)^2. \tag{3.2}$$

*If, in addition, $\Lambda$ is bipartite the canonical flux distribution maximizes* $|\det T|$ *among all flux distributions.*

Before proving the theorem we make a series of remarks:

(i) Unless $|A| = |B|$ in the bipartite case, $D(T) = 0$ and $\det T = 0$ for *every* choice of flux. In the general case, $D(T) = 0$ unless $|\Lambda|$ is even. More generally, we could consider nonbipartite graphs with $T$ of the form $T_K = \begin{pmatrix} 0 & M \\ M^* & K \end{pmatrix}$, with $K$ self-adjoint. This means that edges are added between $B$-vertices but not between $A$-vertices. It is then an easy exercise in linear algebra to prove that $\det T_K = 0$ unless $|B| \geqslant |A|$ and that, if $|B| = |A|$, then $\det T_K$ is independent of $K$, i.e., $\det T_K = \det T_0$. As an example, start with the simple square, i.e., $|\Lambda| = 4$ and (1,2), (2,3), (3,4), and

(4,1) are the edges. Theorem 3.1 says that the determinant is maximized by flux $= \pi$ through the square. Now add a diagonal edge (1,3) with some hopping amplitude $|t_{1,3}|$ on this new edge. We now have a graph that consists of two triangles. The observation just made says that the determinant is independent of the individual fluxes through the two triangles and depends only on their sum. The canonical flux distribution, which is $\pi/2$ in each triangle, is optimal, but is by no means the unique optimizer.

(ii) In the bipartite case, the sign of the determinant as given in (3.2) is correct for any $T$, not just the canonical $T$. This follows from the $\lambda$, $-\lambda$ pairing of the eigenvalues which holds for a bipartite lattice.

We can go further and define the elementary symmetric functions

$$e_k(T) = \sum_{i_1 < i_2 < \cdots < i_k} \prod_{j=1}^{k} \lambda_{i_j}(T) \tag{3.3}$$

for $1 \leqslant k \leqslant |\Lambda|$ and with $e_0(T) = 1$. They are the coefficients in the characteristic polynomial

$$\det(T + z) = \sum_{k=0}^{|\Lambda|} e_k(T) z^{|\Lambda| - k}. \tag{3.4}$$

Here, the $\lambda_i(T)$'s are the eigenvalues of $T$. For the same reason (the $\lambda$, $-\lambda$ pairing) we see that for a bipartite graph

$$\begin{aligned} e_k(T) &= 0, && k \text{ odd} \\ e_k(T) &= (-1)^{k/2} |e_k(T)|, && k \text{ even}. \end{aligned} \tag{3.5}$$

In fact, $e_k(T)$ is a sum of determinants of principal submatrices of $T$ in which $|\Lambda| - k$ columns and corresponding rows are removed. Each such submatrix is naturally associated with a subgraph of $\Lambda$—which is necessarily bipartite as well. Therefore, we can conclude that the sign of the determinant of every principal submatrix of even order $k$ is $(-1)^{k/2}$. For odd $k$, such determinants are always zero. Warning: The canonical flux distribution need not maximize $|e_k(T)|$ for $k \neq |\Lambda|$. The reason is that the canonical flux distribution for a subgraph might differ from the one for the full graph $\Lambda$. See Theorem 5.1, however.

(iii) The canonical flux distribution maximizes det $T$ in the bipartite case. It fails, generally, to do so in the nonbipartite case; nevertheless, it does have a "maximum property" which is given in Theorem A.2 in the appendix.

(iv) In the appendix it is shown that Theorem 3.1 is one of the two key ingredients in a proof of Kasteleyn's theorem.

*Proof.* By definition, the determinant is a sum over permutations of monomials in the matrix elements of $T$, each of which is a product of the kind $\varepsilon(\pi) t_{1,\pi(1)} \cdots t_{|\Lambda|,\pi(|\Lambda|)}$. Here, $\varepsilon(\pi) = \pm 1$ is the signature of $\pi$. Using the cycle decomposition of the permutation $\pi$, we see that the above monomial can be written as $\prod_{j=1}^{k}(-1)^{\ell_j-1} \times \prod_{C_j} T$, where $C_1, \ldots, C_k$ is a family of circuits with the property that every vertex of the graph is in precisely one circuit. Here $\ell_j$ denotes the length of the circuit $C_j$. By the definition of the canonical flux distribution, $\prod_{C_j} T = \prod_{C_j} |t_{xy}| \exp[\pm i\pi f_j/2]$, where $f_j$ is the number of interior triangles of $C_j$, and the sign in the exponent indicates the orientation of $C_j$. Thus, the determinant is now a sum over all circuit decompositions of terms of the form $\prod_{j=1}^{k} \prod_{C_j} |t_{xy}|(-1)^{\ell_j-1} \cos(\pi f_j/2)$. Note that the factor 2 is counted by distinguishing circuits of different orientations.

By Lemma 2.3, $f_j = \ell_j + 2v_j - 2$. Thus, $(-1)^{\ell_j-1}\cos(\pi f_j/2) = (-1)^{\ell_j-1}\cos(\pi\ell_j/2 + \pi v_j - \pi) = (-1)^{\ell_j}\cos(\pi\ell_j/2 + \pi v_j)$. If $\ell_j$ is odd, the cosine vanishes. Hence, *only even circuits contribute to the determinant.* This is a crucial property of the canonical flux distribution! Moreover, since every vertex must belong to a circuit and every circuit has even length, $v_j$ is also even for all $j$ and hence $2v_j \equiv 0 \pmod 4$ and does not contribute to the sign of the monomial. Therefore, the monomial equals $\prod_{j=1}^{k}\cos(\pi\ell_j/2)\prod_{C_j}|t_{xy}| = \prod_{j=1}^{k}(-1)^{\ell_j/2}\prod_{C_j}|t_{xy}| = (-1)^{|\Lambda|/2}\prod_{C_j}|t_{xy}|$ since $\sum_{j=1}^{k}\ell_j = |\Lambda|$. Note that when $|\Lambda|$ is odd, there is at least one circuit of odd length in every circuit decomposition, and hence $\det T = 0$.

The last step is to derive relation (3.2), which is geometrically "obvious". It suffices to note in our case that $D(T)\cdot D(T)$ is a sum of terms, each of which is of the form $\mathscr{D}_1\mathscr{D}_2$, where $\mathscr{D}_1$ (likewise $\mathscr{D}_2$) denotes a single term in (3.1) corresponding to a single dimer covering. If we superimpose the two coverings, we get a collection of disjoint circuits $C_1, \ldots, C_k$ on $\Lambda$. Each site of $\Lambda$ is in exactly one of these circuits. Additionally, each circuit will have an *even* length. This "circuit covering" of $\Lambda$ corresponds to a term in det $T$. Conversely, each term in det $T$ corresponds to a "circuit covering". (Note: it is here that we use the fact that only circuits of even length contribute to det $T$; for otherwise some terms in det $T$ might give rise to "circuit coverings" that contain circuits of odd length.) All that is needed is to check that the weights in $D(T)^2$ correspond to those in det $T$. The weight of a "circuit covering" in det $T$ is $2^n$, where $n \leqslant k$ is the number of circuits whose length exceeds 2. The factor of 2 comes from the two possible orientations of the circuit, or in other words, the contribution of a cyclic permutation and its inverse. The same factor $2^n$ arises in $D(T)^2$ because each circuit can be decomposed into a dimer covering of the circuit in exactly two ways.

The last line of the theorem follows from the observation that, if a graph is bipartite, then det $T$ contains only even length circuits (for *all* fluxes, not just the canonical flux); the canonical flux then makes the contributions of the different circuit coverings add together constructively. ■

**4. Rings with arbitrary weights.** We begin our study of the problem of maximizing eigenvalue sums of $T$ with respect to fluxes by considering the simplest possible case. In the process some notation and identities will be established that will prove useful in later sections of this paper.

A **ring** of $R > 2$ vertices (or $R$ edges) is a graph $\Lambda$ with $|\Lambda| = R$ vertices labeled 1 up to $R$ and with edges $(1, 2), (2, 3), \ldots, (R-1, R), (R, 1)$. The hopping matrix is then determined by $R$ complex numbers $t_{12}, \ldots, t_{R1}$ with magnitudes given a priori as $|t_{i,i+1}|$. Note that $\Lambda$ is not necessarily bipartite, i.e., $R = |\Lambda|$ does not have to be even.

Although the spectrum of $T$ is easy to compute explicitly if $|t_{i,i+1}|$ is independent of $i$, and hence one might think that our main theorem here, Theorem 4.1, is without content, we draw the reader's attention to the fact that we shall consider all possible $T$'s. In other words, we shall be dealing with the "random one-dimensional Laplacian" whose spectrum is the object of much current research. From this point of view, it is somewhat surprising that some physical quantities of this random system can easily be maximized with respect to the flux.

While our goal is to compute $E_0(T)$ in (1.2), we shall consider more general functions of the eigenvalues of $T$. Let $f\colon \mathbb{R}^+ \to \mathbb{R}$ be a real-valued function defined for nonnegative reals and define $F$ by

$$F(T) = \operatorname{Tr} f(T^2) = \sum_{j=1}^{|\Lambda|} f(\lambda_j^2), \tag{4.1}$$

where $\lambda_1 \geqslant \lambda_2 \geqslant \cdots \geqslant \lambda_{|\Lambda|}$ are the eigenvalues of $T$. The $f$ needed for $E_0$ is

$$f_1(x) = \sqrt{x}, \tag{4.2}$$

while for $\ln|\det T|$ it is

$$f_2(x) = \tfrac{1}{2}\ln x. \tag{4.3}$$

Still another physically important function is

$$f_3(x) = \ln\cosh\sqrt{x} \tag{4.4}$$

appropriate to the free energy $\mathscr{F} = -\beta^{-1}\ln \Xi$ in the grand canonical ensemble[1] (1.3):

$$\begin{aligned}\mathscr{F} &= -2\beta^{-1}\operatorname{Tr}\ln(e^{\beta T}+1) = -2\beta^{-1}\operatorname{Tr}\{\ln[\cosh(\beta T/2)] + \ln 2 + \beta T/2\}\\ &= -2\beta^{-1}\operatorname{Tr} f_3(\beta^2 T^2/4) - 2\beta^{-1}|\Lambda|\ln 2\end{aligned} \tag{4.5}$$

since $\operatorname{Tr} T = 0$. Here, $\beta^{-1} =$ (Boltzmann's constant) $\times$ (temperature).

All these functions have the property of being *concave*, i.e., $f(\lambda x + (1-\lambda)y) \geqslant \lambda f(x) + (1-\lambda) f(y)$ for all $x \geqslant 0$, $y \geqslant 0$ and $0 < \lambda < 1$. (In fact they are *strictly* concave, i.e., equality implies that $x = y$.)

[1] In (4.5) we have set the chemical potential $\mu$ equal to zero. For a bipartite lattice this yields an average particle number $M = 2N = |\Lambda|$, which follows from $M = 2\sum_{j=1}^{|\Lambda|}\exp(\beta\lambda_j)[1+\exp(\beta\lambda_j)]^{-1}$ together with the $(\lambda, -\lambda)$ pairing.

These three functions also belong to a more restricted class of functions which we call **integrated Pick functions.** These are functions with the integral representation

$$f(x) = c \ln x + \int_0^\infty \ln\left(1 + \frac{x}{s}\right)\mu(ds) \tag{4.6}$$

where $c \geqslant 0$ and where $\mu$ is a nonnegative measure on $[0, \infty)$ such that this integral is finite. For the functions in (4.2)–(4.4) we have this integral representation with $\mu(ds) = (\text{const.})s^{-1/2}\, ds$ and $c = 0$ for (4.2), $\mu(ds) \equiv 0$ and $c = 1/2$ for (4.3) and $\mu(ds) = \sum_{k=0}^{\infty} \delta(s - [\pi(k + \frac{1}{2})]^2)\, ds$ and $c = 0$ with $\delta =$ Dirac's $\delta$-function for 4.4. See [KL, eqs. (3.12) and (3.16)]. Combining (4.1) with (4.6) yields

$$F(T) = c \ln \det T^2 + \int_0^\infty \ln \det(1 + T^2/s)\mu(ds), \tag{4.7}$$

and we see that the problem of maximizing $F(T)$ is reduced to that of maximizing various determinants with respect to the flux.

The function $G(T)$ given in (1.4) cannot be represented in the form (4.7); nevertheless, we shall also be able to maximize $G(T)$.

4.1. THEOREM (Maximizing flux for the ring). *Consider a hopping matrix $T$ with arbitrary, but fixed, amplitudes $|t_{xy}|$ on a ring of $R$ sites, let $f$ be an integrated Pick function given by (4.6), and let $F(T)$ be as in (4.1). Then the canonical flux $\pi(R + 2)/2$ (mod $2\pi$) maximizes both $F(T)$ and the gap $G(T)$ if $R$ is even. If $R$ is odd, $F(T)$ and $G(T)$ are maximized by both of the choices 0 and $\pi$.*

*Remark.* When $R$ is odd, the canonical flux is always $\pi/2$ or $3\pi/2$ and never 0 or $\pi$.

*Proof.* For $F(T)$ it suffices, by formula (4.7), to show that the flux described above maximizes $\det(c^2 + T^2) = |\det(ic + T)|^2$ for all real numbers $c$.

First, we observe that for the invariants (or elementary symmetric functions) we have that $e_k(T) = 0$ for $k$ odd and $1 \leqslant k \leqslant R - 1$. This follows directly from remark (ii) after Theorem 3.1 when $R$ is even. When $R$ is odd, it also follows from remark (ii) together with the observation that every proper subgraph of a ring is bipartite. We also see from remark (ii) that the sign of $e_{2m}(T)$ is $(-1)^m$.

With this information about the signs of the $e_k$'s, we can write, using (3.4) with $z = ic$,

$$\det(c^2 + T^2) = \begin{cases} \left(\sum_{m=0}^{|\Lambda|/2} |e_{2m}(T)|\, c^{|\Lambda|-2m}\right)^2, & |\Lambda| \text{ even} \\ \left(\sum_{m=0}^{(|\Lambda|-1)/2} |e_{2m}(T)|\, c^{|\Lambda|-2m}\right)^2 + (\det T)^2, & |\Lambda| \text{ odd}. \end{cases} \tag{4.8}$$

For future use, we remark that both parts of (4.8) hold for *any bipartite graph* $\Lambda$, not just a ring with an even number of sites.

Second, we shall show that, in the expression (3.4) for $\det(T + z)$, with $z \in \mathbb{C}$, the invariants $e_k(T)$ are *independent of the flux* if $k < R$. This is true only for a ring. Note that the invariants are real since $T$ is Hermitian. Recall that the number $e_k(T)$ can be computed from $T$ by calculating the subdeterminants of $T$ with any $R - k$ columns and corresponding rows removed, and then summing these numbers over all possible removals. The result is a sum of monomials of the form $\prod_j (-1)^{\ell_j - 1} \prod_{C_j} T$ where the product is taken over all circuits $C_i$ that cover the subgraph obtained by removing $k$ vertices and the corresponding edges. But for $k < R$ the only circuits that cover this subgraph form a dimer covering; their contribution does not depend on the flux but only on the numbers $|t_{i,i+1}|$. Thus the only term in (3.4) that depends on the flux is $e_{|\Lambda|}(T) = \det T$.

In both cases in (4.8), the problem of maximizing $\det(c^2 + T^2)$ is seen to be the same as maximizing $|\det T|$. If $R$ is even, this problem is solved in Theorem 3.1. If $R$ is odd, there are precisely two circuits that contribute to $\det T$. These are the circuits that traverse the entire ring (in either direction) and correspond to an even permutation. Thus, for a ring of odd length

$$\det T = 2\, Re \left\{ \prod_{i=1}^{R} t_{i,i+1} \right\} = 2(\cos \Phi) \prod_{i=1}^{R} |t_{i,i+1}|, \tag{4.9}$$

from which we see that $\Phi = 0$ or $\Phi = \pi$ maximizes $|\det T|$, and hence also $F(T)$. This completes the proof for $F(T)$.

To compute $G(T)$ we return to (4.8) and write $Q_\Phi(\lambda) := \det(T - \lambda) = P(\lambda) + \det T$, with $P$ being a polynomial of order $R$ whose coefficients are *independent* of the flux $\Phi$. $P$ is even if $R$ is even, and $P$ is odd if $R$ is odd. We note that, as $\det T$ varies between its maximum and minimum values, $Q_\Phi$ always has $R$ roots. We leave it to the reader to verify the following with the aid of a graph of $Q_\Phi(\lambda)$. Even $R$: The maximum separation between $\lambda_{R/2}$ and $\lambda_{R/2+1}$ is achieved by making $|Q_\Phi(0)|$ as large as possible. Since $Q_\Phi(0) = \det T$, this means choosing $\Phi$ to make $|\det T|$ as large as possible—as stated in our theorem. Odd $R$: If $\det T = 0$, the eigenvalues of $Q_\Phi$ are paired (because $P$ is odd). Then $-\lambda_{(R-1)/2} = \lambda_{(R+3)/2}$ and $\lambda_{(R+1)/2} = 0$. Thus $G_+ := [\lambda_{(R+1)/2} - \lambda_{(R+3)/2}] = [\lambda_{(R-1)/2} - \lambda_{(R+1)/2}] =: G_-$ when $\det T = 0$. As $|\det T|$ increases, either $G_+$ increases $G_-$ decreases and $\lambda_{(R+1)/2} > 0$ or vice versa and $\lambda_{(R+1)/2} < 0$. Thus $G = \max(G_+, G_-)$, and this increases with $|\det T|$. By (4.9) we see that $\Phi = 0$ or $\Phi = \pi$ maximizes $|\det T|$. ∎

**5. Trees of rings.** Most of the results in Theorem 4.1 for bipartite rings with arbitrary hopping amplitudes $|t_{xy}|$ can be extended to a much larger class of planar graphs. Two special cases of this class are the ladders and the necklaces; they are discussed in detail in the next section because even stronger results can be obtained for them. It was those two classes, in fact, that were the origin of this work and that were reported in [LE].

A planar graph $\Lambda$ is said to be a **tree of rings** if and only if $\Lambda$ has an embedding in $\mathbb{R}^2$ such that every circuit in $\Lambda$ has no interior vertices.

The simplest example consists of two rings which have exactly one vertex in common. Another example consists of two rings that have exactly one edge (i.e., two neighboring vertices) in common. More generally, one can have a "tree of rings" in which two successive rings share either one edge or one vertex. The canonical flux distribution for a tree of rings would have flux $(\pi/2)[(\ell - 2)(\text{mod } 4)]$ in each circuit of length $\ell$.

5.1. THEOREM (Maximizing flux for bipartite trees of rings). *Let $\Lambda$ be a bipartite, planar graph that is a tree of rings and let $|t_{xy}|$ be arbitrary given hopping amplitudes. For $f$ an integrated Pick function, let $F(T)$ be as in (4.1). Then the canonical flux distribution maximizes $F(T)$. Moreover, it also maximizes the magnitude of each elementary symmetric function $e_k(T)$ defined in (3.3).*

*Proof.* As in the proof of Theorem 4.1 (cf. eq. (4.8)), we have that $\det(c^2 + T^2)$ will be maximized if we can *simultaneously* maximize all the $|e_k(T)|$'s and if they all have the sign$(-1)^{k/2}$. The latter question was dealt with in remark (ii) just after Theorem 3.1.

Each $e_k(T)$ can be evaluated as a sum of determinants of principal submatrices of $T$ of order $k$. In terms of graphs, a particular term in the sum is the determinant of $T$ restricted to a subgraph $\Lambda'$ with $|\Lambda'| = k$. The important point is that the circuits of $\Lambda'$: (i) are a subset of the circuits of $\Lambda$ and (ii) have no interior points. The canonical flux distribution for $\Lambda'$ is the same as for $\Lambda$; this means that, if $C$ is a circuit that is both in $\Lambda$ and in $\Lambda'$, then $\Phi_C = \Phi'_C = 0$ or $\pi$ where $\Phi_C$ is the canonical flux through $C$ (in $\Lambda$) and $\Phi'_C$ is the canonical flux (in $\Lambda'$). (Note: The only way in which $\Phi_C$ could differ from $\Phi'_C$ is if $C$ had some interior vertices that were removed in passing from $\Lambda$ to $\Lambda'$. But $C$ had no interior vertices to start with.) Hence each subdeterminant appearing in $e_k(T)$ is maximized (in absolute value) by the original canonical flux distribution in $\Lambda$. Since the signs of all these subdeterminants are the same—in fact they depend only on $k$ (see remark (ii) after Theorem 3.1)—we see that $|e_k(T)|$ is maximized. ■

**6. Ladders and necklaces.** Most, but not all the graphs considered in this section are special cases of those discussed in Section 5. Here we consider certain graphs that are finite subsets of the infinite lattice $\mathbf{Z}^2$, which is the infinite embedded graph whose vertices are points in the plane with integer coordinates and whose edges are the horizontal and vertical line segments joining vertices a unit distance apart. Of particular importance are *boxes*, which are the subgraphs of $\mathbf{Z}^2$ with 4 vertices and 4 edges forming a circuit. In general, our graphs need not be subgraphs of $\mathbf{Z}^2$; i.e., they need not contain *all* the edges of $\mathbf{Z}^2$ that connect the vertices $V$ in our graph. (Example: $\Lambda$ contains the 4 vertices of a box but only 3 of its edges.) Evidently, all our graphs are bipartite—with the $A$-$B$ decomposition of their vertices being the one inherited from $\mathbf{Z}^2$ (see the penultimate paragraph of Section 2). Before describing them in detail, a few remarks are needed.

If $T$ is a hopping matrix of a *bipartite* graph $\Lambda$, the matrix $T^2$ is evidently block diagonal, i.e., $T^2 = \begin{pmatrix} \alpha_T & 0 \\ 0 & \beta_T \end{pmatrix}$, where $\alpha_T$ is $|A| \times |A|$, $\beta_T$ is $|B| \times |B|$ and both are positive semidefinite. Assuming that $v := |B| - |A| \geqslant 0$, we have that the eigenvalues satisfy

$$\operatorname{spec}(\beta_T) = \operatorname{spec}(\alpha_T) \cup \{v \text{ zeros}\}. \tag{6.1}$$

This is a simple consequence of the fact that $T = \begin{pmatrix} 0 & M \\ M^* & 0 \end{pmatrix}$, so that $\alpha_T = MM^*$ and $\beta_T = M^*M$. Since the eigenvalues of $T$ come in pairs and there are $|\Lambda|$ of them, we conclude that

$$\operatorname{spec}(T^2) = \{\operatorname{spec}(\alpha_T) \text{ with double multiplicity}\} \cup \{v \text{ zeros}\}, \tag{6.2}$$

$$\operatorname{spec}(T) = \operatorname{spec}(\sqrt{\alpha_T}) \cup \operatorname{spec}(-\sqrt{\alpha_T}) \cup \{v \text{ zeros}\}. \tag{6.3}$$

Thus spec($T$) is determined by *either* $\alpha_T$ or $\beta_T$ alone.

The matrix $\alpha_T$ has diagonal elements, $(\alpha_T)_{aa} = \sum_{b \in B} |t_{ab}|^2$ for $a \in A$. The off-diagonal elements of $\alpha_T$ can be thought of as a hopping matrix of a new graph $\Lambda_A$, which need not be planar. The vertices of $\Lambda_A$ are the $A$-vertices of $\Lambda$. A necessary condition for the pair $(a, a')$ to be an edge of $\Lambda_A$ is that there is a $B$-vertex $b$ such that $(a, b)$ and $(b, a')$ are edges in $\Lambda$. There may be more than one such $b$ for a given pair $(a, a')$, and it is important to note that, since $(\alpha_T)_{aa'} = \sum_b t_{ab} t_{ba'}$, it can happen that $(\alpha_T)_{aa'} = 0$, in which case $(a, a')$ is *not* an edge of $\Lambda_A$, in conformity with our earlier convention. Similar remarks hold for $(\beta_T)_{bb'}$ and $\Lambda_B$.

There are special edges in $\Lambda_A$ or $\Lambda_B$ which we call **interior diagonals**. These are edges $(a, a')$ or $(b, b')$ in which $a$ and $a'$ (or $b$ and $b'$) belong to some (same) box $S$ that is a subgraph of $\Lambda$. This set of edges is denoted by $D_A$. Since circuits of length 4 can only be the edges of boxes, it follows that the interior diagonals are the *only* edges that can possibly disappear from $\Lambda_A$ because of the equality $\sum_b t_{ab} t_{ba'} = 0$. Likewise for $\Lambda_B$ and $D_B$.

The graphs we shall consider here can be described as follows. Let $\Lambda'_A = \Lambda_A \sim D_A$; i.e., the vertices of $\Lambda'_A$ are those of $\Lambda_A$ but the edges are those of $\Lambda_A$ *without* the interior diagonals. Analogously, $\Lambda'_B$ is defined. We say that $\Lambda$ is a **hidden tree** if either $\Lambda'_A$ or $\Lambda'_B$ is a tree (i.e., does not contain any circuits).

Two important examples introduced in [LE] are **ladders** and **necklaces**. Each is a connected union of $n$ boxes, labeled $1, 2, \ldots, n$, forming a one-dimensional array. In the ladder the boxes are joined along (parallel) edges, with box $j$ connected to $j + 1$. In the necklace, boxes $j$ and $j + 1$ have only a single vertex in common, and these vertices are either all $A$ or all $B$. In both examples, boxes $j$ and $k$ are disjoint if $|j - k| > 1$. With the usual orientation of $\mathbf{Z}^2$, ladders are either horizontal or vertical, while a necklace runs at 45° to either of these directions.

One can generalize the ladder by allowing occasional 90° bends, while still keeping the one-dimensional character. Now, squares $j$ and $k$ can now have a vertex (but not an edge) in common if $|j-k|=2$; squares 1 and $n$ are disjoint. The bends cannot be completely arbitrary, however, because the "hidden tree" condition must be maintained. It can never be maintained for a necklace with 90° bends.

Another not so trivial example is that in which $\Lambda$ is the union of 4 squares, all of which have a vertex in common and which together form a square of side length 2. Here $|A|=4$ and $|B|=5$. Although $\Lambda'_A$ is a tree, $\Lambda'_B$ is *not* a tree. (6.1) notwithstanding, it is somewhat surprising, when viewing the graphs for $\Lambda_A$ and $\Lambda_B$, that they have the same spectrum—except for one zero eigenvalue. Another example in this vein is the $\Lambda$ that resembles a (3,2) Young diagram, i.e., 3 squares in a horizontal row and 2 squares, also in a horizontal row, directly beneath them. The simplest case that is not a hidden tree, and for which none of our theorems apply, is two rows of three squares each.

Finally, we make some remarks about the next theorem.

(i) Ladders and necklaces are trees of rings, but the last three examples (two rows of squares) are not. Thus for ladders and necklaces, the fact that the canonical flux distribution maximizes $\mathrm{Tr}|T|$, $\mathscr{F}$ and $|\det T|$ is already covered by Theorem 5.1. The statement about the gap is new, however, as is the method of proof.

(ii) The concave function $F(T^2)$ is definitely a generalization of the function in (4.1). Not all concave functions (even those that are invariant under unitary transformations) are eigenvalue sums as in (4.1). In particular, the sum of the $k$ lowest eigenvalues of $T^2$ is not such a function, and it is needed, in fact, to prove the theorem about the gap.

6.1. THEOREM (Canonical flux maximizes concave functions on hidden trees). *Let $\Lambda$ be a graph that is a subset of $\mathbf{Z}^2$ and suppose that $\Lambda'_A$ (resp. $\Lambda'_B$) is a tree. Let $F$ be a concave function on the cone of positive definite matrices of order $|A|$ (resp. $|B|$) with the property that $F(U^*PU)=F(P)$ for every $P>0$ and every gauge transformation $U$ (restricted to $\Lambda_A$, of course). Finally, let $\{|t_{xy}|\}$ be unit hopping amplitudes on $\Lambda$, i.e., $|t_{xy}|=1$ if $(x,y)\in E(\Lambda)$.*

*Our conclusion is that among all hopping matrices $T$ with this unit hopping amplitude, $F(\alpha_T)$ (resp. $F(\beta_T)$) is maximized by the canonical flux.*

*Proof.* We shall assume $\Lambda'_A$ is a tree. The proof for $\Lambda'_B$ is similar. Assume $T$ maximizes $F(\alpha_T)$. Let $U=\{u_x\delta_{xy}\}_{x,y\in V(\Lambda)}$ be the following gauge transformation: if $x=(n,m)$ with $n,m\in\mathbf{Z}$, then $u_x=(-1)^n$. Let $Y=U^*TU$. By concavity and gauge invariance, the block diagonal matrix $P^2:=(1/2)(T^2+Y^2)$ satisfies $F(\alpha_P)\geqslant(1/2)F(\alpha_T)+(1/2)F(\alpha_Y)=F(\alpha_T)$. This inequality proves our theorem if we can show that the matrix $\alpha_P$ can be achieved by the canonical flux distribution, i.e., if there is a gauge such that $C:=$ ($T$ with the canonical flux distribution) satisfies $\alpha_C=\alpha_P$.

Now note that $T^2$ and $Y^2 = U^*T^2U$ are related as follows:

$$(Y^2)_{xy} = -(T^2)_{xy} \qquad \text{if } (x, y) \text{ is an interior diagonal;}$$

$$(Y^2)_{xy} = (T^2)_{xy} \qquad \text{otherwise.}$$

Therefore, $(P^2)_{xy} = 0$ for $(x, y) \in D_A \cup D_B$ and $(P^2)_{xy} = (T^2)_{xy}$ otherwise.

For any gauge, $(C^2)_{xy} = 0$ if $(x, y)$ is an interior diagonal. This is so because the flux through each box in $\Lambda$ is $\pi$, and if we label the four vertices 1, 2, 3, 4 (counterclockwise) with 1 and $3 \in V_A$, we have $(C^2)_{13} = C_{12}C_{23} + C_{14}C_{43}$. But $C_{12}C_{23}C_{34}C_{41} = -1$ and $C_{14} = \overline{C}_{41} = 1/C_{41}$, $C_{43} = 1/C_{34}$ (since, e.g., $|C_{14}|^2 = 1$), so $(C^2)_{13} = 0$, as required. As for the diagonal elements, they are clearly equal, i.e., $(C^2)_{xx} = (P^2)_{xx}$.

Finally, we have to compare the other matrix elements of $C^2$ and $P^2$ on $\Lambda_A$. In fact, they are both nonzero only on $\Lambda'_A$, in which case they satisfy $|(C^2)_{aa'}| = |(P^2)_{aa'}| = 1$ because there is precisely one path (i.e., $B$-vertex) between $a$ and $a'$. Therefore $(C^2)_{aa'} = \exp[i\theta(a, a')](P^2)_{aa'}$ for each edge in $\Lambda_{A'}$. The relevant question is then the following: Is there a gauge transformation $U$ such that $(U^*C^2U)_{aa'} = (P^2)_{aa'}$? In other words, can we find $u_x = \exp[i\phi_x]$ such that $\phi_a - \phi_{a'} = \theta(a, a')$ for every edge $(a, a')$ in $\Lambda'_A$? Since $\Lambda'_A$ is a tree, the answer is trivially, yes. All one-forms on a tree are exact. ∎

*Applications.* For a graph with hopping amplitudes satisfying the hypotheses of Theorem 6.1, the canonical flux distribution yields:

(a) the lowest ground state energy $E_0(T)$ and free energy $\mathscr{F}(T)$ for all temperatures;
(b) the largest $|\det T|$ and gap $\tilde{G}(T)$.

The functions $-E_0(T)$, $-\mathscr{F}(T)$ and $\log|\det T|$ are concave since they are integrated Pick functions, which are concave in $T^2$ as mentioned in Section 4. The gap $\tilde{G}(T)$ can be computed from the matrices $\alpha_T$ and $\beta_T$. Assuming that $\nu \equiv |B| - |A| \geqslant 0$, we see that $\tilde{G}(T) = 2(\inf \operatorname{spec} \alpha_T)^{1/2}$ and that $\tilde{G}(T) = 2(\sum_{i=1}^{\nu+1} \gamma_i)^{1/2}$, where $\gamma_i$ denote the eigenvalues of $\beta_T$ arranged in increasing order. Of course, we used that $\gamma_1 = \gamma_2 = \cdots = \gamma_\nu = 0$. Now the sum of the first $k$ eigenvalues of a Hermitian matrix $H$ is a concave function of $H$. Moreover since $x \mapsto \sqrt{x}$ is concave and increasing, we see that $\tilde{G}(T)$ is a concave function of $\alpha_T$ (resp. $\beta_T$). Moreover, it is gauge invariant and hence satisfies the assumptions of Theorem 6.1. Note that we had to discuss the above two formulas for $\tilde{G}(T)$ since both possibilities ($\Lambda'_A$ is a tree or $\Lambda'_B$ is a tree) have to be considered.

In the case of ladders and necklaces the results about $E_0(T)$, $\mathscr{F}(T)$ and $|\det T|$ were covered by Theorem 5.1, but the examples with two rows of boxes, cited above, were not covered. (In fact, the two-rowed examples *cannot* be extended to the full generality of Theorem 5.1; see Section 7B.) For ladders and necklaces, the result about the gap is not covered by Theorem 5.1. In [LE] it was mistakenly asserted in Theorem 1 that (a) and (b) hold for *fully* generalized ladders and necklaces in which *arbitrary* 90° bends are allowed. Indeed, for $E_0(T)$, $\mathscr{F}(T)$ and $|\det T|$ this is correct (by Theorem 5.1). For the gap, however, we must use Theorem 6.1, and this fails for bent necklaces and for ladders with arbitrary bends. It does hold, however, for generalized ladders that are hidden trees.

*Additional remarks and examples.* There are two more cases where the concavity argument in the proof of Theorem 6.1 is applicable but where the graph is not necessarily a hidden tree, or even planar.

(i) *Row of cubes.* Instead of a row of squares as in the ladder, take $\Lambda$ to be a row of cubes joined on their faces, i.e., neighboring cubes have four edges and one face in common. Such a graph is, in fact, bipartite and planar, but it is not a hidden tree. If there are $n$ cubes, then $\Lambda$ is the $4 \times n$ planar, square lattice with "periodic boundary conditions" in one direction. (Indeed, we can even make the row of cubes into a torus—i.e., attach the first cube to the last—which is the same thing as the $4 \times (n + 1)$ planar, square lattice with periodic boundary conditions in both directions; the following argument will continue to work in this case provided $n$ is odd.)

We assume, as in Theorem 6.1, that $|t_{xy}| = 1$ for every edge. The flux in every face can easily be arranged to be $\pi$ in the following way. Start with the face that cube 1 and cube 2 have in common and put flux $\pi$ through it by making $t_{xy} = 1$ on three edges and $-1$ on the fourth. Then use the negative of this on the corresponding edges that cube 2 and cube 3 have in common—and so on alternately. Finally, set $t_{xy} = +1$ on the remaining edges, i.e., those edges that are perpendicular to the faces between the cubes. A first application of the concavity argument shows that we get an upper bound in terms of $T^2$, but without interior diagonals on the faces common to all the cubes. In a second, similar application of the argument, we can get an upper bound in terms of a matrix that has no interior diagonals on any of the other faces of the cubes as well. In fact, the only nonzero elements of $T^2$ that remain will consist of four independent one-dimensional chains (or else, in the case of the torus, four independent rings of length $n + 1$ with zero flux through each ring). Theorem 4.1 says that, if $n + 1 \equiv 2 (\text{mod } 4)$, the optimizing flux for such rings is zero and hence the above concavity argument shows that our choice of fluxes cannot be improved. If $n + 1 \equiv 0 (\text{mod } 4)$, the optimum choice for such a ring is flux $\pi$. This can be achieved by a slight modification of our initial choice of the $t_{xy}$'s along the edges perpendicular to the intercube faces. Initially we chose them all to be $+1$, but now we choose them all to be $+1$ *except* for the $n$th cube. There we choose $t_{xy} = -1$ along the four perpendicular edges.

(ii) *$SU(2)$-valued fields.* We have considered the case that $t_{xy} = |t_{xy}| \exp[i\theta(x, y)]$ with $e^{i\theta}$ the unknown variable. It is also amusing to replace $e^{i\theta}$, which is in $U(1)$ by a $2 \times 2$ matrix $U_{xy}$ in $SU(2)$. In other words, $T$ becomes a $2|\Lambda| \times 2|\Lambda|$ Hermitian matrix in which each $t_{xy}$ (for $x, y \in V(\Lambda)$) equals a given number $|t_{xy}|$ times an $(x, y)$-dependent element of $SU(2)$. We require $t_{xy} = t_{xy}^*$. Theorem 6.1 goes through in this case when $|t_{xy}| = 1$. We do not know whether Theorem 3.1, for instance, can be generalized to the $SU(2)$ case.

There is, however, an interesting special feature of $SU(2)$. For the ladder or row of cubes, we had to break the translation invariance from period one to period two in order to achieve flux $\pi$ in each face; i.e., $T$ could not be made the same in every box but we had to translate by two boxes (or cubes) in order to recover $T$. With $SU(2)$ fields we can achieve the optimal flux distribution with period one. This means the following. We require that the product of the four $t_{xy}$'s around a square face (which is now a matrix product, of course) is the matrix $-I \in SU(2)$. This can be achieved by placing $i\sigma^1$ along all horizontal edges, $i\sigma^2$ along all vertical edges, and (in the case of cubes) $i\sigma^3$ along all the edges in the remaining direction. Here $\sigma^1$, $\sigma^2$, and $\sigma^3$ are the Pauli matrices. We then have $-I$ in every face because, for example, $(i\sigma^1)(i\sigma^2)(-i\sigma^1)(-i\sigma^2) = -I$.

**7. Some conjectures and counterexamples.**

*A. The smallest determinant.* A natural question, to be compared with Theorem 3.1, is "Which flux distribution *minimizes* $|\det T|$ for a bipartite lattice?" Since the canonical flux distribution maximizes $|\det T|$ and since it places flux $\pi$ in each square face (which is the maximum possible flux), it might be supposed that the answer to the question is zero flux, i.e., set $t_{xy} = |t_{xy}|$ for every edge in $E$. In the case where $\Lambda$ is a simply connected net of boxes on $\mathbf{Z}^2$ and $|t_{xy}| = 1$, the determinant was computed by Deift and Tomei [DT] to have the three possible values 0, $-1$, or $+1$. Despite this supportive example, *the above conjecture is wrong.* In the case of two boxes with one common edge and $|t_{xy}| = 1$, the determinant *vanishes* when the flux in each square is $\pi/3$. On the other hand, $\det T = -1$ when the flux is zero.

*B. The smallest energy.* In Section 5 we have seen examples of some graphs whose energy is minimized by the canonical flux distribution *for arbitrary* hopping amplitudes. Moreover, $|\det T|$ is always maximized by that flux distribution for *every* bipartite, planar graph with $|A| = |B|$. For such an arbitrary graph it is therefore natural to conjecture that $E(T)$ is also always minimized by the canonical flux distribution. Alas, this conjecture is also false for arbitrary $|t_{xy}|$, as we now show by an example.

In $\mathbf{Z}^2$ consider the graph consisting of four boxes arranged in a square, i.e., $\Lambda$ has the nine vertices $o = (0, 0)$, $a = (1, 0)$, $b = (0, 1)$, $c = (-1, 0)$, $d = (0, -1)$, and $(\pm 1, \pm 1)$. If the conjecture were true for arbitrary amplitudes, then $E_0(T)$ would always be minimal if the flux in each square is $\pi$. If we now let $|t_{oa}|$, $|t_{ob}|$, $|t_{oc}|$, and $|t_{od}|$ tend to zero, $E_0(T)$ becomes that of a ring of eight sites for which $E_0(T)$ is minimized by flux $\pi$ (not $0 \equiv 4\pi$), as we saw in Theorem 4.1. Indeed, to see that 0 and $\pi$ do not give the same value for $E_0(T)$ in general, for this ring, assume that the $t$'s on the ring all have amplitudes equal to one. If the flux is zero, then $E_0(T) = -\mathrm{Tr}|T|$ is easily computed to be $-4(1 + \sqrt{2})$, and for flux $\pi$ it equals $-2^{9/4}(\sqrt{1 + \sqrt{2}} + \sqrt{\sqrt{2} - 1})$, which is more negative.

**8. The Falicov-Kimball model.** So far, all our models concerned hopping matrices on a graph $\Lambda$. It is likely that the results obtained do not hold if we add a diagonal term to $T$, i.e., if we replace $T$ by the matrix

$$H = -T + 2UW. \tag{8.1}$$

Here $W$ is a real diagonal matrix satisfying $0 \leqslant W_x \leqslant 1$ for each $x$, and $U$ is a given real number called the coupling constant.

The eigenvalues of $H$ can be interpreted as the possible energy levels of a single electron hopping on a graph $\Lambda$ with kinetic energy $-T$ and potential energy $2UW$. This model, with $W_x$ restricted to be 0 or 1, was introduced in [FK] as a model for a semiconductor-metal transition, and it was studied extensively in [KL], where it was called the "static model", and in [BS]. Our generalization to $W_x \in [0, 1]$ for each $x$ is a mild one that we include here primarily because it can be handled without extra complication. The points of view taken in [KL] were different from that in [FK]. In [KL], the model was considered either as a simplified version of the Hubbard model in which electrons of one sign of spin are infinitely massive and therefore do not hop, or else as a model of independent electrons interacting with static nuclei. In the first view, $U > 0$ and $U < 0$ are both relevant. In the second, $U < 0$ is the physically relevant sign, and $W_x = 1$ (resp. 0) denotes the presence (resp. absence) of a nucleus at $x$.

The eigenvalues of $H$ in (8.1) are denoted by $-\nu$ (not $-\lambda$) which, as usual, are ordered $\nu_1 \geqslant \nu_2 \geqslant \cdots \geqslant \nu_{|\Lambda|}$. The ground state energy of a system of $N$ *spinless* electrons, interacting with a magnetic field and with the nuclei, is given, as usual, by

$$E_0^{(N)} = -\sum_{i=1}^{N} \nu_i . \tag{8.2}$$

(There is no 2 here, as in (1.1), because there is no spin.) Notice that the spectrum of (8.1) is still invariant under gauge transformations. Thus the energy $E_0^{(N)}$ depends on $N$, the flux distribution, and of course on $W$.

As mentioned before, minimizing the energy over all fluxes with a fixed $W$ is presumably a hopeless endeavor, but the situation becomes easier if one tries to minimize the energy with respect to the fluxes *and* $W$. It is clear that the minimum is attained when $W = 0$ if $U > 0$ or when $W = I$ if $U < 0$, which is uninteresting both mathematically and physically. If, however, we introduce $N_n := \sum_{x \in \Lambda} W_x =$ (total charge of the static particles), then the half-filled band condition (from the Hubbard model point of view, at least) is $N_n + N = |\Lambda|$. This is the case that parallels the restriction in the previous parts of this paper, since it means that on the average each lattice site is occupied by one particle. In fact, we shall be a bit more general in the $U > 0$ case and will treat a slightly different case when $U < 0$. We shall optimize the energy over all flux distributions, all potentials $W$, and all choices of

$N$ subject to one of the following three constraints:

$$\sum_{x \in \Lambda} W_x + N \leqslant 2|A| \qquad \text{if } U < 0, \tag{8.3a}$$

$$\sum_{x \in \Lambda} W_x + N \leqslant 2|B| \qquad \text{if } U < 0, \tag{8.3b}$$

$$\sum_{x \in \Lambda} W_x + N \geqslant |\Lambda| \qquad \text{if } U > 0. \tag{8.3c}$$

Theorem 2.1 in [KL] says that in these three cases we can easily compute the minimum of $E_0^{(N)}$ with respect to $W$ and $M$—regardless of the flux distribution. This result requires only one particular structure of the graph, namely that it is bipartite. Nothing else is required. As we shall see in Theorem 8.2, the result is that the nuclei want to occupy only the $A$-sites or the $B$-sites in order to minimize the total energy. We emphasize again that this fact is *independent* of the magnetic field.

Lemma 8.1 and Theorem 8.2 are really a transcription to the $W_x \in [0, 1]$ case of Lemma 2.2 and Theorem 2.1 in [KL]. Since they are short, we give them here. It is convenient to introduce the matrix $S = 2W - I$, so that $S$ is diagonal with $S_x \in [-1, 1]$. Thus $H = h + UI$ with

$$h = -T + US. \tag{8.4}$$

The matrix $h$ has eigenvalues $-\mu_1 \leqslant -\mu_2 \leqslant \cdots \leqslant -\mu_{|\Lambda|}$, with $\mu_j = \nu_j + U$.

8.1. Lemma (Maximization with respect to the nuclear configuration). *Let $\Lambda$ be a bipartite (not necessarily planar) graph and $T$ a prescribed hopping matrix on $\Lambda$. We consider all functions $S$ on the vertices of $\Lambda$ satisfying $-1 \leqslant S_x \leqslant 1$ for all $x$. Let $F$ be a concave, nondecreasing function from the set of Hermitian positive semidefinite matrices into the reals. Assume also that $F$ is gauge invariant, $F(U^*PU) = F(P)$ when $U$ is a gauge transformation.*

*Then $F(h^2)$ is maximized with respect to $S$ at $S = V$ and at $S = -V$, where*

$$V_x = \begin{cases} +1 & \textit{for } x \in A \\ -1 & \textit{for } x \in B. \end{cases} \tag{8.5}$$

*If $F$ is strictly concave and strictly increasing, then these are the only maximizers.*

*Proof.* The matrix $V = V^*$ is a gauge transformation, and hence $h$ and $h' := VhV$ satisfy $F(h^2) = F(h'^2)$. Now $h^2 = T^2 + U^2S^2 - U(TS + ST)$ and, since $VTV = -T$ and $VSV = S$, $h'^2 = T^2 + U^2S^2 + U(TS + ST)$. By concavity

$$F(h^2) = \tfrac{1}{2}[F(h^2) + F(h'^2)] \leqslant F(T^2 + U^2S^2) \leqslant F(T^2 + U^2I) \tag{8.6}$$

since $F$ is nondecreasing and $S^2 \leqslant I$. Note that $T^2 + U^2I = h^2$ when $S$ is chosen to be $+V$ or $-V$. If $F$ is strictly increasing and strictly concave, we can have equality in (8.6) only if $TS + ST = 0$ and $S^2 = I$. The former implies that $t_{xy}(S_x + S_y) = 0$, which implies (since $\Lambda$ is connected) that $S = (\text{constant})V$. The latter implies that $(\text{constant}) = \pm 1$. ■

8.2. THEOREM (Energy minima with respect to nuclear configurations). *Let $\Lambda$ be a bipartite graph (not necessarily planar) and let $T$ be a prescribed hopping matrix. We consider functions $W$ on the vertices of $\Lambda$ satisfying $0 \leqslant W_x \leqslant 1$.*

*For the three cases given in (8.3), the minimum value of $E_0^{(N)}$ with respect to $W$ and $N$ is uniquely achieved as follows:*

$$N = |A| \quad and \quad W = W_A := \tfrac{1}{2}(I + V), \qquad U < 0, \tag{8.7a}$$

$$N = |B| \quad and \quad W = W_B := \tfrac{1}{2}(I - V), \qquad U < 0, \tag{8.7b}$$

$$N = |B| \text{ and } W = W_A \quad or \quad N = |A| \text{ and } W = W_B, \qquad U > 0. \tag{8.7c}$$

*For each of these three cases, the minimum $E_0^{(N)}$ is given by*

$$E_0^{(N)} = -\tfrac{1}{2}\operatorname{Tr}|h| + \tfrac{1}{2}\operatorname{Tr} h + UN, \tag{8.8}$$

*where $h = -T + U(2W - I)$.*

*Proof.* By (8.2) and (8.4) we have that $E_0^{(N)} = -\sum_{j=1}^{N} v_j \geqslant -(1/2)\operatorname{Tr}|h| + (1/2)\operatorname{Tr} h + UN =: A(h)$. By Lemma 8.1, $\operatorname{Tr}|h| = \operatorname{Tr}\sqrt{h^2}$ is maximal precisely at $W = W_A$ or $W = W_B$ since $x \to \sqrt{x}$ is a strictly increasing and strictly concave function. Also, $(1/2)\operatorname{Tr} h + UN = U\{\sum_x W_x - |\Lambda|/2 + N\} =: B(h)$. If $U > 0$, $B(h) \geqslant |\Lambda|/2$. If $U < 0$, $B(h) \geqslant U(2|A| - |\Lambda|/2)$ for (8.3a) and $B(h) \geqslant U(2|B| - |\Lambda|/2)$ for (8.3b). These three lower bounds in $B(h)$ are attained (under conditions (8.3)) if $N$ and $W$ satisfy (8.7).

To complete the proof, we have to show that the lower bound on $-(1/2)\operatorname{Tr}|h| =: C(h)$, given in Lemma 8.1, is compatible with the condition on $N$ given in (8.7) (when $W$ is also that given in (8.7)). For example, we have to show that, if $W = W_A$ and $U < 0$, then the sum of the negative eigenvalues of $h$ (namely $-\sum_{\mu_j \geqslant 0} \mu_j = -(1/2)\operatorname{Tr}|h| + (1/2)\operatorname{Tr} h$) equals the sum of the lowest $|A|$ eigenvalues of $h$. In other words, we have to show that $h$ has exactly $|A|$ negative eigenvalues. We do so now with a proof different from the one in [KL]. First note that, for $t \in [0, 1]$, $h_t = -tT + UV$ has no zero eigenvalues when $U \neq 0$ since $h_t^2 = t^2T^2 + U^2I \geqslant U^2I > 0$. Second, the matrix $h_0 = UV$ has precisely $|A|$ negative eigenvalues because $U < 0$. Since the eigenvalues of $h_t$ are continuous functions of $t$ and because no eigenvalue can cross zero, $h$ also has precisely $|A|$ negative eigenvalues. The other two cases (8.7b) and (8.7c) are treated in the same fashion. ■

The next theorem is our main result about the FK model in a magnetic field.

8.3. THEOREM (Canonical flux minimizes energy on trees of rings). *Let $\Lambda$ be a bipartite tree of rings, and $T$ a hopping matrix with arbitrarily prescribed amplitudes, $|t_{xy}|$. As in Theorem 8.2, we consider functions $W$ on the vertices of $\Lambda$ satisfying $0 \leqslant W_x \leqslant 1$.*

*For the three cases given in (8.3), the minimum value of $E_0^{(N)}$ with respect to the flux distribution, $N$, and $W$ is achieved by the canonical flux distribution together with the $N$ and $W$ given by (8.7).*

*Proof.* Minimizing first with respect to $N$ and $W$, we can assume, by Theorem 8.2, that (8.7) is satisfied. In these cases we have that $E_0 = -(1/2)\operatorname{Tr}|h_{A,B}| +$ constant, where the constant depends on the case but *not* on the flux distribution. In each case $W = W_A$ or $W_B$, and we denote the two choices of $h$ by $h_A$ and $h_B$. Note that $N$ no longer enters the discussion. Our only goal now is to maximize $\operatorname{Tr}|h_{A,B}|$ with respect to the flux distribution. But $\operatorname{Tr}|h_{A,B}| = \operatorname{Tr}\sqrt{h_{A,B}^2}$ and $h_{A,B}^2 = T^2 + U^2 I$ since $S_{A,B}^2 = I$. The function $x \mapsto \sqrt{x}$ is an integrated Pick function, i.e., $\sqrt{x} = d\int_0^\infty \ln(1 + x/s)s^{-1/2}\,ds$ for some constant $d > 0$ (see 4.6). Hence maximizing $\operatorname{Tr}|h_{A,B}|$ is reduced to maximizing $\det(c^2 + h_{A,B}^2) = \det(c^2 + U^2 + T^2)$ for all constant $c$. That this is achieved by the canonical flux distribution on trees of rings is precisely the content of Theorem 5.1. ■

## APPENDIX

### Kasteleyn's theorem

We give here a different and, we believe, more transparent proof of a deep theorem due to Kasteleyn [KP], which is one of the main tools for counting dimer configurations on planar graphs. Let us emphasize that $\Lambda$ is now a finite graph that is *not necessarily bipartite.*

Historically, the motivation behind Kasteleyn's theorem was an attempt to calculate efficiently the partition function $D(T)$ in (3.1) for large planar graphs—by reducing the problem to the calculation of a determinant. This was accomplished by Temperley and Fisher [TF] in special cases, but independently and in full generality by Kasteleyn [KP]. The starting point was Pfaff's theorem for an *antisymmetric matrix $A$* (of even order): $\det A = \operatorname{Pf}(A)^2$. Here $\operatorname{Pf}(A)$ is the **Pfaffian** of $A$ (more precisely, the Pfaffian of the upper triangular array of $A = \{a_{xy}\}_{1 \leqslant x < y \leqslant N}$), defined by

$$\operatorname{Pf}(A) = \sum_\pi \varepsilon(\pi) a_{\pi(1),\pi(2)} a_{\pi(3),\pi(4)} \cdots a_{\pi(N-1)}, a_{\pi(N)} \tag{A.1}$$

where the sum is over all permutations $\pi \in S_N$ with $\pi(1) < \pi(3) < \pi(5) < \cdots$ and with $\pi(i) < \pi(i+1)$ for odd $i$. Also, $\varepsilon(\pi)$ is the signature of $\pi$.

Each term in $\operatorname{Pf}(A)$ corresponds to a dimer covering of the graph $\Lambda$ with $N$ vertices and with edges corresponding to the nonzero elements of $A$.

Given $|t_{xy}|$, the Kasteleyn, Temperley-Fisher idea is to set $a_{xy} = \exp[i\theta(x, y)]|t_{xy}|$ for $x < y$, with $\theta(x, y)$ chosen so that all terms in (A.1) have a common argument $\theta$. Then trivially, $D(T) = e^{i\theta}\,\mathrm{Pf}(A)$ for some $\theta$ and $|\det A| = D(T)^2$.

In Theorem 3.1 we solved the $D(T)$ problem, from a different perspective, by using the canonical flux distribution: $|\det T| = D(T)^2$. Moreover, we gave a very simple rule (in the proof of Theorem 3.1 and in the remark following Lemma 2.2) for an explicit construction of $\exp[i(\theta(x, y)]$. We did *not* try to construct an antisymmetric matrix, but it is a fact that among the gauge equivalent $T$'s with canonical flux distribution, there *is* one that is antisymmetric. This is Theorem A.1 below. We note here that $\det T$ is a gauge invariant quantity, but $\mathrm{Pf}(T)$ is not.

Theorem A.1, together with Theorem 3.1, yields *Kasteleyn's theorem*, namely that there is always a real, antisymmetric matrix $A$ such that $|a_{xy}| = |t_{xy}|$ and $\mathrm{Pf}(A) = D(T)$. To see this implication, note that Theorem (3.1) says $\det A = \det(iT) = (-1)^{|\Lambda|/2}(-1)^{|\Lambda|/2}D(T)^2$. On the other hand, $\det A = \mathrm{Pf}(A)^2$. We can then make $\mathrm{Pf}(A) = +D(T)$ by multiplying the first row and column of $A$ by $-1$, if necessary. One way in which our proof is a little simpler than Kasteleyn's is that graphs with cut-points do not require special treatment, in either of Theorems A.1 or 3.1.

Theorem A.2 is another corollary of Theorem A.1. It answers the question: For which class of matrices (in the nonbipartite case) does the canonical flux distribution maximize $|\det T|$? Clearly, the canonical flux cannot maximize $|\det T|$ in general. (A simple counterexample is provided by the triangle $|\Lambda| = 3$, where $|\det T|$ is maximized by flux 0 or $\pi$ (Theorem 4.1) but the canonical flux is $\pi/2$.)

A.1. THEOREM (Canonical flux and antisymmetry). *Let $\Lambda$ be a finite planar graph with given hopping amplitudes $|t_{xy}|$. There exists a gauge such that the Hermitian hopping matrix $T$ defined by the canonical flux distribution has purely imaginary elements, i.e., $T = iA$ with $A^T = -A$ and $A$ real.*

*Proof.* We can assume $\Lambda$ is triangulated and we can start with the remark following Lemma 2.2 which states (indeed, gives an explicit rule) that there is a matrix $T^\circ$ whose fluxes are canonical and such that $t^\circ_{xy} \in \{i, -i, 1, -1\}$ for all $(x, y) \in E$. Edges for which $t^\circ_{xy} = \pm i$ will be called "good", and those for which $t_{xy} = \pm 1$ will be called "bad". Our goal is to find a gauge transformation $U_{xy} = u_x\delta_{xy}$ with $u_x \in \{i, 1\}$ for all $x \in V$, such that $U^*TU$ has no bad edges. We shall do so by showing that, if $T$ is *any* matrix with canonical fluxes and with $t_{xy} \in \{i, -i, 1, -1\}$ and with at least one bad edge, then we can find a gauge transformation $U$ with $u_x \in \{i, 1\}$ such that $U^*TU$ has at least one less bad edge than $T$ has. Since the number of edges of $\Lambda$ is finite, the theorem is proved by induction.

Let $(a, b) \in E$ be a bad edge. Consider the set of sites $S = \{y \in V$: there is a path from $a$ to $y$ whose edges are all good$\}$; by definition $a \in S$. Set $u_x = i$ if $x \in S$ and $u_x = 1$ if $x \notin S$. Clearly, if $(c, d) \in E$ was a good edge for $T$, it remains a good edge for $U^*TU$ (because either $u_c = u_d = 1$ or $u_c = u_d = i$).

To complete the proof we have only to show that the edge $(a, b)$ has become a good edge. Since $a \in S$, we have to show that $b \notin S$. Indeed, suppose $b \in S$. Then there is a circuit $C = a, x_1, x_2, \ldots, x_n, b$, of length greater than 2, such that the edges

$(a, x_1), (x_1, x_2), \ldots, (x_{n-1}, x_n)$ are good while $(x_n, b)$ is bad. We claim that this is impossible; in fact *every* circuit must have an *even* number of bad edges. To see this, use equation (2.2) in Lemma 2.3, which says that $f = \ell$ (mod 2). Here, $f$ is the number of (triangular) faces inside $C$. We have $\Phi_C =$ flux through $C = \pm\pi f/2$ (by the definition of the canonical flux distribution). On the other hand, $\Phi_C = (\pi/2)\sum_G(\pm 1)$. Here, the sum is over the good edges, and $+1$ or $-1$ is taken according to the direction in which the edge is traversed when $C$ is traversed in an anticlockwise sense. In any event, $\Phi_C = \pm(\pi/2)\{|G|(\text{mod } 2)\}$ where $|G|$ is the number of good edges in $C$. Thus $f = |G|(\text{mod } 2)$, which proves our assertion. ■

A.2. Theorem (Canonical flux maximizes antisymmetric determinants). *Let $\Lambda$ be a planar (not necessarily bipartite) graph with hopping amplitudes $|t_{xy}|$ given. The canonical flux distribution maximizes $|\det T|$ among all flux distributions such that $T$ is both Hermitian and antisymmetric.*

*Proof.* If $T$ is antisymmetric, $\det T = \text{Pf}(T)^2$. But as we easily see from definition (3.1), $|\text{Pf}(T)| \leqslant D(T)$. By Theorem 3.1, $|\text{Pf}(T_c)| = D(T)$, where $T_c$ has the canonical flux distribution. By Theorem A.1, the gauge can be chosen so that $T_c$ is antisymmetric. ■

## References


[AM] I. Affleck and J. B. Marston, *Large n-limit of the Heisenberg-Hubbard model: Implications for high-$T_c$ superconductors*, Phys. Rev. B **37** (1988), 3774–3777.

[BBR] A. Barelli, J. Bellissard, and R. Rammal, *Spectrum of 2D Bloch electrons in a periodic magnetic field: algebraic approach*, J. Physique **51** (1990), 2167–2185.

[BR] J. Bellissard and R. Rammal, *An algebraic semiclassical approach to Bloch electrons in a magnetic field*, J. Physique **51** (1990), 1803–1830.

———, *Ground state of the Fermi gas on 2D lattices with a magnetic field*, J. Physique **51** (1990), 2153–2165.

———, *Ground state of the Fermi gas on 2D lattices with a magnetic field: new exact results*, Europhys. Lett. **13** (1990), 205–210.

[BS] U. Brandt and R. Schmidt, *Exact results for the distribution of the f-level ground state occupation in the spinless Falicov-Kimball model*, Z. Phys. B **63** (1986), 45–53.

———, *Ground state properties of a spinless Falicov-Kimball model; Additional features*, Z. Phys. B **67** (1987), 43–51.

[DT] P. A. Deift and C. Tomei, *On the determinant of the adjacency matrix for a planar sublattice*, J. Combin. Theory Ser. B **35** (1983), 278–289.

[FK] L. M. Falicov and J. C. Kimball, *Simple model for semiconductor-metal transitions: $SmB_6$ and transition metal oxides*, Phys. Rev. Lett. **22** (1969), 997–999.

[HD] D. R. Hofstadter, *Energy levels and wave functions of Bloch electrons in a rational or irrational magnetic field*, Phys. Rev. B **14** (1976), 2239–2249.

[HLRW] Y. Hasegawa, P. Lederer, T. M. Rice, and P. B. Wiegmann, *Theory of electronic diamagnetism in two-dimensional lattices*, Phys. Rev. Lett. **63** (1989), 907–910.

[HP] P. G. Harper, *Single band motion of conduction electrons in a uniform magnetic field*, Proc. Phys. Soc. A **68** (1955), 874–878.

———, *The general motion of conduction electrons in a uniform magnetic field, with application to the diamagnetism of metals*, Proc. Phys. Soc. A **68** (1955), 879–892.

[KG] G. Kotliar, *Resonating valence bonds and d-wave superconductivity*, Phys. Rev. B **37** (1988), 3664–3666.

[KP] P. W. Kasteleyn, *The statistics of dimers on a lattice I. The number of dimer arrangements on a quadratic lattice*, Physica **27** (1961), 1209–1225.

———, "Graph theory and crystal physics" in *Graph Theory and Theoretical Physics*, F. Harary ed., Academic Press, London, 1967, 44–110.

[KL] T. Kennedy and E. H. Lieb, *An itinerant electron model with crystalline or magnetic long range order*, Phys. A **138** (1986), 320–358.

[LE] E. H. Lieb, *The flux phase problem on planar lattices*, Helv. Phys. Acta **65** (1992), 247–255.

[RD] D. S. Rokhsar, *Quadratic quantum antiferromagnets in the fermionic large-N limit*, Phys. Rev. B **42** (1990), 2526–2531.

———, *Solitons in chiral-spin liquids*, Phys. Rev. Lett. **65** (1990), 1506–1509.

[TF] H. N. V. Temperley and M. E. Fisher, *Dimer problem in statistical mechanics—An exact result*, Philos. Mag. **6** (1961), 1061–1063.

[WP] P. B. Wiegmann, *Towards a gauge theory of strongly correlated electronic systems*, Physica **153C** (1988), 103–108.

[WWZ] X. G. Wen, F. Wilczek, and A. Zee, *Chiral spin states and superconductivity*, Phys. Rev. B **39** (1989), 11413–11423.

Lieb: Departments of Mathematics and Physics, Princeton University, P.O. Box 708, Princeton, New Jersey 08544-0708, USA

Loss: School of Mathematics, Georgia Institute of Technology, Atlanta, Georgia 30332-0160, USA

[8 9] ... W. Kasteleyn, The statistics of dimers on a lattice I. The number of dimer arrangements on a quadratic lattice, Physica 27 (1961) 1209–1225.

... Harary ed., Academic Press, London, 1967, 43–110.

# Selecta of Elliott H. Lieb

**The Stability of Matter: From Atoms to Stars**
**Selecta of Elliott H. Lieb**
Edited by W. Thirring
With a Preface by F. Dyson

Springer-Verlag Berlin Heidelberg New York 2004
Fourth Edition

**Inequalities**
**Selecta of Elliott H. Lieb**
Edited by M. Loss and M. B. Ruskai

Springer-Verlag Berlin Heidelberg New York 2002

**Condensed Matter Physics and Exactly Soluble Models**
**Selecta of Elliott H. Lieb**
Edited by B. Nachtergaele, J. P. Solovej and J. Yngvason

Springer-Verlag Berlin Heidelberg New York 2004

**Statistical Mechanics**
**Selecta of Elliott H. Lieb**
Edited by B. Nachtergaele, J. P. Solovej and J. Yngvason

Springer-Verlag Berlin Heidelberg New York 2004

# Publications of Elliott H. Lieb

1. Second Order Radiative Corrections to the Magnetic Moment of a Bound Electron, Phil. Mag. Vol. **46**, 311–316 (1955).
2. A Non-Perturbation Method for Non-Linear Field Theories, Proc. Roy. Soc. **241A**, 339–363 (1957).
3. (with K. Yamazaki) Ground State Energy and Effective Mass of the Polaron, Phys. Rev. **111**, 728–733 (1958).
4. (with H. Koppe) Mathematical Analysis of a Simple Model Related to the Stripping Reaction, Phys. Rev. **116**, 367–371 (1959).
5. Hard Sphere Bose Gas – An Exact Momentum Space Formulation, Proc. U.S. Nat. Acad. Sci. **46**, 1000–1002 (1960).
6. Operator Formalism in Statistical Mechanics, J. Math. Phys. **2**, 341–343 (1961).
7. (with D.C. Mattis) Exact Wave Functions in Superconductivity, J. Math. Phys. **2**, 602–609 (1961).

††8. (with T.D. Schultz and D.C. Mattis) Two Soluble Models of an Antiferromagnetic Chain, Annals of Phys. (N.Y.) **16**, 407–466 (1961).

††$9. (with D.C. Mattis) Theory of Ferromagnetism and the Ordering of Electronic Energy Levels, Phys. Rev. **125**, 164–172 (1962).

††$10. (with D.C. Mattis) Ordering Energy Levels of Interacting Spin Systems, J. Math. Phys. **3**, 749–751 (1962).

11. New Method in the Theory of Imperfect Gases and Liquids, J. Math. Phys. **4**, 671–678 (1963).

††12. (with W. Liniger) Exact Analysis of an Interacting Bose Gas. I. The General Solution and the Ground State, Phys. Rev. **130**, 1605–1616 (1963).

††13. Exact Analysis of an Interacting Bose Gas. II. The Excitation Spectrum, Phys. Rev. **130**, 1616–1624 (1963).

††14. Simplified Approach to the Ground State Energy of an Imperfect Bose Gas, Phys. Rev. **130**, 2518–2528 (1963).

---

$*$ means the paper appears in: *Stability of Matter*, ed by W. Thirring, 4th Edition (Springer Berlin Heidelberg 2004)

$ means the paper appears in: *Inequalities*, ed by M. Loss and M.B. Ruskai (Springer Berlin Heidelberg 2002)

†† means the paper appears in: *Condensed Matter Physics and Exactly Soluble Models*, ed by B. Nachtergaele, J.P. Solovej and J. Yngvason (Springer Berlin Heidelberg 2004)

† means the paper appears in: *Statistical Mechanics*, ed by B. Nachtergaele, J.P. Solovej and J. Yngvason (Springer Berlin Heidelberg 2004)

15. (with A. Sakakura) Simplified Approach to the Ground State Energy of an Imperfect Bose Gase. II. The Charged Bose Gas at High Density, Phys. Rev. **133**, A899–A906 (1964).
16. (with W. Liniger) Simplified Approach to the Ground State Energy of an Imperfect Bose Gas. III. Application to the One-Dimensional Model, Phys. Rev. **134**, A312–A315 (1964).
††17. (with T.D. Schultz and D.C. Mattis) Two-Dimensional Ising Model as a Soluble Problem of Many Fermions, Rev. Mod. Phys. **36**, 856–871 (1964).
18. The Bose Fluid, *Lectures in Theoretical Physics, Vol. VIIC,* (Boulder summer school), University of Colorado Press, 175–224 (1965).
††19. (with D.C. Mattis) Exact Solution of a Many-Fermion System and Its Associated Boson Field, J. Math. Phys. **6**, 304–312 (1965).
†20. (with S.Y. Larsen, J.E. Kilpatrick and H.F. Jordan) Suppression at High Temperature of Effects Due to Statistics in the Second Virial Coefficient of a Real Gas, Phys. Rev. **140**, A129–A130 (1965).
21. (with D.C. Mattis) Book *Mathematical Physics in One Dimension*, Academic Press, New York (1966).
$22. Proofs of Some Conjectures on Permanents, J. of Math. and Mech. **16**, 127–139 (1966).
23. Quantum Mechanical Extension of the Lebowitz-Penrose Theorem on the van der Waals Theory, J. Math. Phys. **7**, 1016–1024 (1966).
24. (with D.C. Mattis) Theory of Paramagnetic Impurities in Semiconductors, J. Math. Phys. **7**, 2045–2052 (1966).
25. (with T. Burke and J.L. Lebowitz) Phase Transition in a Model Quantum System: Quantum Corrections to the Location of the Critical Point, Phys. Rev. **149**, 118–122 (1966).
26. Some Comments on the One-Dimensional Many-Body Problem, unpublished Proceedings of Eastern Theoretical Physics Conference, New York (1966).
†27. Calculation of Exchange Second Virial Coefficient of a Hard Sphere Gas by Path Integrals, J. Math. Phys. **8**, 43–52 (1967).
†28. (with Z. Rieder and J.L. Lebowitz) Properties of a Harmonic Crystal in a Stationary Nonequilibrium State, J. Math. Phys. **8**, 1073–1078 (1967).
††29. Exact Solution of the Problem of the Entropy of Two-Dimensional Ice, Phys. Rev. Lett. **18**, 692–694 (1967).
††30. Exact Solution of the *F* Model of an Antiferroelectric, Phys. Rev. Lett. **18**, 1046–1048 (1967).
††31. Exact Solution of the Two-Dimensional Slater KDP Model of a Ferroelectric, Phys. Rev. Lett. **19**, 108–110 (1967).
††32. Residual Entropy of Square Ice, Phys. Rev. **162**, 162–172 (1967).
33. Ice, Ferro- and Antiferroelectrics, in *Methods and Problems in Theoretical Physics, in honour of R.E. Peierls*, Proceedings of the 1967 Birmingham conference, North-Holland, 21–28 (1970).
34. Exactly Soluble Models, in *Mathematical Methods in Solid State and Superfluid Theory*, Proceedings of the 1967 Scottish Universities' Summer School of Physics, Oliver and Boyd, Edinburgh 286–306 (1969).

††35. Solution of the Dimer Problem by the Transfer Matrix Method, J. Math. Phys. **8**, 2339–2341 (1967).
††36. (with M. Flicker) Delta-Function Fermi Gas with Two-Spin Deviates, Phys. Rev. **161**, 179–188 (1967).
$37. Concavity Properties and a Generating Function for Stirling Numbers, J. Combinatorial Theory **5**, 203–206 (1968).
38. A Theorem on Pfaffians, J. Combinatorial Theory **5**, 313–319 (1968).
††39. (with F.Y. Wu) Absence of Mott Transition in an Exact Solution of the Short-Range, One-Band Model in One Dimension, Phys. Rev. Lett. **20**, 1445–1448 (1968).
††40. Two-Dimensional Ferroelectric Models, J. Phys. Soc. (Japan) **26** (supplement), 94–95 (1969).
41. (with W.A. Beyer) Clusters on a Thin Quadratic Lattice, Studies in Appl. Math. **48**, 77–90 (1969).
42. (with C.J. Thompson) Phase Transition in Zero Dimensions: A Remark on the Spherical Model, J. Math. Phys. **10**, 1403–1406 (1969).
†*43. (with J.L. Lebowitz) Existence of Thermodynamics for Real Matter with Coulomb Forces, Phys. Rev. Lett. **22**, 631–634 (1969).
44. Two Dimensional Ice and Ferroelectric Models, in *Lectures in Theoretical Physics, XI D*, (Boulder summer school) Gordon and Breach, 329–354 (1969).
45. Survey of the One Dimensional Many Body Problem and Two Dimensional Ferroelectric Models, in *Contemporary Physics: Trieste Symposium 1968*, International Atomic Energy Agency, Vienna, vol. 1, 163–176 (1969).
46. Models, in *Phase Transitions*, Proceedings of the 14th Solvay Chemistry Conference, May 1969, Interscience, 45–56 (1971).
$47. (with H. Araki) Entropy Inequalities, Commun. Math. Phys. **18**, 160–170 (1970).
††48. (with O.J. Heilmann) Violation of the Noncrossing Rule: The Hubbard Hamiltonian for Benzene, Trans. N.Y. Acad. Sci. **33**, 116–149 (1970). Also in Annals N.Y. Acad. Sci. **172**, 583–617 (1971). (Awarded the 1970 Boris Pregel award for research in chemical physics.)
†49. (with O.J. Heilmann) Monomers and Dimers, Phys. Rev. Lett. **24**, 1412–1414 (1970).
50. Book Review of "Statistical Mechanics" by David Ruelle, Bull. Amer. Math. Soc. **76**, 683–687 (1970).
51. (with J.L. Lebowitz) Thermodynamic Limit for Coulomb Systems, in *Systèmes a un Nombre Infini de Degrés de Liberté*, Colloques Internationaux de Centre National de la Recherche Scientifique **181**, 155–162 (1970).
52. (with D.B. Abraham, T. Oguchi and T. Yamamoto) On the Anomolous Specific Heat of Sodium Trihydrogen Selenite, Progr. Theor. Phys. (Kyoto) **44**, 1114–1115 (1970).

53. (with D.B. Abraham) Anomalous Specific Heat of Sodium Trihydrogen Selenite – An Associated Combinatorial Problem, J. Chem. Phys. **54**, 1446–1450 (1971).
54. (with O.J. Heilmann, D. Kleitman and S. Sherman) Some Positive Definite Functions on Sets and Their Application to the Ising Model, Discrete Math. **1**, 19–27 (1971).
55. (with Th. Niemeijer and G. Vertogen) Models in Statistical Mechanics, in *Statistical Mechanics and Quantum Field Theory*, Proceedings of 1970 Ecole d'Eté de Physique Théorique (Les Houches), Gordon and Breach, 281–326 (1971).
††56. (with H.N.V. Temperley) Relations between the 'Percolation' and 'Colouring' Problem and Other Graph-Theoretical Problems Associated with Regular Planar Lattices: Some Exact Results for the 'Percolation' Problem, Proc. Roy. Soc. **A322**, 251–280 (1971).
57. (with M. de Llano) Some Exact Results in the Hartree-Fock Theory of a Many-Fermion System at High Densities, Phys. Letts. **37B**, 47–49 (1971).
†58. (with J.L. Lebowitz) The Constitution of Matter: Existence of Thermodynamics for Systems Composed of Electrons and Nuclei, Adv. in Math. **9**, 316–398 (1972).
59. (with F.Y. Wu) Two Dimensional Ferroelectric Models, in *Phase Transitions and Critical Phenomena*, C. Domb and M. Green eds., vol. 1, Academic Press 331–490 (1972).
†60. (with D. Ruelle) A Property of Zeros of the Partition Function for Ising Spin Systems, J. Math. Phys. **13**, 781–784 (1972).
†61. (with O.J. Heilmann) Theory of Monomer-Dimer Systems, Commun. Math. Phys. **25**, 190–232 (1972). Errata **27**, 166 (1972).
††62. (with M.L. Glasser and D.B. Abraham) Analytic Properties of the Free Energy for the "Ice" Models, J. Math. Phys. **13**, 887–900 (1972).
†63. (with D.W. Robinson) The Finite Group Velocity of Quantum Spin Systems, Commun. Math. Phys. **28**, 251–257 (1972).
64. (with J.L. Lebowitz) Phase Transition in a Continuum Classical System with Finite Interactions, Phys. Lett. **39A**, 98–100 (1972).
65. (with J.L. Lebowitz) Lectures on the Thermodynamic Limit for Coulomb Systems, in *Statistical Mechanics and Mathematical Problems*, Battelle 1971 Recontres, Springer Lecture Notes in Physics **20**, 136–161 (1973).
66. (with J.L. Lebowitz) Lectures on the Thermodynamic Limit for Coulomb Systems, in *Lectures in Theoretical Physics XIV B*, (Boulder summer school), Colorado Associated University Press, 423–460 (1973).
$67. Convex Trace Functions and the Wigner-Yanase-Dyson Conjecture, Adv. in Math. **11**, 267–288 (1973).
$68. (with M.B. Ruskai) A Fundamental Property of Quantum Mechanical Entropy, Phys. Rev. Lett. **30**, 434–436 (1973).
$69. (with M.B. Ruskai) Proof of the Strong Subadditivity of Quantum-Mechanical Entropy, J. Math. Phys. **14**, 1938–1941 (1973).

70. (with K. Hepp) On the Superradiant Phase Transition for Molecules in a Quantized Radiation Field: The Dicke Maser Model, Annals of Phys. (N.Y.) **76**, 360–404 (1973).
††71. (with K. Hepp) Phase Transitions in Reservoir-Driven Open Systems with Applications to Lasers and Superconductors, Helv. Phys. Acta **46**, 573–602 (1973).
††72. (with K. Hepp) Equilibrium Statistical Mechanics of Matter Interacting with the Quantized Radiation Field, Phys. Rev. **A8**, 2517–2525 (1973).
††73. (with K. Hepp) Constructive Macroscopic Quantum Electrodynamics, in *Constructive Quantum Field Theory*, Proceedings of the 1973 Erice Summer School, G. Velo and A. Wightman, eds., Springer Lecture Notes in Physics **25**, 298–316 (1973).
†$74. The Classical Limit of Quantum Spin Systems, Commun. Math. Phys. **31**, 327–340 (1973).
75. (with B. Simon) Thomas-Fermi Theory Revisited, Phys. Rev. Lett. **31**, 681–683 (1973).
$76. (with M.B. Ruskai) Some Operator Inequalities of the Schwarz Type, Adv. in Math. **12**, 269–273 (1974).
77. Exactly Soluble Models in Statistical Mechanics, lecture given at the 1973 I.U.P.A.P. van der Waals Centennial Conference on Statistical Mechanics, Physica **73**, 226–236 (1974).
78. (with B. Simon) On Solutions to the Hartree-Fock Problem for Atoms and Molecules, J. Chem. Physics **61**, 735–736 (1974).
79. Thomas-Fermi and Hartree-Fock Theory, lecture at 1974 International Congress of Mathematicians, Vancouver. Proceedings, Vol. 2, 383–386 (1975).
$80. Some Convexity and Subadditivity Properties of Entropy, Bull. Amer. Math. Soc. **81**, 1–13 (1975).
$81. (with H.J. Brascamp and J.M. Luttinger) A General Rearrangement Inequality for Multiple Integrals, Jour. Funct. Anal. **17**, 227–237 (1975).
$82. (with H.J. Brascamp) Some Inequalities for Gaussian Measures and the Long-Range Order of the One-Dimensional Plasma, lecture at Conference on Functional Integration, Cumberland Lodge, England. *Functional Integration and its Applications*, A.M. Arthurs ed., Clarendon Press, 1–14 (1975).
††83. (with K. Hepp) The Laser: A Reversible Quantum Dynamical System with Irreversible Classical Macroscopic Motion, in *Dynamical Systems*, Battelle 1974 Rencontres, Springer Lecture Notes in Physics **38**, 178–208 (1975). Also appears in *Melting, Localization and Chaos*, Proc. 9th Midwest Solid State Theory Symposium, 1981, R. Kalia and P. Vashishta eds., North-Holland, 153–177 (1982).
∗84. (with P. Hertel and W. Thirring) Lower Bound to the Energy of Complex Atoms, J. Chem. Phys. **62**, 3355–3356 (1975).
∗85. (with W. Thirring) Bound for the Kinetic Energy of Fermions which Proves the Stability of Matter, Phys. Rev. Lett. **35**, 687–689 (1975). Errata **35**, 1116 (1975).

†86. (with H.J. Brascamp and J.L. Lebowitz) The Statistical Mechanics of Anharmonic Lattices, in the proceedings of the 40th session of the International Statistics Institute, Warsaw, **9**, 393–404 (1975).
$87. (with H.J. Brascamp) Best Constants in Young's Inequality, Its Converse and Its Generalization to More Than Three Functions, Adv. in Math. **20**, 151–172 (1976).
$88. (with H.J. Brascamp) On Extensions of the Brunn-Minkowski and Prékopa-Leindler Theorems, Including Inequalities for Log Concave Functions and with an Application to the Diffusion Equation, J. Funct. Anal. **22**, 366–389 (1976).
89. (with J.F. Barnes and H.J. Brascamp) Lower Bounds for the Ground State Energy of the Schroedinger Equation Using the Sharp Form of Young's Inequality, in *Studies in Mathematical Physics*, Lieb, Simon, Wightman eds., Princeton Press, 83–90 (1976).
$90. Inequalities for Some Operator and Matrix Functions, Adv. in Math. **20**, 174–178 (1976).
*91. (with H. Narnhofer) The Thermodynamic Limit for Jellium, J. Stat. Phys. **12**, 291–310 (1975). Errata J. Stat. Phys. **14**, 465 (1976).
*92. The Stability of Matter, Rev. Mod. Phys. **48**, 553–569 (1976).
93. Bounds on the Eigenvalues of the Laplace and Schroedinger Operators, Bull. Amer. Math. Soc. **82**, 751–753 (1976).
94. (with F.J. Dyson and B. Simon) Phase Transitions in the Quantum Heisenberg Model, Phys. Rev. Lett. **37**, 120–123 (1976). (See no. 104.)
$*95. (with W. Thirring) Inequalities for the Moments of the Eigenvalues of the Schrödinger Hamiltonian and Their Relation to Sobolev Inequalities, in *Studies in Mathematical Physics*, E. Lieb, B. Simon, A. Wightman eds., Princeton University Press, 269–303 (1976).
96. (with B. Simon and A. Wightman) Book *Studies in Mathematical Physics: Essays in Honor of Valentine Bargmann*, Princeton University Press (1976).
97. (with B. Simon) Thomas-Fermi Theory of Atoms, Molecules and Solids, Adv. in Math. **23**, 22–116 (1977).
†98. (with O.E. Lanford and J.L. Lebowitz) Time Evolution of Infinite Anharmonic Systems, J. Stat. Phys. **16**, 453–461 (1977).
99. The Stability of Matter, Proceedings of the Conference on the Fiftieth Anniversary of the Schroedinger equation, Acta Physica Austriaca Suppl. XVII, 181–207 (1977).
$100. Existence and Uniqueness of the Minimizing Solution of Choquard's Non-Linear Equation, Studies in Appl. Math. **57**, 93–105 (1977).
†101. (with J. Fröhlich) Existence of Phase Transitions for Anisotropic Heisenberg Models, Phys. Rev. Lett. **38**, 440–442 (1977).
*102. (with B. Simon) The Hartree-Fock Theory for Coulomb Systems, Commun. Math. Phys. **53**, 185–194 (1977).
103. (with W. Thirring) A Lower Bound for Level Spacings, Annals of Phys. (N.Y.) **103**, 88–96 (1977).

†104. (with F.J. Dyson and B. Simon) Phase Transitions in Quantum Spin Systems with Isotropic and Non-Isotropic Interactions, J. Stat. Phys. **18**, 335–383 (1978).

105. Many Particle Coulomb Systems, lectures given at the 1976 session on statistical mechanics of the International Mathematics Summer Center (C.I.M.E.). In *Statistical Mechanics*, C.I.M.E. 1 Ciclo 1976, G. Gallavotti, ed., Liguore Editore, Naples, 101–166 (1978).

∗106. (with R. Benguria) Many-Body Atomic Potentials in Thomas-Fermi Theory, Annals of Phys. (N.Y.) **110**, 34–45 (1978).

∗107. (with R. Benguria) The Positivity of the Pressure in Thomas-Fermi Theory, Commun. Math. Phys. **63**, 193–218 (1978). Errata **71**, 94 (1980).

108. (with M. de Llano) Solitons and the Delta Function Fermion Gas in Hartree-Fock Theory, J. Math. Phys. **19**, 860–868 (1978).

†109. (with J. Fröhlich) Phase Transitions in Anisotropic Lattice Spin Systems, Commun. Math. Phys. **60**, 233–267 (1978).

†110. (with J. Fröhlich, R. Israel and B. Simon) Phase Transitions and Reflection Positivity. I. General Theory and Long Range Lattice Models, Commun. Math. Phys. **62**, 1–34 (1978). (See no. 124.)

$111. (with M. Aizenman and E.B. Davies) Positive Linear Maps Which are Order Bounded on C* Subalgebras, Adv. in Math. **28**, 84–86 (1978).

$∗112. (with M. Aizenman) On Semi-Classical Bounds for Eigenvalues of Schrödinger Operators, Phys. Lett. **66A**, 427–429 (1978).

113. New Proofs of Long Range Order, in *Proceedings of the International Conference on Mathematical Problems in Theoretical Physics* (June 1977), Springer Lecture Notes in Physics, **80**, 59–67 (1978).

$114. Proof of an Entropy Conjecture of Wehrl, Commun. Math. Phys. **62**, 35–41 (1978).

115. (with B. Simon) Monotonicity of the Electronic Contribution to the Born-Oppenheimer Energy, J. Phys. B. **11**, L537–L542 (1978).

†116. (with O.J. Heilmann) Lattice Models for Liquid Crystals, J. Stat. Phys. **20**, 679–693 (1979).

117. (with H. Brezis) Long Range Atomic Potentials in Thomas-Fermi Theory, Commun. Math. Phys. **65**, 231–246 (1979).

∗118. The $N^{5/3}$ Law for Bosons, Phys. Lett. **70A**, 71–73 (1979).

119. A Lower Bound for Coulomb Energies, Phys. Lett. **70A**, 444–446 (1979).

120. Why Matter is Stable, Kagaku **49**, 301–307 and 385–388 (1979). (In Japanese).

121. The Number of Bound States of One-Body Schrödinger Operators and the Weyl Problem, Symposium of the Research Inst. of Math. Sci., Kyoto University, (1979).

122. Some Open Problems About Coulomb Systems, in *Proceedings of the Lausanne 1979 Conference of the International Association of Mathematical Physics*, Springer Lecture Notes in Physics, **116**, 91–102 (1980).

$∗123. The Number of Bound States of One-Body Schrödinger Operators and the Weyl Problem, *Proceedings of the Amer. Math. Soc. Symposia in Pure Math.*, **36**, 241–252 (1980).

†124. (with J. Fröhlich, R.B. Israel and B. Simon) Phase Transitions and Reflection Positivity. II. Lattice Systems with Short-Range and Coulomb Interactions. J. Stat. Phys. **22**, 297–347 (1980). (See no. 110.)
125. Why Matter is Stable, Chinese Jour. Phys. **17**, 49–62 (1980). (English version of no. 120).
†$126. A Refinement of Simon's Correlation Inequality, Commun. Math. Phys. **77**, 127–135 (1980).
127. (with B. Simon) Pointwise Bounds on Eigenfunctions and Wave Packets in N-Body Quantum Systems. VI. Asymptotics in the Two-Cluster Region, Adv. in Appl. Math. **1**, 324–343 (1980).
128. The Uncertainty Principle, article in *Encyclopedia of Physics*, R. Lerner and G. Trigg eds., Addison Wesley, 1078–1079 (1981).
$∗129. (with S. Oxford) An Improved Lower Bound on the Indirect Coulomb Energy, Int. J. Quant. Chem. **19**, 427–439 (1981).
∗130. (with R. Benguria and H. Brezis) The Thomas-Fermi-von Weizsäcker Theory of Atoms and Molecules, Commun. Math. Phys. **79**, 167–180 (1981).
†131. (with M. Aizenman) The Third Law of Thermodynamics and the Degeneracy of the Ground State for Lattice Systems, J. Stat. Phys. **24**, 279–297 (1981).
†132. (with J.-R. Bricmont, J. Fontaine, J.L. Lebowitz and T. Spencer) Lattice Systems with a Continuous Symmetry III. Low Temperature Asymptotic Expansion for the Plane Rotator Model, Commun. Math. Phys. **78**, 545–566 (1981).
†133. (with A.D. Sokal) A General Lee-Yang Theorem for One-Component and Multicomponent Ferromagnets, Commun. Math. Phys. **80**, 153–179 (1981).
∗134. Variational Principle for Many-Fermion Systems, Phys. Rev. Lett. **46**, 457–459 (1981). Errata **47**, 69 (1981).
135. Thomas-Fermi and Related Theories of Atoms and Molecules, in *Rigorous Atomic and Molecular Physics*, G. Velo and A. Wightman, eds., Plenum Press 213–308 (1981).
∗136. Thomas-Fermi and Related Theories of Atoms and Molecules, Rev. Mod. Phys. *53*, 603–641 (1981). Errata **54**, 311 (1982). (Revised version of no. 135.)
137. Statistical Theories of Large Atoms and Molecules, in *Proceedings of the 1981 Oaxtepec conference on Recent Progress in Many-Body Theories,* Springer Lecture Notes in Physics, **142**, 336–343 (1982).
138. Statistical Theories of Large Atoms and Molecules, Comments Atomic and Mol. Phys. **11**, 147–155 (1982).
∗139. Analysis of the Thomas-Fermi-von Weizsäcker Equation for an Infinite Atom without Electron Repulsion, Commun. Math. Phys. **85**, 15–25 (1982).
140. (with D.A. Liberman) Numerical Calculation of the Thomas-Fermi-von Weizsäcker Function for an Infinite Atom without Electron Repulsion, Los Alamos National Laboratory Report, LA-9186-MS (1982).

∗141. Monotonicity of the Molecular Electronic Energy in the Nuclear Coordinates, J. Phys. B.: At. Mol. Phys. **15**, L63–L66 (1982).
142. Comment on "Approach to Equilibrium of a Boltzmann Equation Solution", Phys. Rev. Lett. **48**, 1057 (1982).
143. Density Functionals for Coulomb Systems, in *Physics as Natural Philosophy: Essays in honor of Laszlo Tisza on his 75th Birthday*, A. Shimony and H. Feshbach eds., M.I.T. Press, 111–149 (1982).
$144. An $L^p$ Bound for the Riesz and Bessel Potentials of Orthonormal Functions, J. Funct. Anal. **51**, 159–165 (1983).
$145. (with H. Brezis) A Relation Between Pointwise Convergence of Functions and Convergence of Functionals, Proc. Amer. Math. Soc. **88**, 486–490 (1983).
∗146. (with R. Benguria) A Proof of the Stability of Highly Negative Ions in the Absence of the Pauli Principle, Phys. Rev. Lett. **50**, 1771–1774 (1983).
$147. Sharp Constants in the Hardy-Littlewood-Sobolev and Related Inequalities, Annals of Math. **118**, 349–374 (1983).
$148. Density Functionals for Coulomb Systems (a revised version of no. 143), Int. Jour. Quant. Chem. **24**, 243–277 (1983). An expanded version appears in *Density Functional Methods in Physics*, R. Dreizler and J. da Providencia eds., Plenum Nato ASI Series **123**, 31–80 (1985).
149. The Significance of the Schrödinger Equation for Atoms, Molecules and Stars, lecture given at the Schrödinger Symposium, Dublin Institute of Advanced Studies, October 1983, unpublished Proceedings.
∗150. (with I. Daubechies) One Electron Relativistic Molecules with Coulomb Interaction, Commun. Math. Phys. **90**, 497–510 (1983).
151. (with I. Daubechies) Relativistic Molecules with Coulomb Interaction, in *Differential Equations, Proc. of the Conference held at the University of Alabama in Birmingham, 1983*, I. Knowles and R. Lewis eds., Math. Studies Series, **92**, 143–148 North-Holland (1984).
152. Some Vector Field Equations, in *Differential Equations, Proc. of the Conference held at the University of Alabama in Birmingham, 1983*, I. Knowles and R. Lewis eds., Math. Studies Series **92**, 403–412 North-Holland (1984).
$153. On the Lowest Eigenvalue of the Laplacian for the Intersection of Two Domains, Inventiones Math. **74**, 441–448 (1983).
154. (with J. Chayes and L. Chayes) The Inverse Problem in Classical Statistical Mechanics, Commun. Math. Phys. **93**, 57–121 (1984).
$155. On Characteristic Exponents in Turbulence, Commun. Math. Phys. **92**, 473–480 (1984).
∗156. Atomic and Molecular Negative Ions, Phys. Rev. Lett. **52**, 315–317 (1984).
∗157. Bound on the Maximum Negative Ionization of Atoms and Molecules, Phys. Rev. **29A**, 3018–3028 (1984).
158. (with W. Thirring) Gravitational Collapse in Quantum Mechanics with Relativistic Kinetic Energy, Annals of Phys. (N.Y.) **155**, 494–512 (1984).
159. (with I.M. Sigal, B. Simon and W. Thirring) Asymptotic Neutrality of Large-Z Ions, Phys. Rev. Lett. **52**, 994–996 (1984). (See no. 185.)

∗160. (with R. Benguria) The Most Negative Ion in the Thomas-Fermi-von Weizsäcker Theory of Atoms and Molecules, J. Phys. B: At. Mol. Phys. **18**, 1045–1059 (1985).

$161. (with H. Brezis) Minimum Action Solutions of Some Vector Field Equations, Commun. Math. Phys. **96**, 97–113 (1984).

$162. (with H. Brezis) Sobolev Inequalities with Remainder Terms, J. Funct. Anal. **62**, 73–86 (1985).

$163. Baryon Mass Inequalities in Quark Models, Phys. Rev. Lett. **54**, 1987–1990 (1985).

∗164. (with J. Fröhlich and M. Loss) Stability of Coulomb Systems with Magnetic Fields I. The One-Electron Atom, Commun. Math. Phys. **104**, 251–270 (1986).

∗165. (with M. Loss) Stability of Coulomb Systems with Magnetic Fields II. The Many-Electron Atom and the One-Electron Molecule, Commun. Math. Phys. **104**, 271–282 (1986).

∗166. (with W. Thirring) Universal Nature of van der Waals Forces for Coulomb Systems, Phys. Rev. A **34**, 40–46 (1986).

167. Some Ginzburg-Landau Type Vector-Field Equations, in *Nonlinear systems of Partial Differential Equations in Applied Mathematics*, B. Nicolaenko, D. Holm and J. Hyman eds., Amer. Math. Soc. Lectures in Appl. Math. **23**, Part 2, 105–107 (1986).

††168. (with I. Affleck) A Proof of Part of Haldane's Conjecture on Spin Chains, Lett. Math. Phys. **12**, 57–69 (1986).

$169. (with H. Brezis and J-M. Coron) Estimations d'Energie pour des Applications de $\mathbf{R}^3$ a Valeurs dans $S^2$, C.R. Acad. Sci. Paris **303** Ser. 1, 207–210 (1986).

170. (with H. Brezis and J-M. Coron) Harmonic Maps with Defects, Commun. Math. Phys. **107**, 649–705 (1986).

171. Some Fundamental Properties of the Ground States of Atoms and Molecules, in *Fundamental Aspects of Quantum Theory*, V. Gorini and A. Frigerio eds., Nato ASI Series B, Vol. 144, 209–214, Plenum Press (1986).

172. (with T. Kennedy) A Model for Crystallization: A Variation on the Hubbard Model, in *Statistical Mechanics and Field Theory: Mathematical Aspects*, Springer Lecture Notes in Physics **257**, 1–9 (1986).

173. (with T. Kennedy) An Itinerant Electron Model with Crystalline or Magnetic Long Range Order, Physica **138A**, 320–358 (1986).

††174. A Model for Crystallization: A Variation on the Hubbard Model, Physica **140A**, 240–250 (1986) (Proceedings of IUPAP Statphys 16, Boston).

††175. (with T. Kennedy) Proof of the Peierls Instability in One Dimension, Phys. Rev. Lett. **59**, 1309–1312 (1987).

††176. (with I. Affleck, T. Kennedy and H. Tasaki) Rigorous Results on Valence-Bond Ground States in Antiferromagnets, Phys. Rev. Lett. **59**, 799–802 (1987).

∗177. (with H.-T. Yau) The Chandrasekhar Theory of Stellar Collapse as the Limit of Quantum Mechanics, Commun. Math. Phys. **112**, 147–174 (1987).

178. (with H.-T. Yau) A Rigorous Examination of the Chandrasekhar Theory of Stellar Collapse, Astrophys. Jour. **323**, 140–144 (1987).

$179. (with F. Almgren) Singularities of Energy Minimizing Maps from the Ball to the Sphere, Bull. Amer. Math. Soc. **17**, 304–306 (1987). (See no. 190.)

180. Bounds on Schrödinger Operators and Generalized Sobolev Type Inequalities, Proceedings of the International Conference on Inequalities, University of Birmingham, England, 1987, Marcel Dekker Lecture Notes in Pure and Appl. Math., W.N. Everitt ed., volume 129, pages 123–133 (1991).

††181. (with I. Affleck, T. Kennedy and H. Tasaki) Valence Bond Ground States in Isotropic Quantum Antiferromagnets, Commun. Math. Phys. **115**, 477–528 (1988).

182. (with T. Kennedy and H. Tasaki) A Two Dimensional Isotropic Quantum Antiferromagnet with Unique Disordered Ground State, J. Stat. Phys. **53**, 383–416 (1988).

†183. (with T. Kennedy and B.S. Shastry) Existence of Néel Order in Some Spin 1/2 Heisenberg Antiferromagnets, J. Stat. Phys. **53**, 1019–1030 (1988).

†184. (with T. Kennedy and B.S. Shastry) The *XY* Model has Long-Range Order for all Spins and all Dimensions Greater than One, Phys. Rev. Lett. **61**, 2582–2584 (1988).

∗185. (with I.M. Sigal, B. Simon and W. Thirring) Approximate Neutrality of Large-Z Ions, Commun. Math. Phys. **116**, 635–644 (1988). (See no. 159.)

∗186. (with H.-T. Yau) The Stability and Instability of Relativistic Matter, Commun. Math. Phys. **118**, 177–213 (1988).

187. (with H.-T. Yau) Many-Body Stability Implies a Bound on the Fine Structure Constant, Phys. Rev. Lett. **61**, 1695–1697 (1988).

∗188. (with J. Conlon and H.-T. Yau) The $N^{7/5}$ Law for Charged Bosons, Commun. Math. Phys. **116**, 417–448 (1988).

$189. (with F. Almgren and W. Browder) Co-area, Liquid Crystals, and Minimal Surfaces, in *Partial Differential Equations*, S.S. Chern ed., Springer Lecture Notes in Math. **1306**, 1–22 (1988).

190. (with F. Almgren) Singularities of Energy Minimizing Maps from the Ball to the Sphere: Examples, Counterexamples and Bounds, Ann. of Math. **128**, 483–530 (1988).

$191. (with F. Almgren) Counting Singularities in Liquid Crystals, in *IXth International Congress on Mathematical Physics*, B. Simon, A. Truman, I.M. Davies eds., Hilger, 396–409 (1989). This also appears in: *Symposia Mathematica, vol. XXX,* Ist. Naz. Alta Matem. Francesco Severi Roma, 103–118, Academic Press (1989); *Variational Methods,* H. Berestycki, J-M. Coron, I. Ekeland eds., Birkhäuser, 17–36 (1990); How many singularities can there be in an energy minimizing map from the ball to the sphere?, in *Ideas and Methods in Mathematical Analysis, Stochastics, and Applications,* S. Albeverio, J.E. Fenstad, H. Holden, T. Lindstrom eds., Cambridge Univ. Press, vol. 1, 394–408 (1992).

$192. (with F. Almgren) Symmetric Decreasing Rearrangement can be Discontinuous, Bull. Amer. Math. Soc. **20**, 177–180 (1989).

§193. (with F. Almgren) Symmetric Decreasing Rearrangement is Sometimes Continuous, Jour. Amer. Math. Soc. **2**, 683–773 (1989). A summary of this work (using 'rectifiable currents') appears as The (Non)continuity of Symmetric Decreasing Rearrangement in *Symposia Mathematica, vol. XXX,* Ist. Naz. Alta Matem. Francesco Severi Roma, 89–102, Academic Press (1989) and in *Variational Methods,* H. Berestycki, J-M. Coron, I. Ekeland eds., Birkhäuser, 3–16 (1990).

††§194. Two Theorems on the Hubbard Model, Phys. Rev. Lett. **62**, 1201–1204 (1989). Errata **62**, 1927 (1989).

195. (with J. Conlon and H.-T. Yau) The Coulomb gas at Low Temperature and Low Density, Commun. Math. Phys. **125**, 153–180 (1989).

§196. Gaussian Kernels have only Gaussian Maximizers, Invent. Math. **102**, 179–208 (1990).

§∗197. Kinetic Energy Bounds and their Application to the Stability of Matter, in *Schrödinger Operators*, Proceedings Sønderborg Denmark 1988, H. Holden and A. Jensen eds., Springer Lecture Notes in Physics **345**, 371–382 (1989). Expanded version of no. 180.

∗198. The Stability of Matter: From Atoms to Stars, 1989 Gibbs Lecture, Bull. Amer. Math. Soc. **22**, 1–49 (1990).

§199. Integral Bounds for Radar Ambiguity Functions and Wigner Distributions, J. Math. Phys. **31**, 594–599 (1990).

200. On the Spectral Radius of the Product of Matrix Exponentials, Linear Alg. and Appl.**141**, 271–273 (1990).

††§201. (with M. Aizenman) Magnetic Properties of Some Itinerant-Electron Systems at $T > 0$, Phys. Rev. Lett. **65**, 1470–1473 (1990).

202. (with H. Siedentop) Convexity and Concavity of Eigenvalue Sums, J. Stat. Phys. **63**, 811–816 (1991).

§203. (with J.P. Solovej) Quantum Coherent Operators: A Generalization of Coherent States, Lett. Math. Phys. **22**, 145–154 (1991).

204. The Flux-Phase Problem on Planar Lattices, Helv. Phys. Acta **65**, 247–255 (1992). Proceedings of the conference "Physics in Two Dimensions", Neuchâtel, August 1991.

205. Atome in starken Magnetfeldern, Physikalische Blätter **48**, 549–552 (1992). Translation by H. Siedentop of the Max-Planck medal lecture (1 April 1992) "Atoms in strong magnetic fields".

206. Absence of Ferromagnetism for One-Dimensional Itinerant Electrons, in *Probabilistic Methods in Mathematical Physics*, Proceedings of the International Workshop Siena, May 1991, F. Guerra, M. Loffredo and C. Marchioro eds., World Scientific pp. 290–294 (1992). A shorter version appears in *Rigorous Results in Quantum Dynamics*, J. Dittrich and P. Exner eds., World Scientific, pp. 243–245 (1991).

207. (with J.P. Solovej and J. Yngvason) Heavy Atoms in the Magnetic Field of a Neutron Star, Phys. Rev. Lett. **69**, 749–752 (1992).

∗208. (with J.P. Solovej) Atoms in the Magnetic Field of a Neutron Star, in *Differential Equations with Applications to Mathematical Physics*, W.F. Ames, J.V. Herod and E.M. Harrell II eds., Academic Press, pages 221–

237 (1993). Also in *Spectral Theory and Scattering Theory and Applications*, K. Yajima, ed., Advanced Studies in Pure Math. **23**, 259–274, Math. Soc. of Japan, Kinokuniya (1994). This is a summary of nos. 215, 216. Earlier summaries also appear in: (a) *Méthodes Semi-Classiques, Colloque international (Nantes 1991)*, Asterisque *210*, 237–246 (1991); (b) *Some New Trends on Fluid Dynamics and Theoretical Physics*, C.C. Lin and N. Hu eds., 149–157, Peking University Press (1993); (c) *Proceedings of the International Symposium on Advanced Topics of Quantum Physics, Shanxi*, J.Q. Lang, M.L. Wang, S.N. Qiao and D.C. Su eds., 5–13, Science Press, Beijing (1993).

209. (with M. Loss and R. McCann) Uniform Density Theorem for the Hubbard Model, J. Math. Phys. **34**, 891–898 (1993).

210. Remarks on the Skyrme Model, in *Proceedings of the Amer. Math. Soc. Symposia in Pure Math.* **54**, part 2, 379–384 (1993). (Proceedings of Summer Research Institute on Differential Geometry at UCLA, July 8–28, 1990.)

$211. (with E. Carlen) Optimal Hypercontractivity for Fermi Fields and Related Noncommutative Integration Inequalities, Commun. Math. Phys. **155**, 27–46 (1993).

212. (with E. Carlen) Optimal Two-Uniform Convexity and Fermion Hypercontractivity, in *Quantum and Non-Commutative Analysis*, Proceedings of June, 1992 Kyoto Conference, H. Araki et.al. eds., Kluwer (1993), pp. 93–111. (Condensed version of no. 211.)

†213. (with M. Loss) Fluxes, Laplacians, and Kasteleyn's Theorem, Duke Math. Journal **71**, 337–363 (1993).

214. (with V. Bach, R. Lewis and H. Siedentop) On the Number of Bound States of a Bosonic $N$-Particle Coulomb System, Zeits. f. Math. **214**, 441–460 (1993).

215. (with J.P. Solovej and J. Yngvason) Asymptotics of Heavy Atoms in High Magnetic Fields: I. Lowest Landau Band Region, Commun. Pure Appl. Math. **47**, 513–591 (1994).

216. (with J.P. Solovej and J. Yngvason) Asymptotics of Heavy Atoms in High Magnetic Fields: II. Semiclassical Regions, Commun. Math. Phys. **161**, 77–124 (1994).

∗217. (with V. Bach, M. Loss and J.P. Solovej) There are No Unfilled Shells in Unrestricted Hartree-Fock Theory, Phys. Rev. Lett. **72**, 2981–2983 (1994).

$218. (with K. Ball and E. Carlen) Sharp Uniform Convexity and Smoothness Inequalities for Trace Norms, Invent. Math. **115**, 463–482 (1994).

$219. Coherent States as a Tool for Obtaining Rigorous Bounds, *Proceedings of the Symposium on Coherent States, past, present and future*, Oak Ridge, D.H. Feng, J. Klauder and M.R. Strayer eds., World Scientific (1994), pages 267–278.

††220. The Hubbard model: Some Rigorous Results and Open Problems, in Proceedings of 1993 conference in honor of G.F. Dell'Antonio, *Advances in Dynamical Systems and Quantum Physics*, S. Albeverio et.al. eds., pp. 173–193, World Scientific (1995). A revised version appears in Pro-

ceedings of 1993 NATO ASI *The Hubbard Model*, D. Baeriswyl et.al. eds., pp. 1–19, Plenum Press (1995). A further revision appears in *Proceedings of the XIth International Congress of Mathematical Physics*, Paris, 1994, D. Iagolnitzer ed., pp. 392–412, International Press (1995).

221. (with V. Bach and J.P. Solovej) Generalized Hartree-Fock Theory of the Hubbard Model, J. Stat. Phys. **76**, 3–90 (1994).

††222. Flux Phase of the Half-Filled Band, Phys. Rev. Lett. **73**, 2158–2161 (1994).

$223. (with M. Loss) Symmetry of the Ginzburg-Landau Minimizer in a Disc, Math. Res. Lett. **1**, 701–715 (1994).

224. (with J.P. Solovej and J. Yngvason) Quantum Dots, in *Proceedings of the Conference on Partial Differential Equations and Mathematical Physics*, University of Alabama, Birmingham, 1994, I. Knowles, ed., International Press (1995), pages 157–172.

*225. (with J.P. Solovej and J. Yngvason) Ground States of Large Quantum Dots in Magnetic Fields, Phys. Rev. B **51**, 10646–10665 (1995).

226. (with J. Freericks) The Ground State of a General Electron-Phonon Hamiltonian is a Spin Singlet, Phys. Rev. B **51**, 2812–2821 (1995).

††227. (with B. Nachtergaele) Stability of the Peierls Instability for Ring-Shaped Molecules, Phys. Rev. B **51**, 4777–4791 (1995).

228. (with B. Nachtergaele) Dimerization in Ring-Shaped Molecules: The Stability of the Peierls Instability in *Proceedings of the XIth International Congress of Mathematical Physics*, Paris, 1994, D. Iagolnitzer ed., pp. 423–431, International Press (1995).

229. (with B. Nachtergaele) Bond Alternation in Ring-Shaped Molecules: The Stability of the Peierls Instability. In Proceedings of the conference *The Chemical Bond*, Copenhagen 1994, Int. J. Quant. Chem. **58**, 699–706 (1996).

230. Fluxes and Dimers in the Hubbard Model, in Proceedings of the International Congress of Mathematicians, Zürich, 1994, S.D. Chatterji ed., vol. 2, pp. 1279–1280, Birkhäuser (1995).

*231. (with M. Loss and J. P. Solovej) Stability of Matter in Magnetic Fields, Phys. Rev. Lett. **75**, 985–989 (1995).

*232. (with O. J. Heilmann) Electron Density near the Nucleus of a large Atom, Phys. Rev A **52**, 3628–3643 (1995).

*233. (with A. Iantchenko and H. Siedentop) Proof of a Conjecture about Atomic and Molecular Cores Related to Scott's Correction, J. reine u. ang. Math. **472**, 177–195 (1996).

††234. (with L.E. Thomas) Exact Ground State Energy of the Strong-Coupling Polaron, Commun. Math. Phys. **183**, 511–519 (1997). Errata **188**, 499–500 (1997).

$235. (with L. Cafarelli and D. Jerison) On the Case of Equality in the Brunn-Minkowski Inequality for Capacity, Adv. in Math. **117**, 193–207 (1996).

*236. (with M. Loss and H. Siedentop) Stability of Relativistic Matter via Thomas-Fermi Theory, Helv. Phys. Acta **69**, 974–984 (1996).

††237. Some of the Early History of Exactly Soluble Models, in *Proceedings of the 1996 Northeastern University conference on Exactly Soluble Models,* Int. Jour. Mod. Phys. B **11**, 3–10 (1997).
∗238. (with H. Siedentop and J. P. Solovej) Stability and Instability of Relativistic Electrons in Magnetic Fields, J. Stat. Phys. **89**, 37–59 (1997).
239. (with H. Siedentop and J-P. Solovej) Stability of Relativistic Matter with Magnetic Fields, Phys. Rev. Lett. **79**, 1785–1788 (1997).
240. Stability of Matter in Magnetic Fields, in *Proceedings of the Conference on Unconventional Quantum Liquids, Evora, Portugal, 1996* Zeits. f. Phys. B **933**, 271–274 (1997).
241. Birmingham in the Good Old Days, in *Proceedings of the Conference on Unconventional Quantum Liquids, Evora, Portugal, 1996* Zeits. f. Phys. B **933**, 125–126 (1997).
242. (with M. Loss) Book *Analysis*, American Mathematical Society (1997).
243. Doing Math with Fred, in *In Memoriam Frederick J. Almgren Jr., 1937–1997*, Experimental Math. **6**, 2–3 (1997).
∗244. (with J.P. Solovej and J. Yngvason) Asymptotics of Natural and Artificial Atoms in Strong Magnetic Fields, in *The Stability of Matter: From Atoms to Stars, Selecta of E. H. Lieb*, W. Thirring ed., second edition, Springer Verlag, pp. 145–167 (1997). This is a summary of nos. 207, 208, 215, 216, 224, 225.
245. Stability and Instability of Relativistic Electrons in Classical Electromagnetic Fields, in *Proceedings of Conference on Partial Differential Eqations and Mathematical Physics, Georgia Inst. of Tech., March, 1997*, Amer. Math. Soc. Contemporary Math. series, E. Carlen, E. Harrell, M. Loss eds., **217**, 99–108 (1998).
∗246. (with J. Yngvason) Ground State Energy of the Low Density Bose Gas, Phys. Rev. Lett. **80**, 2504–2507 (1998). arXiv math-ph/9712138, mp_arc 97-631.
†247. (with J. Yngvason) A Guide to Entropy and the Second Law of Thermodynamics, Notices of the Amer. Math. Soc. **45**, 571–581 (1998). arXiv math-ph/9805005, mp_arc 98-339. http://www.ams.org/notices/199805/lieb.pdf. See no. 266. This paper received the American Mathematical Society 2002 Levi Conant prize for "the best expository paper published in either the Notices of the AMS or the Bulletin of the AMS in the preceding five years".
$248. (with D. Hundertmark and L.E. Thomas) A Sharp Bound for an Eigenvalue Moment of the One-Dimensional Schroedinger Operator, Adv. Theor. Math. Phys. **2**, 719–731 (1998). arXiv math-ph/9806012, mp_arc 98-753.
$249. (with E. Carlen) A Minkowski Type Trace Inequality and Strong Subadditivity of Quantum Entropy, in Amer. Math. Soc. Transl. (2), **189**, 59–69 (1999).
250. (with J. Yngvason) The Physics and Mathematics of the Second Law of Thermodynamics, Physics Reports **310**, 1–96 (1999). arXiv cond-mat/9708200, mp_arc 97-457.
251. Some Problems in Statistical Mechanics that I would like to see Solved, 1998 IUPAP Boltzmann prize lecture, Physica A **263**, 491–499 (1999).

††252. (with P. Schupp) Ground State Properties of a Fully Frustrated Quantum Spin System, Phys. Rev. Lett. **83**, 5362–5365 (1999). arXiv math-ph/9908019, mp_arc 99-304.

253. (with P. Schupp) Singlets and Reflection Symmetric Spin Systems, Physica A **279**, 378–385 (2000). arXiv math-ph/9910037, mp_arc 99-404.

*254. (with R. Seiringer and J.Yngvason) Bosons in a Trap: A Rigorous Derivation of the Gross-Pitaevskii Energy Functional, Phys. Rev A **61**, #043602–1-13 (2000). arXiv math-ph/9908027, mp_arc 99-312.

255. (with J. Yngvason) The Ground State Energy of a Dilute Bose Gas, in *Differential Equations and Mathematical Physics, University of Alabama, Birmingham, 1999*, R. Weikard and G. Weinstein, eds., 295–306, Internat. Press (2000). arXiv math-ph/9910033, mp_arc 99-401.

*256. (with M. Loss) Self-Energy of Electrons in Non-perturbative QED, in *Differential Equations and Mathematical Physics, University of Alabama, Birmingham, 1999*, R. Weikard and G. Weinstein, eds. 279–293, Amer. Math. Soc./Internat. Press (2000). arXiv math-ph/9908020, mp_arc 99-305.

257. (with R. Seiringer and J. Yngvason) The Ground State Energy and Density of Interacting Bosons in a Trap, in *Quantum Theory and Symmetries, Goslar, 1999*, H.-D. Doebner, V.K. Dobrev, J.-D. Hennig and W. Luecke, eds., pp. 101–110, World Scientific (2000). arXiv math-ph/9911026, mp_arc 99-439.

258. (with J. Yngvason) The Ground State Energy of a Dilute Two-dimensional Bose Gas, J. Stat. Phys. **103**, 509–526 (2001). arXiv math-ph/0002014, mp_arc 00-63.

†259. (with J. Yngvason) A Fresh Look at Entropy and the Second Law of Thermodynamics, Physics Today **53**, 32–37 (April 2000). arXiv math-ph/0003028, mp_arc 00-123. See also **53**, 11–14, 106 (October 2000).

260. Lieb-Thirring Inequalities, in *Encyclopaedia of Mathematics, Supplement vol. 2,* pp. 311–313, Kluwer (2000). arXiv math-ph/0003039, mp_arc 00-132.

261. Thomas-Fermi Theory, in *Encyclopaedia of Mathematics, Supplement vol. 2,* pp. 455–457, Kluwer (2000). arXiv math-ph/0003040, mp_arc 00-131.

262. (with H. Siedentop) Renormalization of the Regularized Relativistic Electron-Positron Field, Commun. Math. Phys. **213**, 673–684 (2000). arXiv math-ph/0003001 mp_arc 00-98.

263. (with R. Seiringer and J. Yngvason) A Rigorous Derivation of the Gross-Pitaevskii Energy Functional for a Two-dimensional Bose Gas, Commun. Math. Phys. **224**, 17–31 (2001). arXiv cond-mat/0005026, mp_arc 00-203.

*264. (with M. Griesemer and M. Loss) Ground States in Non-relativistic Quantum Electrodynamics, Invent. Math. **145**, 557–595 (2001). arXiv math-ph/0007014, mp_arc 00-313.

*265. (with J.P. Solovej) Ground State Energy of the One-Component Charged Bose Gas, Commun. Math. Phys. **217**, 127–163 (2001). Errata **225**, 219–221 (2002). arXiv cond-mat/0007425, mp_arc 00-303.

266. (with J. Yngvason) The Mathematics of the Second Law of Thermodynamics, in *Visions in Mathematics, Towards 2000*, A. Alon, J. Bourgain, A.

Connes, M. Gromov and V. Milman, eds., GAFA 2000, no. 1, Birkhauser, p. 334–358 (2000). See no. 247. mp_arc 00-332.

267. The Bose Gas: A Subtle Many-Body Problem, in *Proceedings of the XIII International Congress on Mathematical Physics, London*, A. Fokas, et al. eds. International Press, pp. 91–111, 2001. arXiv math-ph/0009009, mp_arc 00-351.

268. (with J. Freericks and D. Ueltschi) Segregation in the Falicov-Kimball Model, Commun. Math. Phys. **227**, 243–279 (2002). arXiv math-ph/0107003, mp_arc 01-243.

269. (with G.K. Pedersen) Convex Multivariable Trace Functions, Reviews in Math. Phys. **14**, 1–18 (2002). arXiv math.OA/0107062.

††270. (with J. Freericks and D. Ueltschi) Phase Separation due to Quantum Mechanical Correlations, Phys. Rev. Lett. **88**, #106401–1-4 (2002). arXiv cond-mat/0110251.

∗271. (with M. Loss) Stability of a Model of Relativistic Quantum Electrodynamics, Commun. Math. Phys. **228**, 561–588 (2002). arXiv math-ph/0109002, mp_arc 01-315.

∗272. (with M. Loss) A Bound on Binding Energies and Mass Renormalization in Models of Quantum Electrodynamics, J. Stat. Phys. **108**, 1057–1069 (2002). arXiv math-ph/0110027.

∗273. (with R. Seiringer) Proof of Bose-Einstein Condensation for Dilute Trapped Gases, Phys. Rev. Lett. **88**, #170409–1-4 (2002). arXiv math-ph/0112032, mp_arc02-115.

274. (with M. Loss) Stability of Matter in Relativistic Quantum Mechanics, in *Mathematical Results in Quantum Mechanics*, Proceedings of QMath8, Taxco, Amer. Math. Soc. Contemporary Mathematics series, pp. 225–238, 2002.

275. (with J. Yngvason) The Mathematical Structure of the Second Law of Thermodynamics, in *Current Developments in Mathematics 2001*, International Press, de Jong et al eds. 2002, pp. 89–129. arXiv math-ph/0204007.

276. (with R. Seiringer, J. P. Solovej and J. Yngvason) The Ground State of the Bose Gas, in *Current Developments in Mathematics 2001*, de Jong et al eds., International Press, 2002, pp. 131–178. arXiv math-ph/0204027, mp_arc 02-183.

277. (with R. Seiringer and J. Yngvason) Poincaré Inequalities in Punctured Domains, Annals of Math. **158**, 1067–1080 (2003). arXiv math.FA/0205088.

∗278. (with R. Seiringer and J. Yngvason) Superfluidity in Dilute Trapped Bose Gases, Phys. Rev. B **66**, #134529–1-6 (2002). arXiv cond-mat/0205570, mp_arc 02-339.

279. (with F.Y. Wu) The one-dimensional Hubbard model: A reminiscence, Physica A **321**, 1–27 (2003). arXiv cond-mat/0207529.

280. (with E. Eisenberg) Polarization of interacting bosons with spin, Phys. Rev. Lett. **89**, #220403–1-4 (2002), mp_arc 02-446. arXiv cond-mat/0207042.

281. The Stability of Matter and Quantum Electrodynamics, in Proceedings of the Heisenberg symposium, Munich, Dec. 2001, *Fundamental Physics – Heisenberg and Beyond*, G. Buschhorn and J. Wess, eds., pp. 53–68,

Springer (2004). arXiv math-ph/0209034. A modified, up-dated version appears in the Milan Journal of Mathematics **71**, 199–217 (2003). A further modification appears in the Jahresbericht of the German Math. Soc. **106**, 93–110 (2004). arXiv math-ph/0401004.

282. (with R. Seiringer and J. Yngvason) Two-dimensional Gross-Pitaevskii Theory, in: A.G. Litvak (editor). *Progress in Nonlinear Science*, Proceedings of the International Conference Dedicated to the 100th Anniversary of A.A. Andronov, Volume II, pp 582–590, Nizhny Novgorod, Institute of Applied Physics, University of Nizhny Novgorod, 2002.

283. (with R. Seiringer) Equivalent forms of the Bessis-Moussa-Villani Conjecture, J. Stat. Phys. **115**, 185–190 (2004). math-ph/0210027.

∗284. (with R. Seiringer and J. Yngvason) One-dimensional Bosons in Three-dimensional Traps, Phys. Rev. Lett. **91**, #150401–1-4 (2003). arXiv cond-mat/0304071.

285. (with R. Seiringer and J. Yngvason) One-dimensional Behavior of Dilute, Trapped Bose Gases. Commun. Math. Phys. **244**, 347–393 (2004). arXiv math-ph/0305025.

286. (with R. Seiringer) Bose-Einstein Condensation of Dilute Gases in Traps, in *Differential Equations and Mathematical Physics* proceedings of a conference at the Univ. of Birmingham, 2002, Amer. Math. Soc. Contemporary Math. series, Y. Karpeshina, G. Stolz, R. Weikard, Y. Zeng, eds., **327**, 239–250 (2003).

∗287. (with M. Loss) Existence of Atoms and Molecules in Non-Relativistic Quantum Electrodynamics, Adv. Theor. Math. Phys. **7**, 667–710 (2003). arXiv math-ph/0307046.

∗288. (with J.P. Solovej) Ground State Energy of the Two-Component Charged Bose Gas. Commun. Math. Phys. DOI: 10.1007/s00220-004-1144-1 (in press). arXiv math-phys/0311010, mp_arc 03-490.

289. (with J. Yngvason) The Entropy of Classical Thermodynamics, in *Entropy*, the proceedings of a conference in Dresden, 2000, A. Greven, G. Keller and G. Warnecke eds., Princeton University Press, pp. 147–196 (2003).

290. (with M. Loss) A Note on Polarization Vectors in Quantum Electrodynamics. Commun. Math. Phys. (in press) arXiv math-ph/0401016.

††291. (with M. Aizenman, R. Seiringer, J.P. Solovej and J. Yngvason) Bose-Einstein Quantum Phase Transition in an Optical Lattice Model. Phys. Rev. A **70**, #023612–(1-12) (2004). arXiv cond-mat/0403240.

292. (with E. Carlen) Some Matrix Rearrangement Inequalities, Annali Matem. Pura Appl., (in press). arXiv math.OA/0402239.

††293. (with R. Seiringer, J.P. Solovej and J. Yngvason) The Quantum-Mechanical Many-Body Problem: The Bose Gas, in Proceedings of the conference 'Perspectives in Analysis', KTH Stockholm, 2003, M. Benedicks, P. Jones and S. Smirnov eds., Springer (in press). arXiv math-ph/0405004.

294. (with M. Loss) The Thermodynamic Limit for Matter Interacting with Coulomb Forces and with the Quantized Electromagnetic Field: I. The Lower Bound. arXiv math-ph/0408001.

295. (with E. Carlen and M. Loss) A Sharp analog of Young's Inequality on $\mathbf{S}^N$ and Related Entropy Inequalities, Jour. Geom. Anal. **14**, 487–520 (2004). arXiv math.FA/0408030.

MIX
Papier aus verantwortungsvollen Quellen
Paper from responsible sources
FSC® C105338

If you have any concerns about our products,
you can contact us on
**ProductSafety@springernature.com**

In case Publisher is established outside the EU,
the EU authorized representative is:
**Springer Nature Customer Service Center GmbH**
**Europaplatz 3, 69115 Heidelberg, Germany**

Printed by Libri Plureos GmbH
in Hamburg, Germany